SAFEGUARDING THE OZONE LAYER AND THE GLOBAL CLIMATE SYSTEM

This Intergovernmental Panel on Climate Change (IPCC)/Technology and Economic Assessment Panel (TEAP) Special Report provides information relevant to decision-making in regard to safeguarding the ozone layer and the climate system – two global environmental issues involving complex scientific and technical considerations. Scientific evidence linking chlorofluorocarbons (CFCs) and other ozone-depleting substances (ODSs) to global ozone depletion led to the initial control of chemicals under the 1987 Montreal Protocol and to amendments and adjustments in the 1990s that added additional ODSs, agreed phase-outs, and accelerated those phase-outs. As various approaches to the phase-out of ODSs were developed it was realized that some actions taken to reduce future depletion of the ozone layer, in particular the introduction of HFCs and PFCs, could affect global warming. When the Kyoto Protocol was negotiated in 1997, countries had new incentives to take account of how choices among substitutes could affect the objectives of both protocols.

The potential of each ODS substitute to influence the climate system depends not only on the physical and chemical properties of the substance itself, but also on the factors that influence its emission to the atmosphere, such as containment, recycling, destruction and energy efficiency in particular applications. This report provides the scientific context required for consideration of choices among alternatives to ODSs; potential methodologies for assessing options; and technical issues relating to greenhouse gas (GHG) emission-reduction opportunities for each of the sectors involved, including refrigeration, air conditioning, foams, aerosols, fire protection and solvents. The report also addresses the future availability of HFCs. The volume includes a *Summary for Policymakers* approved by governments represented in the IPCC, and a *Technical Summary.*

The IPCC/TEAP Special Report on *Safeguarding the Ozone Layer and the Global Climate System: Issues Related to Hydrofluorocarbons and Perfluorocarbons* provides invaluable information for researchers in environmental science, climatology, and atmospheric chemistry, policy makers in governments and environmental organizations, and scientists and engineers in industry.

IPCC/TEAP Special Report on

Safeguarding the Ozone Layer and the Global Climate System

Issues Related to Hydrofluorocarbons and Perfluorocarbons

Edited by

Bert Metz   Lambert Kuijpers   Susan Solomon

Stephen O. Andersen   Ogunlade Davidson   José Pons

David de Jager   Tahl Kestin   Martin Manning   Leo Meyer

Prepared by Working Groups I and III of the
Intergovernmental Panel on Climate Change, and the
Technical and Economic Assessment Panel

Published for the Intergovernmental Panel on Climate Change

CAMBRIDGE UNIVERSITY PRESS
Cambridge, New York, Melbourne, Madrid, Cape Town, Singapore, São Paulo

Cambridge University Press
40 West 20th Street, New York, NY 10011–4211, USA

Published in the United States of America by Cambridge University Press, New York

www.cambridge.org
Information on this title:www.cambridge.org/9780521863360

First published 2005

Printed in Canada

A catalogue record for this publication is available from the British Library

ISBN-13 978-0-521-86336-0 hardback
ISBN-10 0-521-86336-8 hardback

ISBN-13 978-0-521-68206-0 paperback
ISBN-10 0-521-68206-1 paperback

Please use the following reference to the whole report:
IPCC/TEAP, 2005: IPCC/TEAP Special Report on Safeguarding the Ozone Layer and the Global
Climate System: Issues Related to Hydrofluorocarbons and Perfluorocarbons. Prepared by Working
Group I and III of the Intergovernmental Panel on Climate Change, and the Technology and
Economic Assessment Panel [Metz, B., L. Kuijpers, S. Solomon, S. O. Andersen, O. Davidson,
J. Pons, D. de Jager, T. Kestin, M. Manning, and L. A. Meyer (eds.)]. Cambridge University Press,
Cambridge, United Kingdom and New York, NY, USA, 488 pp.

Cover photo: Nacreous clouds near McMurdo Station in Antarctica. Nacreous clouds are a type of
polar stratospheric cloud (or PSC) that occur when the polar stratosphere is very cold, particularly
during winter or spring. PSCs play a major role in ozone depletion because reactions on the
surfaces of particles in these clouds convert chlorine and bromine into forms that are highly
reactive with ozone. © Seth White.

Contents

Foreword

The Intergovernmental Panel on Climate Change (IPCC) was jointly established by the World Meteorological Organization (WMO) and the United Nations Environment Programme (UNEP) in 1988 to assess available information on the science, the impacts, and the economics of, and the options for mitigating and/or adapting to, climate change. In addition, the IPCC provides, on request, scientific, technical, and socio-economic advice to the Conference of the Parties (COP) to the United Nations Framework Convention on Climate Change (UNFCCC). The IPCC has produced a series of Assessment Reports, Special Reports, Technical Papers, methodologies, and other products that have become standard works of reference and that are widely used by policymakers, scientists, and other experts.

The *Special Report on Safeguarding the Ozone Layer and the Global Climate System* was developed in response to invitations by the United Nations Framework Convention on Climate Change (UNFCCC)[1] and the Montreal Protocol on Substances that Deplete the Ozone Layer[2]. IPCC and the Montreal Protocol's Technology and Economic Assessment Panel (TEAP) were asked to work together to develop a balanced scientific, technical and policy-relevant Special Report. The request covered both a scientific assessment of the interrelations between the ozone layer and climate change and development of user-friendly and policy-neutral information to assist all Parties and stakeholders in making informed decisions when evaluating alternatives to ozone-depleting substances.

Discussions on these topics have a long history, including deliberations at the Fourth Conference of the Parties to the UNFCCC (held in 1998 in Buenos Aires) which invited Parties and all other relevant entities to provide information to the UNFCCC Secretariat on available and potential ways and means of limiting emissions of hydrofluorocarbons (HFCs) and perfluorocarbons (PFCs) when used as replacements for ozone depleting substances. In 1999, an IPCC/TEAP Expert Meeting was organized[3], which addressed the issue and forms important background for the present effort, along with new information on science, technology, and policy needs.

After the decision of the Eighth Conference of the Parties to the UNFCCC[1], the Fourteenth Meeting of the Parties to the Montreal Protocol (Rome, Italy, 25-29 November 2002), welcomed this decision, and requested the TEAP to work with the IPCC in preparing the Special Report, and requested simultaneous submission of the report to the Open Ended Working Group under The Montreal Protocol and the UNFCCC SBSTA. The scope, structure, and outline of the Special Report were approved by the IPCC in plenary meetings during its Twentieth Session in Paris, France, from 19-21 February 2003.

As is usual in the IPCC, success in producing this report has depended first and foremost on the enthusiasm and cooperation of experts worldwide in many related but different disciplines. We would like to express our gratitude to all the Coordinating Lead Authors, Lead Authors, Contributing Authors, Review Editors, and Expert Reviewers. These individuals have devoted enormous time and effort to produce this report and we are extremely grateful for their commitment to the IPCC process.

We would also like to express our sincere thanks to the Steering Committee for this Report, which consisted of co-chairs from both the Technology and Economic Assessment Panel (TEAP) and the Intergovernmental Panel on Climate Change (IPCC):

Stephen Andersen, Lambert Kuijpers, and José Pons
for the Technology and Economic Assessment Panel (TEAP), and

Susan Solomon, Ogunlade Davidson and Bert Metz (chair of the Steering Committee)
for the Intergovernmental Panel on Climate Change (IPCC).

---

[1] Decision 12/CP.8, FCCC/CP/2002/7/Add.1, page 30. Eight Conference of the Parties to the UNFCCC, New Delhi, India, 23 October – 1 November 2002

[2] Decision XIV/10 UNEP/OzL.Pro.14/9, page 42. Fourteenth Meeting of the Parties to the Montreal Protocol, Rome, Italy, 25-29 November 2002

[3] Proceedings of the Joint IPCC/TEAP Expert Meeting on options for the limitation of emissions of HFCs and PFCs, Petten, The Netherlands, 26-28 May 1999, see http://www.ipcc-wg3.org/docs/IPCC-TEAP99/index.html

We are convinced that this Special Report provides a balanced scientific, technical and policy-related assessment that will assist all concerned in taking decisions when considering alternatives to ozone-depleting substances.

Michel Jarraud
Secretary-General,
World Meteorological Organization

Klaus Töpfer
Executive Director,
United Nations Environment Programme and
Director-General,
United Nations Office in Nairobi

Preface

This *Special Report on Safeguarding the Ozone and the Global Climate System* has been developed in response to invitations from Parties to the UNFCCC and the Montreal Protocol. It provides information relevant to decision-making in regard to safeguarding the ozone layer and the global climate system: two global environmental issues involving complex scientific and technical considerations. The scope, structure, and outline of this Special Report were approved by the IPCC at its Twentieth Session in Paris, France, 19–21 February 2003. The responsibility for preparing the report was given jointly to IPCC's Working Groups I and III and the Montreal Protocol's Technology and Economic Assessment Panel (TEAP). A joint IPCC/TEAP Steering Committee (see below) was established to manage preparation of the report following IPCC procedures.

Background

Scientific evidence linking chlorofluorocarbons (CFCs) and other Ozone Depleting Substances (ODSs) to global ozone depletion led to the initial control of chemicals under the 1987 Montreal Protocol and to Amendments and Adjustments in the 1990s that added additional ODSs, agreed phaseouts, and accelerated those phaseouts. This international process has resulted in (i) elimination of production of most CFCs, methyl chloroform, and halons, (ii) the increased use of existing hydrochlorofluorocarbons (HCFCs), (iii) the new production of a wide range of industrial fluorine containing chemicals, including new types of HCFCs, hydrofluorocarbons (HFCs) and perfluorocarbons (PFCs), (iv) use of non-halogenated chemical substitutes such as hydrocarbons, carbon dioxide and ammonia and (v) the development of not-in-kind alternative methods such as water-based cleaning procedures.

The likelihood that CFCs and other ODSs also affect the climate system was first identified in the 1970s, and the global warming effectiveness of halocarbons, including HFCs, has been further elucidated over the past three decades. For example, the 1989 Scientific Assessment of Stratospheric Ozone included a chapter on halocarbon global warming potentials (GWPs) and the 1989 Technology Assessment presented these GWPs in discussions of the importance of energy efficiency in insulating foam, refrigeration, and air conditioning. As various approaches were developed to the phase-out of ODSs under the Montreal Protocol, it was realized that some actions taken to reduce future depletion of the ozone layer, in particular the introduction of HFCs and PFCs, could increase or decrease global warming impact.

This scientific and technical information allowed Parties to the Montreal Protocol to choose options to replace ODSs, mindful of the global warming impact and was reflected in some investment decisions under the Montreal Protocol Multilateral Fund. When the Kyoto Protocol was negotiated in 1997, countries had new incentives to take account of how choices among substitutes could affect the objectives of both protocols. These considerations created a need for more comprehensive information regarding options for ODS replacement that take into account the need of safeguarding the ozone layer as well as the global climate system. In May 1999, the IPCC and TEAP held a joint expert meeting on options for the limitation of Emissions of HFCs and PFCs and in October 1999 TEAP published its report: "The implications to the Montreal Protocol of the Inclusion of HFCs and PFCs in the Kyoto Protocol". This Special Report of 2005 is the latest cooperative effort.

The potential of each ODS substitute to influence the climate system depends not only upon the physical and chemical properties of the substance itself but also upon the factors that influence emissions to the atmosphere, such as containment, recycling, destruction and energy efficiency in particular applications. Gases, applications and sectors considered in the report are those related to emissions of CFCs, HCFCs, HFCs and PFCs, as well as to alternatives for the use of HFCs and PFCs. The report does not consider unrelated industrial or other uses of the same chemicals The report covers chemicals and technologies in use or likely to be used in the next decade.

Organization of the Report

The report provides the scientific context required for consideration of choices among alternatives to ODSs (chapters 1 and 2); potential methodologies for assessing options (chapter 3); and technical issues relating to GHG emission reduction opportunities for each of the sectors involved, including refrigeration, air conditioning, foams, aerosols, fire protection and solvents (chapters 4 to 10). The report also addresses the future availability of HFCs (chapter 11).

Chapters 1 and 2 address linkages between ozone depletion and climate change, and draw from previous international scientific assessments, particularly the periodic assessments conducted under the auspices of WMO, UNEP, and the IPCC (e.g., the most recent *Scientific Assessment of Ozone Depletion, 2002*, the *Assessment of the Environmental Impacts of Ozone Depletion, 2002*, and *Climate Change: The Scientific Basis, 2001*). Chapter 1 covers stratospheric chemistry and dynamics and their coupling to climate change, while chapter 2 covers radiative forcing of each of the relevant gases as well as their roles in tropospheric chemistry and air quality. The present report does not seek to cover the breadth and depth of the more specialized ozone and climate change assessments, but rather to provide a summary of relevant interactions between the two environmental issues to aid the understanding and application of the rest of the report.

Chapter 3 summarizes available methodologies to characterize or compare technologies (such as the lifecycle climate performance parameter, LCCP), particularly those approaches that are applied across the diverse sectors covered in this report.

Chapters 4 through 10 then provide technical descriptions and information for each of the key sectors of halocarbon use: refrigeration (4), residential and commercial air conditioning and heating (5), mobile air conditioning (6), foams (7), medical aerosols (8), fire protection (9), and non-medical aerosols, solvents and HFC-23 byproduct emissions from HCFC-22 production (10). Each chapter provides an overview of its sector, the relevant technologies, information on consumption and emission of relevant gases, and practices and alternative technologies to reduce emissions and net warming impacts. This includes consideration of process improvement in applications, improved containment, recovery and recycling during operation, end-of-life recovery, disposal and destruction. The choices among options within each sector involve detailed consideration of technical factors including performance, environmental health and safety, cost, availability of alternatives, and total energy and resource efficiencies.

Chapter 11 covers both supply and demand issues for HFCs, and integrates emissions estimates across sectors and regions. It aggregates the emissions information for various chemicals from the various sectors, and considers the balance between supply and demand of HFCs.

As in past IPCC reports, this report contains a Summary for Policymakers (SPM) and a Technical Summary (TS), in addition to the main chapters. Each section of the SPM and TS has been referenced to the appropriate section of the relevant chapter, so that material in the SPM and TS can easily be followed up in further detail in the chapters. The report also contains annexes with a the list of Authors and Expert Reviewers, a glossary, a list of acronyms and abbreviations, a list of units and conversion factors, and an overview of major chemical formulae and nomenclature of substances that are considered in this report.

The report was compiled between August 2003 and April 2005 by 145 experts from 35 countries. The draft report was circulated for review by experts, which submitted valuable suggestions for improvement. This was followed by a second review by both governments and experts. In these two review rounds, about 6600 comments were received from about 175 experts, governments and non-governmental organizations.

This review process, and its oversight by Review Editors who are independent of the author teams, is an intrinsic part of any IPCC assessment and is an important part of ensuring the quality and credibility of the product.

The final report was considered by a joint Session of IPCC Working Groups I and III held in Addis Ababa from April 6th to 8th, 2005, where the SPM was approved line-by-line and the underlying report was accepted by the IPCC Panel.

Acknowledgements

The Steering Committee sincerely appreciates all the Coordinating Lead Authors, Lead Authors, and Review Editors whose expertise, diligence, and patience have underpinned the successful completion of this report and who generously contributed substantial amounts of their professional and personal time, and the contribution of the many contributors and reviewers for their valuable and painstaking dedication and work.

We also thank Marco Gonzalez and Megumi Seki from UNEP's Ozone Secretariat, for their co-sponsorship, financial support and commitment to the process that led to this report.

We thank the Governments of the Netherlands, United States of America, Japan, and Argentina for hosting the four report drafting meetings, and the Government of Ethiopia and the United Nations Centre in Addis Ababa for hosting the Joint Working Group I and III Session.

We also thank Renate Christ, Secretary of the IPCC, and the staff of the IPCC Secretariat, who provided logistical support for government liaison and travel of experts from the developing and transitional economy countries.

Finally we also thank the staff of the Working Groups I and III Technical Support Units, for their work in preparing the report, in particular David de Jager (Secretary of the Steering Committee, TSU WG III), Martin Manning (Head TSU WG I), Leo Meyer (Head TSU WG III) for their scientific and management support, and Tahl Kestin, Scott Longmore, Melinda Tignor (WGI), Heleen de Coninck, Anita Meier, Martin Middelburg, Rob Puijk, and Thelma van den Brink (WG III) for their technical and logistic support. Special thanks to Christine Ennis, Dave Thomas and Pete Thomas for their contribution to the copy-editing of the final draft of this report.

The Steering Committee for this Report:
   Stephen O. Andersen, Co-Chair of the TEAP
   Ogunlade Davidson, Co-Chair of IPCC Working Group III
   Lambert Kuijpers, Co-Chair of the TEAP
   Bert Metz, Co-Chair of IPCC Working Group III (Chair of the Steering Committee)
   José Pons, Co-Chair of the TEAP
   Susan Solomon, Co-Chair of IPCC Working Group I

IPCC/TEAP Special Report
Safeguarding the Ozone Layer and the Global Climate System:
Issues Related to Hydrofluorocarbons and Perfluorocarbons

Summary for Policymakers

Contents

1. Introduction

This IPCC Special Report was developed in response to invitations by the *United Nations Framework Convention on Climate Change* (UNFCCC)[1] and the *Montreal Protocol on Substances that Deplete the Ozone Layer*[2] to prepare a balanced scientific, technical and policy relevant report regarding alternatives to ozone-depleting substances (ODSs) that affect the global climate system. It has been prepared by the IPCC and the Technology and Economic Assessment Panel (TEAP) of the Montreal Protocol.

Because ODSs cause depletion of the stratospheric ozone layer[3], their production and consumption are controlled under the Montreal Protocol and consequently are being phased out, with efforts made by both developed and developing country parties to the Montreal Protocol. Both the ODSs and a number of their substitutes are greenhouse gases (GHGs) which contribute to climate change (see Figure SPM-1). Some ODS substitutes, in particular hydrofluorocarbons (HFCs) and perfluorocarbons (PFCs),

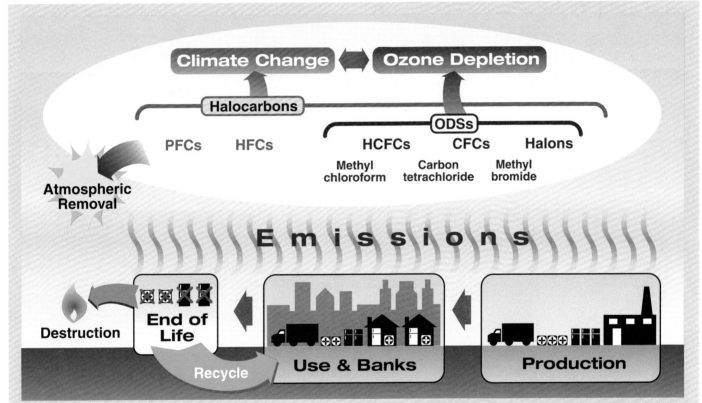

Figure SPM-1. Schematic diagram of major issues addressed by this report. Chlorofluorocarbons (CFCs), halons and hydrochlorofluorocarbons (HCFCs) contribute to ozone depletion and climate change, while hydrofluorocarbons (HFCs) and perfluorocarbons (PFCs) contribute only to climate change and are among possible non-ozone depleting alternatives for ODSs. Red denotes gases included under the Montreal Protocol and its amendments and adjustments[4] while green denotes those included under the UNFCCC and its Kyoto Protocol. Options considered in this report for reducing halocarbon emissions include improved containment, recovery, recycling, destruction of byproducts and existing banks[5], and use of alternative processes, or substances with reduced or negligible global warming potentials.

---

[1] Decision 12/CP.8, FCCC/CP/2002/7/Add.1, page 30.
[2] Decision XIV/10 UNEP/OzL.Pro.14/9, page 42.
[3] Ozone within this report refers to stratospheric ozone unless otherwise noted.
[4] Hereafter referred to as the Montreal Protocol.
[5] Banks are the total amount of substances contained in existing equipment, chemical stockpiles, foams and other products not yet released to the atmosphere.

are covered under the UNFCCC and its Kyoto Protocol. Options chosen to protect the ozone layer could influence climate change. Climate change may also indirectly influence the ozone layer.

This report considers the effects of total emissions of ODSs and their substitutes on the climate system and the ozone layer. In particular, this provides a context for understanding how replacement options could affect global warming. The report does not attempt to cover comprehensively the effect of replacement options on the ozone layer.

The report considers, by sector, options for reducing halocarbon emissions, options involving alternative substances, and technologies, to address greenhouse gas emissions reduction. It considers HFC and PFC emissions insofar as these relate to replacement of ODSs. HFC and PFC emissions from aluminum or semiconductor production or other sectors are not covered.

The major application sectors using ODSs and their HFC/PFC substitutes include refrigeration, air conditioning, foams, aerosols, fire protection and solvents. Emissions of these substances originate from manufacture and any unintended byproduct releases, intentionally emissive applications, evaporation and leakage from banks contained in equipment and products during use, testing and maintenance, and end-of-life practices.

With regard to specific emission reduction options, the report generally limits its coverage to the period up to 2015, for which reliable literature is available on replacement options with significant market potential for these rapidly evolving sectors. Technical performance, potential assessment methodologies and indirect emissions[6] related to energy use are considered, as well as costs, human health and safety, implications for air quality, and future availability issues.

---

[6] It should be noted that the National Inventory Reporting community uses the term 'indirect emissions' to refer specifically to those greenhouse gas emissions which arise from the breakdown of another substance in the environment. This is in contrast to the use of the term in this report, which specifically refers to energy-related CO_2 emissions associated with Life Cycle Assessment (LCA) approaches such as Total Equivalent Warming Impact (TEWI) or Life Cycle Climate Performance (LCCP).

2. Halocarbons, ozone depletion and climate change

2.1 What are the past and present effects of ODSs and their substitutes on the Earth's climate and the ozone layer?

Halocarbons, and in particular ODSs, have contributed to positive direct radiative forcing[7] and associated increases in global average surface temperature (see Figure SPM-2). The total positive direct radiative forcing due to increases in industrially produced ODS and non-ODS halocarbons from 1750 to 2000 is estimated to be 0.33 ± 0.03 W m^{-2}, representing about 13% of the total due to increases in all well-mixed greenhouse gases over that period. Most halocarbon increases have occurred in recent decades. Atmospheric concentrations of CFCs were stable or decreasing in the period 2001–2003 (0 to –3% per year, depending on the specific gas) while the halons and the substitute hydrochlorofluorocarbons (HCFCs) and HFCs increased (+1 to +3% per year, +3 to +7% per year, and +13 to +17% per year, respectively). [1.1, 1.2, 1.5 and 2.3][8]

Stratospheric ozone depletion observed since 1970 is caused primarily by increases in concentrations of reactive chlorine and bromine compounds that are produced by degradation of anthropogenic ODSs, including halons, CFCs, HCFCs, methyl chloroform (CH$_3$CCl$_3$), carbon tetrachloride (CCl$_4$) and methyl bromide (CH$_3$Br). [1.3 and 1.4]

Ozone depletion produces a negative radiative forcing of climate, which is an indirect cooling effect of the ODSs (see Figure SPM-2). Changes in ozone are believed to currently contribute a globally averaged radiative forcing of about –0.15 ± 0.10 W m^{-2}. The large uncertainty in the indirect radiative forcing of ODSs arises mainly because of uncertainties in the detailed vertical distribution of ozone depletion. This negative radiative forcing is *very likely*[10] to be smaller than the positive direct radiative forcing due to ODSs alone (0.32 ± 0.03 W m^{-2}). [1.1, 1.2 and 1.5]

Warming due to ODSs and cooling associated with ozone depletion are two distinct climate forcing mechanisms that do not simply offset one another. The spatial and seasonal distributions of the cooling effect of ozone depletion differ from those of the warming effect. A limited number of global climate modelling and statistical studies suggest that ozone depletion is one mechanism that may affect patterns of climate variability which are important for tropospheric circulation and temperatures in both hemispheres. However, observed changes in these patterns of variability cannot be unambiguously attributed to ozone depletion. [1.3 and 1.5]

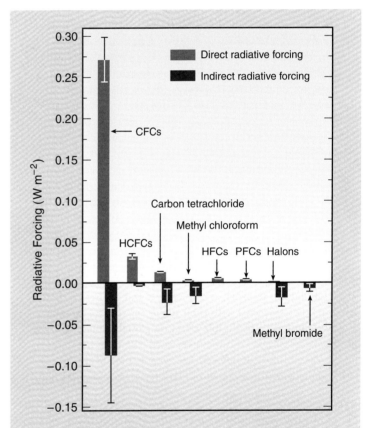

Figure SPM-2. Direct and indirect radiative forcing (RF) due to changes in halocarbons from 1750 to 2000.[9] Error bars denote ±2 standard-deviation uncertainties. [Based on Table 1.1]

---

[7] Radiative forcing is a measure of the influence a factor has in altering the balance of incoming and outgoing energy in the Earth-atmosphere system, and is an index of the importance of the factor as a potential climate change mechanism. It is expressed in watts per square meter (W m^{-2}). A greenhouse gas causes direct radiative forcing through absorption and emission of radiation and may cause indirect radiative forcing through chemical interactions that influence other greenhouse gases or particles.
[8] Numbers in square brackets indicate the sections in the main report where the underlying material and references for the paragraph can be found.
[9] PFCs used as substitutes for ODSs make only a small contribution to the total PFC radiative forcing.
[10] In this Summary for Policymakers, the following words have been used where appropriate to indicate judgmental estimates of confidence: very likely (90–99% chance); *likely* (66–90% chance); *unlikely* (10–33% chance); and *very unlikely* (1–10% chance).

Each type of gas has had different greenhouse warming and ozone depletion effects (see Figure SPM-2) depending mainly on its historic emissions, effectiveness as a greenhouse gas, lifetime and the amount of chlorine and/or bromine in each molecule. Bromine-containing gases currently contribute much more to cooling than to warming, whereas CFCs and HCFCs contribute more to warming than to cooling. HFCs and PFCs contribute only to warming. [1.5 and 2.5]

2.2 How does the phase-out of ODSs affect efforts to address climate change and ozone depletion?

Actions taken under the Montreal Protocol have led to the replacement of CFCs with HCFCs, HFCs, and other substances and processes. Because replacement species generally have lower global warming potentials[11] (GWPs), and because total halocarbon emissions have decreased, their combined CO_2-equivalent (direct GWP-weighted) emission has been reduced. The combined CO_2-equivalent emissions of CFCs, HCFCs and HFCs derived from atmospheric observations decreased from about 7.5 ± 0.4 $GtCO_2$-eq per year around 1990 to 2.5 ± 0.2 $GtCO_2$-eq per year around 2000, equivalent to about 33% and 10%, respectively, of the annual CO_2 emissions due to global fossil fuel burning. Stratospheric chlorine levels have approximately stabilized and may have already started to decline. [1.2, 2.3 and 2.5]

Ammonia and those hydrocarbons (HCs) used as halocarbon substitutes have atmospheric lifetimes ranging from days to months, and the direct and indirect radiative forcings associated with their use as substitutes are *very likely* to have a negligible effect on global climate. Changes in energy-related emissions associated with their use may also need to be considered. (See Section 4 for treatment of comprehensive assessment of ODS replacement options.) [2.5]

Based on the Business-As-Usual scenario developed in this report, the estimated direct radiative forcing of HFCs in 2015 is about 0.030 W m⁻²; based on scenarios from the IPCC Special Report on Emission Scenarios (SRES), the radiative forcing of PFCs[9] in 2015 is about 0.006 W m⁻². Those HFC and PFC radiative forcings correspond to about 1.0% and 0.2%, respectively, of the estimated radiative forcing of all well-mixed greenhouse gases in 2015, with the contribution of ODSs being about 10%. While this report particularly focused on scenarios for the period up to 2015, for the period beyond 2015 the IPCC SRES scenarios were considered but were not re-assessed. These SRES scenarios project significant growth in radiative forcing from HFCs over the following decades, but the estimates are likely to be very uncertain due to growing uncertainties in technological practices and policies. [1.5, 2.5 and 11.5]

Observations and model calculations suggest that the global average amount of ozone depletion has now approximately stabilized (for example, see Figure SPM-3). Although considerable variability in ozone is expected from year to year, including in polar regions where depletion is largest, the ozone layer is expected to begin to recover in coming decades due to declining ODS concentrations, assuming full compliance with the Montreal Protocol. [1.2 and 1.4]

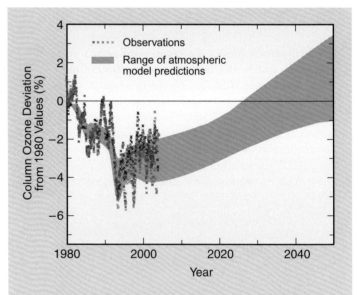

Figure SPM-3. Observed and modelled low- and mid-latitude (60°S–60°N) column ozone amounts as percent deviations from the 1980 values. [Box 1.7]

---

[11] GWPs are indices comparing the climate impact of a pulse emission of a greenhouse gas relative to that of emitting the same amount of CO_2, integrated over a fixed time horizon.

Over the long term, projected increases in other greenhouse gases could increasingly influence the ozone layer by cooling the stratosphere and changing stratospheric circulation. As a result of the cooling effect and of reducing ODS concentrations, ozone is *likely* to increase over much of the stratosphere, but could decrease in some regions, including the Arctic. However, the effects of changes in atmospheric circulation associated with climate change could be larger than these factors, and the net impact on total ozone due to increases in atmospheric concentrations of greenhouse gases is currently uncertain in both magnitude and sign. Based on current models an Arctic 'ozone hole' similar to that presently observed over the Antarctic is *very unlikely* to occur. [1.4]

The relative future warming and cooling effects of emissions of CFCs, HCFCs, HFCs, PFCs and halons vary with gas lifetimes, chemical properties and time of emission (see Table SPM-1). The atmospheric lifetimes range from about a year to two decades for most HFCs and HCFCs, decades to centuries for some HFCs and most halons and CFCs, and 1000 to 50,000 years for PFCs. Direct GWPs for halocarbons range from 5 to over 10,000. ODS indirect cooling is projected to cease upon ozone layer recovery, so that GWPs associated with the indirect cooling effect depend on the year of emission, compliance with the Montreal Protocol and gas lifetimes. These indirect GWPs are subject to much greater uncertainties than direct GWPs. [1.5, 2.2 and 2.5]

2.3 What are the implications of substitution of ODSs for air quality and other environmental issues relating to atmospheric chemistry?

Substitution for ODSs in air conditioning, refrigeration, and foam blowing by HFCs, PFCs, and other gases such as hydrocarbons are not expected to have a significant effect on global tropospheric chemistry. Small but not negligible impacts on air quality could occur near localized emission sources and such impacts may be of some concern, for instance in areas that currently fail to meet local standards. [2.4 and 2.6]

Persistent degradation products (such as trifluoroacetic acid, TFA) of HFCs and HCFCs are removed from the atmosphere via deposition and washout processes. However, existing environmental risk assessment and monitoring studies indicate that these are not expected to result in environmental concentrations capable of causing significant ecosystem damage. Measurements of TFA in sea water indicate that the anthropogenic sources of TFA are smaller than natural sources, but the natural sources are not fully identified. [2.4]

Table SPM-1. GWPs of halocarbons commonly reported under the Montreal Protocol and the UNFCCC and its Kyoto Protocol and assessed in this report relative to CO_2, for a 100-year time horizon, together with their lifetimes and GWPs used for reporting under the UNFCCC. Gases shown in blue (darker shading) are covered under the Montreal Protocol and gases shown in yellow (lighter shading) are covered under the UNFCCC. [Tables 2.6 and 2.7]

Gas	GWP for direct radiative forcing[a]			GWP for indirect radiative forcing (Emission in 2005[b])			Lifetime (years)	UNFCCC Reporting GWP[c]
CFCs								
CFC-12	10,720	±	3750	–1920	±	1630	100	n.a.[d]
CFC-114	9880	±	3460	Not available			300	n.a.[d]
CFC-115	7250	±	2540	Not available			1700	n.a.[d]
CFC-113	6030	±	2110	–2250	±	1890	85	n.a.[d]
CFC-11	4680	±	1640	–3420	±	2710	45	n.a.[d]
HCFCs								
HCFC-142b	2270	±	800	–337	±	237	17.9	n.a.[d]
HCFC-22	1780	±	620	–269	±	183	12	n.a.[d]
HCFC-141b	713	±	250	–631	±	424	9.3	n.a.[d]
HCFC-124	599	±	210	–114	±	76	5.8	n.a.[d]
HCFC-225cb	586	±	205	–148	±	98	5.8	n.a.[d]
HCFC-225ca	120	±	42	–91	±	60	1.9	n.a.[d]
HCFC-123	76	±	27	–82	±	55	1.3	n.a.[d]
HFCs								
HFC-23	14,310	±	5000	~0			270	11,700
HFC-143a	4400	±	1540	~0			52	3800
HFC-125	3450	±	1210	~0			29	2800
HFC-227ea	3140	±	1100	~0			34.2	2900
HFC-43-10mee	1610	±	560	~0			15.9	1300
HFC-134a	1410	±	490	~0			14	1300
HFC-245fa	1020	±	360	~0			7.6	–[e]
HFC-365mfc	782	±	270	~0			8.6	–[e]
HFC-32	670	±	240	~0			4.9	650
HFC-152a	122	±	43	~0			1.4	140
PFCs								
C_2F_6	12,010	±	4200	~0			10,000	9200
C_6F_{14}	9140	±	3200	~0			3200	7400
CF_4	5820	±	2040	~0			50,000	6500
Halons								
Halon-1301	7030	±	2460	–32,900	±	27,100	65	n.a.[d]
Halon-1211	1860	±	650	–28,200	±	19,600	16	n.a.[d]
Halon-2402	1620	±	570	–43,100	±	30,800	20	n.a.[d]
Other Halocarbons								
Carbon tetrachloride (CCl_4)	1380	±	480	–3330	±	2460	26	n.a.[d]
Methyl chloroform (CH_3CCl_3)	144	±	50	–610	±	407	5.0	n.a.[d]
Methyl bromide (CH_3Br)	5	±	2	–1610	±	1070	0.7	n.a.[d]

[a] Uncertainties in GWPs for direct positive radiative forcing are taken to be ±35% (2 standard deviations) (IPCC, 2001).
[b] Uncertainties in GWPs for indirect negative radiative forcing consider estimated uncertainty in the time of recovery of the ozone layer as well as uncertainty in the negative radiative forcing due to ozone depletion.
[c] The UNFCCC reporting guidelines use GWP values from the IPCC Second Assessment Report (see FCCC/SBSTA/2004/8, http://unfccc.int/resource/docs/2004/sbsta/08.pdf).
[d] ODSs are not covered under the UNFCCC.
[e] The IPCC Second Assessment Report does not contain GWP values for HFC-245fa and HFC-365mfc. However, the UNFCCC reporting guidelines contain provisions relating to the reporting of emissions from all greenhouse gases for which IPCC-assessed GWP values exist.

3. Production, banks and emissions

3.1 How are production, banks and emissions related in any particular year?

Current emissions of ODSs and their substitutes are largely determined by historic use patterns. For CFCs and HCFCs, a significant contribution (now and in coming decades) comes from their respective banks. There are no regulatory obligations to restrict these CFC and HCFC emissions either under the Montreal Protocol or the UNFCCC and its Kyoto Protocol, although some countries have effective national policies for this purpose.

Banks are the total amount of substances contained in existing equipment, chemical stockpiles, foams and other products not yet released to the atmosphere (see Figure SPM-1). The build-up of banks of (relatively) new applications of HFCs will – in the absence of additional bank management measures – also significantly determine post 2015 emissions.

3.2 What can observations of atmospheric concentrations tell us about banks and emissions?

Observations of atmospheric concentrations, combined with production and use pattern data, can indicate the significance of banks, but not their exact sizes.
The most accurate estimates of emissions of CFC-11 and CFC-12 are derived from observations of atmospheric concentrations. Those emissions are now larger than estimated releases related to current production, indicating that a substantial fraction of these emissions come from banks built up through past production. Observations of atmospheric concentrations show that global emissions of HFC-134a are presently smaller than reported production, implying that this bank is growing. The total global amount of HFC-134a currently in the atmosphere is believed to be about equal to the amount in banks. [2.5 and 11.3.4]

In the case of CFC-11 and some other gases, the lack of information on use patterns makes it difficult to assess the contribution to observed emissions from current production and use. Further work in this area is required to clarify the sources.

3.3 How are estimated banks and emissions projected to develop in the period 2002 to 2015?

Banks of CFCs, HCFCs, HFCs and PFCs were estimated at about 21 GtCO$_2$-eq in 2002[12,13]. In a Business-As-Usual (BAU) scenario, banks are projected to decline to about 18 GtCO$_2$-eq in 2015[14]. [7, 11.3 and 11.5]

In 2002, CFC, HCFC and HFC banks were about 16, 4 and 1 GtCO$_2$-eq (direct GWP weighted), respectively (see Figure SPM-4). In 2015, the banks are about 8, 5 and 5 GtCO$_2$-eq, respectively, in the BAU scenario. Banks of PFCs used as ODS replacements were about 0.005 GtCO$_2$-eq in 2002.

CFC banks associated with refrigeration, stationary air-conditioning (SAC)[15] and mobile air-conditioning (MAC) equipment are projected to decrease from about 6 to 1 GtCO$_2$-eq over the period 2002 to 2015, mainly due to release to the atmosphere and partly due to end-of-life recovery and destruction. CFC banks in foams are projected to decrease much more slowly over the same period (from 10 to 7 GtCO$_2$-eq), reflecting the much slower release of banked blowing agents from foams when compared with similarly sized banks of refrigerant in the refrigeration and air-conditioning sector.

HFC banks have started to build up and are projected to reach about 5 GtCO$_2$-eq in 2015. Of these, HFCs banked in foams represent only 0.6 GtCO$_2$-eq, but are projected to increase further after 2015.

---

[12] Greenhouse gas (GHG) emissions and banks expressed in terms of CO$_2$-equivalents use GWPs for direct radiative forcing for a 100-year time horizon. Unless stated otherwise, the most recent scientific values for the GWPs are used, as assessed in this report and as presented in Table SPM-1 (Column for 'GWP for direct radiative forcing').

[13] Halons cause much larger negative indirect than positive direct radiative forcing and, in the interest of clarity, their effects are not given here.

[14] In the BAU projections, it is assumed that all existing measures continue, including Montreal Protocol (phase-out) and relevant national policies. The current trends in practices, penetration of alternatives, and emission factors are maintained up to 2015. End-of-life recovery efficiency is assumed not to increase.

[15] In this Summary for Policymakers the 'refrigeration' sector comprises domestic, commercial, industrial (including food processing and cold storage) and transportation refrigeration. [4] 'Stationary air conditioning (SAC)' comprises residential and commercial air conditioning and heating. [5] 'Mobile air conditioning (MAC)' applies to cars, buses and passenger compartments of trucks.

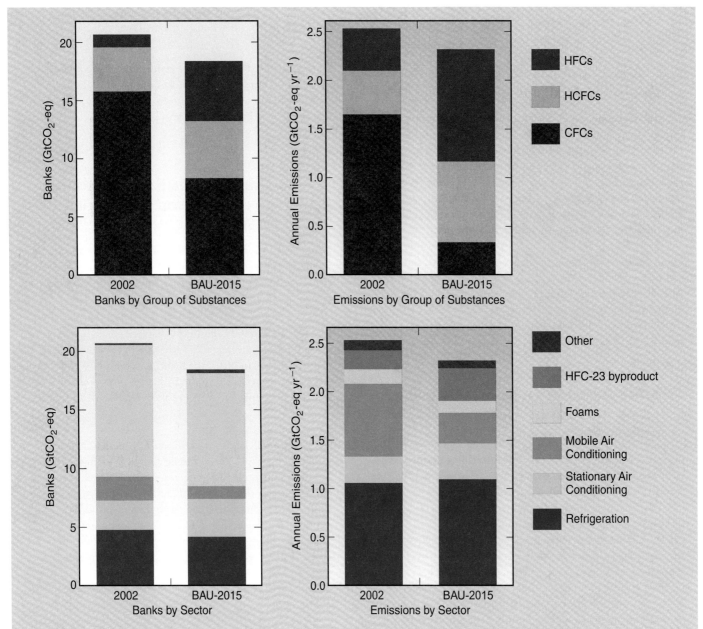

Figure SPM-4. Historic data for 2002 and Business-As-Usual (BAU) projections for 2015 of greenhouse gas CO_2-equivalent banks (left) and direct annual emissions (right), related to the use of CFCs, HCFCs and HFCs. Breakdown per group of greenhouse gases (top) and per emission sector (bottom). 'Other' includes Medical Aerosols, Fire Protection, Non-Medical Aerosols and Solvents. [11.3 and 11.5]

In the BAU scenario, total direct emissions of CFCs, HCFCs, HFCs and PFCs are projected to represent about 2.3 $GtCO_2$-eq per year by 2015 (as compared to about 2.5 $GtCO_2$-eq per year in 2002). CFC and HCFC emissions are together decreasing from 2.1 (2002) to 1.2 $GtCO_2$-eq per year (2015), and emissions of HFCs are increasing from 0.4 (2002) to 1.2 $GtCO_2$-eq per year (2015)[16]. PFC emissions from ODS substitute use are about 0.001 $GtCO_2$-eq per year (2002) and projected to decrease. [11.3 and 11.5]

Figure SPM-4 shows the relative contribution of sectors to global direct greenhouse gas (GHG) emissions that are related to the use of ODSs and their substitutes. Refrigeration applications together with SAC and MAC contribute the bulk of global direct GHG emissions in line with the higher emission rates associated with refrigerant banks. The largest part of GHG emissions from foams is expected to occur after 2015 because most releases occur at end-of-life.

With little new production, total CFC banks will decrease due to release to the atmosphere during operation and disposal. In the absence of additional measures a significant part of the CFC banks will have been emitted by 2015. Consequently, annual CFC emissions are projected to decrease from 1.7 (2002) to 0.3 $GtCO_2$-eq per year (2015).

HCFC emissions are projected to increase from 0.4 (2002) to 0.8 $GtCO_2$-eq per year (2015), owing to a steep increase expected for their use in (commercial) refrigeration and SAC applications.

The projected threefold increase in HFC emissions is the result of increased application of HFCs in the refrigeration, SAC and MAC sectors, and due to byproduct emissions of HFC-23 from increased HCFC-22 production (from 195 $MtCO_2$-eq per year in 2002 to 330 $MtCO_2$-eq per year in 2015 BAU).

Uncertainties in emission estimates are significant. Comparison of results of atmospheric measurements with inventory calculations shows differences per group of substances in the order of 10 to 25%. For individual gases the differences can be much bigger. This is caused by unidentified emissive applications of some substances, not accounted for in inventory calculations, and uncertainties in the geographically distributed datasets of equipment in use. [11.3.4]

The literature does not allow for an estimate of overall indirect[6] GHG emissions related to energy consumption. For individual applications, the relevance of indirect GHG emissions over a life cycle can range from low to high, and for certain applications may be up to an order of magnitude larger than direct GHG emissions. This is highly dependent on the specific sector and product/application characteristics, the carbon-intensity of the consumed electricity and fuels during the complete life cycle of the application, containment during the use-phase, and the end-of-life treatment of the banked substances. [3.2, 4 and 5]

---

[16] For these emission values the most recent scientific values for GWPs are used (see Table SPM-1, second column, 'GWP for direct radiative forcing'). If the UNFCCC GWPs would be used (Table SPM-1, last column, 'UNFCCC Reporting GWP'), reported HFC emissions (expressed in tonnes of CO_2-eq) would be about 15% lower.

4. Options for ODS phase-out and reducing greenhouse gas emissions

4.1 What major opportunities have been identified for reductions of greenhouse gas emissions and how can they be assessed?

Reductions in direct GHG emissions are available for all sectors discussed in this report and can be achieved through:
- **improved containment of substances;**
- **reduced charge of substances in equipment;**
- **end-of-life recovery and recycling or destruction of substances;**
- **increased use of alternative substances with a reduced or negligible global warming potential; and**
- **not-in-kind technologies[17].**

A comprehensive assessment would cover both direct emissions and indirect energy-related emissions, full life-cycle aspects, as well as health, safety and environmental considerations. However, due to limited availability of published data and comparative analyses, such comprehensive assessments are currently almost absent.

Methods for determining which technology option has the highest GHG emission reduction potential address both direct emissions of halocarbons or substitutes and indirect energy-related emissions over the full life cycle. In addition, comprehensive methods[18] assess a wide range of environmental impacts. Other, simplified methods[19] exist to assess life-cycle impacts and commonly provide useful indicators for life-cycle greenhouse gas emissions of an application. Relatively few transparent comparisons applying these methods have been published. The conclusions from these comparisons are sensitive to assumptions about application-specific, and often region- and time-specific parameters (e.g., site-specific situation, prevailing climate, energy system characteristics). [3.5]

Comparative economic analyses are important to identify cost-effective reduction options. However, they require a common set of methods and assumptions (e.g., costing methodology, time-frame, discount rate, future economic conditions, system boundaries). The development of simplified standardized methodologies would enable better comparisons in the future. [3.3]

The risks of health and safety impacts can be assessed in most cases using standardized methods. [3.4 and 3.5]

GHG emissions related to energy consumption can be significant over the lifetime of appliances considered in this report. Energy efficiency improvements can thus lead to reductions in indirect emissions from these appliances, depending on the particular energy source used and other circumstances, and produce net cost reductions, particularly where the use-phase of the application is long (e.g., in refrigeration and SAC).
The assessed literature did not allow for a global estimate of this reduction potential, although several case studies at technology and country level illustrate this point.

Through application of current best practices[20] and recovery methods, there is potential to halve (1.2 $GtCO_2$-eq per year reduction) the BAU direct emissions from ODSs and their GHG substitutes by 2015[21]. About 60% of this potential concerns HFC emissions, 30% HCFCs and 10% CFCs.
The estimates are based on a Mitigation Scenario[22] which makes regionally differentiated assumptions on best practices in production, use, substitution, recovery and destruction of these substances. Sectoral contributions are shown in Figure SPM-5. [11.5]

---

[17] Not-in-kind technologies achieve the same product objective without the use of halocarbons, typically with an alternative approach or unconventional technique. Examples include the use of stick or spray pump deodorants to replace CFC-12 aerosol deodorants; the use of mineral wool to replace CFC, HFC or HCFC insulating foam; and the use of dry powder inhalers (DPIs) to replace CFC or HFC metered dose inhalers (MDIs).

[18] Comprehensive methods, e.g. Life Cycle Assessment (LCA), cover all phases of the life cycle for a number of environmental impact categories. The respective methodologies are detailed in international ISO standards ISO 14040:1997, ISO 14041:1998, ISO 14042:2000, and ISO 14043:2000.

[19] Typical simplified methods include Total Equivalent Warming Impact (TEWI), which assesses direct and indirect greenhouse emissions connected only with the use-phase and disposal; and Life Cycle Climate Performance (LCCP), which also includes direct and indirect greenhouse emissions from the manufacture of the active substances.

[20] For this Report, best practice is considered the lowest achievable value of halocarbon emission at a given date, using commercially proven technologies in the production, use, substitution, recovery and destruction of halocarbon or halocarbon-based products (for specific numbers, see Table TS-6).

[21] For comparison, CO_2 emissions related to fossil fuel combustion and cement production were about 24 $GtCO_2$ per year in 2000.

[22] The Mitigation Scenario used in this Report, projects the future up to 2015 for the reduction of halocarbon emissions, based on regionally differentiated assumptions of best practices.

Of the bank-related emissions that can be prevented in the period until 2015, the bulk is in refrigerant-based applications where business-as-usual emission rates are considerably more significant than they are for foams during the period in question. With earlier action, such as recovery/destruction and improved containment, more of the emissions from CFC banks can be captured.

4.2 What are the sectoral emission reduction potentials in 2015 and what are associated costs?

In refrigeration applications direct GHG emissions can be reduced by 10% to 30%. For the refrigeration sector as a whole, the Mitigation Scenario shows an overall direct emission reduction of about 490 MtCO$_2$-eq per year by 2015, with about 400 MtCO$_2$-eq per year predicted for commercial refrigeration. Specific costs are in the range of 10 to 300 US$/tCO$_2$-eq[23,24]. Improved system energy efficiencies can also significantly reduce indirect GHG emissions.

In full supermarket systems, up to 60% lower LCCP[19] values can be obtained by using alternative refrigerants, improved containment, distributed systems, indirect systems or cascade systems. Refrigerant specific emission abatement costs range for the commercial refrigeration sector from 20 to 280 US$/tCO$_2$-eq.

In food processing, cold storage and industrial refrigeration, ammonia is forecast for increased use in the future, with HFCs replacing HCFC-22 and CFCs. Industrial refrigeration refrigerant specific emissions abatement costs were determined to be in the range from 27 to 37 US/tCO$_2$-eq. In transport refrigeration, lower GWP alternatives, such as ammonia, hydrocarbons and ammonia/carbon dioxide have been commercialized.

The emission reduction potential in domestic refrigeration is relatively small, with specific costs in the range of 0 to 130 US$/tCO$_2$-eq. Indirect emissions of systems using either HFC-134a or HC-600a (isobutane) dominate total emissions, for different carbon intensities of electric power generation. The difference between the LCCP[19] of HFC-134a and isobutane systems is small and end-of-life recovery, at a

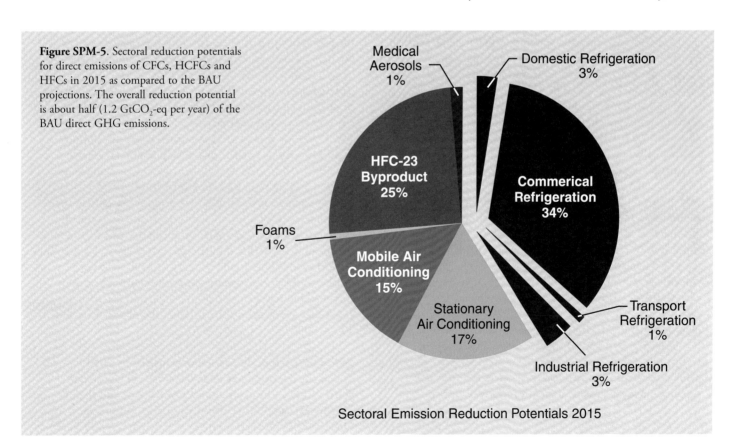

Figure SPM-5. Sectoral reduction potentials for direct emissions of CFCs, HCFCs and HFCs in 2015 as compared to the BAU projections. The overall reduction potential is about half (1.2 GtCO$_2$-eq per year) of the BAU direct GHG emissions.

Sectoral Emission Reduction Potentials 2015

[23] The presented cost data concern direct emission reductions only. Taking into account energy efficiency improvements may result in even net negative specific costs (savings).
[24] Costs in this report are given in 2002 US dollars unless otherwise stated.

certain cost increase, can further reduce the magnitude of the difference. [4]

Direct GHG emissions of residential and commercial air-conditioning and heating equipment (SAC) can be reduced by about 200 MtCO$_2$-eq per year by 2015 relative to the BAU scenario. Specific costs range from −3 to 170 US$/tCO$_2$-eq[23]. When combined with improvements in system energy efficiencies, which reduce indirect GHG emissions, in many cases net financial benefits accrue. Opportunities to reduce direct GHG (i.e., refrigerant) emissions can be found in (i) more efficient recovery of refrigerant at end-of-life (in the Mitigation Scenario assumed to be 50% and 80% for developing and developed countries, respectively); (ii) refrigerant charge reduction (up to 20%); (iii) better containment and (iv) the use of refrigerants with reduced or negligible GWPs in suitable applications.

Improving the integrity of the building envelope (reduced heat gain or loss) can have a significant impact on indirect emissions.

HFC mixtures and hydrocarbons (HCs) (for small systems) are used as alternatives for HCFC-22 in developed countries. For those applications where HCs can be safely applied, the energy efficiency is comparable to fluorocarbon refrigerants. Future technical developments could reduce refrigerant charge, expanding the applicability of HCs. [5]

In mobile air conditioning, a reduction potential of 180 MtCO$_2$-eq per year by 2015 could be achieved at a cost of 20 to 250 US$/tCO$_2$-eq[23]. Specific costs differ per region and per solution.
Improved containment, and end-of-life recovery (both of CFC-12 and HFC-134a) and recycling (of HFC-134a) could reduce direct GHG emissions by up to 50%, and total (direct and indirect) GHG emissions of the MAC unit by 30 to 40% at a financial benefit to vehicle owners. New systems with either CO$_2$ or HFC-152a, with equivalent LCCP, are likely to enter the market, leading to total GHG system emission reductions estimated at 50 to 70% in 2015 at an estimated added specific cost of 50 to 180 US$ per vehicle.

Hydrocarbons and hydrocarbon blends, which have been used to a limited extent, present suitable thermodynamic properties and permit high energy efficiency. However, the safety and liability concerns identified by vehicle manufacturers and suppliers limit the possible use of hydrocarbons in new vehicles. [6.4.4]

Due to the long life-span of most foam applications, by 2015 a limited emission reduction of 15 to 20 MtCO$_2$-eq per year is projected at specific costs ranging from 10 to 100 US$/tCO$_2$-eq[23]. The potential for emission reduction

increases in following decades.
Several short-term emission reduction steps, such as the planned elimination of HFC use in emissive one-component foams in Europe, are already in progress and are considered as part of the BAU. Two further key areas of potential emission reduction exist in the foams sector. The first is a potential reduction in halocarbon use in newly manufactured foams. However, the enhanced use of blends and the further phase-out of fluorocarbon use both depend on further technology development and market acceptance. Actions to reduce HFC use by 50% between 2010 and 2015, would result in emission reduction of about 10 MtCO$_2$-eq per year, at a specific cost of 15 to 100 US$/tCO$_2$-eq, with further reductions thereafter[23].

The second opportunity for emission reduction can be found in the worldwide banks of halocarbons contained in insulating foams in existing buildings and appliances (about 9 and 1 GtCO$_2$-eq for CFC and HCFC, respectively in 2002). Although recovery effectiveness is yet to be proven, and there is little experience to date, particularly in the buildings sector, commercial operations are already recovering halocarbons from appliances at 10 to 50 US$/tCO$_2$-eq[23]. Emission reductions may be about 7 MtCO$_2$-eq per year in 2015. However, this potential could increase significantly in the period between 2030 and 2050, when large quantities of building insulation foams will be decommissioned. [7]

The reduction potential for medical aerosols is limited due to medical constraints, the relatively low emission level and the higher costs of alternatives. The major contribution (14 MtCO$_2$-eq per year by 2015 compared to a BAU emission of 40 MtCO$_2$-eq per year) to a reduction of GHG emissions for metered dose inhalers (MDIs) would be the completion of the transition from CFC to HFC MDIs beyond what was already assumed as BAU. The health and safety of the patient is considered to be of paramount importance in treatment decisions, and there are significant medical constraints to limit the use of HFC MDIs. If salbutamol MDIs (approximately 50% of total MDIs) would be replaced by dry powder inhalers (which is not assumed in the Mitigation Scenario) this would result in an annual emission reduction of about 10 MtCO$_2$-eq per year by 2015 at projected costs in the range of 150 to 300 US$/tCO$_2$-eq. [8]

In fire protection, the reduction potential by 2015 is small due to the relatively low emission level, the significant shifts to not-in-kind alternatives in the past and the lengthy procedures for introducing new equipment.
Direct GHG emissions for the sector are estimated at about 5 MtCO$_2$-eq per year in 2015 (BAU). Seventy five percent of original halon use has been shifted to agents with no climate impact. Four percent of the original halon applications continue to employ halons. The remaining 21% has been

shifted to HFCs with a small number of applications shifted to HCFCs and to PFCs. PFCs are no longer needed for new fixed systems and are limited to use as a propellant in one manufacturer's portable extinguisher agent blend. Due to the lengthy process of testing, approval and market acceptance of new fire protection equipment types and agents, no additional options will likely have appreciable impact by 2015. With the introduction of a fluoroketone (FK) in 2002, additional reductions at an increased cost are possible in this sector through 2015. Currently those reductions are estimated to be small compared to other sectors. [9]

For non-medical aerosols and solvents there are several reduction opportunities, but the reduction potentials are likely to be rather small because most remaining uses are critical to performance or safety. The projected BAU emissions by 2015 for solvents and aerosols are about 14 and 23 $MtCO_2$-eq per year, respectively. Substitution of HFC-134a by HFC-152a in technical aerosol dusters is a leading option for reducing GHG emissions. For contact cleaners and plastic casting mould release agents, the substitution of HCFCs by hydrofluoroethers (HFEs) and HFCs with lower GWPs offers an opportunity. Some countries have banned HFC use in cosmetic, convenience and novelty aerosol products, although HFC-134a continues to be used in many countries for safety reasons.

A variety of organic solvents can replace HFCs, PFCs and ODSs in many applications. These alternative fluids include lower GWP compounds such as traditional chlorinated solvents, HFEs, HCs and oxygenated solvents. Many not-in-kind technologies, including no-clean and aqueous cleaning processes, are also viable alternatives. [10]

Destruction of byproduct emissions of HFC-23 from HCFC-22 production has a reduction potential of up to 300 $MtCO_2$-eq per year by 2015 and specific costs below 0.2 US$/t$CO_2$-eq according to two European studies in 2000.
Reduction of HCFC-22 production due to market forces or national policies, or improvements in facility design and construction also could reduce HFC-23 emissions. [10.4]

4.3 What are the current policies, measures and instruments?

A variety of policies, measures and instruments have been implemented in reducing the use or emissions of ODSs and their substitutes, such as HFCs and PFCs. These include regulations, economic instruments, voluntary agreements and international cooperation. Furthermore, general energy or climate policies affect the indirect GHG emissions of applications with ODSs, their substitutes, or not-in-kind alternatives.

This report contains information on policies and approaches in place in some countries (mainly developed) for reducing the use or emissions of ODSs and their substitutes. Those relevant policies and approaches include:
- Regulations (e.g., performance standards, certification, restrictions, end-of-life management)
- Economic instruments (e.g., taxation, emissions trading, financial incentives and deposit refunds)
- Voluntary agreements (e.g., voluntary reductions in use and emissions, industry partnerships and implementation of good practice guidelines)
- International cooperation (e.g., Clean Development Mechanism)

It should be noted that policy considerations are dependent on specific applications, national circumstances and other factors.

4.4 What can be said about availability of HFCs/PFCs in the future for use in developing countries?

No published data are available to project future production capacity. However, as there are no technical or legal limits to HFC and PFC production, it can be assumed that the global production capacity will generally continue to satisfy or exceed demand. Future production is therefore estimated in this report by aggregating sectoral demand.

In the BAU scenario, global production capacity is expected to expand with additions taking place mainly in developing countries and through joint ventures. Global production capacity of HFCs and PFCs most often exceeds current demand. There are a number of HFC-134a plants in developed countries and one plant in a developing country with others planned; the few plants for other HFCs are almost exclusively in developed countries. The proposed European Community phase-out of HFC-134a in mobile air conditioners in new cars and the industry voluntary programme to reduce their HFC-134a emissions by 50% will affect demand and production capacity and output. Rapidly expanding markets in developing countries, in particular for replacements for CFCs, is resulting in new capacity for fluorinated gases which is at present being satisfied through the expansion of HCFC-22 and 141b capacity. [11]

Safeguarding the Ozone Layer and the Global Climate System: Issues Related to Hydrofluorocarbons and Perfluorocarbons

Technical Summary

Coordinating Lead Authors

David de Jager (The Netherlands), Martin Manning (USA), Lambert Kuijpers (The Netherlands)

Lead Authors

Stephen O. Andersen (USA), Paul Ashford (UK), Paul Atkins (USA), Nick Campbell (France), Denis Clodic (France), Sukumar Devotta (India), Dave Godwin (USA), Jochen Harnisch (Germany), Malcolm Ko (USA), Suzanne Kocchi (USA), Sasha Madronich (USA), Bert Metz (The Netherlands), Leo Meyer (The Netherlands), José Roberto Moreira (Brazil), John Owens (USA), Roberto Peixoto (Brazil), José Pons (Venezuela), John Pyle (UK), Sally Rand (USA), Rajendra Shende (India), Theodore Shepherd (Canada), Stephan Sicars (Canada), Susan Solomon (USA), Guus Velders (The Netherlands), Dan Verdonik (USA), Robert Wickham (USA), Ashley Woodcock (UK), Paul Wright (UK) and Masaaki Yamabe (Japan)

Review Editors

Ogunlade Davidson (Sierra Leone), Mack McFarland (USA), Pauline Midgley (Germany)

Contents

1. Introduction

This IPCC Special Report was developed in response to invitations by the *United Nations Framework Convention on Climate Change* (UNFCCC)[1,2] and the *Montreal Protocol on Substances that Deplete the Ozone Layer*[3] to prepare a balanced scientific, technical and policy relevant report regarding alternatives to ozone-depleting substances (ODSs) that affect the global climate system. It has been prepared by the IPCC and the Technology and Economic Assessment Panel (TEAP) of the Montreal Protocol.

Because ODSs cause depletion of the stratospheric ozone layer[4], their production and consumption are controlled under the Montreal Protocol and consequently are being phased out, with efforts made by both developed and developing country parties to the Montreal Protocol. Both the ODSs and a number of their substitutes are greenhouse gases (GHGs) which contribute to climate change (see Figure TS-1). Some ODS substitutes, in particular hydrofluorocarbons (HFCs) and perfluorocarbons (PFCs), are covered under the UNFCCC and its Kyoto Protocol. Options

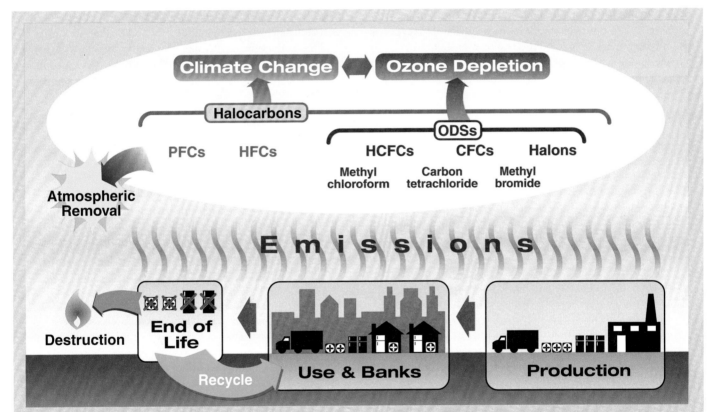

Figure TS-1. Schematic diagram of major issues addressed by this report. Chlorofluorocarbons (CFCs), halons and hydrochlorofluorocarbons (HCFCs) contribute to ozone depletion and climate change, while hydrofluorocarbons (HFCs) and perfluorocarbons (PFCs) contribute only to climate change and are among possible non-ozone depleting alternatives for ODSs. Red denotes gases included under the Montreal Protocol and its amendments and adjustments[5] while green denotes those included under the UNFCCC and its Kyoto Protocol. Options considered in this report for reducing halocarbon emissions include improved containment, recovery, recycling, destruction of byproducts and existing banks[6], and use of alternative processes, or substances with reduced or negligible global warming potentials.

---

[1] Decision 12/CP.8, FCCC/CP/2002/7/Add.1, page 30.
[2] Terms defined in the Glossary are highlighted in colour the first time they are used in this Technical Summary.
[3] Decision XIV/10 UNEP/OzL.Pro.14/9, page 42.
[4] Ozone within this report refers to stratospheric ozone unless otherwise noted.
[5] Hereafter referred to as the Montreal Protocol.
[6] Banks are the total amount of substances contained in existing equipment, chemical stockpiles, foams and other products not yet released to the atmosphere.

chosen to protect the ozone layer could influence climate change. Climate change may also indirectly influence the ozone layer.

This report considers the effects of total emissions of ODSs and their substitutes on the climate system and the ozone layer. In particular, this provides a context for understanding how replacement options could affect global warming. The report does not attempt to cover comprehensively the effect of replacement options on the ozone layer.

Hydrochlorofluorocarbons (HCFCs) have been used to replace chlorofluorocarbons (CFCs) in several applications as they have shorter lifetimes in the atmosphere and consequently cause less ozone depletion. HFCs and PFCs have been identified as potential long-term replacements for ODSs because they contain neither bromine nor chlorine and do not cause any significant ozone depletion. However, all of these species are also GHGs and so contribute to climate change in varying degrees. Additional alternatives for halocarbon use include ammonia and organic substances, the direct emissions of which have a very small effect on climate although indirect emissions may be important.

The report considers, by sector, options for reducing halocarbon emissions, options involving alternative substances, and technologies, to address greenhouse gas emissions reduction. It considers HFC and PFC emissions insofar as these relate to replacement of ODSs. HFC and PFC emissions from aluminum or semiconductor production or other sectors where ODS replacement is not involved are not covered.

The major application sectors using ODSs and their HFC/PFC substitutes include refrigeration, air conditioning, foams, aerosols, fire protection and solvents. Emissions of these substances originate from manufacture and any unintended byproduct releases, intentionally emissive applications, evaporation and leakage from banks contained in equipment and products during use, testing and maintenance, and end-of-life practices.

With regard to specific emission reduction options, the report generally limits its coverage to the period up to 2015, for which reliable literature is available on replacement options with significant market potential for these rapidly evolving sectors. Technical performance, potential assessment methodologies and indirect emissions[7] related to energy use are considered, as well as costs, human health and safety, implications for air quality, and future availability issues.

The Technical Summary (TS) brings together key information from the underlying report and follows to some extent the report structure, which is in three parts. The first part describes scientific links between stratospheric ozone depletion and climate change and provides relevant information on radiative forcing, observations of changes in forcing agents and emissions (section 2 of the TS). It addresses how the phase-out of ODSs is affecting both stratospheric ozone and climate change as well as the implications for air quality and local environmental issues. The report does not seek to cover the breadth and depth of other specialized assessments of ozone depletion and climate change, but rather to assess relevant interactions between the two environmental issues pertinent to the consideration of replacement options.

The second part assesses options to replace ODSs, including environmental, health, safety, availability and technical performance issues (section 3 and 4 of the TS). The report assesses practices and alternative technologies to reduce emissions and net warming impacts within each use sector, including consideration of process improvement in applications, improved containment, end-of-life recovery, recycling, disposal and destruction as well as relevant policies and measures.

The third part of the report covers supply and demand issues. The report aggregates available information on emissions from the various sectors and regions and considers the balance between supply and demand, bearing in mind those issues relevant to developing countries (section 3.8 in the TS).

---

[7] It should be noted that the National Inventory Reporting community uses the term 'indirect emissions' to refer specifically to those greenhouse gas emissions which arise from the breakdown of another substance in the environment. This is in contrast to the use of the term in this report, which specifically refers to energy-related CO_2 emissions associated with Life Cycle Assessment (LCA) approaches such as Total Equivalent Warming Impact (TEWI) or Life Cycle Climate Performance (LCCP).

2. Halocarbons, ozone depletion and climate change

2.1 How do the CFCs and their replacements contribute to the radiative forcing of the climate system?

Many halocarbons, including CFCs, PFCs, HFCs and HCFCs, are effective GHGs because they absorb Earth's outgoing infrared radiation in a spectral range where energy is not removed by CO_2 or water vapour (sometimes referred to as the *atmospheric window*, see Figure TS-2). Halocarbon molecules can be many thousand times more efficient at absorbing radiant energy emitted from the Earth than a molecule of CO_2, and small amounts of these gases can contribute significantly to radiative forcing[8] of the climate system. [1.1][9]

Radiative efficiencies ($W\ m^{-2}\ ppb^{-1}$) for the halocarbons and other well-mixed GHGs that are reported under the Montreal and Kyoto Protocols are given in Table TS-1. For most of the species considered here, the magnitude of the direct radiative forcing generated by a gas is given by the product of its mixing ratio (in parts per billion, ppb) and radiative efficiency. For the more abundant greenhouse gases – CO_2, methane and nitrous oxide – there is a nonlinear relationship between mixing ratio and radiative forcing. [1.1]

The primary radiative effect of CO_2 and water vapour is to warm the surface climate but cool the stratosphere. However, due to their absorption in the atmospheric window, the direct radiative effect of halocarbons is to warm both the troposphere and stratosphere. [1.2 and Box 1.4]

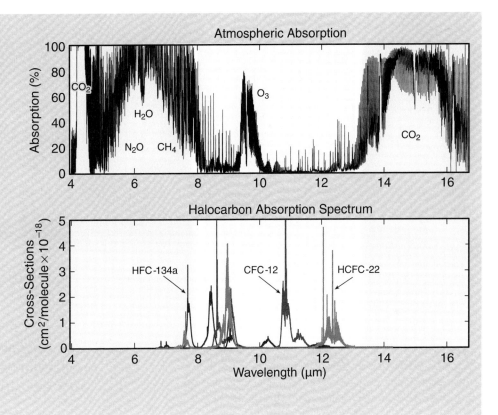

Figure TS-2. Top panel: Infrared atmospheric absorption (0 represents no absorption and 100% represents complete absorption of radiation) as derived from the space borne IMG/ADEOS radiance measurements (3 April 1997, 9.5°W, 38.4°N). Bottom panel: Absorption cross-sections for halocarbons (HCFC-22, CFC-12, HFC-134a) in the infrared atmospheric window, which lies between the nearly opaque regions due to strong absorptions by CO_2, H_2O, O_3, CH_4 and N_2O. [Figure 2.6]

[8] Radiative forcing is a measure of the influence a factor has in altering the balance of incoming and outgoing energy in the Earth-atmosphere system, and is an index of the importance of the factor as a potential climate change mechanism. It is expressed in watts per square meter ($W\ m^{-2}$). A greenhouse gas causes direct radiative forcing through absorption and emission of radiation and may cause indirect radiative forcing through chemical interactions that influence other greenhouse gases or particles.
[9] Numbers in square brackets indicate the sections in the main report where the underlying material and references for the paragraph can be found.

Table TS-1. Radiative efficiencies, lifetimes and positive direct radiative forcing for the well-mixed GHGs and halocarbons normally reported under the Montreal and Kyoto Protocols due to their increases between 1750 and 2000 and between 1970 and 2000. See Section 1.1, especially Table 1.1, and Section 2.2, especially Table 2.6, for details.

Gas species	Radiative efficiency (W m^{-2} ppb^{-1})	Lifetime (years)	Radiative forcing (W m^{-2}) 1750–2000	1970–2000
CO$_2$	1.55×10^{-5} [a]	-[b]	1.50	0.67
CH$_4$	3.7×10^{-4}	12[c]	0.49	0.13
N$_2$O	3.1×10^{-3}	114[c]	0.15	0.068
CFC-11	0.25	45	0.066	0.053
CFC-12	0.32	100	0.173	0.136
CFC-113	0.3	85	0.025	0.023
CFC-114	0.31	300	0.005	0.003
CFC-115	0.18	1700	0.002	0.002
HCFC-22	0.20	12	0.0283	0.0263
HCFC-123	0.14	1.3	0.0000	0.0000
HCFC-124	0.22	5.8	0.0003	0.0003
HCFC-141b	0.14	9.3	0.0018	0.0018
HCFC-142b	0.2	17.9	0.0024	0.0024
HCFC-225ca	0.2	1.9	0.0000	0.0000
HCFC-225cb	0.32	5.8	0.0000	0.0000
HFC-23	0.19	270	0.0029	0.0029
HFC-32	0.11	4.9	0.0000	0.0000
HFC-125	0.23	29	0.0003	0.0003
HFC-134a	0.16	14	0.0024	0.0024
HFC-152a	0.09	1.4	0.0002	0.0002
HFC-227ea	0.26	34.2	0.0000	0.0000
Halon-1211	0.3	16	0.0012	0.0012
Halon-1301	0.32	65	0.0009	0.0009
Halon-2402	0.33	20	0.0001	0.0001
CCl$_4$	0.13	26	0.0127	0.0029
CH$_3$Br	0.01	0.7	0.0001	0.0000
CH$_3$CCl$_3$	0.06	5	0.0028	0.0018
CF$_4$	0.08	50,000	0.0029	0.0029
C$_2$F$_6$	0.26	10,000	0.0006	0.0006
C$_3$F$_8$	0.26	2600	0.0001	0.0001
Ethane	0.0032	0.21	-	-
Pentane	0.0046	0.010	-	-

Notes

[a] The radiative efficiency of CO$_2$ decreases as its concentration increases.

[b] Removal of CO$_2$ from the atmosphere involves many different processes and its rate cannot be expressed accurately with a single lifetime. However, carbon cycle models typically estimate that 30 to 50% of CO$_2$ emissions remain in the atmosphere for more than 100 years.

[c] The lifetimes of CH$_4$ and N$_2$O incorporate indirect effects of the emission of each gas on its own lifetime.

Apart from their direct effect on climate, some gases have an indirect effect either from radiative forcing caused by their degradation products or through their influences on atmospheric chemistry. Halocarbons containing chlorine and bromine are ODSs and, because ozone is a strong absorber of UV radiation, they have an indirect cooling effect which is significant compared to their direct warming effect. For hydrocarbons, indirect effects associated with tropospheric ozone production may be significantly greater than direct effects. [1.1, 1.5, Box 1.3, 2.2 and 2.5]

2.2 How long do CFCs and their replacements remain in the atmosphere after they are emitted?

Gases with longer lifetimes remain in the atmosphere, and consequently may affect climate, for a longer time. The lifetimes of several halocarbons and replacement species are given in Table TS-1. Most CFCs are removed from the atmosphere on time scales ranging from about 50 to 100 years. With the exception of HFC-23, which has a lifetime of 270 years, HCFCs and HFCs are removed efficiently in the troposphere through atmospheric chemistry oxidation processes. As a result, they have lifetimes ranging from about one year to a few decades. PFCs are very inert molecules and their emissions contribute to warming of the climate system over timescales that can exceed 1000 years. [2.2]

Most halocarbon gases are sufficiently long-lived that they become mixed throughout the atmosphere before they are destroyed. They therefore have nearly constant mixing ratios throughout the troposphere. In contrast, ammonia and organic compounds have lifetimes that typically range from days to weeks, thereby making their distributions variable both spatially and temporally. [2.2]

2.3 How are the atmospheric concentrations and emissions of CFCs, halons, HCFCs, HFCs and PFCs changing?

Atmospheric observations, as seen for example in Figure TS-3, show that global concentrations of the CFCs increased largely during the period from the 1970s to the 1990s, so that their contributions to radiative forcing grew mainly during this period, while the concentrations of replacement species are now increasing. HCFC-22 is the most abundant HCFC, with a current concentration of 160 ppt. Its concentration began to increase in the early 1970s concurrently with those of the CFCs, while the concentrations of other significant HCFCs

increased mainly in the 1990s, as did those of the HFCs. [1.2, 2.3]

Currently observed rates of change differ among ODSs, depending mainly upon their emissions and atmospheric lifetimes (see Table TS-2). Following the Montreal Protocol and its subsequent amendments, large reductions in ODS production and emission have occurred. The concentrations of some CFCs have peaked, while others are expected to decline in the future. In the complete absence of emissions, the concentrations of these gases would decrease, but at a slow rate that is determined by their atmospheric lifetimes of decades to centuries. Continuing emissions will cause increases in the concentrations of these gases or diminish their rates of decrease. Observations of annual changes in atmospheric concentrations provide the most reliable estimates of total global emissions for long-lived gases. For example, CFC-11 is observed to be decreasing at a rate about 60% slower than would occur in the absence of emissions, while CFC-12 is still increasing slightly, indicating continuing emissions of both species. CFC-113, however, is decreasing at a rate close to that expected in the absence of emissions. Table TS-2 shows observed global concentrations and growth or decay rates of the major CFCs, halons, HCFCs, HFCs and PFCs, together with the estimated emissions to the atmosphere required to explain currently observed trends. [2.3, 2.5]

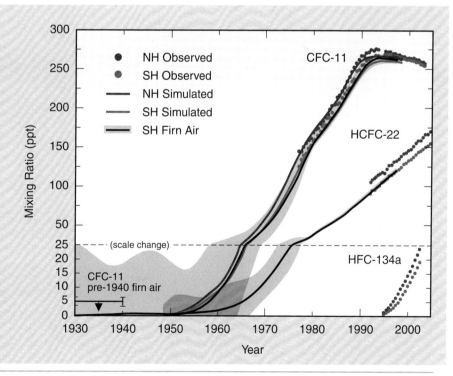

Figure TS-3. Estimated global tropospheric mixing ratios (ppt) for CFC-11, HCFC-22 and HFC-134a shown separately for the Northern and Southern Hemisphere. Red and blue symbols show measurements from the AGAGE (Advanced Global Atmospheric Gases Experiment) and CMDL (Climate Monitoring and Diagnostics Laboratory) networks, while red and blue lines show simulated CFC-11 concentrations based on estimates of emissions and atmospheric lifetimes. Black lines and the shaded area show estimates and uncertainty bands for CFC-11 and HCFC-22 derived by synthesis inversion of Antarctic firn air measurements and *in situ* Cape Grim atmospheric measurements. The thick black horizontal line with arrow and error bars shows a separate upper bound estimate of pre-1940 CFC-11 concentrations based on South Pole firn air measurements. Note that the gases shown here are used in different applications and are presented for illustrative purposes only. [Figure 1.8]

Table TS-2. Observed global concentrations and trends for some of the most abundant CFCs, HCFCs, HFCs and PFCs near the year 2003, together with the total global emissions required to explain these trends. For comparison estimated emissions are shown for the year 1990, a little after the peak in total ODS emissions. See Section 2.3, especially Table 2.1, and Figure 2.4 for details.

Species	Concentration in 2003[a] (ppt)	Trend between 2001 and 2003[a] (ppt yr^{-1})	Trend as percentage of concentration %	Estimated emission in 2002[b] (kt yr^{-1})	Estimated emission in 2002[b] (GtCO$_2$-eq yr^{-1})[d]	Estimated emissions in 1990[c] (kt yr^{-1})	Estimated emissions in 1990[c] (GtCO$_2$-eq yr^{-1})[d]
CFC-11	256	-1.9 – -2.7	-0.7 – -1.1	70 – 90	0.33 – 0.42	280 – 300	1.31 – 1.40
CFC-12	538	+0.2 – +0.8	+0.04 – +0.16	110 – 130	1.2 – 1.4	400 – 430	4.29 – 4.61
CFC-113	80	-0.6 – -0.7	-0.8 – -1.0	5 – 12	0.03 – 0.07	180 – 230	1.09 – 1.39
Total major CFCs				***185 – 232***	***1.54 – 1.89***	***860 – 960***	***6.68 – 7.40***
HCFC-22	157	+4.5 – +5.4	+2.8 – +3.4	240 – 260	0.43 – 0.46	185 – 205	0.33 – 0.36
HCFC-123	0.03	0	0	n.a.[e]			0
HCFC-141b	16	+1.0 – +1.2	+6.3 – +7.5	55 – 58	0.04		0
HCFC-142b	14	+0.7 – +0.8	+4.3 – +5.7	25	0.06	10 – 20	0.02 – 0.05
Total major HCFCs				***320 – 343***	***0.53 – 0.56***	***195 – 225***	***0.35 – 0.41***
HFC-23	17.5	+0.58	+3.3	13	0.19	6.4	0.09
HFC-125	2.7	+0.46	+17	9 – 10	0.03	0	0
HFC-134a	26	+3.8 – +4.1	+15 – +16	96 – 98	0.14	0	0
HFC-152a	2.6	+0.34	+13	21 – 22	0.003	0	0
Total major HFCs				***139 – 143***	***0.36***	***6.4***	***0.09***
Halon-1211	4.3	+0.04 – +0.09	+0.9 – +2.8	7 – 8	0.013 – 0.015	11.5	0.02
Halon-1301	2.9	+0.04 – +0.08	+1.4 – +2.8	1 – 2	0.007 – 0.014	5.1	0.04
CCl$_4$	95	-0.9 – -1.0	-1.0 – -1.1	64 – 76	0.09 – 0.10	120 – 130	0.17 – 0.18
CH$_3$CCl$_3$	27	-5.6 – -5.8	-21 – -23	15 – 17	0.002	646	0.09
CF$_4$	76	n.a.[e]	n.a.[e]	n.a.[e]			
C$_2$F$_6$	2.9	+0.1	+3.4				
C$_3$F$_8$	0.26	n.a.[e]	n.a.[e]				

[a] Average of tropospheric concentrations and range of trends from different monitoring networks.

[b] Estimated from current concentrations and trends.

[c] Estimated emissions for 1990 based on Chapter 2, Figure 2.4, except for HFC-23 which is taken from Chapter 10.

[d] GWP-weighted emissions using direct GWPs from this report.

[e] n.a. – not available. Insufficient data to determine reliable trend or emission value.

For comparison, earlier estimates of emissions (WMO, 2003)[10] are also shown for the year 1990, which is slightly after the peak in ODS emissions. Emissions of CFC-113 and CH_3CCl_3, both used largely as solvents and having no accumulated banks, decreased by more than a factor of 10 from 1990 to 2000. Stratospheric chlorine levels have approximately stabilized and may have already started to decline. [2.3]

The estimated current emissions of CFC-11 and CFC-12 shown in Table TS-2 are larger than estimates of new production, indicating that a substantial fraction of these emissions originates from banks of these chemicals built up from past production. Such banks include material contained in foams, air conditioning, refrigeration and other applications. In contrast, production is currently greater than emission for nearly all of the HCFCs and HFCs, implying that banks of these chemicals are currently accumulating and could contribute to future radiative forcing. One measure of the relevance of such banks is the ratio of bank size to the amount already in the atmosphere. For example, in the case of HFC-134a these amounts are estimated to be about equal. [2.3, 2.5 and 11.3]

Continuing atmospheric observations of CFCs and other ODSs now enable improved validation of estimates for the lag between production and emission to the atmosphere. This provides new insight into the overall significance of banks and of end-of-life options which are relevant to the future use of HCFC and HFC substitutes. [2.5]

For some gases, there are now sufficient atmospheric observations to constrain not just global but also regional emissions in certain areas. For example, atmospheric measurements suggest sharp increases in European emissions of HFC-134a over the period 1995–1998 and in HFC-152a over the period 1996–2000, with some subsequent levelling off through to 2003. [2.3]

2.4 How much do the halocarbon gases and their replacements contribute to positive radiative forcing of the climate system relative to pre-industrial times? What about relative to 1970?

The contributions to direct radiative forcing due to increases in halocarbon concentrations from 1750 to 2000 and from 1970 to 2000 are summarized in Table TS-1 and Figure

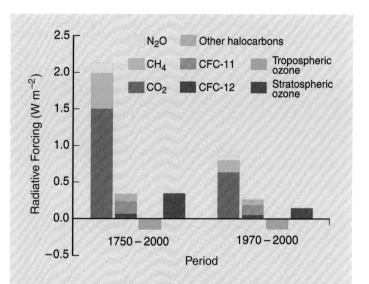

Figure TS-4. Radiative forcing (W m⁻²) due to changes in well mixed GHGs and ozone over the time periods 1750 – 2000 and 1970 – 2000. The negative radiative forcing from stratospheric ozone is due to stratospheric ozone depletion resulting from halocarbon emissions between 1970 and 2000. In contrast, the tropospheric ozone radiative forcing is largely independent of the halocarbon radiative forcing. [Figure 1.3]

TS-4. The direct radiative forcing due to these increases from 1750 to 2000 is estimated to be 0.33 ± 0.03 W m⁻², which represents about 13% of the total due to increases in all of the well-mixed GHGs over that period. The contributions of CFCs, HCFCs and HFCs are about 0.27 W m⁻², 0.033 W m⁻² and 0.006 W m⁻² respectively. [1.1 and 1.5]

Because increases in halocarbon concentrations occurred mainly during the last three decades, their relative contribution to total radiative forcing is larger during this period. The direct radiative forcing due to increases in halocarbons from 1970 to 2000 was 0.27 ± 0.03 W m⁻², which represents about 23% of that due to increases in all of the well-mixed GHGs. The contribution to direct radiative forcing due to HCFCs is presently dominated by HCFC-22, while that due to HFCs is dominated by HFC-134a and HFC-23, with the latter being emitted mainly as a byproduct of manufacture of HCFC-22. [1.1, 1.5]

---

[10] WMO, 2003: Scientific Assessment of Ozone Depletion: 2002. Global Ozone Research and Monitoring Project – Report No. 47, World Meteorological Organization, Geneva, 498 pp.

2.5 How has stratospheric ozone changed in recent decades and why?

As shown in Figure TS-5, the amount of stratospheric ozone has decreased over the past few decades, particularly in the Antarctic. The largest decreases since 1980 have been observed over the Antarctic during the spring (the *Antarctic ozone hole*), with the monthly total column ozone amounts in September and October being about 40–50% below pre-ozone-hole values. [1.2, 1.3 and 1.4]

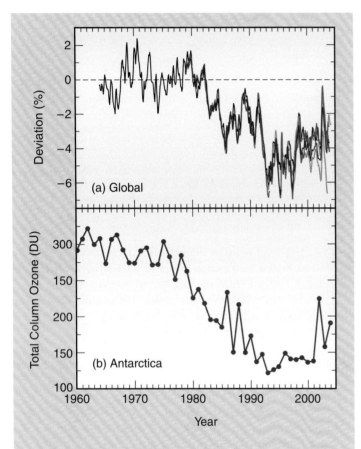

Figure TS-5. Top: Time-series of de-seasonalized global mean column ozone anomalies estimated from five different data sets, based on ground-based (black line) and satellite measurements (colored lines). Anomalies are expressed as percentages of the time average for the period 1964–1980. Bottom: October mean total column ozone measurements from the Dobson spectrophotometer at Halley, Antarctica (73.5°S, 26.7°W). [Figures 1.4 and 1.5]

Arctic ozone loss in any given year depends strongly on the meteorological conditions. Arctic ozone has been chemically depleted by up to 30% in recent cold years, but the losses observed in warm years have been very small. Globally averaged ozone has decreased by roughly 3% since 1980. The ozone column decreased by about 6% over the mid-latitudes (35°–60°) in the Southern Hemisphere and 3% in the Northern Hemisphere. Significant long-term changes in column ozone have not been observed in the tropics. Observations and model calculations suggest that the global average amount of ozone depletion has now approximately stabilized (see Figure TS-5). [1.2]

The observed ozone changes are caused by both chemical and dynamical factors, with the chemical factors being dominant. This ozone depletion is caused primarily by increases in concentrations of reactive chlorine and bromine compounds that are produced by degradation of anthropogenic ODSs, including halons, CFCs, HCFCs, methyl chloroform (CH_3CCl_3), carbon tetrachloride (CCl_4) and methyl bromide (CH_3Br). Human activities have increased the amount of chlorine in the stratosphere, relative to estimated natural background levels, by a factor of about 5 since 1970. CFCs are the primary source of this change, while HCFCs currently contribute about 5% to the total stratospheric chlorine loading. [1.2, 1.3 and 1.4]

2.6 How has ozone depletion affected the radiative forcing of the climate system?

Observations and modelling show that ozone depletion has acted to cool the stratosphere, which in turn can contribute to cooling of the troposphere and surface. The warming of the climate by ODSs and the cooling associated with ozone depletion are two distinct mechanisms that are governed by different physical processes and feedbacks and for which there are quite different levels of scientific understanding. For the purposes of this report, we follow IPCC (2001)[11] and assume that the observed depletion is caused entirely by ODSs and that the ozone radiative forcing can be considered to be an indirect effect due to ODSs. Gases containing bromine (such as halons) are particularly effective ozone depleters and have a larger contribution to the indirect effect on a per-molecule basis than other ozone-depleting gases such as the CFCs. [1.1, 1.2, 1.3 and 1.5]

---

[11] IPCC, 2001: Climate Change 2001: The Scientific Basis. Contribution of Working Group I to the Third Assessment Report of the Intergovernmental Panel on Climate Change [Houghton, J. T., Y. Ding, D. J. Griggs, M. Noguer, P. J. van der Linden, X. Dai, K. Maskell, and C. A. Johnson (eds.)]. Cambridge University Press, Cambridge, United Kingdom, and New York, NY, USA, 944 pp.

The best estimate of the negative indirect radiative forcing associated with ozone depletion over the period 1970–2000 is –0.15 ± 0.10 W m^{-2}, where the large uncertainty is determined by the range of model estimates and arises mainly because of uncertainties in the detailed vertical distribution of ozone depletion. This indirect effect is *very likely*[12] to be smaller in magnitude than the positive direct radiative forcing due to ODSs alone (0.32 ± 0.03 W m^{-2}) which is far better understood . If some fraction of the observed global ozone changes were not attributable to ODSs the magnitude of this indirect effect would be reduced. [1.5]

The relative contributions of different types of gas to positive direct and negative indirect radiative forcing are shown in Figure TS-6. However, the warming and cooling effects produced by direct and indirect radiative forcing do not simply offset one another because the spatial and seasonal distributions of the effects on surface climate are different. [1.2, 1.5 and Box 1.4]

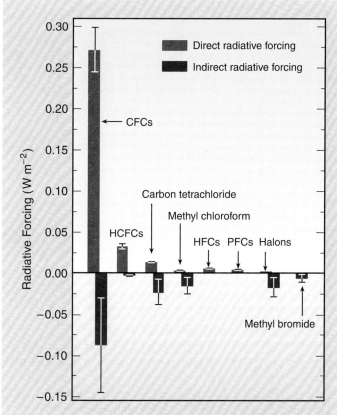

Figure TS-6. Radiative forcing due to changes in halocarbons from 1750 to 2000. [Based on Table 1.1]

A limited number of global climate modelling and statistical studies suggest that ozone depletion is one mechanism that may affect patterns of climate variability which are important for tropospheric circulation and temperatures in both hemispheres. In particular, the significant depletion of stratospheric ozone occurring in the Antarctic region is *likely* to have affected stratospheric circulation, and consequently the troposphere. There are indications that the Antarctic ozone hole has contributed to the cooling observed over the Antarctic plateau and to warming in the region of the Antarctic Peninsula. [1.3]

2.7 What factors are expected to control ozone in the coming century? Will it 'recover' to pre-ozone-hole values? Has the recovery already begun?

Global ozone recovery is expected to follow decreases in chlorine and bromine loading in the stratosphere as ODS concentrations decline due to reductions in their emissions. While this is expected to be the dominant factor in ozone recovery, emissions of other GHGs (such as CO_2, methane and nitrous oxide) can affect both tropospheric and stratospheric chemistry and climate, and will have some effect on ozone recovery. [1.3 and 1.4]

Figure TS-7 shows a range of predictions for changes in stratospheric ozone for the latitude range 60°S–60°N from two-dimensional photochemical models together with comparable ground-based and satellite measurements taken up to 2003. Such computer simulations show global ozone recovery occurring gradually as halogen gas emissions and concentrations decrease. However, the time of recovery varies significantly, depending on assumptions made about future climate and future composition of the atmosphere, and consequently remains quite uncertain. [1.4 and Box 1.7]

Future temperature changes related to GHG emissions are expected to enhance stratospheric ozone depletion in some parts of the stratosphere and decrease it in others. Increases in CO_2 concentration are expected to cool the stratosphere, which is known to reduce the rates of gas-phase ozone destruction in much of the stratosphere and thereby increase ozone concentrations at altitudes above about 25 km. In contrast, lower temperatures could decrease ozone concentrations at lower altitudes. While the latter effect is expected to be most important in the Arctic in late winter to early spring, it may be small compared with other

---

[12] In this report the following words have been used where appropriate to indicate judgmental estimates of confidence: *very likely* (90–99% chance); *likely* (66–90% chance); *unlikely* (10–33% chance); and *very unlikely* (1–10% chance).

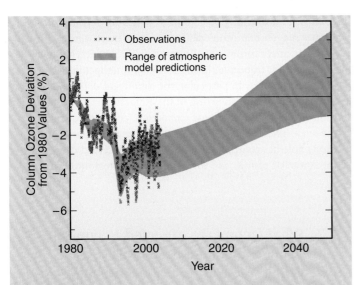

Figure TS-7. Observed and modelled changes in low- and mid-latitude (60°S – 60°N) de-seasonalised column ozone relative to 1980. The black symbols indicate ground-based measurements, and the coloured symbols various satellite-based data sets. The range of model predictions comes from the use of several different two-dimensional photochemistry models forced with the same halocarbon scenario; some models also allowed for the effect of changing CO_2 amounts on stratospheric temperatures. The measurements show that column ozone values between 60°S and 60°N decreased beginning in the early 1980s and the models capture the timing and extent of this decrease quite well. Modelled halogen source gas concentrations decrease in the early 21st century in response to the Montreal Protocol, so that simulated ozone values increase and recover towards pre-1980 values. [Box 1.7]

processes and will slowly decrease with time as the chlorine and bromine loadings decrease. Changes in stratospheric circulation may also occur in association with increases in GHGs and these could either increase or decrease future mid-latitude and polar ozone. The net result of future GHG emissions on global ozone depends upon the combination of these effects and its magnitude and direction are poorly quantified at present. [1.3 and 1.4]

As can be seen from the measurements shown in Figure TS-7, detection of the beginning of ozone recovery is difficult because of the high variability in ozone levels. This variability is due to both meteorological variability and the confounding influence of volcanic eruptions on the ozone layer. As a result, it is not yet possible to state that the beginning of ozone recovery has been unambiguously identified. [1.2, 1.4 and Box 1.7]

Models suggest that minimum levels of Antarctic ozone may already have occurred or should occur within the next few years. Predictions of the timing of a minimum in Arctic ozone

are more uncertain due to far greater natural variability in this region, but models suggest that it should occur within the next two decades. An Arctic 'ozone hole' similar to that currently observed over the Antarctic is *very unlikely* to occur. [1.3 and 1.4]

2.8 How much are CFCs, HCFCs and their possible replacements expected to affect the future radiative forcing of the climate system?

The estimated radiative forcing of HFCs in 2015 is in the range 0.022–0.025 W m⁻² based on the SRES emission scenarios and in the range 0.019–0.030 W m⁻² based on scenarios from Chapter 11 of this report. The radiative forcing of PFCs in 2015 is about 0.006 W m⁻² based on SRES scenarios. These HFC and PFC radiative forcings correspond to about 6–10% and 2% respectively of the total estimated radiative forcing due to CFCs and HCFCs in 2015 (estimated to be 0.297 W m⁻² for the baseline scenario). [2.5 and 11.5]

Scenario-based projections of future radiative forcing are shown in Figure TS-8. Such projections over longer time scales become more uncertain due to the growing influences of uncertainties in future technological practices and policies, but the contribution of HFCs may be 0.1 to 0.25 W m⁻² by 2100 based upon the range of SRES emission scenarios, while that of PFCs may be 0.02 to 0.04 W m⁻² by 2100. [1.5 and 2.5]

Figure TS-8 shows estimates of the separate contributions to future halocarbon radiative forcing from: past emissions (i.e. what is currently in the atmosphere); future emissions of new production assuming no change in current practices (based on the WMO Ab scenario); and emissions from current banks of halocarbons. Two different estimates of this last component are shown, one based on the WMO (2003) Ozone Assessment and the other based on later sections of this report. Although the magnitude of current ODS banks remains uncertain, their contribution to radiative forcing is expected to be comparable with that of HFC emissions to the atmosphere in the next few decades. This indicates that choices with respect to end-of-life options, such as the destruction of currently banked material, can provide significant benefits for the climate system. End-of-life recovery, recycling and destruction practices could also reduce emissions of all newly produced halocarbons and their contribution to radiative forcing shown in Figure TS-8. [1.5]

In addition to reducing atmospheric chlorine loading, actions taken under the Montreal Protocol and its adjustments and amendments have also acted to reduce total CO_2-equivalent

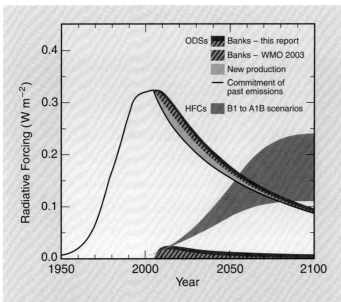

Figure TS-8. Direct radiative forcing of all ODSs compared with that of SRES projections for HFCs. The direct radiative forcing is split up into contributions from the commitment of past emissions (solid black line), release of allowed new production under the Montreal Protocol (grey shaded area), and release from ODS banks existing in 2004. Two estimates are given for these latter emissions – see Chapter 2. Radiative forcing due to HFCs are shown for the SRES B1 and A1B scenarios (boundaries of the purple shaded area). The contribution due to the delayed release of ODSs in banks is shown separately and is comparable with that projected due to HFCs for the next two decades. ODSs also have other indirect effects on radiative forcing. [Figure 1.19]

emissions. This has occurred because the replacement species generally have lower global warming potentials (GWPs) and because total halocarbon emissions have been reduced. The reduction can be seen by comparing emissions in 1990 and in 2000 given in Table TS-2, and is shown more generally in Figure TS-9. Thus the combined CO_2-equivalent emissions of CFCs, HCFCs and HFCs have decreased from a peak of 7.5 $\pm$ 0.4 GtCO$_2$-eq yr^{-1} around 1990 to 2.5 $\pm$ 0.2 GtCO$_2$-eq yr^{-1} around 2000, which is equivalent to about 10% of the annual contribution due to global fossil fuel burning in that year. [2.3 and 2.5]

Ammonia and those hydrocarbons used as halocarbon substitutes have atmospheric lifetimes ranging from days to months, and the direct and indirect radiative forcings associated with their use as substitutes are *very likely* to have a negligible effect on global climate.

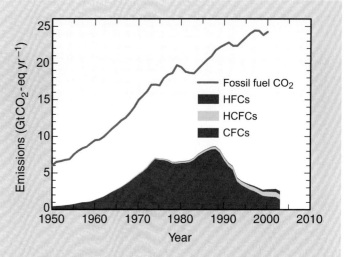

Figure TS-9. Direct GWP-weighted emissions (100-yr time horizon) for CFCs and HCFCs and HFCs compared with total CO_2 emissions due to fossil fuel combustion and cement production. [Figure 2.11]

2.9 What is the global warming potential and how is it used?

The GWP is a measure of the future radiative effect of an emission of a substance relative to the emission of the same amount of CO_2 integrated over a chosen time horizon. The GWP value for a species is proportional to its radiative efficiency and increases with its lifetime in the atmosphere. GWPs are most useful as relative measures of the climate response due to direct radiative forcing of well-mixed greenhouse gases whose atmospheric lifetimes are controlled by similar processes, which includes most of the halocarbons. [2.5 and Box 2.4]

The choice of time horizon is a policy consideration and makes a compromise between the relative weighting of short-term and long-term effects. Current practice most often uses GWPs calculated with a 100-year time horizon. This does not take into account the full effect of very long-lived gases with high radiative efficiency, such as PFCs, that persist in the atmosphere for much longer than 100 years. Similarly, integrating over 100 years reduces the contribution of short-lived species that last for only part of that period, relative to that of CO_2 which continues to contribute to radiative forcing throughout the 100-year time horizon and beyond. [2.5]

Table TS-3. GWPs of halocarbons commonly reported under the Montreal Protocol and the UNFCCC and its Kyoto Protocol and assessed in this report relative to CO_2, for a 100-year time horizon, together with their lifetimes and GWPs used for reporting under the UNFCCC. Gases shown in blue (darker shading) are covered under the Montreal Protocol and gases shown in yellow (lighter shading) are covered under the UNFCCC. [Tables 2.6 and 2.7]

Gas	GWP for direct radiative forcing[a]			GWP for indirect radiative forcing (Emission in 2005[b])			Lifetime (years)	UNFCCC Reporting GWP[c]
CFCs								
CFC-12	10,720	±	3750	−1920	±	1630	100	n.a.[d]
CFC-114	9880	±	3460	Not available			300	n.a.[d]
CFC-115	7250	±	2540	Not available			1700	n.a.[d]
CFC-113	6030	±	2110	−2250	±	1890	85	n.a.[d]
CFC-11	4680	±	1640	−3420	±	2710	45	n.a.[d]
HCFCs								
HCFC-142b	2270	±	800	−337	±	237	17.9	n.a.[d]
HCFC-22	1780	±	620	−269	±	183	12	n.a.[d]
HCFC-141b	713	±	250	−631	±	424	9.3	n.a.[d]
HCFC-124	599	±	210	−114	±	76	5.8	n.a.[d]
HCFC-225cb	586	±	205	−148	±	98	5.8	n.a.[d]
HCFC-225ca	120	±	42	−91	±	60	1.9	n.a.[d]
HCFC-123	76	±	27	−82	±	55	1.3	n.a.[d]
HFCs								
HFC-23	14,310	±	5000	~0			270	11,700
HFC-143a	4400	±	1540	~0			52	3800
HFC-125	3450	±	1210	~0			29	2800
HFC-227ea	3140	±	1100	~0			34.2	2900
HFC-43-10mee	1610	±	560	~0			15.9	1300
HFC-134a	1410	±	490	~0			14	1300
HFC-245fa	1020	±	360	~0			7.6	−[e]
HFC-365mfc	782	±	270	~0			8.6	−[e]
HFC-32	670	±	240	~0			4.9	650
HFC-152a	122	±	43	~0			1.4	140
PFCs								
C_2F_6	12,010	±	4200	~0			10,000	9200
C_6F_{14}	9140	±	3200	~0			3200	7400
CF_4	5820	±	2040	~0			50,000	6500
Halons								
Halon-1301	7030	±	2460	−32,900	±	27,100	65	n.a.[d]
Halon-1211	1860	±	650	−28,200	±	19,600	16	n.a.[d]
Halon-2402	1620	±	570	−43,100	±	30,800	20	n.a.[d]
Other Halocarbons								
Carbon tetrachloride (CCl_4)	1380	±	480	−3330	±	2460	26	n.a.[d]
Methyl chloroform (CH_3CCl_3)	144	±	50	−610	±	407	5.0	n.a.[d]
Methyl bromide (CH_3Br)	5	±	2	−1610	±	1070	0.7	n.a.[d]

[a] Uncertainties in GWPs for direct positive radiative forcing are taken to be ±35% (2 standard deviations) (IPCC, 2001).

[b] Uncertainties in GWPs for indirect negative radiative forcing consider estimated uncertainty in the time of recovery of the ozone layer as well as uncertainty in the negative radiative forcing due to ozone depletion.

[c] The UNFCCC reporting guidelines use GWP values from the IPCC Second Assessment Report (see FCCC/SBSTA/2004/8, http://unfccc.int/resource/docs/2004/sbsta/08.pdf).

[d] ODSs are not covered under the UNFCCC.

[e] The IPCC Second Assessment Report does not contain GWP values for HFC-245fa and HFC-365mfc. However, the UNFCCC reporting guidelines contain provisions relating to the reporting of emissions from all greenhouse gases for which IPCC-assessed GWP values exist.

Direct GWP values are given in Table TS-3. These were changed in the IPCC Third Assessment Report (IPCC, 2001) relative to the previous assessment (IPCC, 1996)[13] due to revisions to the radiative efficiency of CO_2 and to some lifetimes and radiative efficiencies for other species. Revisions to GWP values given for some species in this report are due mainly to the use of updated lifetimes as described in section 2.5.4. [2.5]

The indirect GWPs associated with ozone depletion caused by the different ODSs are given in Table TS-3. As ODS indirect cooling effects are projected to cease upon ozone layer recovery, their duration depends not only on gas lifetime but also on the time of ozone recovery. Estimates of indirect GWPs incorporate this latter dependence by setting all indirect effects to zero after the time at which equivalent effective stratospheric chlorine (EESC) is estimated to return to its pre-1980 values. Indirect GWPs therefore depend on the year of emission and have large uncertainties arising from: uncertainty in the radiative forcing caused by ozone depletion; uncertainties in the amount of EESC attributed to each species; and uncertainty in the time at which EESC returns to its pre-1980 values. [1.5 and 2.5]

Given the very different levels of scientific understanding and relative uncertainties associated with direct and indirect radiative forcing of ODSs, the lack of cancellation in their effects on surface climate and the dependence of indirect GWPs on the year of emission, this report does not consider the use of net GWPs combining direct and indirect effects. Where direct GWPs are used with ODS emissions, or to construct CO_2-equivalent values, it should be recognized that there are also indirect effects that may be significant over the next several decades. [1.2, 1.5, Box 1.4 and 2.5]

2.10 Are HCFCs, HFCs or their replacements expected to have other effects on future environmental chemistry?

The emissions of organic gases (including HCFCs, HFCs, PFCs and hydrocarbons) and ammonia due to the replacement of ODSs in refrigeration and air conditioning are not expected to have significant large-scale impacts on air quality. The local impact of hydrocarbon and ammonia substitutes can be estimated by comparing the anticipated emission to local emissions from all sources. Small but not negligible impacts could occur near localized emission sources and such increases may be of concern, for instance, in areas that currently fail to meet local standards. [2.4 and 2.6]

Persistent degradation products of HFCs and HCFCs (such as trifluoroacetic acid, TFA) are removed from the atmosphere via deposition and washout processes and TFA is toxic to some aquatic life at concentrations approaching 1 mg L^{-1}. However, degradation of identified sources cannot account for the observed TFA abundances in the oceans, surface waters and atmosphere, indicating that there are larger natural sources of TFA. Current observations show typical concentrations in the ocean of about 0.2 µg L^{-1}, while concentrations as high as 40 µg L^{-1} have been observed in the Dead Sea and Nevada lakes, suggesting a linkage to salt chemistry. Calculations based on projected HCFC and HFC emissions suggest that the concentration of TFA in rain-water due to their degradation is expected to be between 0.1 µg L^{-1} and 0.5 µg L^{-1} in the year 2010. Thus increases to toxic levels of 1 mg L^{-1} in specific ecosystems resulting from use of halocarbons are not supported by current studies. [2.4]

---

[13] IPCC, 1996: Climate Change 1995: The Science of Climate Change. Contribution of Working Group I to the Second Assessment Report of the Intergovernmental Panel on Climate Change [Houghton, J. T., L. G. Meira Filho, B. A. Callander, N. Harris, A. Kattenberg, and K. Maskell (eds.)]. Cambridge University Press, Cambridge, United Kingdom, and New York, NY, USA, 572 pp.

3. Options for ODS phase-out and reducing GHG emissions

3.1 Where do GHG emissions occur that are related to the use of ODSs and their substitutes?

Ozone-depleting substances (ODSs) and their substitutes are being used in a wide range of products and processes. Many of these substances (or byproducts released during manufacture) are greenhouse gases (GHGs), the emissions of which will result in a contribution to the direct positive forcing of the climate. Direct emissions of GHGs may occur during the manufacture of these substances, during their use in products and processes and at the end of the life of these products (see Figure TS-1). Banks are the total amount of substances contained in existing equipment, chemical stockpiles, foams and other products, which are not yet released to the atmosphere.

The indirect GHG emissions of applications of ODSs and their replacements are the GHG emissions related to the energy consumption (fuels and electricity) during the entire life cycle of the application[7]. This effect is different from the indirect negative radiative forcing of ODSs discussed in preceding sections.

The UNFCCC addresses anthropogenic emissions by sources and the removals by sinks of all greenhouse gases not controlled by the Montreal Protocol, and its Kyoto protocol regulates emissions of carbon dioxide (CO_2), methane (CH_4), nitrous oxide (N_2O), HFCs, PFCs and sulphur hexafluoride (SF_6). The Montreal Protocol, on the other hand, controls not the emission but rather the production and consumption of ODSs. Thus, the emissions due to releases of CFCs and HCFCs present in banks (e.g. refrigeration equipment, foams) are not covered by either the Montreal Protocol or Climate Convention and Kyoto Protocol. These emissions could make a significant future contribution to global warming.

3.2 How are estimated banks and emissions projected to develop during the period 2002 to 2015?

Current banks and emissions

Current emission profiles of ODSs and their substitutes are largely determined by historical use patterns, resulting in a relatively high contribution (at the present time and in the coming decades) from the CFCs and HCFCs banked in equipment and foams. Annual emissions of CFCs, HCFCs, HFCs and PFCs[14] in 2002 were about 2.5 $GtCO_2$-eq yr^{-1} (see Table TS-4[15,16]). Refrigeration applications together with stationary air conditioning (SAC) and mobile air conditioning (MAC) contribute the bulk of global direct GHG emissions. About 80% of the 2002 emissions are CFCs and HCFCs.

The banks stored in equipment and foams may leak during the use phase of the products they are part of and at the end of the product life cycle (if they are not recovered or destructed). The bank-turnover varies significantly from application to application: from months (e.g. solvents) to several years (refrigeration applications) to over half a century (foam insulation).

Banks of CFCs, HCFCs, HFCs and PFCs were estimated at about 21 $GtCO_2$-eq (2002). CFCs, HCFCs and HFCs contribute about 16, 4 and 1 $GtCO_2$-eq, respectively (see Table TS-5), while banks of PFCs used as ODS replacements contribute only about 0.005 GtCO2-eq. The build-up of the banks of (relatively) new applications of HFCs will significantly determine future (>2015) emissions without additional bank management measures.

---

[14] This concerns only emissions of HFCs and PFCs that result from their use as ODS-substitutes. Total emissions of HFCs and notably PFCs are higher because of emissions from other applications that are not within the scope of this report (e.g. emissions from aluminum production and the semiconductor industry).
[15] Greenhouse gas (GHG) emissions and banks expressed in terms of CO_2-equivalents use GWPs for direct radiative forcing for a 100-year time horizon. Unless stated otherwise, the most recent scientific values for the GWPs are used, as assessed in this report and as presented in Table TS-3.
[16] Halons cause much larger negative indirect radiative forcing than positive direct radiative forcing and, in the interest of clarity, their effects are not included in estimates of total emissions and banks expressed in $MtCO_2$-equivalents.

Table TS-4. Greenhouse gas CO_2-equivalent (GWP-weighted) annual emissions of halons, CFCs, HCFCs and HFCs and of PFCs that are used as ODS substitutes: Breakdown per group of GHGs and per emission sector. Historical data for 2002, and business-as-usual (BAU) projections for 2015 emissions and emissions under a mitigation scenario (MIT) in 2015. The reduction potential is the difference between 2015 BAU and mitigation projections.

Note: Direct GWPs for a 100-year time horizon were used from IPCC (2001) and WMO (2003) (as listed in Table TS-3) 'Total' may not add up, due to rounding.

2002	Annual emissions ($MtCO_2$-eq yr^{-1})								
	Refrige-ration[a]	SAC[b]	MAC[c]	Foams	Medical aerosols	Fire protection	HFC-23 byproduct	Other[d]	Total
Halons[e]	-	-	-	-	-	[47][e]	-	-	[47][e]
CFCs	726	99	641	117	69	0	-	0	1651
HCFCs	232	164	15	32	-	0.1	-	6	447
HFCs	102	9	93	3	6	1	195	25	434
PFCs	0	0	0	0	-	0.1	-	1	1
Total[e]	1060	271	749	152	75	1	195	32	2534

2015 BAU scenario	Annual emissions ($MtCO_2$-eq yr^{-1})								
	Refrige-ration	SAC	MAC	Foams	Medical aerosols	Fire protection	HFC-23 byproduct	Other	Total
Halons	-	-	-	-	-	[12][e]	-	-	[12][e]
CFCs	136	50	49	85	17	0	-	0	338
HCFCs	570	210	19	20	-	0.1	-	9	828
HFCs	391	109	247	18	23	4	332	27	1153
PFCs	0	0	0	0	-	0.1	-	0.1	0.2
Total[e]	1097	370	315	124	40	5	332	37	2319

2015 Mitigation scenario	Annual emissions ($MtCO_2$-eq yr^{-1})								
	Refrige-ration	SAC	MAC	Foams	Medical aerosols	Fire protection	HFC-23 byproduct	Other	Total
Halons	-	-	-	-	-	[12][e]	-	-	[12][e]
CFCs	84	24	32	81	0	0	-	0	221
HCFCs	359	86	12	17	-	0.1	-	9	484
HFCs	164	60	92	9	26	4	33	27	416
PFCs	0	0	0	0	-	0.1	-	0.1	0.2
Total[e]	607	170	136	107	26	5	33	37	1121

2015 Reduction potential	Emissions reduction ($MtCO_2$-eq yr^{-1})								
	Refrige-ration	SAC	MAC	Foams	Medical aerosols	Fire protection	HFC-23 byproduct	Other	Total
Halons	-	-	-	-	-	n.q.	-	-	-
CFCs	53	26	17	4	17	-	-	-	117
HCFCs	210	124	7	3	-	n.q.	-	n.q.	344
HFCs	227	49	155	10	-3	n.q.	299	n.q.	737
PFCs	-	-	-	-	-	-	-	-	0
Total[e]	490	200	179	17	14	n.q.	299	n.q.	1198

Notes:

n.q. Not quantified

[a] 'Refrigeration' comprises domestic, commercial, industrial (including food processing and cold storage) and transportation refrigeration.

[b] 'SAC' (stationary air conditioning) comprises residential and commercial air conditioning and heating.

[c] 'MAC'(mobile air conditioning) applies to cars, buses and passenger compartments of trucks.

[d] 'Other' includes non-medical aerosols and solvents

[e] Halons cause much larger negative indirect than positive direct radiative forcing and, in the interest of clarity, their effects are not included in the totals and are shown in brackets in the table.

Table TS-5. Greenhouse gas CO_2-equivalent (GWP-weighted) banks of halons, CFCs, HCFCs and HFCs and of PFCs that are used as ODS substitutes: Breakdown per group of GHGs and per emission sector. Historical data for 2002 and BAU and MIT projections for 2015. Note: Direct GWPs for a 100-year time horizon were used from IPCC (2001) and WMO (2003) (as listed in Table TS-3). 'Total' may not add up, due to rounding.

2002	Banks ($MtCO_2$-eq yr^{-1})							
	Refrigeration[a]	SAC[b]	MAC[c]	Foams	Medical aerosols[f]	Fire protection	Other[d,f]	Total
Halons[e]	-	-	-	-	-	[531][e]	-	[531][e]
CFCs	3423	631	1600	10,026	69	0	0	15,749
HCFCs	810	1755	36	1229	-	5	6	3841
HFCs	518	123	350	16	6	65	25	1103
PFCs	0	0	0	0	-	4	1	5
Total[e]	**4751**	**2509**	**1987**	**11,270**	**75**	**74**	**32**	**20,698**

2015 BAU scenario	Banks ($MtCO_2$-eq yr^{-1})							
	Refrigeration[a]	SAC[b]	MAC[c]	Foams	Medical aerosols[f]	Fire protection	Other[d,f]	Total
Halons	-	-	-	-	-	[206][e]	-	[206][e]
CFCs	653	208	138	7286	17	0	0	8302
HCFCs	1582	1536	42	1696		6	9	4871
HFCs	1922	1488	896	644	23	226	27	5227
PFCs				0		4	0.1	4
Total[e]	**4157**	**3232**	**1076**	**9626**	**40**	**236**	**37**	**18,404**

2015 Mitigation	Banks ($MtCO_2$-eq yr^{-1})							
	Refrigeration[a]	SAC[b]	MAC[c]	Foams	Medical aerosols[f]	Fire protection	Other[d,f]	Total
Halons	-	-	-	-	-	[206][e]	-	[206][e]
CFCs	627	208	138	7286	0	0	0	8258
HCFCs	1466	1134	41	1696		6	9	4352
HFCs	1455	1586	712	494	26	226	27	4527
PFCs				0		4	0.1	4
Total[e]	**3548**	**2928**	**891**	**9475**	**26**	**236**	**37**	**17,141**

Notes

[1] 'Refrigeration' comprises domestic, commercial, industrial (including food processing and cold storage) and transportation refrigeration.

[b] 'SAC' (stationary air conditioning) comprises residential and commercial air conditioning and heating.

[c] 'MAC' (mobile air conditioning) applies to cars, buses and passenger compartments of trucks.

[d] 'Other' includes non-medical aerosols and solvents.

[e] Halons cause much larger negative indirect than positive direct radiative forcing and, in the interest of clarity, their effects are not included in the totals and are shown in brackets in the table.

[f] Emissive use applications are assumed to have banks that are equal to annual emissions.

2015 Business-as-usual projections

The sector chapters have developed business-as-usual (BAU) projections for the use and emissions of CFCs, HCFCs, halons, HFCs and some PFCs (where these are used as replacements for ODSs). These projections have assumed that all existing measures will continue, including the Montreal Protocol (phase-out) and relevant national regulations.

The usual practices and emission rates are kept unchanged up to 2015. End-of-life recovery efficiency is assumed not to increase. An overview of key assumptions for the BAU projections of 2015 is given in Table TS-6.

Table TS-6. Key assumptions in the business-as-usual (BAU) and mitigation (MIT) scenarios.

| Sector | Annual market growth 2002-2015 (both in BAU and MIT)[1] (% yr⁻¹) | | | | Best-practice assumptions | | | | | | | | |
Refrigeration SAC and MAC	EU % yr⁻¹	USA % yr⁻¹	Japan % yr⁻¹	DCs[1] % yr⁻¹	Type of reduction	EU BAU	EU MIT	USA BAU	USA MIT	Japan BAU	Japan MIT	DCs[1] BAU	DCs[1] MIT
Domestic refrigeration	1	2.2	1.6	2–4.8	Substance	HFC-134a / HC-600a	HC-600a	HFC-134a	HFC-134a / HC-600a (50%)	HFC-134a	HC-600a	CFC-12 / HFC-134a	Plus HC-600a (50% in 2010)
					Recovery	0%	80%	0%	80%	0%	80%	0%	50%
Commercial refrigeration	1.8	2.7	1.8	2.6–5.2	Substance	R-404A / R 410A	R-404A / R-404A (50%)	HCFC-22 / R-404A /	R-404A / R 410A (50%)	HCFC / R-404A	R-404A R-410A (50%)	CFC / HCFC	R-404A / R 410A (50%)
					Recovery	50%	90%	50%	90%	50%	90%	25%	30%
					Charge		–30%		–30%		–30%		–10%
Industrial refrigeration	1	1	1	3.6–4	Substance	HFC-NH₃ (35%)	HFC-NH₃ (70%)	HCFC / HFC-NH₃ (60%)	HCFC / HFC-NH₃ (80%)	HCFC / HFC-NH₃ (35%)	HCFC / HFC-NH₃ (70%)	CFC / HCFC-22	NH₃ (40–70%)
					Recovery	50%	90%	50%	90%	50%	90%	15–25%	50%
					Charge		–40%		–40%		–40%		–10%
Transport refrigeration	2	3	1	3.3–5.2	Substance	HFCs	HFCs	HCFCs / HFCs	HCFCs / HFCs	HCFCs / HFCs	HCFCs / HFCs	CFC / HCFC-22	Plus HFCs, up tp 30% 20–30%
					Recovery	50%	80%	50%	70%	50%	70%	0%	50%
Stationary AC	3.8	3	1	5.4–6	Substance	HFCs	HFCs	HCFCs / HFCs	HCFCs / HFCs	HCFCs / HFCs	HCFCs / HFCs	CFC / HCFC-22	CFC / HCFC-22 (HFCs 30% in some DCs)
					Recovery	50%	80%	30%	80%	30%	80%	0%	50%
					Charge				–20%		–20%		
Mobile AC	4	4	1	6–8	Substance	HFC-134a / CO₂ (10%) as of 2008	HFC-134a / CO₂ (50%) as of 2008	HFC-134a	HFC-134a / CO₂ (30%) as of 2008	HFC-134a	HFC-134a / CO₂ (30%) as of 2008	CFC / HCFC-134a	HFC-134a
					Recovery	50%	80%	0%	70%	0%	70%	0%	50%
					Charge	700 g	500 g	900 g	700 g	750 g	500 g	750–900 g	750–900 g

Table TS-6. (continued)

Sector	Annual market growth 2002-2015 (both in BAU and MIT)[1] (% yr⁻¹)		Best-practice assumptions
Foams	About 2% yr⁻¹	BAU	Assumptions on substance use (see Chapter 7)
		MIT	HFC consumption reduction: A linear decrease in use of HFCS between 2010 and 2015 leading to 50% reduction by 2015. Production / installation improvements: The adoption of production emission reduction strategies from 2005 for all block foams and from 2008 in other foam sub-sectors. End-of-life management options: The extension of existing end-of-life measures to all appliances and steel-faced panels by 2010 together with a 20% recovery rate from other building-based foams from 2010.
Medical aerosols	1.5–3% yr⁻¹	BAU	Partial phase-out of CFCs
		MIT	Complete phase-out of CFCs
Fire protection	–4.5% yr⁻¹ (all substances) +0.4% yr⁻¹ (HCFCs/HFCs/PFCs)	BAU	Phase-out of halons
		MIT	Not quantifiable
HFC-23 byproduct	2.5% yr⁻¹	BAU	HFC-23 emissions of existing production capacity: 2% of HCFC-22 production (in kt) HFC-23 emissions of new production capacity: 4% of HCFC-22 production (in kt)
		MIT	100% implementation of reduction options (90% emission reduction)
Non-medical aerosols	16% increase period in total CO-weighted emissions over 2002–2015	BAU	See Chapter 10
		MIT	Not quantifiable

[1] BAU: Business-As-Usual Scenario; MIT: Mitigation Scenario; DCs: developing countries

The activities underlying the emissions of fluorocarbons are expected to grow significantly between 2002 and 2015. These activities and services (such as refrigeration, air conditioning and insulation) will be provided by a number of technologies and substances, including CFCs and HCFCs. In industrialized countries, the use and emissions of CFCs and HCFCs will decline following the Montreal Protocol phase-out requirement as obsolete equipment is retired. In developing countries, the production of HCFCs can continue until 2040, and significant increase in their production is expected. These changes, and their impacts, are reflected in the data in Table TS-4. [11.6]

The decline in CFC emissions is not accompanied by a similar increase in emissions of HFCs because of continuing trends towards non-HFC technology and substitutes with lower GWPs. In addition, but not included in the BAU scenario, the capture and safe disposal of substances that were emitted in the past are likely to increase with respect to HFCs since these substances are controlled under the Kyoto Protocol. The BAU case assumes the continuing application of all existing measures, and the mitigation scenario embodies improvements that could be implemented assuming global application of current best-practice emission reduction techniques.

In the BAU scenario, banks are projected to decline to 18 $GtCO_2$-eq in 2015. CFC banks associated with refrigeration, SAC[17] and MAC equipment are expected to fall from about 6 $GtCO_2$-eq in 2002 to 1 $GtCO_2$-eq by 2015, mainly due to release to the atmosphere and partly due to end-of-life recovery and destruction. CFC banks in foams remain significant (decreasing from 10 to 7 $GtCO_2$-eq over the same period). HCFC banks will increase from about 4 to 5 $GtCO_2$-eq, primarily due to the projected increase of HCFC-22 use in commercial refrigeration. Total HFC banks will start to build up to 5 $GtCO_2$-eq in 2015. HFC banks in foams represent only 0.6 $GtCO_2$-eq and are projected to increase further after 2015. [11.4 and 11.6]

In the BAU scenario, total direct emissions of CFCs, HCFCs, HFCs and PFCs are projected to represent about 2.3 $GtCO_2$-eq yr^{-1} by 2015 (as compared to about 2.5 $GtCO_2$-eq yr^{-1} in 2002)[16]: Combined CFC and HCFC emissions are decreasing from 2.1 (2002) to 1.2 $GtCO_2$-eq yr^{-1} (2015), and emissions

of HFCs are increasing from 0.4 (2002) to 1.2 $GtCO_2$-eq yr^{-1} (2015)[18]. PFC emissions from ODS substitute use are about 0.001 $GtCO_2$-eq yr^{-1} (2002) and projected to decrease.

Table TS-4 shows the relative contribution of sectors to global direct GHG emissions that are related to the use of ODSs and their substitutes. Refrigeration applications together with SAC and MAC contribute the bulk (77% in 2015 BAU) of global direct GHG emissions, which is in line with the higher emission rates associated with refrigerant banks. The largest part of GHG emissions from foams are expected to occur after 2015 because most releases occur at end-of-life. HFC-23 byproduct emissions account for 14% of all direct GHG emissions (2015 BAU).

Due to the leakage of CFCs from banks to the atmosphere, emissions of CFCs will decrease from 1.7 (2002) to 0.3 $GtCO_2$-eq (2015). HCFC emissions are projected to increase from 0.4 (2002) to 0.8 $GtCO_2$-eq yr^{-1} (2015), owing to a steep increase in their use in (commercial) refrigeration and SAC applications. The projected threefold increase in HFC emissions is the result of the increased application of HFCs in the refrigeration, SAC and MAC sectors and due to byproduct emissions of HFC-23 from increased HCFC-22 production. HCFC-22 production is projected to increase by about 40% over the 2002–2015 period. [11.4 and 11.6]

The literature does not allow for an estimate of overall indirect GHG emissions related to energy consumption. For individual applications, the relevance of indirect GHG emissions over a life cycle can range from low to high, and for certain applications may be up to an order of magnitude larger than direct GHG emissions. This is highly dependent on the specific sector and product/application characteristics, the carbon-intensity of the consumed electricity and fuels during the complete life cycle of the application, containment during the use-phase and the end-of-life treatment of the banked substances. Table TS-7 presents examples of the ranges found in the literature with respect to the proportion of direct emissions to total GHG emissions for applications using HFCs. For applications using other substances, these proportions may differ significantly. The relatively old vintage stock of refrigeration equipment using CFCs may, in particular, provide a larger share of direct emissions. [3.2, 4 and 5]

---

[17] In this Technical Summary, the 'refrigeration' sector comprises domestic, commercial, industrial (including food processing and cold storage) and transportation refrigeration. [4] 'Stationary air conditioning (SAC)' comprises residential and commercial air conditioning and heating. [5] 'Mobile air conditioning (MAC)' applies to cars, buses and passenger compartments of trucks.

[18] For these emission values the most recent scientific values for GWPs were used (see Table TS-3). If the UNFCCC GWPs would be used, reported HFC emissions (expressed in tonnes of CO_2-eq) would be about 15% lower.

Table TS-7. Percentage contribution of direct emissions to total lifetime greenhouse gas emissions in various applications (emissions associated to functional unit) – selected indicative examples for applications using HFCs.

Application sector	Method applied	Percentage of HFC emissions of lifetime system greenhouse emissions (using GWP-100)	Characterization of system and key assumptions
MAC	TEWI	**40–60%** – Current systems (gasoline engine) **50–70%** – Current systems (diesel engine)	Passenger vehicle; HFC-134a Sevilla (Spain)
Commercial refrigeration	LCCP	**20–50%** – For a wide range of sensitivity tests on leakage rate, energy efficiency and energy supply	Direct expansion refrigeration unit; supermarket (1000 m²); R-404A; Germany
Domestic refrigeration	TEWI	**2–3%** – No recovery at end-of-life	European standard domestic refrigerator; HFC-134a; world average electricity mix
Insulation foam of domestic refrigerators	LCCP	**6%** – With 90% blowing agent recovered at disposal **17%** – With 50% blowing agent recovered at disposal	HFC-24fa; Europe
PU insulation foam in refrigerated truck	LCCP	**2%** – With full recovery of HFC at disposal **13%** – without recovery of HFC at disposal	Refrigerated diesel truck; Germany
PU spray foam industrial flat warm roof	LCA	**13%** – With full recovery of HFC at disposal **20%** – without recovery of HFC at disposal	4 cm thickness; HFC-365mfc; Germany
PU boardstock in private building cavity wall	LCA	**4%** – With full recovery of HFC at disposal **17%** – Without recovery of HFC at disposal	5 cm thickness; HFC-365mfc; Germany
PU boardstock in private building pitched warm roof	LCA	**10%** – With full recovery of HFC at disposal **33%** – Without recovery of HFC at disposal	10 cm thickness; HFC-365mfc; Germany

2015 mitigation-scenario projections

Mitigation options are identified and described for each sector in the respective sector chapters. In Section 4 of this Technical Summary, more detailed information on sectoral reduction opportunities is given. On a more aggregated level, overall sector emission reduction potentials are determined for 2015 relative to the BAU scenario. The estimates are based on a mitigation scenario, which assumes the global application of best practices in use, recovery and destruction of ODSs and ODS-substitutes. The scenario assumptions are presented in Table TS-6, and the sectoral GHG emissions in the mitigation scenario in 2015 are shown in Table TS-4.

Through global application of best practices and recovery methods, about 1.2 $GtCO_2$-eq yr^{-1} of direct GHG emissions can be reduced by 2015, as compared with the BAU scenario. About 60% of this potential is HFC emission reduction; HCFCs and CFCs contribute about 30% and 10%, respectively. Almost 75% of the reduction potential can be found in the refrigeration, SAC and MAC sectors, and about 25% can be found in the destruction of HFC-23 byproduct emissions from HCFC-22 production. This latter option represents about 40% of the HFC-reduction potential. [11.6]

Of the bank-related emissions that can be prevented during the period preceding 2015, the bulk is in refrigerant-based applications where BAU emission rates are considerably more significant than they are for foams during the same period. With earlier action, more of the CFC banks can be captured.

Most indirect energy-related GHG emissions occur during the use-phase of the applications and, in many cases, energy efficiency improvements can result in significant reductions in GHG emissions, particularly where the use-phase is long. Energy efficiency improvements can be profitable and reduce the net costs of the emission reduction options, although the reduction potential is again highly dependent on the specific circumstances. While the assessed literature did not allow for a global estimate of this reduction potential, several case studies at the technology and country level illustrate this point.

Uncertainties

Uncertainties in emission projections are significant. A comparison of atmospheric measurements with inventory calculations shows differences per group of substances (CFCs, HCFCs, HFCs and PFCs) in the order of 10–25%. For individual gases, the differences can be much bigger. These differences are caused by unidentified emissive applications of some substances (e.g. CFC-11, HCFC-141b, HCFC-142b) that are not accounted for in inventory calculations and by uncertainties in the geographically distributed data sets of equipment in use. [11.3.4]

3.3 Which options are available for reducing GHG emissions?

In general, technical options to reduce direct GHG emissions can be achieved through:

- improved containment of substances;
- reduced charge of substances in equipment and products;
- end-of-life recovery and recycling or destruction of substances;
- increased use of alternative substances with a reduced or negligible global-warming potential;
- not-in-kind technologies.

Reductions of indirect GHG emissions can be achieved by improving the energy efficiency of products and processes (and by reducing the specific GHG emissions of the energy system). To determine which technology option has the highest GHG emission reduction potential, both direct and indirect emissions have to be assessed. The comparison of technology options is not a straightforward exercise as even within one technological application significant variations in direct and indirect emissions may occur. Figure TS-10 shows the distribution of direct emissions from mobile air conditioners in a fleet of passenger cars and of indirect energy-related emissions from supermarket refrigerators. The graphs show that even within a single technology class considerable differences in potentials usually exist to reduce direct and/or indirect emissions. The proper monitoring,

benchmarking and understanding of system performance will be a crucial first step in facilitating short- and mid-term emission reductions. However, to achieve major emission reductions in the long run, choices have to be made between the different established technology options and, consequently, there is a need to be prepared and informed by standardized environmental assessment tools.

3.4 Which assessment tools can provide information on technology choices?

The protection of the stratospheric ozone layer and climate requires the selection of technologies which differ with respect to their impacts on climate, on health, safety and other environmental endpoints and on their private and social costs. Analyses of these various impacts can assist decision-makers to choose among competing replacement technologies. However, the results of such analyses can vary, depending upon which of the many factors that are not intrinsic to the technologies are taken into consideration; these include the analytical approach (e.g. top-down compared with bottom-up), degree of product or process optimization, service and disposal practices, regional circumstances and a wealth of other inputs and assumptions. To make intelligent choices, decision-makers must therefore be aware of the sensitivities, uncertainties and limitations inherent in each type of analysis and must also be able to evaluate whether the approach and assumptions used in an analysis are reasonable for the regions

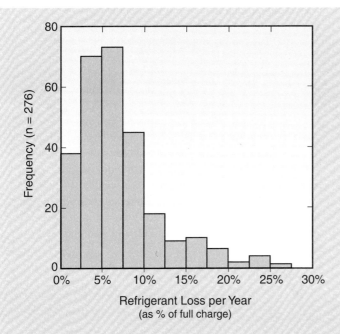
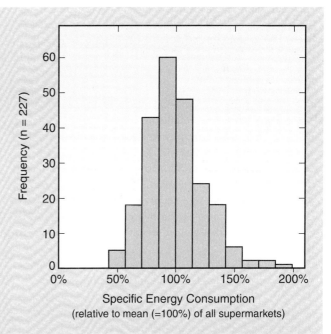

Figure TS-10. Variance of direct and indirect emissions within equipment populations. Annual leakage rates of mobile air conditioning systems in a fleet of passenger vehicles (left panel, n=276). Specific energy consumptions expressed as percentage of the mean for a group of standard layout supermarket refrigeration units (right panel, n=227).

and time periods in which the competing technologies would be applied.

To face such challenges, well-established and clearly described methodologies are needed. This report provides an overview of the different types of analyses and provides concise guidance on how to evaluate and apply them. For each type of analysis, the most important analytical approaches and variables are discussed, along with their sensitivities, uncertainties and limitations.

Any assessment of technical options to reduce the impacts of emissions involves the evaluation of multiple factors. These include the influence of direct and indirect emissions on radiative forcing, costs, health, safety and environmental impacts for each application and compound; service or design practices that can lead to reduced leakage; the effects of recovery and recycling. These factors may be subject to large uncertainties and differ from application to application and region to region. Non-technical factors also need to be considered, such as differing regulatory and management environments in different parts of the world, the availability of infrastructure, investment financing and historical factors. In many cases, only incomplete information on these factors is available, thereby limiting the comprehensiveness of the assessment.

This report deals with direct emissions of halocarbons that are associated with their production, use and decommissioning as well as indirect emissions due to energy requirements. Such emissions are quantified and costs associated with their reduction are evaluated according to methodologies presented herein. For a comprehensive assessment, not only private costs need to be quantified: also external costs that are not paid by the private sector but which are a burden to society should be accounted for.

The purpose of the methodologies chapter in this report (Chapter 3) is to describe procedures for quantifying both ODS and GHG emissions and the costs associated with reducing these emissions. A broad spectrum of assessment tools exists, ranging from tools with very well-established procedures applicable at global level to those loosely defined and not applied in a consistent manner. Table TS-8 provides an overview of methodologies identified as being relevant to this report.

In assessing environmental and climate change impacts, decision-makers prefer to have a comprehensive picture of all relevant environmental aspects. However, information other than that on direct and indirect GHG emissions is often difficult to quantify. Consequently, Total Equivalent Warming Impact (TEWI, a measure of GHG emissions during the use-phase and disposal) and Life Cycle Climate

Performance (LCCP, which also includes the direct GHG emissions during manufacture and the indirect GHG emissions that are associated with the embodied energy of the substances of concern) have more practical value than more encompassing methodologies like Life Cycle Assessment (LCA), Environmental Burden and others. It is worth noting that there is no scientifically established basis for reducing multiple impact results (like LCA) to a single overall score or number.

In the past, little attention was paid to ensure the comparability of results from different technology assessments. There is a wide range of available results on GHG performance from different assessments – often not comparable. The treatment of uncertainties is often incomplete, and resulting recommendations are not sufficiently robust to be compared across sectors. In the light of the many assumptions and different methodologies, an important role has been identified for technology comparisons under agreed conditions using a common set of methods and assumptions. The development of simple and pragmatic standard methodologies and their respective quality criteria is recommended. Future work will need to bridge the gap between the application of specific comparisons and sufficiently robust results that can be used for policy design in entire subsectors.

Analyses of sectoral and global emissions and emission reduction potentials are based on extensive databases on equipment populations and other product distributions that comprise field data on substance emissions and energy consumption. These databases should ideally be consistent and compatible with national GHG emission inventories. Information on fluid sales to the different parties involved in the subsector will need to be made available. The improvement of these fairly comprehensive data sets for analysis in support of robust sectoral policies requires significant resources and results in a number of confidentiality issues which need to be addressed cautiously. To achieve acceptability across subsectors with respect to future developments, decision-makers could consider paying special attention to increasing the involvement of relevant stakeholders and to introducing additional measures in order to increase transparency for outside users through more extensive documentation of methods and assumptions.

Table TS-8. Overview of assessment methodologies reviewed in this report.

Methodology to assess	Overview
Direct emissions	
Production	1) Identification of all feedstock requirements and chemical processing stages necessary to transform feedstock to intermediaries and final product. 2) Accounting for all emissions occurring at each stage through measurement and/or modelling.
Use	1) Measurements to estimate losses of fluids during lifetime of equipment, which yield direct emissions. 2) Transportation and distribution losses are included as direct use emissions.
Decommissioning	1) Emissions are accounted for based on the final destination of the products. 2) If no recovery, all remaining fluids are assumed to be direct emissions. 3) If recovered, emissions may also occur during fluid reprocessing.
Indirect emissions	
Production	1) Modelling and/or measurement of built-in energy in equipment used for feedstock sourcing, preprocessing, transport and transformation to final product in the plants. 2) Modelling and/or measurents of total energy consumption to produce one unit of a particular fluid from feedstocks. 3) Relating energy consumption to emission of GHGs through region- or country-specific data.

Use	1) Energy consumption evaluation during equipment lifetime.
	   a) Refrigeration, air conditioning and heating applications: modelling and/or measuring equipment energy consumption.
	   b) Insulating foams:
	      b1) Thickness compensation modelling and/or
	      b2) measuring energy consumption for baseline and for foam insulated application.
	2) Relating energy consumption to emission of GHGs through region- or country-specific data.

Decommissioning	1) Energy consumption required for product recycling or destruction at end-of-life. 2) Relating energy consumption to emission of GHGs through region- or country-specific data.
Costs	
Private costs	Costs are calculated based on expertise of private companies. These mainly include capital costs, labour, land, materials, maintenance and administrative costs.
Social costs	Costs usually charged to society, such as air and water pollution, associated with production of goods by private sectors. Costs are estimated through the quantification of benefits and damages induced by final product.
Discount rates	Use this to consider different time distribution of costs incurred. Present value or levelized cost is evaluated taking into consideration market or social discount rates for private or social costs, respectively.
Sustainable development	Quantification, or at least qualification, of positive and negative impacts caused on a society's well-being by changes in production and use of fluids.
Health and safety	
Health and safety	1) Focus on risk assessment of chemicals and then minimising the negative health and safety impacts through risk management of systems. 2) For relevant substances use the references to existing databases and data sources.
Environmental impacts	
TEWI (Total Equivalent Warming Impact) – Accounts for GHGs from direct emissions of operating fluids together with the energy-related CO_2	$TEWI = \sum(a_i * b_i + a_i * c_i) + d$ $a_i =$ GWP of gas i emitted; $b_i =$ mass of gas i released during the operating lifetime of the system; $c_i =$ mass of gas i released when the system is decommissioned at end-of-life; $d =$ emission of CO_2 resulting from the energy used to operate the system (for its entire lifetime).
LCCP (Life Cycle Climate Performance) – Accounts for TEWI factors plus the fugitive emissions arising during manufacture (of the operating fluids only) and the CO_2 associated with their embodied energy.	$LCCP = TEWI$ (calculated value as above described) $+ \sum x_i * y_i + z$ $x_i =$ GWP of gas i emitted during the manufacture of fluid; $y_i =$ mass of gas i released during the manufacture of fluid; $z =$ embodied energy of all material used for the manufacture of fluid (the specific energy used for the manufacture of one unit mass of each material multiplied by the total mass emitted) expressed in CO_2-eq.
LCA (Life Cycle Assessment) – Describes the environmental impacts of product systems from raw material acquisition to final disposal	1) Describe the system in terms of its unit processes and their interrelationships. 2) Compile an inventory of relevant inputs and outputs of each unit process and of the systems that are involved in those inputs and outputs (Life Cycle Inventory Analysis). 3) Evaluate potential environmental impacts of these inputs and outputs. The most obvious environmental impact categories are climate change and ozone depletion, but some or all of the environmental categories may be important.

3.5   What are regional differences?

Difference in economic development is a very important factor explaining regional differences in assessments. Key assumptions on technical performance indicators, such as equipment lifetime and operational cost of equipment and products, vary greatly from developed to developing countries in ways that seem to be related with the cost of capital.

In developing countries, traditional private costs are usually accounted for, while other hidden private costs (e.g. for Research and Development (R&D), training, environmental liability) are seldom considered. External costs are generally poorly accounted for because there are usually no regulations in place to deal with externality issues and the awareness amongst the population is low. On-site health and safety regulations are usually taken into consideration due to the existence and enforcement of labour laws in most developing countries.

Between developed and developing countries significant variations can also be found with respect to the uncertainty range of their emission estimates (both direct and indirect emissions), which in turn are used as input for further analysis. In the case of direct emissions, some progress has been made in the diffusion of emission inventory methodologies under the Montreal Protocol Multilateral Fund investments in developing countries. However, for both developed and developing countries, uncertainties are generally significant. Improving the quantification of indirect GHG emissions, which are significant relative to total GHG emissions, remains a challenge for all countries. Difficulties centre around such issues as data availability on energy consumption, determination of the carbon intensity of the energy consumed and the GHG emission estimates related to the embodied energy in production inputs. These challenges point to the necessity for a concerted global effort if decision-makers are to be provided with the information needed to make decisions supportive of global ozone layer and climate policies.

The above challenges are compounded by the fact that each technology that results in direct or indirect emissions has unique data requirements for determining their climate and ozone impacts. This situation raises issues of capacity, standards, policy and regulation, for which developed countries have established better, though not comprehensive, frameworks within which to respond. This emphasizes the need to develop simple, standard methodologies and the respective quality criteria, as recommended in the report. [3.6]

3.6   What major opportunities have been identified for reducing GHG emissions associated with the use of ODSs, their substitutes and related indirect emissions?

The major opportunities for reducing direct emissions in terms per sector and per substance group are quantified in Table TS-4. Table TS-9 summarizes the sector characteristics and specific mitigation opportunities. In Section 4 of this Technical Summary, the opportunities for reducing GHG emissions are discussed in more detail.

Table TS-9. Overview of sector- and application-specific findings.

Sector	Description and status of sector	Emission status and BAU trends and opportunities for emissions reduction
Domestic refrigeration	• Manufacturing transitions from CFC-12 are complete in developed countries and in progress in developing countries	Emission status and BAU trends • Refrigerant leakage rates derived with a bottom-up approach suggest a global annual emissions rate of 6% of banked system charge, caused by the significant bank of CFCs in old equipment. New non-CFC systems have typical leakage rates of about 1% yr⁻¹. • The projected emissions in this subsector by 2015 are about 65 MtCO$_2$-eq yr⁻¹ in the BAU-scenario. Opportunities for emissions reduction • HFC-134a and isobutane (HC-600a) are the primary alternative refrigerants for the previously used CFC-12. Each has demonstrated a mass production capability for safe, efficient, reliable and economic use. The choice of HFC-134a or HC-600a varies per region and is strongly influenced by the regulatory environment and liability. • For both refrigerants, indirect emissions dominate total emissions, almost without taking the carbon intensity of electric power generation into consideration. At equal energy efficiencies, HC-600a domestic refrigerators have a better LCCP, with or without end-of-life recovery. The difference with HFC-134a is small, and end-of-life recovery can further reduce the magnitude of the difference. • State-of-the-art refrigeration products are at least 50% more energy-efficient than the 20-year-old units they typically replace.
Commercial refrigeration	• Commercial refrigeration comprises three main types of equipment: stand-alone equipment, condensing units and full supermarket systems. • The most used refrigerants in this sector are HCFC-22, R-404A and HFC-134a.	Emissions status and BAU trends • Refrigerant leakage rates derived with a bottom-up approach suggest a global annual emissions rate of 30% of banked system charge, meaning that the refrigerant emissions typically represent 60% of the total emissions of GHGs resulting from the system operation. • Refrigerant leakage rate data from over 1700 full supermarket systems in USA and Europe were in the range of 3–22%, with an average of 18%. • In 2002, all refrigerant types banked in commercial refrigeration equipment – predominantly CFCs, HCFCs and HFCs – represented 606 ktonnes of a total of 2691 ktonnes for all refrigerating and air conditioning (AC) systems and all refrigerant types, which is 22.5% of the total refrigeration and AC bank. • The projected emissions by 2015 in this subsector are about 902 MtCO$_2$-eq yr⁻¹ in the BAU-scenario. Opportunities for emissions reduction • Significant total emission reductions, that is improved LCCP, can be achieved by using refrigerants HFCs, hydrocarbons, ammonia or CO$_2$, employing charge reduction, more efficient containment and overall improved energy efficiency through new system designs. • In full supermarket systems, LCCP values up to 60% lower than centralized direct systems of traditional design can be obtained by applying direct systems using alternative refrigerants, better containment, distributed systems, indirect systems or cascade systems. Published results show alternatives to have 0–35% higher initial cost and an energy usage 0–20% higher than that of current systems. • Refrigerant emissions abatement cost was found to be in the range of 20–280 US$ per tCO$_2$-eq (10% yr⁻¹ discount rate). Development work on new systems is continuing to reduce cost and energy usage in these systems. This will further reduce the abatement costs. Taking a possible increase in energy efficiency into consideration may also result in negative abatement costs (savings). • For small commercial units, namely stand-alone equipment and condensing units (vending machines, ice-cream freezers, walk-in coolers, etc.), global companies have begun employing low- or zero-GWP alternatives to HFCs (hydrocarbons and CO$_2$) and alternative technologies, which promise reduced direct and comparable or lower indirect emissions.

Table TS-9. (2) Overview of sector- and application-specific findings.

Sector	Description and status of sector	Emission status and BAU trends and opportunities for emissions reduction
Food processing, cold storage and industrial refrigeration	• This broad sector comprises refrigeration equipment for chilled and frozen food processing and storage, plus industrial applications in the chemical, oil and gas industries, air liquefaction and industrial and recreational-facility ice making. • Refrigerants predominantly used in this sector are ammonia and HCFC-22, with smaller amounts of CFCs and HFCs; hydrocarbons are used in the petrochemical industry.	<u>Emissions status and BAU trends</u> • In 2002, all refrigerant types banked in this sector amounted to 298 ktonnes (35% ammonia and 43% HCFC-22). Annual refrigerant emissions were 17% of banked system charge. • The projected emissions in this subsector by 2015 are about 104 MtCO$_2$-eq yr^{-1} in the BAU-scenario. <u>Opportunities for emissions reduction</u> • Ammonia is forecast for increased use in the future, with HFCs 404A, 507A, 410A and 134a replacing HCFC-22 and CFCs. CO$_2$ and ammonia/CO$_2$ cascade systems are beginning to be used in applications having evaporator temperatures of –40°C and lower. • Significant total emission reductions can be achieved by using lower GWP refrigerants, employing system refrigerant charge reduction, more efficient containment, improved refrigerant recovery and overall improved energy efficiency through new system designs. LCCP calculations are used to optimize refrigerant choice and system design for the lowest environmental impact. The abatement costs of refrigerant emissions from industrial refrigeration were determined to be in the range of 27–37 US\$ (2002) per tCO$_2$-eq (8% yr^{-1} discount rate).
Transport refrigeration	• The transport refrigeration sector consists of systems for transporting chilled or frozen goods by road, rail, sea and air. Current systems use refrigerants CFC-12, R-502 (a CFC/HCFC blend), HCFC-22, HFCs (HFC-134a, R-404A, R-507A, R-410A, and R-407C) and smaller amounts of ammonia, hydrocarbons and CO$_2$ in vapour compression systems. Ice and liquid or solid CO$_2$ are also used for refrigeration in this sector. • Several types of refrigeration configurations are used, such as shipboard systems, containers with individual refrigeration units which can be transported by sea, rail or road, and refrigerated trucks and railcars. Air transport refrigeration is mainly with ice or solid CO$_2$.	<u>Emissions status and BAU trends</u> • Relatively severe operating environments such as exposure to outdoor conditions of high or low temperatures, high humidity, salt-water corrosion, road vibrations and container-handling shocks lead to refrigerant leakage rates of 25–35% yr^{-1}. • The projected emissions in this subsector by 2015 are about 22 MtCO$_2$-eq yr^{-1} in the BAU-scenario. <u>Opportunities for emissions reduction</u> • Better containment of refrigerants by improved system design, one example being the recent application of low-leakage hermetic compressor systems for road transport and container refrigeration. • Reduced energy consumption via improved insulation, compressor motor frequency control for partial load conditions, water-cooled condensers for shipboard systems and preventive maintenance to reduce heat exchanger fouling. • Use refrigerants with lower GWP, examples being ammonia or ammonia/CO$_2$ systems for ship refrigeration and hydrocarbon or CO$_2$ vapour-compression systems for road, rail and container refrigeration. Considerations for using these refrigerants compared with fluorocarbon systems include safety requirements, system costs of energy efficiency, and the status of commercialization. CO$_2$ systems are still in the testing and demonstration stages.
Stationary air conditioning and heat-pumps	• 'Stationary air conditioning (SAC) and heat pumps' comprises unitary air conditioners (window-mounted, split systems), cold-water centralized AC systems (chillers) and water-heating heat pumps. • HCFC-22 is the most widely used refrigerant for air-to-air systems. HFC blends were used for the first time on a significant scale in Europe and Japan. Globally, 90% of air conditioners are still produced with HCFC-22. In the past 5 years, China has become the largest producer and consumer of AC units in the world. Chinese production is approximately equal to the annual global production in the rest of the world. • Currently, most of centrifugal chillers sold use HCFC-123 and HFC-134a.	<u>Emissions status and BAU trends</u> • Smaller chillers using positive displacement compressors generally employed HCFC-22 as the refrigerant. This refrigerant is being replaced by HFC-134a, HFC blends and, to a lesser extent, by ammonia and hydrocarbons. • The projected emissions in this sector by 2015 are about 370 MtCO$_2$-eq yr^{-} in the BAU-scenario. <u>Opportunities for emissions reduction</u> • Improving the integrity of the building envelope (reduced heat gain or loss) can have a very significant impact on indirect emissions. • HFC mixtures (R-407C and R-410A) and hydrocarbons (for small systems, mainly portable, in Europe) are used as alternatives for HCFC-22 in developed countries. For those applications for which hydrocarbons can be safely applied, the energy efficiency is comparable to HCFC-22 and R-410A. Future technical developments could reduce refrigerant charge, thereby expanding the applicability of hydrocarbons.

Table TS-9. (3) Overview of sector- and application-specific findings.

Sector	Description and status of sector	Emission status and BAU trends and opportunities for emissions reduction
Stationary air conditioning and heat-pumps (cont.)	• CFC (centrifugal) chiller manufacture was halted globally in 1993, but about 50% of centrifugal units still use CFC-11 and 12 due to the long lifetime of the equipment. • Commercial and residential AC and heating consumes significant quantities of electrical power with associated significant GHG emissions and a use-pattern that usually coincides with typical electricity peak demand periods. They account for more than 50% of building energy use in some tropical climates. In most cases, the indirect energy related GHG emissions far outweigh the direct emissions of the refrigerant.	Opportunities for emissions reduction (cont.) • The application limits for hydrocarbons are defined by national and international standards, regulations and building codes. • Residential heat pumps represent significant opportunities to lower energy consumption for building heating. CO_2 has particular advantages for tap-water-heating heat pumps because it benefits from the use of counter-flow heat exchange and high operating temperatures. • High-efficiency equipment is available in markets where both AC/chiller market volumes and electricity prices are high. Compared to the average installed base, substantive improvements can therefore be attained: for example, up to 33% energy reduction. • Specific costs of abatement options range from –3 to 170 US$ per tCO_2-eq. Improved system energy efficiencies can significantly reduce indirect GHG emissions, leading in some cases to overall costs of –75 US$ per tCO_2-eq.

Mobile Air Conditioning	• Mobile Air Conditioning (MAC) systems have been mass-produced in the US since the early 1960s and since the 1970s in Japan. The main refrigerant was CFC-12. The significant increase in the numbers of air-conditioned cars in Europe began later, around 1995, with the introduction of HFC-134a use.	Emissions status and BAU trends
		• The projected emissions in this subsector by 2015 are about 315 $MtCO_2$-eq yr^{-1} in the BAU scenario.
		Opportunities for emissions reduction
		• Options to reduce direct GHG impacts of MACs are: (1) switch to low-GWP refrigerants; (2) better containment of HFC-134a; (3) increase efficiency and reduce cooling load.
		• HFC-152a and CO_2 (R-744) are the two main options to replace HFC-134a. HFC-152a, apart from its flammability, is largely similar to existing HFC-134a technology. CO_2 systems require newly-developed components and technology. No motor vehicle manufacturer is considering hydrocarbons as an option for new vehicles, but hydrocarbons are in use as service refrigerants in several countries against manufacturers' recommendations (and often against regulations).
		• Improved HFC-134a systems show a reduction of direct GHG emissions by 50%, HFC-152a systems by 92% and CO_2 systems by nearly 100% on a CO_2-equivalent basis relative to the current HFC-134a systems.
		• It is technically and economically feasible to reduce indirect emissions through higher system energy efficiency and reduced heat load, regardless of the refrigerant chosen.
		• There is currently no significant difference between the technically achievable LCCP for HFC-152a and CO_2 systems.
		• Barriers to commercialization of HFC-152a and CO_2 are the required resolution of:
		  o the flammability risk and assurance of commercial supply of the refrigerant for HFC-152a;
		  o the suffocation risk;
		  o the residual technical and cost issues for CO_2 technology.
		• Reference cost of a typical European HFC-134a system is evaluated at about US$ 215 with an internal control compressor. Additional cost of a CO_2 system is evaluated between US$ 48 and US$ 180. Additional cost of a HFC-152a system is evaluated at US$ 48 for an added safety system.

Table TS-9. (4) Overview of sector- and application-specific findings.

Sector	Description and status of sector	Emission status and BAU trends and opportunities for emissions reduction
Foams	• Foams fall into two main categories – insulating (buildings, appliances, cold storage, etc.) and non-insulating (bedding, furniture, packaging, safety, etc.) foams. • Not-in-kind materials such as mineral fibre have held a majority share of the thermal insulation market for the past 40 years. However, foam-based insulation solutions have gained a market share over the past 15 years driven, in part, by increasing trends towards prefabrication, where structural integrity and lightweight characteristics are important. Longevity of thermal performance is also of increasing importance. However, investment cost and fire performance continue to support mineral fibre as the major component of most thermal insulation markets.	<u>Emissions status and BAU trends</u> • Manufacturing safety considerations result in the use of HFC blowing agents for smaller companies, where safety investments have a proportionally higher impact on product cost. Product safety can lead to HFC choice for certain applications in buildings, particularly where insurance concerns exist. • For non-insulating rigid and flexible foams, non-halocarbon blowing agents are now widely used. Hydrocarbon and CO_2 (both liquid- and water-based) technologies have been demonstrated to be as technically viable in a variety of foam categories, implying very little continuing global halocarbon consumption for this category. • The projected emissions by 2015 in this sector are about 124 $MtCO_2$-eq yr^{-1} in the BAU-scenario. <u>Opportunities for emissions reduction</u> • Insulating foams are expected to contribute significantly to CO_2 emission reductions in buildings and appliances as energy-efficiency improvements are demanded. Blowing-agent selection can affect thermal performance significantly. • LCCP analyses can provide insight when comparing insulation types. However, calculations are very sensitive to the carbon intensity of energy used, the product lifetime assumed, the thickness of insulation and the degree of end-of-life recovery/destruction. • By 2015, hydrocarbons are projected to be the major blowing agent in use within the rigid foam sector, when they are expected to account for nearly 60% of total consumption. Other more modest uses will be HFCs (24%) and HCFCs (16%). The HCFC use will be confined to developing countries where most use will be in appliances. Current HFC estimates of future use are lower than previously predicted, primarily because of high HFC costs. Co-blowing with CO_2 has emerged as an important means of limiting HFC use in some key applications. • Actions to reduce HFC use by 50% between 2010 and 2015 would result in an emission reduction of about 10 $MtCO_2$-eq yr^{-1}, with further increases thereafter, at a cost of 15–100 US\$ per tCO_2-eq • Although the effectiveness of recovery has yet to be proven, particularly in the buildings sector, commercial operations are already recovering at 10–50 US\$ per tCO_2-eq for appliances. Emission reductions may be about 7 $MtCO_2$-eq yr^{-1} in 2015. However, this potential could increase significantly in the period between 2030 and 2050, when large quantities of building insulation foams will be decommissioned.
Medical aerosols	• Asthma and chronic obstructive pulmonary disease (COPD) are major illnesses affecting over 300 million people worldwide. Metered dose inhalers (MDIs) are the dominant treatment. Dry powder inhalers (DPIs) which do not contain propellants have become more widely available, but are not suitable for all patients and are more expensive.	<u>Emissions status and BAU trends</u> • The projected emissions by 2015 in this sector are about 40 $MtCO_2$-eq yr^{-1} in the BAU scenario. • No major breakthroughs for inhaled drug delivery are anticipated in the next 10–15 years given the current status of technologies and the development time scales involved. <u>Opportunities for emissions reduction</u> • The major impact in reducing GHG emissions with respect to MDIs is the completion of the transition from CFC to HFC MDIs beyond the BAU trend (17 $MtCO_2$-eq yr^{-1} by 2015). • The health and safety of the patient is of paramount importance in treatment decisions and in policymaking that might impact those decisions. This may constrain the use of DPIs. • Based on the hypothetical case of switching the most widely used inhaled medicine (salbutamol) from HFC MDIs to DPI, which would lead to a modest reduction of about 10 $MtCO_2$-eq yr^{-1}, the projected recurring annual costs would be in the order of \$1.7 billion with an effective mitigation cost of 150–300 US\$ per tCO_2-eq.

Table TS-9. (5) Overview of sector- and application-specific findings.

Sector	Description and status of sector	Emission status and BAU trends and opportunities for emissions reduction
Fire protection	• 75% of original halon use has been switched to agents with no climate impact. 4% of the original halon applications continue to employ halons. The remaining 21% has been switched to HFCs and a small number of applications have been switched to PFCs and HCFCs. • HFCs and inert gases have evolved as the most commonly used gaseous agents and have achieved a degree of equilibrium in terms of market applications and share. • A new FK with nearly zero climate impact has been commercialized, but there is no basis for quantifying its market acceptance.	Emissions status and BAU trends • The projected emissions in this sector by 2015 are about 5 MtCO$_2$-eq yr^{-1} in the BAU-scenario (excluding halons). • Cost remains the main factor in limiting the market acceptance of HFCs, HCFCs and PFCs in portable extinguishers when compared to alternative extinguishers employing more traditional extinguishing agents, such as CO$_2$, dry chemicals and water. PFC use is currently limited to a minor component in one HCFC-containing blend. Opportunities for emissions reduction • Halons and HFCs are the only available alternatives in a number of fixed-system applications when safety, space and weight, cost, speed of extinguishment and special capabilities, such as those that operate in very cold conditions and on board military ships and vehicles, are taken into account. PFCs and HCFCs offer no advantage over other alternatives. A new low-GWP fluoroketone (FK), not yet tested in specialized applications, will provide additional choices in the future with lower climate impact at an additional cost. Due to the lengthy process of testing, approval and market acceptance of new fire protection equipment types and agents, no additional options are likely to have appreciable impact by 2015. • When possible, the use of agents with no climate effect can reduce GHG emissions from this sector, provided that their use meets the requirements of the specific fire-protection application in a cost-effective manner. • Management of halon, HFC, HCFC and PFC banks requires special attention so that economic incentives are created which ensure that policy intention (e.g. mandatory decommissioning) is achieved. Implementing responsible-agent management practices will reduce annual emissions from fixed system banks to 2 ± 1% and from portable extinguisher banks to 4 ± 2%.
Non-medical aerosol products	• This sector includes technical, safety, consumer and novelty aerosols. • More than 98% of non-medical aerosols in developed countries have converted from CFCs to ozone- and climate-safe alternatives. • The largest single use of HFCs in non-medical aerosol products is for 'dusters', where pressurized gas is used to blow particles from work surfaces and devices.	Emissions status and BAU trends • The projected emissions in this sector by 2015 are about 23 MtCO$_2$-eq yr^{-1} in the BAU-scenario. Opportunities for emissions reduction • The reduction potential is uncertain but estimated to be rather small. • Substitution of HFC-134a by HFC-152a in technical aerosol dusters is a leading option for reducing GHG emissions. For contact cleaners and plastic-casting mould release agents, the substitution of HCFCs by HFEs and HFCs with lower GWPs offers an opportunity for emission reduction. Safety aspects, where flammable propellants and ingredients cannot be used, continue to rely on HFC-134a for its non-flammability. Some countries have banned HFC use in novelty aerosol products. HFC-134a continues to be used in many countries for safety reasons.

Table TS-9. (6) Overview of sector- and application-specific findings.

Sector	Description and status of sector	Emission status and BAU trends and opportunities for emissions reduction
Solvents	• Prior to the Montreal Protocol, CFC-113 and methyl chloroform were widely used as cleaning solvents for metals, electronics, precision and fabric applications. ODS use in these applications has been eliminated or dramatically reduced. Most cleaning solvent applications now rely on not-in-kind substitutes. A small percentage have or are expected to transition to HFCs or HFEs. PFC use is declining and expected to be eliminated by 2025.	Emissions status and BAU trends • The projected emissions in this sector by 2015 are about 14 MtCO$_2$-eq yr^{-1} in the BAU-scenario. Opportunities for emissions reduction • A variety of organic solvents can replace HFCs, PFCs and ODSs in many applications. • These alternative fluids include lower GWP compounds such as traditional chlorinated solvents, hydrofluoroethers (HFEs), and n-propyl bromide. Numerous not-in-kind technologies, including hydrocarbon and oxygenated solvents, are also viable alternatives in some applications. • Safety, especially toxicity, plays a key role in the choice of solvents. Caution is warranted prior to adoption of any alternatives whose toxicity profile is not complete. • HFC solvents are primarily used in technically-demanding specialty applications and then only in selected countries. Uses tend to be focused in critical applications where there are no other substitutes. Consumption may decline in the future. • Use of PFC solvent is constrained to a few niche applications due to the limited solvency, high cost and substitution with lower GWP solvents. • Improved containment is important in existing uses because optimized equipment can reduce solvent consumption by as much as 80% in some applications. Due to their high cost and ease of purification during recycling, fluorinated solvents can be and often are recovered and reused.
HFC-23 byproduct	• HFC-23 is a byproduct of HCFC-22 manufacturing • Although the production of HCFC-22 for direct use is ending in developed countries and will eventually end in developing countries, its production as a feedstock is projected to continue to grow.	Emissions status and BAU trends • The projected emissions by 2015 in this sector are about 332 MtCO$_2$-eq yr^{-1} in the BAU-scenario. • The upper bound of HFC-23 emissions is in the order of 4% of HCFC-22 production. Process optimization can reduce average emissions to 2% or less. However, the actual reduction achieved is facility-specific. Opportunities for emissions reduction • Capture and destruction of HFC-23 by thermal oxidation is a highly effective option for reducing emissions at specific costs below 0.2 US$ per tCO$_2$-eq. Emissions can be reduced by more than 90%.

3.7 Which policy instruments are available to achieve the reductions in the emissions of GHGs addressed in this report?

As discussed in the IPCC Third Assessment Report, a wide range of policies, measures and instruments can be used to reduce GHG emissions. These include:

- regulations (e.g. mandatory technology and performance standards; product bans);
- financial incentives (e.g. taxes on emissions, production, import or consumption; subsidies and direct government spending and investment; deposit-refund systems; tradable and non-tradable permits);
- voluntary agreements.

Almost all of the policy instrument categories mentioned above have been considered for, or implemented in, reducing the use or emissions of ODSs and their substitutes, such as HFCs and PFCs. Furthermore, general energy or climate policies affect the indirect GHG emissions of applications with ODSs, their substitutes or not-in-kind alternatives. In addition, specific policies for reducing the GHG emissions of ODS substitutes (i.e. HFCs and PFCs) have been formulated. Examples are given in Table TS-10.

3.8 What can be said about the future availability of HFCs/PFCs for use in developing countries?

No published data are available to project future production capacity. However, as there are no technical or legal limits to HFC and PFC production, it can be assumed that the global production capacity will generally continue to satisfy or exceed demand. Future production is therefore estimated in this report by aggregating sectoral demand.

In the BAU scenario, global production capacity is expected to expand with additions taking place mainly in developing countries and through joint ventures. Global production capacity of HFCs and PFCs most often exceeds current demand. There are a number of HFC-134a plants in developed countries and one plant in a developing country with others planned; the few plants for other HFCs are almost exclusively in developed countries. The proposed European Community phase-out of HFC-134a in mobile air conditioners in new cars and the industry voluntary programme to reduce their HFC-134a emissions by 50% will affect demand and production capacity and output. Rapidly expanding markets in developing countries, in particular for replacements for CFCs, is resulting in new capacity for fluorinated gases which is at present being satisfied through the expansion of HCFC-22 and 141b capacity. [11]

Table TS-10. Policy instruments for reducing GHG emissions of ODS-substitutes.

Type of instrument	Specific instrument examples
Regulations	Mandatory performance standards: Energy-efficiency performance standards and leak-tightness performance standards are in place in several countries, notably for refrigeration and cooling applications. Regulations that prohibit venting and require recycling have been enacted in many countries but are often difficult to enforce. Effective application requires complementary compliance programmes and mandatory training of technicians. Vehicle emission performance standards (e.g. restriction of the amount of GHGs emitted by cars, including both fuel and MAC emissions) are being considered (e.g. by the State of California).
	Obligation to use certified companies for servicing of installations and equipment (e.g. 'STEK' programme in the Netherlands).
	Bans and restrictions on the use of specific substances for certain applications: HFCs phase-out schemes are in place or proposed in several countries (e.g. Austria, Denmark, Switzerland). The proposed EU Directive aims to amend the existing EU Vehicle Type Approval legislation which will introduce a GWP threshold for HFCs used in MAC.
	End-of-life management measures, such as mandatory recycling and bans on venting
Financial incentives and market mechanisms	The relative costs of HFCs/PFCs and other alternatives to ODSs will impact the choices of both the users and the producers of these substances. HFCs and PFCs are complex chemicals, and they tend to be more expensive than the ODSs they replace, thus further encouraging the substitution to not-in-kind alternatives. Financial incentives can further shape this cost differential between substances and technologies.
	Deposits or taxes on HFC import and production are collected by several countries. Deposits and taxes raise the cost of HFCs, encouraging containment and making recycling more attractive.
	Tax rebates for delivery of used HFCs and PFCs to destruction facilities provide incentives to minimize emissions. In Norway rebates are NKr 183 (26 US$) per tCO$_2$-eq
	Subsidies: In addition to the international grants from the Multilateral Fund under the Montreal Protocol and, for example, the Global Environment Facility, national governments subsidize research and development as well as the implementation of new, low-GHG emission technologies. Subsidy (50%) on the cost of collection and destruction of halons and CFCs was provided in the Netherlands to discourage venting, just before stocks of these substances became illegal in 2004.
	Emission reduction of HFCs and PFCs can be funded by the Clean Development Mechanism (CDM) of the Kyoto Protocol. CDM projects exist for Korea and China and are under development for India and Mexico.
	Large point sources of HFCs may be included in emission trading schemes. To date, these sources (e.g. the HFC-23 byproduct emissions of HCFC-22 production) have not usually been included in emissions trading schemes, although the UK scheme is a notable exception. Monitoring the use and emissions of such substances may be less practical for the more diffuse emission sources.
Voluntary agreements	There are several industry- and government-sponsored emission reduction and responsible use programmes. Adherence to responsible use principles can lead to reductions in HFC emissions beyond current projections. Principles for responsible use include: • Use HFCs only in applications where they provide safety, energy efficiency, environmental or critical economic or public health advantage; • Limit emissions of HFCs to the lowest practical level during manufacture, use and disposal of equipment and products; • If HFCs are to be used, select the compound or system with the smaller climate impact that satisfies the application requirements.
	Good practice guidelines have been developed regarding the selection and maintenance of equipment, including improved containment of substances and recovery during servicing and at the end-of-life.

4. Important findings for sectors using ODSs and their alternatives

4.1 What are the most important findings for the refrigeration sector?

Refrigerants are by far the largest contributors to direct emissions of GHGs. In this report, the refrigeration sector is classified into the following subsectors: domestic refrigeration, commercial refrigeration, industrial refrigeration and food processing and cold storage and transport refrigeration. The sectors residential and commercial air conditioning and heating (SAC) and mobile air conditioning (MAC) are presented in separate sections (4.2 and 4.3) in this Technical Summary. Table TS-11 presents a detailed breakdown of banks and direct emissions of GHGs over all these sectors utilizing refrigerants.

The five general options to reduce direct GHG emissions for the refrigeration sector can be specified as follows:

- improved containment: leak-tight systems;
- recovery, recycling and the destruction of refrigerants during servicing and at the end-of-life of the equipment;
- application of reduced charge systems:
  - lower refrigerant charge per unit of cooling capacity;
  - reduced refrigeration capacity demand;
- use of alternative refrigerants with a reduced or negligible global-warming potential (e.g. hydrocarbons (HCs), carbon dioxide (CO_2), ammonia, etc.);
- not-in-kind technologies.

These principles equally apply for the SAC and MAC sectors.

4.1.1 Domestic refrigeration

Domestic refrigerators and freezers are used for food storage in dwelling units and in non-commercial areas such as offices throughout the world. More than 80 million units are produced annually with internal storage capacities ranging from 20 litres to more than 850 litres. With an estimated unit average lifespan of 20 years, this results in an installed inventory of approximately 1500 million units. As a result of the Montreal Protocol, manufacturers initiated the transition from CFC refrigerant applications during the early 1990s. This transition is complete in developed countries and has made significant progress in developing countries. However, the typical lifespan of domestic refrigerators results in units manufactured using CFC-12 refrigerant still comprising the majority of the installed base. This in turn significantly retards the rate of reduction in the demand for CFC-12 refrigerant in the servicing sector.

Isobutane (HC-600a) and HFC-134a are the major alternative refrigerants replacing CFC-12 in new domestic refrigeration equipment (see Table TS-12). Each of these has demonstrated mass production capability for safe, efficient, reliable and economic use. Similar product efficiencies result from the use of either refrigerant. Independent studies have concluded that application design parameters introduce more efficiency variation than is presented by the refrigerant choice. Comprehensive refrigerant selection criteria include safety, environmental, functional, cost and performance requirements. The choice of refrigerant can be strongly influenced by local regulatory and litigation environments. Each refrigerator typically contains 50–250 g of refrigerant enclosed in a factory-sealed hermetic system. A simplified summary of relative technical considerations for these two refrigerants is:

- HC-600a uses historically familiar mineral oil as the lubricant in the hermetic system. Manufacturing processes and designs must properly deal with the flammable nature of this refrigerant. Among these are the need for proper factory ventilation and appropriate electrical equipment, prevention of refrigerant leakage and access to electrical components, use of sealed or non-sparking electrical components when there is accessibility to leaking refrigerant and the use of proper brazing techniques or, preferably, avoidance of brazing operations on charged systems. Field service procedures must also properly accommodate the refrigerant flammability.
- HFC-134a uses moisture-sensitive polyolester oil as the lubricant in the hermetic system. Manufacturing processes and service procedures must take care to properly maintain low moisture levels. Long-term reliability requires a more careful avoidance of contaminants during production or servicing compared to either CFC-12 or HC-600a practices.

The use of the hydrocarbon blend propane (HC-290)/isobutane (HC-600a) allows the matching of CFC-12 volumetric capacity and avoids capital expense for retooling compressors. These blends introduce manufacturing complexities and require the use of charging techniques suitable for refrigerant blends having components with different boiling points. Application of these blends in Europe during the 1990s was an interim step towards the transition to HC-600a using retooled compressors. The safety considerations for hydrocarbon blends are consistent with those for HC-600a.

Table TS-11. Refrigerant bank and direct emissions of CFCs, HCFCs, HFCs and other substances (hydrocarbons, ammonia and carbon dioxide) in 2002, the 2015 business-as-usual scenario and the 2015 mitigation scenario, for the refrigeration sector, the residential and commercial air-conditioning and heating sector ('stationary air conditioning') and the mobile air-conditioning sector.

	Banks (kt)					Emissions (kt yr⁻¹)					Emissions (MtCO$_2$-eq yr⁻¹) SAR/TAR[b]	Emissions (MtCO$_2$-eq yr⁻¹) This Report[c]
	CFCs	HCFCs	HFCs	Other	Total	CFCs	HCFCs	HFCs	Other	Total		
2002												
Refrigeration	330	461	180	108	1079	71	132	29	18	250	848	1060
- Domestic refrigeration	107	-	50	3	160	8	-	0.5	0.04	9	69	91
- Commercial refrigeration	187	316	104	-	606	55	107	23	-	185	669	837
- Industrial refrigeration[a]	34	142	16	105	298	7	24	2	18	50	92	110
- Transport refrigeration	2	4	10	-	16	1	1	3	-	6	19	22
Stationary Air Conditioning	84	1028	81	1	1194	13	96	6	0.2	115	222	271
Mobile Air Conditioning	149	20	249	-	418	60	8	66	-	134	583	749
Total 2002	**563**	**1509**	**509**	**109**	**2691**	**144**	**236**	**100**	**18**	**499**	**1653**	**2080**
2015 BAU												
Refrigeration	64	891	720	136	1811	13	321	115	21	471	919	1097
- Domestic refrigeration[a]	37	-	189	13	239	5	-	8	1	13	51	65
- Commercial refrigeration	6	762	425	-	1193	5	299	89	-	393	758	902
- Industrial refrigeration[a]	21	126	85	123	356	4	21	11	21	56	88	104
- Transport refrigeration	0.1	2.8	20.3	-	23.2	0.1	1.3	7.4	-	9	22	26
Stationary Air Conditioning	27	878	951	2	1858	7	124	68	0	199	314	370
Mobile Air Conditioning	13	23	635	4	676	5	11	175	1	191	281	315
Total 2015-BAU	**104**	**1792**	**2306**	**143**	**4345**	**25**	**455**	**359**	**23**	**861**	**1514**	**1782**
2015 Mitigation												
Refrigeration	62	825	568	186	1641	8	202	52	15	278	508	607
- Domestic refrigeration	35	-	105	60	200	3	-	3	1	6	27	35
- Commercial refrigeration	6	703	378	-	1087	3	188	40	-	230	414	494
- Industrial refrigeration[a]	21	120	65	126	331	3	13	5	14	36	53	63
- Transport refrigeration	0.1	2.8	20.3	-	23.2	0.0	0.9	4.3	-	5	13	15
Stationary Air Conditioning	27	644	1018	2	1691	3	50	38	0	91	145	170
Mobile Air Conditioning	13	23	505	70	611	3	7	65	7	82	119	136
Total 2015 Mitigation	**102**	**1493**	**2090**	**259**	**3943**	**14**	**259**	**155**	**22**	**451**	**772**	**914**

a Including food processing/cold storage
b Greenhouse gas CO$_2$-equivalent (GWP-weighted) emissions, using direct GWPs, taken from IPCC (1996 and 2001) (SAR/TAR)
c Greenhouse gas CO$_2$-equivalent (GWP-weighted) emissions, using direct GWPs, taken from Chapter 2 in this report

Table TS-12. Domestic refrigeration, current status and abatement options.

Product Configuration		Cold Wall	Open Evaporator Roll Bond	No-Frost
Cooling capacity	From	60 W	60 W	120 W
	To	140 W	140 W	250 W
Refrigerant charge (HFC)	From	40 g	40 g	120 g
	To	170 g	170 g	180 g
Approximate percentage of sector refrigerant bank (160 kt) in configuration		20 units @ 100 g average 18% of 160 kt	15 units @ 100 g average 14% of 160 kt	50 units @ 150 g average 68% of 160 kt
Approximate percentage of sector refrigerant emissions (8950 tonnes) in subsector		18% of 8950 tonnes	14% of 8950 tonnes	68% of 8950 tonnes
Predominant technology		HC-600a	HFC-134a	HFC-134a
Other commercialized technologies		HFC-134a, CFC-12	HC-600a, CFC-12	HC-600a, CFC-12
Low GWP technologies with fair or better than fair potential for replacement of HCFC/HFC in the markets		R-600a	HC-600a	HC-600a
Status of alternatives		Fully developed and in production	Fully developed and in production	Fully developed and in production
HC-600a Mfg. Cost Premium		No Premium	3–5 US$	8–30 US$
Capital Investment		0	45–75 million US$	400–1500 million US$
Emission reduction		1432 tonnes	1253 tonnes	6086 tonnes

Alternative refrigeration technologies such as the Stirling cycle, absorption cycle and thermoelectric, thermionic and thermoacoustic systems continue to be pursued for special applications or situations with primary drivers different from conventional domestic refrigerators. These technology options are not expected to significantly alter the position of vapour compression technology as the technology of choice for domestic refrigeration in the foreseeable future.

Vapour compression technology is established and readily available worldwide. Current technology designs, based on HC-600a or HFC-134a, typically use less than one-half the electrical energy required by the units they replace. This reliable performance is provided without resorting to higher cost or more complex designs. Continued incremental improvements in unit performance and/or energy efficiency are anticipated. Government regulations and voluntary agreements on energy efficiency and labelling programmes have demonstrated their effectiveness in driving improved efficiency product offerings in several countries.

Good design, manufacturing and service practices will minimize refrigerant emissions during refrigerator production and use phases; however, special attention must be given to the retirement of the large number of units containing CFC-12. With a typical 20-year lifespan, refrigerator end-of-life retirement and disposal occurs at a frequency of about 5% of the installed base each year. This means approximately 75 million refrigerators containing 100 g per unit, or a total of 7500 tonnes of refrigerant, are disposed of annually. For at least another 10 years, this refrigerant will predominantly be CFC-12. The small refrigerant charge per unit detracts from the economic justification for refrigerant recovery. Regulating agencies around the world have provided incentives or non-compliance penalties to promote recovery of this ODS.

The current (2002 data) annualized HFC-134a emissions rate from domestic refrigerators is 1.0% during product usage. HFC emissions for domestic refrigeration are estimated to be 480 tonnes in 2002, increasing to 7800 tonnes by 2015 in a BAU scenario. In the mitigation-scenario, emissions in 2015 are 2800 tonnes due to improved refrigerant containment and recovery. Table TS-12 summarizes emission abatement opportunities with increased application of HC-600a refrigerant. Similarly, the manufacturing cost premiums and capital investments and development costs required for implementation are tabulated for the three most common refrigerator configurations.

4.1.2 Commercial refrigeration

Commercial refrigeration comprises three main types of equipment: stand-alone equipment, condensing units and full supermarket systems. A wide variety of refrigeration systems fall within the subsector of commercial refrigeration, from ice-cream freezers, with a cooling capacity of about 200 W and not varying greatly from domestic freezers, up to machinery rooms containing multiple compressor racks that consume several hundreds of kilowatts. The most commonly used refrigerants in this subsector are HCFC-22, R-404A and HFC-134a.

In 2002, all refrigerant types banked in commercial refrigeration, predominantly equipment containing CFCs, HCFCs and HFCs, represented 605 kt of a total of 2690 kt for all refrigerating and air conditioning (AC) systems and all refrigerant types; this represents 22% of the total refrigeration and AC bank.

On a global basis, commercial refrigeration is the refrigeration subsector with the largest refrigerant emissions, calculated as CO_2-equivalents, representing 40% of the total refrigerant emissions (in refrigeration, SAC and MAC). The emission levels, including fugitive emissions, ruptures and emissions during servicing and at end-of-life, are generally very high, especially for supermarkets and hypermarkets. The larger the charge, the larger the average emission rate, which results from very long pipes, large numbers of fittings and valves and very high emissions when ruptures occur.

Refrigerant leakage rates derived with a bottom-up approach suggest a global annual emissions rate of 30% of banked system charge. Refrigerant emissions typically represent 60% of the total emissions of GHGs resulting from the system operation, with the remainder being indirect emissions caused by power production. These percentages indicate how important emission reductions from this sector are.

Annual refrigerant leakage rate data from over 1700 full supermarket systems in the USA and Europe are in the range of 3–22% of system charge, with an average of 18%. It may be concluded that if the emission estimate of 30% on a global basis is correct, the values of 3–22% must represent selected company data from countries with a strong emphasis on emission reductions.

Projections of future halocarbon refrigerant emissions according to different scenarios are given in Table TS-11. The high economic growth rates of some of the developing countries will have a very significant impact on the growth of the refrigerant bank and emissions.

Significant total emission reductions, that is improved LCCP, can be achieved by using refrigerants like HFCs, hydrocarbons, ammonia or CO_2, by employing charge reduction and more efficient containment and by overall improved energy efficiency through new system designs. This is summarized in Table TS-13. Safety issues have to be taken into account if toxic or flammable refrigerants are used; these depend on national and, occasionally, on local regulations, which may limit the degree to which some of these refrigerants can be applied.

In full supermarket systems, up to 60% lower LCCP values than in centralized direct systems of traditional design can be obtained by applying direct systems using alternative refrigerants, better containment, distributed systems, indirect systems or cascade systems.

Published results show that alternative systems have a 0–35% higher initial cost and a 0–20% higher energy usage than current systems.

Refrigerant emissions abatement cost is in the range of 20–280 US$/t$CO_2$-eq[19,20]. Development work on new systems is continuing in order to reduce cost and energy usage in these systems, which will further reduce abatement costs. Taking into consideration possible increases in energy efficiency may also result in negative abatement costs (savings).

For small commercial units, namely stand-alone equipment and condensing units (vending machines, ice-cream freezers, walk-in coolers, etc.), global companies have begun employing low- or zero-GWP alternatives to HFCs (hydrocarbons and CO_2) and alternative technologies. Both of these options promise reduced direct and comparable or lower indirect emissions.

4.1.3 Food processing, cold storage and industrial refrigeration

Food processing and cold storage is one of the important applications of refrigeration; the preservation and distribution of food while keeping nutrients intact. This application is very significant in size and economic importance in all countries, including developing ones. The application includes both cold storage (at temperatures from –1°C to 10°C), freezing (–30°C to –35°C) and the long-term storage of frozen products (–20°C to –30°C). The amount of chilled food is about 10 to 12 times greater than the amount of frozen products.

---

[19] Costs in this report are given in US dollars in 2002 unless otherwise stated.

[20] The presented cost data are for direct emission reductions only. The taking into account of energy efficiency improvements may result in even net negative specific costs (savings).

Table TS-13. Sector summary for commercial refrigeration – current status and abatement options.

Subsector		Stand-alone Equipment	Condensing Units	Full supermarket system			
				Direct Centralized	Indirect Centralized	Distributed	Hybrids
Cooling capacity	From	0.2 kW	2 kW	20 kW			
	To	3 kW	30 kW	>1000 kW			
Refrigerant charge	From	0.5 kg	1 kg	100 kg	20	*	*
	To	~2 kg	15 kg	2000 kg	500 kg	*	*
Approximate percentage of sector refrigerant bank in subsector		11% of 606 kt	46% of 606 kt	43% of 606 kt			
Approximate percentage of sector refrigerant emissions in subsector		3% of 185 kt	50% of 185 kt	47% of 185 kt			
2002 Refrigerant bank, percentage by weight		CFCs 33%, HCFCs 53%, HFCs 14%					
Typical annual average charge emission rate		30%					

Subsector	Stand-alone Equipment	Condensing Units	Full supermarket system			
			Direct Centralized	Indirect Centralized	Distributed	Hybrids
Technologies with reduced LCCP	**Improved HFC** SDNA	**Improved HFC** SDNA	**Improved HFC** EmR 30% ChEU 0% ChCst 0 ±10%	**Ammonia** EmR 100% ChEU 0–20% ChCst 20–30%	**HFC** EmR 75% ChEU 0–10% ChCst 0–10%	**Cascade-HFC/CO$_2$** EmR 50–90% ChEU 0%
	HC SDNA	**R-410A** SDNA	**CO$_2$ (all-CO$_2$)** EmR 100% ChEU 0 ±10% ChCst 0±10%	**HC** EmR 100% ChEU 0–20 % ChCst 20–30%	**Economized-HFC-404A** SDNA	**Cascade-Ammonia/CO$_2$** SDNA
	CO$_2$ SDNA	**HC** SDNA		**HFC** EmR 50–90% ChEU 0–20% ChCst 10–25%	**Economized-HFC-410A** SDNA	**Cascade-HC/CO$_2$** SDNA
		CO$_2$ SDNA			**CO$_2$** SDNA	
LCCP reduction potential (world avg. emission factor for power production)	SDNA		35–60%			
Abatement cost estimates (10-yr lifetime, 10% interest rate)	SDNA		20–280 US$ per tonne CO$_2$ mitigated			

Notes:

* Alternatives in these categories have been commercialized, but since the current number of systems are limited, they are only referenced as options below

EmR (%):   Direct Emission Reduction (compared with installed systems)
ChEU (%):   Change in Energy Usage (+/–) (compared with state of the art)
ChCst (%):   Change in Cost (+/–) (compared with state of the art)
SDNA: Sufficient data on emission reduction, energy usage and change in cost not available from literature

The majority of refrigerating systems for food processing and cold storage are based on reciprocating and screw compressors. Ammonia, HCFC-22, R-502 and CFC-12 are the historically used refrigerants. HFC refrigerants are currently being used in place of CFC-12, R-502 and HCFC-22 in certain regions. The preferred HFCs are HFC-134a and HFC blends with a small temperature glide such as R-404A, R-507A and R-410A. Ammonia/CO$_2$ cascade systems are also being used, as are hydrocarbons as primary refrigerants in indirect systems.

Limited data are available on TEWI/LCCP for this category. A recent study of system performance and LCCP calculations for an 11-kW refrigeration system operating with R-404A, R-410A and HC-290 showed negligible differences in LCCP, based on the assumptions used in the calculations.

Additional information on refrigerant leakage and emissions reduction options for the total sector of food processing, cold storage and industrial refrigeration is included at the end of the following section.

Industrial refrigeration includes a wide range of cooling and freezing applications, such as those in the chemical, oil and gas industries, industrial and recreational ice making and air liquefaction. Most systems are vapour compression cycles, with evaporator temperatures ranging from 15°C down to –70°C. Cryogenic applications operate at even lower temperatures. The capacity of the units varies from 25 kW to 30 MW, with systems often custom-made and erected on-site. The refrigerants used are preferably single component or azeotropes because many of the systems use flooded evaporators to achieve high efficiency. Some designs use indirect systems with heat transfer fluids to reduce refrigerant charge size and/or to minimize the risk of direct contact with the refrigerant.

These refrigeration systems are normally located in industrial areas with limited public access. Ammonia is the predominant and increasingly used refrigerant, with the second most common refrigerant in terms of volume use being HCFC-22, although the use of HCFC-22 in new systems is forbidden by European regulations since January 2001 for all types of refrigerating equipment. The smaller volume CFC refrigerants CFC-12 and R-502 are being replaced by HFC-134a and by R-404A, R-507A and R-410A. CFC-13 and R-503 are being replaced by HFC-23 and by R-508A or R-508B. HCFC-22 is being replaced by R-410A, as the energy efficiency of R-410A systems can be slightly higher than that of HCFC-22, and R-410A is similar to ammonia with respect to evaporation

temperatures down to –40°C. Hydrocarbon refrigerants historically have been used in large refrigeration plants within the oil and gas industry.

CO_2 is beginning to find application in this subsector as a low-temperature refrigerant and in cascade systems, with ammonia in the high stage and CO_2 in the low stage. The energy efficiency of CO_2 systems can be similar to that of HCFC-22, ammonia and R-410A in the evaporator range of –40°C to –50°C. CO_2 is also being used as an indirect system heat transfer fluid.

Annual refrigerant leakage rates of industrial refrigeration systems have been estimated to be in the range of 7–10%, while those for the combined sector of food processing, cold storage and industrial refrigeration were reported to be 17% of the total systems refrigerant bank in 2002. The refrigerant bank consisted of 35% ammonia and 43% HCFC-22 by weight, with the remainder being CFCs, HFCs and hydrocarbons. The 2002 distribution of the major refrigerants and emissions in this total sector are shown in Table TS-14.

Emission reduction options are to use refrigerants with lower GWP, to improve design of equipment and operations resulting in a reduced system refrigerant charge, to improve refrigerant containment and recovery and to improve energy efficiency through new system designs. LCCP calculations should be used in optimizing refrigerant choice and system design for the lowest environmental impact. Industrial refrigeration refrigerant emissions abatement cost was determined to be in the range of 27–37 US$/tCO$_2$-eq (8% yr^{-1} discount rate).

4.1.4 Transport refrigeration

The transport refrigeration subsector consists of refrigeration systems for transporting chilled or frozen goods by road, rail, air and sea. Several types of refrigeration configurations are used, such as shipboard systems, containers with individual refrigeration units which can be transported by sea, rail or road and refrigerated trucks and railcars. The transport subsector also covers the use of refrigeration in fishing

Table TS-14. Food processing, cold storage and industrial refrigeration (2002).

	CFCs (CFC-12 and R-502)	HCFC-22	NH$_3$	HFCs (HFC-134a, R-404A, R-507A, R-410A)
Cooling Capacity	25 kW–1000 kW	25 kW–30 MW	25 kW–30 MW	25 kW–1000 kW
Emissions, t yr^{-1}	9500	23,500	17,700	1900
Refrigerant in bank, tonnes	48,500	127,500	105,300	16,200
Emissions % yr^{-1}	14%	18%	17%	12%

Table TS-15. Transport refrigeration, characteristics and alternatives.

Subsector		Sea Transport & Fishing	Road Transport	Rail Transport	Container Transport
Cooling capacity	From	5 kW	2 kW	10 kW	Approx. 5 kW
	To	1400 kW	30 kW	30 kW	
Refrigerant charge	From	1 kg	1 kg	10 kg	Approx. 5 kg
	To	Several tonnes	20 kg	20 kg	
Approximate percentage of sector refrigerant bank in subsector		52% of 15,900 tonnes	27% of 15,900 tonnes	5% of 15,900 tonnes	16% of 15,900 tonnes
Approximate percentage of sector refrigerant emissions in subsector		46% of 6000 tonnes	30% of 6000 tonnes	6% of 6000 tonnes	18% of 6000 tonnes
Predominant technology		HCFC-22	HFC-134a, HFC-404A, HFC-410A	HFC-134a, HFC-404A, HFC-410A	HFC-404A
Other commercialized technologies		Various HFCs, ammonia, ammonia, CO_2/ammonia for low temperatures; hydrocarbon systems for gas tankers; sorption systems for part of the cooling load	Hydrocarbons, liquid or solid CO_2, ice slurry, eutectic plates	Solid CO_2	HFC-134a, HCFC-22
Low GWP technologies with fair or better than fair potential for replacement of HCFC/HFC in the markets		Ammonia, CO_2/ammonia for low temperatures	Hydrocarbons, CO_2 compression systems; for short hauls a combination of stationary hydrocarbon or ammonia with liquid CO_2, ice slurry or eutectic plates	Hydrocarbons, CO_2 compression systems; for specific transports (certain fruits) a combination of stationary hydrocarbon or ammonia with liquid CO_2, ice slurry or eutectic plates	CO_2 compression system
Status of alternatives		Fully developed. Some cost issues related to additional safety for ammonia plants on ships. Hydrocarbon practical mainly for ships which are built according to explosion-proof standards (e.g. gas carriers)	Hydrocarbon mini-series successfully field tested, lack of demand and additional requirements on utilization (driver training, parking). Liquid CO_2 systems commercialized. CO_2 compression tested in proto-types, but open-drive compressor needed for most systems in combination with leakage is an issue	Solid CO_2 is standard use, but not very energy efficient, difficult to handle and high infrastructure requirements, therefore presently being phased out. Increasingly use of systems designed for trailer use with optimization for rail requirements (shock resistance)	Under development – prototype testing; might be available in the near future if demanded

vessels where the refrigeration systems are used for both food processing and storage.

Technical requirements for transport refrigeration units are more stringent than for many other applications of refrigeration. The equipment has to operate in a wide range of ambient temperatures and under extremely variable weather conditions (solar radiation, rain, etc). The transport equipment must be able to carry a highly diverse range of cargoes with differing temperature requirements and must be robust and reliable in the often severe transport environment. Despite the robust and sturdy design of transport refrigeration units, leaks within the refrigeration system can occur as a result of vibrations, sudden shocks, collisions with other

objects and salt-water corrosion. Ensuring the safe operation with all working fluids is essential if – for example, in the case of ships – there are no easy options for evacuation in the case of refrigerant leakage. Safety must be inherent through the choice of fluids or ensured through a number of technical measures. There is also a need for continuity with respect to equipment maintenance, as the transport equipment can require servicing in many locations around the world.

Refrigeration systems are typically using CFC-12, R-502, HCFC-22, HFC-134a, R-404A, R-507A, R-410A and R-407C. Lower GWP alternatives, such as ammonia, hydrocarbons and ammonia/CO_2, have been commercialized in some vapour compression applications. Ice and liquid or solid CO_2 are used in some sectors of transport refrigeration such as road, rail and air. An overview of the different applications currently used as well as the state of development of alternatives can be found in Table TS-15. The refrigerant bank is presently estimated to be 3300 tonnes of CFCs, 3200 tonnes of HCFC-22 and 9500 tonnes of HFCs and HFC mixtures; the total bank is expected to increase from 16,000 tonnes at the present time to 23,200 tonnes in 2015 (BAU scenario). The expectation is that present-day combined refrigerant emissions of 6000 tonnes annually will increase to 8700 tonnes annually by 2015 for a BAU scenario or will be 5250 tonnes annually following significantly increased efforts in refrigerant recovery and recycling and better containment such as the use of hermetically sealed compressors. These latter options would significantly lower CO_2-equivalent emissions, as would the replacement of fluorocarbons with lower GWP alternatives.

There are lower-GWP refrigerant replacement options available for all transport refrigeration applications where CFCs, HCFCs or HFCs are presently being used; see Table TS-15. In several cases, these options might increase the costs of the refrigeration system due to equipment- and safety-related costs. It must be remembered that for the owners of transport equipment, and in the absence of external incentives, the initial cost of the transport system and refrigeration plant is far more important than the running costs of the installation.

Due to refrigerant leakage rates of 25–35%, the change from an HFC such as R-404A to a lower GWP alternative will usually lead to a reduction of the TEWI, if the energy consumption is not substantially higher than in today's systems. In several applications, the reduction of TEWI could be very significant.

There are many opportunities for reducing the energy consumption of transport refrigeration systems, including improved insulation to reduce cooling losses and load, compressor frequency control for partial load conditions, water-cooled condensers for shipboard systems and preventive maintenance to reduce heat exchanger fouling.

4.2 What are the most important findings for residential and commercial air conditioning and heating?

The applications, equipment and products that are included in the sector of residential and commercial air conditioning and heating can be classified into three groups: stationary air conditioners (including both equipment that cools air and heat pumps that directly heat air), chillers and water-heating heat pumps.

Stationary air conditioners generally fall into six distinct categories: (1) window-mounted and through-the-wall; (2) non-ducted split residential and commercial; (3) non-ducted single-package; (4) non-ducted water-source; (5) ducted residential split and single packaged; (6) ducted commercial split and packaged. Water chillers coupled with air handling and distribution systems commonly provide comfort air conditioning in large commercial buildings. Water-heating heat pumps are manufactured using various heat sources: air, water from ponds and rivers and the ground.

Which refrigerants were used in the past?

- HCFC-22 in unitary air conditioners;
- HCFC-22 and R-502 in water-heating heat pumps;
- CFC-11, CFC-12, HCFC-22 and R-500 in centrifugal water chillers;
- HCFC-22 and CFC-12 (to a much lower extent) in positive displacement water chillers.

Stationary air conditioners: The vast majority of stationary air conditioners (and air-heating heat pumps) use the vapour-compression cycle technology with HCFC-22 refrigerant. Nearly all air-cooled air conditioners manufactured prior to 2000 use this refrigerant as their working fluid.

Water chillers: Chillers employing screw, scroll and reciprocating compressors generally employ HCFC-22. Some of the smaller reciprocating chillers (under 100 kW) were offered with CFC-12 as the refrigerant. Centrifugal chillers are manufactured in the United States, Asia and Europe. Prior to 1993, these chillers were offered with CFC-11, CFC-12, R-500 and HCFC-22 as refrigerants.

Water-heating heat pumps: In the past, the most common refrigerants for vapour compression heat pumps have been R-502 and HCFC-22.

What are current practices?

Stationary air conditioners: A rough estimate would indicate that globally more than 90% of the air-cooled air conditioner (and heat pump) units currently being produced still use HCFC-22 as the refrigerant. This refrigerant is beginning to be phased out in some countries ahead of the schedule dictated by the Montreal Protocol. The refrigerant options being used to replace HCFC-22 are the same for all of the stationary air conditioner categories: HFC-134a, HFC blends and hydrocarbons. CO_2 is also being considered for this application. At present, HFC blends are being used in the vast majority of non-ODS systems, with hydrocarbons being used in a small percentage of low charge systems.

<u>Water chillers:</u> HFCs and HFC blends (particularly R-407C and R-410A) are beginning to replace HCFC-22 unit sales in new positive displacement chillers. Larger water-cooled screw chillers (e.g. above 350 kW) have been developed to use HFC-134a rather than HCFC-22. Ammonia is used in some chillers in Europe, and a few small chillers using hydrocarbon refrigerants are also produced there each year. HCFC-123 and HFC-134a have replaced CFC-11 and CFC-12 in centrifugal chillers produced since 1993.

<u>Water-heating heat pumps:</u> In developed countries, HCFC-22 is still the most commonly used refrigerant, but HFC alternatives are being introduced. In developing countries, CFC-12 is also still used to a limited extent. Alternatives to HFCs in small residential and commercial water-heating systems include hydrocarbons and CO_2. Hydrocarbons are being used in Europe and CO_2 is being used in Europe and Asia.

What are possible future trends?

Options to reduce GHG emissions in residential and commercial air conditioning and heating equipment involve containment in CFC/HCFC/HFC vapour-compression systems that are applicable worldwide and for all equipment, and the use of non-CFC/HCFC/HFC systems. This latter option is not applicable in all instances due to economic, safety and energy efficiency considerations.

Containment can be achieved through the improved design, installation and maintenance of systems to reduce leakage and minimize refrigerant charge quantities in the systems, and the recovery, recycling and reclaiming of refrigerant during servicing and at equipment disposal. In order to minimize installation, service and disposal emissions, a trained labour force using special equipment is required. The baseline emissions and the emissions in a mitigation scenario for 2015 are presented in Table TS-11.

What are alternative low GWP-refrigerants?

Depending on the application, alternative refrigerants to HFCs in residential and commercial air conditioning and heating equipment can include hydrocarbons, ammonia, water and CO_2.

<u>Stationary air conditioners:</u> The use of hydrocarbons in air-cooled air conditioning products having refrigerant charge levels greater than 1 kg has been the focus of considerable risk analysis and safety standards development activities (e.g. European Standard EN 378). The most significant issue that

will confront a manufacturer when considering applying hydrocarbon refrigerants is the determination of an acceptable level of risk and the associated liability.

CO_2 offers a number of desirable properties as a refrigerant: availability, low toxicity, low direct GWP and low cost. CO_2 systems are also likely to be smaller than those using common refrigerants. These benefits are offset by the fact that the use of CO_2 in air conditioning applications requires high operating pressures and results in low operating efficiencies, thereby contributing to increased indirect CO_2 emissions through higher energy consumption. Actual system tests of non-optimized air conditioning systems have demonstrated COPs[21] up to 2.25 at air inlet temperatures of 35°C compared to COPs up to 4.0 for typical HCFC-22 equipment.

<u>Water chillers:</u> Positive displacement chillers using ammonia as the refrigerant are available in the capacity range from 100 to 2000 kW with a few being larger. Recommended practice guidelines limit the use of large ammonia systems in public buildings to situations in which the ammonia is confined to the machine room where alarms, venting devices and scrubbers can enhance safety. Guidelines are available for the safe design and application of ammonia systems. Modern, compact factory-built units contain the ammonia much more effectively than old ammonia plants.

Hydrocarbon refrigerants have a long history of application in industrial chillers in petrochemical plants. Before 1997, they were not used in comfort air conditioning chiller applications due to reservations about systems safety. European manufacturers now offer a range of positive displacement hydrocarbon chillers. Unit sales of hydrocarbon chillers amount to about 100 to 150 annually, primarily in northern Europe. This is a small number compared with the installed base of more than 77,000 HCFC and HFC chillers in Europe. Typical safety measures include proper placement and/or gas-tight enclosure of the chiller, application of the low-charge system design, fail-safe ventilation systems, and gas detector alarm-activating systems. Efficiency is similar to that of equivalent HCFC-22 products. The cost of hydrocarbon chillers is higher than that of HCFC or HFC equivalents.

CO_2 is being investigated for a wide range of potential applications. However, CO_2 does not match cycle energy efficiencies of fluorocarbon refrigerants for typical water chilling applications and, consequently, there is no environmental incentive to use CO_2 in chillers in place of HFCs. There has been no commercial application of CO_2 in chillers to date.

---

[21] COP stands for Coefficient of Performance, a measure of the energy efficiency of a refrigerating system.

Water is an attractive alternative because it is non-toxic and non-flammable. However, it is a very low-pressure refrigerant. The low pressures and very high volumetric flow rates required in water vapour-compression systems necessitate compressor designs that are uncommon in the air conditioning field. The few applications that exist use water as a refrigerant to chill water or to produce ice slurries by direct evaporation from a pool of water. These systems currently carry a cost premium of more than 50% above conventional systems. The higher costs are inherent and are associated with the large physical size of water vapour chillers and the complexity of their compressor technology.

Water-heating heat pumps: Some European manufacturers are using propane (HC-290) or propylene (HC-1270) as refrigerants in small low-charge residential and commercial water-heating heat pumps. The hydrocarbon circuit is typically located outdoors or in a ventilated indoor space and uses ambient air, earth or ground water sources.

In water-heating applications, propane will yield the same or slightly higher energy efficiency as HCFC-22. When designing new heat pump systems with propane or other flammable refrigerants, adequate safety precautions must be taken to ensure safe operation and maintenance. Several standards that regulate the use of hydrocarbons in heat pumps exist or are being developed in Europe, Australia and New Zealand. An example of a standard under development is an update of European Standard EN 378.

The transcritical CO_2 cycle exhibits a significant temperature glide on the high temperature side. Such a glide can be of benefit in a counter-flow heat exchanger. Heat pumps using CO_2 as the refrigerant can generate water temperatures up to 90°C and have been developed in Japan for home use. A typical heating capacity is 4.5 kW.

Ammonia has been applied in medium-size and large-capacity heat pumps, mainly in Scandinavia, Germany, Switzerland and the Netherlands. System safety requirements for ammonia heat pumps are similar to those for ammonia chillers.

What are alternative not-in-kind technologies?

A number of other non-traditional technologies have been examined for their potential to reduce the consumption and emission of HFCs. These include desiccant systems, Stirling cycle systems, thermoelectrics, thermoacoustics and magnetic refrigeration. With the exception of the Stirling cycle and desiccants, these all suffer such large efficiency penalties that the consequent indirect effects would overwhelm any direct emission reduction benefit. Despite receiving research interest, the Stirling cycle has remained limited to niche applications and has never been commercialized for air conditioning. In

high latent load applications, desiccant systems have been applied to supplement the latent performance of traditional mechanical air conditioning.

What is the overall reduction potential?

Direct GHG emissions of residential and commercial air conditioning and heating equipment can be reduced by about 200 MtCO$_2$-eq yr^{-1} relative to the BAU scenario (2015). Specific costs range from –3 to 170 US$/tCO$_2$-eq. Improved system energy efficiencies can significantly reduce indirect GHG emissions, leading in some cases to overall savings of 75 US$/tCO$_2$-eq. Opportunities to reduce direct GHG (i.e. refrigerant) emissions can be found in (1) a more efficient recovery of refrigerant at end-of-life (in the mitigation scenario, assumed to be up to 50% and 80% for developing and developed countries, respectively); (2) charge reduction (up to 20%); (3) better containment; (4) the use of non-fluorocarbon refrigerants in suitable applications.

4.3   What are the most important findings for mobile air conditioning?

What are past and current trends in MAC?

MAC systems have been mass produced in the USA since the early 1960s and in Japan since the 1970s. The significant increase in the numbers of air-conditioned cars in Europe and also in developing countries began later, around 1995.

As indicated in Table TS-16, the global CFC-12 fleet has decreased from approximately 212 million vehicles in 1990 to 119 million vehicles in 2003, while the HFC-134a fleet has increased from fewer than 1 million in 1992 up to 338 million in 2003.

On the basis of a BAU scenario, and including high economic growth of rapidly-developing countries, Figure TS-11 shows a projected increase in the air-conditioned fleet that reaches approximately 965 million air-conditioned vehicles by 2015.

What are current emissions and projections?

Emissions from vehicles continuing to use MAC systems with CFC-12 are about 531 g yr^{-1} vehicle^{-1} when all types of emissions are included (fugitive emissions represent 220 g yr^{-1} vehicle^{-1}). Recovery and recycling are performed for CFC-12 in end-of-life of vehicles in some countries, but the CFC-12 will still eventually be released into the atmosphere following these practices unless it is destroyed. The annual emissions (2002) from the global fleet of air-conditioned CFC-12-based vehicles are about 514 MtCO$_2$-eq yr^{-1} (fugitive emissions represent 213 MtCO$_2$-eq yr^{-1}). Projections suggest

Table TS-16. MAC fleet evolution and refrigerant choice from 1990 to 2003.

Year	AC vehicle fleet (million)	
	CFC-12	HFC-134a
1990	212	-
1991	220	-
1992	229	0.7
1993	229	10
1994	222	27
1995	215	49
1996	206	74
1997	197	100
1998	186	128
1999	175	161
2000	163	198
2001	149	238
2002	134	285
2003	119	338

that the effect of the MAC subsector on radiative forcing of the climate system will be dominated by emissions of CFC-12 through to 2006 at the earliest.

The direct emissions from the global fleet of HFC-134a-based MAC systems are estimated to be about 220 g yr^{-1} vehicle^{-1} including fugitive emissions that account for 130 g yr^{-1} vehicle^{-1} or, when expressed as CO_2-equivalents, about 96 $MtCO_2$-eq yr^{-1} including fugitive emissions that account for 56 $MtCO_2$-eq yr^{-1}. Current 'do-it-yourself' kits for recharging air conditioners that use disposable cans of fluids lead to at least a twofold increase in emissions to the atmosphere

compared with professional servicing that uses more efficient fluid-delivery components.

Table TS-11 emphasizes the rapid change in refrigerant choice that has resulted from implementation of the Montreal Protocol.

What are the indirect energy-related CO_2 emissions related to MAC operation?

The operation of MAC systems worldwide leads to a substantial indirect effect through increased fuel use and associated CO_2 emissions. Current fuel consumption tests and standards in the motor vehicle industry do not explicitly evaluate this effect, which varies by climate zone. Depending on climate conditions, it is estimated that MAC systems represent 2.5–7.5% additional fuel consumption, or about 126 kg (Tokyo) to 369 (Phoenix) kg of CO_2 yr^{-1} vehicle^{-1}. With the total number of air-conditioned vehicles estimated to be 457 million (in 2003), the indirect effect corresponds to as much as 114 $MtCO_2$-eq yr^{-1} (when averaged across the global fleet and assuming an average value of 250 kg CO_2 yr^{-1} vehicle^{-1}) relative to 750 $MtCO_2$-eq yr^{-1} of direct emissions.

What are possible future trends in MAC?

The following options exist for reducing GHG emissions: (1) enhance the current HFC-134a systems; (2) move to lower GWP refrigerants, either HFC-152a or CO_2. Hydrocarbons, even if they are low GWP refrigerants and efficient when properly used, are not seen as suitable options by car makers and suppliers due to safety concerns.

'Improved' HFC-134a systems are being introduced progressively onto the market, at an additional cost varying from 24–36 US$ per system. These employ tighter hoses, compressors and service valves, all of which reduce leakage. Recent studies suggest that improved HFC-134a systems and improved servicing could result in emissions of about 70 g yr^{-1} vehicle^{-1}, which is about 60% lower than those from current HFC-134a systems. Improvements in recovery practices and service training could further reduce emissions. Significant energy savings are related to the use of variable volume compressors with an external control, which are also being progressively introduced in the market. Additional savings are related to a MAC system design that integrates energy efficiency constraints.

Several recent studies suggest that improvements in energy efficiency, through measures such as on/off controls instead of standard continuous operation, insulation of doors and roofs and so forth, could reduce these emissions by about 30–40%, which represents 30–40 $MtCO_2$-eq yr^{-1}.

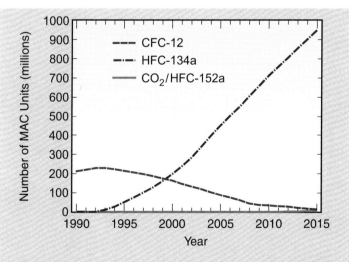

Figure TS-11. MAC fleet evolution from 1990 to 2015 in the BAU scenario.

Table TS-17. Comparison of MAC options.

	HFC-134a (reference)	Improved HFC-134a	CFC-12 (old type) development)	CO$_2$ (under development)	HFC-152a (direct system, under devolpment)
Substance characteristics					
Radiative efficiency (W m^{-2} ppb^{-1})	0.16	0.16	0.32	See Ch. 2	0.09
Atmospheric lifetime (yr)	14	14	100	See Ch. 2	1.4
Direct GWP (100-yr time horizon)					
- This report	1410	1410	10,720	1	122
- UNFCCC[a]	1300	1300	8100	1	140
Technical data					
Stage of development	Commercial	Near commercial	Commercial	Demonstration	Demonstration
System lifetime	12–16	12–16	12–16	12–16	12–16
Cooling capacity (kW)	6	5	6	6	6
Charge (kg/system)					
- range	0.7–0.9	0.6–0.75	1–1.2	0.5–0.7	0.45–0.55
- relative figures (%)	100	80	125	70	70
# Charges over lifetime	2–3	1–2	4	2–4	1–2
Coefficient of Performance (COP)	0.9–1.6	1.2–2.5	0.9–1.2	0.9–2.0	1.2–2.0
Energy consumption (relative figures)	100	80	130	70	70
Emissions per functional unit					
Direct emissions					
- in % of charge yr^{-1}	15	7	20	15	7
- in kg CO$_2$-eq yr^{-1}	166	64	1782	0.09	4.9
- relative figures (%)	100	40	1043	0.05	2.9
Indirect CO$_2$-emissions (kg CO$_2$ yr^{-1})					
- Sevilla	184	147	239	129	129
- Tokyo	126	101	163	88	88
- Phoenix	369	295	480	258	258
End-of-life emissions recovery efficiency[b]	0	50	0	0	50
TEWI (kg CO$_2$-eq 14 years)					
- Sevilla	4900	2954	28,294	1807	1875
- Tokyo	4088	2310	27,230	1233	1301
- Phoenix (without recovery)[b]	7490	5026	31,668	3613	3681
Costs per functional unit					
Investment costs (US$)	215	24–36	n.a.	48–180	48

Notes:

[a] The GWP values used for the calculations are the UNFCCC GWPs.

[b] Due to large uncertainties in the effectiveness of recovery, the TEWI calculations have not taken recovery into account and so the average direct emission per year for 'improved HFC-134a systems' is 100 g yr^{-1}.

MAC systems using CO_2 have been successfully demonstrated in passenger vehicles and commercial buses. CO_2 has a GWP that is 1300-fold smaller than HFC-134a, thereby limiting the direct effects per charge. However, CO_2 systems operate at pressures about eightfold higher than those of CFC-12 and HFC-134a (discharge pressures in the order of 12 MPa), and since leak flow rates are related to the square of the pressure, CO_2 systems imply a much more difficult containment. To date, CO_2 systems have shown an energy efficiency comparable to or better than improved HFC-134a systems in cooler ambient climates, but they are likely to be less efficient in warmer climates. Nevertheless, as indicated in Table TS-17, due to its nearly negligible direct effect, the TEWI of CO_2 systems is significantly better than that of improved HFC-134a systems. Barriers to commercialization include the resolution of additional costs and safety issues associated with the release of CO_2 within the passenger cabin and maintenance issues as well as the conversion cost of the service system.

HFC-152a has also been successfully demonstrated in MAC systems. While HFC-152a systems can use the same components as HFC-134a systems, the former require an added safety system because HFC-152a is flammable while HFC-134a is non-flammable. Direct emissions (in CO_2-equivalents) are very low (92% reduction, referred to as the HFC-134a baseline). HFC-152a systems have so far shown an energy efficiency comparable with or better than improved HFC-134a systems, but the energy gain could be lost if a secondary loop system required by safety considerations is used. However, its overall climate impact, expressed in TEWI, is still significantly lower than that of HFC-134a and in the same order of magnitude of CO_2 systems. The principle barriers to commercialization at the present time are the resolution of the flammability risk and the assurance of a commercial global availability of HFC-152a.

Table TS-17 presents a comparison of the primary MAC systems currently in use, under development or demonstrated. For each option, issues of relative cost are indicated, along with points that need to be considered to evaluate the effects on the radiative forcing of the climate system, including the indirect effects. It has to be emphasized that the choice of a given technical option in a given year will have only a limited effect in the first years of introduction due to all the refrigerant banked in – and emitted by – the current fleet.

4.4 What are the most important findings for foams?

What applications are foams currently used for and why?

Foamed (or cellular) polymers have been used historically in a variety of applications that utilize the potential for creating either flexible or rigid structures. Flexible foams continue to be used effectively for furniture cushioning, packaging and impact management (safety) foams. Rigid foams are used primarily for thermal insulation applications such as those required for appliances, transport and in buildings. In addition, rigid foams are used to provide structural integrity and buoyancy.

For thermal insulation applications (the majority of rigid foam use), mineral fibre alternatives (e.g. glass fibre and mineral wool) have been, and continue to be, major not-in-kind alternatives. Table TS-18 illustrates the major benefits and limitations of both approaches.

The implications of these relative benefits and limitations vary substantially, both between products within a category and between applications. This makes a generic conclusion about

Table TS-18. Benefits and limitations of the use of both mineral fibres and cellular polymers in thermal insulation applications.

	Mineral Fibre	*Cellular Polymers*
Benefits	• Initial cost • Availability • High max. temperature • Fire performance	• Blowing-agent-based thermal properties • Moisture resistance • Structural integrity • Lightweight
Limitations	• Air-based thermal properties • Moisture resistance[a] • Low structural integrity	• Fire performance (organic) • Limited max. temperature • Initial cost (in some cases)

Notes:
[a] Potential effect on long-term thermal performance

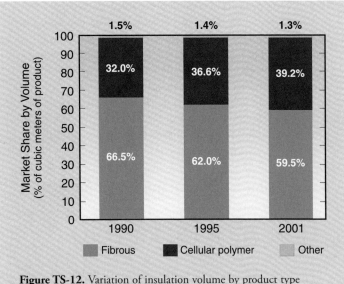

Figure TS-12. Variation of insulation volume by product type (1990–2001) in Europe.

preferences impossible. The current thermal insulation market supports a variety of solutions (at least 15 major product types), and this reflects the range of requirements demanded for the applications served. Unfortunately, there are only limited data available on the thermal insulation market at global and regional levels. One of the complexities of global market analysis is the difference in building practices around the world, which often respond to material availability and climatic conditions.

Purely as an example, a systematic and periodic analysis of the European thermal insulation market has enabled the trends over the period from 1990–2001 to be identified (Figure TS-12). This analysis indicates a growing reliance on foamed products in thermal insulation applications that has been driven in part by the increased use of metal-faced panels in Europe, which in turn relies increasingly on foamed cores. However, the harmonization of fire classifications in Europe over the coming 5 years may cause the trend to be arrested or even reversed. Such is the volatility of these markets and the importance of maintaining ranges of product types.

In reviewing the not-in-kind options, it is important to acknowledge continuing development. For example, it seems likely that the use of vacuum insulation panels (evacuated and sealed foam matrices) in domestic refrigerators and freezers will increase. In fact, most Japanese units already contain at least one such panel in strategic design positions. Other opportunities include multilayer reflective foils but the thermal efficiency of these is, as yet, far from proven.

The relationship between foam products, processes for manufacture and applications is complex. Table TS-19 summarizes the main interrelationships between generic product types and applications for both non-insulating and insulating foams, while the main chapter deals with the additional overlay of processes for manufacture.

Table TS-19. Main interrelationships between generic product types and applications for both non-insulating and insulating foams.

Foam type (insulating)	Application Area							
	Refrigeration and transport			**Buildings and building services**				
	Domestic appliances	Other appliances	Reefers and transport	Wall insulation	Roof insulation	Floor insulation	Pipe insulation	Cold stores
Polyurethane	√	√	√	√	√	√	√	√
Extruded polystyrene			√	√	√	√	√	√
Phenolic				√	√		√	√
Polyethylene						√	√	

Foam type (non-insulating)	Application area					
	Transport		**Comfort**		**Packaging**	**Buoyancy**
	Seating	Safety	Bedding	Furniture	Food and other	Marine and leisure
Polyurethane	√	√	√	√	√	√
Extruded polystyrene					√	√
Polyethylene					√	√

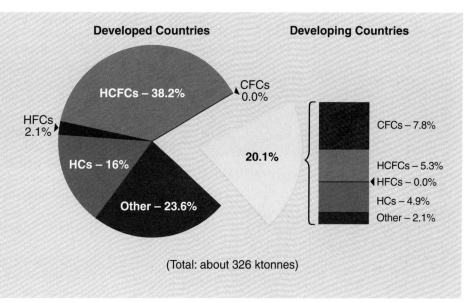

Figure TS-13. Breakdown of blowing-agent use by type and country type (2001).

Developed Countries

Developing Countries

HCFCs – 38.2%

CFCs 0.0%

HFCs 2.1%

20.1%

HCs – 16%

Other – 23.6%

CFCs – 7.8%

HCFCs – 5.3%

HFCs – 0.0%

HCs – 4.9%

Other – 2.1%

(Total: about 326 ktonnes)

What blowing agents have been used historically and what are the trends for the future?

At the point of discovery of the ozone hole in the early 1980s, virtually all of the applications and product types listed above used CFCs as either a primary blowing agent (rigid foams) or auxiliary blowing agent (flexible foams). The total consumption of CFCs by the foam sector in 1986 was approximately 250 ktonnes (165 ktonnes rigid; 85 ktonnes flexible). Blowing-agent use overall grew by a further 30% over the next 15 years, despite improved blowing efficiencies and reduced losses. In the meantime, however, a variety

of alternative blowing agents have been evaluated and adopted. These include HCFCs (as transitional substances), hydrocarbons, HFCs, methylene chloride (for flexible foams) and various forms of CO_2. Figure TS-13 summarizes the situation in 2001.

Figure TS-14 illustrates the projected growth of blowing-agent consumption in the rigid foam sector for the period to 2015.

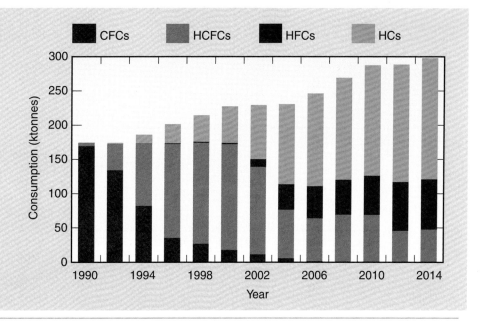

Figure TS-14. Projected growth in blowing-agent consumption in rigid foams – beyond 2000.

CFCs   HCFCs   HFCs   HCs

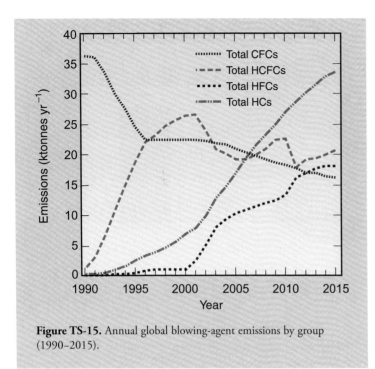

Figure TS-15. Annual global blowing-agent emissions by group (1990–2015).

What are the implications of past, present and future use patterns?

Past, present and future use patterns have impacts on both emissions and the accrual of banked blowing agents. Figure TS-15 illustrates the projected emissions between 1990 and 2015 based on the historical and future use of blowing agents in foam. The graph shows the projected annual emission of all blowing-agent types up to 2015. Table TS-20 assesses the projected development of banks by region and application area. It is clear that much of the emissions from the foams used in buildings has yet to take place.

What drives the selection of blowing agents?

Thermal conductivity
The ability to retain blowing agents within foams provides opportunities to improve the efficiency of the thermal insulation relative to air-filled products. However, such benefits can only be achieved when the thermal conductivities of the retained blowing agents are lower than that of air. This is the case for all blowing agents highlighted in the Figure

Table TS-20. Total global accrued banks of blowing-agent types by group (1990–2015).

Application area	Blowing agent	1990 (tonnes)		2000 (tonnes)		2015 (tonnes)	
		Developed Countries	Developing Countries	Developed Countries	Developing Countries	Developed Countries	Developing Countries
Appliances and transport	CFC	378,000	108,000	238,000	222,000	450	15,500
	HCFC	0	0	177,000	32,100	75,700	265,000
	HFC	0	0	1150	0	154,000	0
	HC	0	0	87,100	31,600	354,000	329,000
	All agents	378,000 (24.6%)	108,000 (53.7%)	503,250 (20.1%)	285,700 (58.1%)	584,150 (17.2%)	609,500 (58.5%)
Polyurethane panel subtotal	CFC	233,000	34,300	283,000	70,500	262,000	75,100
	HCFC	0	0	96,000	3700	142,000	94,800
	HFC	0	0	2150	0	135,000	0
	HC	0	0	43,800	250	238,000	0
	All agents	233,000 (15.1%)	34,300 (17.1%)	424,950 (16.9%)	74,450 (15.1%)	777,000 (22.9%)	169,900 (16.3%)
Buildings and other subtotal	CFC	921,000	58,800	964,000	127,300	769,000	106,000
	HCFC	5200	0	568,000	4650	683,000	156,000
	HFC	0	0	200	0	269,000	150
	HC	1150	0	47,500	50	311,000	0
	All agents	927,350 (60.3%)	58,800 (29.2%)	1,579,700 (63.0%)	132,000 (26.8%)	2,032,000 (59.9%)	262,150 (25.2%)
Total	CFC	1,532,000	201,100	1,485,000	419,800	1,031,450	196,600
	HCFC	5200	0	841,000	40,450	900,700	515,800
	HFC	0	0	3500	0	558,000	150
	HC	1,150	0	178,400	31,900	903,000	329,000
	All agents	1,538,350	201,100	2,507,900	492,150	3,393,150	1,041,550

TS-15. However, the relative performance of the various blowing-agent types does vary with temperature; for example, the comparative advantages of HFCs over hydrocarbons are greater in refrigerators (average temperature of 5°C) than in water heaters (average temperature of 40°C). In addition, the size and shape of the cells also affects the overall performance of the foams and, therefore, product comparisons are not always straightforward.

Flammability (product and process)
The overall flammability of a foam product is influenced by the choice of polymeric matrix and facing material as well by the choice of blowing agent. However, it is often the case that the contribution of the blowing agent can shift the classification of a product or can change the attitude of insurers to the risk posed. In addition, the handling of certain flammable blowing agents can create fundamental challenges in some foam processes. This is particularly the case for small and medium enterprises (SMEs), for which economies of scale do not exist and discontinuous processes dominate. In the case of flammable blowing agents, investment criteria vary considerably depending on whether the investment relates to a new plant or modification of an existing plant. In the latter case, costs can often be prohibitive if the plant is old or if it is owned by a small or medium enterprise. Employer liability issues can also be a cause for concern in some regions with a strong litigious heritage.

What has already been done to minimize use?

When the three aspects presented above are taken into consideration, HFCs emerge as the preferred option in several key sectors, although the progress made in developing alternative technologies has ensured that uptake has been constrained. An example of this is the continuing extension of the application of hydrocarbon technologies. However, even in cases where HFCs have been adopted, there are two additional points to consider:

(1) How much HFC is required in the formulation to achieve the required performance?
(2) Which HFC should be chosen?

The cost of HFCs can be a general constraint on uptake. Blowing-agent costs typically represent a significant element of overall variable costs. Accordingly, any substantial elevation in blowing-agent costs can influence variable costs by up to 15%. In a highly competitive market, such increases are unsustainable and prevent selection, unless formulation changes can be made to reduce dependence on the more expensive blowing agent. An example of this is the co-blowing of HFC-based polyurethane foams with CO_2 generated from the reaction of isocyanate and water.

Nonetheless, the overall decision is a complex one based on combinations of direct blowing-agent cost, related formulation issues (e.g. the use of increased flame retardant or foam density penalties), product performance, process safety and capital costs. As noted previously, the latter are particularly relevant to SMEs and other small volume consumers.

The net effect of the above selection of considerations on HFC demand has been to reduce previous (1999) projections of 115 ktonnes consumption in 2010 to 60 ktonnes in 2010 in this current report. This can already be considered as a reduction facilitated by the application of responsible use principles by the foam industry.

Lowest GWP selection
Since the two prime liquid HFCs (HFC-245fa and HFC-365mfc) have similar 100-year GWPs, the choice between the two has been driven more by a consideration of respective boiling points and blowing efficiencies. For gaseous blowing agents, HFC-152a has a much lower 100-year GWP than HFC-134a. However, HFC-152a is more flammable than HFC-134a and is also emitted much faster from some types of foam (e.g. extruded polystyrene). This can mean that short-term impacts of HFC-152a use can be as significant as those of HFC-134a. In addition, the thermal insulating benefit of HFC-152a can be short-lived. All of these factors have to be evaluated during the selection of the appropriate blowing agent.

What additional measures can reduce future emissions further and what actions are required?

Further substitution
While the adoption of responsible use criteria in HFC selection has successfully reduced the consumption of HFCs in the foam sector by nearly 50% over that predicted in 1999, there are several areas in which further substitution may be possible over the next 5–10 years. For example:

• wider hydrocarbon use in polyurethane spray foam;
• wider CO_2 use in extruded polystyrene (XPS);
• wider hydrocarbon use in appliance foams;
• changes in the attitude of insurers to hydrocarbons in panels.

Although the impacts of each of these trends can be modelled individually, the uncertainties are too great to be meaningful. Accordingly, in this assessment, two high-level mitigation scenarios are presented to evaluate the impact of pursuing such options.

Process of good practice
Work has already been initiated on establishing procedures for the identification and minimization of process losses. While this work is important in setting the right tone for the handling of HFCs in foam processes, the potential savings are unlikely to reach more than 2–3% of total life-cycle emissions, since most processes are already well enclosed. One

exception might be in the emissions during polyurethane spray application, where further effort is still required to quantify losses; future efforts may lead to improvements in spray head design.

Waste management
Waste minimization is a clear objective of all businesses. However, foam manufacturers face specific challenges:
• product proliferation requiring more versatile production processes;
• intrinsic fabrication losses (e.g. cutting pipe sections from block foam).
The management of this waste is therefore a key issue in minimizing emissions. The mitigation-scenario models assess the effects of a combination of process and waste management improvements.

Bank management (refrigerators)
The size of existing and future banks of blowing agent in the appliance and transport sectors has been estimated. The baseline scenario already takes into account the recovery activity occurring in Europe and Japan, so bank sizes do not automatically equate to future emissions. With proven technology and the cost of recovery from refrigerators currently estimated at 10–50 US$/tCO$_2$-eq, it would appear reasonable to assume that all refrigerator foams could be managed at end-of-life by 2015, if the investment in plants to do so were appropriately dispersed geographically. This would, however, involve investment in developing as well as developed countries. One scenario evaluated in this report examines the potential impact of all appliances being processed at end-of-life with anticipated recovery levels in excess of 80% of the original blowing-agent loading.

Bank management (buildings)
For the building sector, the technical feasibility and economic viability of blowing-agent recovery is less well established. Activities such as the Japan Testing Center for Construction Materials (JTCCM) project in Japan are assisting the development of further knowledge in this area. At present, however, the general consensus is that recovery will be considerably more expensive than from appliances because of the lower yield (caused by losses in the use and recovery phases) and the additional costs of demolition waste separation.

One exception to this trend is in the metal-faced panel market where blowing-agent retention and ease of dismantling may allow recovery through existing refrigerator plants. The bank available from polyurethane panels has been assessed and is expected to exceed 700 ktonnes in fluorocarbon blowing agents by 2015. Recovery costs are expected to be in the same range as for appliances, but work is continuing to confirm this. Both scenarios have been modelled, but with a more modest expectation of 20% recovery from traditional building sources.

What is the significance of these potential scenarios to wider climate and ozone strategies?

The baseline emissions for the BAU scenario are shown in Figure TS-16. Because the life cycles of foams are so significant, all graphs in this section illustrate the likely impact of possible emission reduction scenarios up to 2100. The baseline assumption is a freeze at 2015 consumption levels for both HCFCs and HFCs. It is assumed that HCFCs are phased-out linearly between 2030 and 2040. Bearing in mind that technology developments are likely to continue in the foams sector, reliance on HFCs is not expected beyond 2030, and a linear decline is assumed from 2020. As a further reference point, the ongoing emissions from the banks already established at 2015 are also shown. The three primary elements of the mitigation scenario can be summarized as follows:
• a linear decrease in the use of HFCs between 2010 and 2015, leading to a 50% reduction by 2015;
• the adoption of production emission reduction strategies from 2005 onwards for all block foams and from 2008 onwards in other foam subsectors;
• the extension of existing end-of-life measures to all appliances and steel-faced panels by 2010 together with a 20% recovery rate from other building-based foams from 2010.

The resulting impacts of these three measures are as shown in Figure TS-17 and Table TS-21.

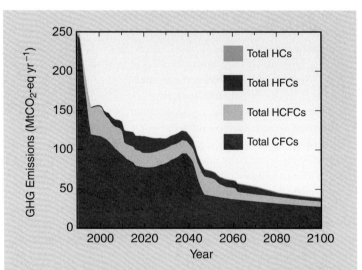

Figure TS-16. GWP-weighted blowing-agent emissions by group (1990–2100) – business-as-usual scenario.

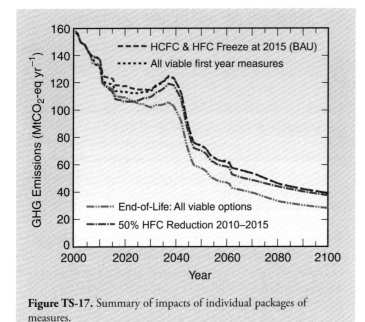

Figure TS-17. Summary of impacts of individual packages of measures.

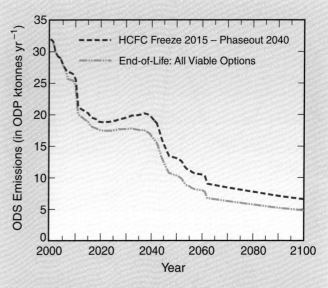

Figure TS-18. Impact of all viable end-of-life measures on ODS emission reduction.

It can be seen that focusing on reducing HFC consumption provides the most significant saving in the period up to 2015 and, on the basis that any such reduction is extrapolated out to use patterns beyond 2015, this focus offers the greatest 'HFC-specific' benefit to 2100 as well. In contrast, end-of-life measures deliver lower savings during the period up to 2015, but they do have the potential to deliver more overall savings in the period up to 2100 if all blowing-agent types are considered. The value is particularly significant for CFCs, for which the GWPs are high and there is an incremental effect of ozone depletion.

The potential savings in ODSs emissions from all viable end-of-life strategies is shown n Figure TS-18 based in ODP tonnes. It can be seen that year-on-year savings in the order of 2000–3000 ODP tonnes will accrue for the period to 2100.

Table TS-21. Summary of impacts of individual packages of measures by blowing agent type: cumulative emission reductions resulting under each scenario assessed.

Measure	Year	Cumulative Emission Reductions			
		CFCs (tonnes)	HCFCs (tonnes)	HFCs (tonnes)	CO_2-equivalents (MtCO$_2$-eq)
HFC consumption reduction (2010–2015)	2015	0	0	31,775	36
	2050	0	0	225,950	259
	2100	0	0	352,350	411
Production/installation improvements	2015	78	14,450	16,700	36
	2050	58	31,700	32,700	68
	2100	47	24,350	26,500	55
End-of-life management options	2015	8545	16,375	105	52
	2050	64,150	144,650	88,540	540
	2100	137,700	358,300	194,800	1200

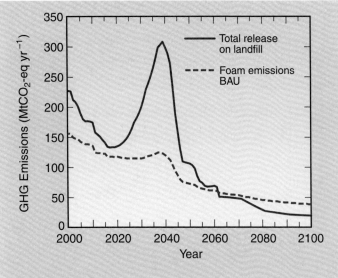

Figure TS-19. The effects of applying different initial landfill emission assumptions.

The estimates for reductions in both GHG and ODS emissions identified in this analysis from end-of-life measures are relatively conservative, since they are measured against a baseline where only 10–20% of blowing loss is accounted for when the foam is land-filled. This partially explains why there are significant emissions after 2065. Effectively, the landfill sites have become banks of their own. If a more aggressive assumption were to be applied to foams destined for landfill (i.e. 100% emission at the point of landfill), Figure TS-19 illustrates the consequence in GHG emission terms.

While it is recognized that the truth probably lies somewhere between the two extremes, the potential release of such significant amounts of blowing agent over a relatively short period (2030–2050) draws attention to the potential incremental value of end-of-life management as a mitigation option.

4.5 What are the most important findings for medical aerosols?

Medical aerosols are important in the treatment of asthma and COPD

Asthma and chronic obstructive pulmonary disease (COPD) are the most common chronic diseases of the air passages (airways or bronchi) of the lungs and are estimated to affect over 300 million people worldwide. These illnesses account for high health care expenditure and cause significant loss of time from work and school and, in addition, COPD is responsible for premature death.

Asthma is a chronic condition with two main components, airway inflammation and airway narrowing. Most asthma patients have symptoms every day, with more severe attacks intermittently. Asthma most often starts in childhood, and prevalence varies from approximately 1% in some countries such as Indonesia to over 30% in children in New Zealand and Australia.

COPD is a condition typified by narrowing and inflammation of the airways in conjunction with damage to the lung tissue. It is caused primarily by cigarette smoking, with environmental air pollution as a potential cofactor, and ultimately leads to permanent disability and death. The prevalence of COPD in many developed countries is between 4–17% in adults over 40 years old. Data are less certain in developing countries, but figures as high as 26% have been quoted. Smoking is declining in some developed countries, but in developing countries both smoking and COPD are increasing.

Inhaled therapy is currently and likely to remain the gold standard for treatment. Inhalation of aerosol medication of a specific particle size (1–5 microns) optimizes the local clinical effect in the airways where it is needed, with minimal side effects. Inhalation aerosols have been the subject of significant investment in research and development, in response to both therapeutic and environmental needs. Currently the two main methods of delivering respiratory drugs for most patients are the metered dose inhaler (MDI) and the dry powder inhaler (DPI).

What is a MDI?

MDIs are the dominant form of treatment for asthma and COPD worldwide. The MDI was introduced in the mid-1950s with CFC-11 and CFC-12 as propellants; CFC-114 was introduced later. In order to accomplish the phase-out of CFCs under the Montreal Protocol, the MDI industry undertook an exhaustive search for an appropriate alternative aerosol propellant. A medical propellant must be safe for human use and meet several additional strict criteria related to safety and efficacy: (1) liquefied gas with appropriate vapour pressure, (2) low toxicity, (3) non-flammable, (4) chemically stable, (5) acceptable to patients (in terms of taste and smell), (6) appropriate solvency characteristics and (7) appropriate density. It was extremely difficult to identify compounds fulfilling all of these criteria and in the end only two HFCs – HFC-134a and HFC-227ea – emerged as viable alternatives to CFCs.

The components and formulations of CFC-based MDIs had to be substantially modified to use the new HFC propellants. As drug devices, MDIs are subject to extensive regulation by national health authorities to ensure product safety, product efficacy and manufacturing quality. Therefore, the process for developing HFC MDIs is essentially the same as the development of a wholly new drug product in that it involves full clinical trials for each reformulated MDI. The development (technical, pharmaceutical, clinical) costs for the CFC-to-HFC transition were estimated to be approximately US$ 1 billion in 1999 and hence now will be significantly higher. Similar costs would be expected for de-novo DPI development programmes to replace existing molecules in MDIs.

What is a DPI?

DPIs deliver powdered medication of specific particle size, do not use propellants and have no impact on the ozone layer or climate. Delivery of the active drug in powder form is technically difficult. For example, particles of respirable size tend to have poor flow characteristics due to adhesive interparticle forces. Additionally, most DPI formulations are sensitive to moisture during processing, storage and in use, thereby limiting their utility in humid climates.

Early DPIs providing single premeasured doses had limited use in the 1960s and 1970s. Significant technical progress has led to patient-friendly multidose DPIs becoming more widely available in the past decade, and this has mitigated the increase in MDI use. DPIs have been formulated successfully for many inhaled drugs and are now widely available in many but not all countries. However, they are not an alternative to pressurized MDIs for all patients or for all drugs.

The relative cost of DPIs is high, especially compared with MDIs containing salbutamol, which still account for approximately 50% of the MDIs prescribed worldwide. In a study conducted to compare the costs across seven European countries, the salbutamol DPIs were found to cost on average 2.6 times more than MDIs.

What factors influence treatment choice?

Primary prevention of asthma is not yet feasible, while primary prevention of COPD entails not commencing tobacco smoking. The prevalence of asthma and COPD is likely to continue to increase.

The choice of the most suitable drugs and inhaler are decided by physician and patient based on many factors including disease and severity, compliance, ease of use, cost, availability and patient preference. Inhaler devices are only effective if used correctly. Patients will often be able to use one device correctly but not another. Both MDIs and DPIs have an important role in treatment, and no single delivery system is universally acceptable for all patients. It is critical to preserve the range of therapeutic options.

MDIs are the dominant form of treatment for asthma and COPD worldwide. In developed countries, the proportion of current MDI to DPI use varies substantially between countries: in the USA, 9:1 (MDI:DPI); in the UK, 7:3; in Sweden, 2:8. This relates to a number of factors, including availability (e.g. multidose DPIs are only recently available in the USA compared with a local company with a long tradition of DPI manufacture in Sweden) and affordability.

What are the future technical developments?

Annual growth in the global market in inhaled asthma/COPD medication up to 2015 is projected at approximately 1.5–3% yr^{-1}. A large portion of CFCs are being replaced by HFCs (approximately 90% HFC-134a and 10% HFC-227ea), and all MDI use in the developed world will be HFC by 2010. From a peak annual CFC use of over 15,000 tonnes in 1987–2000, CFC use in MDIs has fallen to an estimated 8000 tonnes, with HFC accounting for 3000–4000 tonnes, in the period 2001–2004, and by 2015, HFC use is estimated to rise to 13,000–15,000 tonnes. The lower use of HFCs compared to peak CFC use is partly due to increased use of DPIs and partly because some HFC MDIs use less propellant per actuation.

No major technical breakthroughs in device technology are expected in the short-term. Research and development for a new inhalation product is a lengthy, technically challenging and expensive process and typically takes over 10 years to reach the market. Future inhalation devices such as nebulizers and DPIs with a power source to make them independent of a patient's breath or small aqueous multidose devices will probably be more expensive than present-day DPIs and will therefore be even more expensive than HFC MDIs

In developing countries, inhaled therapy is almost exclusively with pressurized MDIs, either from multinational or local manufacturers. Improved economic circumstances together with the adoption of international treatment guidelines will likely substantially increase inhaled therapy. Affordable and less complex DPIs are technically feasible and could be manufactured locally in developing countries. There would be significant pharmaceutical difficulties in hot and humid climates, and they would remain more expensive than MDIs on a cost-per-dose basis. If these became available and achieved a significant market share, they could mitigate the future increases in volumes of HFC needed for MDIs.

What would be the cost of a complete switch from HFC MDIs to DPIs?

Typically, newer multidose DPIs contain more expensive drugs, whereas about 50% of MDIs contain salbutamol, which is less expensive and off patent. This explains part of the difference in individual inhaler cost. It has been estimated that by 2015 there could be as many as 340 million HFC MDI units containing salbutamol. Switching these to an equivalent salbutamol DPI would incur significant costs to health care systems. Hypothetical estimates of the cost of switching completely from HFC MDIs to DPIs (assuming a minimal twofold increase in price) would be on the order of an additional and recurrent US$ 1.7–3.4 billion yr^{-1} (150–300 US$/tCO$_2$-eq). The emission reduction achieved would be about 10 MtCO$_2$-eq yr^{-1} by 2015. This additional cost would significantly impact on patient care.

Would there be any medical constraints for a switch from HFC MDIs to DPIs?

Switching patients from reliable and effective medications has significant implications for patient health and safety, and the provision of a range of safe alternatives is critical before enforcing change on environmental grounds. Any future environmental policy measures that could impact on patient use of HFC MDIs would require careful consideration and consultation with physicians, patients, national health authorities and other health care experts.

What the are key conclusions?

- The major impact in reducing GWP with respect to MDIs is the completion of the transition from CFC to HFC MDIs.
- No major breakthroughs for inhaled drug delivery are anticipated in the next 10–15 years given the current status of technologies and the development time scales involved.
- The health and safety of the patient is of paramount importance in treatment decisions and in policy making that might impact those decisions.
- Based on the hypothetical case of switching the most widely used inhaled medicine (salbutamol) from HFC MDIs to DPI, the projected recurring annual costs would be in the order of US$ 1.7 billion with an effective mitigation cost of 150–300 US$/tCO$_2$-eq for a reduction of about 10 MtCO$_2$-eq yr^{-1} by 2015.

4.6 What are the most important findings for fire protection?

What are past and current trends in fire protection?

Halons are gases that display exceptional safety, efficiency and cleanliness in fire-fighting applications. These gases were widely used worldwide in both fixed and portable fire-extinguishing equipment beginning in the early 1960s. Because of their high ODPs, governments and fire protection professionals led the first sector-wide phase-out under the Montreal Protocol. This led to the development of a range of effective alternatives for new systems. Since fire protection is a highly regulated sector, adopting alternatives requires extensive changes in local, national and international standards, practices and technology. These changes have reduced unnecessary emissions from the halon bank and are also being applied to the halon alternatives.

There are two categories of applications that can require halon or an alternative: fixed systems and portable extinguishers. Halon 1301 dominated the market in fixed systems prior to the Montreal Protocol, and its remaining bank was about 45 ktonnes in 2000. Halon 1211 was primarily used in portable extinguishers, and the bank in 2000 was estimated at about 154 ktonnes. Halon 2402 was used predominantly in the former Soviet Union, and no information on banks or emissions is available in the literature. One estimate of 2000 emissions is 2.3 ktonnes for halon 1301 and 17.8 ktonnes for halon 1211, or about 5% and 11% of the bank per year, respectively. One study suggests that the emission rate for halon 1301 in fixed systems, excluding ships, aircraft and military systems, is as little as 0.12% yr^{-1} when an exceptional level of diligence is made in tracking and keeping the halon in place for providing critical fire protection. An extremely low (0.12% yr^{-1}) rate of emissions has been achieved in one region, primarily due to unique cultural factors together with unusually strong enforcement action, both of which may be difficult to replicate in other regions. On average, emission rates for fixed systems are about 2 ± 1% yr^{-1} and about twice that for portable extinguishers, that is, 4 ± 2% yr^{-1} of the bank (installed base including stocks for recharge).

Fire protection is strictly regulated in most countries. New agents/techniques can only be used following a demonstration of acceptable safety and fire-extinguishment performance according to specific protocols. It is important that countries without national standards strive to adopt the practices recommended in international standards in order to protect against the introduction of unsafe or ineffective alternatives.

Selecting an alternative to halon involves evaluating a wide range of factors. These include space and weight, cost, safety, requirements for 'cleanliness' (i.e. without residue or damage such as in the storage of records or cultural heritage buildings), environmental performance, effectiveness against a specific fire threat (fires in solid materials ('Class A' fires), flammable liquids ('Class B' fires) and energized electrical equipment ('Class C' fires) and special circumstances (e.g. very cold conditions).

Halon is no longer necessary in most (>95%) new installations that would have used halons in pre-Montreal Protocol times. The remaining new installations still using halons are principally in commercial aircraft and some military applications for which an effective alternative to halons has yet to be found. Among the applications formerly protected by halons, about half of today's new installations are based upon non-gaseous alternatives, such as water and dry powders, while the other half make use of in-kind gaseous agents, including a range of halocarbons and inert gases.

In fixed systems where a clean agent is necessary, the alternatives currently available are CO_2, inert gases (such as nitrogen and argon), HFCs, PFCs, HCFCs and, more recently, a fluoroketone (FK). Some of these alternatives have no significant effect on the climate system, while others have substantial GWPs. Only the HCFCs are also ozone-depleting. PFCs and HCFCs were used in the early stages of implementation of the Montreal Protocol but do not provide any advantage over other halocarbon clean agents. New PFC systems are no longer being produced due to the environmental impacts of these gases upon the climate relative to other alternatives with similar capabilities and costs. CO_2 systems may be appropriate for some applications but are lethal at concentrations necessary to extinguish fires, while inert gas systems may also be appropriate for use in some applications but have significant weight and volume impacts and are not recommended when speed of fire suppression is an issue, due to a discharge rate that is five to six times slower than that of halocarbon systems.

Table TS-22 presents a comparison of the primary systems currently in use, under development or demonstratied for clean, fixed systems of fire extinguishment suitable for occupied spaces (typically replacing halon 1301). For each option, issues of relative cost are indicated, along with considerations needed to evaluate the effects on the radiative forcing of the climate system and practical concerns such as system weight and space requirements, special capabilities and availability.

Figure TS-20 is an illustration of the production, emission levels and resultant bank sizes of halon 1301 and its HFC/ PFC/HCFC/FK alternatives for the period 1965–2015. The bank of Halon 1301 in 2002 is projected to be 42,434 tonnes with 2052 tonnes of emissions, which agrees well with atmospheric measurements indicating 1000–2000 tonnes of emissions. The combined bank of all HFC/PFC/HCFC/ FK alternatives in fixed systems in 2004 is estimated to be approximately 26,700 tonnes. PFCs make up about 2.5% of that total. By one estimate, the HCFC portion can be as high as approximately 3600 tonnes (about 13%). Studies suggest that emission rates of $2 \pm 1\%$ year^{-1} are now practical in these systems. At a 2% emission rate, the 2004 emissions represent 1.4 million tonnes of CO_2-equivalent (MtCO$_2$-eq).

For portable extinguishers, fire codes and costs are the primary drivers in choosing an alternative (typically replacing Halon 1211). Portable extinguishers employing HFCs and PFCs have found limited market acceptance due primarily to their high costs compared to more traditional extinguishing agents such as CO_2, dry chemicals and water. HCFC acceptance has been greater but is also limited by its high cost relative to those of more traditional agents. Dry chemical agents are between about six and sixteen times cheaper than the clean agents and are most effective in terms of fire ratings, but they have the disadvantage of agent residue. HFCs, PFCs and HCFCs are the most expensive and least effective in terms of fire-extinguishing performance (i.e. their fire ratings). Table TS-23 presents a comparison of the alternative portable fire extinguishers. For each option, relative cost and climate considerations are indicated as well as practical concerns such as weights and dimensions. Users of Halon 1211 portable extinguishers in the past currently have three choices: a single HFC/HCFC extinguisher at increased cost, a single dry chemical extinguisher if the residue can be tolerated or two extinguishers – one of water for ordinary combustibles, and a second of CO_2 for use in flammable liquid fires or near electrically energized equipment. Local and national regulations often dictate the choice of portable extinguishers.

Figure TS-21 is an illustration of the production, emission levels and resultant bank sizes of Halon 1211 and its HFC/ PFC/HCFC alternatives for the period 1965–2015. The Halon 1211 bank in 2002 is projected to be 124,843 tonnes with 17,319 tonnes of emissions. This is approximately twice the 7000–8000 tonnes of emissions one would expect based on atmospheric measurements. While no data are available in the literature, information provided by a producer combined with modelling, estimates the portable extinguisher bank of HCFCs, HFCs and PFCs at approximately 1471 tonnes at the end of 2002 with 0.12 MtCO$_2$-eq of emissions. The estimate for 2004 is approximately 1852 tonnes with emissions of 0.16 MtCO$_2$-eq at a 4% emission rate. Approximately 68% are HCFCs, 30% are HFCs and 2% are PFCs.

Table TS-22. Comparison table – clean-agent systems suitable for occupied spaces.

Fixed systems	Halon 1301 (reference)	HFC-23	HFC-227ea	HFC-125[1]	FK- 5-1-12	Inert Gas
Substance characteristics						
Radiative efficiency (W m^{-2} ppb^{-1})	0.32	0.19	0.26	0.23	0.3	n.a.
Atmospheric lifetime (yr)	65	270	34.2	29	0.038	n.a.
Direct GWP (100-yr time horizon)						
- This report	7030	14,310	3140	3450	not	n.a.
- IPCC (1996)	5400	11,700	2900	2800	available[2]	
Ozone depletion potential	12	~0	-	~0	-	n.a.
Technical data						
Demonstrated special capabilities	yes	yes[3]	yes[4]	yes[4]	note[6]	no
Weight (kg m^{-3})[a]	0.8	2.3	1.1	1.1	1.2	4.3
Area (10^4 m^2/m^3)[b]	5.8	12.0	6.8	7.4	7.3	28.2
Volume (10^4 m^3/m^3)[c]	8.6	18.0	13.1	14.4	13.8	56.6
Emission rate [d]	2 ± 1%	2 ± 1%	2 ± 1%	2 ± 1%	2 ± 1%	2 ± 1%
Costs						
Investment cost (relative to Halon 1301)	100%	535%	377%	355%	484%	458%
Additional service costs (US\$ kg^{-1})[e]	0.15	0.43	0.60	0.53	0.72	0.31
Additional recovery costs at end-of-life (US\$ kg^{-1}) [f] ( ) indicates income	(3.85)	(10.75)	(15.07)	(13.20)	(18.00)	0.00
HFC abatement costs (US\$ per tCO$_2$-eq)[g]	-	-	-	-	21–22	14–27
Commercial considerations						
Multiple agent manufacturers	-	yes	yes	yes	no[7]	yes

Notes:

[a] Average weight of the agent storage containers and contents in kilogrammes per cubic metre of space protected.

[b] Average area of a square or rectangle circumscribing the agent cylinder bank expressed in square metres × 10^4 per cubic metre of volume protected.

[c] Average volume is the area multiplied by the height of the cylinders measured to the top of the valves expressed in cubic metres × 10^4 per cubic metre of volume protected.

[d] Total average in-service-life annual emissions rate including system discharges for fire and inadvertent discharges.

[e] Additional annual service costs are based on the replacement of 2% of the agent charge emitted per year.

[f] For the halocarbon agents, the end-of-life agent value is positive and represents a cost recovery equivalent to 50% of the initial cost of the agent as the agent is recovered, recycled and resold for use in either new systems or for the replenishment of existing systems.

[g] HFC abatement costs for FK-5-1-12 and inert gas are based on HFC-227ea, the predominant HFC, as the reference. The lower value reflects the cost in US\$ per tonne of CO$_2$-equivalent at a discount rate of 4% and tax rate of 0%. The range includes both the lowest and highest of costs for the USA, non-USA Annex 1 and non-Annex 1 countries.

Explanation of special capabilities:

1. In some jurisdictions HFC-125 is not allowed for use in occupied spaces while in other jurisdictions that use is permitted under certain conditions.

2. Due to the short atmospheric lifetime, no GWP can be given. It is expected to be negligible for all practical purposes (Taniguchi *et al.*, 2003). See Section 2.5.3.3 'Very short-lived hydrocarbons' for additional information.

3. HFC-23 is effective at low temperatures (cold climates) and in large volumes due to its high vapour pressure.

4. HFC-227ea is effective in shipboard and vehicle applications due to extensive large-scale testing that has established the use parameters and demonstrated its specialized capabilities in these applications.

5. HFC-125 is effective in vehicle and aircraft engine applications as a result of extensive large-scale testing that has established the use parameters and demonstrated its specialized capabilities in these applications.

6. FK-5-1-12 is in the early stages of its product life cycle and has yet to be tested for special applications beyond those achieved through the conventional approval testing of the requirements in ISO and NFPA type standards.

7. While the agent FK-5-1-12 is a proprietary product of a single agent-manufacturer, the agent is available from multiple systems manufacturers.

Figure TS-20. Halon 1301 and halocarbon alternatives time-series for fixed fire-extinguishing systems.

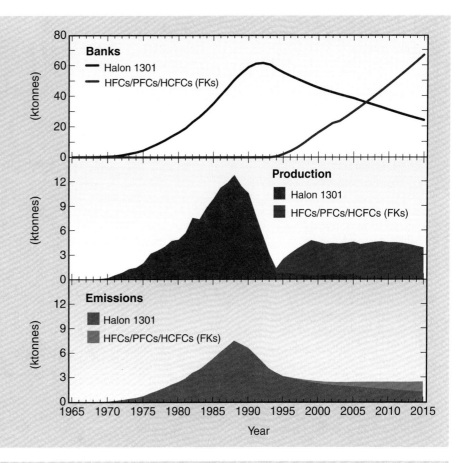

Figure TS-21. Halon 1211 time-series for portable fire-extinguishers.

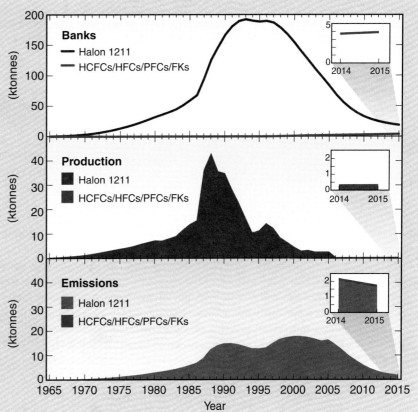

Table TS-23. Comparison table – extinguishing agents for portable fire extinguishers.

Portable systems	Halon 1211 (reference)	HCFC Blend B	HFC-236fa	Carbon Dioxide	Dry Chemical	Water
Substance characteristics						
Radiative efficiency (W m^{-2} ppb^{-1})	0.3	Note[a]	0.28	See Ch. 2	-	-
Atmospheric lifetime (yr)	16	Note[a]	240	See Ch. 2	-	-
Direct GWP (100-yr time horizon)						
- This report	1860	<650[a]	9500	1	-	-
- IPCC (1996)	not given	<730[a]	6300	1	-	-
Ozone depletion potential	5.3	<0.02[a]	-	-	-	-
Technical data						
Agent residue after discharge	no	no	no	no	yes	yes
Suitable for Class A fires	yes	yes	yes	no	yes	yes
Suitable for Class B fires	yes	yes	yes	yes	yes	no
Suitable for energized electrical	yes	yes	yes	yes	yes	no
Extinguisher fire rating [b]	2-A:40-B:C	2-A:10-B:C	2-A:10-B:C	10-B:C	3-A:40-B:C	2-A
Agent charge (kg)	6.4	7.0	6.0	4.5	2.3	9.5
Extinguisher charged weight (kg)	9.9	12.5	11.6	15.4	4.15	13.1
Extinguisher height (mm)	489	546	572	591	432	629
Extinguisher width (mm)	229	241	241	276	216	229
Emission rate [c]	4 ± 2%	4 ± 2%	4 ± 2%	4 ± 2%	4 ± 2%	4 ± 2%
Costs						
Investment costs (relative to Halon 1211)	100%	186%	221%	78%	14%	28%
Additional service costs (US$ kg^{-1})	-[d]	-[d]	-[d]	-[d]	-[d]	-[d]
Additional recovery costs at end-of-life (US$ kg^{-1})	-[d]	-[d]	-[d]	0.00	0.00	0.00

Notes:

[a] HCFC Blend B is a mixture of HCFC-123, CF$_4$ and argon. While the ratio of the components is considered proprietary by the manufacturer, two sources report that HCFC-123 represents over 90% of the blend on a weight basis, with CF$_4$ and argon accounting for the remainder. The atmospheric lifetime of HCFC-123 is 1.3 years; this figure is 50,000 years for CF$_4$.

[b] Fire extinguisher rating in accordance with the requirements of Underwriters Laboratories, Inc. The higher the number, the more effective the extinguisher.

[c] This value is the total average in-service-life annual emissions rate, including both intentional discharges for fire and inadvertent discharges.

[d] This information is neither in the literature nor available from other sources, as it is considered confidential.

What are possible future trends in fire protection?

In 2010 and 2015, emissions of the halocarbon alternatives in fixed fire-extinguishing systems are estimated by modelling to be as high as 2.74 and 3.72 MtCO$_2$-eq, respectively. In portable extinguishers, these emissions are estimated by modelling to be a high as 0.25 and 0.34 MtCO$_2$-eq, respectively. These estimates are based on an emission rate of 2% of the fixed system bank and 4% of the portable extinguisher bank per year and assume a 3% growth rate per year. Efforts to reduce further unnecessary emissions in fire-extinguishing systems could reduce these values by about 50%, while less comprehensive care in emission reductions would likely increase them by 50%. This puts the total emissions from fixed fire-extinguishing systems in the range of 2 ± 1% yr^{-1} and from portable extinguishers in the range of 4 ± 2% yr^{-1}.

While several alternatives to HFCs have been proposed for fire protection, including inert gases, fluoroketones (FKs) and water mist, HFCs and inert gases have become, and appear likely to remain, the most commonly used clean agents and have achieved equilibrium in the market. Due to the lengthy process of testing and approving new fire protection equipment types, no additional options are likely to have an appreciable impact by 2015. FK 5-1-12 has been commercialized and is now available, but there is no basis for predicting its rate of market acceptance or its effect on the already established equilibrium. There is currently no basis for estimating any reduction in the use or emissions of HFC/PFC/HCFCs in fire protection by 2015. In addition, a relationship exists between the halon bank and the use of HFCs. Reductions in the use of halons will result in an increased use of HFCs (and other alternatives) to meet fire protection requirements. Care must continue in the management of the halon bank to ensure an appropriate availability of halons. Therefore, clean agent demand will be influenced by economic growth and by decisions that regulators and halon owners make with respect to the disposition of agents from decommissioned systems.

In 2010 and 2015, banks of 31 and 24 ktonnes of Halon 1301, respectively and 33 and 19 ktonnes of Halon 1211, respectively, are projected to remain, but the emissions and the bank sizes will depend upon the effectiveness of practices to control leakage and to handle recovery at end-of-life. Banks of HFC/PFC/HCFC/FK in fixed systems at a 2% emission rate are projected to be 44 ktonnes in 2010 and 63 ktonnes in 2015, of which approximately 3.6 ktonnes consists of HCFCs. The portable extinguisher bank at a 4% emission rate is projected to be 3.0 ktonnes in 2010 and 3.9 ktonnes in 2015, assuming a 3% growth rate, of which it is estimated that approximately 68% are HCFCs, 30% are HFCs and 2% are PFCs. Looking to the future, GHG emissions from halocarbon-based clean agent systems may either increase or decrease depending on future market acceptance of the alternatives to halons. As research into new fire protection technologies continues, additional replacement options will likely emerge post-2015.

4.7 What are the most important findings for non-medical aerosols and solvents and for HFC-23 emissions?

What are past and current trends?

Prior to the Montreal Protocol, ODSs were widely used as a cleaning solvent for metals, electronics and precision applications, and in consumer and technical and safety aerosols as propellants or solvents. ODS use in these applications has been eliminated or reduced dramatically. Most solvent cleaning applications now rely on not-in-kind substitutes. A small percentage have or are expected to transition to HFCs or hydrofluoroethers (HFEs). PFC use is declining and expected to be eliminated by 2025.

Non-medical aerosols
Aerosol products use gas pressure to propel liquid, paste or powder active ingredients in precise spray patterns with controlled droplet size and quantity. They can also be made into products that use the gas only. In developed countries, 98% of non-medical aerosols now use non-ozone-depleting, very low GWP propellants (hydrocarbons, dimethylether, CO_2 or nitrogen). These substitutions led to a total reduction of GHG emissions from non-medical aerosol origin by over 99% between 1977 and 2001. The remaining aerosol products using either HCFCs (in developing countries where HCFC consumption is allowed until 2040) or HFCs

(HFC-152a and HFC–134a) do so because these propellants provide a safety, functional or health benefit for the users. Additionally, the use of HFCs in non-medical aerosols is further limited by cost. HFCs are between five and eight times more expensive than hydrocarbons. In 2003, HFC use in aerosols represented total emissions of about 22 $MtCO_2$-eq.

Solvents
It is estimated that by 1999, 90% of the ODS solvent use had been reduced through conservation and substitution with not-in-kind technologies (no-clean flux, aqueous or semi-aqueous cleaning and hydrocarbon solvents). The remaining 10% of solvent use is shared by several organic solvent alternatives. The in-kind substitutes for CFC-113 and CFC-11 include HCFCs, PFCs, HFCs and HFEs. The only HCFC solvents currently used are HCFC-141b and HCFC-225ca/cb. Most HCFC-141b use is for foam blowing; solvent applications represented less than 10% of its global use in 2002. The use of HCFC-141b is banned in the EU and is rapidly declining in other developed countries. In developing countries, the use of HCFC-141b is still increasing, especially in China, India and Brazil, as economic growth rates are high. HCFC-225ca/cb use is directed to niche applications, and because of its ODP and phase-out schedule, it is being gradually replaced by HFC, HFE and not-in-kind alternatives.

Production byproducts and fugitives
Emissions of ODSs, HFCs and PFCs also occur during the production of fluorocarbons, either as undesired byproducts or as losses of useful material as fugitive emissions. Fugitive losses are small and generally represent less than 1% of total production. The most significant of the byproducts is HFC-23 (fluoroform), which is generated during the manufacture of HCFC-22. While the Montreal Protocol will eventually phase out the direct use of HCFC-22, its use as a feedstock is permitted to continue indefinitely because it does not involve the release of HCFC-22 to the atmosphere. Global feedstock demand has been increasing and is expected to continue to grow beyond 2015. HCFC-22 production is growing rapidly in developing countries, especially China and India. Commercial (non-feedstock) uses will end by 2020 in developed countries and by 2040 in developing countries.

HFC-23 generation ranges from 1.4–4% of total HCFC-22 production, depending on production management and process circumstances. HFC-23 is the most potent (GWP of 14,310) and persistent (atmospheric life 270 years) of the HFCs. Global emissions of HFC-23 increased by an estimated 12% between 1990 and 1995 as a result of a similar increase in the global production of HCFC-22. However, due to the widespread implementation of process optimization and thermal destruction in developed countries, this trend has not continued and since 1995 has become smaller than the increase in production.

Table TS-24. Overview of non-medical aerosol propellant alternatives.

	HCFC-22	HFC-134a	HFC-152a	Dimethylether	Isobutane[a]
Substance characteristics					
Radiative efficiency (W m^{-2} ppb^{-1})	0.20	0.16	0.09	0.02	0.0047
Atmospheric lifetime (yr)	12	14	1.4	0.015	0.019
GWP (100-yr time horizon)					
- This report	1780	1410	122	1	n/a
- IPCC (1996)	1500	1300	140	1	
ODP	0.05	~0	-	-	-
Ground-level ozone impact					
- MIR[2] (g-O1/g-substance)	<0.1	<0.1	<0.1	0.93	1.34
- POCP[3] (relative units)	0.1	0.1	1	17	31
Flammability (based on flashpoint)	None	None	Flammable	Flammable	Flammable
Technical data					
Stage of development	Commercial	Commercial	Commercial	Commercial	Commercial
Type of application:					
- Technical aerosols	X	X	X	X	X
- Safety aerosols	X	X			
- Consumer products	Phased out in industrialized countries		X	X	X
Emissions	Use totally emissive in all cases				
Costs					
Additional investment costs			Special safety required at filling plant	Special safety required at filling plant	Special safety required at filling plant

Notes:

[a] Listed values refer to isobutane only. Additional hydrocarbon aerosol propellants are used in non-medical aerosol applications as indicated in Chapter 10.

What emission reduction options are available?

Non-medical aerosols

Although there are no technical barriers for the transition out of CFCs to alternatives for non-medical aerosol products, in 2001, there was still an estimated 4300 tonnes of CFCs used in developing countries and countries with economies in transition (CEIT). Technical aerosols are pressurized gas products used to clean, maintain, fix, test, manufacture or disinfect various types of equipment or used in a number of processes. The largest use of HFCs in technical aerosols occurs in dusters where the substitution of HFC-134a by HFC-152a is a leading factor in reducing GHG emissions. For cleaners (contact cleaners, flux removers) and mould release agents, the substitution of HCFC-141b by HFEs and HFCs with lower GWP offers the opportunity for additional emission reduction with no substantial technical issues. Safety aerosols (safety signal horns, tire inflators) and insecticides for planes and restricted areas continue to rely on HFC-134a due to its non-flammability. Cosmetic, convenience and novelty aerosol products include artificial snow, silly string and noise-makers (horns). The majority of noise-makers (>80%) use hydrocarbons; artificial snow and string novelties originally transitioned to hydrocarbons but, after highly publicized safety incidents, were reformulated to HFC-134a. HFC use in cosmetic, convenience and novelty aerosols is being banned in the EU.

Table TS-24 presents a comparison of non-medical aerosol alternatives.

Solvents

Although HFCs are available in all regions, their use as solvents has been primarily in developed countries due to high costs and the concentration for applications in high-tech industries. With increasing concern about climate protection, HFC uses tend to be focused in critical applications with no other substitutes. Current use in developed countries is considered to have peaked and may even decline in the future.

PFC solvents are no longer considered technically necessary for most applications, and their use is constrained to a few

Table TS-25. Overview of HFCs, PFCs and HCFCs in solvent applications.

	HCFC-141b	HCFC-225ca/cb	HFC-43-10mee	HFC-365mfc	PFC–51-14 (C_6F_{14})
Substance characteristics					
Radiative efficiency (W m^{-2} ppb^{-1})	0.14	0.2/0.32	0.4	0.21	0.49
Atmospheric lifetime (yr)	9.3	1.9/5.8	15.9	8.6	3,200
GWP (100-yr time horizon)					
- This report	713	120/586	1,610	782	9,140
- IPCC (1996, 2001a)	600	180/620	1,300	890	7,400
ODP	0.12	0.02/0.03	-	-	-
Ground-level ozone impact					
- MIR (g-O_1/g-substance)	<0.1	<0.1	n/a	n/a	n/a
- POCP (relative units)	0.1	0.2/0.1	n/a	n/a	n/a
Ground-level ozone impact	None	None	None	None	None
Flammability (based on flashpoint)	None	None	None	Flammable	None
Technical data					
Stage of development	Commercial	Commercial	Commercial	Commercial	Commercial
Type of application:					
- Electronics cleaning	X	X	X		
- Precision cleaning	X	X	X	X	X
- Metal cleaning	X	X	X		
- Drying	X	X	X	X	
- Carrier solvent	X	X	X		X

niche applications due to very limited performance and high cost. Volumes are known to have decreased since the mid-1990s as a result of replacement with lower GWP solvents.

Emission reduction options in solvent applications fall into two categories:

(1) Improved containment in existing uses. New and retrofitted equipment can significantly reduce emissions of all solvents. Optimized equipment can reduce solvent consumption by as much as 80% in some applications. Due to their high cost and ease of recycling, the fluorinated solvents are generally recovered and reused by the end-users or their suppliers.

(2) Alternative fluids and technologies. A variety of organic solvents can replace HFCs, PFCs and ODSs in many applications. These alternative fluids include lower GWP compounds such as traditional chlorinated solvents, HFEs and n-propyl bromide. The numerous not-in-kind technologies, including hydrocarbon and oxygenated solvents, are also viable alternatives in some applications. Caution is warranted prior to adoption of any alternatives whose toxicity profile is not complete. In a limited number of applications, no substitutes are available due to the unique performance characteristics of the HFC or PFC in that case.

Tables TS-25 and TS-26 present comparisons of solvent alternatives.

Production byproducts and fugitives

It is technically feasible to reduce future emissions of HFC-23 from HCFC-22 by over 90% (or by a factor of ten) through capture and destruction of the HFC-23 byproduct. However, emissions of HFC-23 could grow by as much as 60% between now and 2015, from about 15 ktonnes yr^{-1} to 23 ktonnes yr^{-1} due to anticipated growth in HCFC-22 production. The upper bound of HFC-23 emissions is in the order of 3–4% of HCFC-22 production, but the actual quantity of HFC-23 produced depends in part on how the process is operated at each facility.

Techniques and procedures to reduce the generation of HFC-23 through process optimization can reduce average emissions to 2% or less of production. However, actual achievements vary for each facility, and it is not possible to eliminate HFC-23 emission by this means. Capture and destruction of HFC-23 by thermal oxidation is a highly effective option to reduce emissions. Destruction efficiency can be more than 99.0%, but the impact of 'down time' of thermal oxidation units on emissions needs to be taken into account. Assuming a technological lifetime of 15 years, specific abatement costs of less than 0.2 US\$/tCO_2-eq can be calculated.

Table TS-26. Overview of alternative fluids and not-in-kind technologies in solvent applications.

	CH_2Cl_2[a]	HFE-449s1[b]	n-propyl bromide	No Clean	Hydro-carbon / oxygenated	Aqueous / semi-aqueous
Substance characteristics						
Radiative efficiency (W m^{-2} ppb^{-1})	0.03	0.31	0.3	n.a.		n.a.
Atmospheric lifetime (yr)	0.38	5	0.04	n.a.		n.a.
GWP (100-yr time horizon)						
- This report	10	397	n/a	n.a.		n.a.
- IPCC (1996)	9	not given				
ODP	-	-	-	-	-	-
Ground-level ozone impact						
- MIR^3 (g-O$_1$/g-substance)	0.07	n.a.	n.a.			
- POCP^4 (relative units)	7	n.a.	n.a.			
Ground-level ozone impact	Low to moderate	None	Low to moderate	None	Low to moderate	None
Flammability (based on flashpoint)	None	None	None	n.a.	Flammable	n.a.
Technical data						
Stage of development	Commercial	Commercial	Commercial	Commercial	Commercial	Commercial
Type of application:						
- Electronics cleaning		X	X	X	X	X
- Precision cleaning		X	X		X	X
- Metal cleaning	X	X	X		X	X
- Drying					X	
- Carrier solvent	X	X	X		X	

Notes:
[a] The listed values refer to CH_2Cl_2 only. Additional chlorinated solvents are used in these applications as indicated in Chapter 10.
[b] The listed values refer to HFE-449s1 only. Additional HFE solvents are used in these applications as indicated in Chapter 10.

The calculation of HFC-23 emissions requires data not only on the quantities of HCFC-22 produced (the activity) and the rate of emission (which is influenced by process design and operating culture) but also on the extent to which emissions are abated. This has a particular influence on the uncertainty of HFC-23 estimates of future emissions.

Table TS-27 presents a comparison of the process optimization and thermal oxidation as reduction options for HFC-23 byproduct emissions.

What are possible future trends?

Non-medical aerosols
HFC emissions in non-medical aerosols are estimated at 23 MtCO$_2$-eq in 2010. Low growth is projected for this sector through to 2015. While there are no technical barriers to formulate consumer products without HFCs, the use of HFC-152a in some products such as hairspray and deodorants will increase in the USA due to the implementation of regulations to control ground level ozone formation from hydrocarbon emission. Current volatile organic compound (VOC) controls in Europe do not exempt

HFCs because of the broad definition of VOC (boiling point <250°C under standard pressure/temperature conditions). No other VOC regulations identified elsewhere in the world restrict the use of hydrocarbons in non-medical aerosols.

Solvents
Most solvent uses are emissive in nature with a short inventory period of a few months to 2 years. Although used solvents can and are distilled and recycled on site, essentially all quantities sold are eventually emitted. The distinction between consumption and emission (i.e. banking) is not significant for these applications. Projected global emissions of HFCs and PFCs from solvent uses are 4.2 MtCO$_2$-eq in 2010 and 4.4 MtCO$_2$-eq in 2015. Emissions of PFCs are assumed to decline linearly until they are no longer used in solvent applications by 2025.

Table TS-27. Comparison of HFC-23 byproduct from HCFC-22 production reduction options: process optimization and thermal oxidation.

	HCFC-22	HFC-23		
Substance characteristics				
Radiative efficiency (W m^{-2} ppb^{-1})	0.20	0.19		
Atmospheric lifetime (yr)	12	270		
GWP (100-yr time horizon)				
- This report	1780	14,310		
- IPCC (1996)	1500	11,700		
ODP	0.05	~0		
Flammability	None	None		
HFC-23 emission reduction options		**No Optimization**	**Process Optimization**	**Thermal Oxidation**
Stage of development		Commercial	Commercial	Commercial
Direct emissions		3–4% HCFC-22 produced	2–3% HCFC-22 produced	<1% HCFC-22 produced
Additional costs		Reference	Dependent on process and market can range from marginal saving to significant penalty	US$ 2–8 million total installed capital costs, with US$ 189–350 thousand annual operating costs

Byproduct emissions of HFC-23

The quantity of HFC-23 produced (and that, potentially, may be emitted) is directly related to the production of HCFC-22 and, as a result, emission forecasts require a scenario for future HCFC-22 production volumes. These will depend on the consumption of HCFC-22 in developed countries, which is declining, and the consumption in developing countries and global demand for fluoropolymers feedstock, both of which are increasing.

Based on a BAU scenario that exactly follows the requirements of the Montreal Protocol, consumption and production of non-feedstock HCFC-22 will fall by a factor of ten by 2015 from the average level in 2000–2003 in developed countries. In the same countries, growth in the demand for fluoropolymer feedstock is projected to continue increasing linearly, leading to the feedstock demand for HCFC-22 doubling there by 2015. In developing countries, the production of HCFC-22 for both feedstock and non-feedstock uses has grown rapidly in recent years; over the period 1997–2001, production for commercial (or non-feedstock) uses grew linearly at 20 ktonnes yr^{-1} and feedstock use grew at 4.1 ktonnes yr^{-1}. Projected at these rates until 2015, the total global requirement for HCFC-22 would become about 730 ktonnes yr^{-1} – about 40% of which would be for feedstock – compared with a total of 470 ktonnes yr^{-1} in the year 2000. (Table TS-28).

In the BAU case to 2015, it has been assumed that emissions from existing capacity (in both developed and developing countries) will continue at 2% of HCFC-22 production and that new capacity (mainly in developing countries) will emit HFC-23 at a rate of 4%. Consequently, emissions of HFC-23 could grow by 60% between now and 2015 – from about 15 ktonnes yr^{-1} in 2003 to 23 ktonnes yr^{-1} (Table TS-28).

In a variation of this scenario, the current best-practice technology comprising capture and thermal oxidation of the 'vent gases' is progressively introduced into all facilities, commencing in 2005. Destruction technology is assumed to be 100% efficient and to operate for 90% of the on-line time of the HCFC-22 plant. Reduced emissions were calculated from the same activity (in the form of assumed future HCFC-

Table TS-28. Historical and future emissions of HFC-23.

Year	HCFC-22 Production scenario (kt)	HFC-23 BAU emissions (kt)	HFC-23 Current best practice emissions (kt)
1990	341	6.4	6.4
1995	385	7.3	7.3
2000	491	11.5	11.5
2005	550	15.2	13.8
2010	622	19.0	8.8
2015	707	23.2	2.3

22 production) as the BAU case. The difference between the two HFC-23 forecasts is therefore solely due to the extent to which destruction technology is deployed. The forecasts represent potential extreme cases, and future changes in activity will tend to increase the probability of one or the other.

Destruction of byproduct emissions of HFC-23 from HCFC-22 production has a reduction potential of up to 300 $MtCO_2$-eq per year by 2015 and specific costs below 0.2 $US\$/tCO_2$-eq according to two European studies in 2000. Reduction of HCFC-22 production due to market forces or national policies, or improvements in facility design and construction also could reduce HFC-23 emissions. [10.4]

1

Ozone and Climate:
A Review of Interconnections

Coordinating Lead Authors
John Pyle (UK), Theodore Shepherd (Canada)

Lead Authors
Gregory Bodeker (New Zealand), Pablo Canziani (Argentina), Martin Dameris (Germany), Piers Forster (UK), Aleksandr Gruzdev (Russia), Rolf Müller (Germany), Nzioka John Muthama (Kenya), Giovanni Pitari (Italy), William Randel (USA)

Contributing Authors
Vitali Fioletov (Canada), Jens-Uwe Grooß (Germany), Stephen Montzka (USA), Paul Newman (USA), Larry Thomason (USA), Guus Velders (The Netherlands)

Review Editors
Mack McFarland (USA)

Contents

EXECUTIVE SUMMARY

Stratospheric ozone has been depleted over the last 25 years following anthropogenic emissions of a number of chlorine- and bromine-containing compounds (ozone-depleting substances, ODSs), which are now regulated under the Montreal Protocol. The Protocol has been effective in controlling the net growth of these compounds in the atmosphere. As chlorine and bromine slowly decrease in the future, ozone levels are expected to increase in the coming decades, although the evolution will also depend on the changing climate system.

- *Emissions of chlorofluorocarbons (CFCs) and hydrochlorofluorocarbons (HCFCs) have depleted stratospheric ozone.* Globally, ozone has decreased by roughly 3% since 1980. The largest decreases since 1980 have been observed over the Antarctic in spring (the 'ozone hole'), where the monthly column-ozone amounts in September and October have been about 40–50% below pre-ozone-hole values. Arctic ozone shows high year-to-year variability during winter-spring due to variability in dynamical transport and chemical loss (which act in concert). Chemical losses of up to 30% are observed during cold winters, whereas they are small for dynamically active warm years.

- *Due to the control of ODSs under the Montreal Protocol and its Amendments and Adjustments, the abundance of anthropogenic chlorine in the troposphere has peaked and is now declining.* The tropospheric abundance of anthropogenic bromine began decreasing in the late 1990s. Stratospheric concentrations of ODSs lag those in the troposphere by several years. Stratospheric chlorine levels have approximately stabilized and may have already started to decline.

- *Changing ODS concentrations will be a dominant factor controlling changes in stratospheric ozone for the next few decades.* Stratospheric ozone depletion is likely to be near its maximum, but ozone abundance is subject to considerable interannual variability. Assuming full compliance with measures adopted under the Montreal Protocol, ozone should slowly recover. Currently, there is no unequivocal evidence from measurements in the atmosphere that the onset of ozone recovery has begun.

- *Models suggest that in Antarctica the peak in springtime ozone depletion may already have occurred or should occur within the next few years.* Because of greater variability, predicting the timing of the peak in springtime Arctic ozone depletion is more uncertain but models suggest it should occur within the first two decades of the 21st century. Based on these model calculations, we do not expect an Arctic 'ozone hole' similar to that observed over the Antarctic.

- *The ozone layer will not necessarily return to its pre-depleted state, even when the abundance of stratospheric chlorine and bromine returns to previous levels.* 'Recovery' of the ozone layer is a complex issue: it depends not just on the extent to which the ODSs are replaced by non-ODSs, but also on emissions of gases (including ODS substitutes) that affect the climate system directly.

There are many complex two-way interactions between stratospheric ozone and climate. Changes in stratospheric temperature and transport affect the concentration and distribution of stratospheric ozone; changes in tropospheric climate affect stratospheric circulation; changes in stratospheric ozone influence the radiative forcing of the atmosphere, and hence surface climate, as well as the chemistry of the troposphere. While our understanding is still far from complete, new evidence about some of these interactions has emerged in recent years.

- *Stratospheric ozone depletion has led to a cooling of the stratosphere.* A significant annual-mean cooling of the lower stratosphere over the past two decades (of approximately 0.6 K per decade) has been found over the mid-latitudes of both hemispheres. Modelling studies indicate that changes in stratospheric ozone, well-mixed greenhouse gases and stratospheric water vapour could all have contributed to these observed temperature changes. In the upper stratosphere the annual mean temperature trends are larger, with an approximately uniform cooling over 1979–1998 of roughly 2 K per decade at an altitude of around 50 km. Model studies indicate that stratospheric ozone changes and carbon dioxide changes are each responsible for about 50% of the upper stratospheric temperature trend.

- *The southern polar vortex, which creates the dynamical setting for the Antarctic ozone hole, tends to persist longer now than in the decades before the appearance of the ozone hole.* During the period 1990–2000, the vortex break-up in the late spring to early summer has been delayed by two to three weeks relative to the 1970–1980 period.

- *Future temperature changes in the stratosphere could either enhance or reduce stratospheric ozone depletion, depending on the region.* Increases in the concentrations of well-mixed greenhouse gases, which are expected to cool the stratosphere, could reduce the rate of gas-phase ozone destruction in much of the stratosphere and hence reduce stratospheric ozone depletion. However ozone depletion in the polar lower stratosphere depends on the low temperatures there. Until stratospheric ODS abundances return to pre-ozone-hole levels, temperature reductions in polar latitudes could enhance polar ozone depletion; this effect is expected to be most important in the Arctic late winter to spring where a small decrease in temperature could have a large effect on ozone.

- *Ozone abundance in Northern Hemisphere mid-latitudes and the Arctic is particularly sensitive to dynamical effects (transport and temperature)*. The dynamical feedbacks from greenhouse-gas increases could either enhance or reduce ozone abundance in these regions; currently, not even the sign of the feedback is known. Furthermore, dynamical variability can affect ozone in these regions on decadal time scales.

- *Statistical and modelling studies suggest that changes in stratospheric circulation regimes (e.g., between strong and weak polar vortices) can have an impact on surface climate patterns.* Because future changes in ozone will have an impact on the circulation of the stratosphere, these changes could also influence the troposphere. In particular there are indications that the Antarctic ozone hole has led to a change in surface temperature and circulation – including the cooling that has been observed over the Antarctic continent, except over the Antarctic Peninsula where a significant warming has been observed.

Halocarbons are particularly strong greenhouse gases (gases that absorb and emit thermal infrared radiation); a halocarbon molecule can be many thousands of times more efficient at absorbing radiant energy emitted from the Earth than a molecule of carbon dioxide. Ozone is a greenhouse gas that also strongly absorbs ultraviolet and visible radiation. Changes in these gases have affected the Earth's radiative balance in the past and will continue to affect it in the future.

- *Over the period 1750–2000 halocarbon gases have contributed a positive direct radiative forcing of about 0.33 W m^{-2}, which represents about 13% of the total well-mixed greenhouse gas radiative forcing over this period.* Over the period 1970–2000 halocarbon gases represent about 23% of the total well-mixed greenhouse gas radiative forcing. Over the same period, only about 5% of the total halocarbon radiative forcing is from the ODS substitutes; ODSs themselves account for 95%.

- *Stratospheric ozone depletion has led to a negative radiative forcing of about 0.15 W m^{-2} with a range of ±0.1 W m^{-2} from different model estimates.* Therefore the positive radiative forcing due to the combined effect of all ODSs (0.32 W m^{-2}) over the period 1750–2000 is larger than the negative forcing due to stratospheric ozone depletion.

- *Assuming that the observed changes in ozone are caused entirely by ODSs, the ozone radiative forcing can be considered an indirect forcing from the ODSs.* However, there is evidence that some fraction of the observed global ozone changes cannot be attributed to ODSs, in which case the indirect forcing by ODSs would be weaker.

- *The relative contributions of various halocarbons to positive direct and negative indirect forcing differ from gas to gas.* Gases containing bromine (such as halons) are particularly effective ozone depleters, so their relative contribution to negative forcing is greater than that of other ozone-depleting gases, such as the CFCs.

- *The same magnitudes of positive direct radiative forcing from halocarbons and of negative indirect forcing from ozone depletion are highly likely to have a different magnitude of impact on global mean surface temperature.* The climate responses will also have different spatial patterns. As a consequence, the two forcings do not simply offset one another.

- *Both the direct and indirect radiative forcings of ODSs will be reduced as a result of the Montreal Protocol.* Nevertheless, these forcings are expected to remain significant during the next several decades. The precise change of each of the forcings will be affected by several factors, including (1) the way climate change may affect stratospheric ozone chemistry and hence the ozone indirect radiative forcing; (2) future emissions of ODSs and their substitutes (see Chapter 2); and (3) continuing emissions from 'ODS banks' (ODSs that have already been manufactured but have not yet been released into the atmosphere).

1.1   Introduction

1.1.1   *Purpose and scope of this chapter*

Ozone absorbs solar ultraviolet (UV) radiation (thereby warming the stratosphere) and is a greenhouse gas (thereby warming the troposphere). It thus plays an important role in the climate system. Furthermore, absorption of potentially damaging UV radiation by the stratospheric ozone layer protects life at the Earth's surface. Stratospheric ozone amounts have been depleted in recent decades, following emission into the atmosphere of various ozone-depleting substances (ODSs), most of which are also greenhouse gases. The CFCs (chlorofluorocarbons) and halons, the major anthropogenic depleters of stratospheric ozone, are now controlled under the Montreal Protocol and its Amendments and Adjustments. They are being replaced by a variety of substances with lower ozone depletion potentials (ODPs) but that are generally still greenhouse gases, often with large global warming potentials (GWPs). The scientific connection between the different environmental problems of climate change and ozone depletion, and between the provisions of the Montreal Protocol and the Kyoto Protocol (and the United Nations Framework Convention on Climate Change, UNFCCC), is the subject of this chapter.

The scientific issues are complex and the connections between the two problems more intricate than suggested above. For example, not only is ozone a greenhouse gas, so that any change in ozone abundance could have an impact on climate, but changes in climate could affect ozone in a number of different ways. So, the ozone distribution in the future will depend, among other things, on the emission and impact of other greenhouse gases and not just on those that deplete ozone. For this reason, ozone recovery following the reduction in atmospheric abundances of the ODSs will not be a simple return to the pre-ozone-hole atmosphere (WMO, 1999, Chapter 12).

Climate change and ozone depletion have both been the subject of recent assessments (IPCC, 2001; WMO, 2003) describing the advances in understanding made in recent years. It is now well established, for example, that observed polar and mid-latitude ozone depletion is a consequence of the increase in stratospheric chlorine and bromine concentrations. The chemical processes involving these compounds and leading to polar ozone depletion are well understood. Similarly, the important role of ozone and the ODSs in the climate system has been documented. Stratospheric ozone depletion during recent decades has represented a negative radiative forcing of the climate system; in contrast, the increase in ODSs has been a positive radiative forcing. These earlier reports form an important basis for this special report, which aims to address specifically the coupling between the chemistry and climate problems. In particular, this report will explore the scientific issues arising from the phase-out and replacement of ODSs as they affect stratospheric composition and climate change. The earlier reports did not have this particular emphasis (although the issues were addressed by a European Commission report (EC, 2003)).

1.1.2   *Ozone in the atmosphere and its role in climate*

Figure 1.1 illustrates a number of important concepts concerning the ozone layer (Figure 1.1a) and its role in the stratosphere (the region between approximately 15 and 50 km altitude in which temperature rises with altitude, Figure 1.1b). Ozone, like water vapour and carbon dioxide, is an important and naturally occurring greenhouse gas; that is, it absorbs and emits radiation in the thermal infrared (Figure 1.1c), trapping heat to warm the Earth's surface (Figure 1.1d). In contrast to the so-called well-mixed greenhouse gases (WMGHGs), stratospheric ozone has two distinguishing properties. First, its relatively short chemical lifetime means that it is not uniformly mixed throughout the atmosphere and therefore its distribution is controlled by both dynamical and chemical processes (Section 1.3). In fact, unlike the WMGHGs, ozone is produced entirely within the atmosphere rather than being emitted into it. Second (Figure 1.1c), it is a very strong absorber of short wavelength UV radiation (it is also a weak absorber of visible radiation). The ozone layer's absorption of this UV radiation leads to the characteristic increase of temperature with altitude in the stratosphere and, in consequence, to a strong resistance to vertical motion. As well as ozone's role in climate it also has more direct links to humans: its absorption of UV radiation protects much of Earth's biota from this potentially damaging short wavelength radiation. In contrast to the benefits of stratospheric ozone, high surface ozone values are detrimental to human health.

The distribution of ozone in the atmosphere is maintained by a balance between photochemical production and loss, and by transport between regions of net production and net loss. A number of different chemical regimes can be identified for ozone. In the upper stratosphere, the ozone distribution arises from a balance between production following photolysis of molecular oxygen and destruction via a number of catalytic cycles involving hydrogen, nitrogen and halogen radical species (see Box 1.1). The halogens arise mainly from anthropogenic ODSs (the CFCs, HCFCs and halons). In the upper stratosphere, the rates of ozone destruction depend on temperature (a reduction in temperature slows the destruction of ozone, see Section 1.3.2) and on the concentrations of the radical species. In the lower stratosphere, reactions on aerosols become important. The distribution of the radicals (and the partitioning of the nitrogen, hydrogen and halogen species between radicals and 'reservoirs' that do not destroy ozone) can be affected by heterogeneous and multiphase chemistry acting on condensed matter (liquid and solid particles). At the low temperature of the wintertime polar lower stratosphere, this is the chemistry that leads to the ozone hole (see Box 1.1).

The large-scale circulation of the stratosphere, known as the Brewer-Dobson circulation (see Box 1.2), systematically transports ozone poleward and downward (figure in Box 1.2). Because ozone photochemical reactions proceed quickly in the sunlit upper stratosphere, this transport has little effect on the ozone distribution there as ozone removal by transport is quickly replenished by photochemical production. However, this trans-

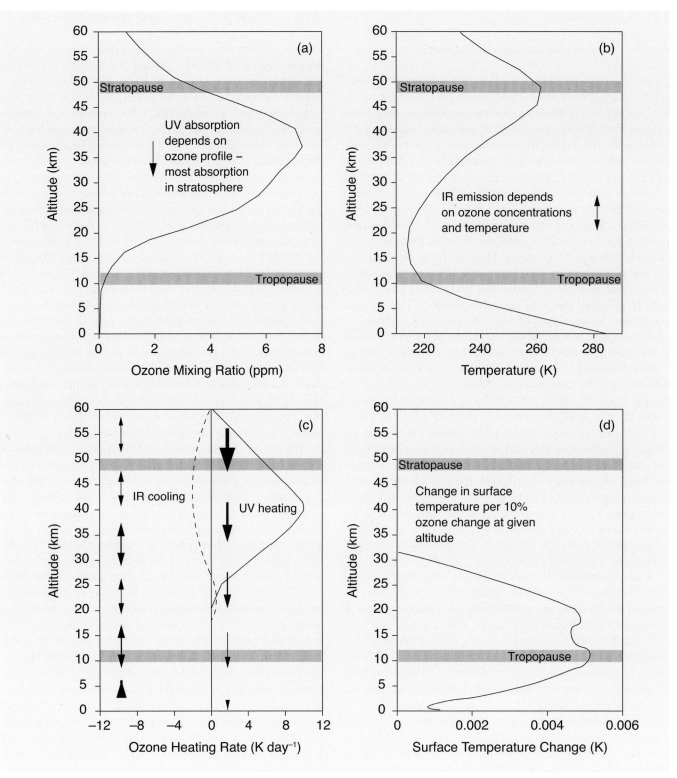

Figure 1.1. Vertical profiles of ozone-related quantities. (a) Typical mid-latitude ozone mixing ratio profile, based on an update of Fortuin and Langematz (1994); (b) atmospheric temperature profile, based on Fleming *et al.* (1990), showing the stratosphere bounded by the tropopause below and the stratopause above; (c) schematic showing the ultraviolet (UV) radiative flux through the atmosphere (single-headed arrows) and infrared (IR) emission at around 9.6 μm (the ozone absorption band, double-headed arrows), and the heating in the ultraviolet (solid curve) and infrared (dashed curve) associated with these fluxes; (d) schematic of the change in surface temperature due to a 10% change in ozone concentration at different altitudes (based on Figure 6.1 of IPCC, 2001). Figure provided by Peter Braesicke, University of Cambridge.

Box 1.1. Ozone chemistry

Stratospheric ozone is produced naturally by photolysis of molecular oxygen (O_2) at ultraviolet wavelengths below 242 nm,

$$O_2 + h\nu \rightarrow O + O \tag{1}$$

The atomic oxygen produced in this reaction reacts rapidly with O_2 to form ozone (O_3),

$$O + O_2 + M \rightarrow O_3 + M \tag{2}$$

where M denotes a collision partner, not affected by the reaction. O_3 itself is photolyzed rapidly,

$$O_3 + h\nu \rightarrow O + O_2 \tag{3}$$

O_3 and O establish a rapid photochemical equilibrium through Reactions [2] and [3], and together are called 'odd oxygen'. Finally, in this sequence of reactions (known as the Chapman reactions), ozone is removed by

$$O + O_3 \rightarrow 2O_2 \tag{4}$$

Destruction by Reaction [4] alone cannot explain observed ozone abundances in the stratosphere and it is now known that, away from polar latitudes, the ozone production through Reaction [1] is largely balanced by destruction in catalytic cycles of the form

$$XO + O \rightarrow X + O_2$$
$$X + O_3 \rightarrow XO + O_2$$
$$\text{Net: } O + O_3 \rightarrow 2O_2 \tag{Cycle 1}$$

The net reaction is equivalent to Reaction [4]. Note that, because O and O_3 are in rapid photochemical equilibrium, the loss of one oxygen atom also effectively implies the loss of an ozone molecule, so that the cycle destroys two molecules of 'odd oxygen'. Notice also that the catalyst, X, is not used up in the reaction cycle. The most important cycles of this type in the stratosphere involve reactive nitrogen (X = NO), halogen (X = Cl) and hydrogen (X = H, OH) radicals. In the lower stratosphere, cycles catalyzed by Br also contribute to the ozone loss. Owing to the large increase of O with altitude, the rates of these cycles increase substantially between 25 and 40 km, as does the rate of ozone production.

In polar regions, the abundance of ClO is greatly enhanced during winter as a result of reactions on the surfaces of polar stratospheric cloud particles that form at the low temperatures found there. However, atomic oxygen, O, has very low concentrations there, which limits the efficiency of Cycle 1. In that case, two other catalytic cycles become the dominant reaction mechanisms for polar ozone loss. The first, the so-called ClO dimer cycle, is initiated by the reaction of ClO with another ClO,

$$ClO + ClO + M \rightarrow (ClO)_2 + M$$
$$(ClO)_2 + h\nu \rightarrow ClOO + Cl$$
$$ClOO + M \rightarrow Cl + O_2 + M$$
$$2 (Cl + O_3 \rightarrow ClO + O_2)$$
$$\text{Net: } 2O_3 \rightarrow 3O_2 \tag{Cycle 2}$$

and the second, the ClO-BrO cycle, is initiated by the reaction of ClO with BrO:

$$ClO + BrO \rightarrow Cl + Br + O_2$$
$$Cl + O_3 \rightarrow ClO + O_2$$
$$Br + O_3 \rightarrow BrO + O_2$$
$$\text{Net: } 2O_3 \rightarrow 3O_2 \tag{Cycle 3}$$

The net result of both Cycle 2 and Cycle 3 is to destroy two ozone molecules and to produce three oxygen molecules. Both cycles are catalytic, as chlorine (Cl) and bromine (Br) are not lost in the cycles. Sunlight is required to complete the cycles and to help maintain the large ClO abundance. Cycles 2 and 3 account for most of the ozone loss observed in the late winter-spring season in the Arctic and Antarctic stratosphere. At high ClO abundances, the rate of ozone destruction can reach 2 to 3% per day in late winter-spring. Outside the polar regions the ClO-BrO cycle is of minor importance because of much lower ClO concentrations, and the effect of the ClO dimer cycle is negligible as the cycle is only effective at the low polar temperatures.

Box 1.2. Stratospheric transport, planetary waves and the Brewer-Dobson circulation

The stratospheric Brewer-Dobson circulation consists of rising motion in the tropics and sinking motion in the extratropics, together with an associated poleward mass flux. Its effect on chemical species is complemented by mixing (mainly quasi-horizontal) that acts to transport air parcels both poleward and equatorward. Both processes are primarily driven by mechanical forcing (wave drag) arising from the dissipation of planetary-scale waves in the stratosphere. These planetary waves are generated in the troposphere by topographic and thermal forcing, and by synoptic meteorological activity, and propagate vertically into the stratosphere. Because of filtering by the large-scale stratospheric winds, vertical propagation of planetary waves into the stratosphere occurs primarily during winter, and this seasonality in wave forcing accounts for the winter maximum in the Brewer-Dobson circulation. In the case of ozone, the Brewer-Dobson circulation (together with the associated horizontal mixing) transports ozone poleward and downward and leads to a springtime maximum in extratropical ozone abundance (see figure).

Because air enters the stratosphere primarily in the tropics, the physical and chemical characteristics of air near the tropical tropopause behave as boundary conditions for the global stratosphere. For example, dehydration of air near the cold tropical tropopause accounts for the extreme dryness of the global stratosphere (Brewer, 1949). Overall, the region of the tropical atmosphere between about 12 km and the altitude of the tropopause (about 17 km) has characteristics intermediate to those of the troposphere and stratosphere, and is referred to as the *tropical tropopause layer* (TTL).

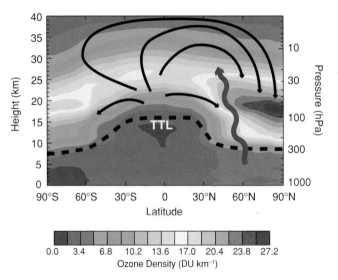

Box 1.2, Figure. Meridional cross-section of the atmosphere showing ozone density (colour contours; in Dobson units (DU) per km) during Northern Hemisphere (NH) winter (January to March), from the climatology of Fortuin and Kelder (1998). The dashed line denotes the tropopause, and TTL stands for tropical tropopause layer (see text). The black arrows indicate the Brewer-Dobson circulation during NH winter, and the wiggly red arrow represents planetary waves that propagate from the troposphere into the winter stratosphere.

port leads to significant variations of ozone in the extra-tropical lower stratosphere, where the photochemical relaxation time is very long (several months or longer) and ozone can accumulate on seasonal time scales. Due to the seasonality of the Brewer-Dobson circulation (maximum during winter and spring), ozone builds up in the extra-tropical lower stratosphere during winter and spring through transport, and then decays photochemically during the summer when transport is weaker. The column-ozone distribution (measured in Dobson units, DU) is dominated by its distribution in the lower stratosphere and reflects this seasonality (Figure 1.2). Furthermore, planetary waves are stronger (and more variable) in the Northern Hemisphere

(NH) than in the Southern Hemisphere (SH), because of the asymmetric distribution of the surface features (topography and land-sea thermal contrasts) that, in combination with surface winds, force the waves. Accordingly, the stratospheric Brewer-Dobson circulation is stronger during the NH winter, and the resulting extra-tropical build-up of ozone during the winter and spring is greater in the NH than in the SH (Andrews *et al.,* 1987; Figure 1.2).

Variations in the Brewer-Dobson circulation also influence polar temperatures in the lower stratosphere (via the vertical motions); stronger wave forcing coincides with enhanced circulation and higher polar temperatures (and more ozone trans-

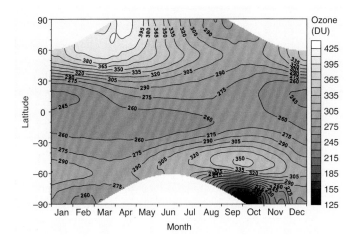

Figure 1.2. A climatology of total column ozone plotted as a function of latitude and month. Version 8 Total Ozone Mapping Spectrometer (TOMS) data were used together with version 3.1 Global Ozone Monitoring Experiment (GOME) data over the period 1994 to 2003. Updated from Bodeker *et al.* (2001).

port). Since temperature affects ozone chemistry, dynamical and chemical effects on column ozone thus tend to act in concert and are coupled.

Human activities have led to changes in the atmospheric concentrations of several greenhouse gases, including tropospheric and stratospheric ozone and ODSs and their substitutes. Changes to the concentrations of these gases alter the radiative balance of the Earth's atmosphere by changing the balance between incoming solar radiation and outgoing infrared radiation. Such an alteration in the Earth's radiative balance is called a *radiative forcing*. This report, past IPCC reports and climate change protocols have universally adopted the concept of radiative forcing as a tool to gauge and contrast *surface* climate change caused by different mechanisms (see Box 1.3). Positive radiative forcings are expected to warm the Earth's surface and negative radiative forcings are expected to cool it. The radiative forcings from the principal greenhouse gases are summarized in Figure 1.3 and Table 1.1 (in Section 1.5.2). Other significant contributors to radiative forcing that are not shown in Figure 1.3 include tropospheric aerosol changes and changes in the Sun's output. Changes in carbon dioxide (CO_2) provide the largest radiative forcing term and are expected to be the largest overall contributor to climate change. In contrast with the positive radiative forcings due to increases in other greenhouse gases, the radiative forcing due to stratospheric ozone depletion is negative.

Halocarbons are particularly effective greenhouse gases in part because they absorb the Earth's outgoing infrared radiation in a spectral range where energy is not removed by carbon dioxide or water vapour (sometimes referred to as the *atmospheric window*). Halocarbon molecules can be many thousands of times more efficient at absorbing the radiant energy emitted from the Earth than a molecule of carbon dioxide, which ex-

plains why relatively small amounts of these gases can contribute significantly to radiative forcing of the climate system. Because halocarbons have low concentrations and absorb in the atmospheric window, the magnitude of the direct radiative forcing from a halocarbon is given by the product of its tropospheric mixing ratio and its radiative efficiency. In contrast, for the more abundant greenhouse gases (carbon dioxide, methane and nitrous oxide) there is a nonlinear relationship between the mixing ratio and the radiative forcing.

Since 1970 the growth in halocarbon concentrations and the changes in ozone concentrations (depletion in the stratosphere and increases in the troposphere) have been very significant contributors to the total radiative forcing of the Earth's atmosphere. Because halocarbons have likely caused most of the stratospheric ozone loss (see Section 1.4.1), there is the possibility of a partial offset between the positive forcing of halocarbon that are ODSs and the negative forcing from stratospheric ozone loss. This is discussed further in Section 1.5.

The climate impacts of ozone changes are not confined to the surface: stratospheric ozone changes are probably responsible for a significant fraction of the observed cooling in the lower stratosphere over the last two decades (Section 1.2.4) and may alter atmospheric dynamics and chemistry (Section 1.3.6). Further, it was predicted that depletion of stratospheric ozone would lead to a global increase in erythemal UV radiation (the radiation that causes sunburn) at the surface of about 3%, with much larger increases at high latitudes; these predicted high lat-

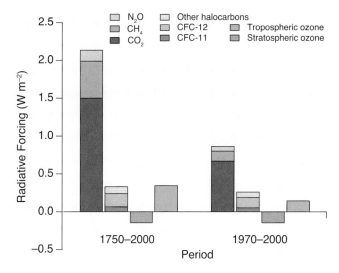

Figure 1.3. Changes in radiative forcing (in W m⁻²) from WMGHGs and ozone over two periods: 1750–2000 and 1970–2000. Numbers are taken from estimates in Tables 6.11 and 6.13 of IPCC (2001). The halocarbon direct radiative forcing is shown separately from the other WMGHG forcing. The negative forcing from stratospheric ozone is largely a result of the stratospheric ozone depletion, which resulted from halocarbon emissions between 1970 and 2000; it can therefore be considered an indirect radiative forcing from halocarbons, which offsets their positive direct forcing. In contrast, the tropospheric ozone radiative forcing is largely independent of the halocarbon forcing.

Box 1.3. Radiative forcing and climate sensitivity

Radiative forcing is a useful tool for estimating the relative climate impacts, in the global mean, of different climate change mechanisms. Radiative forcing uses the notion that the globally averaged radiative forcing (ΔF) is related to the globally averaged equilibrium surface temperature change (ΔT_s) through the climate sensitivity (λ):

$$\Delta T_s = \lambda \Delta F$$

The reason surface temperature changes in different models are not usually compared directly is that the climate sensitivity is poorly known and varies by a factor of three between different climate models (IPCC, 2001, Chapter 9). Further, climate model studies have shown that, for many forcing mechanisms, λ in an individual climate model is more or less independent of the mechanism. These two factors enable the climate change potential of different mechanisms to be quantitatively contrasted through their radiative forcings, while they remain difficult to compare through their predicted surface temperature changes. Unfortunately, certain radiative forcings, including ozone changes, have been shown to have different climate sensitivities compared with carbon dioxide changes (IPCC, 2001, Chapter 6; WMO, 2003, Chapter 4). One of the reasons for this difference is that some aspects of the radiative forcing definition, such as tropopause height, have a large impact on the ozone radiative forcing (see below). These factors are the main reasons for the large uncertainty in the ozone radiative forcing and the global mean climate response.

Why is the radiative forcing of stratospheric ozone negative?

The radiative forcing is defined as the change in net radiative flux at the tropopause after allowing for stratospheric temperatures to readjust to radiative equilibrium (IPCC, 2001, Chapter 6). The details of this definition are crucial for stratospheric ozone, and are explained in the figure.

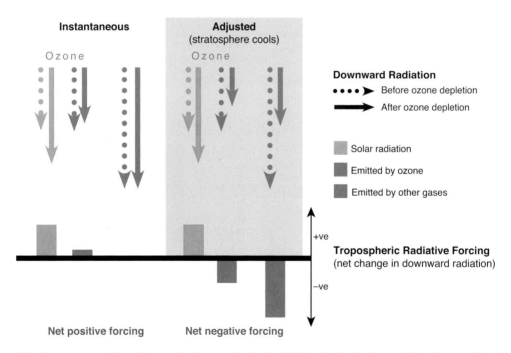

Box 1.3, Figure. The instantaneous effect of stratospheric ozone depletion (left-hand side of schematic) is to increase the shortwave radiation from the Sun reaching the tropopause (because there is less ozone to absorb it), and to slightly reduce the downward longwave radiation from the stratosphere, as there is less ozone in the stratosphere to emit radiation. This gives an instantaneous net positive radiative forcing. However, in response to less absorption of both shortwave and longwave radiation in the stratosphere, the region cools, which leads to an overall reduction of thermal radiation emitted downward from the stratosphere (right-hand side of schematic). The size of this adjustment term depends on the vertical profile of ozone change and is largest for changes near the tropopause. For the observed stratospheric ozone changes the adjustment term is larger than the positive instantaneous term, thus the stratospheric ozone radiative forcing is negative.

itude increases in surface UV dose have indeed been observed (WMO, 2003, Chapter 5).

1.1.3 Chapter outline

Our aim in this chapter is to review the scientific understanding of the interactions between ozone and climate. We interpret the term 'climate' broadly, to include stratospheric temperature and circulation, and refer to tropospheric effects as 'tropospheric climate'. This is a broad topic; our review will take a more restricted view concentrating on the interactions as they relate to stratospheric ozone and the role of the ODSs and their substitutes. For this reason the role of tropospheric ozone in the climate system is mentioned only briefly. Similarly, some broader issues, including possible changes in stratospheric water vapour (where the role of the ODSs and their substitutes should be minor) and climate-dependent changes in biogenic emissions (whose first order effect is likely to be in the troposphere), will not be discussed in any detail. The relationship between ozone and the solar cycle is well established, and the solar cycle has been used as an explanatory variable in ozone trend analysis. The connection between ozone and the solar cycle was assessed most recently in Chapter 12 of WMO (1999) and will not be discussed further here.

Section 1.2 presents an update on stratospheric observations of ozone, ozone-related species and temperature. Section 1.3 considers, at a process level, the various feedbacks connecting stratospheric ozone and the climate system. The understanding of these processes leads, in Section 1.4, to a discussion of the attribution of past changes in ozone, the prediction of future changes, and the connection of both with climate. Finally, Section 1.5 reviews the trade-offs between ozone depletion and radiative forcing, focusing on ODSs and their substitutes.

1.2 Observed changes in the stratosphere

1.2.1 Observed changes in stratospheric ozone

Global and hemispheric-scale variations in stratospheric ozone can be quantified from extensive observational records covering the past 20 to 30 years. There are numerous ways to measure ozone in the atmosphere, but they fall broadly into two categories: measurements of column ozone (the vertically integrated amount of ozone above the surface), and measurements of the vertical profile of ozone. Approximately 90% of the vertically integrated ozone column resides in the stratosphere. There are more independent measurements, longer time-series, and better global coverage for column ozone. Regular measurements of column ozone are available from a network of surface stations, mostly in the mid-latitude NH, with reasonable coverage extending back to the 1960s. Near-global, continuous column ozone data are available from satellite measurements beginning in 1979. The different observational data sets can be used to estimate past ozone changes, and the differences between data sets provide a lower bound of overall uncertainty. The differ-

ences indicate good overall agreement between different data sources for changes in column ozone, and thus we have reasonable confidence in describing the spatial and temporal characteristics of past changes (WMO, 2003, Chapter 4).

Five data sets of zonal and monthly mean column ozone values developed by different scientific teams were used to quantify past ozone changes in Chapter 4 of WMO (2003); they include ground-based measurements covering 1964–2001, and several different satellite data sets extending in time over 1979–2001. Figure 1.4a shows globally averaged column ozone changes derived from each of these data sets, updated through 2004. The analyses first remove the seasonal cycle from each

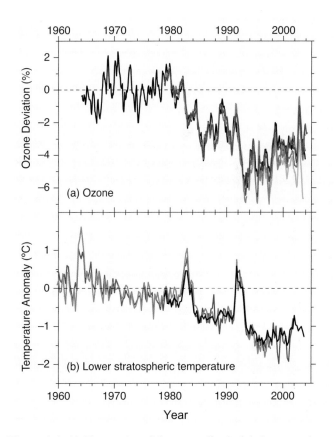

Figure 1.4. (a) Time-series of de-seasonalized global mean column ozone anomalies estimated from five different data sets, including ground-based (black line) and satellite measurements (colour lines). Anomalies are expressed as percentage differences from the 1964–1980 average, and the seasonal component of the linear trend has been removed. Updated from Fioletov *et al.* (2002). (b) Time-series of de-seasonalized global mean lower-stratospheric temperature anomalies estimated from radiosondes (colour lines) and satellite data (black line). Anomalies (in °C) are calculated with respect to the 1960–1980 average. The radiosonde data represent mean temperature in the 100–50 hPa layer (about 16–22 km), and are derived from the HadRT and LKS data sets described in Seidel *et al.* (2004). Satellite temperatures (available for 1979–2003) are from the Microwave Sounding Unit Channel 4, and represent mean temperature in a layer over about 13–22 km. For direct comparison, the satellite data have been normalized to equal the radiosonde time means over 1979–1997. Panel (b) is discussed in Section 1.2.4.

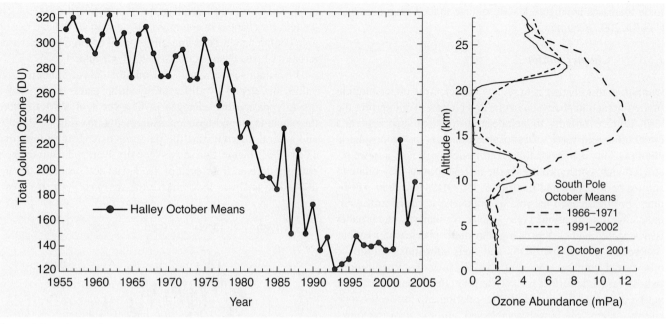

Figure 1.5. Left panel: October mean total column ozone measurements from the Dobson spectrophotometer at Halley, Antarctica (73.5°S, 26.7°W). Right panel: Vertical ozone profiles measured by ozonesondes at South Pole station, Antarctica (90°S). These data are from the WOUDC (World Ozone and UV Data Centre) and NDSC (Network for Detection of Stratospheric Change) databases. The 1966–1971 October mean profile is shown using a long-dashed line, the 1991–2002 October mean profile as a short-dashed line, and the single ozonesonde flight on 2 October 2001 as a solid line.

data set (mean and linear trend), and the deviations are area weighted and expressed as anomalies with respect to the period 1964–1980. The global ozone amount shows decreasing values between the late 1970s and the early 1990s, a relative minimum during 1992–1994, and slightly increasing values during the late 1990s. Global ozone for the period 1997–2004 was approximately 3% below the 1964–1980 average values. Since systematic global observations began in the mid-1960s, the lowest annually averaged global ozone occurred during 1992–1993 (5% below the 1964–1980 average). These changes are evident in each of the available global data sets.

The global ozone changes shown in Figure 1.4a occur mainly in the extratropics of both hemispheres (poleward of 25°–35°). No significant long-term changes in column ozone have been observed in the tropics (25°N to 25°S). Column ozone changes averaged over mid-latitudes (from 35° to 60° latitude) are significantly larger in the SH than in the NH; averaged for the period 1997–2001, SH values are 6% below pre-1980 values, whereas NH values are 3% lower. Also, there is significant seasonality to the NH mid-latitude losses (4% losses in the winter-spring period and 2% in summer), whereas long-term SH losses are about 6% year round (WMO, 2003, Chapter 4).

The most dramatic changes in ozone have occurred during the spring season over Antarctica, with the development during the 1980s of a phenomenon known as the *ozone hole* (Figure 1.5). The ozone hole now recurs every spring, with some inter-annual variability and occasional extreme behaviour (see Box 1.6 in Section 1.4.1). In most years the ozone concentration

is reduced to nearly zero over a layer several kilometres deep within the lower stratosphere in the Antarctic polar vortex. Since the early 1990s, the average October column ozone poleward of

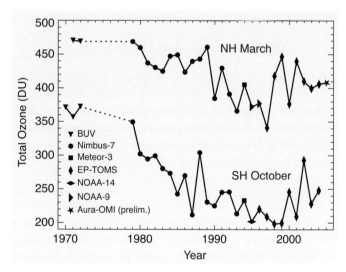

Figure 1.6. Average column ozone poleward of 63° latitude in the springtime of each hemisphere (March for the NH and October for the SH), in Dobson units, based on data from various satellite instruments as indicated. The data point from the Ozone Monitoring Instrument (OMI) is preliminary. Updated from Newman *et al.* (1997), courtesy of NIVR (Netherlands), KNMI (Netherlands), FMI (Finland), and NASA (USA).

63°S has been more than 100 DU below pre-ozone-hole values (a 40 to 50% decrease), with up to a 70% local decrease for periods of a week or so (WMO, 2003, Chapter 3; Figure 1.6).

Compared with the Antarctic, Arctic ozone abundance in the winter and spring is highly variable (Figure 1.6), because of interannual variability in chemical loss and in dynamical transport. Dynamical variability within the winter stratosphere leads to changes in ozone transport to high latitudes, and these transport changes are correlated with polar temperature variability – with less ozone transport being associated with lower temperatures. Low temperatures favour halogen-induced chemical ozone loss. Thus, in recent decades, halogen-induced polar ozone chemistry has acted in concert with the dynamically induced ozone variability, and has led to Arctic column ozone losses of up to 30% in particularly cold winters (WMO, 2003, Chapter 3). In particularly dynamically active, warm winters,

however, the estimated chemical ozone loss has been very small.

Changes in the vertical profile of ozone are derived primarily from satellites, ground-based measurements and balloon-borne ozonesondes. Long records from ground-based and balloon data are available mainly for stations over NH mid-latitudes. Ozone profile changes over NH mid-latitudes exhibit two maxima, with decreasing trends in the upper stratosphere (about 7% per decade at 35–45 km) and in the lower stratosphere (about 5% per decade at 15–25 km) during the period 1979–2000. Ozone profile trends show a minimum (about 2% per decade decrease) near 30 km. The vertically integrated profile trends are in agreement with the measured changes in column ozone (WMO, 2003, Chapter 4).

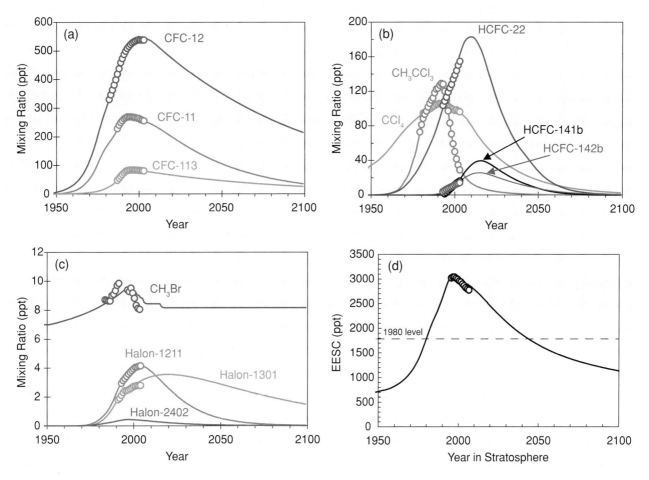

Figure 1.7. Global atmospheric mixing ratios (in ppt) in the WMO (2003) baseline scenario Ab (lines) of the most abundant CFCs, HCFCs, chlorinated solvents and brominated gases at the Earth's surface. Also shown are measured annual means (circles) that are based on results from two independent, global sampling networks (an update of Figure 1-22 of WMO, 2003) for all gases except CH_3Br. Global surface means for CH_3Br are taken from Khalil *et al.* (1993) for the years before 1992, and from Montzka *et al.* (2003) for 1996–2003. Panel (d) shows past and potential future changes in the total ozone-depleting halogen burden of the stratosphere, estimated as equivalent effective stratospheric chlorine (EESC). The EESC values are derived from the baseline scenario Ab (line) or surface measurements (circles). EESC is derived from measured surface mixing ratios of the most abundant CFCs, HCFCs, chlorinated solvents, halons and methyl bromide; it is based on the assumption of a 3-year time lag between the troposphere and mid-stratosphere and a factor of 45 to account for the enhanced reactivity of Br relative to Cl in depleting stratospheric ozone (Daniel *et al.*, 1999).

1.2.2 Observed changes in ODSs

As a result of reduced emissions because of the Montreal Protocol and its Amendments and Adjustments, mixing ratios for most ODSs have stopped increasing near the Earth's surface (Montzka *et al.*, 1999; Prinn *et al.*, 2000). The response to the Protocol, however, is reflected in quite different observed behaviour for different substances. By 2003, the mixing ratios for CFC-12 were close to their peak, CFC-11 had clearly decreased, while methyl chloroform (CH_3CCl_3) had dropped by 80% from its maximum (Figure 1.7; see also Chapter 1 of WMO, 2003).

Halons and HCFCs are among the few ODSs whose mixing ratios were still increasing in 2000 (Montzka *et al.*, 2003; O'Doherty *et al.*, 2004). Halons contain bromine, which is on average 40 to 50 times more efficient on a per-atom basis at destroying stratospheric ozone than chlorine. However, growth rates for most halons have steadily decreased during recent years (Fraser *et al.*, 1999; Montzka *et al.*, 2003). Furthermore, increases in tropospheric bromine from halons have been offset by the decline observed for methyl bromide (CH_3Br) since 1998

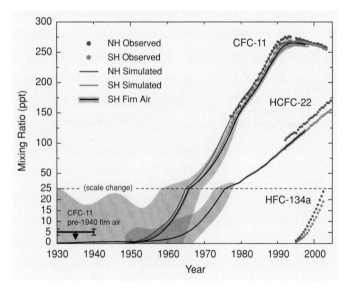

Figure 1.8. Estimated global atmospheric mixing ratios (in ppt) of CFC-11, HCFC-22 and HFC-134a, shown separately for the NH (red) and SH (blue). Circles show measurements from the AGAGE (Advanced Global Atmospheric Gases Experiment) and CMDL (Climate Monitoring and Diagnostics Laboratory) networks, while colour lines show simulated CFC-11 concentrations based on estimates of emissions and atmospheric lifetimes (updated from Walker *et al.*, 2000). The black lines and the shaded areas show estimates and uncertainty bands, respectively, for CFC-11 and HCFC-22 that were derived by synthesis inversion of Antarctic firn air measurements and *in situ* Cape Grim atmospheric measurements (Sturrock *et al.*, 2002). The thick black horizontal line with an arrow and an error bar shows a separate upper-bound estimate of pre-1940 CFC-11 concentrations based on South Pole firn air measurements (see Battle and Butler, 'Determining the atmospheric mixing ratio of CFC-11 from firn air samples', which is available at ftp://ftp.cmdl.noaa.gov/hats/firnair/ and is based on the analysis of Butler *et al.*, 1999).

(Yokouchi *et al.*, 2002; Montzka *et al.*, 2003; Simmonds *et al.*, 2004).

Atmospheric amounts of HCFCs continue to increase because of their use as CFC substitutes (Figure 1.7b). Chlorine from HCFCs has increased at a fairly constant rate of 10 ppt Cl yr[-1] since 1996, although HCFCs accounted for only 5% of all chlorine from long-lived gases in the atmosphere by 2000 (WMO, 2003, Chapter 1). The ODPs of the most abundant HCFCs are only about 5–10% of those of the CFCs (Table 1.2 in Section 1.5.3).

Figure 1.8 contrasts the observed atmospheric abundances of CFC-11 with those of HCFC-22 and HFC-134a. The behaviour of CFC-11 is representative of the behaviour of CFCs in general, and the behaviour of HCFC-22 and HFC-134a is representative of HCFCs (as well as HFCs) in general (see Figure 2.3 in Chapter 2). The most rapid growth rates of CFC-11 occurred in the 1970s and 1980s. The largest emissions were in the NH, and concentrations in the SH lagged behind those in the NH, consistent with an inter-hemispheric mixing time scale of about 1 to 2 years. In recent years, following the implementation of the Montreal Protocol, the observed growth rate has declined, concentrations appear to be at their peak and, as emissions have declined, the inter-hemispheric gradient has almost disappeared. In contrast, HCFC-22 and HFC-134a concentrations are still growing rapidly and there is a marked inter-hemispheric gradient.

Ground-based observations suggest that by 2003 the cumulative totals of both chlorine- and bromine-containing gases regulated by the Montreal Protocol were decreasing in the lower atmosphere (Montzka *et al.*, 2003). Although tropospheric chlorine levels peaked in the early 1990s and have since declined, atmospheric bromine began decreasing in 1998 (Montzka *et al.*, 2003).

The net effect of changes in the abundance of both chlorine and bromine on the total ozone-depleting halogens in the stratosphere is estimated roughly by calculating the equivalent effective stratospheric chlorine (EESC; Daniel *et al.*, 1995) (Figure 1.7d). The calculation of EESC (see Box 1.8 in Section 1.5.3) includes consideration of the total amount of chlorine and bromine accounted for by long-lived halocarbons, how rapidly these halocarbons degrade and release their halogen in the stratosphere and a nominal lag time of three years to allow for transport from the troposphere into the stratosphere. The tropospheric observational data suggest that EESC peaked in the mid-1990s and has been decreasing at a mean rate of 22 ppt yr[-1] (0.7% yr[-1]) over the past eight years (Chapter 1 of WMO, 2003). Direct stratospheric measurements show that stratospheric chlorine reached a broad plateau after 1996, characterized by variability (Rinsland *et al.*, 2003).

1.2.3 Observed changes in stratospheric aerosols, water vapour, methane and nitrous oxide

In addition to ODSs, stratospheric ozone is influenced by the abundance of stratospheric aerosols, water vapour, methane

(CH$_4$) and nitrous oxide (N$_2$O). Observed variations in these constituents are summarized in this section.

During the past three decades, aerosol loading in the stratosphere has primarily reflected the effects of a few volcanic eruptions that inject aerosol and its gaseous precursors (primarily sulphur dioxide, SO$_2$) into the stratosphere. The most noteworthy of these eruptions are El Chichón (1982) and Pinatubo (1991). The 1991 Pinatubo eruption likely had the largest impact of any event in the 20th century (McCormick *et al.*, 1995), producing about 30 Tg of aerosol (compared with approximately 12 Tg from El Chichón) that persisted into at least the late 1990s. Current aerosol loading, which is at the lowest observed levels, is less than 0.5 Tg, so the Pinatubo event represents nearly a factor of 100 enhancement relative to non-volcanic levels. The source of the non-volcanic stratospheric aerosol is primarily carbonyl sulfide (OCS), and there is general agreement between the aerosols estimated by modelling the transformation of observed OCS to sulphate aerosols and observed aerosols. However, there is a significant dearth of SO$_2$ measurements, and the role of tropospheric SO$_2$ in the stratospheric aerosol budget – while significant – remains a matter of some uncertainty. Because of the high variability of stratospheric aerosol loading it is difficult to detect trends in the non-volcanic aerosol component. Trends derived from the late 1970s to the current period are likely to encompass a value of zero.

The recent Stratospheric Processes And their Role in Climate (SPARC) Assessment of Upper Tropospheric and Stratospheric Water Vapour (SPARC, 2000) provided an extensive review of data sources and quality for stratospheric water vapour, together with detailed analyses of observed seasonal and interannual variability. The longest continuous reliable data set is at a single location (Boulder, Colorado, USA), is based on balloon-borne frost-point hygrometer measurements (approximately one per month), and dates back to 1980. Over the period 1980–2000, a statistically significant positive trend of approximately 1% yr^{-1} is observed at all levels between about 15 and 26 km in altitude (SPARC, 2000; Oltmans *et al.*, 2000). However, although a linear trend can be fitted to these data, there is a high degree of variability in the infrequent sampling, and the increases are neither continuous nor steady (Figure 1.9). Long-term increases in stratospheric water vapour are also inferred from a number of other ground-based, balloon, aircraft and satellite data sets spanning approximately 1980–2000 (Rosenlof *et al.*, 2001), although the time records are short and the sampling uncertainty is high in many cases.

Global stratospheric water vapour measurements have been made by the Halogen Occultation Experiment (HALOE) satellite instrument for more than a decade (late 1991 to 2004). Interannual changes in water vapour derived from HALOE data show excellent agreement with Polar Ozone and Aerosol Measurement (POAM) satellite data, and also exhibit strong coherence with tropical tropopause temperature changes (Randel *et al., 2004*). An updated comparison of the HALOE measurements with the Boulder balloon data for the period 1992–2004 is shown in Figure 1.9. The Boulder and HALOE data show

reasonable agreement for the early part of the record (1992–1996), but there is an offset after 1997, with the Boulder data showing higher values than HALOE measurements. As a result, linear trends derived from these two data sets for the (short) period 1992–2004 show very different results (increases for the Boulder data, but not for HALOE). The reason for the differences between the balloon and satellite data (for the same time period and location) is unclear at present, but the discrepancy calls into question interpretation of water vapour trends derived from short or infrequently sampled data records. It will be important to reconcile these differences because these data sets are the two longest and most continuous data records available for stratospheric water vapour.

It is a challenge to explain the magnitude of the water vapour increases seen in the Boulder frost-point data. Somewhat less than half the observed increase of about 10% per decade (through 2001) can be explained as a result of increasing tropospheric methane (transported to the stratosphere and oxidized to form water vapour). The remaining increase could be reconciled with a warming of the tropical tropopause of approximately 1 K per decade, assuming that air entered the stratosphere with water vapour in equilibrium with ice. However, observations suggest that the tropical tropopause has cooled slightly for this time period, by approximately 0.5 K per decade, and risen slightly in altitude by about 20 m per decade (e.g., Seidel *et al.*, 2001; Zhou *et al.*, 2001). Although regional-scale processes may also influence stratospheric water vapour (such as summer monsoon circulations, e.g., Potter and Holton, 1995), there is no evidence

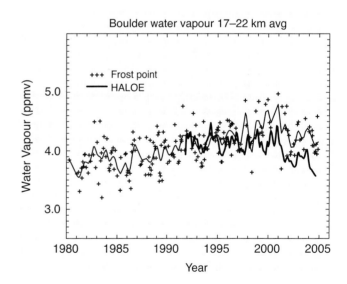

Figure 1.9. Time-series of stratospheric water vapour. Symbols show frost-point hygrometer measurements (averaged over 17–22 km) from Boulder, Colorado covering 1980–2004. The thin line shows a smooth fit through the measurements, using a running Gaussian window with a half-width of three months. The heavy line shows HALOE satellite water vapour data during 1992–2004 for the same altitude region, and using measurements near Boulder (over latitudes 35°N–45°N and longitudes 80°W–130°W). Note the difference between the two data sets after about 1997. Updated from Randel *et al.* (2004).

for increases in tropopause temperature in these regions either (Seidel *et al.*, 2001). From this perspective, the extent of the decadal water vapour increases inferred from the Boulder measurements is inconsistent with the observed tropical tropopause cooling, and this inconsistency limits confidence in predicting the future evolution of stratospheric water vapour.

The atmospheric abundance of methane has increased by a factor of about 2.5 since the pre-industrial era (IPCC, 2001, Chapter 4). Measurements of methane from a global monitoring network showed increasing values through the 1990s, but approximately constant values during 1999–2002 (Dlugokencky *et al.*, 2003). Changes in stratospheric methane have been monitored on a global scale using HALOE satellite measurements since 1991. The HALOE data show increases in lower stratospheric methane during 1992–1997 that are in reasonable agreement with tropospheric increases during this time (Randel *et al.*, 1999). In the upper stratosphere the HALOE data show an overall decrease in methane since 1991, which is likely attributable to a combination of chemical and dynamical influences (Nedoluha *et al.*, 1998; Considine *et al.*, 2001; Röckmann *et al.*, 2004).

Measurements of tropospheric N_2O show a consistent increase of about 3% per decade (IPCC, 2001, Chapter 4). Because tropospheric air is transported into the stratosphere, these positive N_2O trends produce increases in stratospheric reactive nitrogen (NO_y), which plays a key role in ozone photochemistry. Measurements of stratospheric column NO_2 (a main component of NO_y) from the SH (1980–2000) and the NH (1985–2000) mid-latitudes (about 45°) show long-term increases of approximately 6% per decade (Liley *et al.*, 2000; WMO, 2003, Chapter 4), and these are reconciled with the N_2O changes by considering effects of changing levels of stratospheric ozone, water vapour and halogens (Fish *et al.*, 2000; McLinden *et al.*, 2001).

1.2.4   Observed temperature changes in the stratosphere

There is strong evidence of a large and significant cooling in most of the stratosphere since 1980. Recent updates to the observed changes in stratospheric temperature and to the understanding of those changes have been presented in Chapters 3 and 4 of WMO (2003). Current long-term monitoring of stratospheric temperature relies on satellite instruments and radiosonde analyses. The Microwave Sounding Unit (MSU) and Stratospheric Sounding Unit (SSU) instruments record temperatures in several 10–15 km thick layers between 17 and 50 km in altitude. Radiosonde trend analyses are available up to altitudes of roughly 25 km. Determining accurate trends with these data sets is difficult. In particular the radiosonde coverage is not global and suffers from data quality concerns, whereas the satellite trend data is a result of merging data sets from several different instruments. Figure 1.4b shows global temperature time-series in the lower and middle stratosphere derived from satellites and radiosondes. It reveals a strong imprint of 1 to 2 years of warming following the volcanic eruptions of Agung (1963), El Chichón (1982) and Mt. Pinatubo (1991).

When these years are excluded from long-term trend analyses, a significant global cooling is seen in both the radiosonde and the satellite records over the last few decades. This cooling is significant (at the 97% level or greater) at all levels of the stratosphere except the 30 km (10 hPa) level in the SSU record (Figure 1.10). The largest global cooling is found in the upper stratosphere, where it is fairly uniform in time at a rate of about 2 K per decade (for 1979–1997). In the lower stratosphere (below 25 km) this long-term global cooling manifests itself as more of a step-like change following the volcanic warming events (Figure 1.4b). The cooling also varies with latitude. The lower stratosphere extratropics (from 20° to 60° latitude) show a cooling of 0.4–0.8 K per decade in both hemispheres, which remains roughly constant throughout the year. At high latitudes most of the cooling, up to 2 K per decade for both hemispheres, occurs during spring.

Much recent progress has been made in modelling these temperature trends (see Shine *et al.*, 2003; WMO, 2003, Chapters 3 and 4). Models range from one-dimensional (1-D) fixed dynamical heating rate (FDH) calculations to three-dimensional (3-D) coupled chemistry-climate models; many of their findings are presented later in this chapter. FDH is a simple way of determining the radiative response to an imposed change whilst 'fixing' the background dynamical heating to its climatological value; this makes the calculation of temperature change simpler, as only a radiation model needs to be used. Changes in dynamical circulations simulated by models can lead to effects over latitudinal bands, which vary between models. However, dynamics cannot easily produce a global mean temperature change. Because of this, global mean temperature is radiatively controlled and provides an important focus for attribution (Figure 1.10). For the global mean in the upper stratosphere the models suggest roughly equal contributions to the cooling from ozone decreases and carbon dioxide increases. In the global mean mid-stratosphere there appears to be some discrepancy between the SSU and modelled trends near 30 km: models predict a definite radiative cooling from carbon dioxide at these altitudes, which is not evident in the SSU record. In the lower stratosphere the cooling from carbon dioxide is much smaller than higher in the stratosphere. Although ozone depletion probably accounts for up to half of the observed cooling trend, it appears that a significant cooling from another mechanism may be needed to account for the rest of the observed cooling. One possible cause of this extra cooling could be stratospheric water vapour increases. However, stratospheric water vapour changes are currently too uncertain (see Section 1.2.3) to pinpoint their precise role. Tropospheric ozone increases have probably contributed slightly to lower stratospheric cooling by reducing upwelling thermal radiation; one estimate suggests a cooling of 0.05 K per decade at 50 hPa and a total cooling of up to 0.5 K over the last century (Sexton *et al.*, 2003). The springtime cooling in the Antarctic lower stratosphere is almost certainly nearly all caused by stratospheric ozone depletion. However, the similar magnitude of cooling in the Arctic spring does not seem to be solely caused by ozone changes, which are much smaller

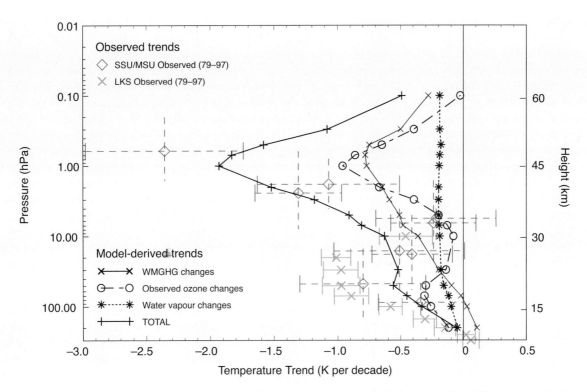

Figure 1.10. Observed and model-derived global and annual mean temperature trends in the stratosphere. The model-derived trends show the contributions of observed changes in WMGHGs, ozone and water vapour to the total temperature trend, based on an average of several model results for ozone and WMGHGs and a single model result for stratospheric water vapour, which is based on satellite-derived trends. The satellite temperature trends are from MSU and SSU (WMO, 2003, Chapter 4) and the radiosonde trends are from the LKS data set. The horizontal bars show the 2σ errors in the observations; the vertical bars give the approximate altitude range sensed by the particular satellite channel. After Shine *et al.* (2003).

than in the Antarctic; interannual variability may contribute substantially to the cooling observed in the Arctic wintertime (see Section 1.4.1).

One mechanism for altering temperatures in the upper troposphere and lower stratosphere comes from the direct radiative effects of halocarbons (WMO, 1986; Ramaswamy *et al.*, 1996; Hansen *et al.*, 1997). In contrast with the role of carbon dioxide, halocarbons can actually warm the upper tropospheric and lower stratospheric region (see Box 1.4).

In summary, stratospheric temperature changes over the past few decades are significant (they are, in fact, substantially larger than those seen in the surface temperature record over the same period), and there are clear quantifiable features of the contributions from ozone, carbon dioxide and volcanism in the past stratospheric temperature record. More definite attribution of the causes of these trends is limited by the short time-series. Future increases of carbon dioxide can be expected to substantially cool the upper stratosphere. However, this cooling could be partially offset by any future ozone increase (Section 1.4.2). In the lower stratosphere, both ozone recovery and halocarbon increases would warm this region compared with the present. Any changes in stratospheric water vapour or changes in tropospheric conditions, such as high-cloud properties and tropospheric ozone, would also affect future temperatures in

the lower stratosphere. Furthermore, circulation changes can affect temperatures over sub-global scales, especially at mid- and polar latitudes. Several studies have modelled parts of these expected temperature changes, and are discussed in the rest of this chapter (see especially Section 1.4.2).

1.3 Stratospheric ozone and climate feedback processes

The distribution of ozone depends on a balance between chemical processes, which can be affected by changes in the concentration of the ODSs, and transport processes, which can be affected by climate change. Climate change, in its broadest sense, can also affect ozone chemistry directly by modifying the rates of temperature-dependent reactions. These interactions are discussed in Sections 1.3.1 to 1.3.4. In these sections we also discuss the possible impact on ozone of changing stratospheric water vapour abundances, which may be regulated by climate.

Changes in stratospheric ozone can also affect climate. The impact of ozone changes on stratospheric temperatures has already been discussed in Section 1.2.4. As well as the direct impact of ozone as an absorber of UV radiation and as a greenhouse gas, there are other effects that could be important. For example, changes in ozone could affect the lifetime of reactive

Box 1.4. Halocarbons and the tropical tropopause

Halocarbons and carbon dioxide are greenhouse gases: they trap longwave radiation to warm the Earth's surface. However, compared with carbon dioxide halocarbons have quite different spectral absorption characteristics and they interact very differently with the Earth's radiation field. The very strong 15 μm band dominates the role of carbon dioxide, whereas the halocarbons tend to absorb weakly in the 8–13 μm atmospheric window, a region of the spectrum where other gases have only a small effect on outgoing longwave radiation. These absorption properties, combined with a typical vertical temperature profile, means that halocarbons usually warm the atmosphere locally, whereas carbon dioxide generally cools it (the atmosphere only warms as a response to the induced surface warming). Further, this effect is largest at the tropical tropopause (about 17 km altitude), where temperatures are most different from those of the underlying surface (e.g., Dickinson, 1978; Wang *et al.*, 1991). The tropical tropopause can be defined as the height at which the coldest temperatures are found in a vertical temperature profile (see Figure 1.1b). The effects of halocarbon, carbon dioxide and ozone changes are contrasted in the figure in this box.

Panel (a) shows that halocarbons may have warmed the tropical upper troposphere and lower stratosphere by as much as 0.3 K, which is locally larger than the cooling effects of carbon dioxide. For the calculation of a globally averaged temperature change this warming can be thought of as partially cancelling out the cooling effect of ozone depletion in the extra-tropical lower stratosphere. However, the patterns are quite distinct (compare panels (a) and (b)), and as a result the equator-to-pole temperature gradient in the lower stratosphere, where temperature increases towards higher latitudes, would be reduced.

Although halocarbons are potentially important, because of their coupling to water vapour, it is unlikely that halocarbons dominate the response of the tropical tropopause to changing greenhouse gases, as there is some observational evidence for a general cooling of the tropical tropopause over the last few decades (Seidel *et al.*, 2001; Zhou *et al.*, 2001).

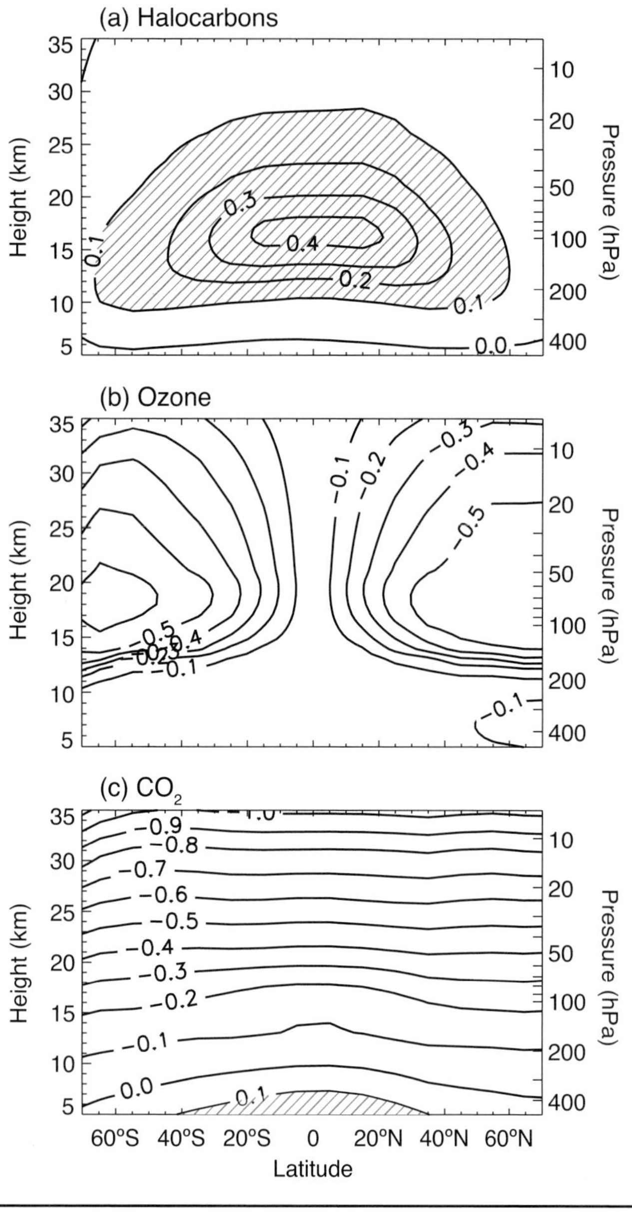

Box 1.4, Figure. The expected contribution (in kelvin) to the upper-tropospheric and lower-stratospheric temperature changes over 1970–2000 from changes in (a) halocarbons, (b) ozone and (c) carbon dioxide. Results are based on FDH model calculations. The hatched area denotes positive changes greater than 0.1 K. After Forster and Joshi (2005).

greenhouse gases, by changing the penetration of UV radiation. Changes in the structure of the stratosphere, caused by ozone changes, could alter the interaction between the troposphere and stratosphere and lead to further changes in stratospheric ozone. These latter interactions are discussed in Sections 1.3.5 and 1.3.6.

Improved knowledge of these various feedback processes is essential for informing the numerical models used to predict future chemistry-climate interactions; these models are discussed in Section 1.4.

Note that other factors, which are not discussed in detail here, can also affect the interaction between ozone and climate. Perhaps the most obvious is the impact of major volcanic eruptions. These can lead to an increase in volcanic aerosol in the stratosphere, which can influence both climate and the chemical processes controlling the ozone layer.

1.3.1   Impact of ODSs on stratospheric ozone

The abundance of ozone in the stratosphere at a particular location is governed by three processes: photochemical production, destruction by catalytic cycles, and transport processes. Photochemical production in the stratosphere occurs mostly through the photolysis of O_2, with loss via catalytic cycles involving hydrogen (HO_x), nitrogen (NO_x), and halogen (ClO_x, BrO_x) radicals (see Box 1.1). The relative importance of the various loss cycles in the stratosphere varies substantially with altitude (Figure 1.11). Above about 45 km, loss through HO_x dominates, while below this altitude NO_x-catalyzed ozone loss is most important. The importance of the ClO_x-catalyzed ozone loss cycle, which varies with chlorine loading, peaks at about 40 km. Below about 25 km HO_x-driven ozone loss cycles dominate again.

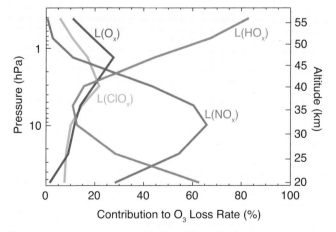

Figure 1.11. The vertical distribution of the relative importance of the individual contributions to ozone loss by the HO_x, ClO_x and NO_x cycles as well as the Chapman loss cycle (O_x). The calculations are based on HALOE (V19) satellite measurements and are for overhead sun (23°S, January) and for total inorganic chlorine (Cl_y) in the stratosphere corresponding to 1994 conditions. Reaction rate constants were taken from DeMore *et al.* (1997). Updated and extended in altitude range from Grooß *et al.* (1999) and Chapter 6 of WMO (1999).

1.3.1.1   Upper stratosphere

In the upper stratosphere the ozone budget is largely understood, although uncertainties remain regarding the rate constants of key radical reactions. In particular, strong evidence has been accumulated that the observed ozone depletion in the upper stratosphere is caused by increased levels of stratospheric chlorine (WMO, 1999, Chapter 6; WMO, 2003, Chapter 4), as originally proposed by Molina and Rowland (1974) and Crutzen (1974). Because of the direct correspondence between the stratospheric chlorine abundance and ozone depletion in the upper stratosphere, it has been suggested (e.g., WMO, 1999, Chapter 12) that the response of stratospheric ozone to the declining stratospheric chlorine levels might be first detectable in the upper stratosphere (see also Box 1.7 in Section 1.4.2).

1.3.1.2   Polar regions

In recent decades stratospheric ozone losses have been most pronounced in polar regions during winter and spring (WMO 2003, Chapter 3; see also Figures 1.5 and 1.6). These losses are determined largely by three chemical factors: (1) the conversion of chlorine reservoirs into active, ozone-destroying forms through heterogeneous reactions on the surfaces of polar stratospheric cloud particles; (2) the availability of sunlight that drives the catalytic photochemical cycles that destroy ozone; and (3) the timing of the deactivation of chlorine (i.e., the timing of the conversion of active chlorine back to the reservoir species). Temperature controls the formation and destruction of polar stratospheric clouds (PSCs) and thus the timing of activation (1) and deactivation (3) of chlorine. Furthermore, temperature controls the efficiency of the catalytic cycles (especially the ClO dimer cycle; see Box 1.1) that destroy ozone in the presence of sunlight (2).

However, polar ozone loss would not occur without a prominent dynamical feature of the stratosphere in winter and spring: the polar vortex. In both hemispheres, the polar vortex separates polar air from mid-latitude air to a large extent, and within the vortex the low-temperature conditions that develop are the key factor for polar ozone loss. These two factors are dynamically related: a strong vortex is generally also a colder vortex. The two crucial questions for future polar ozone are whether, in an increased greenhouse-gas climate, the region of low stratospheric temperatures will increase in area and whether it will persist for longer in any given year (see Section 1.3.2).

When anthropogenically emitted ODSs are eventually removed from the stratosphere, the stratospheric halogen burden will be much lower than it is today and will be controlled by the naturally occurring source gases methyl chloride (CH_3Cl) and methyl bromide (CH_3Br). Under such conditions, dramatic losses of polar ozone in winter and spring as we see today are not expected to occur. However, the rate of removal of anthropogenic halogens from the atmosphere depends on the atmospheric lifetimes of CFCs (typically 50–100 years) and is considerably slower than the rate at which halogens have been increasing in the decades prior to approximately the year 2000, when the stratospheric halogen loading peaked (Figure 1.7). A complete removal of anthropogenic halogens from the atmosphere will take more than a century. During this period of enhanced levels of halogens caused by past anthropogenic emissions, the polar stratosphere will remain vulnerable to climate perturbations, such as increasing water vapour or a cooling of the stratosphere, that lead to enhanced ozone destruction.

1.3.1.3 Lower-stratospheric mid-latitudes

It is well established that ozone in the lower stratosphere at mid-latitudes has been decreasing for a few decades; both measurements of ozone locally in the lower stratosphere and measurements of column ozone (a quantity that is dominated by the amount of ozone in the lower stratosphere) show a clear decline (WMO, 2003, Chapter 4). However, because mid-latitude ozone loss is much less severe than polar ozone loss, it cannot be identified in measurements in any one year but is rather detected as a downward trend in statistical analyses of longer time-series. It is clear that chemical loss driven by halogens is very important, but other possible effects that may contribute to the observed mid-latitude trends have also been identified. A definitive quantitative attribution of the trends to particular mechanisms has not yet been achieved. Chapter 4 of WMO (2003) reviewed this issue most recently; we will not repeat that detailed analysis here but instead provide a brief summary.

The chemical processes that may affect trends of mid-latitude ozone are essentially related to the ODS trends that are known to be responsible for the observed ozone loss in the upper stratosphere and in the polar regions. (Transport effects are discussed in Section 1.3.4.) Halogen chemistry may lead to ozone depletion in the mid-latitude lower stratosphere through a number of possible mechanisms, including:

1. Export of air that has encountered ozone destruction during the winter from the polar vortex (e.g., Prather *et al.*, 1990).
2. Export of air with enhanced levels of active chlorine from the polar vortex (e.g., Prather and Jaffe, 1990; Norton and Chipperfield, 1995).
3. *In situ* activation of chlorine either on cold liquid sulfate aerosol particles (e.g., Keim *et al.*, 1996) or on ice particles (e.g., Borrmann *et al.*, 1997; Bregman *et al.*, 2002). Further, the reaction of N_2O_5 with water on liquid aerosol particles at higher temperatures indirectly enhances the concentrations of ClO at mid-latitudes (McElroy *et al.*, 1992).
4. Ozone depletion due to elevated levels of BrO in the lower stratosphere, possibly caused by transport of very short-lived halogen-containing compounds or BrO across the tropopause (WMO, 2003, Chapter 2).

The first mechanism results from transport combined with polar ozone loss, whereas the remaining three mechanisms all involve *in situ* chemical destruction of ozone at mid-latitudes. All four mechanisms listed above are ultimately driven by the increase of halogens in the atmosphere over the past decades. Thus, while Millard *et al.* (2002) emphasized the strong interannual variability in the relative contributions of the different mechanisms to the seasonal mid-latitude ozone loss during the winter-spring period, they nonetheless showed that ozone loss driven by catalytic cycles involving halogens was always an important contributor (40–70%) to the simulated mid-latitude ozone loss in the five winters in the 1990s that they studied. Similarly, Chipperfield (2003) found that the observed mid-latitude column ozone decrease from 1980 to the early 1990s could be reproduced in long-term simulations in a 3-D chemistry-transport model (CTM). The modelled ozone is affected by dynamical interannual variability, but the overall decreases are dominated by halogen trends; and about 30–50% of the modelled halogen-induced change is a result of high latitude processing on PSCs.

Under climate change the strength of the polar vortex may change (see Section 1.3.2), but the sign of this change is uncertain. A change in the strength and temperature of the vortex will affect chlorine activation and ozone loss there and, through mechanisms (1) and (2), ozone loss in mid-latitudes.

The possibility that chlorine might be activated on cirrus clouds or on cold liquid aerosol particles in the lowermost stratosphere (mechanism (3)) was revisited recently by Bregman *et al.* (2002). Based on their model results it seems unlikely that this process is the main mechanism for the observed long-term decline of ozone in the mid-latitude lower stratosphere.

Very short-lived organic chlorine-, bromine-, and iodine-containing compounds possess a potential to deplete stratospheric ozone. However a quantitative assessment of their impact on stratospheric ozone is made difficult by their short lifetime, so there is need to consider the transport pathways from the troposphere to the stratosphere of these compounds in detail (WMO, 2003, Chapter 2). An upper limit for total stratospheric iodine, I_y, of 0.10 ± 0.02 ppt was recently reported (for below 20 km) by Bösch *et al.* (2003). The impact of this magnitude of iodine loading on stratospheric ozone is negligible.

The disturbance of the mid-latitude ozone budget caused by anthropogenic emissions of ODSs will ultimately cease when the stratospheric halogen burden has reached low enough levels (see Section 1.2.2 and Box 1.7 in Section 1.4.2). However, like polar ozone, mid-latitude ozone will for many decades remain vulnerable to an enhancement of halogen-catalyzed ozone loss caused by climate change and by natural phenomena such as volcanic eruptions.

1.3.2 Impact of temperature changes on ozone chemistry

The increasing abundance of WMGHGs in the atmosphere is expected to lead to an increase in temperature in the troposphere. Furthermore, increasing concentrations of most of these gases, notably CO_2, N_2O and CH_4, are expected to lead to a temperature decrease in the stratosphere. By far the strongest contribution to stratospheric cooling from the WMGHGs comes from CO_2 (Section 1.2.4). As noted in Section 1.2.4, temperatures in the stratosphere are also believed to have decreased in part because of the observed reductions in ozone concentrations; any such cooling would have a feedback on the ozone changes.

Decreasing stratospheric temperatures lead to a reduction of

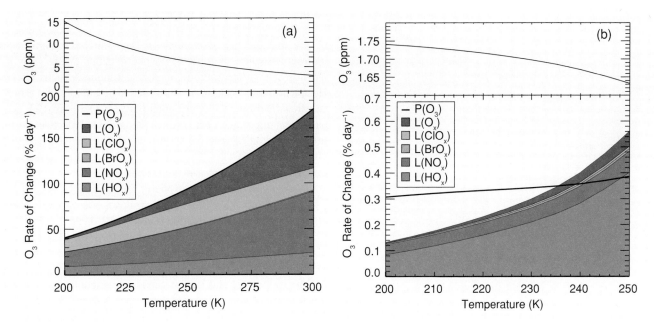

Figure 1.12. An estimate of the contribution of the various ozone loss cycles to the ozone loss rate, together with the ozone production rate P(O$_3$), at 45°N for equinox conditions (end of March) as a function of temperature (bottom panels). The estimates were obtained from short (20-day) runs of a chemical box model, starting from climatological values of ozone; the ozone mixing ratios at the end of the run are shown in the top panels. Panels (a) show conditions for 40 km altitude (2.5 hPa), where the climatological temperature is about 250 K; panels (b) show conditions for 20 km (55 hPa), where the climatological temperature is about 215 K. Reaction rate constants were taken from Sander *et al.* (2003). The production and loss rates are for the simulated ozone value. Note that at 40 km ozone is in steady-state (i.e., production equals the sum of all loss terms) and thus ozone is under photochemical control. At 20 km ozone production does not equal loss, implying that transport also has a strong influence on ozone concentrations. At 20 km the simulated ozone value after 20 days remains close to its initial climatological value of 1.70 ppm.

the ozone loss rate in the upper stratosphere, thereby indirectly leading to more ozone in this region. This reduction of ozone loss is caused by the very strong positive temperature dependence of the ozone loss rate, mainly owing to the Chapman reactions and the NO$_x$ cycle (Figure 1.12). This inverse relationship between ozone and temperature changes in the upper stratosphere has been known for many years (Barnett *et al.*, 1975). More recently, differences in temperature between the two hemispheres have been identified (Li *et al.*, 2002) as a cause of inter-hemispheric differences in both the seasonal cycle of upper-stratospheric ozone abundances and in the upper-stratospheric ozone trend deduced from satellite measurements.

In the mid-latitude lower stratosphere, one mechanism that leads to ozone loss and is directly sensitive to temperature changes involves heterogeneous reactions on the surfaces of cloud and cold aerosol particles (see mechanism (3) in Section 1.3.1.3). The rates of many of these reactions increase strongly with decreasing temperature. Similarly, the reaction rates increase with increasing water-vapour concentrations, so that future increases in water, should they occur, would also have an impact (see Sections 1.2.3 and 1.3.3). Note, however, that in the lower-stratospheric mid-latitudes the most important heterogeneous reactions are hydrolysis of N$_2$O$_5$ and bromine nitrate, which are relatively insensitive to temperature and water vapour concentrations. Therefore, the expected impact of cli-

mate change on the chemical mechanism (3) is expected to be relatively small. In addition to this impact on heterogeneous chemistry, Zeng and Pyle (2003) have argued that a reduction in temperature in the lower stratosphere can slow the rate of HO$_x$-driven ozone destruction.

Polar ozone loss occurs when temperatures in a large enough region sink below the threshold temperature for the existence of PSCs (approximately 195 K), because chlorine is activated by heterogeneous reactions on the surfaces of PSCs. Therefore, a cooling of the stratosphere enhances Arctic ozone loss if the volume of polar air with temperatures below the PSC threshold value increases. (The Antarctic in winter and spring is already consistently below this threshold.) Moreover, a stronger PSC activity is expected to lead to a greater denitrification of the Arctic vortex, hence a slower deactivation of chlorine and, consequently, a greater chemical ozone loss (Waibel *et al.*, 1999; Tabazadeh *et al.*, 2000). Rex *et al.* (2004) have deduced an empirical relation between observed temperatures and observed winter-spring chemical loss of Arctic ozone (Figure 1.13). Based on this relation, and for current levels of chlorine, about 15 DU additional loss of total ozone is expected for each 1 K cooling of the Arctic lower stratosphere.

Although the lower stratosphere is expected to generally cool with increasing greenhouse-gas concentrations (Section 1.2.4), the temperature changes in the lower stratosphere dur-

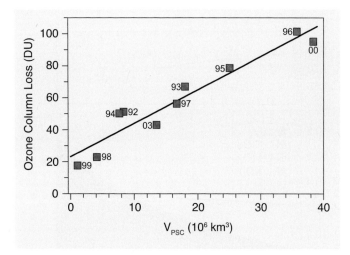

Figure 1.13. The overall chemical loss in average polar column ozone during a given Arctic winter versus the winter average of the stratospheric volume where conditions were cold enough (based on ECMWF temperature data) for the existence of PSCs (V_{PSC}). Also shown is the linear fit to the data. The calculations could not be performed for 2001 or 2002. Adapted from Rex *et al.* (2004).

ing the polar wintertime will be sensitive to any change in circulation associated with dynamical feedbacks, principally from changes in planetary wave drag (see Section 1.3.4.1). In principle, these feedbacks could be of either sign and thus could lead to enhanced cooling or even to warming. An early model study found enhanced cooling, to the extent that polar ozone loss would be expected to increase during the next 10 to 15 years even while halogen levels decreased (Shindell *et al.*, 1998). More recent studies with higher-resolution models have found a much less dramatic dynamical feedback, with some models showing an increase and some a decrease in planetary wave drag, and with all models predicting a relatively small change in Arctic ozone over the next few decades (Austin *et al.*, 2003). However, it should be noted that model results for the Arctic are difficult to assess because the processes leading to polar ozone depletion show so much natural variability that the atmosphere may evolve anywhere within (or even outside) the envelope provided by an ensemble of model simulations (see Section 3.5 of WMO, 2003).

In any event, it is the Arctic that is most sensitive to the effects caused by climate change that are discussed above. Present-day temperatures in the Arctic lie close to the threshold value for the onset of heterogeneous chlorine activation and thus close to the threshold value for the onset of rapid ozone loss chemistry; in the Antarctic temperatures are much lower and thus ozone loss is not as sensitive to temperature changes.

1.3.3 Impact of methane and water vapour changes on ozone chemistry

With the exception of the high latitudes in winter, ozone in the upper stratosphere is in a photochemical steady-state; photo-

chemical reactions are sufficiently fast that ozone concentrations are determined by a local balance between photochemical production and loss. Nonetheless, transport has a significant indirect influence on upper stratospheric ozone insofar as it determines the concentrations of trace compounds such as CH_4, H_2O, CFCs and N_2O, all of which act as precursors of the radicals that determine the ozone chemistry. Furthermore, CH_4 is of particular importance because it is the primary mechanism for the conversion of reactive Cl to the unreactive HCl reservoir via the reaction $CH_4 + Cl \rightarrow HCl + CH_3$, and thus affects the efficiency of the chlorine-driven ozone loss.

Changes in upper stratospheric CH_4 (Section 1.2.3) have important implications for upper stratospheric ozone. Siskind *et al.* (1998) found that for the period 1992–1995, the increase in active chlorine (Cl and ClO) resulting from the CH_4 decrease was the largest contributor to the ozone changes occurring over that time. Indeed, a measured increase in upper-stratospheric ClO between 1991 and 1997, which is significantly greater than that expected from the increase in the chlorine source gases alone, may be explained by the observed concurrent decrease of CH_4 (Froidevaux *et al.*, 2000). Further, Li *et al.* (2002) find that inter-hemispheric differences in CH_4 are partly responsible (together with inter-hemispheric differences in temperature) for observed differences in upper stratospheric ozone trends between the two hemispheres (from the SAGE I and SAGE II satellite experiments for 1979–1997).

Stratospheric water vapour is the primary source of the HO_x radicals that drive the dominant ozone loss cycles in the upper stratosphere (Figure 1.11). An increase in stratospheric water vapour is therefore expected to lead to a greater chemical loss of ozone in the upper stratosphere (e.g., Siskind *et al.*, 1998). This effect has been investigated quantitatively in model simulations. In a study of conditions for the year 2010, Jucks and Salawitch (2000) find that above 45 km an increase of 1% in the water vapour mixing ratio would completely negate the increase of ozone driven by the 15% decrease of inorganic chlorine that is expected by the year 2010. At 40 km the increase of ozone would still be reduced by about 50%. Shindell (2001) conducted a general circulation model (GCM) study for the period 1979–1996. The simulations show a significant chemical effect of water vapour increases on ozone concentrations, with a reduction of more than 1% between 45 and 55 km, and a maximum impact of about 4% at 50 km. Further, Li *et al.* (2002) found that the annually averaged (downward) ozone trend at 45°S and 1.8 hPa (45 km) increases by 1% per decade for a water vapour increase of 1% yr⁻¹. These model results point to an anticorrelation between ozone and water vapour in the upper stratosphere and lower mesosphere. However, such an anticorrelation is not seen in observations and ozone in this region varies much less than predicted by models (Siskind *et al.*, 2002).

In the mid-latitude lower stratosphere, any increase in water vapour would also be expected to lead to ozone loss because of an intensified HO_x ozone loss cycle (see Box 1.1). The model results of Dvortsov and Solomon (2001) predict that an increase in stratospheric water vapour of 1% yr⁻¹ translates to an additional depletion of mid-latitude column ozone by 0.3% per decade.

In the polar lower stratosphere, the rates of the heterogeneous reactions that are responsible for the activation of chlorine (which eventually leads to chemical ozone destruction) increase with increasing water vapour concentrations. An increase in stratospheric water vapour essentially means that the temperature threshold at which PSCs form, and thus heterogeneous reactions rates begin to become significant for chlorine activation, is shifted to higher values. If stratospheric water vapour were to increase it could lead to a substantially enhanced Arctic ozone loss in the future (e.g., Kirk-Davidoff *et al.*, 1999).

In considering the above discussion concerning stratospheric water vapour, it needs to be borne in mind that there is considerable uncertainty about the sign of future water vapour changes in light of the puzzling past record (Section 1.2.3).

1.3.4 *The role of transport for ozone changes*

Transport is a key factor influencing the seasonal and interannual variability of stratospheric ozone. Seasonal variations in transport force the large winter-spring build-up of extra-tropical total ozone in both hemispheres, and inter-hemispheric differences in transport (larger in the NH) cause corresponding differences in extra-tropical total ozone (see Box 1.2). In both hemispheres, mid-latitude ozone decreases during summer and returns to approximate photochemical balance by autumn, and there is a strong persistence of the dynamically forced anomalies throughout summer (Fioletov and Shepherd, 2003; Weber *et al.*, 2003). The interannual variability in the winter-spring build-up is greater in the NH, reflecting the greater dynamical variability of the NH stratosphere.

1.3.4.1 *Stratospheric planetary-wave-induced transport and mixing*

Large-scale transport of ozone is a result of advection by the Brewer-Dobson circulation and of eddy transport effects; both of these mechanisms are first-order terms in the zonal mean ozone transport equation (Andrews *et al.*, 1987). The strength of the Brewer-Dobson circulation is directly tied to dissipating planetary waves forced from the troposphere, and eddy transports of ozone are also linked to planetary wave activity (although this latter linkage is difficult to quantify), so that net ozone transport is tied to the variability of forced planetary waves. The amount of dissipating wave activity (also called planetary wave drag, PWD) within the stratosphere is related to the vertical component of wave activity entering the lower stratosphere (the so-called Eliassen-Palm (EP) flux). This quantity can be derived from conventional meteorological analyses and is a convenient proxy used to quantify PWD. A significant correlation has been found between interannual changes in PWD and total ozone build-up during winter and spring (Fusco and Salby, 1999; Randel *et al.*, 2002). The fact that the effect of PWD on ozone is seasonal and has essentially no interannual memory suggests that long-term changes in PWD can be

expected to lead to long-term changes in winter-spring ozone, all else being equal. However, although the basic physics of the ozone-PWD connection is well understood, its quantification via correlations is at best crude, and this limits our ability to attribute changes in ozone to changes in PWD.

In the NH, there have been interannual variations in various meteorological parameters during the period 1980–2000 that together paint a fairly consistent, albeit incomplete, picture. During the mid-1990s the NH exhibited a number of years when the Arctic wintertime vortex was colder and stronger (Graf *et al.*, 1995; Pawson and Naujokat, 1999) and more persistent (Waugh *et al.*, 1999). Any dynamically induced component of these changes requires a weakened Brewer-Dobson circulation, which in turn requires a decrease in PWD. Such a decrease in PWD during this period has been documented (Newman and Nash, 2000), although the results were sensitive to which months and time periods were considered. Randel *et al.* (2002) show that for the period 1979–2000, PWD in the NH increased during early winter (November to December) and decreased during mid-winter (January to February). This seasonal variation is consistent with the Arctic early winter warming and late winter cooling seen over the same time period at 100 hPa (Langematz *et al.*, 2003; see also Figure 1.16 in Section 1.4.1.2). The weakened Brewer-Dobson circulation during mid-winter implies a decrease in the winter build-up of mid-latitude ozone, and Randel *et al.* (2002) estimated that the decreased wave driving may account for about 20–30% of the observed changes in ozone in the January to March period. Changes in SH dynamics are not as clear as in the NH, primarily because meteorological reanalysis data sets are less well constrained by observations.

Planetary waves, in addition to affecting the temperature and chemistry of the polar stratosphere through the processes described earlier, can also displace the centre of the polar vortex off the pole. This has important implications for ozone and NO_x chemistry because air parcel trajectories within the vortex are then no longer confined to the polar night but experience short periods of sunlight (Solomon *et al.*, 1993). Polar ozone loss processes are usually limited by the poleward retreat of the terminator (the boundary that delineates polar night). Planetary wave distortion of the vortex can expose deeper vortex air to sunlight (Lee *et al.*, 2000) and cause ozone depletion chemistry to start earlier than it would have otherwise. In this way, wave-induced displacements of the vortex can drive a mid-winter start to Antarctic ozone depletion (Roscoe *et al.*, 1997; Bodeker *et al.*, 2001).

Because wave-induced forcing in the stratosphere is believed to come primarily from planetary-scale Rossby waves that are generated in the troposphere during wintertime, future changes in the generation of tropospheric waves may influence polar ozone abundance. However there is as yet no consensus from models on the sign of this change (Austin *et al.*, 2003).

1.3.4.2 *Tropopause variations and ozone mini-holes*
Tropospheric circulation and tropopause height variations also

affect the mid-latitude distribution of column ozone. The relationship between local mid-latitude tropopause height and column ozone on day-to-day time scales is well documented (e.g., Dobson, 1963; Bojkov *et al.*, 1993). Day-to-day changes in tropopause height are associated with the passage of synoptic-scale disturbances in the upper troposphere and lower stratosphere, which affect ozone in the lower stratosphere through transport (Salby and Callaghan, 1993) and can result in large local changes in column ozone, particularly in the vicinity of storm tracks (James, 1998). In extreme situations, they can lead to so-called ozone mini-holes, which occur over both hemispheres and have the lowest column ozone levels observed outside the polar vortices, sometimes well under 220 DU (Allaart *et al.*, 2000; Hood *et al.*, 2001; Teitelbaum *et al.*, 2001; Canziani *et al.*, 2002). Ozone mini-holes do not primarily entail a destruction of ozone, but rather its re-distribution. Changes in the spatial and temporal occurrence of synoptic-scale processes and Rossby wave breaking (e.g., storm track displacements) that are induced by climate change or natural variability, could lead to changes in the distribution or frequency of the occurrence of mini-hole and low-ozone events, and thus to regional changes in the mean column ozone (Hood *et al.*, 1999; Reid *et al.*, 2000; Orsolini and Limpasuvan, 2001).

Observations have shown that over the NH the altitude of the extra-tropical tropopause has generally increased over recent decades. Radiosonde measurements over both Europe and Canada show an increase in altitude of about 300–600 m over the past 30 years, with the precise amount depending on location (Forster and Tourpali, 2001; Steinbrecht *et al.*, 2001). Consistent regional increases are also seen in meteorological reanalyses over both hemispheres (Hoinka, 1999; Thompson *et al.*, 2000), although reanalysis trend studies should be viewed with care (Bengtsson *et al.*, 2004). Spatial patterns show increases over the NH and SH beginning at mid-latitudes and reaching a maximum towards the poles. The magnitude of the changes can also depend on the longitude.

However, the effects of tropospheric circulation changes on ozone and the relation of both to tropopause height changes remain poorly understood at present. In particular, the relationship between tropopause height and ozone mentioned earlier applies to single stations; there is no reason to expect it to apply in the zonal mean, or on longer time scales (e.g., seasonal or interannual) over which ozone transport is irreversible. Because of our poor understanding of what controls the zonal-mean mid-latitude tropopause height, or whether it is possible to consider such processes in a zonal-mean approach, it is not clear that the ozone-tropopause height correlations (derived from daily or monthly statistics) can be extended to decadal time scales in order to estimate changes in ozone. Some recent model studies suggest that lower stratospheric cooling caused by ozone depletion, acting together with tropospheric warming due to WMGHGs, can be a contributor to tropopause height changes (Santer *et al.*, 2003a,b), in which case the tropopause height changes cannot be entirely considered as a cause of the ozone changes. One source of uncertainty in these model calculations

is that most climate models cannot resolve the observed tropopause height changes, so these must be inferred by interpolation; another is that the observed ozone changes are not well quantified close to the tropopause (WMO, 2003, Chapter 4). Furthermore there remain considerable differences between the various reanalysis products available, as well as between the model results.

1.3.4.3   Stratosphere-troposphere exchange

Stratosphere-troposphere exchange (STE) processes also affect the ozone distribution and have the capability to affect tropospheric chemistry. STE is a two-way process that encompasses transport from the troposphere into the stratosphere (TST) and from the stratosphere into the troposphere (STT) through a variety of processes (see Holton *et al.*, 1995, and references therein). Some aspects of STE are implicit in the preceding discussions. The net mass flux from the troposphere into the stratosphere and back again is driven by PWD through the Brewer-Dobson circulation (Section 1.1.2; Holton *et al.*, 1995). Although the global approach can explain the net global flux, local and regional processes need to be identified and assessed to fully understand the distribution of trace species being exchanged. Such understanding is relevant both for chemical budgets and for climate change and variability studies. Although there is a net TST in the tropics and a net STT in the extratropics, a number of mechanisms that lead to STT in the tropics and TST in the extratropics have been identified (e.g., Appenzeller and Davies, 1992; Poulida *et al.*, 1996; Hintsa *et al.*, 1998; Ray *et al.*, 1999; Lelieveld and Dentener, 2000; Stohl *et al.*, 2003). In the extratropics TST can significantly affect the composition of the lowermost stratosphere, even if the exchange cannot reach higher into the stratosphere, by introducing, during deep TST events, near-surface pollutants, and ozone-poor and humid air. Similarly, deep STT events can transport ozone-rich air into the mid- and lower troposphere. It should be noted, however, that so-called shallow events, with exchanges near the tropopause, remain the main feature in the extratropics. Regions of occurrence of such exchange processes, at least in the NH (Sprenger and Wernli, 2003), appear to be linked to the storm tracks and their variability, that is, in the same region where mini-holes are most frequent.

The past and future variability of STE remains to be assessed. It is clear that changes in planetary wave activity will modify the Brewer-Dobson circulation and hence affect global transport processes (Rind *et al.*, 1990; Butchart and Scaife, 2001). As for regional mechanisms, seasonal and interannual variability, driven, for example, by the North Atlantic Oscillation (NAO) and the El Niño-Southern Oscillation (ENSO), have already been observed (James *et al.*, 2003; Sprenger and Wernli, 2003). Changes in the occurrence of regional and local weather phenomena could also modify STE. Fifteen-year studies with reanalysis products have not yielded any distinct trends (Sprenger and Wernli, 2003). Given the discrepancies between different approaches to evaluate STE and inhomogeneities in meteorological analyses and reanalyses, consistent trend studies are not available as of yet.

1.3.4.4 The tropical tropopause layer

Air enters the stratosphere primarily in the tropics, and hence the physical and chemical characteristics of air near the tropical tropopause behave as boundary conditions for the global stratosphere. The tropical tropopause is relatively high, near 17 km. The tropospheric lapse rate (up to 12–14 km) is determined by radiative-convective equilibrium, whereas the thermal structure above 14 km is primarily in radiative balance, which is characteristic of the stratosphere (Thuburn and Craig, 2002). Overall, the region of the tropical atmosphere between about 12 km and the tropopause has characteristics intermediate to those of the troposphere and stratosphere, and is referred to as the tropical tropopause layer (TTL) (Highwood and Hoskins, 1998). Thin (sometimes subvisible) cirrus clouds are observed over large areas of the TTL (Wang *et al.*, 1996; Winker and Trepte, 1998), although their formation mechanism(s) and effects on large-scale circulation are poorly known. In the tropics, the background clear-sky radiative balance shifts from cooling in the troposphere to heating in the stratosphere, with the transition occurring at around 15 km. The region of heating above about 15 km is linked to mean upward motion into the lower stratosphere, and this region marks the base of the stratospheric Brewer-Dobson circulation.

The TTL is coupled to stratospheric ozone through its control of stratospheric water vapour and by the transport of tropospheric source gases and ODSs into the lower stratosphere, so changes in the TTL could affect the stratospheric ozone layer. If the residence time of air in the TTL before it is transported into the stratosphere is short, then it may be possible that short-lived halogen compounds (natural or anthropogenic replacements for the CFCs) or their degradation products could enter the lower stratosphere and contribute to ozone loss there (WMO, 2003, Chapter 2).

Air entering the stratosphere is dehydrated as it passes through the cold tropical tropopause, and this drying accounts for the extreme aridity of the global stratosphere (Brewer, 1949). Furthermore, the seasonal cycle in tropopause temperature imparts a strong seasonal variation in stratospheric water vapour, which then propagates with the mean stratospheric transport circulation (Mote *et al.*, 1996). Year-to-year variations in tropical tropopause temperatures are also highly correlated with global stratospheric water vapour anomalies (Randel *et al.*, 2004), although there remain some issues concerning the consistency of decadal-scale changes in water vapour (Section 1.2.3). However, while there is strong empirical coupling between tropopause temperatures and stratospheric water vapour, details of the dehydration mechanism(s) within the TTL are still a topic of scientific debate (Holton and Gettelman, 2001; Sherwood and Dessler, 2001). One critical, unanswered question is whether, and how, the TTL will change in response to climate change and what will be the resulting impact on stratospheric ozone.

1.3.5 Stratosphere-troposphere dynamical coupling

Analysis of observational data shows that atmospheric circulation tends to maintain spatially coherent, large-scale patterns for extended periods of time, and then to shift to similar patterns of opposite phase. These patterns represent preferred modes of variability of the coupled atmosphere-ocean-land-sea-ice system. They fluctuate on intra-seasonal, seasonal, interannual and decadal time scales, and are influenced by externally and anthropogenically caused climate variability. Some of the patterns exhibit seesaw-like behaviour and are usually called oscillations. A few circulation modes, listed below, have been implicated in stratosphere-troposphere dynamical coupling, and thus may provide a coupling between stratospheric ozone depletion and tropospheric climate.

- *Northern Annular Mode (NAM)* (Thompson and Wallace, 1998, 2000; Baldwin and Dunkerton, 1999): The NAM, also referred to as the Arctic Oscillation (AO), is a hemisphere-wide annular atmospheric circulation pattern in which atmospheric pressure over the northern polar region varies out of phase with pressure over northern mid-latitudes (around 45°N), on time scales ranging from weeks to decades.

- *Southern Annular Mode (SAM)* (Gong and Wang, 1999; Thompson and Wallace, 2000): The SAM, also referred to as the Antarctic Oscillation, is the SH analogue of the NAM. The SAM exhibits a large-scale alternation of pressure and temperature between the mid-latitudes and the polar region.

- *North Atlantic Oscillation (NAO)* (Walker and Bliss, 1932; Hurrell, 1995): The NAO was originally identified as a seesaw of sea-level pressure between the Icelandic Low and the Azores High, but an associated circulation (and pressure) pattern is also exhibited well above in the troposphere. The NAO is the dominant regional pattern of wintertime atmospheric circulation variability over the extra-tropical North Atlantic, and has exhibited variability and trends over long time periods (Appenzeller *et al.*, 2000). The relationship between the NAO and the NAM remains a matter of debate (Wallace, 2000; Ambaum *et al.*, 2001; Rogers and McHugh, 2002).

The NAM extends through the depth of the troposphere. During the cold season (January to March), when the stratosphere has large-amplitude disturbances, the NAM also has a strong signature in the stratosphere, where it is associated with variations in the strength of the westerly vortex that encircles the Arctic polar stratosphere; this signature suggests a coupling between the stratosphere and the troposphere (Perlwitz and Graf, 1995; Thompson and Wallace, 1998; Baldwin and Dunkerton, 1999). During winters when the stratospheric vortex is stronger than normal, the NAM (and NAO) tends to be in a positive phase.

Circulation modes can affect the ozone distribution directly (in the troposphere and the lowermost stratosphere; see Lamarque and Hess, 2004), and indirectly by influencing propagation of planetary waves from the troposphere into the middle atmosphere (Ambaum and Hoskins, 2002). Therefore, changes in circulation modes can produce changes in ozone distribution. It has been shown that because the NAO has a large vertical extent during the winter and because it modulates the tropopause height, it can explain much of the spatial pattern in column ozone trends in the North Atlantic over the past 30 years (Appenzeller *et al.*, 2000).

Stratospheric changes may feed back onto changes in circulation modes. Observations suggest that at least in some cases the large amplitude NAM anomalies tend to propagate from the stratosphere to the troposphere on time scales of weeks to a few months (Baldwin and Dunkerton, 1999; Christiansen, 2001). Furthermore, because of the strong coupling between the stratospheric vortex and the NAM, the recent trend in the NAM and NAO has been associated with processes that are known to affect the strength of the stratospheric polar vortex, such as tropical volcanic eruptions (Kodera, 1994; Kelly *et al.*, 1996; Rozanov *et al.*, 2002; Stenchikov *et al.*, 2002), ozone depletion (Graf *et al.*, 1998; Shindell *et al.*, 2001a), and anthropogenic changes in greenhouse gas concentrations (Shindell *et al.*, 2001a; Gillett *et al.*, 2002a). However, some modelling studies have shown that a simulated trend in the tropospheric NAM and SAM does not necessarily depend on stratospheric involvement (Fyfe *et al.*, 1999; Gillett *et al.*, 2002b). Global climate modelling simulations that include interactive stratospheric chemistry suggest that one mechanism by which solar variability may affect tropospheric climate is through solar-forced changes in upper-stratospheric ozone that induce changes in the leading mode of variability of the coupled troposphere-stratosphere circulation (Shindell *et al.*, 2001b).

A number of modelling studies have examined the effect of increased concentrations of greenhouse gases on the annular modes (e.g., Perlwitz *et al.*, 2000; Shindell *et al.*, 2001a; Gillett *et al.*, 2002b; Rauthe *et al.*, 2004). Most coupled atmosphere-ocean climate models agree in finding a positive NAM trend under increasing greenhouse-gas concentrations, a trend which is qualitatively consistent with the observed positive NAM trend. A more positive NAM is consistent with a stronger stratospheric vortex. On the other hand, an intensified polar vortex is related to changes in planetary- and synoptic-scale wave characteristics, and may produce tropospheric circulation anomalies similar to the positive phase of the NAM. These potential feedbacks obscure cause and effect. For example, whereas Rind *et al.* (2002) and Sigmond *et al.* (2004) found that in the middle stratosphere the perturbation to the NAM from a $2 \times CO_2$ climate depends on modelled changes in sea-surface temperatures (SSTs), Sigmond *et al.* (2004) suggested that perturbations in the zonal wind near the surface (which will affect the response to SSTs) are mainly caused by doubling of stratospheric CO_2.

The stratosphere-troposphere dynamical coupling through the circulation modes discussed in this section suggests that stratospheric dynamics should be accounted for in the problem of detection and prediction of future tropospheric climate change. Because increased concentrations of greenhouse gases may cause changes in stratospheric dynamics (and in ozone,

which in turn cause changes in stratospheric dynamics), greenhouse-gas changes may induce changes in surface climate through stratosphere-troposphere dynamical coupling, in addition to radiative forcing.

1.3.6   *Possible dynamical feedbacks of ozone changes*

The effect of ozone loss on polar stratospheric temperatures results in concomitant changes in stratospheric zonal winds and polar vortex structure, and also possible changes in planetary-wave behaviour. The strongest effect is seen in the Antarctic, where a large cooling of the lower stratosphere, which is associated with the ozone hole (Randel and Wu, 1999), has resulted in an intensified and more persistent springtime polar vortex. Waugh *et al.* (1999) show that since the ozone hole developed, the break-up date of the Antarctic vortex occurs two to three weeks later (moving from mid-November to early December) than before. Recent research has shown that trends in surface temperatures over Antarctica (cooling over the interior and part of the warming over the peninsula) may be in part traceable, at least over the period 1980–2000, to trends in the lower stratospheric polar vortex, which are largely caused by the ozone hole (Thompson and Solomon, 2002). It has been suggested that during the early summer the strengthening of the westerly flow extends all the way to the surface (Thompson and Solomon, 2002). The role of this mechanism – which has yet to be elucidated – has been highlighted by the modelling study of Gillett and Thompson (2003), who prescribed ozone depletion in an atmospheric climate model. They found that the seasonality, structure and amplitude of the modelled changes in 500 hPa geopotential height and near-surface air temperature in the Antarctic had similar spatial patterns to observations, which were a cooling over the Antarctic interior and a warming over the peninsula and South America (Figure 1.14). This result suggests that anthropogenic emissions of ozone-depleting gases have had a distinct impact on climate not only in the stratosphere but also at the Earth's surface. These surface changes appear to act in the same direction as changes resulting from increases in greenhouse gases (Shindell and Schmidt, 2004).

An increase in the strength of the Antarctic polar vortex by stratospheric ozone depletion does not affect only surface winds and temperatures. Using a 15,000-year integration of a coupled ocean-atmosphere model, Hall and Visbeck (2002) showed that fluctuations of the mid-latitude westerly winds generate ocean circulation and sea-ice variations on interannual to centennial time scales.

Cause and effect are more difficult to separate in the Arctic stratosphere than in the Antarctic stratosphere because of higher 'natural' meteorological variability, relatively smaller ozone losses and a less clear separation between transport and chemical effects on ozone. Furthermore, temperatures need to be sufficiently low in any given year to initiate polar ozone chemistry. Nonetheless, in several years during the mid-1990s the Arctic experienced low ozone, low temperatures in late spring and enhanced vortex persistence (Randel and Wu, 1999; Waugh *et al.*,

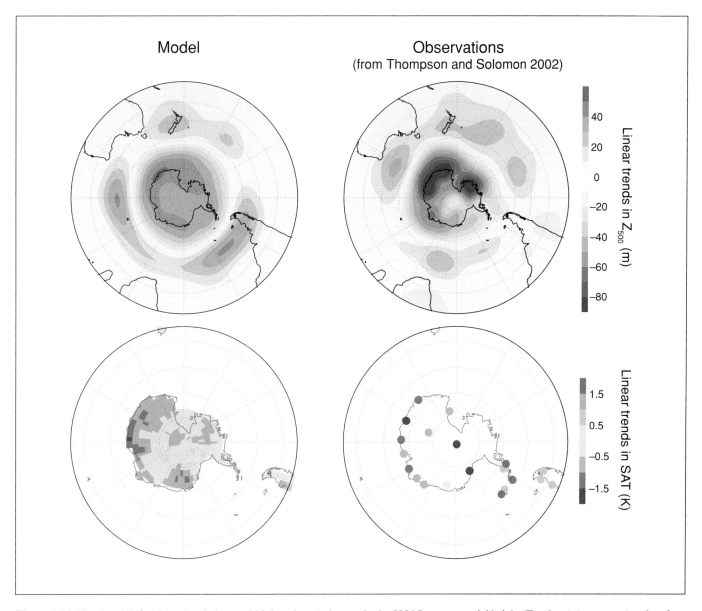

Figure 1.14. Simulated (left column) and observed (right column) changes in the 500 hPa geopotential height (Z_{500}, in m) (upper row) and surface air temperature (SAT, in K) (lower row). Observed changes are linear trends over 22 years (1979–2000) in the 500 hPa geopotential height, and over 32 years (1969–2000) in surface air temperature, averaged over December to May. Simulated changes are differences between the perturbed and control integrations in the 500 hPa geopotential height and the land surface temperature averaged over December to February. Adapted from Gillett and Thompson (2003).

1999), which resembled the coupled changes observed in the Antarctic.

A crucial issue is the potential feedback between changes in stratospheric climate and planetary wave forcing (through both modification of tropospheric planetary wave sources and changes in wave propagation into and within the stratosphere). However, this topic is poorly understood at present. Chen and Robinson (1992) and Limpasuvan and Hartmann (2000) have argued that stronger vertical shear of the zonal wind at high latitudes will reduce the strength of stratospheric PWD, whereas

Hu and Tung (2002) have argued for precisely the opposite effect. Many climate model simulations (e.g., Rind *et al.*, 1990; Butchart and Scaife, 2001) found enhanced PWD in an atmosphere with elevated WMGHG concentrations, which leads to a stronger Brewer-Dobson circulation; however models were not unanimous on this result (Shindell *et al.*, 2001a). Because planetary wave forcing is a primary driver of the stratospheric circulation, improved understanding of this coupling will be necessary to predict future ozone changes.

1.4   Past and future stratospheric ozone changes (attribution and prediction)

1.4.1   *Current understanding of past ozone changes*

In Section 1.3 a number of processes were described that can influence the distribution of and changes in stratospheric ozone. Here we discuss the current understanding of past ozone changes, based in part on numerical models that attempt to include these processes. A hierarchy of models of increasing sophistication is used (see Box 1.5). Some models include relatively few processes (e.g., box models, which include only chemical processes, but perhaps in great detail), whereas others attempt to include many more (e.g., the chemistry-climate models, CCMs). The models have their different uses, and strengths, some of which are described in the remainder of Section 1.4.

The understanding of polar and mid-latitude ozone decline has been assessed most recently in Chapters 3 and 4 of WMO (2003). The material here relies heavily on that report, which deals with these topics in much greater detail.

1.4.1.1   *Mid-latitude ozone depletion*

In the upper stratosphere the observed ozone depletion over the last 25 years is statistically robust in the sense that its value is not sensitive to small differences in the choice of the time period analyzed (WMO, 2003, Chapter 4). This behaviour accords with the fact that in this region of the atmosphere, ozone is under photochemical control and thus only weakly affected by dynamical variability. We therefore expect upper-stratospheric ozone abundance to reflect long-term changes in temperature and in the abundance of species (principally halogens) that react chemically with ozone.

Over the past 25 years there have been large changes in the abundance of halogen compounds, and the extent of the observed ozone decrease is consistent with the observed increase in anthropogenic chlorine, as originally predicted by Molina and Rowland (1974) and Crutzen (1974). In particular, the vertical and latitudinal profiles of ozone trends in the upper stratosphere are reproduced by 2-D photochemistry models. In the upper stratosphere the attribution of ozone loss in the last couple of decades to anthropogenic halogens is clear-cut. The 2-D models indicate that changes in halogens make the largest contribution to the observed loss of about 7% per decade (WMO, 2003, Chapter 4). The observed cooling of the upper stratosphere (of perhaps 2 K per decade, see Figure 1.10) will have reduced the rate of ozone destruction in this region. If we take the observed variation of ozone with temperature in the upper stratosphere (e.g., Barnett *et al.*, 1975; Froidevaux *et al.*, 1989), then a cooling of 2 K per decade should have led to an increase in ozone of about 2% per decade, partially compensating the loss caused by halogens. Note, however, that in the upper stratosphere radiative-transfer models suggest roughly equal contributions to the cooling from ozone decreases and carbon dioxide increases (Figure 1.10); the changes in the ozone-temperature system are nonlinear. Changes in CH_4 and N_2O will have also played a

small role here. Note also that although 2-D models have been widely used in many ozone assessments, they have important limitations in terms of, for example, their ability to include the full range of feedbacks between chemistry and dynamics, or their ability to reproduce the polar vortex (which will then influence the treatment of mid-latitudes). They are expected to be most accurate in the upper stratosphere, where dynamical effects on ozone are weakest.

However, the changes in upper-stratospheric ozone represent only a small contribution to the total changes in column ozone observed over the last 25 years, except in the tropics where there is no statistically significant trend in column ozone (WMO, 2003, Chapter 4). Most of the column ozone depletion in the extratropics occurs in the lower stratosphere, where the photochemical time scale for ozone becomes long and the ozone distribution is sensitive to dynamical variability as well as to chemical processes. This complicates the problem of attribution of the observed ozone decline. Strong ozone variability and complex coupled interactions between dynamics and chemistry, which are not separable in a simple manner, make attribution particularly difficult.

A number of 2-D models contributed to Chapter 4 of WMO (2003), and a schematic showing their simulations of column ozone between 60°S and 60°N from 1980 to 2050 is provided in the figure in Box 1.7 (in Section 1.4.2). The symbols in this figure show the observed changes relative to 1980. As there are no significant trends in tropical ozone, the observed and modelled changes are attributable to mid-latitudes. The shaded area shows the results of 2-D photochemistry models forced by observed changes in halocarbons, other source gases and aerosols (and in some cases, temperatures) from 1980 to 2000. Overall, the models broadly reproduce the long-term changes in mid-latitude column ozone for 1980–2000, within the range of uncertainties of the observations and the model range. The spread in the model results comes mainly from their large spread over the SH mid-latitudes, and is at least partly a result of their treatment of the Antarctic ozone hole (which cannot be well represented in a 2-D model). In addition, the agreement between models and observations over 60°S to 60°N hides some important disagreements within each hemisphere. In particular, models suggest that the chemical signal of ozone loss following the major eruption of the Mt. Pinatubo volcano in the early 1990s should have been symmetric between the hemispheres, but observations show a large degree of inter-hemispheric asymmetry in mid-latitudes (see the more detailed discussion in Section 4.6 of WMO, 2003).

As discussed in Section 1.3.4, changes in atmospheric dynamics can also have a significant influence on NH mid-latitude column ozone on decadal time scales. Natural variability, changes in greenhouse gases and changes in column ozone itself are all likely to contribute to these dynamical changes. Furthermore, because chemical and dynamical processes are coupled, their contributions to ozone changes cannot be considered in isolation. This coupling is especially complex and difficult to understand with regard to dynamical changes in the

---

Box 1.5. Atmospheric models

The chemical composition and the thermal and dynamical structure of the atmosphere are determined by a large number of simultaneously operating and interacting processes. Therefore, numerical mathematical models have become indispensable tools to study these complex atmospheric interactions. Scientific progress is achieved in part by understanding the discrepancies between atmospheric observations and results from the models. Furthermore, numerical models allow us to make *predictions* about the future development of the atmosphere. Today, a hierarchy of atmospheric models of increasing complexity is employed for the investigation of the Earth's atmosphere. Models range from simple (with perhaps one process or one spatial dimension considered) to the most complex 3-D and interactive models. In the following list, a brief summary of the most important types of models is given:

- *Fixed dynamical-heating (FDH) model:* Calculation of stratospheric temperature changes and radiative forcing with a radiation scheme, assuming that the stratosphere is in an equilibrium state and no dynamical changes occur in the atmosphere.
- *Trajectory (Lagrangian) model:* Simulation of air-parcel movement through the atmosphere based on meteorological analyses.
- *Box-trajectory model:* Simulation of chemical processes within a parcel of air that moves through the atmosphere.
- *Mesoscale (regional) model:* Analysis and forecast of medium-scale (a few tens of kilometers) radiative, dynamical and chemical structures in the atmosphere; investigation of transport and exchange processes.
- *Contour-advection model:* Simulation of highly resolved specific two-dimensional (2-D) fluid-dynamical processes, such as processes at transport barriers in the atmosphere.
- *Mechanistic circulation model:* Simplified 3-D atmospheric circulation model that allows for the investigation of specific dynamical processes.
- *Two-dimensional photochemistry model:* Zonally averaged representation of the middle atmosphere, with detailed chemistry but highly simplified transport and mixing.
- *General-circulation climate model (GCM):* Three-dimensional simulation of large-scale radiative and dynamical processes (spatial resolution of a few hundred kilometers) in the atmosphere over years and decades; investigation of the climate effects of atmospheric trace gases (greenhouse gases); investigations of the interaction of the atmosphere with the biosphere and oceans.
- *Chemistry-transport model (CTM):* Three-dimensional (or 2-D latitude-longitude) simulation of chemical processes in the atmosphere employing meteorological analyses derived from observations or GCMs; simulation of spatial and temporal development and distribution of chemical species.
- *Chemistry-climate model (CCM):* Interactively coupled 3-D GCM with chemistry; investigation of the interaction of radiative, dynamical, physical and chemical processes of the atmosphere; assessment of future development of chemical composition and climate.

Several decades ago, when numerical models of the atmosphere were first developed, much less computational power was available than today. Early studies with numerical models focused on the simulation of individual radiative, dynamical or chemical processes of the atmosphere. These early and rather simple models have evolved today into very complex tools, although, for reasons of computational efficiency, simplifying assumptions (parametrizations) must still be made. For example, the atmosphere in a global model requires discretization, which is generally done by decomposing it into boxes of a specific size. Processes acting on smaller scales than the boxes cannot be treated individually and must be parametrized, that is, their effects must be prescribed by functional dependencies of resolved quantities.

---

tropopause region. There is an observed relationship between column ozone and several tropospheric circulation indices, including tropopause height (Section 1.3.4.2). Over time scales of up to about one month, it is the dynamical changes that cause the ozone changes (Randel and Cobb, 1994), whereas on longer time scales feedbacks occur and the causality in the relationship becomes unclear. Thus, although various tropospheric circulation indices (including tropopause height) have changed over the last 20 years in the NH in such a way as to imply a decrease

in column ozone, this inference is based on an extrapolation of short-time-scale correlations to longer time scales, which may not be valid (Section 4.6 of WMO, 2003).

Above the tropopause, stratospheric PWD drives the seasonal winter-spring ozone build-up in the extratropics, and has essentially no interannual memory (Fioletov and Shepherd, 2003). It follows that the observed decrease in NH PWD in the late winter and spring has likely contributed to the observed decrease in NH column ozone over the last 20 to 25 years (Fusco and Salby, 1999; Randel *et al.*, 2002). The effect of changes in PWD on the ozone distribution is understood in general terms, but its quantification in observations is relatively crude.

The seasonality of the long-term changes in mid-latitude column ozone differs between hemispheres. In the NH, the maximum decrease is found in spring, and it decays through to late autumn. Fioletov and Shepherd (2003) have shown that in the NH the ozone decreases in summer and early autumn are the photochemically damped signal of winter-spring losses, and thus arise from the winter-spring losses (however they occur) without any need for perturbed chemistry in the summertime. The same seasonality is not enough for explaining the summer and early autumn ozone decreases observed in the SH mid-latitudes (Fioletov and Shepherd, 2003) because they are comparable with the winter-spring losses and therefore point to the influence of transport of ozone-depleted air into mid-latitudes following the break-up of the ozone hole.

In summary, the vertical, latitudinal and seasonal characteristics of past changes in mid-latitude ozone are broadly consistent with the understanding that halogens are the primary cause of these changes. However, to account for decadal variations it is necessary to include consideration of the interplay between dynamical and chemical effects as well as the impact of variations in aerosol loading (WMO, 2003, Chapter 4); our inadequate quantitative understanding of these processes limits our predictive capability.

1.4.1.2 Winter-spring polar depletion

Polar ozone depletion in the winter-spring period is generally considered separately from mid-latitude depletion, because of the extremely severe depletion that can occur in polar regions from heterogeneous chemistry on PSCs (Section 1.3.1). We consider first the Antarctic, and then the Arctic.

The Antarctic ozone hole represents the most striking example of ozone depletion in the atmosphere. It developed through the 1980s as chlorine loading increased, and has recurred every year since then (Figure 1.5). The Antarctic ozone hole has been clearly attributed to anthropogenic chlorine through field campaigns and photochemical modelling (WMO, 1999, Chapter 7). An ozone hole now occurs every year because, in addition to the availability of anthropogenic chlorine, wintertime temperatures in the Antarctic lower stratosphere are always low enough for chlorine activation to occur on PSCs prior to the return of sunlight to the vortex in the spring, and the air is always sufficiently isolated within the vortex for it to become strongly depleted of ozone (see Box 1.1). There is nevertheless dynamically induced variability in the extent and severity of the ozone hole, as seen in Figure 1.5. The most dramatic instance of such variability occurred in 2002, when the Antarctic stratosphere experienced its first observed sudden warming (see Box 1.6). The sudden warming split the vortex in two (figure in Box 1.6) and halted the development of the ozone hole that year. Such events occur commonly in the NH, as a result of the stronger planetary-wave forcing in the NH, but had never before been seen in the SH, although there had been previous instances of disturbed winters

(e.g., 1988). Although unprecedented, this event resulted from dynamical variability and was not indicative of ozone recovery; indeed, the ozone hole in 2003 was back to a severity characteristic of the 1990s (Figure 1.5 and figure in Box 1.6).

Given the fairly predictable nature of the Antarctic ozone hole, its simulation constitutes a basic test for models. Two-dimensional models are not expected to represent the ozone hole well because they cannot properly represent isolation of polar vortex air (WMO, 2003, Chapter 4), and their estimates of Antarctic ozone loss vary greatly. However, an emerging tool for attribution is the 3-D chemistry-climate model (CCM). These models include an on-line feedback of chemical composition to the dynamical and radiative components of the model. The CCMs currently available have been developed with an emphasis on the troposphere and stratosphere. They consider a wide range of chemical, dynamical and radiative feedback processes and can be used to address the question of why changes in stratospheric dynamics and chemistry may have occurred as a result of anthropogenic forcing. The models specify WMGHGs, source gases, aerosols and often SSTs, but otherwise run freely and exhibit considerable interannual variability (in contrast with the 2-D models). Thus in addition to issues of model accuracy, it is necessary to consider how representative are the model simulations.

In recent years a number of CCMs have been employed to examine the effects of climate change on ozone (Austin *et al.*, 2003). They have been used for long-term simulations to try to reproduce observed past changes, such as ozone and temperature trends. The models have been run either with fixed forcings to investigate the subsequent 'equilibrium climate' (so-called time-slice runs) or with time-varying forcings (so-called transient runs). The current CCMs can reproduce qualitatively the most important atmospheric features (e.g., stratospheric winds and temperatures, and ozone columns and profiles) with respect to the mean conditions and the seasonal and interannual variability; the long-term changes in the dynamical and chemical composition of the upper troposphere and the stratosphere are also in reasonable agreement with observations. But a reasonable reproduction of the timing of the ozone hole and an adequate description of total column ozone do not necessarily mean that all processes in the CCMs are correctly captured. There are obvious discrepancies between the models themselves, and between model results and detailed observations. Some of the differences among the models are caused by the fact that the specific model systems employed differ considerably, not only in complexity (e.g., the number of chemical reactions considered and the parametrization of sub-grid-scale processes), but also in the vertical extent of the model domain and in the horizontal and vertical resolutions.

An important difference between models and observations originates to a large extent from the cold bias (the 'cold-pole problem') that is found in the high latitudes of many CCMs, particularly near the tropopause and in the lower stratosphere. The low-temperature bias of the models is generally largest in winter over the South Pole, where it is of the order of 5–10 K;

Box 1.6. The 2002 Antarctic ozone hole

The unusual behaviour of the Antarctic ozone hole in 2002 highlights the important role that tropospheric climate and dynamics play in ozone depletion processes. Year-to-year changes in ozone hole area (or depth) are more likely to reflect changes in stratospheric dynamics that influence ozone loss over the short term, rather than long-term issues such as changes in stratospheric concentrations of ODSs controlled by the Montreal Protocol. During the early winter of 2002, the Antarctic ozone hole was unusually disturbed (Allen *et al.*, 2003; Stolarski *et al.*, 2005) as a result of a series of wave events (Newman and Nash, 2005). The anomalously high stratospheric wave activity was forced by strong levels of planetary wave 1 (a single wave encircling the globe) in the mid-latitude lower troposphere and a tropopause wave propagation regime that favoured transmission of these waves to the lower stratosphere. The wave events occurred irregularly over the course of the winter and preconditioned the polar vortex for an extremely large wave event on 22 September. This large wave event resulted in the first ever major stratospheric sudden warming to be observed in the SH, which split the Antarctic ozone hole in two (see figure). Such a splitting of the vortex in two is a characteristic signature of wave-2 sudden warmings in the NH and is well understood dynamically.

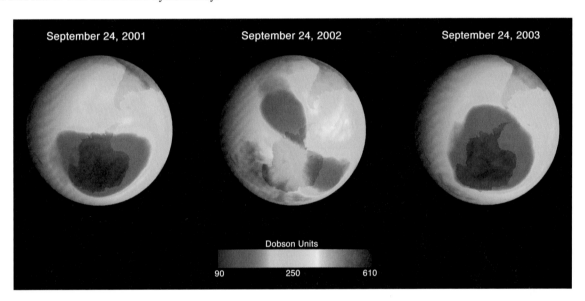

Box 1.6, Figure. Comparison of the first split ozone hole on record (middle, 2002) and the Antarctic ozone hole at the same time one year earlier (left, 2001) and one year later (right, 2003). The hole is dark blue. In 2001, the area of the ozone layer thinning over Antarctica reached 26.5 million km², larger than the size of the entire North American continent. Because of higher Antarctic winter temperatures, the 2002 'hole' appears to be about 40% smaller than in 2001. In 2003, Antarctic winter temperatures returned to normal and the ozone hole returned to its usual state. Figure provided by Stuart Snodgrass, NASA/Goddard Space Flight Center Scientific Visualization Studio (SVS).

this magnitude of bias could be significant in controlling planetary-wave dynamics and restricting the interannual variability of the models. A low-temperature bias in the lower stratosphere of a model has a significant impact not only on model heterogeneous chemistry (i.e., leading to enhanced ozone destruction via enhanced occurrence of PSCs), but also on the transport of chemical species and its potential change due to changes in circulation. Moreover, changes in stratospheric dynamics alter the conditions for wave forcing and wave propagation (of small-scale gravity waves as well as large-scale planetary waves), which in turn influence the seasonal and interannual variability of the atmosphere. The reasons for the cold bias in the models are still unknown. This bias has been reduced in some models, for example by considering non-orographic gravity-wave drag

schemes (Manzini and McFarlane, 1998), but the problem is not yet solved. Without its resolution, the reliability of these models for attribution and prediction is reduced.

Despite these potential problems, CCMs simulate the development of the Antarctic ozone hole reasonably well (Figure 1.15). The models examined in the intercomparison of Austin *et al.* (2003) agree with observations of minimum Antarctic springtime column ozone over 1980–2000 within the model variability (Figure 1.15a), confirming that the Antarctic ozone hole is indeed a robust response to anthropogenic chlorine. The cold-pole biases of some models seem not to affect minimum column ozone too much, although some modelled minima are significantly lower than observed. None of the models exhibited a sudden warming as seen in 2002, although several exhibit

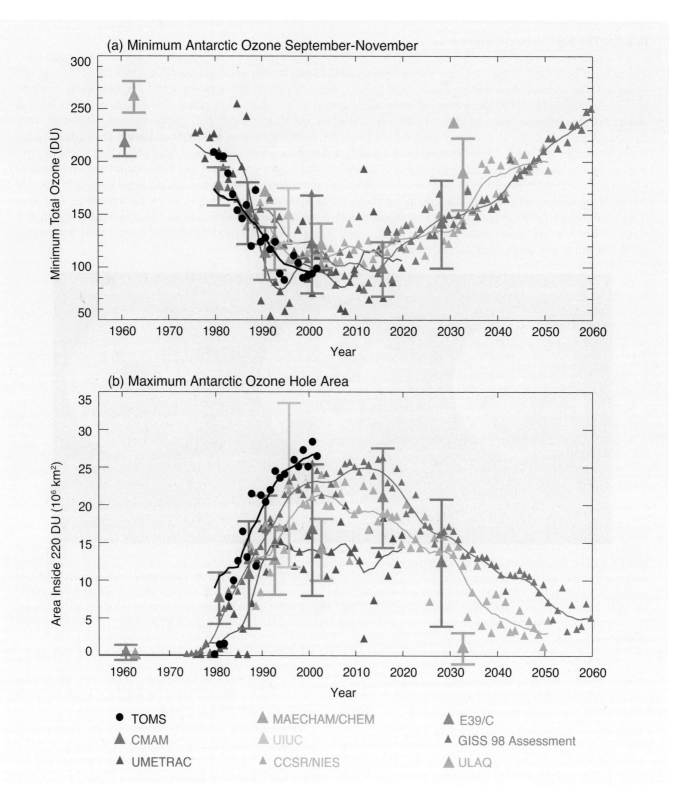

Figure 1.15. (a) Time development of the Antarctic ozone column: measurements (TOMS, black dots) and CCM model calculations (coloured triangles) of minimum total ozone from 60°S to 90°S for September to November. For the transient simulations (small triangles), the solid lines show the results of a Gaussian smoothing applied to the annual results. For the time-slice simulations, the mean and twice the standard deviation of the values within each model sample are indicated by the large coloured triangles and the error bars, respectively. For the MAECHEM/CHEM model only, the values have been plotted two years late for clarity, and a standard tropospheric ozone column of 40 DU has been added to the computed stratospheric columns. (b) As (a), but for the time development of the maximum area of the ozone hole, defined by the 220 DU contour, for September to November. Adapted from WMO (2003) based on Austin *et al.* (2003).

years with a disturbed vortex, as seen in their interannual variability in minimum column ozone. (The models with cold-pole biases appear to underestimate the observed variability of the ozone hole, as would be expected.) However, minimum total ozone is not the only diagnostic of the ozone hole. Figure 1.15b shows the modelled ozone hole area in comparison with observations. Many models significantly underestimate the area of the ozone hole.

In contrast with the Antarctic, winter-spring ozone abundance in the Arctic exhibits significant year-to-year variability, as seen in Figure 1.6. This variability arises from the highly disturbed nature of Arctic stratospheric dynamics, including relatively frequent sudden warmings (every 2 to 4 years). Because dynamical variability affects both transport and temperature, dynamical and chemical effects on ozone are coupled. The dynamical variability is controlled by stratospheric PWD; years with strong PWD, compared with years with weak PWD, have stronger downwelling over the pole, and thus higher temperatures, a weaker vortex and more ozone transport (Newman *et al.*, 2001). Because low temperatures and a stronger vortex tend to favour chemical ozone loss in the presence of elevated halogen levels (Section 1.3.2), it follows that dynamical and chemical effects tend to act in concert. In the warmest years (e.g., 1999) there is essentially no chemical ozone loss and ozone levels are similar to those seen pre-1980. The chemical ozone loss was calculated for cold years using various methods in Section 3.3 of WMO (2003), and the agreement between the methods provided confidence in the estimates. The calculations showed a roughly linear relationship between chemical ozone loss and temperature (see also Figure 1.13, which shows the relationship of ozone loss to V_{PSC}, which itself varies linearly with temperature (Rex *et al.*, 2004)). Furthermore, the *in situ* chemical ozone loss was found to account for roughly one half of the observed ozone decrease between cold years and warm years. Given the high confidence in the estimates of chemical ozone loss, the remaining half can be attributed to reduced ozone transport, as is expected in cold years with weak PWD, a strong vortex, and weak downwelling over the pole. From this it can be concluded that in cold years, chlorine chemistry doubled the ozone decrease that would have occurred from transport alone (WMO, 2003, Chapter 3).

Compared with other recent decades, the 1990s had an unusually high number of cold years, and these led to low values of Arctic ozone (Figure 1.6). In more recent years, Arctic ozone has been generally higher, although still apparently below pre-1980 values. This behaviour does not reflect the time evolution of stratospheric chlorine loading (Figure 1.7d). Rather, it reflects the meteorological variability in the presence of chlorine loading. In this respect the Arctic ozone record needs to be interpreted in the context of the meteorology of a given year, far more than is generally the case in the Antarctic.

This interpretation of Arctic ozone changes is consistent with the fact that the observed decrease in Arctic stratospheric temperature over 1980–2000 cannot be explained from direct radiative forcing due to ozone depletion or changes in green-

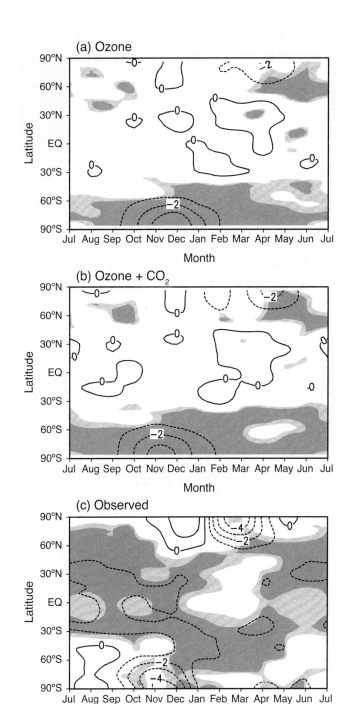

Figure 1.16. Latitude-month distribution of the zonal-mean temperature trend (in K per decade) at 100 hPa over the period 1980–2000: (a) modelled trend obtained by imposing the observed ozone trend over the same period (note that this was not done with a CCM), (b) modelled trend obtained by imposing the observed ozone trend plus the observed CO_2 trend, and (c) observed trend derived from NCEP reanalysis data. The contour interval is 1 K per decade. Dark (light) shaded areas denote regions where the trend is significant at the 99% (95%) confidence level. From Langematz *et al.* (2003).

house gases alone (Shine *et al.*, 2003), although they do make a contribution. The inference is that the observed springtime cooling was mainly the result of decadal meteorological variability in PWD. This is in contrast with the Antarctic, where the observed cooling in November and the prolonged persistence of the vortex have been shown to be the result of the ozone hole (Randel and Wu, 1999; Waugh *et al.*, 1999). These results (for both hemispheres) are illustrated by Figure 1.16. There is seen to be little impact from CO_2 at these altitudes over this time period. In the Arctic, the cooling induced by ozone loss is too small and too late in the season to account for the observed cooling. Rather, as discussed above, the observed cooling is required in order to initiate severe Arctic ozone depletion.

It is not known what has caused the recent decadal variations in Arctic temperature. Rex *et al.* (2004) have argued that the value of V_{PSC} in cold years has systematically increased since the 1960s. However, the Arctic exhibits significant decadal variability (Scaife *et al.*, 2000) and it is not possible to exclude natural variability as the cause of these changes.

These considerations have implications for the attribution of past changes. No matter how good a CCM is and how well its climate-change experiments are characterized, it cannot exactly reproduce the real atmosphere because the real atmosphere is only one possible realization of a chaotic system. The best one can expect, even for a perfect model (that is, a model that correctly considers all relevant processes), is that the observations fall within the range of model-predicted behaviours, according to appropriate statistical criteria. Whether this permits a meaningful prediction depends on the relative magnitudes and time scales of the forced signal and the natural noise. Whereas the evolution of Antarctic ozone is expected to be fairly predictable over decadal time scales (past and future), it is not at all clear whether this is the case in the Arctic. Thus, it is not a priori obvious that even a perfect CCM would reproduce the decreases in Arctic ozone observed over the past 20 years, for example.

Bearing these caveats in mind, the simulations of past Arctic minimum ozone from the CCMs considered in Austin *et al.* (2003) are shown in Figure 1.17. (Because of the large degree of scatter, the results are shown separately for the transient and time-slice simulations.) The models seem generally to have a positive bias with respect to the observations over the same period. None of the CCMs achieve column ozone values as low as those observed, and the modelled ozone trends are generally smaller than the observed trend (Austin *et al.*, 2003). However, the range of variability exhibited by each of the models is considerable, and it is therefore difficult to say that the models are definitely deficient on the basis of the ozone behaviour alone; more detailed diagnostics are required – such as comparison with the observed relationship shown in Figure 1.13 and with observations of other chemical species.

In summary, the development of the Antarctic ozone hole through the 1980s was a direct response to increasing chlorine loading, and the severity of the ozone decrease has not changed since the early 1990s, although there are year-to-year variations (especially in 2002, which was a surprising anomaly). In the

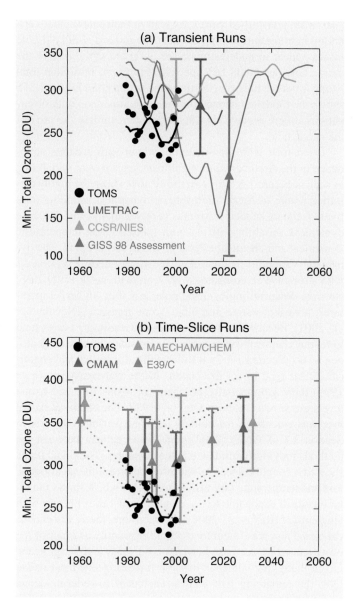

Figure 1.17. Development of the Arctic ozone column: measurements (TOMS, black dots) and model calculations of minimum total ozone from 60°N to 90°N for March and April. For the transient simulations (panel (a)), the solid lines show the results of a Gaussian smoothing applied to the annual results, and the error bars denote twice the standard deviation of the annual values from the smoothed curve (plotted just once for clarity). For the time-slice simulations (panel (b)), the mean and twice the standard deviation of the values within each model sample are indicated by the coloured triangles and the error bars, respectively. For the MAECHEM/CHEM model only, the values have been plotted two years late for clarity, and a standard tropospheric ozone column of 100 DU has been added to the computed stratospheric columns. Adapted from Austin *et al.* (2003).

Arctic, the extent of chemical ozone loss due to chlorine depends on the meteorology of a given winter. In winters that are cold enough for the existence of PSCs, chemical ozone loss has been identified and acts in concert with reduced transport (the effects being of a similar magnitude in terms of column ozone loss) to give Arctic ozone depletion. There was particularly severe Arctic ozone depletion in many years of the 1990s (and in 2000), as a result of a series of particularly cold winters. It is not known whether this period of low Arctic winter temperatures is just natural variability or a response to changes in greenhouse gases.

1.4.2 The Montreal Protocol, future ozone changes and their links to climate

The halogen loading of the stratosphere increased rapidly in the 1970s and 1980s. As a result of the Montreal Protocol and its Amendments and Adjustments, the stratospheric loading of chlorine and bromine is expected to decrease slowly in the coming decades, reaching pre-1980 levels some time around 2050 (WMO, 2003, Chapter 1). If chlorine and bromine were the only factors affecting stratosphere ozone, we would then expect stratospheric ozone to 'recover' at about the same time. Over this long (about 50-year) time scale, the state of the stratosphere may well change because of other anthropogenic effects, in ways that affect ozone abundance. For example, increasing concentrations of CO_2 are expected to further cool the stratosphere, and therefore to influence the rates of ozone destruction. Any changes in stratospheric water vapour, CH_4 and N_2O, all of which are difficult to predict quantitatively, will also affect stratospheric chemistry and radiation. In addition, natural climate variability including, volcanic eruptions, can affect ozone on decadal time scales. For these reasons, 'recovery' of stratospheric ozone is a complicated issue (see Box 1.7). A number of model calculations to investigate recovery were reported in Chapters 3 and 4 of WMO (2003), using both 2-D and 3-D models. The main results from that assessment, and new studies reported since then, are summarized here. As with the discussion of past ozone changes, we first discuss mid-latitude (or global) changes and then polar changes.

1.4.2.1 Mid-latitude ozone

Predictions of future mid-latitude ozone change in response to expected decreasing halogen levels have been extensively discussed in Chapter 4 of WMO (2003), based primarily on simulations with eight separate 2-D photochemistry models that incorporated predicted future changes in halogen loading. Several scenarios for future changes in trace climate gases (CO_2, N_2O and CH_4) were also incorporated in these simulations (based on IPCC, 2001, Chapter 4), although the ozone results were not particularly sensitive to which scenario was employed. Results of the simulations for the period 1980–2050 are illustrated in the figure in Box 1.7, which shows the range of near-global (60°N to 60°S) mean column-ozone anomalies derived from the set of 2-D models. The models generally predict a minimum in global column ozone in 1992–1993 following the eruption of Mt. Pinatubo, followed by steady increases (although some models have a secondary minimum in about 2000). The latter evolution is primarily determined by changes in atmospheric

chlorine and bromine loading, which reaches a maximum in approximately 1995 and then slowly decreases (Figure 1.7d). There is, however, a large spread in the times the models predict that global ozone will return to 1980 levels, ranging from 2025 to after 2050. The spread of results arises in part from the differences between the models used.

Two of the 2-D models used in the simulations included interactive temperature changes caused by increasing greenhouse gases, which result in a long-term cooling of the stratosphere (one model includes this feedback throughout the entire stratosphere, and the other only above about 30 km). Cooling in the upper stratosphere results in slowing of the gas-phase chemical cycles that destroy ozone there (Section 1.3.2), and consequently these two models show a greater increase with time of total ozone (defining the upper range in the middle 21st century in the figure in Box 1.7) than the other models. An important caveat is that most of the past (1980–2000) column ozone change has occurred in the lower stratosphere, and future temperature feedback effects are more complicated and uncertain in the lower stratosphere. Furthermore, these effects involve changes in transport, heterogeneous chemistry and polar processes, which are not accurately simulated in 2-D models. (Note, however, that at least one CCM (Pitari *et al.*, 2002) also predicts a similar fast recovery.) At present there is considerable uncertainty regarding the details of temperature feedbacks on future mid-latitude column ozone changes.

Because 2-D models can be expected to capture the main chemical processes involved in ozone recovery, reliable predictions of future changes in the coupled ozone-climate system require the use of CCMs, since they consider possible changes in climate feedback mechanisms. Despite some present limitations, these models have been employed in sensitivity studies to assess the global future development of the chemical composition of the stratosphere and climate. The results have been compared with each other to assess the uncertainties of such predictions (Austin *et al.*, 2003), and have been documented in international assessment reports (WMO, 2003, Chapter 3; EC, 2003, Chapter 3). However most of the attention has been focused on polar ozone.

As discussed in Section 1.3.4, two dynamical influences appear to be related to NH mid-latitude ozone decreases over 1980–2000: a decrease in PWD and an increase in tropopause height. These mechanisms likewise have the potential to influence future ozone. If the past dynamical changes represent natural variability, then one cannot extrapolate past dynamical trends, and ozone recovery could be either hastened or delayed by dynamical variability. If, on the other hand, the past dynamical changes represent the dynamical response to WMGHG-induced climate change, then one might expect these changes to increase in magnitude in the future. This would decrease future ozone levels and delay ozone recovery. In the case of tropopause height changes, to the extent that they are themselves caused by ozone depletion, the changes should reverse in the future.

Another potential dynamical influence on ozone arises via

Box 1.7. Stratospheric ozone recovery

It is well established that the depletion of the ozone layer, both globally and in the polar regions, is attributable to an atmospheric chlorine burden that has been strongly enhanced, compared with natural levels, by anthropogenic emissions of halogen-containing compounds. Today, the production of such compounds has largely stopped following the regulations of the Montreal Protocol. However, owing to the long lifetime of the most important halogen source gases, the removal of these gases from the atmosphere will take many decades. The expectation that they will be ultimately removed is based on the assumption that global compliance with the Montreal Protocol will continue.

As a consequence, it might be expected that in about 50 years ozone depletion as it is observed today will have disappeared or, in other words, that the ozone layer will have 'recovered' (see figure). However, because of global climate change the state of the atmosphere has changed in recent decades and is expected to change further over the coming decades, so that a recovery to precisely the pre-1980 (that is, to the pre-'ozone hole') conditions will not occur.

Therefore, it is not obvious what should be considered as a detection of the *onset* of a 'recovery' of the ozone layer. A true detection of the onset of recovery requires more than the observation of a slow down or even a reversal of a downward ozone trend in a particular region, even if known natural periodic signals, such as the quasi-biennial oscillation or the 11-year solar cycle, have been accounted for. Rather, a detection of the onset of recovery requires that any temporal change in ozone in a particular region of the atmosphere can be *attributed* to the reduction of the atmospheric concentrations of anthropogenic halogen compounds as a result of the Montreal Protocol. Although one recent paper (Newchurch *et al.*, 2003) has reported that the first stage of a recovery of the ozone layer has been detected in observations of upper-stratospheric ozone, there is currently no consensus in the scientific community that the recovery of stratospheric ozone has been unequivocally established.

1. The detection of the onset of the recovery of stratospheric ozone will result from careful comparisons of the latest ozone measurements with past values. It is expected that ozone recovery will be first noticeable as a change in particular features, including: An increase in global column ozone towards values observed before 1980, when halogen source-gas abundances were much lower than they are today.
2. A sustained reduction in the maximum size of the Antarctic ozone hole, and an increase in the minimum value of column ozone in the hole.
3. Less ozone depletion in those Arctic winters in which temperatures fall below the threshold for the existence of polar stratospheric clouds (195 K).

As the ozone layer approaches full recovery, we expect to observe changes in all these features.

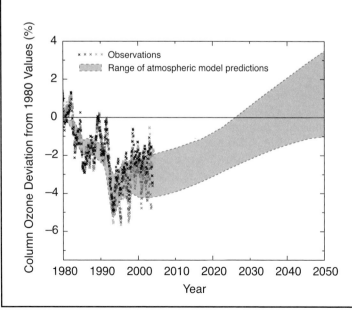

Box 1.7, Figure. Observed and modelled changes in low and middle latitude (60°S–60°N) de-seasonalized column ozone relative to 1980. The black symbols indicate ground-based measurements, and the coloured symbols various satellite-based data sets, updated from Fioletov *et al.* (2002). The range of model predictions comes from the use of several different 2-D photochemistry models forced with the same halocarbon scenario (WMO, 2003, Chapter 4); some models also allowed for the effect on stratospheric temperatures of changing CO_2 amounts. The measurements show that 60°S–60°N column ozone values decreased beginning in the early 1980s, and the models capture the timing and extent of this decrease quite well. Modelled halogen source-gas concentrations decrease in the early 21st century in response to the Montreal Protocol (Figure 1.7), so the simulated ozone values increase and recover towards pre-1980 values. Adapted from Figure Q20-1 of WMO (2003).

water vapour. Stratospheric water vapour is controlled by two processes: methane oxidation within the stratosphere and the transport of water vapour into the stratosphere. The latter depends in large part on the temperature of the tropical tropopause, although the precise details of this relationship remain unclear (see Section 1.3.4.4). In the future, changes in tropical tropopause temperature could conceivably affect the water vapour content of the stratosphere.

Unfortunately, our ability to predict future dynamical influences on ozone is very poor, for two reasons. First, dynamical processes affecting ozone exhibit significant temporal variability and are highly sensitive to other aspects of the atmospheric circulation. This means that the WMGHG-induced signal is inherently difficult to isolate or to represent accurately in CCMs, whereas the climate noise is relatively high (especially in the NH). Second, there are still uncertainties in the performance of the CCMs, which are the tools needed to address this question. As discussed in Section 1.4.1, predictions of WMGHG-induced dynamical changes by CCMs do not even agree on the sign of the changes: some models predict a weakened PWD, which would decrease mid-latitude (and polar) ozone by weakening transport, whereas others predict a strengthened PWD. Note that these dynamical changes would also affect the lifetime of stratospheric pollutants, with stronger PWD leading to shorter lifetimes (Rind *et al.*, 1990; Butchart and Scaife, 2001). (The latter effect acts on the several-year time scale associated with the Brewer-Dobson circulation, in contrast with the essentially instantaneous effect of PWD on ozone transport.) Thus, the direct effect of PWD change on ozone transport (no matter what its sign) would be reinforced over decadal time scales by altered chemical ozone loss arising from PWD-induced changes in stratospheric chlorine loading.

To summarize the results from 2-D models, upper-stratospheric ozone in the mid-latitudes will recover to pre-1980 levels well before stratospheric chlorine loading returns to pre-1980 levels, and could even overshoot pre-1980 levels because of CO_2-induced upper-stratospheric cooling (see, for example, WMO, 2003, Chapter 4). However, details of this future evolution are sensitive to many factors (uncertain emission scenarios, future climate change, changes in water vapour, volcanic eruptions, etc.) and there is significant divergence between different models. In terms of mid-latitude column ozone, where changes are expected to be dominated by ozone in the lower stratosphere, 2-D models all predict a steadily increasing ozone abundance as halogen levels decrease. While these models usually include a detailed treatment of polar chemistry, they do not include a detailed treatment of polar dynamics and mixing to mid-latitudes, or of dynamical aspects of climate change. Any future changes in stratospheric circulation and transport, or a large Pinatubo-like volcanic eruption, could have the potential to affect global column ozone, both directly and indirectly via chemical processes. Quantitative prediction is clearly difficult. At this stage, CCM simulations provide sensitivity calculations, which allow for the exploration of some examples of possible future evolution.

1.4.2.2 Polar ozone

As noted earlier, 2-D models do not provide a realistic treatment of polar processes, so predictions of polar ozone rely principally on CCMs. An intensive intercomparison of CCM predictions, the first of its kind, was performed for Chapter 3 of WMO (2003) (see Austin *et al.*, 2003). It is important to recognize that, apart from model deficiencies, CCM predictions are themselves subject to model variability and thus must be viewed in a statistical sense. Because of computer limitations (CPU time and mass storage), a large number of simulations with a single CCM cannot be carried out, so the models cannot yet be generally employed for ensemble runs. The approach taken in Austin *et al.* (2003) was to regard the collection of different CCMs as representing an ensemble. The collection included, moreover, a mixture of transient and time-slice runs (see Section 1.4.1).

For the Antarctic, where the CCMs all reproduce the development of the ozone hole in a reasonably realistic manner, the predicted future evolution is shown in Figure 1.15. Considering all the models, there is an overall consensus that the recovery of the Antarctic ozone hole will essentially follow the stratospheric chlorine loading. There is a hint in Figure 1.15 of a slight delay in the recovery compared with the peak in chlorine loading, presumably because of a cooling arising from the specified increase in WMGHG concentrations. Austin *et al.* (2003) estimated that the recovery of the Antarctic ozone layer can be expected to begin any year within the range 2001 to 2008. This would mean that Antarctic ozone depletion could slightly increase within the next few years despite a decrease in stratospheric chlorine loading, because lower temperatures would increase chlorine activation. However there is considerable natural variability, so it is also quite possible that the most severe Antarctic ozone hole has already occurred.

In the Arctic, where ozone depletion is more sensitive to meteorological conditions, the picture drawn by the CCMs is not consistent (Figure 1.17). Whereas most CCMs also predict a delayed start of Arctic ozone recovery (according to Austin *et al.* (2003), any year within the range 2004 to 2019), others indicate a different development: they simulate an enhanced PWD that produces a more disturbed and warmer NH stratospheric vortex in the future. (Enhanced PWD would also lead to a stronger meridional circulation and faster removal of CFCs.) This 'dynamical heating' more than compensates for the radiative cooling due to enhanced greenhouse-gas concentrations. Under these circumstances, and in combination with reduced stratospheric chlorine concentrations, polar ozone in these CCMs recovers within the next decade to values measured before the start of stratospheric ozone depletion. However, none of the current models suggest that an 'ozone hole', similar to that observed in the Antarctic, will occur over the Arctic (Figure 1.17). In this respect, the earlier GISS result (Shindell *et al.*, 1998), which is also depicted in Figure 1.17a, has not been supported by the more recent models (Austin *et al.*, 2003; WMO, 2003, Chapter 3). Analyses of model results show that the choice of the prescribed SSTs seems to play a critical role in the planetary-wave forcing (Schnadt and Dameris, 2003). Predicted SSTs (derived from coupled atmosphere-ocean models) vary in response to the same forcings that are already included in CCMs. Realistic predictions of future SSTs are therefore a necessary prerequisite for stratospheric circulation predictions, at least in the NH extratropics.

The differences between the currently available CCM results clearly indicate the uncertainties in the assessment of the future development of polar ozone. In the NH the differences are more pronounced than in the SH. Part of these differences arises from the highly variable nature of the NH circulation, as is reflected in the past record (Section 1.4.1). Until now, the different transient model predictions have been based on single realizations, not ensembles, so it is not yet possible to determine whether the differences between the models – with the sole exception of the GISS results shown in Figure 1.17, which are significantly different from the other model results – are statistically significant. The time-slice simulations give some indication of the range of natural variability that is possible, and allow for a wide range of future possibilities. Nevertheless there are also significant uncertainties arising from the performance of the underlying dynamical models. Cold biases have been found in the stratosphere of many of the models, with obvious direct consequences for chemistry. In consequence, at the current stage of model development, uncertainties in the details of PSC formation and sedimentation (denitrification) might be less important than model temperature biases for simulating accurate ozone amounts, although it is possible that the models underestimate the sensitivity of chemical ozone loss to temperature changes. However, because it has been shown that denitrification does contribute significantly to Arctic ozone loss and is very sensitive to temperature, both problems have to be solved before a reliable estimate of future Arctic ozone losses is possible. Further investigations and model developments, combined with data analysis, are needed in order to reduce or eliminate these various model deficiencies and quantify the natural variability.

In summary, the Antarctic ozone hole is expected to recover more or less following the decrease in chlorine loading, and return to 1980 levels in the 2045 to 2055 time frame. There may be a slight delay arising from WMGHG-induced cooling, but natural variability in the extent of the ozone hole is sufficiently large that the most severe ozone hole may have already occurred, or may occur in the next five years or so. With regard to the Arctic, the future evolution of ozone is potentially sensitive to climate change and to natural variability, and will not necessarily follow strictly the chlorine loading. There is uncertainty in even the sign of the dynamical feedback to WMGHG changes. Numerical models like CCMs are needed to make sensitivity studies to estimate possible future changes, although they are currently not fully evaluated and their deficiencies are obvious. Therefore, the interpretation of such 'predictions' must be performed with care. Progress will result from further development of CCMs and from comparisons of results between models and with observations. This will help to get a better understanding of potential feedbacks in the atmosphere, thereby leading to more reliable estimates of future changes.

1.5 Climate change from ODSs, their substitutes and ozone depletion

Previous sections in this chapter have primarily been concerned with complex interactions between stratospheric ozone loss and the climate of the stratosphere, including past and future behaviour. In this section we discuss and use the concept of radiative forcing to quantify the effect of ODSs and their substitutes on *surface* climate. Radiative forcing is used to quantify the direct role of these compounds, as well as their indirect role through stratospheric ozone depletion. Section 1.5.1 discusses the applicability of radiative forcing, Section 1.5.2 examines the direct radiative forcing from ODSs and their substitutes, and Section 1.5.3 goes on to discuss their indirect radiative forcing. Lastly, Section 1.5.4 presents an example scenario for the net (direct plus indirect) radiative forcing from the ODSs and their substitutes, placing it within our general understanding of climate change.

1.5.1 *Radiative forcing and climate sensitivity*

Sections 1.5.2 to 1.5.4 compare the radiative forcings from halocarbons and ozone to assess their roles in climate change. To make this comparison we have to assume that (a) radiative forcings can be compared between different mechanisms and (b) radiative forcing is related to the climate change issue of interest. The first assumption is equivalent to saying that the climate sensitivity (λ) is constant between different mechanisms (see Box 1.3), and is implicit whenever direct and indirect radiative forcings are compared (e.g., Section 1.5.4).

However, several climate modelling studies have found that for many climate mechanisms λ varies with the latitude of the imposed forcing, and is higher for changes in the extratropics than for changes in the tropics (IPCC, 2001, Chapter 6; Joshi *et al.*, 2003). Additionally, several climate-modelling studies have compared the climate sensitivity for stratospheric ozone increases (WMO, 2003, Chapter 4; Joshi *et al.*, 2003) and generally found that global stratospheric ozone increases have a 20–80% higher climate sensitivity than carbon dioxide. This finding has been attributed to an additional positive feedback that results from an increase in stratospheric water vapour, which in turn arises from a warmer tropical tropopause (Stuber *et al.*, 2001; Joshi *et al.*, 2003). However, the latter studies all used idealized stratospheric ozone changes, and to date no study has performed similar analyses with realistic ozone changes.

Climate models also typically have different responses to equivalent forcings from carbon dioxide and from other WMGHGs (Wang *et al.*, 1992; Govindasamy *et al.*, 2001); these results indicate possible differences in climate sensitivity for CO_2 and halocarbons. The two studies that have examined this directly reached contradictory conclusions. Hansen *et al.* (1997) found halocarbons to have about a 20% larger climate sensitivity than carbon dioxide, mainly because of a stronger positive cloud feedback in their halocarbon experiments. In contrast, Forster and Joshi (2005) found that, because halocarbons preferentially heat the tropical tropopause region rather than the surface (see Box 1.4), climate sensitivity for halocarbon changes was 6% smaller than for carbon dioxide changes.

The trade-off or partial cancellation between direct and indirect radiative forcing discussed in Section 1.5.4 only occurs in the global mean: the latitudinal forcing patterns actually complement each other. Both stratospheric ozone depletion and increases in halocarbons realize more positive forcing in the tropics and more negative forcing at higher latitudes (IPCC, 2001, Chapter 6). Therefore, the variation in climate sensitivity with latitude of any applied forcing would likely mean that, even if the halocarbon and ozone forcing cancelled each other out in the global mean (to give a net forcing of zero), an overall global cooling would result. Further, the patterns of surface temperature response may well be distinct.

Radiative forcing estimates, such as those shown in Figure 1.3, are measured as changes from pre-industrial times and are indicative of the equilibrium surface temperature change one might have expected since then. Radiative forcing is not necessarily indicative of patterns of temperature change, transient temperature changes or other metrics. In reality, only 60–80% of the temperature changes associated with the radiative forcing shown in Figure 1.3 would have already been realized in the observed surface warming of the last 150+ years (IPCC, 2001, Chapter 6). From any time period, such as the present, future temperature changes will come from the combination of temperatures continuing to respond to past radiative forcing and any new changes to the radiative forcing. The rate of change of radiative forcing is potentially more important for evaluating short-term climate change and possibilities for mitigation than the total radiative forcing. It is therefore useful to examine the rate, as well as the total radiative forcing (Solomon and Daniel, 1996).

In summary, adopting radiative forcing and examining its rate of change are among the best methods for comparing the climate roles of carbon dioxide, ozone-depleting gases, their substitutes and ozone. However, radiative forcing gives an estimate only of the equilibrium globally averaged surface temperature response, and could lead to errors of around 50% (given the range in model results discussed above) when comparing the predicted global mean temperature change from ozone and the halocarbons with that from carbon dioxide.

1.5.2 Direct radiative forcing of ODSs and their substitutes

The direct radiative forcing from the ozone-depleting gases and their substitutes are relatively well known, and have been comprehensively assessed in Chapter 1 of WMO (2003) and Chapter 6 of IPCC (2001). The radiative forcing of the substitute gases is re-assessed in Chapter 2 of this report. The radiative forcings of the ODSs have individual uncertainties of about 10%, and together they have contributed about 0.26 W m^{-2} (22%) to the total WMGHG radiative forcing since 1970 (Figure 1.3; Table 1.1).

Table 1.1. Positive direct radiative forcing by WMGHGs, including halocarbon gases, since 1750 and since 1970, and negative indirect radiative forcing since 1980 (no indirect forcing occurred prior to 1980). Based on data in Chapter 6 of IPCC (2001) and Chapter 1 of WMO (2003). Totals may not sum up consistently because of rounding.

Species	Radiative Forcing (W m^{-2})		
	Direct		Indirect
	1750–2000	1970–2000	1980–2000
CFCs			
CFC-11	0.066	0.053	−0.043
CFC-12	0.17	0.14	−0.034
CFC-13	0.001	0.001	n.a.
CFC-113	0.025	0.023	−0.010
CFC-114	0.005	0.003	n.a.
CFC-115	0.002	0.002	n.a.
Total	**0.27**	**0.22**	**−0.088**
HCFCs			
HCFC-22	0.028	0.026	−0.002
HCFC-141b	0.0018	0.0018	0
HCFC-142b	0.0024	0.0024	0
HCFC-124	0.0005	0.0005	n.a.
Total	**0.033**	**0.031**	**−0.002**
Halons and methyl bromide			
Halon-1211	0.0012	0.0012	−0.010
Halon-1301	0.0009	0.0009	−0.004
Halon-2402	0.0001	0.0001	−0.002
CH_3Br	0.0001	0.0001	−0.006
Total	**0.0022**	**0.0022**	**−0.023**
Chlorocarbons			
CH_3Cl	0.0007	0.0001	~0
CCl_4	0.0127	0.0029	−0.023
CH_3CCl_3	0.0028	0.0018	−0.015
Total	**0.021**	**0.0048**	**−0.038**
Total ODSs	*0.32*	*0.26*	*−0.150*
HFCs			
HFC-23	0.0029	0.0029	
HFC-134a	0.0024	0.0024	
HFC-125	0.0003	0.0003	
HFC-152a	0.0002	0.0002	
Total	**0.0058**	**0.0058**	
PFCs			
CF4	0.0029	0.0029	
C2F6	0.0006	0.0006	
C3F8	0.0001	0.0001	
C4F8	0.0003	0.0003	
Total	**0.0039**	**0.0039**	
Total halocarbons	*0.33*	*0.26*	*−0.150*
Other WMGHGs			
CO_2	1.50	0.67	
Methane	0.49	0.13	
Nitrous oxide	0.15	0.068	
SF_6	0.0024	0.0024	
Total	**2.14**	**0.87**	
Total WMGHGs	*2.48*	*1.14*	*−0.150*

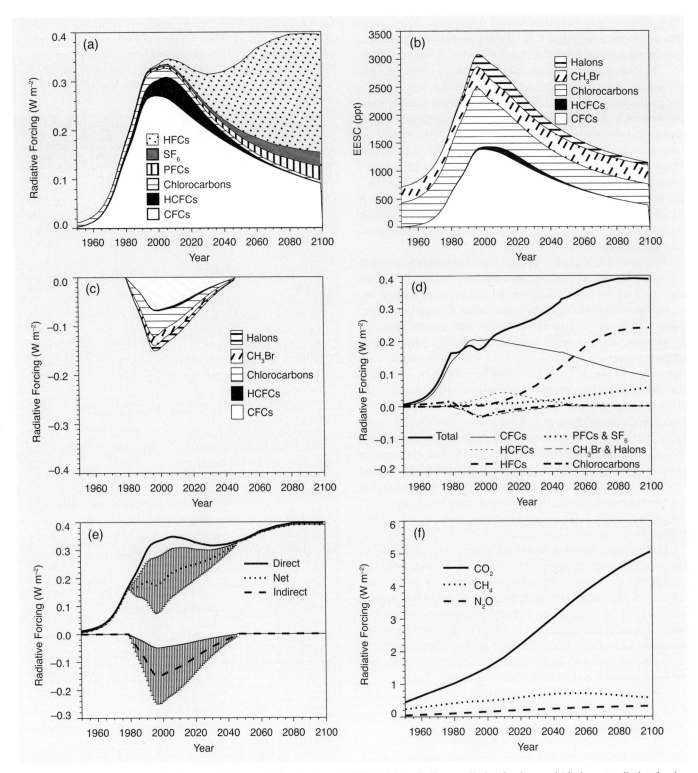

Figure 1.18. The breakdown of (a) the total direct radiative forcing, (b) EESC, (c) the indirect radiative forcing, and (d) the net radiative forcing time-series for individual halocarbon groups. In (e) the total direct, indirect and net radiative forcings are shown, with error bars arising from uncertainties in the stratospheric ozone radiative forcing from Chapter 6 of IPCC (2001). In (f) the radiative forcing for CO_2, N_2O and CH_4 is shown for comparison with the halocarbons. For (a), (b) and (c) the individual forcings have been stacked to show the total forcing. These time-series are derived from data in scenario Ab from Chapter 1 of WMO (2003) for the ozone-depleting halocarbons and scenario SRES A1B from Chapter 6 of IPCC (2001) for the other gases. The EESC has been calculated using data in Chapter 1 of WMO (2003), assuming that bromine is 45 times more effective on a per-atom basis at destroying ozone than chlorine. The change in EESC between 1979 and 1997 is assumed to give an indirect radiative forcing of –0.15 W m^{-2} (IPCC, 2001, Chapter 6). This radiative forcing scales with EESC amounts above the 1979 threshold. From Forster and Joshi (2005).

Section 1.2.2 discussed past changes and a future scenario for the abundances of the ozone-depleting gases. Chapter 2 discusses a range of future scenarios for the substitute gases. In this section we examine scenarios for future radiative forcing of the ODSs themselves. One possible future scenario is the A1B scenario from the Special Report on Emission Scenarios (SRES; IPCC, 2000), which for the ODSs is consistent with the Ab scenario from WMO (2003). Figure 1.18a shows a time-series of radiative forcing for the different groups of gases for this scenario. The figure shows that the current radiative forcing from ODSs is beginning to decline. However, because of their long lifetime in the atmosphere and their continued emission from the 'ODS bank' (namely those ODSs that have already been manufactured but have not yet been released into the atmosphere), they could dominate the radiative forcing for the next four to five decades. In this scenario HFCs dominate the radiative forcing by the end of the century, but this is only one example of several possible scenarios (see Chapter 2).

As the production of ODSs has now essentially ceased (WMO, 2003, Chapter 1), the principal issue for the emissions of ODSs concerns the treatment of the ODS bank. The scenario for ODSs in Figure 1.18a includes continued emissions of ODSs from the ODS bank, and if these emissions from the bank could be cut by recovering the ODSs, the radiative forcing time-series could be somewhat altered. Figure 1.19 illustrates the role of post-2004 releases from the ODS bank in continuing the ODS radiative forcing, and compares it with the HFC radiative forcing. If emissions of ODSs from the bank continue at their current rate, ODS radiative forcing will still decrease. However, for at least the next two decades, the continued emission from the ODS bank is expected to have a comparable contribution to the total radiative forcing with that of HFC emissions. This calculation does not take into account the effect that ODSs have on the stratospheric ozone radiative forcing, which is discussed next.

1.5.3 *Indirect radiative forcing of ODSs*

IPCC (2001) gives the value for the radiative forcing of stratospheric ozone as -0.15 W m^{-2}, with a range of ± 0.1 W m^{-2}. This radiative forcing primarily arises from mid-and high-latitude ozone depletion in the lower stratosphere (see Box 1.3). Past calculations of this radiative forcing have extensively employed observed trends in ozone. It is generally assumed that these observed ozone trends have been caused entirely by emissions of ODSs, so the negative radiative forcing of stratospheric ozone can be thought of as 'indirect'. For the ODSs it is possible to calculate EESC values and use them to scale the ozone radiative forcing (see Box 1.8). EESC values for different ODSs are shown in Figure 1.18b. The indirect radiative forcing of each ODS is shown in Figure 1.18c. These values have been calculated by assuming that EESC values above a 1979 background level give a forcing that scales with the -0.15 W m^{-2} best-estimate forcing from Chapter 6 of IPCC (2001) over 1979–1997. This calculation follows the approaches of Daniel *et al.* (1999)

and Forster and Joshi (2005). It results in a value of the indirect forcing that is purely from stratospheric ozone loss and ignores any possible changes to tropospheric chemistry and climate, which are discussed in Chapter 2. Table 1.1 shows the indirect radiative forcing of stratospheric ozone broken down by species.

Uncertainties in the indirect radiative forcing of stratospheric ozone arise from many different factors:

- The quoted IPCC (2001, Chapter 6) range of ± 0.1 W m^{-2} for the radiative forcing of stratospheric ozone is based on differing model results and arises primarily from the underlying uncertainty in the vertical distribution of the ozone trend relative to the tropopause. The quoted range can be used to scale the indirect radiative forcing in Figure 1.18c, as shown with the error bars in Figure 1.18e.

- The future radiative forcing from ozone is in fact more uncertain than the range quoted in IPCC (2001) because there is a lack of confidence in the details of future ozone changes, particularly near the tropopause, and many different scenarios are possible. These scenarios are discussed further in Section 1.5.4.

- Although most of the ozone changes can be attributed to ODSs, a sizeable fraction of the NH changes could have oth-

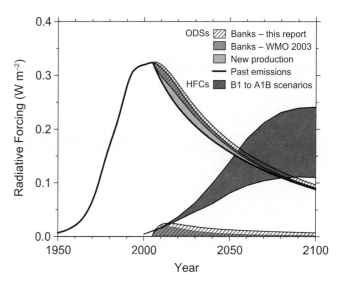

Figure 1.19. Direct radiative forcing of all ODSs compared with that of SRES projections for HFCs. The direct radiative forcing is split up into contributions from the commitment of past emissions (solid black line), release of allowed new production under the Montreal Protocol (grey shaded area), and release from ODS banks existing in 2004. Two estimates are given for these latter emissions, one based on bank estimates from WMO (2003), and one based on estimates in Ashford *et al.* (2004) – see Chapter 11. Radiative forcing due to HFCs is shown for the SRES B1 and A1B scenarios (boundaries of the red shaded area). The contribution due to the delayed release of ODSs in banks is shown separately and is comparable to that projected due to HFCs for the next two decades. ODSs also have other effects on radiative forcing.

Box 1.8. Ozone depletion potentials (ODPs) and equivalent effective stratospheric chlorine (EESC)

Ozone depletion potentials (ODPs) are used as a simple measure for quantifying the effect of various ozone-depleting compounds on the ozone layer, and have proved to be an important quantity for the formulation of the Montreal Protocol and its Amendments. The ODP is defined as the integrated change in total ozone per unit mass emission of a specific ozone-depleting substance relative to the integrated change in total ozone per unit mass emission of CFC-11 (WMO, 1995; WMO, 2003).

For the calculation of the ODP of an ozone-depleting substance, the change in total ozone per unit mass of emission of this substance may be determined with numerical models. As an alternative, Solomon *et al.* (1992) have formulated a semi-empirical approach for determining ODPs based mainly on observations rather than on models.

The quantities required for the semi-empirical approach are the physical properties of the halocarbons, their lifetimes, the so-called fractional release factor Φ – a factor representing the factor of inorganic halogen release in the stratosphere from observations relative to CFC-11 – and in the case of bromine, a quantification of the catalytic efficiency of ozone destruction relative to chlorine. For long-lived gases that are well mixed in the troposphere the definition of the semi-empirical ODP of a particular halogen compound is (Solomon *et al.*, 1992; Chapter 1 of WMO, 2003)

$$\text{ODP} = \Phi \cdot \alpha \cdot \frac{\tau_x}{\tau_{\text{CFC-11}}} \frac{M_{\text{CFC-11}}}{M_x} \frac{n_x}{3}$$

Here, τ is the global lifetime of the long-lived gas, M is the molecular weight, n is the number of halogen atoms and α is the relative efficiency of any halogen compound compared with chlorine. For bromine, $\alpha = 45$ (Daniel *et al.*, 1999). The current best estimates of ODPs by both the model and the semi-empirical methods are listed in Table 1.2 together with the ODP values originally adopted for the formulation of the Montreal Protocol.

The equivalent effective stratospheric chlorine (ESSC) is an index that is similar to an ODP. It relates the total stratospheric chlorine and bromine levels to the tropospheric release of halocarbons. It was defined as a mixing ratio by Daniel *et al.* (1995):

$$\text{EESC} = \left(\underset{\substack{\text{chlorine-} \\ \text{containing} \\ \text{compounds}}}{\sum} n_x Cl_{\text{trop}}^{t-3} \Phi + \underset{\substack{\text{bromine-} \\ \text{containing} \\ \text{compounds}}}{\sum} n_x \alpha Br_{\text{trop}}^{t-3} \Phi \right) F_{\text{CFC-11}}$$

where Cl_{trop}^{t-3} represents the stratospheric halocarbon mixing ratio at time t, and accounts for the approximately three years it takes to travel from the source of emission into the lower stratosphere; $F_{\text{CFC-11}}$ is the fractional release of halogens into the stratosphere from CFC-11, which is considered to be proportional to the change in column ozone.

er causes, especially those that occur close to the tropopause (Section 1.4.1). This means that a small but possibly significant part of the radiative forcing of stratospheric ozone may not be attributable to ODSs (Forster and Tourpali, 2001).

- The simple linear scaling between EESC values, above a 1979 background level, and the stratospheric ozone radiative forcing is used for obtaining an approximation to the ozone radiative forcing time-series. It was first proposed by Daniel *et al.* (1999). For EESC values below the 1979 level there is assumed to be no ozone depletion and no indirect radiative forcing. This simplistic approximation does not

take into account the many chemistry and climate feedbacks discussed in this chapter (see also Daniel *et al.*, 1999).

1.5.4   *Net radiative forcing*

Figures 1.18d–f show the direct, indirect and net radiative forcing time-series for the ODSs and other WMGHGs, based on the data in Figures 1.18a–c. These time-series depend critically on which scenario is used, which in the case of Figure 1.18 is the A1B SRES scenario from IPCC (2000). Nevertheless the figure gives one estimate of the possible global mean trade-offs involved in assessing the role of halocarbons in climate. Direct radiative forcing from halocarbons has risen sharply since 1980. However, ozone depletion has contributed a negative indirect forcing. Uncertainties in the magnitude of the indirect forcing are sufficiently large that the net radiative forcing since 1980 may have either increased or decreased (Figure 1.18e). The best estimate is that the indirect forcing has largely offset the increase in the direct radiative forcing. This suggests that surface temperatures over the last few decades would have increased even more rapidly if stratospheric ozone depletion had not occurred. However, it is important to note that ozone depletion since 1980 is the result, in part, of ODS increases prior to 1980.

The balance between direct and indirect radiative forcing varies substantially between different classes of ODSs (Table 1.1). For the halons, the negative indirect forcing dominates their very small positive direct effect, causing their net forcing to be negative. In contrast, the CFCs and HCFCs have a net positive radiative forcing, despite their associated ozone loss.

The scenario shown in Figure 1.18 assumes that concentrations of ODS substitutes will increase rapidly in the future while those of the ODSs will decay. In that case, even though the direct radiative forcing of the halocarbons would decrease during the next few decades, the net forcing would actually increase because of ozone recovery. (Note that the future positive forcing from ozone recovery is not included in the SRES scenarios.) However, this increase in net forcing depends on which scenario is used for the ODS substitutes. The detailed quantification of the radiative forcing from the ODS substitutes is discussed in Chapter 2.

Table 1.2. Steady-state ozone depletion potentials (ODPs) for long-lived halocarbons (normalized to unity for CFC-11).

Halocarbon	Steady-State ODPs		
	WMO (1999) Model	WMO (2003) Semi-Empirical	Montreal Protocol
CFC-11		1	1
CFC-12	0.82	1.0	1.0
CFC-113	0.90	1.0	0.8
CFC-114	0.94[a]		1.0
CFC-115	0.44[a]		0.6
Halon-1301	12	12	10.0
Halon-1211	5.1	5.3[b]	3.0
Halon-2402		<8.6	6.0
Halon-1202		1.3	
CCl$_4$	1.20	0.73	1.1
CH$_3$CCl$_3$	0.11	0.12	0.1
HCFC-22	0.034	0.05	0.055
HCFC-123	0.012	0.02	0.02
HCFC-124	0.026	0.02	0.022
HCFC-141b	0.086	0.12	0.11
HCFC-142b	0.043	0.07	0.065
HCFC-225ca	0.017	0.02	0.025
HCFC-225cb	0.017	0.03	0.033
CH$_3$Cl		0.02	
CH$_3$Br	0.37	0.38	0.6

Upper limits for selected hydrofluorocarbons

HFC-134a	$<1.5 \times 10^{-5}$
HFC-23	$<4 \times 10^{-4}$
HFC-125	$<3 \times 10^{-5}$

[a] Updated in WMO (2003).
[b] Value was erroneously reported in Table 1-5 of WMO (2003) as 6.0.

References

Allaart, M., P. Valks, R. van der A, A. Piters, H. Kelder, and P. F. J. van Velthoven, 2000: Ozone mini-hole observed over Europe, influence of low stratospheric temperature on observations. *Geophysical Research Letters,* **27,** 4089–4092.

Allen, D. R., R. M. Bevilacqua, G. E. Nedoluha, C. E. Randall, and G. L. Manney, 2003: Unusual stratospheric transport and mixing during the 2002 Antarctic winter. *Geophysical Research Letters,* **30**(12), 1599.

Ambaum, M. H. P. and B. J. Hoskins, 2002: The NAO troposphere-stratosphere connection. *Journal of Climate,* **15,** 1969–1978.

Ambaum, M. H. P., B. J. Hoskins, and D. B. Stephenson, 2001: Arctic Oscillation or North Atlantic Oscillation? *Journal of Climate,* **14**(16), 3495–3507.

Andrews, D. G., J. R. Holton, and C. B. Leovy, 1987: *Middle Atmosphere Dynamics.* Academic Press, Orlando, 489 pp.

Appenzeller, C. and H. C. Davies, 1992: Structure of stratospheric intrusions into the troposphere. *Nature,* **358,** 570–572.

Appenzeller, C., A. K. Weiss, and J. Staehelin, 2000: North Atlantic oscillation modulates total ozone winter trends. *Geophysical Research Letters,* **27**(8), 1131–1134.

Ashford, P., D. Clodic, A. McCulloch, L. Kuijpers, 2004: Emission profiles from the foam and refrigeration sectors compared with atmospheric concentrations, Part 2 – Results and discussion. *International Journal of Refrigeration,* **27**(7), 701–716.

Austin, J., D. Shindell, S. R. Beagley, C. Brühl, M. Dameris, E. Manzini, T. Nagashima, P. Newman, S. Pawson, G. Pitari, E. Rozanov, C. Schnadt, and T. G. Shepherd, 2003: Uncertainties and assessments of chemistry-climate models of the stratosphere. *Atmospheric Chemistry and Physics,* **3,** 1–27.

Baldwin, M. P. and T. J. Dunkerton, 1999: Propagation of the Arctic Oscillation from the stratosphere to the troposphere. *Journal of Geophysical Research – Atmospheres,* **104**(D24), 30,937–30,946.

Barnett, J. J., J. T. Houghton, and J. A. Pyle, 1975: The temperature dependence of the ozone concentration near the stratosphere. *Quarterly Journal of the Royal Meteorological Society,* **101,** 245–257.

Bengtsson, L., S. Hagemann, and K. I. Hodges, 2004: Can climate trends be calculated from reanalysis data? *Journal of Geophysical Research – Atmospheres,* **109**(D11), doi:10.1029/2004JD004536.

Bodeker, G. E., J. C. Scott, K. Kreher, and R. L. McKenzie, 2001: Global ozone trends in potential vorticity coordinates using TOMS and GOME intercompared against the Dobson network: 1978–1998. *Journal of Geophysical Research – Atmospheres,* **106**(D19), 23,029–23,042.

Bojkov, R. D., C. S. Zerefos, D. S. Balis, I. C. Ziomas, and A. F. Bais, 1993: Record low total ozone during northern winters of 1992 and 1993. *Geophysical Research Letters,* **20**(13), 1351–1354.

Borrmann, S., S. Solomon, J. E. Dye, D. Baumgardner, K. K. Kelly, and K. R. Chan, 1997: Heterogeneous reactions on stratospheric background aerosols, volcanic sulfuric acid droplets, and type I polar stratospheric clouds: Effects of temperature fluctuations and differences in particle phase. *Journal of Geophysical Research – Atmospheres,* **102**(D3), 3639–3648.

Bösch, H., C. Camy-Peyret, M. P. Chipperfield, R. Fitzenberger, H. Harder, U. Platt, and K. Pfeilsticker, 2003: Upper limits of stratospheric IO and OIO inferred from center-to-limb-darkening-corrected balloon-borne solar occultation visible spectra: Implications for total gaseous iodine and stratospheric ozone. *Journal of Geophysical Research – Atmospheres,* **108**(D15), 4455, doi:10.1029/2002JD003078.

Bregman, B., P. H. Wang, and J. Lelieveld, 2002: Chemical ozone loss in the tropopause region on subvisible ice clouds, calculated with a chemistry-transport model. *Journal of Geophysical Research – Atmospheres,* **107**(D3), doi:10.1029/2001JD000761.

Brewer, A.M., 1949: Evidence of a world circulation provided by the measurement of helium and water distribution in the stratosphere. *Quarterly Journal of the Royal Meteorological Society,* **75,** 351–363.

Butchart, N. and A. A. Scaife, 2001: Removal of chlorofluorocarbons by increased mass exchange between the stratosphere and troposphere in a changing climate. *Nature,* **410**(6830), 799–802.

Butler, J. H., M. Battle, M. Bender, S. A. Montzka, A. D. Clarke, E. S. Saltzman, C. Sucher, J. Severinghaus, and J. W. Elkins, 1999: A twentieth century record of atmospheric halocarbons in polar firn air. *Nature,* **399,** 749–755.

Canziani, P. O., R. H. Compagnucci, S. A. Bischoff, and W. E. Legnani, 2002: A study of impacts of tropospheric synoptic processes on the genesis and evolution of extreme total ozone anomalies over southern South America. *Journal of Geophysical Research – Atmospheres,* **107**(D24), 4741, doi:10.1029/2001JD000965.

Chen, P. and W. A. Robinson, 1992: Propagation of planetary waves between the troposphere and stratosphere. *Journal of the Atmospheric Sciences,* **49,** 2533–2545.

Chipperfield, M. P., 2003: A three-dimensional model study of long-term mid-high latitude lower stratosphere ozone changes. *Atmospheric Chemistry and Physics,* **3,** 1253–1265.

Christiansen, B., 2001: Downward propagation of zonal mean zonal wind anomalies from the stratosphere to the troposphere: Model and reanalysis. *Journal of Geophysical Research – Atmospheres,* **106**(D21), 27,307–27,322.

Considine, D. B., J. E. Rosenfield, and E. L. Fleming, 2001: An interactive model study of the influence of the Mount Pinatubo aerosol on stratospheric methane and water trends. *Journal of Geophysical Research – Atmospheres,* **106**(D21), 27,711–27,727.

Crutzen, P. J., 1974: Estimates of possible future ozone reductions from continued use of fluorochloro-methanes (CF_2Cl_2, $CFCl_3$). *Geophysical Research Letters,* **1,** 205–208.

Daniel, J. S., S. Solomon, and D. L. Albritton, 1995: On the evaluation of halocarbon radiative forcing and global warming potentials. *Journal of Geophysical Research – Atmospheres,* **100**(D1), 1271–1285.

Daniel, J. S., S. Solomon, R. W. Portmann, and R. R. Garcia, 1999: Stratospheric ozone destruction: The importance of bromine relative to chlorine. *Journal of Geophysical Research – Atmospheres,* **104,** 23,871–23,880.

DeMore, W. B., C. J. Howard, S. P. Sander, A. R. Ravishankara, D. M. Golden, C. E. Kolb, R. F. Hampson, M. J. Molina, and M. J. Kurylo, 1997: Chemical Kinetics and Photochemical

Data for Use in Stratospheric Modeling. Evaluation Number 12. JPL Publication No. 97-4, National Aeronautics and Space Administration, Jet Propulsion Laboratory, California Institute of Technology, Pasadena, California. (Available at http://jpldataeval. jpl.nasa.gov/pdf/Atmos97_Anotated.pdf)

Dickinson, R. E., 1978: Effect of chlorofluromethane infrared radiation on zonal atmospheric temperatures. *Journal of the Atmospheric Sciences*, **35**(11), 2142–2152.

Dlugokencky, E. J., S. Houweling, L. Bruhwiler, K. A. Masarie, P. M. Lang, J. B. Miller, and P. P. Tans, 2003: Atmospheric methane levels off: Temporary pause or a new steady-state? *Geophysical Research Letters*, **30**(19), 1992, doi:10.1029/2003GL018126.

Dobson, G. M. B., 1963: *Exploring the Atmosphere*. Clarendon Press, London.

Dvortsov, V. L. and S. Solomon, 2001: Response of the stratospheric temperatures and ozone to past and future increases in stratospheric humidity. *Journal of Geophysical Research – Atmospheres*, **106**(D7), 7505–7514.

EC, 2003: *Ozone-Climate Interactions*. Air Pollution Research Report No. 81, EUR 20623, European Commission, Office for Official Publications of the European Communities, Luxembourg, 143 pp.

Fioletov, V. E. and T. G. Shepherd, 2003: Seasonal persistence of midlatitude total ozone anomalies. *Geophysical Research Letters*, **30**(7), 1417, doi:10.1029/2002GL016739.

Fioletov, V. E., G. E. Bodeker, A. J. Miller, R. D. McPeters, and R. Stolarski, 2002: Global and zonal total ozone variations estimated from ground-based and satellite measurements: 1964–2000. *Journal of Geophysical Research – Atmospheres*, **107**(D22), 4647, doi:10.1029/2001JD001350.

Fish, D. J., H. K. Roscoe, and P. V. Johnston, 2000: Possible causes of stratospheric NO_2 trends observed at Lauder, New Zealand. *Geophysical Research Letters*, **27**(20), 3313–3316.

Fleming, E. L., S. Chandra, J. J. Barnett, and M. Corney, 1990: Zonal mean temperature, pressure, zonal wind and geopotential height as functions of latitude. *Advances in Space Research*, **12**, 1211–1259.

Forster, P. M. d. F. and M. Joshi, 2005: The role of halocarbons in the climate change of the troposphere and stratosphere. *Climatic Change* (in press).

Forster, P. M. d. F. and K. Tourpali, 2001: Effect of tropopause height changes on the calculation of ozone trends and their radiative forcing. *Journal of Geophysical Research – Atmospheres*, **106**(D11), 12,241–12,251.

Fortuin, J. P. F. and H. Kelder, 1998: An ozone climatology based on ozonesonde and satellite measurements. *Journal of Geophysical Research – Atmospheres*, **103**(D24), 31,709–31,734.

Fortuin, J. P. F. and U. Langematz, 1994: An update on the global ozone climatology and on concurrent ozone and temperature trends. *SPIE, Atmospheric Sensing and Modeling*, **2311**, 207–216.

Fraser, P. J., D. E. Oram, C. E. Reeves, S. A. Penkett, and A. McCulloch, 1999: Southern Hemispheric halon trends (1978–1998) and global halon emissions. *Journal of Geophysical Research – Atmospheres*, **104**, 15,985–15,999.

Froidevaux, L., M. Allen, S. Berman, and A. Daughton, 1989: The mean ozone profile and its temperature sensitivity in the upper stratosphere and lower mesosphere. *Journal of Geophysical Research*, **94**, 6389–6417.

Froidevaux, L., J. W. Waters, W. G. Read, P. S. Connell, D. E. Kinnison, and J. M. Russell, 2000: Variations in the free chlorine content of the stratosphere (1991–1997): Anthropogenic, volcanic, and methane influences. *Journal of Geophysical Research – Atmospheres*, **105**(D4), 4471–4481.

Fusco, A. C. and M. L. Salby, 1999: Interannual variations of total ozone and their relationship to variations of planetary wave activity. *Journal of Climate*, **12**(6), 1619–1629.

Fyfe, J. C., G. J. Boer, and G. M. Flato, 1999: The Arctic and Antarctic oscillations and their projected changes under global warming. *Geophysical Research Letters*, **26**(11), 1601–1604.

Gillett, N. P. and D. W. J. Thompson, 2003: Simulation of recent Southern Hemisphere climate change. *Science*, **302**(5643), 273–275.

Gillett, N. P., M. R. Allen, R. E. McDonald, C. A. Senior, D. T. Shindell, and G. A. Schmidt, 2002a: How linear is the Arctic Oscillation response to greenhouse gases? *Journal of Geophysical Research*, **107**(D3), 10.1029/2001JD000589.

Gillett, N. P., M. R. Allen, and K. D. Williams, 2002b: The role of stratospheric resolution in simulating the Arctic Oscillation response to greenhouse gases. *Geophysical Research Leters*, **29**(10), doi:10.1029/2001GL014444.

Gong, D. Y. and S. W. Wang, 1999: Definition of Antarctic Oscillation Index. *Geophysical Research Letters*, **26**(4), 459–462.

Govindasamy, B., K. E. Taylor, P. B. Duffy, B. D. Santer, A. S. Grossman, and K. E. Grant, 2001: Limitations of the equivalent CO_2 approximation in climate change simulations. *Journal of Geophysical Research – Atmospheres*, **106**, 22,593–22,603.

Graf, H. F., J. Perlwitz, I. Kirchner, and I. Schult, 1995: Recent Northern Hemisphere climate trends, ozone changes, and increased greenhouse gas forcing. *Beiträge zur Physik der Atmosphäre*, **68**, 233–248.

Graf, H. F., I. Kirchner, and J. Perlwitz, 1998: Changing lower stratospheric circulation: The role of ozone and greenhouse gases. *Journal of Geophysical Research – Atmospheres*, **103**(D10), 11,251–11,261.

Grooß, J. U., R. Müller, G. Becker, D. S. McKenna, and P. J. Crutzen, 1999: The upper stratospheric ozone budget: An update of calculations based on HALOE data. *Journal of Atmospheric Chemistry*, **34**(2), 171–183.

Hall, A. and M. Visbeck, 2002: Synchronous variability in the Southern Hemisphere atmosphere, sea ice, and ocean resulting from the annular mode. *Journal of Climate*, **15**(21), 3043–3057.

Hansen, J., M. Sato, and R. Ruedy, 1997: Radiative forcing and climate response. *Journal of Geophysical Research – Atmospheres*, **102**(D6), 6831–6864.

Highwood, E. J. and B. J. Hoskins, 1998: The tropical tropopause. *Quarterly Journal of the Royal Meteorological Society*, **124**, 1579–1604.

Hintsa, E. J., K. A. Boering, E. M. Weinstock, J. G. Anderson, B. L. Gary, L. Pfister, B. C. Daube, S. C. Wofsy, M. Loewenstein, J.

R. Podolske, J. J. Margitan, and T. P. Bui, 1998: Troposphere-to-stratosphere transport in the lowermost stratosphere from measurements of H_2O, CO_2, N_2O and O_3. *Geophysical Research Letters,* **25**, 2655–2658.

Hoinka, K. P., 1999: Temperature, humidity, and wind at the global tropopause. *Monthly Weather Review,* **127**, 2248–2265.

Holton, J. R. and A. Gettelman, 2001: Horizontal transport and the dehydration of the stratosphere. *Geophysical Research Letters,* **28**(14), 2799–2802.

Holton, J. R., P. H. Haynes, M. E. McIntyre, A. R. Douglass, R. B. Rood, and L. Pfister, 1995: Stratosphere-troposphere exchange. *Reviews of Geophysics,* **33**(4), 403–439.

Hood, L., S. Rossi, and M. Beulen, 1999: Trends in lower stratospheric zonal winds, Rossby wave breaking behavior, and column ozone at northern midlatitudes. *Journal of Geophysical Research – Atmospheres,* **104**(D20), 24,321–24,339.

Hood, L. L., B. E. Soukharev, M. Fromm, and J. P. McCormack, 2001: Origin of extreme ozone minima at middle to high northern latitudes. *Journal of Geophysical Research – Atmospheres,* **106**(D18), 20,925–20,940.

Hu, Y. and K.-K. Tung, 2002: Interannual and decadal variations of planetary wave activity, stratospheric cooling, and Northern Hemisphere annular mode. *Journal of Climate,* **15**, 1659–1673.

Hurrell, J. W., 1995: Decadal trends in the North Atlantic Oscillation – Regional temperatures and precipitation. *Science,* **269**(5224), 676–679.

IPCC, 2000: *Emissions Scenarios.* Special Report of the Intergovernmental Panel on Climate Change [Nakicenovic, N. and R. Swart (eds.)]. Cambridge University Press, Cambridge, United Kingdom, and New York, NY, USA, 570 pp.

IPCC, 2001: *Climate Change 2001: The Scientific Basis. Contribution of Working Group 1 to the Third Assessment Report of the Intergovernmental Panel on Climate Change* [Houghton, J. T., Y. Ding, D. J. Griggs, M. Noguer, P. J. van der Linden, X. Dai, K. Maskell, and C. A. Johnson (eds.)]. Cambridge University Press, Cambridge, United Kingdom, and New York, NY, USA, 944 pp.

James, P. M., 1998: A climatology of ozone mini-holes over the Northern Hemisphere. *International Journal of Climatology,* **18**(12), 1287–1303.

James, P., A. Stohl, C. Forster, S. Eckhardt, P. Seibert, and A. Frank, 2003: A 15-year climatology of stratosphere-troposphere exchange with a Lagrangian particle dispersion model – 2. Mean climate and seasonal variability. *Journal of Geophysical Research – Atmospheres,* **108**(D12), 8522, doi:10.1029/2002JD002639.

Joshi, M., K. Shine, M. Ponater, N. Stuber, R. Sausen, and L. Li, 2003: A comparison of climate response to different radiative forcings in three general circulation models: Towards an improved metric of climate change. *Climate Dynamics,* **20**(7–8), 843–854.

Jucks, K. W. and R. J. Salawitch, 2000: Future changes in upper stratospheric ozone. In: *Atmospheric Science Across the Stratopause* [Siskind, D. E., S. D. Eckermann, and M. E. Summers (eds.)] Geophysical Monograph Series No. 123, American Geophysical Union, pp. 241–255.

Keim, E. R., D. W. Fahey, L. A. Del Negro, E. L. Woodbridge, R. S. Gao, P. O. Wennberg, R. C. Cohen, R. M. Stimpfle, K. K. Kelly, E.

J. Hintsa, J. C. Wilson, H. H. Jonsson, J. E. Dye, D. Baumgardner, S. R. Kawa, R. J. Salawitch, M. H. Proffitt, M. Loewenstein, J. R. Podolske, and K. R. Chan, 1996: Observations of large reductions in the NO/NO_y ratio near the mid-latitude tropopause and the role of heterogeneous chemistry. *Geophysical Research Letters,* **23**(22), 3223–3226.

Kelly, P. M., P. D. Jones, and P. Q. Jia, 1996: The spatial response of the climate system to explosive volcanic eruptions. *International Journal of Climatology,* **16**(5), 537–550.

Khalil, M. A. K., R. A. Rasmussen, and R. Gunawardena, 1993: Atmospheric methyl bromide: Trends and global mass balance. *Journal of Geophysical Research – Atmospheres,* **98**, 2887–2896.

Kirk-Davidoff, D. B., E. J. Hintsa, J. G. Anderson, and D. W. Keith, 1999: The effect of climate change on ozone depletion through changes in stratospheric water vapour. *Nature,* **402**(6760), 399–401.

Kodera, K., 1994: Influence of volcanic eruptions on the troposphere through stratospheric dynamical processes in the Northern Hemisphere winter. *Journal of Geophysical Research – Atmospheres,* **99**(D1), 1273–1282.

Lamarque, J.-F. and P. G. Hess, 2004: Arctic Oscillation modulation of the Northern Hemisphere spring tropospheric ozone. *Geophysical Research Letters,* **31**, L06127, doi:10.1029/2003GL019116.

Langematz, U., M. Kunze, K. Krüger, K. Labitzke, and G. L. Roff, 2003: Thermal and dynamical changes of the stratosphere since 1979 and their link to ozone and CO_2 changes. *Journal of Geophysical Research – Atmospheres,* **108**(D1), 4027, doi:10.1029/2002JD002069.

Lee, A. M., H. K. Roscoe, and S. Oltmans, 2000: Model and measurements show Antarctic ozone loss follows edge of polar night. *Geophysical Research Letters,* **27**(23), 3845–3848.

Lelieveld, J. and F. J. Dentener, 2000: What controls tropospheric ozone? *Journal of Geophysical Research – Atmospheres,* **105**, 3531–3551.

Li, J. L., D. M. Cunnold, H. J. Wang, E. S. Yang, and M. J. Newchurch, 2002: A discussion of upper stratospheric ozone asymmetries and SAGE trends. *Journal of Geophysical Research – Atmospheres,* **107**(D23), 4705, doi:10.1029/2001JD001398.

Liley, J. B., P. V. Johnston, R. L. McKenzie, A. J. Thomas, and I. S. Boyd, 2000: Stratospheric NO_2 variations from a long time series at Lauder, New Zealand. *Journal of Geophysical Research – Atmospheres,* **105**(D9), 11,633–11,640.

Limpasuvan, V. and D. L. Hartmann, 2000: Wave-maintained annular modes of climate variability. *Journal of Climate,* **13**, 4414–4429.

Manzini, E. and N. A. McFarlane, 1998: The effect of varying the source spectrum of a gravity wave parameterization in a middle atmosphere general circulation model. *Journal of Geophysical Research – Atmospheres,* **103**(D24), 31,523–31,539.

McCormick, M. P., L. W. Thomason, and C. R. Trepte, 1995: Atmospheric effects of the Mount Pinatubo eruption. *Nature,* **373**, 399–404.

McElroy, M. B., R. J. Salawitch, and K. Minschwaner, 1992: The changing stratosphere. *Planetary and Space Science,* **40**, 373–401.

McLinden, C. A., S. C. Olsen, M. J. Prather, and J. B. Liley, 2001: Understanding trends in stratospheric NO_y and NO_2. *Journal of Geophysical Research – Atmospheres,* **106**(D21), 27,787–27,793.

Millard, G. A., A. M. Lee, and J. A. Pyle, 2002: A model study of the connection between polar and midlatitude ozone loss in the Northern Hemisphere lower stratosphere. *Journal of Geophysical Research – Atmospheres,* **107**(D5), 8323, doi:10.1029/2001JD000899.

Molina, M. J. and F. S. Rowland, 1974: Stratospheric sink for chlorofluoromethanes: Chlorine atom catalysed destruction of ozone. *Nature,* **249**, 810–812.

Montzka, S. A., J. H. Butler, J. W. Elkins, T. M. Thompson, A. D. Clarke, and L. T. Lock, 1999: Present and future trends in the atmospheric burden of ozone-depleting halogens. *Nature,* **398**, 690–694.

Montzka, S. A., J. H. Butler, B. D. Hall, D. J. Mondeel, and J. W. Elkins, 2003: A decline in tropospheric organic bromine. *Geophysical Research Letters,* **30**(15), 1826, doi:10.1029/2003GL017745.

Mote, P. W., K. H. Rosenlof, M. E. McIntyre, E. S. Carr, J. C. Gille, J. R. Holton, J. S. Kinnersley, H. C. Pumphrey, J. M. Russell III, and J. C. Waters, 1996: An atmospheric tape recorder: The imprint of tropical tropopause temperatures on stratospheric water vapor. *Journal of Geophysical Research – Atmospheres,* **101**, 3989–4006.

Nedoluha, G. E., D. E. Siskind, J. T. Bacmeister, R. M. Bevilacqua, and J. M. Russell, 1998: Changes in upper stratospheric CH_4 and NO_2 as measured by HALOE and implications for changes in transport. *Geophysical Research Letters,* **25**(7), 987–990.

Newchurch, M. J., E. S. Yang, D. M. Cunnold, G. C. Reinsel, J. M. Zawodny, and J. M. Russell, 2003: Evidence for slowdown in stratospheric ozone loss: First stage of ozone recovery. *Journal of Geophysical Research – Atmospheres,* **108**(D16), 4507, doi:10.1029/2003JD003471.

Newman, P. A. and E. R. Nash, 2000: Quantifying the wave driving of the stratosphere. *Journal of Geophysical Research – Atmospheres,* **105**(D10), 12,485–12,497.

Newman, P. A. and E. R. Nash, 2005. The unusual Southern Hemisphere stratosphere winter of 2002. *Journal of the Atmospheric Sciences,* **62**(3), 614–628.

Newman, P. A., J. F. Gleason, R. D. McPeters, and R. S. Stolarski, 1997: Anomalously low ozone over the Arctic. *Geophysical Research Letters,* **24**(22), 2689–2692.

Newman, P. A., E. R. Nash, and J. E. Rosenfield, 2001: What controls the temperature of the Arctic stratosphere during the spring? *Journal of Geophysical Research – Atmospheres,* **106**(D17), 19,999–20,010.

Norton, W. A. and M. P. Chipperfield, 1995: Quantification of the transport of chemically activated air from the Northern Hemisphere polar vortex. *Journal of Geophysical Research – Atmospheres,* **100**(D12), 25,817–25,840.

O'Doherty, S., D. M. Cunnold, A. Manning, B. R. Miller, R. H. J. Wang, P. B. Krummel, P. J. Fraser, P. G. Simmonds, A. McCulloch, R. F. Weiss, P. Salameh, L. W. Porter, R. G. Prinn, J. Huang, G. Sturrock, D. Ryall, R. G. Derwent, and S. A. Montzka, 2004: Rapid growth of hydrofluorocarbon 134a, and hydrochlorofluorocarbons

141b, 142b and 22 from Advanced Global Atmospheric Gases Experiment (AGAGE) observations at Cape Grim, Tasmania and Mace Head, Ireland. *Journal of Geophysical Research – Atmospheres,* **109**, D06310, doi:10.1029/2003JD004277.

Oltmans, S. J., H. Vömel, D. J. Hofmann, K. H. Rosenlof, and D. Kley, 2000: The increase in stratospheric water vapour from balloonborne, frostpoint hygrometer measurements at Washington, D.C. and Boulder, Colorado. *Geophysical Research Letters,* **27**, 3453–3456.

Orsolini, Y. J. and V. Limpasuvan, 2001: The North Atlantic Oscillation and the occurrences of ozone miniholes. *Geophysical Research Letters,* **28**(21), 4099–4102.

Pawson, S. and B. Naujokat, 1999: The cold winters of the middle 1990s in the northern lower stratosphere. *Journal of Geophysical Research – Atmospheres,* **104**(D12), 14,209–14,222.

Perlwitz, J. and H. F. Graf, 1995: The statistical connection between tropospheric and stratospheric circulation of the Northern Hemisphere in winter. *Journal of Climate,* **8**(10), 2281–2295.

Perlwitz, J., H. F. Graf, and R. Voss, 2000: The leading variability mode of the coupled troposphere-stratosphere winter circulation in different climate regimes. *Journal of Geophysical Research – Atmospheres,* **105**(D5), 6915–6926.

Pitari, G., E. Mancini, V. Rizi, and D. T. Shindell, 2002: Impact of future climate and emission changes on stratospheric aerosols and ozone. *Journal of the Atmospheric Sciences,* **59**(3), 414–440.

Potter, B. E. and J. R. Holton, 1995: The role of monsoon convection in the dehydration of the lower tropical stratosphere. *Journal of the Atmospheric Sciences,* **52**, 1034–1050.

Poulida, O., R. R. Dickerson, and A. Heymsfield, 1996: Stratosphere-troposphere exchange in a midlatitude mesoscale convective complex: 1. Observations. *Journal of Geophysical Research – Atmospheres,* **101**, 6823–6836.

Prather, M. and A. H. Jaffe, 1990: Global impact of the Antarctic ozone hole: Chemical propagation. *Journal of Geophysical Research – Atmospheres,* **95**(D4), 3473–3492.

Prather, M., M. M. Garcia, R. Suozzo, and D. Rind, 1990: Global impact of the Antarctic ozone hole: Dynamic dilution with a three-dimensional chemical transport model. *Journal of Geophysical Research – Atmospheres,* **95**(D4), 3449–3471.

Prinn, R. G., R. F. Weiss, P. J. Fraser, P. G. Simmonds, D. M. Cunnold, F. N. Alyea, S. O'Doherty, P. Salameh, B. R. Miller, J. Huang, R. H. J. Wang, D. E. Hartley, C. Harth, L. P. Steele, G. Sturrock, P. M. Midgley, and A. McCulloch, 2000: A history of chemically and radiatively important gases in air deduced from ALE/GAGE/AGAGE. *Journal of Geophysical Research – Atmospheres,* **105**, 17,751–17,792.

Ramaswamy, V., M. D. Schwarzkopf, and W. J. Randel, 1996: Fingerprint of ozone depletion in the spatial and temporal pattern of recent lower-stratospheric cooling. *Nature,* **382**(6592), 616–618.

Randel, W. J. and J. B. Cobb, 1994: Coherent variations of monthly mean total ozone and lower stratospheric temperature. *Journal of Geophysical Research,* **99**, 5433–5447.

Randel, W. J. and F. Wu, 1999: Cooling of the Arctic and Antarctic polar stratospheres due to ozone depletion. *Journal of Climate,* **12**(5), 1467–1479.

Randel, W. J., F. Wu, J. M. Russell, and J. Waters, 1999: Space-time patterns of trends in stratospheric constituents derived from UARS measurements. *Journal of Geophysical Research – Atmospheres,* **104**(D3), 3711–3727.

Randel, W. J., F. Wu, and R. Stolarski, 2002: Changes in column ozone correlated with the stratospheric EP flux. *Journal of the Meteorological Society of Japan,* **80**(4B), 849–862.

Randel, W. J., F. Wu, S. J. Oltmans, K. Rosenlof, and G. E. Nedoluha, 2004: Interannual changes of stratospheric water vapor and correlations with tropical tropopause temperatures. *Journal of the Atmospheric Sciences,* **61**, 2133–2148.

Rauthe, M., A. Hense, and H. Paeth, 2004: A model intercomparison study of climate change signals in extratropical circulation. *International Journal of Climatology,* **24**, 643–662.

Ray, E. A., F. L. Moore, J. W. Elkins, G. S. Dutton, D. W. Fahey, H. Vömel, S. J. Oltmans, and K. H. Rosenlof, 1999: Transport into the Northern Hemisphere lowermost stratosphere revealed by in situ tracer measurements. *Journal of Geophysical Research – Atmospheres,* **104**(D21), 26,565–26,580.

Reid, S. J., A. F. Tuck, and G. Kiladis, 2000: On the changing abundance of ozone minima at northern midlatitudes. *Journal of Geophysical Research – Atmospheres,* **105**(D10), 12,169–12,180.

Rex, M., R. J. Salawitch, P. von der Gathen, N. Harris, M. P. Chipperfield, and B. Naujokat, 2004: Arctic ozone loss and climate change. *Geophysical Research Letters,* **31**(4), L04116, doi:10.1029/2003GL018844.

Rind, D., R. Suozzo, N. K. Balachandran, and M. J. Prather, 1990: Climate change and the middle atmosphere. 1. The doubled CO_2 climate. *Journal of the Atmospheric Sciences,* **47**(4), 475–494.

Rind, D., J. Lerner, J. Perlwitz, C. McLinden, and M. Prather, 2002: Sensitivity of tracer transports and stratospheric ozone to sea surface temperature patterns in the doubled CO_2 climate. *Journal of Geophysical Research – Atmospheres,* **107**(D24), 4800, doi:10.1029/2002JD002483.

Rinsland, C. P., E. Mahieu, R. Zander, N. B. Jones, M. P. Chipperfield, A. Goldman, J. Anderson, J. M. Russell III, P. Demoulin, J. Notholt, G. C. Toon, J.-F. Blavier, B. Sen, R. Sussmann, S. W. Wood, A. Meier, D. W. T. Griffith, L. S. Chiou, F. J. Murcray, T. M. Stephen, F. Hase, S. Mikuteit, A. Schulz, and T. Blumenstock, 2003: Long-term trends of inorganic chlorine from ground-based infrared solar spectra: Past increases and evidence for stabilization. *Journal of Geophysical Research – Atmospheres,* **108**(D8), 4252, doi:10.1029/2002JD003001.

Röckmann, T., J.-U. Grooß, and R. Müller, 2004: The impact of anthropogenic chlorine emissions, stratospheric ozone change and chemical feedbacks on stratospheric water. *Atmospheric Chemistry and Physics,* **4**, 693–699.

Rogers, J. C. and M. J. McHugh, 2002: On the separability of the North Atlantic Oscillation and Arctic Oscillation. *Climate Dynamics,* **19**, 599–608.

Roscoe, H. K., A. E. Jones, and A. M. Lee, 1997: Midwinter start to Antarctic ozone depletion: Evidence from observations and models. *Science,* **278**(5335), 93–96.

Rosenlof, K. H., S. J. Oltmans, D. Kley, J. M. Russell, E. W. Chiou, W. P. Chu, D. G. Johnson, K. K. Kelly, H. A. Michelsen, G. E.

Nedoluha, E. E. Remsberg, G. C. Toon, and M. P. McCormick, 2001: Stratospheric water vapor increases over the past half-century. *Geophysical Research Letters,* **28**(7), 1195–1198.

Rozanov, E. V., M. E. Schlesinger, N. G. Andronova, F. Yang, S. L. Malyshev, V. A. Zubov, T. A. Egorova, and B. Li, 2002: Climate/chemistry effects of the Pinatubo volcanic eruption simulated by the UIUC stratosphere/troposphere GCM with interactive photochemistry. *Journal of Geophysical Research – Atmospheres,* **107**(D21), 4594, doi:10.1029/2001JD000974.

Salby, M. and P. F. Callaghan, 1993: Fluctuations of total ozone and their relationship to stratospheric air motions. *Journal of Geophysical Research – Atmospheres,* **98**, 2715–2727.

Sander, S. P., R. R. Friedl, A. R. Ravishankara, D. M. Golden, C. E. Kolb, M. J. Kurylo, R. E. Huie, V. L. Orkin, M. J. Molina, G. K. Moortgat, and B. J. Finlayson-Pitts, 2003: Chemical Kinetics and Photochemical Data for Use in Atmospheric Studies. Evaluation Number 14. JPL Publication No. 02-25, National Aeronautics and Space Administration, Jet Propulsion Laboratory, California Institute of Technology, Pasadena, CA, USA. (Available at http://jpldataeval.jpl.nasa.gov/pdf/JPL_02-25_rev02.pdf)

Santer, B. D., M. F. Wehner, T. M. L. Wigley, R. Sausen, G. A. Meehl, K. E. Taylor, C. Ammann, J. Arblaster, W. M. Washington, J. S. Boyle, and W. Bruggemann, 2003a: Contributions of anthropogenic and natural forcing to recent tropopause height changes. *Science,* **301**(5632), 479–483.

Santer, B. D., R. Sausen, T. M. L. Wigley, J. S. Boyle, K. Achuta-Rao, C. Doutriaux, J. E. Hansen, G. A. Meehl, E. Roeckner, R. Ruedy, G. Schmidt, and K. E. Taylor, 2003b: Behavior of tropopause height and atmospheric temperature in models, reanalyses, and observations: Decadal changes. *Journal of Geophysical Research – Atmospheres,* **108**(D1), 4002, doi:10.1029/2002JD002258.

Scaife, A. A., J. Austin, N. Butchart, S. Pawson, M. Keil, J. Nash, and I. N. James, 2000: Seasonal and interannual variability of the stratosphere diagnosed from UKMO TOVS analyses. *Quarterly Journal of the Royal Meteorological Society,* **126**(568), 2585–2604.

Schnadt, C. and M. Dameris, 2003: Relationship between North Atlantic Oscillation changes and stratospheric ozone recovery in the Northern Hemisphere in a chemistry-climate model. *Geophysical Research Letters,* **30**(9), 1487, doi:10.1029/2003GL017006.

Seidel, D. J., R. J. Ross, J. K. Angell and G. C. Reid, 2001: Climatological characteristics of the tropical tropopause as revealed by radiosondes. *Journal of Geophysical Research – Atmospheres,* **106**, 7857–7878.

Seidel, D. J., J. K. Angell, M. Free, J. Christy, R. Spencer, S. A. Klein, J. R. Lanzante, C. Mears, M. Schabel, F. Wentz, D. Parker, P. Thorne, and A. Sterin, 2004: Uncertainties in signals of large-scale climate variations in radiosonde and satellite upper-air temperature data sets. *Journal of Climate,* **17**(11), 2225–2240.

Sexton, D. M. H., H. Grubb, K. P. Shine, and C. K. Folland, 2003: Design and analysis of climate model experiments for the efficient estimation of anthropogenic signals. *Journal of Climate,* **16**(9), 1320–1336.

Sherwood, S. C., and A. E. Dessler, 2001: A model for transport across the tropical tropopause. *Journal of the Atmospheric Sciences,* **58**, 765–779.

Shindell, D. T., 2001: Climate and ozone response to increased stratospheric water vapor. *Geophysical Research Letters,* **28**(8), 1551–1554.

Shindell, D. T. and G. A. Schmidt, 2004: Southern Hemisphere climate response to ozone changes and greenhouse gas increases. *Geophysical Research Letters,* **31**, L18209, doi:10.1029/2004GL020724.

Shindell, D. T., D. Rind, and P. Lonergan, 1998: Increased polar stratospheric ozone losses and delayed eventual recovery owing to increasing greenhouse-gas concentrations. *Nature,* **392**(6676), 589–592.

Shindell, D. T., G. A. Schmidt, R. L. Miller, and D. Rind, 2001a: Northern Hemisphere winter climate response to greenhouse gas, ozone, solar, and volcanic forcing. *Journal of Geophysical Research – Atmospheres,* **106**(D7), 7193–7210.

Shindell, D. T., G. A. Schmidt, M. E. Mann, D. Rind, and A. Waple, 2001b: Solar forcing of regional climate during the Maunder Minimum. *Science,* **294**, 2149–2152.

Shine, K. P., M. S. Bourqui, P. M. D. Forster, S. H. E. Hare, U. Langematz, P. Braesicke, V. Grewe, M. Ponater, C. Schnadt, C. A. Smith, J. D. Haigh, J. Austin, N. Butchart, D. T. Shindell, W. J. Randel, T. Nagashima, R. W. Portmann, S. Solomon, D. J. Seidel, J. Lanzante, S. Klein, V. Ramaswamy, and M. D. Schwarzkopf, 2003: A comparison of model-simulated trends in stratospheric temperatures. *Quarterly Journal of the Royal Meteorological Society,* **129**(590), 1565–1588.

Sigmond, M., P. C. Siegmund, E. Manzini, and H. Kelder, 2004: A simulation of the separate climate effects of middle atmospheric and tropospheric CO_2 doubling. *Journal of Climate,* **17**(12), 2352–2367.

Simmonds, P. G., R. G. Derwent, A. J. Manning, P. J. Fraser, P. B. Krummel, S. O'Doherty, R. G. Prinn, D. M. Cunnold, B. R. Miller, H. J. Wang, D. B. Ryall, L. W. Porter, R. F. Weiss, and P. K. Salameh, 2004: AGAGE observations of methyl bromide and methyl chloride at Mace Head, Ireland, and Cape Grim, Tasmania, 1998–2001. *Journal of Atmospheric Chemistry,* **47**, 243–269.

Siskind, D. E., L. Froidevaux, J. M. Russell, and J. Lean, 1998: Implications of upper stratospheric trace constituent changes observed by HALOE for O_3 and ClO from 1992 to 1995. *Geophysical Research Letters,* **25**(18), 3513–3516.

Siskind, D. E., G. E. Nedoluha, M. E. Summers, and J. M. Russell, 2002: A search for an anticorrelation between H_2O and O_3 in the lower mesosphere. *Journal of Geophysical Research – Atmospheres,* **107**(D20), 4435, doi:10.1029/2001JD001276.

Solomon, S. and J. S. Daniel, 1996: Impact of the Montreal Protocol and its amendments on the rate of change of global radiative forcing. *Climatic Change,* **32**(1), 7–17.

Solomon, S., M. Mills, L. E. Height, W. H. Pollock, and A. F. Tuck, 1992: On the evaluation of ozone depletion potentials. *Journal of Geophysical Research – Atmospheres,* **97**, 825–842.

Solomon, S., J. P. Smith, R. W. Sanders, L. Perliski, H. L. Miller, G. H. Mount, J. G. Keys, and A. L. Schmeltekopf, 1993: Visible and near-ultraviolet spectroscopy at McMurdo Station, Antarctica. 8. Observations of nighttime NO_2 and NO_3 from April to October 1991. *Journal of Geophysical Research – Atmospheres,* **98**(D1), 993–1000.

SPARC, 2000: *SPARC Assessment of Upper Tropospheric and Stratospheric Water Vapour* [Kley, D., J. M. Russell III, and C. Phillips (eds.)]. WMO-TD No. 1043, WCRP Series Report No. 113, SPARC Report No. 2, Stratospheric Processes and Their Role in Climate (SPARC). (Available at http://www.atmosp.physics.utoronto.ca/SPARC/index.html)

Sprenger, M. and H. Wernli, 2003: A northern hemispheric climatology of cross-tropopause exchange for the ERA15 time period (1979–1993). *Journal of Geophysical Research – Atmospheres,* **108**(D12), 8521, doi:10.1029/2002JD002636.

Steinbrecht, W., H. Claude, U. Kohler, and P. Winkler, 2001: Interannual changes of total ozone and Northern Hemisphere circulation patterns. *Geophysical Research Letters,* **26**, 1191–1194.

Stenchikov, G., A. Robock, V. Ramaswamy, M. D. Schwarzkopf, K. Hamilton, and S. Ramachandran, 2002: Arctic Oscillation response to the 1991 Mount Pinatubo eruption: Effects of volcanic aerosols and ozone depletion. *Journal of Geophysical Research – Atmospheres,* **107**(D24), 4803, doi:10.1029/2002JD002090.

Stohl, A., P. Bonasoni, P. Cristofanelli, W. Collins, J. Feichter, A. Frank, C. Forster, E. Gerasopoulos, H. Gaggeler, P. James, T. Kentarchos, H. Kromp-Kolb, B. Kruger, C. Land, J. Meloen, A. Papayannis, A. Priller, P. Seibert, M. Sprenger, G. J. Roelofs, H. E. Scheel, C. Schnabel, P. Siegmund, L. Tobler, T. Trickl, H. Wernli, V. Wirth, P. Zanis, and C. Zerefos, 2003: Stratosphere-troposphere exchange: A review, and what we have learned from STACCATO. *Journal of Geophysical Research – Atmospheres,* **108**(D12), 8516, doi:10.1029/2002JD002490.

Stolarski, R. J., R. D. McPeters, and P. A. Newman, 2005: The ozone hole of 2002 as measured by TOMS. *Journal of the Atmospheric Sciences,* **62**(3), 716–720.

Stuber, N., M. Ponater, and R. Sausen, 2001: Is the climate sensitivity to ozone perturbations enhanced by stratospheric water vapor feedback? *Geophysical Research Letters,* **28**(15), 2887–2890.

Sturrock, G. A., D. M. Etheridge, C. M. Trudinger, P. J. Fraser, and A. M. Smith, 2002: Atmospheric histories of halocarbons from analysis of Antarctic firn air: Major Montreal Protocol species. *Journal of Geophysical Research – Atmospheres,* **107**(D24), 4765, doi:10.1029/2002JD002548.

Tabazadeh, A., M. L. Santee, M. Y. Danilin, H. C. Pumphrey, P. A. Newman, P. J. Hamill, and J. L. Mergenthaler, 2000: Quantifying denitrification and its effect on ozone recovery. *Science,* **288**(5470), 1407–1411.

Teitelbaum, H., M. Moustaoui, and M. Fromm, 2001: Exploring polar stratospheric cloud and ozone minihole formation: The primary importance of synoptic-scale flow perturbations. *Journal of Geophysical Research – Atmospheres,* **106**(D22), 28,173–28,188.

Thompson, D. W. J. and S. Solomon, 2002: Interpretation of recent Southern Hemisphere climate change. *Science,* **296**(5569), 895–899.

Thompson, D. W. J. and J. M. Wallace, 1998: The Arctic Oscillation signature in the wintertime geopotential height and temperature fields. *Geophysical Research Letters,* **25**(9), 1297–1300.

Thompson, D. W. J. and J. M. Wallace, 2000: Annular modes in the extratropical circulation, part I: Month-to-month variability. *Journal of Climate,* **13**, 1000–1016.

Thompson, D. W. J., J. M. Wallace, and G. C. Hegerl, 2000: Annular modes in the extratropical circulation. Part II: Trends. *Journal of Climate,* **13**(5), 1018–1036.

Thuburn, J. and G. C. Craig, 2002: On the temperature structure of the tropical substratosphere. *Journal of Geophysical Research – Atmospheres,* **107**(D1–D2), 4017.

Waibel, A. E., T. Peter, K. S. Carslaw, H. Oelhaf, G. Wetzel, P. J. Crutzen, U. Poschl, A. Tsias, E. Reimer, and H. Fischer, 1999: Arctic ozone loss due to denitrification. *Science,* **283**(5410), 2064–2069.

Walker, G. T. and E. W. Bliss, 1932: World weather V. *Memoirs of the Royal Meteorological Society,* **4,** 53–84.

Walker, S. J., R. F. Weiss, and P. K. Salameh, 2000: Reconstructed histories of the annual mean atmospheric mole fractions for the halocarbons CFC-11, CFC-12, CFC-113, and carbon tetrachloride. *Journal of Geophysical Research – Oceans,* **105**(C6), 14,285–14,296.

Wallace, J. M., 2000: North Atlantic Oscillation/annular mode: Two paradigms – one phenomenon. *Quarterly Journal of the Royal Meteorological Society,* **126**(564), 791–805.

Wang, P.-H., P. Minnis, M. P. McCormick, G. S. Kent, and K. M. Skeens, 1996: A 6-year climatology of cloud occurrence frequency from Stratospheric Aerosol and Gas Experiment II observations (1985–1990). *Journal of Geophysical Research – Atmospheres,* **101**(D23), 29,407–29,429.

Wang, W.-C., M. P. Dudek, X.-Z. Liang, and J. T. Kiehl, 1991: Inadequacy of effective CO_2 as a proxy in simulating the greenhouse effect of the other radiatively active gases. *Nature,* **350,** 573–577.

Wang, W.-C., M. P. Dudek, and X.-Z. Liang, 1992: Inadequacy of effective CO_2 as a proxy in assessing the regional climate change due to other radiatively active gases. *Geophysical Research Letters,* **19,** 1375–1378.

Waugh, D. W., W. J. Randel, S. Pawson, P. A. Newman, and E. R. Nash, 1999: Persistence of the lower stratospheric polar vortices. *Journal of Geophysical Research – Atmospheres,* **104**(D22), 27,191–27,201.

Weber, M., S. Dhomse, F. Wittrock, A. Richter, B. M. Sinnhuber, and J. P. Burrows, 2003: Dynamical control of NH and SH winter/spring total ozone from GOME observations in 1995–2002. *Geophysical Research Letters,* **30**(11), 1583.

Winker, D. M. and C. R. Trepte, 1998: Laminar cirrus observed near the tropical tropopause by LITE. *Geophysical Research Letters,* **25**(17), 3351–3354.

WMO, 1986: *Atmospheric Ozone: 1985.* Global Ozone Research and Monitoring Project – Report No. 16, World Meteorological Organization (WMO), Geneva.

WMO, 1995: *Scientific Assessment of Ozone Depletion: 1994.* Global Ozone Research and Monitoring Project – Report No. 37, World Meteorological Organization (WMO), Geneva.

WMO, 1999: *Scientific Assessment of Ozone Depletion: 1998.* Global Ozone Research and Monitoring Project – Report No. 44, World Meteorological Organization (WMO), Geneva.

WMO, 2003: *Scientific Assessment of Ozone Depletion: 2002.* Global Ozone Research and Monitoring Project – Report No. 47, World Meteorological Organization, Geneva, 498 pp.

Yokouchi, Y., D. Toom-Sauntry, K. Yazawa, T. Inagaki, and T. Tamaru, 2002: Recent decline of methyl bromide in the troposphere. *Atmospheric Environment,* **36**(32), 4985–4989.

Zeng, G. and J. A. Pyle, 2003: Changes in tropospheric ozone between 2000 and 2100 modelled in a chemistry-climate model. *Geophysical Research Letters,* **30**(7), 1392, doi:10.1029/2002GL016708.

Zhou, X. L., M. A. Geller, and M. H. Zhang, 2001: Cooling trend of the tropical cold point tropopause temperatures and its implications. *Journal of Geophysical Research – Atmospheres,* **106**(D2), 1511–1522.

2

Chemical and Radiative Effects of Halocarbons and Their Replacement Compounds

Coordinating Lead Authors
Guus J.M. Velders (The Netherlands), Sasha Madronich (USA)

Lead Authors
Cathy Clerbaux (France), Richard Derwent (UK), Michel Grutter (Mexico), Didier Hauglustaine (France), Selahattin Incecik (Turkey), Malcolm Ko (USA), Jean-Marie Libre (France), Ole John Nielsen (Denmark), Frode Stordal (Norway), Tong Zhu (China)

Contributing Authors
Donald Blake (USA), Derek Cunnold (USA), John Daniel (USA), Piers Forster (UK), Paul Fraser (Australia), Paul Krummel (Australia), Alistair Manning (UK), Steve Montzka (USA), Gunnar Myhre (Norway), Simon O'Doherty (UK), David Oram (UK), Michael Prather (USA), Ronald Prinn (USA), Stefan Reimann (Switzerland), Peter Simmonds (UK), Tim Wallington (USA), Ray Weiss (USA)

Review Editors
Ivar S.A. Isaksen (Norway), Bubu P. Jallow (Gambia)

Contents

EXECUTIVE SUMMARY

- Human-made halocarbons, including chlorofluorocarbons (CFCs), perfluorocarbons (PFCs), hydrofluorocarbons (HFCs) and hydrochlorofluorocarbons (HCFCs), are effective absorbers of infrared radiation, so that even small amounts of these gases contribute to the radiative forcing of the climate system. Some of these gases and their replacements also contribute to stratospheric ozone depletion and to the deterioration of local air quality.

Concentrations and radiative forcing of fluorinated gases and replacement chemicals

- HCFCs of industrial importance have lifetimes in the range 1.3–20 years. Their global-mean tropospheric concentrations in 2003, expressed as molar mixing ratios (parts per trillion, ppt, 10^{-12}), were 157 ppt for HCFC-22, 16 ppt for HCFC-141b, and 14 ppt for HCFC-142b.
- HFCs of industrial importance have lifetimes in the range 1.4–270 years. In 2003 their observed atmospheric concentrations were 17.5 ppt for HFC-23, 2.7 ppt for HFC-125, 26 ppt for HFC-134a, and 2.6 ppt for HFC-152a.
- The observed atmospheric concentrations of these HCFCs and HFCs can be explained by anthropogenic emissions, within the range of uncertainties in calibration and emission estimates.
- PFCs and SF_6 have lifetimes in the range 1000–50,000 years and make an essentially permanent contribution to radiative forcing. Concentrations in 2003 were 76 ppt for CF_4, 2.9 ppt for C_2F_6, and 5 ppt for SF_6. Both anthropogenic and natural sources of CF_4 are needed to explain its observed atmospheric abundance.
- Global and regional emissions of CFCs, HCFCs and HFCs have been derived from observed concentrations and can be used to check emission inventory estimates. Global emissions of HCFC-22 have risen steadily over the period 1975–2000, whereas those of HCFC-141b, HCFC-142b and HFC-134a started to increase quickly in the early 1990s. In Europe, sharp increases in emissions occurred for HFC-134a over 1995–1998 and for HFC-152a over 1996–2000, with some levelling off of both through 2003.
- Volatile organic compounds (VOCs) and ammonia (NH_3), which are considered as replacement species for refrigerants or foam blowing agents, have lifetimes of several months or less, hence their distributions are spatially and temporally variable. It is therefore difficult to quantify their climate impacts with single globally averaged numbers. The direct and indirect radiative forcings associated with their use as substitutes are very likely to have a negligible effect on global climate because of their small contribution to existing VOC and NH_3 abundances in the atmosphere.
- The lifetimes of many halocarbons depend on the concentration of the hydroxyl radical (OH), which might change in the 21st century as a result of changes in emissions of carbon monoxide, natural and anthropogenic VOCs, and nitrogen oxides, and as a result of climate change and changes in tropospheric ultraviolet radiation. The sign and magnitude of the future change in OH concentrations are uncertain, and range from –18 to +5% for the year 2100, depending on the emission scenario. For the same emissions of halocarbons, a decrease in future OH would lead to higher concentrations of these halocarbons and hence would increase their radiative forcing, whereas an increase in future OH would lead to lower concentrations and smaller radiative forcing.
- The ODSs have contributed 0.32 W m^{-2} (13%) to the direct radiative forcing of all well-mixed greenhouse gases, relative to pre-industrial times. This contribution is dominated by CFCs. The current (2003) contributions of HFCs and PFCs to the total radiative forcing by long-lived greenhouse gases are 0.0083 W m^{-2} (0.31%) and 0.0038 W m^{-2} (0.15%), respectively.
- Based on the emissions reported in Chapter 11 of this report, the estimated radiative forcing of HFCs in 2015 is in the range of 0.019–0.030 W m^{-2} compared with the range 0.022–0.025 W m^{-2} based on projections in the IPCC Special Report on Emission Scenarios (SRES). Based on SRES projections, the radiative forcing of HFCs and PFCs corresponds to about 6–10% and 2%, respectively, of the total estimated radiative forcing due to CFCs and HCFCs in 2015.
- Projections of emissions over longer time scales become more uncertain because of the growing influence of uncertainties in technological practices and policies. However, based on the range of SRES emission scenarios, the contribution of HFCs to radiative forcing in 2100 could range from 0.1 to 0.25 W m^{-2} and of from PFCs 0.02 to 0.04 W m^{-2}. In comparison, the forcing from carbon dioxide (CO_2) in these scenarios ranges from about 4.0 to 6.7 W m^{-2} in 2100.
- Actions taken under the Montreal Protocol and its Adjustments and Amendments have led to the replacement of CFCs with HCFCs, HFCs, and other substances, and to changing of industrial processes. These actions have begun to reduce atmospheric chlorine loading. Because replacement species generally have lower global warming potentials (GWPs), and because total halocarbon emissions have decreased, the total CO_2-equivalent emission (direct GWP-weighted using a 100-year time horizon) of all halocarbons has also been reduced. The combined CO_2-equivalent emissions of CFCs, HCFCs and HFCs decreased from a peak of about 7.5 ± 0.4 GtCO$_2$-eq yr^{-1} around 1990 to about 2.5 ± 0.2 GtCO$_2$-eq yr^{-1} in the year 2000, which corresponds to about 10% of that year's CO_2 emissions from global fossil-fuel burning.

Banks

- Continuing observations of CFCs and other ODSs in the atmosphere enable improved validation of estimates of the lag between production and emission to the atmosphere, and of the associated banks of these gases. Because new production of ODSs has been greatly reduced, annual changes in their concentrations are increasingly dominated by releases from existing banks. The decrease in the atmospheric concentrations of several gases is placing tighter constraints on their derived emissions and banks. This information provides new insights into the significance of banks and end-of-life options for applications using HCFCs and HFCs as well.

- Current banked halocarbons will make a substantial contribution to future radiative forcing of climate for many decades unless a large proportion of these banks is destroyed. Large portions of the global inventories of CFCs, HCFCs and HFCs currently reside in banks. For example, it is estimated that about 50% of the total global inventory of HFC-134a currently resides in the atmosphere and 50% resides in banks. One top-down estimate based on past reported production and on detailed comparison with atmospheric-concentration changes suggests that if the current banks of CFCs, HCFCs and HFCs were to be released to the atmosphere, their corresponding GWP-weighted emissions (using a 100-year time horizon) would represent about 2.6, 3.4 and 1.2 $GtCO_2$-eq, respectively.

- If all the ODSs in banks in 2004 were not released to the atmosphere, the direct positive radiative forcing could be reduced by about 0.018–0.025 W m^{-2} by 2015. Over the next two decades this positive radiative forcing change is expected to be about 4–5% of that due to carbon dioxide emissions over the same period.

Global warming potentials (GWPs)

- The climate impact of a given mass of a halocarbon emitted to the atmosphere depends on its radiative properties and atmospheric lifetime. The two can be combined to compute the global warming potential (GWP), which is a proxy for the climate effect of a gas relative to the emission of a pulse of an equal mass of CO_2. Multiplying emissions of a gas by its GWP gives the CO_2-equivalent emission of that gas over a given time horizon.

- The GWP value of a gas is subject to change as better estimates of the radiative forcing and the lifetime associated with the gas or with the reference gas (CO_2) become available. The GWP values (over 100 years) recommended in this report (Table 2.6) have been modified, depending on the gas, by –16% to +51%, compared with values reported in IPCC (1996), which were adopted by the United Nations Framework Convention on Climate Change (UNFCCC) for use in national emission inventories. Some of these changes have already been reported in IPCC (2001).

- HFC-23 is a byproduct of the manufacturing of HCFC-22 and has a substantial radiative efficiency. Because of its long lifetime (270 years) its GWP is the largest of the HFCs, at 14,310 for a 100-year time horizon.

Degradation products

- The intermediate degradation products of most long-lived CFCs, HCFCs and HFCs have shorter lifetimes than the source gases, and therefore have lower atmospheric concentrations and smaller radiative forcing. Intermediate products and final products are removed from the atmosphere via deposition and washout processes and may accumulate in oceans, lakes and other reservoirs.

- Trifluoroacetic acid (TFA) is a persistent degradation product of some HFCs and HCFCs, with yields that are known from laboratory studies. TFA is removed from the atmosphere mainly by wet deposition. TFA is toxic to some aquatic life forms at concentrations at concentrations approaching 1 mg L^{-1}. Current observations show that the typical concentration of TFA in the oceans is 0.2 µg L^{-1}, but concentrations as high as 40 µg L^{-1} have been observed in the Dead Sea and Nevada lakes.

- New studies based on measured concentrations of TFA in sea water provide stronger evidence that the cumulative source of TFA from the degradation of HFCs is smaller than natural sources, which, however, have not been fully identified. The available environmental risk assessment and monitoring data indicate that the source of TFA from the degradation of HFCs is not expected to result in environmental concentrations capable of significant ecosystem damage.

Air-quality effects

- HFCs, PFCs, other replacement gases such as organic compounds, and their degradation intermediates are not expected to have a significant effect on global concentrations of OH radicals (which essentially determine the tropospheric self-cleaning capacity), because global OH concentrations are much more strongly influenced by carbon monoxide, methane and natural hydrocarbons.

- The local impact of hydrocarbon and ammonia substitutes for air-conditioning, refrigeration and foam-blowing applications can be estimated by comparing their anticipated emissions to local pollutant emissions from all sources. Small but not negligible impacts could occur near highly localized emission sources. However, even small impacts may be of some concern in urban areas that currently fail to meet air quality standards.

2.1 Introduction

The depletion of ozone in the stratosphere and the radiative forcing of climate change are caused by different processes. Chlorine and bromine released from chlorofluorocarbons (CFCs), halons and other halogen-containing species emitted in the last few decades are primarily responsible for the depletion of the ozone layer in the stratosphere. Anthropogenic emissions of carbon dioxide, methane, nitrous oxide and other greenhouse gases are, together with aerosols, the main contributors to the change in radiative forcing of the climate system over the past 150 years. The possible adverse effects of both global environmental issues are being dealt with in different political arenas. The Vienna Convention and the Montreal Protocol are designed to protect the ozone layer, whereas the United Nations Framework Convention on Climate Change (UNFCCC) and the Kyoto Protocol play the same role for the climate system. However these two environmental issues interact both with respect to the trade-off in emissions of species as technology evolves, and with respect to the chemical and physical processes in the atmosphere. The chemical and physical processes of these interactions and their effects on ozone are discussed in Chapter 1. This chapter focuses on the atmospheric abundance and radiative properties of CFCs, hydrochlorofluorocarbons (HCFCs), hydrofluorocarbons (HFCs), perfluorocarbons (PFCs), and their possible replacements, as well as on their degradation products, and on their effects on global, regional and local air quality.

The total amount, or burden, of a gas in the atmosphere is determined by the competing processes of its emission into and its removal from the atmosphere. Whereas emissions of most of the gases considered here are primarily a result of human activities, their rates of removal are largely determined by complex natural processes that vary both spatially and temporally, and may differ from gas to gas. Rapid removal implies a short atmospheric lifetime, whereas slow removal and therefore a longer lifetime means that a given emission rate will result in a larger atmospheric burden. The concept of the atmospheric lifetime of a gas, which summarizes its destruction patterns in the atmosphere, is therefore central to understanding its atmospheric accumulation. Section 2.2 discusses atmospheric lifetimes, how they are calculated, their influence on the global distribution of halocarbons, and their use in evaluating the effects of various species on the ozone layer and climate change under different future scenarios. Section 2.3 discusses the observed atmospheric abundance of CFCs, HCFCs, HFCs and PFCs, as well as emissions derived from these observations. Section 2.4 discusses the decomposition products (e.g., trifluoroacetic acid, TFA) of these species and their possible effects on ecosystems.

Halocarbons, as well as carbon dioxide and water vapour, are greenhouse gases; that is, they allow incoming solar shortwave radiation to reach the Earth's surface, while absorbing outgoing longwave radiation. The radiative forcing of a gas quantifies its ability to perturb the Earth's radiative energy budget, and can be either positive (warming) or negative (cooling). The global warming potential (GWP) is an index used to compare the climate impact of a pulse emission of a greenhouse gas, over time, relative to the same mass emission of carbon dioxide. Radiative forcings and GWPs are discussed in Section 2.5. The concentrations of emitted species and their degradation products determine to a large extent their atmospheric impacts. Some of these concentrations are currently low but might increase in the future to levels that could cause environmental impacts.

Some of the species being considered as replacements for ozone-depleting substances (ODSs), in particular hydrocarbons (HCs) and ammonia (NH_3), are known to cause deterioration of tropospheric air quality in urban and industrialized continental areas, are involved in regulation of acidity and precipitation (ammonia), and could perturb the general self-cleaning (oxidizing) capacity of the global troposphere. These issues are considered in Section 2.6, where potential increments in reactivity from the replacement compounds are compared with the reactivity of current emissions from other sources.

2.2 Atmospheric lifetimes and removal processes

2.2.1 Calculation of atmospheric lifetimes of halocarbons and replacement compounds

To evaluate the environmental impact of a given gas molecule, one needs to know, first, how long it remains in the atmosphere. CFCs, HCFCs, HFCs, PFCs, hydrofluoroethers (HFEs), HCs and other replacement species for ODSs are chemically transformed or physically removed from the atmosphere after their release. Known sinks for halocarbons and replacement gases include photolysis; reaction with the hydroxyl radical (OH), the electronically excited atomic oxygen ($O(^1D)$) and atomic chlorine (Cl); uptake in oceanic surface waters through dissolution; chemical and biological degradation processes; biological degradation in soils; and possibly surface reactions on minerals. For other trace gases, such as NH_3, uptake by aerosols, cloud removal, aqueous-phase chemistry, and surface deposition also provide important sink processes. A schematic of the various removal processes affecting the atmospheric concentration of the considered species is provided in Figure 2.1.

Individual removal processes have various impacts on different gases in different regions of the atmosphere. The dominant process (the one with the fastest removal rate) controls the local loss of a gas. For example, reaction with OH is the dominant removal process for many hydrogenated halocarbons in both the troposphere and stratosphere. For those same gases, reactions with $O(^1D)$ and Cl play a large role only in the stratosphere. The determination of the spatial distribution of the sink strength of a trace gas is of interest because it is relevant to both the calculation of the atmospheric lifetime and the environmental impact of the substance. The lifetimes of atmospheric trace gases have been recently re-assessed by the IPCC (2001, Chapter 4) and by the WMO Scientific Assessment of Ozone Depletion (WMO, 2003, Chapter 1). Below is an updated summary of key features and methods used to determine the lifetimes.

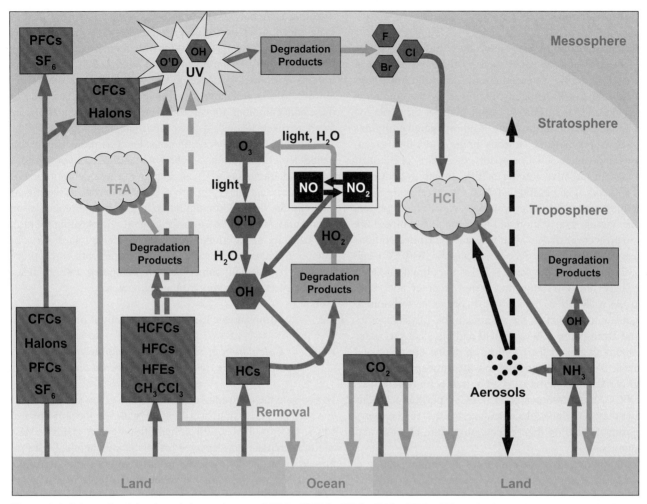

Figure 2.1. Schematic of the degradation pathways of CFCs, HCFCs, HFCs, PFCs, HFEs, HCs and other replacements species in the various atmospheric reservoirs. Colour coding: brown denotes halogens and replacement species emitted and orange their degradation products in the atmosphere; red denote radicals; green arrows denote interaction with soils and the biosphere; blue arrows denote interaction with the hydrological cycle; and black the aerosol phase.

For a given trace gas, each relevant sink process, such as photodissociation or oxidation by OH, contributes to the additive first-order total loss frequency, l, which is variable in space and time. A local lifetime τ_{local} can be defined as the inverse of l evaluated at a point in space (x, y, z) and time (t):

$$\tau_{local} = 1/l(x, y, z, t) \qquad (2.1)$$

The global instantaneous atmospheric lifetime of the gas is obtained by integrating l over the considered atmospheric domain. The integral must be weighted by the distribution of the trace gas on which the sink processes act. If $C(x, y, z, t)$ is the distribution of the trace gas , then the global instantaneous lifetime derived from the budget can be defined as:

$$\tau_{global} = \int C \, dv \, / \int C \, l \, dv \qquad (2.2)$$

where dv is an atmospheric volume element. This expression

can be averaged over a year to determine the global and annually averaged lifetime. The sinks that dominate the atmospheric lifetime are those that have significant strength on a mass or molecule basis, that is, where loss frequency and atmospheric abundance are correlated. Because the total loss frequency l is the sum of the individual sink-process frequencies, τ_{global} can also be expressed in terms of process lifetimes:

$$1/\tau_{global} = 1/\tau_{tropospheric\ OH} + 1/\tau_{photolysis} + 1/\tau_{ocean} \\ + 1/\tau_{scavenging} + 1/\tau_{other\ processes} \qquad (2.3)$$

It is convenient to consider lifetime with respect to individual sink processes limited to specific regions, for example, reaction with OH in the troposphere. However, the associated burden must always be global and include all communicating reservoirs in order for Equation (2.3) to remain valid. In Equation (2.2) the numerator is therefore integrated over the whole atmospheric domain and the denominator is integrated over the domain in which the individual sink process is considered. In the case of $\tau_{\text{tropospheric OH}}$, the convention is that integration is performed over the tropospheric domain. The use of different domains or different definitions for the troposphere can lead to differences of

10% in the calculated value (Lawrence *et al.*, 2001).

A more detailed discussion of lifetimes is given in Box 2.1; the reader is referred to the cited literature for further information.

Box 2.1. Global atmospheric, steady-state and perturbation lifetimes

The conservation equation of a given trace gas in the atmosphere is given by:

$$dB(t)/dt = E(t) - L(B, t) \tag{1}$$

where the time evolution of the burden, $B(t)$ [kg] $= \int C \, dv$, is given by the emissions, $E(t)$ [kg s^{-1}], minus the loss, $L(B, t)$ [kg s^{-1}] $= \int C \, l \, dv$. The emission is an external parameter. The loss of the gas may depend on its concentration as well as that of other gases. The global instantaneous atmospheric lifetime (also called 'burden lifetime' or 'turnover lifetime'), defined in Equation (2.2), is given by the ratio $\tau_{\text{global}} = B(t)/L(t)$. If this quantity is constant in time, then the conservation equation takes on a much simpler form:

$$dB(t)/dt = E(t) - B(t)/\tau_{\text{global}} \tag{2}$$

Equation (2) is for instance valid for a gas whose local chemical lifetime is constant in space and time, such as the noble gas radon (Rn), whose lifetime is fixed by the rate of its radioactive disintegration. In such a case the mean atmospheric lifetime equals the local lifetime: the lifetime that relates source strength to global burden is exactly the decay time of a perturbation. The global atmospheric lifetime characterizes the time to achieve an e-fold (63.2%, see 'lifetime' in glossary) decrease of the global atmospheric burden. Unfortunately, τ_{global} is truly a constant only in very limited circumstances because both C and l change with time. Note that when in steady-state (i.e., with unchanging burden), Equation (1) implies that the source strength is equal to the sink strength. In this case, the steady-state lifetime ($\tau^{SS}_{\text{global}}$) is a scale factor relating constant emissions to the steady-state burden of a gas ($B^{SS} = \tau^{SS}_{\text{global}} E$).

In the case that the loss rate depends on the burden, Equation (2) must be modified. In general, a more rigorous (but still approximate) solution can be achieved by linearizing the problem. The perturbation or pulse decay lifetime (τ_{pert}) is defined simply as:

$$1/\tau_{\text{pert}} = \partial L(B, t)/\partial B \tag{3}$$

and the correct form of the conservation equation is (Prather, 2002):

$$d\Delta B(t)/dt = \Delta E(t) - \Delta B(t)/\tau_{\text{pert}} \tag{4}$$

where $\Delta B(t)$ is the perturbation to the steady-state background $B(t)$ and $\Delta E(t)$ is an emission perturbation (pulse or constant increment). Equation (4) is particularly useful for determining how a one-time pulse emission may decay with time, which is a quantity needed for GWP calculation. For some gases, the perturbation lifetime is considerably different from the steady-state lifetime (i.e., the lifetime that relates source strength to steady-state global burden is not exactly equal to the decay time of a perturbation). For example, if the abundance of CH_4 is increased above its present-day value by a one-time emission, the time it would take for CH_4 to decay back to its background value is longer than its global unperturbed atmospheric lifetime. This occurs because the added CH_4 causes a suppression of OH, which in turn increases the background abundance of CH_4. Such feedbacks cause the time scale of a perturbation (τ_{pert}) to differ from the steady-state lifetime ($\tau^{SS}_{\text{global}}$). In the limit of small perturbations, the relation between the perturbation lifetime of a gas and its global atmospheric lifetime can be derived from the simple budget relationship $\tau_{\text{pert}} = \tau^{SS}_{\text{global}}/(1 - f)$, where $f = d\ln(\tau_{\text{global}}^{SS})/d\ln(B)$ is the sensitivity coefficient (without a feedback on lifetime, $f = 0$ and τ_{pert} is identical to $\tau^{SS}_{\text{global}}$). The feedback of CH_4 on tropospheric OH and its own lifetime has been re-evaluated with contemporary global chemical-transport models (CTMs) as part of an IPCC intercomparison exercise (IPCC, 2001, Chapter 4). The calculated OH feedback, $d\ln(OH)/d\ln(CH_4)$, was consistent among the models, indicating that tropospheric OH abundances decline by 0.32% for every 1% increase in CH_4. When other loss processes for CH_4 (loss to stratosphere, soil uptake) were included, the feedback factor reduced to 0.28 and the ratio $\tau_{\text{pert}}/\tau^{SS}_{\text{global}}$ was 1.4. For a single pulse, this 40% increase in the integrated effect of a CH_4 perturbation does not translate to a 40% larger burden in the perturbation but rather to a lengthening of the duration of the perturbation. If the increased emissions are maintained to steady-state, then the 40% increase does translate to a larger burden (Isaksen and Hov, 1987).

The pulse lifetime is the lifetime that should be used in the GWP calculation (in particular in the case of CH_4). For the long-lived halocarbons, l is approximately constant in time and C can be approximated by the steady-state distribution in order to calculate the steady-state lifetimes reported in Table 2.6. In this case, Equation (2) is approximately valid and the decay lifetime of a pulse is taken to be the same as the steady-state global atmospheric lifetime.

In the case of a constant increment in emissions, at steady-state, Equation (4) shows that steady-state lifetime of the perturbation (τ^{SS}_{pert}) is a scaling factor relating the steady-state perturbation burden (ΔB^{SS}) to the change in emission ($\Delta B^{SS} = \tau^{SS}_{pert}$ ΔE). Furthermore, Prather (1996, 2002) has also shown that the integrated atmospheric abundance following a single pulse emission is equal to the product of the amount emitted and the perturbation steady-state lifetime for that emission pattern. The steady-state lifetime of the perturbation is then a scaling factor relating the emission pulse to the time-integrated burden of that pulse.

Lifetimes can be determined using global tropospheric models by simulating the injection of a pulse of a given gas and watching the decay of this added amount. This decay can be represented by a sum of exponential functions, each with its own decay time. These exponential functions are the chemical modes of the linearized chemistry-transport equations of a global model (Prather, 1996). In the case of a CH_4 addition, the longest-lived mode has an e-fold time of 12 years, which is very close to the steady-state perturbation lifetime of CH_4 and carries most of the added burden. In the case of a carbon monoxide (CO), HCFCs or HCs addition, this mode is also excited, but at a much-reduced amplitude, which depends on the amount of gas added (Prather, 1996; Daniel and Solomon, 1998). The pulse of added CO, HCFCs or HCs causes a build-up of CH_4 while the added burden of the gas persists, by causing the concentration of OH to decrease and thus the lifetime of CH_4 to increase temporarily. After the initial period defined by the photochemical lifetime of the injected trace gas, this built-up CH_4 decays in the same manner as would a direct pulse of CH_4. Thus, changes in the emissions of short-lived gases can generate long-lived perturbations, a result which is also shown in global models (Wild *et al.*, 2001; Derwent *et al.*, 2001).

Figure 2.2 shows the time evolution of the remaining fraction of several constituents in the atmosphere after their pulse emission at time $t = 0$. This figure illustrates the impact of the lifetime of the constituent on the decay of the perturbation. The half-life is the time required to remove 50% of the initial mass injected and the e-fold time is the time required to remove 63.2%. As mentioned earlier, a pulse emission of most gases into the atmosphere decays on a range of time scales until atmospheric transport and chemistry redistribute the gas into its longest-lived decay pattern. Most of this adjustment occurs within 1 to 2 years as the gas mixes throughout the atmosphere. The final e-fold decay occurs on a time scale very close, but not exactly equal, to the steady-state lifetime used to prepare Figure 2.2 and Table 2.6. Note that the removal of CO_2 from the atmosphere cannot be adequately described by a single, simple exponential lifetime (see IPCC, 1994, and IPCC, 2001, for a discussion).

The general applicability of atmospheric lifetimes breaks down for gases and pollutants whose chemical losses or local lifetimes vary in space and time and the average duration of the lifetimes is weeks rather than years or months. For these gases the value of the global atmospheric lifetime is not unique and depends on the location (and season) and the magnitude of the emission (see Box 2.1). The majority of halogen-containing species and ODS-replacement species considered here have atmospheric lifetimes greater than two years, much longer than tropospheric mixing times; hence their lifetimes are not signifi-

cantly altered by the location of sources within the troposphere. When lifetimes are reported for gases in Table 2.6, it is assumed that the gases are uniformly mixed throughout the troposphere. This assumption is less accurate for gases with lifetimes shorter than one year. For such short-lived gases (e.g., HCs, NH_3), reported values for a single global lifetime, ozone depletion potential (ODP) or GWP become inappropriate.

2.2.2   *Oxidation by OH in the troposphere*

The hydroxyl radical (OH) is the primary cleansing agent of the lower atmosphere, and in particular it provides the dominant sink for HCFCs, HFCs, HCs and many chlorinated hydrocarbons. The steady-state lifetimes of these trace gases are determined by the morphology of the species' distribution, the kinetics of the reaction with OH, and the OH distribution. The local abundance of OH is mainly controlled by the local abundances of nitrogen oxides ($NO_x = NO + NO_2$), CO, CH_4 and higher hydrocarbons, O_3, and water vapour, as well as by the intensity of solar ultraviolet radiation. The primary source of tropospheric OH is a pair of reactions that start with the photodissociation of O_3 by solar ultraviolet radiation:

$$O_3 + h\nu \rightarrow O(^1D) + O_2 \qquad [2.1]$$

$$O(^1D) + H_2O \rightarrow OH + OH \qquad [2.2]$$

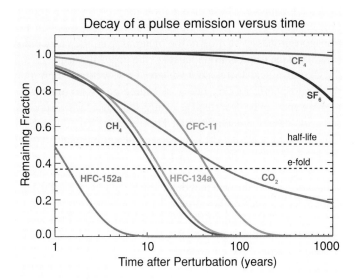

Figure 2.2. Decay of a pulse emission, released into the atmosphere at time $t = 0$, of various gases with atmospheric lifetimes spanning 1.4 years (HFC-152a) to 50,000 years (CF_4). The CO_2 curve is based on an analytical fit to the CO_2 response function (WMO, 1999, Chapter 10).

In polluted regions and in the upper troposphere, photodissociation of other trace gases, such as peroxides, acetone and formaldehyde (Singh *et al.*, 1995; Arnold *et al.*, 1997), may provide the dominant source of OH (e.g., Folkins *et al.*, 1997; Prather and Jacob, 1997; Müller and Brasseur, 1999; Wennberg *et al.*, 1998). OH reacts with many atmospheric trace gases, in most cases as the first and rate-determining step of a reaction chain that leads to more or less complete oxidation of the compound. These chains often lead to formation of an HO_2 radical, which then reacts with O_3 or NO to recycle back to OH. Tropospheric OH and HO_2 are lost through radical-radical reactions that lead to the formation of peroxides, or by reaction with NO_2 to form HNO_3. The sources and sinks of OH involve most of the fast photochemistry of the troposphere.

The global distribution of OH radicals cannot be observed directly because of the difficulty in measuring its small concentrations (of about 10^6 OH molecules per cm^3 on average during daylight in the lower free troposphere) and because of high variability of OH with geographical location, time of day and season. However, indirect estimates of the average OH concentration can be obtained from observations of atmospheric concentrations of trace gases, such methyl chloroform (CH_3CCl_3), that are removed mostly by reaction with OH (with rate constants known from laboratory studies) and whose emission history is relatively well known. The lifetime of CH_3CCl_3 is often used as a reference number to derive the lifetime of other species (see Section 2.2.5) and, by convention, provides a measure of the global OH burden. Observations over a long time (years to decades) can provide estimates of long-term OH trends that would change the lifetimes, ODPs and GWPs of some halocarbons, and increase or decrease their impacts relative to the

current values.

There is an ongoing debate in the literature about the emissions and lifetime of CH_3CCl_3 and hence on the deduced variability of OH concentrations over the last 20 years. Prinn *et al.* (2001) analyzed a 22-year record of global CH_3CCl_3 measurements and emission estimates and suggested that the trend in global OH over that period was $-0.66 \pm 0.57\%$ yr^{-1}. Prinn and Huang (2001) analyzed the record only from 1978 to 1993 and deduced a trend of $+0.3\%$ yr^{-1} using the same emissions as Krol *et al.* (1998). These results through 1993 are essentially consistent with the conclusions of Krol *et al.* (1998, 2001), who used the same measurements and emission record but an independent calculation technique to infer a trend of $+0.46 \pm 0.6\%$ yr^{-1} between 1978 and 1994. However, Prinn *et al.* (2001) inferred a larger interannual and inter-decadal variability in global OH than Krol *et al.* (1998, 2001). Prinn *et al.* (2001) also inferred that OH concentrations in the late 1990s were lower than those in the late 1970s to early 1980s, in agreement with the longer CH_3CCl_3 lifetime reported by Montzka *et al.* (2000) for 1998–1999 relative to the Prinn *et al.* (2001) value for the full-period average. A recent study by Krol and Lelieveld (2003) indicated a larger variation of OH of $+12\%$ during 1978–1990, followed by a decrease slightly larger than 12% in the decade 1991–2000. Over the entire 1978–2000 period, the study found that the overall change was close to zero.

As discussed by Prinn *et al.* (1995, 2001), Krol and Lelieveld (2003) and Krol *et al.* (2003), inferences regarding CH_3CCl_3 lifetimes or trends in OH are sensitive to errors in the absolute magnitude of estimated emissions and to the estimates of other sinks (stratospheric loss, ocean sink). The errors on emissions could be significant and enhanced during the late 1990s because the annual emissions of CH_3CCl_3 were dropping precipitously at the time. These difficulties in the use of CH_3CCl_3 to infer the trend in OH have also been pointed out by Jöckel *et al.* (2003), who suggested the use of other dedicated tracers to estimate the global OH distribution.

The fluctuations in global OH derived from CH_3CCl_3 measurements are in conflict with observed CH_4 growth rates and with model calculations. Karlsdottir and Isaksen (2000) and Dentener *et al.* (2003a,b) present multi-dimensional model results for the period 1980–1996 and 1979–1993, respectively. These studies produce increases in mean global tropospheric OH levels of respectively 0.41% yr^{-1} and 0.25% yr^{-1} over the considered periods. These changes are driven largely by increases in low-latitude emissions of NO_x and CO, in the case of Karlsdottir and Isaksen (2000), and also by changes in stratospheric ozone and meteorological variability, in the case of Dentener *et al.* (2003a,b). The modelling study by Warwick *et al.* (2002) also stressed the importance of meteorological interannual variability on global OH and on the growth rate of CH_4 in the atmosphere. Wang *et al.* (2004) also derived a positive trend in OH of $+0.63\%$ yr^{-1} over the period 1988–1997. Their calculated trend in OH is primarily associated with the negative trend in overhead column ozone. We note however that their forward simulations did not account for the interannual variability of all the variables that affect OH and were conditional on their assumed emissions being correct.

Because of its dependence on CH_4 and other pollutants, the concentration of tropospheric OH is likely to have changed since the pre-industrial era and is expected to change in the future. Pre-industrial OH is likely to have been different than it is today, but because of the counteracting effects of higher concentrations of CO and CH_4 (which decrease OH) and higher concentrations of NO_x and O_3 (which increase OH) there is little consensus on the magnitude or even the sign of this change. Several model studies have suggested that weighted global mean OH has decreased from pre-industrial time to the present day by less than 10% (Berntsen *et al.*, 1997; Wang and Jacob, 1998; Shindell *et al.*, 2001; Lelieveld *et al.*, 2002). Other studies have reported larger decreases in global OH of 16% (Mickley *et al.*, 1999) and 33% (Hauglustaine and Brasseur, 2001). The model study by Lelieveld *et al.* (2002) suggests that during the past century OH concentrations decreased substantially in the marine troposphere by reaction with CH_4 and CO; however, on a global scale, this decrease has been offset by an increase over the continents associated with large emissions of NO_x.

As for future changes in OH, the IPCC (2001, Chapter 4) used scenarios reported in the IPCC Special Report on Emissions Scenarios (SRES, IPCC, 2000) and a comparison of results from 14 models to predict that global OH could decrease by 10% to 18% by 2100 for five emission scenarios, and increase by 5% for one scenario that assumed large decreases in CH_4 and other ozone precursor emissions. Based on a different emission scenario Wang and Prinn (1999) projected a decrease in OH concentrations of $16 \pm 3\%$. In addition to emission changes, future increases in direct and indirect greenhouse gases could also induce changes in OH through direct participation in OH-controlling chemistry, indirectly through stratospheric ozone changes that could change solar ultraviolet in the troposphere, and potentially through climate change effects on biogenic emissions, temperature, humidity and clouds. Changes in tropospheric water could have important chemical repercussions, because the reaction between water vapour and electronically excited oxygen atoms constitutes the major source of tropospheric OH (Reaction [2.1]). So in a warmer and potentially wetter climate, the abundance of OH is expected to increase.

2.2.3   *Removal processes in the stratosphere*

Stratospheric *in situ* sinks for halocarbons and ODS replacements include photolysis and homogeneous gas-phase reactions with OH, Cl and $O(^1D)$. Because about 90% of the burden of well-mixed gases resides in the troposphere, stratospheric removal does not contribute much to the atmospheric lifetimes of gases that are removed efficiently in the troposphere. For most of the HCFCs and HFCs considered in this report, stratospheric removal typically accounts for less than 10% of the total loss. However, stratospheric removal is important for determining the spatial distributions of a source gas and its degradation products

in the stratosphere. These distributions depend on the competition between local photochemical removal processes and the transport processes that carry the material from the entry point (mainly at the tropical tropopause) to the upper stratosphere and the extra-tropical lower stratosphere. Observations show that the stratospheric mixing ratios of source gases decrease with altitude and can be described at steady-state by a local exponential scale height at each latitude. Theoretical calculations show that the local scale height is proportional to the square root of the local lifetime (Ehhalt *et al.*, 1998).

Previous studies (Hansen *et al.*, 1997; Christidis *et al.*, 1997; Jain *et al.*, 2000) showed that depending on the values of the assumed scale height in the stratosphere, the calculated radiative forcing for CFC-11 can differ by as much as 30%. Thus, accurate determination of the GWP of a gas also requires knowing its scale height in the stratosphere. Again, observations will be useful for verifying model-calculated scale heights. Other diagnostics, such as the *age of air* (see glossary), will improve our confidence in the models' ability to simulate the transport in the atmosphere and accurately predict the scale heights appropriate for radiative forcing calculations.

Perfluorinated compounds, such as PFCs, SF_6 and SF_5CF_3, have limited use as ODS replacements is limited (see Chapter 10). The carbon-fluorine bond is remarkably strong and resistant to chemical attack. Atmospheric removal processes for PFCs are extremely slow and these compounds have lifetimes measured in thousands of years. Photolysis at short wavelengths (e.g., Lyman-α at 121.6 nm in the mesosphere) was first suggested to be a possible degradation pathway for CF_4 (Cicerone, 1979). Other possible reactions with $O(^1D)$, H atoms and OH radicals, and combustion in high-temperature systems (e.g., incinerators, engines) were considered (Ravishankara *et al.*, 1993). Reactions with electrons in the mesosphere and with ions were further considered by Morris *et al.* (1995), and a review of the importance of the different processes has been carried out (WMO, 1995). More recently, the degradation processes of SF_5CF_3 were studied (Takahashi *et al.*, 2002). The rate constant for its reaction with OH was found to be less than 10^{-18} cm^3 molecule^{-1} s^{-1} and can be neglected. The main degradation process for CF_4 and C_2F_6 is probably the reaction with O^+ in the mesosphere, but destruction in high-temperature combustion systems remains the principal near-surface removal process. In the case of c-C_4F_8, SF_6 and SF_5CF_3, the main degradation processes are reaction with electrons and photolysis, whereas C_6F_{14} degradation would mainly occur by photolysis at 121.6 nm.

2.2.4   *Other sinks*

For several species other sink processes are also important in determining their global lifetime in the atmosphere. One such process is wet deposition (scavenging by atmospheric hydrometeors including cloud and fog drops, rain and snow), which is an important sink for NH_3. In general, scavenging by large-scale and convective precipitation has the potential to limit the upward transport of gases and aerosols from source regions.

This effect is largely controlled by the solubility of a species in water and its uptake in ice. Crutzen and Lawrence (2000) noted that the solubilities of most of the HCFCs, HFCs and HCs considered in this chapter are too low for significant scavenging to occur. However, because NH_3 is largely removed by liquid-phase scavenging at pH lower than about 7, its lifetime is controlled by uptake on aerosol and cloud drops.

Irreversible deposition is facilitated by the dynamics of tropospheric mixing, which expose tropospheric air to contact with the surface. Irreversible deposition can occur through organisms in ocean surface waters that can both consume and produce halocarbons; chemical degradation of dissolved halocarbons through hydrolysis; and physical dissolution of halocarbons into ocean waters, which does not represent a significant sink for most halocarbons. These processes are highly variable in the ocean, and depend on physical processes of the ocean mixed layer, temperature, productivity, surface saturation and other variables. Determining a net global sink through observation is a difficult task. Yvon-Lewis and Butler (2002) have constructed a high-resolution model of the ocean surface layer, which included its interaction with the atmosphere, and physical, chemical and biological ocean processes. They examined ocean uptake for a range of halocarbons, using known solubilities and chemical and biological degradation rates. Their results show that lifetimes of atmospheric HCFCs and HFCs with respect to hydrolysis in sea water are very long, and range from hundreds to thousands of years. Therefore, they found that for most HFCs and HCFCs the ocean sink was insignificant compared with *in situ* atmospheric sinks. However, both atmospheric CH_3CCl_3 and CCl_4 have shorter (and coincidentally the same) oceanic-loss lifetimes of 94 years, which must be included in determining the total lifetimes of these compounds.

Dry deposition, which is the transfer of trace gases and aerosols from the atmosphere onto surfaces in the absence of precipitation, is also important to consider for some species, particularly NH_3. Dry deposition is governed by the level of turbulence in the atmosphere, the solubility and reactivity of the species, and the nature of the surface itself.

2.2.5 *Halogenated trace gas steady-state lifetimes*

The steady-state lifetimes used in this report and reported in Table 2.6 are taken mainly from the work of WMO (2003, Chapter 1). Chemical reaction coefficients and photodissociation rates used by WMO (2003, Chapter 1) to calculate atmospheric lifetimes for gases destroyed by tropospheric OH are mainly from the latest NASA/JPL evaluations (Sander *et al.*, 2000, 2003). These rate coefficients are sensitive to atmospheric temperature and can be significantly faster near the surface than in the upper troposphere. The global mean abundance of OH cannot be directly measured, but a weighted average of the OH sink for certain synthetic trace gases (whose budgets are well established and whose total atmospheric sinks are essentially controlled by OH) can be derived. The ratio of the atmospheric lifetimes against tropospheric OH loss for a gas is scaled

to that of CH_3CCl_3 by the inverse ratio of their OH reaction rate coefficients at an appropriate scaling temperature of 272 K (Spivakovsky *et al.*, 2000; WMO, 2003, Chapter 1; IPCC, 2001, Chapter 4). Stratospheric losses for all gases considered by WMO (2003, Chapter 1) were taken from published values (WMO, 1999, Chapters 1 and 2; Ko *et al.*, 1999) or calculated as 8% of the tropospheric loss (with a minimum lifetime of 30 years).

WMO (2003, Chapter 1) used Equation (2.3) to determine that the value of τ_{OH} for CH_3CCl_3 is 6.1 years, by using the following values for the other components of the equation: τ_{global} (= 5.0 yr) was inferred from direct observations of CH_3CCl_3 and estimates of emissions; $\tau_{photolysis}$ (= 38–41 yr) arises from loss in the stratosphere and was inferred from observed stratospheric correlations among CH_3CCl_3, CFC-11 and the observed stratospheric age of the air mass; τ_{ocean} (= 94 yr) was derived from a model of the oceanic loss process and has already been used by WMO (2003, Chapter 1); and τ_{others} was taken to be zero.

2.3 Concentrations of CFCs, HCFCs, HFCs and PFCs

2.3.1 *Measurements of surface concentrations and growth rates*

Observations of concentrations of several halocarbons (CFCs, HCFCs, HFCs and PFCs) in surface air have been made by global networks (Atmospheric Lifetime Experiment, ALE; Global Atmospheric Gases Experiment, GAGE; Advanced GAGE, AGAGE; National Oceanic and Atmospheric Administration Climate Monitoring and Diagnostics Laboratory, NOAA/CMDL; and University of California at Irvine, UCI) and as atmospheric columns (Network for Detection of Stratospheric Change, NDSC). The observed concentrations are usually expressed as mole fractions: as ppt, parts in 10^{12} (parts per trillion), or ppb, parts in 10^9 (parts per billion).

Tropospheric concentrations and their growth rates were given by WMO (2003, Chapter 1, Tables 1-1 and 1-12) for a range of halocarbons measured in global networks through 2000. Readers are referred to WMO (2003, Chapter 1) for a more detailed discussion of the observed trends. This chapter provides in Table 2.1 data (with some exceptions) for concentrations through 2003 and for growth rates averaged over the period 2001–2003. Historical data are available on several web sites (e.g., NOAA/CMDL at http://www.cmdl.noaa.gov/hats/index.html and AGAGE at http://cdiac.ornl.gov/ftp/ale_gage_Agage/AGAGE) and are shown here in Figure 2.3.

Many of the species for which data are tabulated here and in WMO (2003, Chapter 1) are regulated in the Montreal Protocol. The concentrations of two of the more abundant CFCs, CFC-11 and CFC-113, peaked around 1996 and have decreased since then. For CFC-12, the concentrations have continued increasing up to 2002, but the rate of increase is now close to zero. The concentrations of the less abundant CFC-114 and CFC-115 (with lifetimes longer than 150 yr) are relatively stable at present. The

Table 2.1. Mole fractions (atmospheric abundance) and growth rates for selected CFCs, halons, HCFCs, HFCs and PFCs. Global mole fractions are for the year 2003 and growth rates averages are for the period 2001–2003, unless mentioned otherwise.

Species	Chemical Formula	Tropospheric Abundance (2003) (ppt)	Growth Rate (2001–2003) (ppt yr^{-1})	Notes[a]
CFCs				
CFC-12	CCl_2F_2	544.4	0.2	AGAGE, *in situ*
		535.4	0.6	CMDL, *in situ*
		535.7	0.8	CMDL, flasks
		538.5	0.6	UCI, flasks
CFC-11	CCl_3F	255.2	−1.9	AGAGE, *in situ*
		257.7	−2.0	CMDL, *in situ*
		256.0	−2.7	CMDL, flasks
		256.5	−2.2	SOGE, Europe, *in situ*
		255.6	−1.9	UCI, flasks
CFC-113	CCl_2FCClF_2	79.5	−0.7	AGAGE, *in situ*
		81.8	−0.7	CMDL, *in situ*
		80.5	−0.6	CMDL, flasks
		79.9	−0.7	SOGE, Europe, *in situ*
		79.5	−0.6	UCI, flasks
CFC-114	$CClF_2CClF_2$	16.4	−0.02	UEA, Cape Grim, flasks
		17.2	−0.1	AGAGE, *in situ*
		17.0	−0.1	SOGE, Europe, *in situ*
CFC-115	$CClF_2CF_3$	8.6	0.07	UEA, Cape Grim, flasks
		8.1	0.16	AGAGE, *in situ*
		8.2	0.03	SOGE, Europe, *in situ*
Halons				
Halon-1211	$CBrClF_2$	4.3	0.04	AGAGE, *in situ*
		4.2	0.09	CMDL, *in situ*
		4.1	0.05	CMDL, flasks
		4.5	0.05	SOGE, Europe, *in situ*
		4.2	0.06	UCI, flasks
		4.6	0.07	UEA, Cape Grim, flasks
Halon-1301	$CBrF_3$	3.1	0.04	AGAGE, *in situ*
		2.6[b]	0.01[b]	CMDL, flasks
		3.2	0.08	SOGE, Europe, *in situ*
		2.4	0.04	UEA, Cape Grim, flasks
Chlorocarbons				
Carbon tetrachloride	CCl_4	93.6	−0.9	AGAGE, *in situ*
		97.1	−1.0	CMDL, *in situ*
		95.5	−0.3	SOGE, Europe, *in situ*
		96.0	−1.0	UCI, flasks
Methyl chloroform	CH_3CCl_3	26.6	−5.8	AGAGE, *in situ*
		27.0	−5.7	CMDL, *in situ*
		26.5	−5.8	CMDL, flasks
		27.5	−5.4	SOGE, Europe, *in situ*
		28.3	−5.8	UCI, flasks
HCFCs				
HCFC-22	$CHClF_2$	156.6	4.5	AGAGE, *in situ*
		158.1	5.4	CMDL, flasks
		156.0[b]	6.9[b]	CMDL, *in situ*
HCFC-141b	CH_3CCl_2F	15.4	1.1	AGAGE, *in situ*
		16.6	1.2	CMDL, flasks
		19.0	1.0	SOGE, Europe, *in situ*
HCFC-142b	CH_3CClF_2	14.7	0.7	AGAGE, *in situ*
		14.0	0.7	CMDL, flasks
		14.1	0.8	CMDL, *in situ*
HCFC-123	$CHCl_2CF_3$	0.03 [96]	0 [96]	UEA, SH, flasks
HCFC-124	$CHClFCF_3$	1.34	0.35	AGAGE, *in situ*
		1.67	0.06	SOGE, Europe, *in situ*

Table 2.1. (continued)

Species	Chemical Formula	Tropospheric Abundance (2003) (ppt)	Growth Rate (2001–2003) (ppt yr^{-1})	Notes[a]
HFCs				
HFC-23	CHF_3	17.5	0.58	UEA, Cape Grim, flasks
HFC-125	CHF_2CF_3	2.7	0.46	AGAGE, *in situ*
		3.2	0.56	SOGE, Europe, *in situ*
		2.6	0.43	UEA, Cape Grim, flasks
HFC-134a	CH_2FCF_3	25.7	3.8	AGAGE, *in situ*
		25.5	4.1	CMDL, flasks
		30.6	4.3	SOGE, Europe, *in situ*
HFC-143a	CH_3CF_3	3.3	0.50	UEA, Cape Grim, flasks
HFC-152a	CH_3CHF_2	2.6	0.34	AGAGE, *in situ*
		4.1	0.60	SOGE, Europe, *in situ*
PFCs				
PFC-14	CF_4	76 [98]		MPAE, NH, flasks
PFC-116	C_2F_6	2.9	0.10	UEA, Cape Grim, flasks
PFC-218	C_3F_8	0.2 [97]		Culbertson *et al.* (2004)
		0.22	0.02	UEA, Cape Grim, flasks
Fluorinated species				
SF_6		3.9 [98]		UH, SH, flasks
		5.1	0.2	UEA, Cape Grim, flasks
		5.2	0.23	CMDL, flasks
		5.2	0.21	CMDL, *in situ*
SF_5CF_3		0.15	0.006	UEA, Cape Grim, flasks

[a] Data sources:
- SH stands for Southern Hemisphere, and NH stands for Northern Hemisphere.
- AGAGE: Advanced Global Atmospheric Gases Experiment. Observed data were provided by D. Cunnold, and were processed through a 12-box model (Prinn *et al.*, 2000). Tropospheric abundances for 2003 are 12-month averages. Growth rates for 2001–2003 are from a linear regression through the 36 monthly averages.
- CMDL: National Oceanic and Atmospheric Administration (NOAA)/Climate Monitoring and Diagnostics Laboratory. Global mean data, downloaded from http://cmdl.noaa.gov/. Tropospheric abundances for 2003 are 12-month averages. Growth rates for 2001–2003 are from a linear regression through the 36 monthly averages. Data for 2003 are preliminary and will be subject to recalibration.
- SOGE: System for Observation of Halogenated Greenhouse Gases in Europe. Values based on observations at Mace Head (Ireland), Jungfraujoch (Switzerland) and Ny-Ålesund (Norway) (Stordal *et al.*, 2002; Reimann *et al.*, 2004), using the same calibration and data-analysis system as AGAGE (Prinn *et al.*, 2000).
- UEA: University of East Anglia, UK. (See references in WMO, 2003, Chapter 1, and Oram *et al.*, 1998).
- UCI: University of California at Irvine, USA. Values based on samples at latitudes between 71°N and 47°S (see references in WMO, 2003, Chapter 1).
- MPAE: Max Planck Institute for Aeronomy (now MPS: Max Planck Institute for Solar System Research), Katlenburg-Lindau, Germany (Harnisch *et al.*, 1999).
- UH: University of Heidelberg, Germany (Maiss *et al.*, 1998)
- Culbertson *et al.* (2004), based on samples from Cape Meares, Oregon; Point Barrow, Alaska; and Palmer Station, Antarctica.

[b] For Halon-1301 and HCFC-22, the tropospheric-abundance data from CMDL is given for year 2002 and growth rates for the period 2001–2002.

[98] Data for the year 1998.

[97] Data for the year 1997.

[96] Data for the year 1996.

two most abundant halons, Halon-1211 and Halon-1301, are still increasing but at a reduced rate. The concentration of CCl_4 has declined since 1990 as a consequence of reduced emissions. The atmospheric abundance of CH_3CCl_3 has declined rapidly because of large reductions in its emissions and its relatively short lifetime. The three most abundant HCFCs, HCFC-22, HCFC-141b and HCFC-142b, increased significantly during the 2000s (at a mean rate of between 4 and 8% yr^{-1}), but the rates of increase for HCFC-141b and HCFC-142b have slowed

somewhat from those in the 1990s. The concentration of SF_6 is growing, but at a slightly reduced rate over the last few years, suggesting that its emissions may be slowing.

Uncertainties in the measurement of absolute concentrations are largely associated with calibration procedures and are species dependent. Calibration differences between the different reporting networks are of the order of 1–2% for CFC-11 and CFC-12, and slightly lower for CFC-113; <3% for CH_3CCl_3; 3–4% for CCl_4; 5% or less for HCFC-22, HCFC-141b and

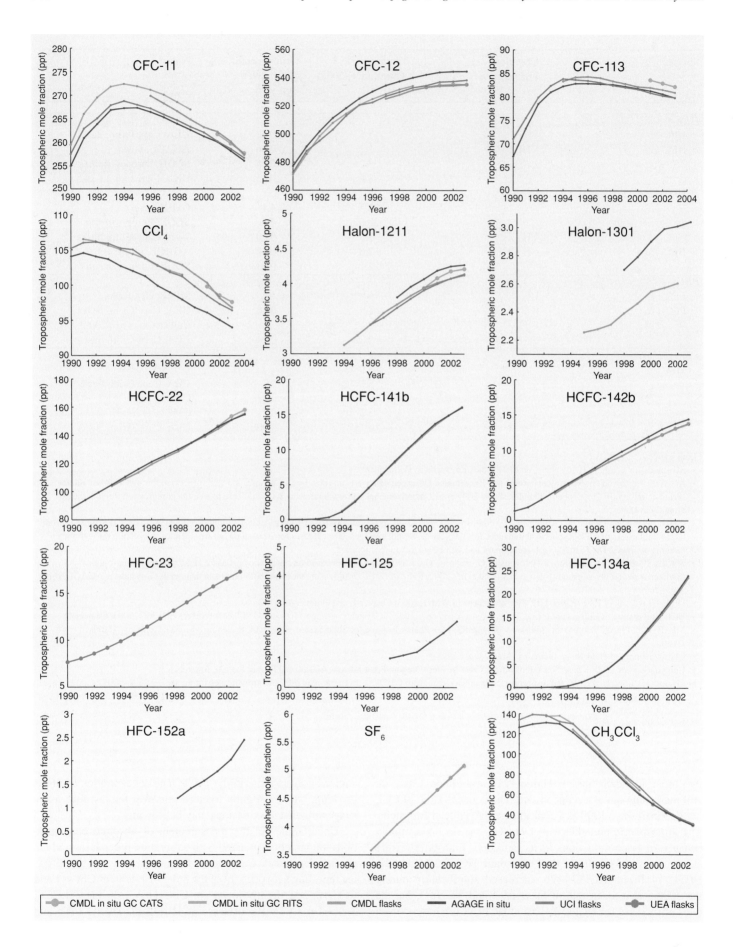

HCFC-142b; 10–15% for Halon-1211; and 25% for Halon-1301 (WMO, 2003, Chapter 1).

HFCs were developed as replacements for CFCs and have a relatively short emission history. Their concentrations are currently significantly lower than those of the most abundant CFCs, but are increasing rapidly. HFC-134a and HFC-23 are the most abundant species, with HFC-134a growing most rapidly at a mean rate of almost 20% yr^{-1} since 2000.

Observations of PFCs are sparse and growth rates are not reported here. CF_4 is by far the most abundant species in this group. About half of the current abundance of CF_4 arises from aluminium production and the electronics industry. Measurements of CF_4 in ice cores have revealed a natural source, which accounts for the other half of current abundances (Harnisch *et al.*, 1996). The level of SF_6 continues to rise, at a rate of about 5% yr^{-1}.

Further, the atmospheric histories of a group of halocarbons have been reconstructed from analyses of air trapped in firn, or snow above glaciers (Butler *et al.*, 1999; Sturges *et al.*, 2001; Sturrock *et al.*, 2002; Trudinger *et al.*, 2002). The results show that concentrations of CFCs, halons and HCFCs at the beginning of the 20th century were generally less than 2% of the current concentrations.

2.3.2 Deriving global emissions from observed concentrations and trends

Emissions of halocarbons can be inferred from observations of their concentrations in the atmosphere when their loss rates are accurately known. Such estimates can be used to validate and verify emission-inventory data produced by industries and reported by parties of international regulatory conventions. The atmospheric emissions of halocarbons estimated in emission inventories are based on compilations of global halocarbon production (AFEAS, 2004), sales into each end-use and the time schedule for atmospheric emission from each end-use (McCulloch *et al.*, 2001, 2003). Uncertainties arise if the global production figures do not cover all manufacturing countries and if there are variations in the behaviour of the end-use categories. For some halocarbons, there are no global emission inventory estimates. Observed concentrations are often the only means of verifying halocarbon emission inventories and the emissions from the 'banks' of each halocarbon.

In principle, estimates of the emissions of halocarbons can be obtained from industrial production figures, provided that sufficient information is available concerning the end-uses of halocarbons and the extent of the banking of the unreleased halocarbon. Industrial production data for the halocarbons have

been compiled in the Alternative Fluorocarbons Environmental Acceptability Study (AFEAS, 2004). AFEAS compiles data for a halocarbon if three or more companies worldwide produced more than 1 kt yr^{-1} each of the halocarbon. These figures have been converted into time histories of global atmospheric release rates for CFC-11 (McCulloch *et al.*, 2001), and for CFC-12, HCFC-22 and HFC-134a (McCulloch *et al.*, 2003). The Emission Database for Global Atmospheric Research (EDGAR) has included emissions of several HFCs and PFCs, and of SF_6 based on various industry estimates (Olivier and Berdowski, 2001; Olivier, 2002; http://arch.rivm.nl/env/int/coredata/edgar/index.html).

Global emissions based on AGAGE and CMDL data have been estimated using a 2-D 12-box model and an optimal linear least-squares Kalman filter. Emissions of HFC-134a, HCFC-141b and HCFC-142b over the period 1992–2000 were estimated by Huang and Prinn (2002) using AGAGE and NOAA-CMDL measurements. They compared them to industry (AFEAS) estimates and concluded that there are significant differences for HCFC-141b and HCFC-142b, but not for HFC-134a. Later O'Doherty *et al.* (2004) used AGAGE data and the same model to infer global emissions of the same species plus HCFC-22. They found a fair agreement between emission estimates based on consumption and on their measurements for HCFC-22, with the former exceeding the latter by about 10% during parts of the 1990s. Emissions for CFC-11, CFC-12, CFC-113, CCl_4 and HCFC-22 have been computed from global observations and compared with industry estimates by Cunnold *et al.* (1997) and Prinn *et al.* (2000), with updates by WMO (2003, Chapter 1). The general conclusion was that emissions estimated in this way are with few exceptions consistent with expectations from the Montreal Protocol.

Estimation of the global emissions (E) from observed trends is based on Equation (2) in Box 2.1 rewritten below as:

$$E(t) = dB(t)/dt + B(t)/\tau^{SS}_{global} \qquad (2.4)$$

where B is the global halocarbon burden and τ^{SS}_{global} is the steady-state lifetime from Table 2.6. The first term on the right-hand side represents the trend in the global burden and the second term represents the decay in the global burden due to atmospheric loss processes. This approach is only valid for long-lived well-mixed gases, and it is subject to uncertainties. First, the estimation of a global burden ($B(t)$ in the second term) from a limited number of surface observation stations is uncertain because it involves variability in both the horizontal and the vertical distributions of the gases. Second, some of the es-

---

Figure 2.3. Global annually averaged tropospheric mole fractions. *In situ* abundances from AGAGE and CMDL (measured with the CATS or RITS gas chromatographs) and flask samples from CMDL are included. The AGAGE abundances are global lower-troposphere averages processed through the AGAGE 12-box model (Prinn *et al.*, 2000; updates provided by D. Cunnold). The CMDL abundances are area-weighted global means for the lower troposphere (from http://www.cmdl.noaa.gov/hats/index.html), estimated as 12-month averages centred around 1 January each year (e.g., 2002.0 = 1 January 2002). CMDL data for 2003 are preliminary and will be subject to recalibration. The UCI data are based on samples at latitudes between 71°N and 47°S. The UEA data (HFC-23) are from Cape Grim, Tasmania, and are represented here by a fit to a series of Legendre polynomials (up to third degree).

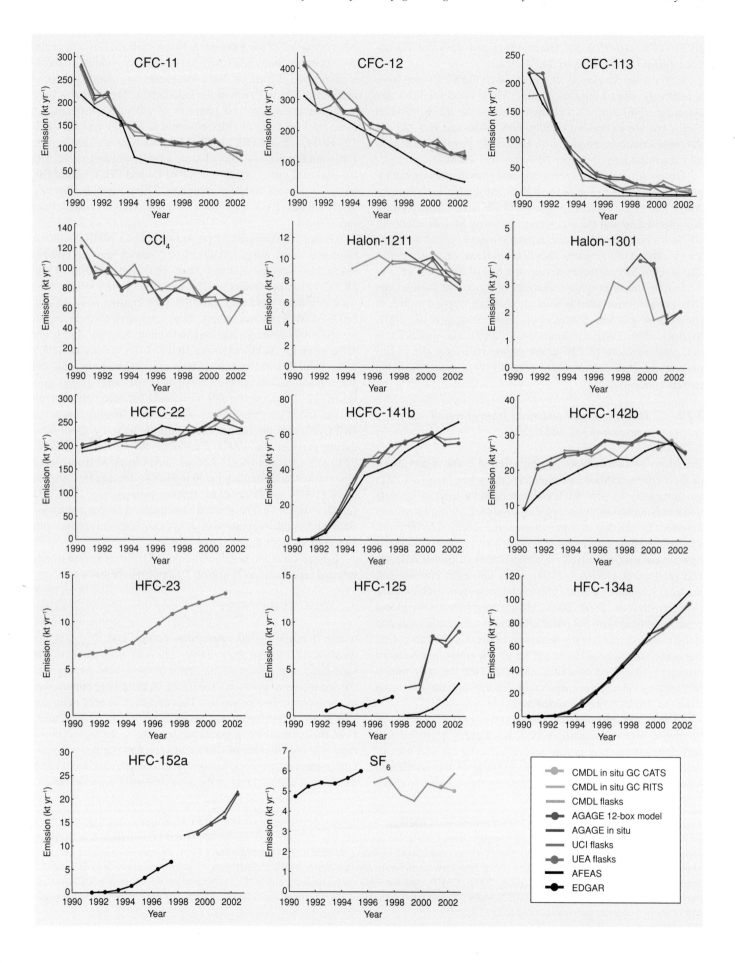

timates presented in this report are based on a 1-box model of the atmosphere, in which case the extrapolation from a limited number of surface observations to the global burden introduces some further uncertainty. Additional uncertainties in the estimated emissions arise from the assumption that the loss can be approximated by $B(t)/\tau^{SS}_{global}$ and from uncertainties in the absolute calibration of the observations.

This method has been employed to infer global emissions in a range of studies. For example, Höhne and Harnisch (2002) found that the calculated emissions of HFC-134a were in good agreement with reported emissions in the 1990–1995 period, whereas the emissions inferred for SF_6 were considerably lower than those reported to the UNFCCC. In a recent paper by Culbertson *et al.* (2004) the 1-box model was used to infer emissions for several CF_3-containing compounds, including CFC-115, Halon-1301, HFC-23, HFC-143a and HFC-134a, using flask samples from Oregon, Alaska and Antarctica. The emissions calculated in their study generally agreed well with emissions from other studies. They also provided first estimates of emissions of some rarer gases, such as CFC-13 and C_3F_8.

In Figures 2.3 and 2.4 we present the concentrations and inferred emissions of several species since 1990. We have used *in situ* data from AGAGE and CMDL, as well as flask samples from, for example, CMDL. The inferred emissions are compared with emissions from AFEAS and EDGAR when those were available. The range in the observation-based emissions in Figure 2.4 includes some but not all of the uncertainties discussed earlier in this section.

In general the results presented in Figures 2.3 and 2.4 confirm the findings of previous studies. The global emission estimates presented in Figure 2.4 show a clear downward trend for most compounds regulated by the Montreal Protocol. Inferred emissions for the HCFCs have been rising strongly since 1990 and those of HCFC-141b and HCFC-142b have levelled off from 2000 onwards. However, emissions of the HFCs have been growing in most cases, most noticeably for HFC-134a, HFC-125 and HFC-152a. In contrast to the other HFCs, emissions of HFC-23 began several decades before the others (Oram *et al.*, 1998) because of its release as a byproduct of HCFC-22 production. Its emissions have increased slightly over the last decade.

Since the mid-1990s, the emissions derived for CFC-11 and CFC-12 (Figure 2.4) have been larger than the best estimate of emissions based on global inventories. The inferred emissions for CFC-12 have decreased strongly, but have gradually become larger than the inventories since the mid-1990s. The inferred emissions of HFC-134a, which have increased strongly since the early 1990s, were in good agreement with the inven-

tories of EDGAR and AFEAS until 2000, after which they were somewhat lower than the AFEAS values.

2.3.3 Deriving regional emissions from observed concentrations

Formally, the application of halocarbon observations to the determination of regional emissions is an inverse problem. Some investigations have treated the problem in this way and used observations to constrain global or regional sources or sinks. In other approaches, the problem has been solved in a direct manner from emissions to mixing ratios, and iteration or extrapolation has been used to work back to the emissions estimated to support and explain the observations. Both approaches have been applied to halocarbons and both have advantages and disadvantages. The focus here is primarily on studies that directly answer questions concerning the regional emissions of halocarbons and their replacements, rather than on documenting the wide range of inverse methods available. The aim is to provide verification or validation of the available emission inventories.

High-frequency *in situ* observations of the anthropogenic halocarbons and greenhouse gases have been made continuously at the Adrigole and Mace Head Atmospheric Observatories on the Atlantic Ocean coast of Ireland since 1987 as part of the ALE/GAGE/AGAGE programmes. To obtain global baseline trends, the European regional pollution events, which affect the Mace Head site about 30% of the time, were removed from the data records (Prinn *et al.*, 2000). Prather (1985) used the polluted data, with its short time-scale variations over 1- to 10-day periods, to determine the relative magnitudes of European continental sources of halocarbons and to quantify previously unrecognized European sources of trace gases such as nitrous oxide (N_2O) and CCl_4.

The European emission source strengths estimated to support the ALE/GAGE/AGAGE observations at Mace Head have been determined initially using a simple long-range transport model (Simmonds *et al.*, 1996) and more recently a Lagrangian dispersion model (Ryall *et al.*, 2001), and are presented in Table 2.2. These studies have demonstrated that although the releases of CFC-11 and CFC-12 have been dramatically reduced, by factors of at least 20, European emissions are still continuing at readily detectable and non-negligible levels. Table 2.2 also provides cogent evidence of the phase-out of the emissions of CCl_4, CH_3CCl_3 and CFC-113 under the provisions of the Montreal Protocol and its Amendments during the 1990s. The contributions of source regions to the observations at Mace Head fall off rapidly to the east and south, with contributions from the eastern Mediterranean being more than three orders of

---

Figure 2.4. Global annual emissions (in kt yr^{-1}) inferred from the mole fractions in Figure 2.3. Emissions are estimated using a 1-box model. In addition, the AGAGE 12-box model has been used to infer emissions from the AGAGE network (Prinn et al., 2000; updates provided by D. Cunnold). A time-dependent scaling for each component, taking into account the vertical distribution in the troposphere and the stratosphere, has been adopted in all the estimates. These scaling factors are taken from the AGAGE 12-box model. Emissions of HFC-23 are based on a 2-D model (Oram *et al.*, 1998, with updates). The inferred emissions are compared with emissions from AFEAS and EDGAR (see references in the text) when those were available.

Table 2.2. European emission source strengths (in kt yr^{-1}) estimated to support ALE/GAGE/AGAGE observations of each halocarbon at Mace Head, Ireland, over the period 1987–2000. A simple long-range transport model (Simmonds *et al.*, 1996) was used for the 1987–1994 period, and a Lagrangian dispersion model (Ryall *et al.*, 2001) was used for the 1995–2000 period.

Year	CFC-11	CFC-12	CFC-113	Methyl Chloroform	Carbon Tetrachloride	Chloroform
1987	215	153	68	207	27	
1988	132	98	43	115	20	
1989	94	87	63	159	26	
1990	67	64	63	153	12	
1991	52	59	44	108	14	
1992	27	32	31	88	9	
1993	20	22	21	75	6	
1994	25	31	20	115	5	
1995	9	15	5	31	3	20
1996	8	14	4	19	4	15
1997	9	13	2	8	3	19
1998	10	13	2	2	3	20
1999	9	11	1	<1	2	19
2000	6	6	1	<1	1	16
2000–2001[a]				>20		
2000–2002[b]				0.3–3.4		

[a] Krol *et al.* (2003). Numbers are based on MINOS and EXPORT aircraft campaigns.
[b] Reimann *et al.* (2004). Numbers are based on observations from Mace Head and Jungfraujoch.

magnitude smaller than contributions from closer sources, as a result of both fewer transport events to Mace Head and greater dilution during transport (Ryall *et al.*, 2001).

Table 2.2 shows that from 1987 to 2000 European emissions inferred from the Mace Head observations of CH$_3$CCl$_3$ appear to have declined from over 200 kt yr^{-1} to under 1 kt yr^{-1}. This sharp decline reflects the influence of its phase-out under the Montreal Protocol and its Amendments, and its use largely as a solvent with minimal 'banking'. In contrast, Krol *et al.* (2003) have estimated European emissions of CH$_3$CCl$_3$ to have been greater than 20 kt yr^{-1} during 2000, based on aircraft sampling during the EXPORT (European Export of Precursors and Ozone by Long-Range Transport) campaign over central Europe. Furthermore, substantial emissions were found during the MINOS (Mediterranean Intensive Oxidant Study) experiment in southeast Europe during 2001. A more detailed reanalysis of the observations at Mace Head and Jungfraujoch, Switzerland, gave emissions in the range 0.3–3.4 kt yr^{-1} for 2000–2002 (Reimann *et al.*, 2004). The reasons for the significant differences between the estimates of European emissions based on long-term observations and those from the EXPORT campaign are unclear. Data from the EXPORT campaign were restricted to four days in the summer of 2000, so they were more prone to the potential influence of regional events of limited duration compared with the long-term observations.

European source strengths of a number of halons and CFC-replacement halocarbons have been determined (Manning *et al.* 2003) using the AGAGE *in situ* high-frequency gas chromatog-

raphy-mass spectrometric GC-MS observations made alongside the AGAGE electron capture detector (ECD) measurements (Simmonds *et al.*, 1998) at Mace Head, and are shown in Table 2.3. The emissions of HCFC-22, HCFC-141b and HCFC-142b appear to have reached a peak during the 1997–2001 period, and have begun to decline rapidly from about 1998–2000 onwards. The corresponding emissions of HFC-134a rose quickly during the late 1990s, with some levelling off through 2003. The emissions of HFC-152a have only recently, from about 1998 onwards, begun to rise and have reached about 2 kt yr^{-1} each by 2002–2003.

In situ GC-MS observations of HFC-134a, HFC-125, HFC-152a and HCFC-141b have been reported from the high-altitude research station in Jungfraujoch, Switzerland, for the 2000–2002 period by Reimann *et al.* (2004). Using mole fraction ratios relative to CO, together with a European CO emission inventory, Reimann *et al.* (2004) estimated European emissions of HFC-134a, HFC-125, HFC-152a and HCFC-141b to be 23.6, 2.2, 0.8 and 9.0 kt yr^{-1}, respectively. These estimated emissions of HFC-134a are twice those from the Mace Head observations in Table 2.3. Reimann *et al.* (2004) explain this difference by the close proximity of the Jungfraujoch station to a potent source region in northern Italy. The estimated European emissions of HFC-125 and of HCFC-141b from Jungfraujoch and Mace Head agree closely. For HFC-152a, the estimated emissions from Jungfraujoch appear to be about half of those from Mace Head.

In addition to the application of long-range transport models, other techniques can be used to determine regional emis-

Table 2.3. European emission source strengths (in kt yr^{-1}) estimated to support the AGAGE GC-MS observations (O'Doherty *et al.*, 2004) of each halon, HCFC and HFC species at Mace Head, Ireland, based on the methodology in Manning *et al.* (2003) and updated to 2003.

Period	Halon-1211	Halon-1301	HCFC-22	HCFC-124	HCFC-141b	HCFC-142b	HFC-125	HFC-134a	HFC-152a
1995–1996					6.9	5.3		3.7	0.5
1996–1997					8.0	6.2		5.8	0.5
1997–1998					12.8	9.3		11.3	0.8
1998–1999	0.7	0.4	18.9	1.1	10.6	6.7	1.5	10.0	1.0
1999–2000	0.7	0.5	22.4	0.5	8.0	5.8	1.3	9.4	1.2
2000–2001	0.6	0.4	33.6	0.6	10.1	4.0	1.6	10.2	2.0
2001–2002	0.6	0.5	21.5	0.6	9.1	3.1	2.1	14.9	2.1
2002–2003	0.8	0.6	13.7	0.61	5.9	1.7	2.0	12.4	2.0

sions from trace-gas observations. Concurrent measurement of the trace gas concentrations with those of ^{222}Rn, a radioactive noble gas with a short half-life that is emitted by soils, has been used to determine continental-scale emission source strengths (Thom *et al.*, 1993). Using the ^{222}Rn method for the year 1996, Biraud *et al.* (2000) determined European emissions of CFC-11 to be 1.8–2.5 kg km^{-2} yr^{-1} and of CFC-12 to be 2.9–4.2 kg km^{-2} yr^{-1}. Schmidt *et al.* (2001) reported a mean continental N$_2$O flux of 42 μg m^{-2} h^{-1} (580 kg km^{-2} yr^{-1}) for Western Europe. These estimates of European CFC emissions compare closely with the estimates from Ryall *et al.* (2001) in Table 2.2, which use the Lagrangian dispersion model method, when they are scaled up with a European surface area of the order of 1.45 × 10^7 km^2.

Estimates of halocarbon emissions for North America have been made by Bakwin *et al.* (1997) and Hurst *et al.* (1998) using the simultaneous high-frequency measurements of perchloroethylene (C$_2$Cl$_4$) and a range of halocarbons made on a 610 m tall tower in North Carolina. The North American halocarbon source strengths for 1995 were 12.9 kt yr^{-1} for CFC-11, 49 kt yr^{-1} for CFC-12, 3.9 kt yr^{-1} for CFC-113, 47.9 kt yr^{-1} for CH$_3$CCl$_3$, and 2.2 kt yr^{-1} for CCl$_4$, as reported by Bakwin *et al.* (1997), and these estimates are significantly lower than the emission inventory estimates of McCulloch *et al.* (2001). Downwards trends are reported for North American emissions of the major anthropogenic halocarbons (Hurst *et al.*, 1998).

Barnes *et al.* (2003a,b) found contrasting results when they used a ratio technique involving CO, C$_2$Cl$_4$ and the halocarbon observations at Harvard Forest, Massachusetts, USA, to estimate emissions. They found that, of all the ODSs emitted from the New York City-Washington D.C. corridor during 1996–1998, only the emissions of CFC-12 and CH$_3$CCl$_3$ showed a detectable decline, which approached a factor of three. In contrast, the regional emissions of CFC-11 appeared to rise slightly, by about 6%, over this period. The emissions of CFC-113 and Halon-1211 did not show any distinguishable trend pattern.

2.4 Decomposition and degradation products from HCFCs, HFCs, HFEs, PFCs and NH$_3$

This section discusses the degradation products of HCFCs, HFCs, HFEs, PFCs and NH$_3$ and their impacts on local and regional air quality, human and ecosystem health, radiative forcing of climate change, and stratospheric ozone depletion. The concentration of degradation products is a key factor in quantifying their impacts. A method to estimate the concentrations of the degradation products from emissions of the parent compounds is discussed in Section 2.4.1.2. Although each chemical has to be evaluated on its own, three general remarks are useful in framing the following discussion. First, fluorine-containing radicals produced in the atmospheric degradation of HFCs, HFEs and PFCs do not participate in catalytic ozone destruction cycles (Ravishankara *et al.*, 1994; Wallington *et al.*, 1995). Hence, the ozone depletion potentials (ODPs) of HFCs, HFEs, and PFCs are essentially zero. In contrast, HCFCs contain chlorine and consequently have non-zero ODPs. Second, the emissions of HCFCs, HFCs and HFEs are very small compared with the mass of hydrocarbons released into the atmosphere, and the atmospheric lifetime of most HCFCs, HFCs, HFEs and PFCs allows their effective dispersal. As a result the concentrations of the degradation products of HCFCs, HFCs, HFEs and PFCs are small and their impact on local and regional air quality (i.e., on tropospheric ozone) is negligible (Hayman and Derwent, 1997). Third, the ultimate atmospheric fate of all HCFCs, HFCs, HFEs and PFCs is oxidation to halogenated carbonyl compounds and HF, which are transferred via dry deposition and wet deposition (rain-out or washout) from the atmosphere to the hydrosphere. The impact of these compounds on terrestrial ecosystems and the hydrosphere (e.g., lakes, oceans) needs to be considered.

2.4.1 Degradation products from HCFCs, HFCs and HFEs

2.4.1.1 Chemical degradation mechanisms

The atmospheric chemistry of HCFCs and HFCs is, in general, well established (see WMO, 1999; WMO, 2003). PFCs degrade extremely slowly and persist for thousands of years (Ravishankara *et al.*, 1993). Hydrofluoroethers (HFEs) have been considered recently as possible CFC replacements. Kinetic and mechanistic data for the tropospheric degradation of a number of HFEs have become available over the past five years.

The environmental impact of HCFCs and HFCs is determined mainly by the tropospheric lifetimes and emission rates of the parent compounds and by the halogenated carbonyl species formed as oxidation products in the atmosphere. The general scheme for the tropospheric degradation of HCFCs and HFCs into halogenated carbonyl compounds is outlined below. Figure 2.5a shows the degradation mechanism for a generic two-carbon HCFC or HFC. Figure 2.5b shows similar data for $C_4F_9OCH_3$ as an example HFE. The oxidation products of HCFCs, HFCs, HFEs and PFCs are not routinely measured in the atmosphere so their concentrations need to be estimated. There are some measurements of the concentrations of the degradation product trifluoroacetic acid (TFA, $CF_3C(O)OH$) (Martin *et al.*, 2003).

Degradation is initiated by the gas-phase reaction with hydroxyl (OH) radicals. This process, which involves either H-atom abstraction from C-H groups, or addition to unsaturated >C=C< groups, is the slowest step in the atmospheric degradation process. There is a large database of rate coefficients for reactions of OH radicals with halogenated compounds (Sander *et al.* 2000, 2003), which provides a means of estimating tropospheric lifetimes for these compounds. Moreover, this database allows rate coefficients to be estimated (typically within a factor of about two) for compounds for which no experimental data are available. Reaction with OH generates a radical, which adds O_2 rapidly (within 10^{-6} s) to give a peroxy radical.

The lifetime of peroxy radicals with respect to their reaction with NO is approximately 1 to 10 minutes. The reaction gives an alkoxy radical, CX_3CXYO, which will either decompose or react with O_2 on a time scale of typically 10^{-3} to 10^{-6} s (Wallington *et al.*, 1994). Decomposition can occur either by C-C bond fission or Cl-atom elimination. Reaction with O_2 is only possible when an α-H atom is available (e.g., in CF_3CFHO). In the case of the alkoxy radicals derived from HFC-32, HFC-125 and HCFC-22, only one reaction pathway is available. Hence, CHF_2O radicals react with O_2 to give $C(O)F_2$, CF_3CF_2O radicals decompose to give CF_3 radicals and $C(O)F_2$, and CF_2ClO radicals eliminate a Cl atom to give $C(O)F_2$. The alkoxy radicals derived from HFC-143a, HCFC-123, HCFC-124, HCFC-141b and HCFC-142b have two or more possible fates, but one loss mechanism dominates in the atmosphere. For HCFC-123 and HCFC-124 the dominant process is elimination of a Cl atom to give $CF_3C(O)Cl$ and $CF_3C(O)F$, respectively. For HFC-143a, HCFC-141b and HCFC-142b reaction with O_2 dominates, giving CF_3CHO, $CFCl_2CHO$ and CF_2ClCHO respectively. The case of HFC-134a is the most complex. Under atmospheric conditions, the alkoxy radical derived from HFC-134a, CF_3CFHO, decomposes (to give CF_3 radicals and $HC(O)F$) and reacts with O_2 (to give $CF_3C(O)F$ and HO_2 radicals) at comparable rates. In the atmosphere 7–20% of the CF_3CFHO radicals formed in the CF_3CFHO_2 + NO reaction react with O_2 to form $CF_3C(O)F$, whereas the remainder decompose to give CF_3 radicals and $HC(O)F$ (Wallington *et al.*, 1996).

The carbonyl products (e.g., $HC(O)F$, $C(O)F_2$, $CF_3C(O)F$) have atmospheric lifetimes measured in days. Incorporation into water droplets followed by hydrolysis plays an important role in the removal of halogenated carbonyl compounds (DeBruyn

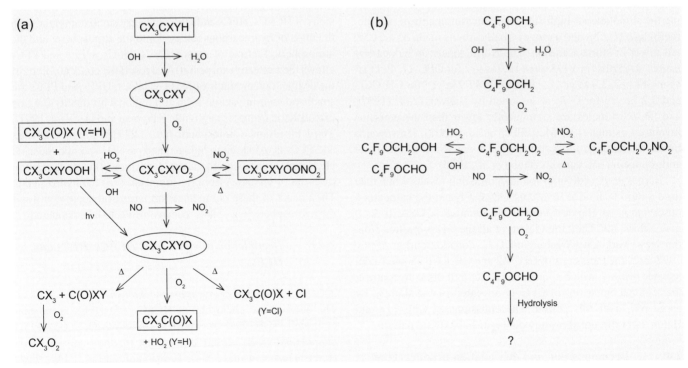

Figure 2.5. (a) Generalized scheme for the atmospheric oxidation of a halogenated organic compound, CX_3CXYH (X, Y = H, Cl or F). Transient radical intermediates are enclosed in ellipses, products with less transitory existence are given in the boxes. (b) Degradation scheme for $C_4F_9OCH_3$.

Table 2.4. Gas-phase atmospheric degradation products of HFCs, HCFCs and HFEs.

Species	Chemical Formula	Degradation Products
HFC-23	CF_3H	COF_2, CF_3OH
HFC-32	CH_2F_2	COF_2
HFC-41	CH_3F	HCOF
HFC-125	CF_3CF_2H	COF_2, CF_3OH
HFC-134a	CF_3CFH_2	HCOF, CF_3OH, COF_2, CF_3COF
HFC-143a	CF_3CH_3	CF_3COH, CF_3OH, COF_2, CO_2
HFC-152a	CF_2HCH_3	COF_2
HFC-161	CH_2FCH_3	HCOF, CH_3COF
HFC-227ca	$CF_3CF_2CHF_2$	COF_2, CF_3OH
HFC-227ea	CF_3CHFCF_3	COF_2, CF_3OH, CF_3COF
HFC-236cb	$CF_3CF_2CH_2F$	HCOF, CF_3OH, COF_2, C_2F_5COF
HFC-236fa	$CF_3CH_2CF_3$	CF_3COCF_3
HFC-245fa	$CF_3CH_2CHF_2$	COF_2, CF_3COH, CF_3OH
HCFC-123	CF_3CCl_2H	$CF_3C(O)Cl$, CF_3OH, $C(O)F_2$, CO
HCFC-124	CF_3CFClH	$CF_3C(O)F$
HCFC-141b	$CFCl_2CH_3$	$CFCl_2CHO$, $C(O)FCl$, CO, CO_2
HCFC-142b	CF_2ClCH_3	CF_2ClCHO, $C(O)F_2$, CO, CO_2
HFE-125	CF_3OCF_2H	COF_2, CF_3OH
HFE-143a	CF_3OCH_3	CF_3OCHO
HFE-449sl	$C_4F_9OCH_3$	C_4F_9OCHO
HFE-569sf2	$C_4F_9OC_2H_5$	C_4F_9OCOH, HCHO
H-Galden 1040X	$CHF_2OCF_2OC_2F_4OCHF_2$	COF_2

et al., 1992). In the case of HC(O)F, the reactions of $C(O)F_2$, FC(O)Cl and $CF_3C(O)F$ with OH radicals (Wallington and Hurley, 1993) and photolysis (Nölle *et al.*, 1992) are too slow to be of any significance. These compounds are removed entirely by incorporation into water droplets. Following uptake in clouds or surface water, the halogenated acetyl halogens ($CF_3C(O)Cl$ and $CF_3C(O)F$) are hydrolyzed to TFA. Degradation products are summarized in Table 2.4.

Although there have been no studies of the degradation products of PFCs, the atmospheric oxidation of PFCs will give essentially the same fluorinated radical species that are formed during the oxidation of HFCs, HFEs and HCFCs. Based on our knowledge of the atmospheric degradation products of HFCs, HFEs and HCFCs it can be stated with high confidence that PFC degradation products will have short atmospheric lifetimes and negligible GWPs.

2.4.1.2 Atmospheric concentrations of degradation products

Using a mass balance argument, the global production rate of a degradation product should be equal to the yield of the product times the emission rate of the source gas. If the local lifetime of the degradation product is sufficiently long that its concentration is approximately uniform in the troposphere or stratosphere, the production rate can then be used to estimate the average con-

centration of the degradation product in either compartment. In more general cases, the concentration of degradation products is determined by (1) the local concentration and local lifetime of the parent compound, (2) the product yields, and (3) the local lifetime of the degradation products. The evaluation of the degradation products for shorter-lived species will be discussed in Sections 2.5.3 and 2.6.

Removal of HCFCs and HFCs is initiated by gas-phase reactions with local lifetimes that are typically of the order of years, whereas the first generation degradation products are removed by rain and clouds, with local lifetimes that are typically of the order of days. Thus, the local concentration of the degradation products will be much smaller (approximately 100 to 1000 times) than the source gas. Short-lived intermediate degradation products such as halogenated carbonyl compounds will be present at extremely low atmospheric concentrations.

2.4.1.3 Concentrations in other environmental compartments

Halogenated carbonyl degradation products (e.g., CF_3COF, COF_2) that resist gas-phase reactions are removed from the atmosphere by rain-out, wet deposition, and dry deposition. An interesting quantity is the concentration of these species in surface water. Using production of TFA from HFC-134a as an example Rodriguez *et al.* (1993) calculated that the steady-state

Table 2.5. Sources of TFA.

Species	Chemical Formula	Molar Yield of TFA	Lifetime (yr)	Global TFA Production (t yr^{-1})
Halothane	$CF_3CHClBr$	0.6	1.2	520[a]
Isoflurane	$CF_3CHClOCHF_2$	0.6	5	280[a]
HCFC-123	CF_3CHCl2	0.6	1.3	266[b]
HCFC-124	CF_3CHFCl	1.0	5.8	4440[b]
HFC-134a	CF_3CH_2F	0.13	14	4560[b]
Fluoropolymers				200[c]
TFA (lab use etc)				Negligible
Total				**10,266**

[a] Tang *et al.* (1998).
[b] Based on the atmospheric burden derived from data in Table 2.1.
[c] Jordan and Frank (1999).

global averaged rain-water concentration of TFA is of the order of 1 µg L^{-1} for an annual emission of 1000 kt yr^{-1} of HFC-134a. Subsequent calculations using a 3-D model (Kotamarthi *et al.*, 1998) showed that the local rain-water concentration averaged over 10-degree latitude and longitude bands would typically deviate from the global average number by a factor of two. Variation can be much larger on a local scale.

2.4.2 *Trifluoroacetic acid (TFA, CF$_3$C(O)OH))*

TFA is produced during the atmospheric degradation of several CFC replacements, and partitions into the aqueous compartments of the environment (Bowden *et al.*, 1996). Reaction with OH radicals in the gas phase accounts for 10–20% of the loss of TFA (Møgelberg *et al.*, 1994). The major fate of TFA is rain-out. The environmental impact of TFA has been studied thoroughly and results have been reviewed in the UNEP effects-assessment reports of 1998 (Tang *et al.*, 1998) and 2002 (Solomon *et al.*, 2003). TFA is highly soluble in water, is a strong acid with Pka of 0.23 and the logarithm of its n-octanol/water partition coefficient is –0.2, which indicates that it will essentially partition in the water compartment and will not bioaccumulate in animals. Little accumulation (concentration factor of approximately 10) occurs in plants exposed to TFA. In a long-term study (90 weeks) (Kim *et al.*, 2000) it was shown that TFA could be biodegraded in an engineered anaerobic waste-water treatment system at a TFA concentration up to 30 mg L^{-1} (as fluoride). However, it is unclear how this result should be extrapolated to natural environments, where concentrations of TFA and nutrients will be different. Ellis *et al.* (2001) observed no significant degradation during a 1-year mesocosm study. In the water compartment no significant abiotic degradation process has been identified (Boutonnet *et al.*, 1999). TFA is a persistent substance. Aquatic ecotoxicity studies showed that the most sensitive standard algae species was the algae *Selenastrum capricornutum* with a no-effect concentration of 0.10 mg L^{-1} (0.12 mg L^{-1} for the sodium salt NaTFA) (Berends *et al.*, 1999). Thus, TFA con-

centrations approaching a milligram per litre may be toxic to some aquatic life forms. The TFA concentration in rain water resulting from HFC and HCFC degradation for the year 2010 is expected to be approximately 100 to 160 ng L^{-1} (Kotamarthi *et al.*, 1998), which is approximately 1000 times smaller than the no-observed-effect concentration of *S. Capricornutum*. This result and the absence of significant bioaccumulation in biota indicate that no adverse effect on the environment is expected (Tang *et al.*, 1998; Solomon *et al.*, 2003). TFA is not metabolized in mammalian systems. Toxicity studies indicate that TFA will have biological effects similar to other strong acids (Tang *et al.*, 1998). In conclusion, no adverse effects on human or ecosystem health are expected from the TFA produced by atmospheric degradation of CFC substitutes.

Atmospheric concentrations of HFC-134a, HCFC-124, and HCFC-123 have been measured at 25.5 ppt, 1.34 ppt, and 0.03 ppt, respectively (see Table 2.1). The corresponding atmospheric burdens are 439.0 kt, 30.84 kt, and 0.774 kt, respectively. Degradation fluxes can be calculated using lifetime values from WMO (2003) and TFA yields, and TFA fluxes can be calculated at 10,266 t yr^{-1} of TFA for the year 2000 (Table 2.5).

TFA has been observed in varying concentrations in surface waters (oceans, rivers and lakes) and in fog, snow and rain-water samples around the globe, for example, in the USA, Germany, Israel, Ireland, France, Switzerland, Austria, Russia, South Africa and Finland (see references in Nielsen *et al.*, 2001). TFA appears to be a ubiquitous component of the contemporary hydrosphere. TFA is reported in concentrations of about 200 ng L^{-1} in ocean water down to depths of several thousand meters (Frank *et al.*, 2002). If 200 ng L^{-1} is the average concentration of TFA in all ocean water, the oceans contain around 3×10^8 t of TFA. With an anthropogenic contribution of 10,266 t yr^{-1} it would have taken approximately 3400 years to achieve the present TFA concentration in ocean waters; therefore, industrial sources cannot explain the observed abundance of TFA in ocean water. High TFA concentrations in the Dead Sea and Nevada lakes, of 6400 ng L^{-1} and 40,000 ng L^{-1}, respectively, suggest long-term accumulation over centuries and the existence of pre-industrial sources of TFA (Boutonnet *et al.*, 1999). However, TFA was not found in pre-industrial (>2000 years old) fresh water taken from Greenland and Denmark (Nielsen *et al.*, 2001). Therefore, although it appears that there is a significant natural source of TFA, the identity of this source is unknown.

Because of the persistence of TFA, it has been suggested that it could accumulate in aquatic ecosystems like vernal pools or seasonal wetlands, which dry out periodically, are replenished by rainfall and are presumed to have little or no seepage (Tromp *et al.*, 1995). A sensitivity study (Tromp *et al.*, 1995) based on mathematical modelling suggested that such accumulation could take place if a series of conditions (a polluted area with high atmospheric concentrations of the precursors and the OH radical; pollution and rainfall events that occur at the same time; and little, or no, seepage) could be maintained simultaneously for several decades. An example cited by the authors indicated that if the concentration of TFA in rain water was assumed to be 1 μg L^{-1}, and the loss frequencies were 5 yr^{-1} for evaporation and of 0.1 yr^{-1} for seepage, the TFA concentration could reach 100 μg L^{-1} (which corresponds to the no-effect concentration of the most sensitive aquatic species) in 30 years. The probability that such a combination of events would be maintained over several decades appears to be rather low (Solomon *et al.*, 2003). Studies of pond water (Tang *et al.*, 1998) confirmed TFA's persistent behaviour but did not find significant accumulation. TFA evapoconcentration was observed during two years in 1998 and 1999 in vernal pools in California (Cahill *et al.*, 2001). Some TFA retention was also observed between the years 1998 and 1999 but was not easily quantified. Cahill *et al.* (2001) suggested that in very wet years surface-water export may occur and may limit long-term TFA accumulation. Concentrations observed at the beginning of the study in January 1998 were about 130 ng L^{-1} (in the range of expected rain-water concentrations) and suggest that accumulation was not maintained in previous years. Although some accumulation is likely in such systems because of TFA persistence, observations indicate that the specific conditions required for accumulation in seasonal wetlands are unlikely to be maintained for several decades.

2.4.3 Other halogenated acids in the environment

Long-chain perfluorinated carboxylic acids (PFCAs, $C_nF_{2n+1}COOH$, where $n = 6$–12) have been observed in biota, and in surface and ground water (Moody and Field, 1999; Moody *et al.*, 2001; Moody *et al.*, 2002; Martin *et al.*, 2004). These PFCAs have no known natural sources, are bioaccumulative, and have no known loss mechanisms in the environment. The health effects from exposure to perfluorooctanoic acid are the subject of a present risk assessment (US EPA, 2003).

PFCAs are not generally used directly in industrial materials or consumer products. The observation of PFCAs in remote locations presumably reflects their formation as degradation products of precursor chemicals in the atmosphere. It has been suggested that degradation of fluorotelomer alcohols,

$F(CF_2CF_2)_nCH_2CH_2OH$ ($n = 3$–6), is a likely source of PFCAs observed in remote locations (Andersen *et al.*, 2003; Ellis *et al.*, 2004), but the importance of this source is unclear. Further studies are needed to quantify the sources of PFCAs in the environment. HFCs, HCFCs and HFEs used as CFC replacements generally have short-chain fluorinated alkyl substituents and so will not contribute to long-chain PFCA pollution.

Several haloacetic acids (HAA) have been detected in environmental samples. Besides TFA, monochloroacetic acid (MCA), dichloroacetic acid (DCA), and trichloroacetic acid (TCA) (Scott *et al.*, 2000) and chlorodifluoroacetic acid (CDFA) (Martin *et al.*, 2000) have been measured in, for example, rain, snow and lake samples. The contribution of HCFCs to chlorofluoro-substituted carboxylic acids is unclear.

2.4.4 Ammonia (NH_3)

It is widely recognized that the bulk of the atmospheric emissions of ammonia (NH_3) is removed from the atmosphere by dry and wet deposition processes and by reaction with strong acids to form ammonium compounds. The latter are important aerosol components with direct and indirect effects on radiative forcing. Some of the NH_3 emitted at the surface survives and is carried into the free troposphere above the atmospheric boundary layer. There, the main removal process for NH_3 is by the reaction with hydroxyl OH radicals:

$$OH + NH_3 \rightarrow H_2O + NH_2 \qquad [2.3]$$

The lifetime of NH_3 in the troposphere attributed to Reaction [2.3] has been estimated to range from 72 to 109 days (Finlayson-Pitts and Pitts, 2000; Warneck, 1999). The main reaction product of NH_3 degradation is the amidogen NH_2 radical, whose main fate is to react with NO_x and ozone and thereby act as a source or a sink for NO_x and as a source of nitrous oxide (N_2O), which is a long-lived well-mixed greenhouse gas:

$$NH_2 + NO \rightarrow N_2 + H_2O \qquad [2.4]$$

$$NH_2 + NO_2 \rightarrow N_2O + H_2O \qquad [2.5]$$

$$NH_2 + O_3 \rightarrow NH_2O + O_2 \qquad [2.6]$$

In a 3-D chemistry-transport model study, Dentener and Crutzen (1994) suggested that the degradation of NH_3 may generate up to about 1 TgN yr^{-1} of N_2O from a global NH_3 source strength of 45 Tg yr^{-1}. It should be noted that 1 TgN yr^{-1} of N_2O would represent about 6% of the global source strength of N_2O (as N) inferred by IPCC (2001, Chapter 4). However, because there is some question whether Reaction [2.5] occurs in the atmosphere as written here, the conversion of NH_3 to N_2O may be much smaller (Warneck, 1999).

2.5   Radiative properties and global warming potentials

Halocarbons released into the atmosphere can affect climate in several ways. These molecules mostly absorb radiation in a spectral window region of the outgoing thermal longwave radiation (see Figure 2.6) and hence are efficient greenhouse gases. Halocarbons can also have indirect effects on climate by causing destruction of ozone (see Chapter 1) or through alterations to tropospheric chemistry (see Section 2.5.3).

Global climate models are the tools that are used to understand and predict how current and future human emissions of halocarbons contribute to climate forcing. The concepts of radiative forcing and global warming potential (GWP) (see Box 2.2 for definitions) were introduced in 1990 (IPCC, 1990; Fisher *et al.*, 1990). They are still widely used as convenient ways to

compare and quantify the relative contribution of equal-mass emissions of different gases to climate forcing, even if it is acknowledged that the use of these simplified formulations has its limitations. Recent publications highlight that different calculations are possible to derive radiative forcings (Hansen *et al.*, 2002; Gregory *et al.*, 2004) and that other possible tools can be constructed to compare the climate impacts of greenhouse gases (Shine *et al.*, 2003; Shine *et al.*, 2005).

2.5.1   Calculation of GWPs

The contribution of a gas to the warming of the atmosphere is a function of its ability to absorb the longwave infrared radiation over a specified period of time, which in turn depends on its concentration, atmospheric steady-state lifetime, and infrared absorption properties (see Figure 2.7).

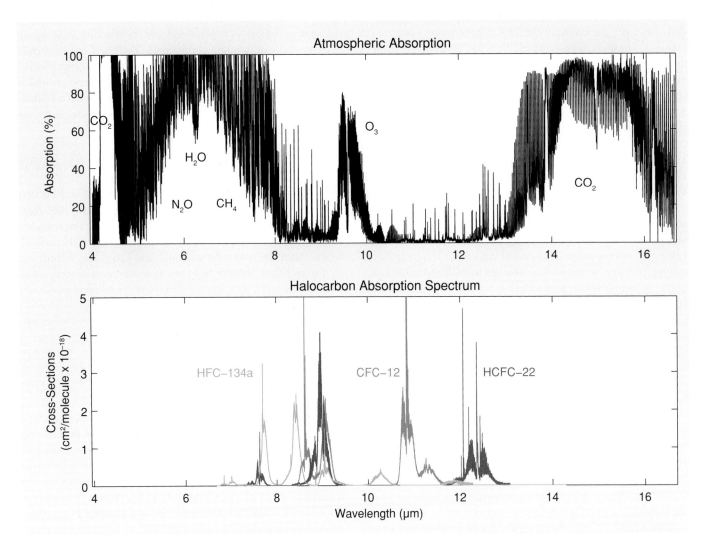

Figure 2.6. Top panel: Atmospheric absorption of infrared radiation (0 is for no absorption and 100% is for full absorption) as derived from the space-borne IMG/ADEOS radiance measurements (3 April 1997; 9.5°W, 38.4°N). Bottom panel: Halocarbons (HCFC-22, CFC-12, HFC-134a) absorption cross-sections in the infrared atmospheric window, which lies between the nearly opaque regions caused by strong absorption by CO_2, H_2O, O_3, CH_4 and N_2O.

Box 2.2. Definitions of radiative forcing and global warming potential (GWP)

Radiative forcing
- The radiative forcing quantifies the ability of a gas to perturb the Earth's radiative energy budget.
- *Definition:* The radiative forcing of the surface-troposphere system due to the perturbation in or the introduction of gas is the change in net (down minus up) irradiance (solar plus longwave, in W m^{-2}) at the tropopause, after allowing for stratospheric temperatures to re-adjust to radiative equilibrium, but with surface and tropospheric temperatures and other state variables (clouds, water) held fixed at the unperturbed values.

Global warming potential (GWP)
- The global warming potential is a relative index used to compare the climate impact of an emitted greenhouse gas, relative to an equal amount of carbon dioxide.
- *Definition:* The global warming potential is the ratio of the time-integrated radiative forcing from a pulse emission of 1 kg of a substance, relative to that of 1 kg of carbon dioxide, over a fixed horizon period.

The direct GWP of a gas x is calculated as the ratio of the time-integrated radiative forcing from a pulse emission of 1 kg of that gas relative to that of 1 kg of a reference gas

$$\text{GWP}_x(TH) = \int_0^{TH} \Delta F_x \cdot dt \Big/ \int_0^{TH} \Delta F_r \cdot dt \qquad (2.5)$$

where *TH* is the integration time (the time horizon) over which the calculation is performed. The gas chosen as reference is generally CO_2, although its atmospheric decay function is subject to substantial scientific uncertainties (IPCC, 1994).

The numerator of Equation (2.5) is the absolute GWP (AGWP) of a gas. In practice, the AGWP is calculated using the following procedure:

$$\int_0^{TH} \Delta F_x \cdot dt \cong a_x \int_0^{TH} \Delta B_x(t) \cdot dt \cong a_x \Delta B_x(0) \tau_x (1 - \exp^{(-TH/\tau_x)})$$

$$(2.6)$$

where, a_x is the radiative forcing due to a unit mass increase of the gas x distributed according to its expected steady-state distribution for continuous emission, $\Delta B_x(t)$ is the change in burden due to the pulse emission, and τ_x is the lifetime of the perturbation of species x.

The calculation is performed over a finite period of time to facilitate policy considerations. The integration time ranges from 20 to 50 years if atmospheric response (e.g., surface temperature change) is of interest, or from 100 to 500 years if a long-term effect (such as sea-level rise) is to be considered.

Substances with very long lifetimes, such as PFCs and SF_6, have contributions that may exceed these time scales, and hence the use of GWP may be inadequate for these species. Conversely, one should be cautious about using the derived GWP values for gases with lifetimes shorter than 5 years that may not be uniformly mixed in the atmosphere, although the concept may be used to some extent for short-lived species and can be applied to the calculation of indirect contributions as addressed in Section 2.5.3.

2.5.2   *Calculation of radiative forcing*

The radiative forcing due to a change in the abundance of a greenhouse gas is the net (down minus up) irradiance change (in watts per square meter, W m^{-2}) at the tropopause induced by this perturbation. It is calculated using a forward radiative-transfer code that computes the irradiance at different atmospheric levels, with the infrared absorption spectrum of the molecule as a key input. Calculation can be performed using simplified assumptions (clear-sky instantaneous forcings) or using improved schemes that take into account cloud coverage and allow stratospheric temperatures to re-adjust to radiative equilibrium (ad-

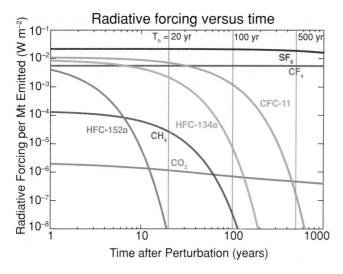

Figure 2.7. The time evolution of the radiative forcing (in W m^{-2}) associated with the decay of a pulse emission, released into the atmosphere at time $t = 0$, of various gases with an atmospheric lifetime spanning 1.4 years (HFC-152a) to 50,000 years (CF4) (see also Figure 2.2). The 20-, 100-, and 500-year time horizons (TH) used for GWP calculations are shown by vertical lines.

justed cloud forcings). Because of the current low abundances of the halocarbons in the atmosphere, their radiative forcing is proportional to their atmospheric abundance. This is to be contrasted with abundant gases, such as CO_2, for which the effect is nonlinear because of saturation of absorption.

Halocarbons exhibit strong absorption bands in the thermal infrared atmospheric window (see Figure 2.6) because of their various vibrational modes. Absorption spectra of CFCs and the majority of proposed substitutes are characterized by absorption bands rather than lines, because individual spectral lines for these heavy molecules are overlapping and not resolved at tropospheric pressures and temperatures. For each gas, the absorption cross-sections (in cm^2 per molecule) are derived from laboratory measurements for specific atmospheric conditions. Several independent measurements of absorption cross-sections for halocarbons have been reported in the literature (e.g., McDaniel *et al.*, 1991; Clerbaux *et al.*, 1993; Clerbaux and Colin, 1994; Varanasi *et al.*, 1994; Pinnock *et al.*, 1995; Barry *et al.*, 1997; Christidis *et al.*, 1997, Ko *et al.*, 1999; Sihra *et al.*, 2001; Orkin *et al.*, 2003; Nemtchinov and Varanasi, 2004; Hurley *et al.*, 2005[1]) and some of these data sets are available through the widely used molecular spectroscopic databases, such as HITRAN (Rothman *et al.*, 2003) and GEISA (Jacquinet-Husson *et al.*, 1999). The intercomparison of measured cross-sections and integrated absorption intensities performed on the same molecule by different groups provides an estimate of the discrepancies among the obtained results, and hence of the measurement accuracy (e.g., Ballard *et al.*, 2000; Forster *et al.*, 2005). Although the discrepancies between different cross-section measurements can reach 40% (Hurley *et al.*, 2005), it is recognized that the typical uncertainties associated with the measured cross-section when integrated over the infrared spectral range are less than 10%.

The measured spectroscopic data are then implemented into a radiative transfer model to compute the radiative fluxes. Several radiative-transfer models are available, ranging from line-by-line to wide-bands models, depending on the spectral interval over which the calculation is performed. Apart from the uncertainties stemming from the cross-sections themselves, differences in the flux calculation can arise from the spectral resolution used; tropopause heights; vertical, spatial and seasonal distributions of the gases; cloud cover; and how stratospheric temperature adjustments are performed. The impact of these parameters on radiative forcing calculations is discussed in comprehensive studies such as Christidis *et al.* (1997), Hansen *et al.* (1997), Myhre and Stordal (1997), Freckleton *et al.* (1998), Highwood and Shine (2000), Naik *et al.* (2000), Jain *et al.* (2000), Sihra *et al.* (2001) and Forster *et al.* (2005). The discrepancy in the radiative forcing calculation for different halocarbons, which is associated with different assumptions used in the radiative transfer calculation or with use of different

cross-section sets, can reach 40% (Gohar *et al.*, 2004). Recent studies performed for the more abundant HFCs (HFC-23, HFC-32, HFC-134a and HFC-227ea) report that an agreement better than 12% can be reached when the calculation conditions are better constrained (Gohar *et al.*, 2004; Forster *et al.*, 2005).

Recommended radiative forcings and associated GWP values, gathered from previous reports and recently published data, are provided in Section 2.5.4.

2.5.3   *Other aspects affecting GWP calculations*

The discussion in the preceding sections describes methods to evaluate the direct radiative impact of emission of a source gas that has a sufficiently long lifetime so that it has a uniform mixing ratio in the troposphere. This section considers two aspects of GWP calculations that are not covered by these methods. First, the direct impact refers only to the greenhouse warming arising from the accumulation of the source gas in the atmosphere, but does not consider its other radiatively important byproducts. There could be indirect effects if the degradation products of the gas also behaved like greenhouse gases (Section 2.5.3.1), or if the presence of the emitted gas and its degradation products affected the distribution of other greenhouse gases (Section 2.5.3.2). Second, the special case of very short-lived source gases with non-uniform mixing ratios in the troposphere will be discussed (Section 2.5.3.3).

2.5.3.1   *Direct effects from decomposition products*
The indirect radiative effects of the source gas include the direct effects from its degradation products. As was discussed in Section 2.4.1 the local concentrations of most degradation products, with very few exceptions, are a factor of 100 (or more) smaller than that of the source gas. Assuming that the values for the radiative forcing of the source gas and its degradation products are similar, the indirect GWP from the degradation products should be much smaller than the direct GWP of the source gas except for cases where the local lifetime of the degradation product is longer than that of the source gas. CO_2 is one of the final degradation products of organic compounds. However the magnitude of the indirect GWP attributed to CO_2 is equal to the number of carbons atoms in the original molecule, and is small compared with the direct GWP values of most of the halocarbons given in Table 2.6.

2.5.3.2   *Indirect effects from influences on other atmospheric constituents*
As discussed in Chapter 1, chlorine and bromine atoms released by halocarbons deplete ozone in the stratosphere. Because ozone is a greenhouse gas, its depletion leads to cooling of the climate system. This cooling is a key indirect effect of halocarbons on radiative forcing and is particularly significant for gases containing bromine.

Daniel *et al.* (1995) described a method for characterizing the net radiative forcing from halocarbons emitted in the year 1990. Although the method and the values are often discussed

---

[1] This most recently published work was not available in time to be fully considered by the authors.

in the context of net GWP, it was clearly explained in Daniel *et al.* (1995) that they differ from direct GWP in the following ways:

1. The net GWP is most useful for comparing the relative radiative effects of different ODSs when emitted at the same time. In the type of calculations that may be performed in the following chapters, one may come across a situation where the GWP weighted emissions from several well-mixed greenhouse gases and ODSs add to zero. It is important to recognize that the result does not guarantee zero climate impact.

2. The amount of ozone depletion caused by an incremental emission of an ODS depends on the equivalent effective stratospheric chlorine (EESC) loading in the atmosphere. The concept of EESC has long been used to assess the effect of halocarbon emissions on the ozone-depleting effect of halogens in the stratosphere (Prather and Watson, 1990; Chapter 1 of WMO, 2003 and references therein; and Figure 1.18 in Chapter 1 of this report). Ozone depletion first began to be observable around 1980, and the date when EESC is projected to fall below its 1980 value can provide a rough estimate of the effects of emissions on the time scale for ozone recovery, assuming no changes in other parameters, such as atmospheric circulation or chemical composition. Chapter 1 of this report assesses in detail the factors that are likely to influence the date of ozone recovery, intercompares model estimates of the date, and discusses uncertainties in its absolute value. EESC is used here to indicate the relative effect of several different future emission scenarios on the date of ozone recovery, but not to identify the absolute value of the date. Current projections based on emission scenarios indicate that the threshold values for EESC will be achieved between 2043 and 2046, when there will be no further ozone depletion for incremental ODS emission. Because the indirect forcing of ODSs becomes zero after the EESC threshold is reached, indirect GWPs for emissions prior to that time decrease as they occur closer to the recovery threshold. Indirect GWPs for emissions that occur after the threshold is reached are zero.

The method described in Daniel *et al.* (1995) assumes an EESC-loading curve for the future based on an emission scenario in compliance with the Montreal Protocol and its subsequent updates. Values given in Daniel *et al.* (1995) are for several emission scenarios (future EESC curves) appropriate for emissions in 1990. Subsequent updates in IPCC (1994), WMO (1999) and IPCC (2001) use a single emission scenario for 1990 emission. The values in WMO (2003) are for an emission in the year 2002.

HCFCs, HFCs and some of their replacement gases (hydrocarbons and ammonia) also have the potential to alter the radiative balance in the atmosphere indirectly by influencing the sources or sinks of greenhouse gases and aerosols. Oxidation by hydroxyl (OH) radicals in the troposphere is the main removal process for the organic compounds emitted by human activi-

ties. This oxidation acts as a source of ozone and as a removal process for hydroxyl radicals, thereby reducing the efficiency of methane oxidation and promoting the build-up of methane. Therefore, emissions of organic compounds, including HCFCs, HFCs and some of their replacement gases, may lead to the build-up of two important greenhouse trace gases, methane and ozone, and consequently may cause an indirect radiative forcing (e.g., Lelieveld *et al.*, 1998; Johnson and Derwent, 1996; Fuglestvedt *et al.*, 1999). The main factors influencing the magnitudes of such indirect radiative impacts were found to be their spatial emission patterns, chemical reactivity and transport, molecular complexity, and the oxidation products formed (Collins *et al.*, 2002).

Another possible indirect effect may result from the reaction of the degradation products of HCFCs and HFCs with OH or O_3, which may modify the OH or O_3 distributions. However, it is unlikely that degradation products of the long-lived HCFCs and HFCs have a large impact on a global scale, because the effect would occur after the emission is distributed around the globe, which will result in relatively low concentrations.

2.5.3.3 *Very short-lived hydrocarbons*

Assigning GWP values to very short-lived (VSL) species (species with a lifetime of a month or shorter) presents a special challenge (WMO, 2003, Chapter 2). Because of their short lifetimes VSL species are not uniformly distributed in the troposphere. Their distributions depend on where and when (during the year) they are emitted. Thus, it is not possible to assign a single steady-state change in burden in the troposphere per unit mass emission. In calculating the local change in radiative forcing, one would have to use their actual three-dimensional distributions. Furthermore, it is not obvious how one would estimate the change in surface temperature from the local changes in forcing. In a sense, the issue is similar to the situation for aerosol forcing. In such cases, the notion of GWP may prove less useful and one would have to examine the local climate impact directly, along with other indirect effects, such as ozone and aerosol formation.

Given the infrared absorption cross-section of a VSL species, it is possible to use current tools to compute a local radiative forcing per unit burden change for a local column. This value can be used to obtain estimates for a GWP-like quantity by examining the decay rate of a pulse emission. In most cases this value will be small compared with the GWP value of the long-lived HCFCs and HFCs, and may indicate that the global impact from VSL species used as halocarbon substitutes is small. However, it is not clear how such forcing may produce impacts on a smaller (e.g., regional) scale.

2.5.4 *Reported values*

The lifetimes, the radiative efficiencies (radiative forcing per concentration unit) and the GWPs of the substances controlled by the Montreal Protocol and of their replacements are given in Table 2.6. The GWP for a gas is the ratio of the absolute global

Table 2.6. Lifetimes, radiative efficiencies, and direct global warming potentials (GWPs) relative to carbon dioxide, for the ODSs and their replacements.

Industrial Designation or Common Name	Chemical Formula	Other Name	Lifetime[a] (yr)	Radiative Efficiency[a] (W m^{-2} ppb^{-1})	Global Warming Potential for a Given Time Horizon			
					IPCC (1996)[b] 100 yr	IPCC (2001) & WMO (2003) 20 yr	100 yr	500 yr
Carbon dioxide	CO_2			See text	1	1	1	1
Methane	CH_4		12.0[c]	3.7×10^{-4}	21	63[c]	23[c]	7[c]
Substances controlled by the Montreal Protocol								
CFC-11	CCl_3F	Trichlorofluoromethane	45	0.25	3800	6330	4680	1630
CFC-12	CCl_2F_2	Dichlorodifluoromethane	100	0.32	8100	10,340	10,720	5230
CFC-113	CCl_2FCClF_2	1,1,2-Trichlorotrifluoroethane	85	0.3	4800	6150	6030	2700
CFC-114	$CClF_2CClF_2$	Dichlorotetrafluoroethane	300	0.31		7560	9880	8780
CFC-115	$CClF_2CF_3$	Monochloropentafluoroethane	1700	0.18		4990	7250	10,040
Halon-1301	$CBrF_3$	Bromotrifluoromethane	65	0.32	5400	7970	7030	2780
Halon-1211	$CBrClF_2$	Bromochlorodifluoromethane	16[d]	0.3		4460	1860	578
Halon-2402	$CBrF_2CBrF_2$	Dibromotetrafluoroethane	20[d]	0.33[d]		3460[d]	1620[d]	505[d]
Carbon tetrachloride	CCl_4		26[d]	0.13	1400	2540[d]	1380[d]	437[d]
Methyl bromide	CH_3Br		0.7	0.01		16	5	1
Bromochloromethane	CH_2BrCl		0.37[d]					
Methyl chloroform	CH_3CCl_3	1,1,1-Trichloroethane	5.0[d]	0.06		476[d]	144[d]	45[d]
HCFC-22	$CHClF_2$	Chlorodifluoromethane	12[d]	0.20	1500	4850[d]	1780[d]	552[d]
HCFC-123	$CHCl_2CF_3$	Dichlorotrifluoroethane	1.3[d]	0.14[d]	90	257[d]	76[d]	24[d]
HCFC-124	$CHClFCF_3$	Chlorotetrafluoroethane	5.8[d]	0.22	470	1950[d]	599[d]	186[d]
HCFC-141b	CH_3CCl_2F	Dichlorofluoroethane	9.3	0.14		2120	713	222
HCFC-142b	CH_3CClF_2	Chlorodifluoroethane	17.9[c]	0.2	1800	5170	2270	709
HCFC-225ca	$CHCl_2CF_2CF_3$	Dichloropentafluoropropane	1.9[d]	0.2[d]		404[d]	120[d]	37[d]
HCFC-225cb	$CHClFCF_2CClF_2$	Dichloropentafluoropropane	5.8[d]	0.32		1910[d]	586[d]	182[d]
Hydrofluorocarbons								
HFC-23	CHF_3	Trifluoromethane	270[d]	0.19[e]	11,700	11,100[f]	14,310[f]	12,100[f]
HFC-32	CH_2F_2	Difluoromethane	4.9[d]	0.11[e]	650	2220[f]	670[f]	210[f]
HFC-125	CHF_2CF_3	Pentafluoroethane	29	0.23	2800	5970	3450	1110
HFC-134a	CH_2FCF_3	1,1,1,2-Tetrafluoroethane	14[d]	0.16[e]	1300	3590[f]	1410[f]	440[f]
HFC-143a	CH_3CF_3	1,1,1-Trifluoroethane	52	0.13	3800	5540	4400	1600
HFC-152a	CH_3CHF_2	1,1-Difluoroethane	1.4	0.09	140	411	122	38
HFC-227ea	CF_3CHFCF_3	1,1,1,2,3,3,3-Heptafluoropropane	34.2[d]	0.26[e]	2900	4930[f]	3140[f]	1030[f]
HFC-236fa	$CF_3CH_2CF_3$	1,1,1,3,3,3-Hexafluoropropane	240[d]	0.28	6300	7620[d]	9500[d]	7700[d]
HFC-245fa	$CHF_2CH_2CF_3$	1,1,1,3,3-Pentafluoropropane	7.6[d]	0.28		3180[d]	1020[d]	316[d]
HFC-365mfc	$CH_3CF_2CH_2CF_3$	1,1,1,3,3-Pentafluorobutane	8.6[d]	0.21		2370[d]	782[d]	243[d]
HFC-43-10mee	$CF_3CHFCHFCF_2CF_3$	1,1,1,2,3,4,4,5,5,5-Decafluoropentane	15.9[d]	0.4	1300	3890[d]	1610[d]	502[d]
Perfluorinated compounds								
Sulphur hexafluoride	SF_6		3200	0.52	23,900	15,290	22,450	32,780
Nitrogen trifluoride	NF_3		740	0.13		7780	10,970	13,240
PFC-14	CF_4	Carbon tetrafluoride	50,000	0.08	6500	3920	5820	9000
PFC-116	C_2F_6	Perfluoroethane	10,000	0.26	9200	8110	12,010	18,280
PFC-218	C_3F_8	Perfluoropropane	2600	0.26	7000	5940	8690	12,520
PFC-318	$c\text{-}C_4F_8$	Perfluorocyclobutane	3200	0.32	8700	6870	10,090	14,740

Table 2.6. (continued)

Industrial Designation or Common Name	Chemical Formula	Other Name	Lifetime[a] (yr)	Radiative Efficiency[a] ($W\ m^{-2}\ ppb^{-1}$)	Global Warming Potential for a Given Time Horizon			
					IPCC (1996)[b] 100 yr	IPCC (2001) & WMO (2003) 20 yr	100 yr	500 yr
Perfluorinated compounds								
PFC-3-1-10	C_4F_{10}	Perfluorobutane	2600	0.33	7000	5950	8710	12,550
PFC-5-1-14	C_6F_{14}	Perfluorohexane	3200	0.49	7400	6230	9140	13,350
Fluorinated ethers								
HFE-449sl	$CH_3O(CF_2)_3CF_3$		5	0.31		1310	397	123
HFE-569sf2	$CH_3CH_2O(CF_2)_3CF_3$		0.77	0.3		189	56	17
HFE-347pcf2[g]	$CF_3CH_2OCF_2CHF_2$		7.1	0.25		1800	540	170
Hydrocarbons and other compounds[h]								
Ethane	C_2H_6	(R-170)	0.21[i]	0.0032[j]				
Cyclopropane	$c\text{-}C_3H_6$	(C-270)	0.44[i]					
Propane	C_3H_8	(R-290)	0.041[i]	0.0031[j]				
n-Butane	$CH_3(CH_2)_2CH_3$	(R-600)	0.018[i]	0.0047[j]				
Isobutane	$(CH_3)_2CHCH_3$	(R-600a) 2-Methylpropane	0.019[i]	0.0047[j]				
Pentane	$CH_3(CH_2)_3CH_3$	(R-601)	0.010[i]	0.0046[i]				
Isopentane	$(CH_3)_2CHCH_2CH_3$	(R-601a) 2-Methylbutane	0.010[i]					
Cyclopentane	$c\text{-}C_5H_{10}$		0.008[i]					
Ethylene	CH_2CH_2	(R-1150) Ethene	0.004[i]	0.035[j]				
Propylene	CH_3CHCH_2	(R-1270) 1-Propene	0.001[i]					
Ammonia	NH_3	(R-717)	a few days					
Dimethylether	CH_3OCH_3		0.015	0.02		1[a]	1[a]	<<1[a]
Methylene chloride	CH_2Cl_2	Dichloromethane	0.38[i]	0.03	9	35[a]	10[a]	3[a]
Methyl chloride	CH_3Cl	Chloromethane	1.3	0.01		55[a]	16[a]	5[a]
Ethyl chloride	CH_3CH_2Cl	Chloroethane	0.11[d]					
Methyl formate	$C_2H_4O_2$		0.16[i]					
Isopropanol	$CH_3CHOHCH_3$	Isopropyl alcohol	0.013[j]					
Trichloroethylene	CCl_2CHCl	Trichlorethene	0.013[l]					
FK-5-1-12	$CF_3CF_2C(O)CF(CF_3)_2$		0.038[k]	0.3[l]				
n-Propyl bromide	$CH_3CH_2CH_2Br$	1-Bromopropane, n-PB	0.04	0.3[l]				

[a] From IPCC (2001, Chapter 6).
[b] Values adopted under the UNFCCC for the national inventories.
[c] The lifetime of methane includes feedbacks on emissions (IPCC, 2001, Chapter 6), and GWPs include indirect effects (see Section 2.5.3.2).
[d] Updated in WMO (2003, Chapter 1).
[e] Updated from two averaged model results in Gohar *et al.* (2004) and rounded for consistency.
[f] Scaled for the updated radiative efficiency noted in (e).
[g] From original paper by Tokuhashi *et al.* (2000). IPCC (2001) erroneously referred to this compounds as HFE-374pcf2.
[h] From direct effects only. Some values for indirect effects are given in Table 2.8.
[i] Global lifetime estimated from a process lifetime, with respect to tropospheric OH calculated relative to 6.1 years for CH_3CCl_3, assuming an average temperature of 272 K.
[j] Highwood *et al.* (1999).
[k] Upper value reported by Taniguchi *et al.* (2003).
[l] Suggested as upper limit.

Table 2.7. Direct and indirect GWPs of ODSs for a 100-year time horizon. The indirect GWP values are estimated from observed ozone depletion between 1980 and 1990 for 2005 emissions.

Species	Direct GWP(100 yr)		Indirect GWP[a] (2005 emission, 100 yr)	
	Best Estimate	Uncertainty[b]	Best Estimate	Uncertainty[c]
CFC-11	4680	±1640	−3420	±2710
CFC-12	10720	±3750	−1920	±1630
CFC-113	6030	±2110	−2250	±1890
HCFC-22	1780	±620	−269	±183
HCFC-123	76	±27	−82	±55
HCFC-124	599	±210	−114	±76
HCFC-141b	713	±250	−631	±424
HCFC-142b	2270	±800	−337	±237
HCFC-225ca	120	±42	−91	±60
HCFC-225cb	586	±205	−148	±98
CH_3CCl_3	144	±50	−610	±407
CCl_4	1380	±480	−3330	±2460
CH_3Br	5	±2	−1610	±1070
Halon-1211	1860	±650	−28,200	±19,600
Halon-1301	7030	±2460	−32,900	±27,100
Halon-2402	1620	±570	−43,100	±30,800

[a] The bromine release factors for CH_3Br, Halon-1211, Halon-1301 and Halon-2402 have been updated to be consistent with the factors in WMO (2003, Chapter 1, Table 1-4).

[b] Uncertainties in GWPs for direct positive radiative forcing are taken to be ±35% (2-σ) (IPCC, 2001).

[c] Uncertainties in GWPs for indirect negative radiative forcings are calculated using the same assumptions as in WMO (2003, Table 1-8 on page 1.35), except that an uncertainty of ±10 years (1-σ) in the time at which ozone depletion no longer occurs is included here, and that an estimated radiative forcing between 1980 and 1990 of −0.1 ± 0.07 (2-σ) W m^{-2} is used. The latter is derived from the updated radiative forcings from IPCC (2001) using the Daniel *et al.* (1995) formalism.

warming potential (AGWP) of the gas to that of the reference gas CO_2. GWP values are subject to change if the radiative efficiency or the lifetime of the gas are updated, or if the AGWP for CO_2 changes. The radiative efficiency of a gas can change if there is an update to its absorption cross-section, or if there is a change in the background atmosphere. IPCC (2001) used 364 ppm for the background mixing ratio of CO_2 and 1.548×10^{-5} W m^{-2} ppb^{-1} for its radiative efficiency for GWP calculations. The corresponding AGWP values for the 20-, 100- and 500-year time horizons were 0.207, 0.696 and 2.241 W m^{-2} yr^{-1} ppm^{-1}, respectively. Most of the GWP values in Table 2.6 are taken from the WMO (2003) report, but they were updated if more recent published data were available (Gohar *et al.*, 2004; Forster *et al.*, 2005). Current values differ from the previously reported values (IPCC, 2001) by amounts ranging from −37% (HFC-123) to +43% (Halon-1211), mostly because of lifetime updates. For completeness, the GWP values as reported in IPCC (1996) – that is, the values adopted under the UNFCCC for the national inventories – are also provided for the 100-year time horizon. Relative to the values reported in IPCC (1996), the recommended GWPs (for the 100-year time horizon) have been modified from −16% (HCFC-123) to +51% (HFC-236fa), depending on the gas, with an average change of +15%. Most of these changes occur because of the updated AGWP of CO_2 used in IPCC (2001).

The absolute accuracy in the GWP calculations is subject to the uncertainties in estimating the atmospheric steady-state lifetimes and radiative efficiencies of the individual gases and of the CO_2 reference. These uncertainties have been described to some extent in the earlier sections. It was shown in the previous paragraph that the discrepancies between the direct reported GWPs and the values from IPCC (1996) can be as high as ±50%.

The direct radiative forcings caused by a 1-ppb increase (the radiative efficiency) of some non-methane hydrocarbons (NMHCs) are also included in Table 2.6. It has been estimated that the global mean direct radiative forcing attributed to anthropogenic emissions of NMHCs in the present-day atmosphere is unlikely to be more than 0.015 W m^{-2} higher than in the pre-industrial atmosphere (Highwood *et al.*, 1999). This value is highly uncertain because of the large dependence on the vertical profiles of these short-lived gases, the natural contribution to the burdens considered and the area-weighted distributions of the mixing-ratio scenarios. The corresponding direct GWPs of the NMHCs are probably insignificant, because NMHCs have much shorter lifetimes than halocarbons and other greenhouse gases.

The indirect GWPs calculated from the cooling that results from stratospheric ozone depletion are discussed in Section 2.5.3.2. Table 2.7 presents direct and indirect GWPs for a 100-year horizon for ODSs. The indirect GWPs are estimated from observed ozone depletion between 1980 and 1990 and adapted

Table 2.8. Global warming potentials (GWPs) for a 100-year time horizon, and lifetimes for several hydrocarbons estimated from the indirect effects by Collins *et al.* (2002).

Species	Chemical Formula	Other Name	Lifetime[a] (yr)	Indirect GWP (100 yr)
Ethane	C_2H_6	(R-170)	0.214	8.4
Propane	C_3H_8	(R-290)	0.041	6.3
Butane	C_4H_{10}	(R-600)	0.018	7
Ethylene	CH_2CH_2	(R-1150) Ethene	0.004	6.8
Propylene	CH_3CHCH_2	(R-1270) 1-Propene	0.001	4.9

[a] Global lifetime estimated from a process lifetime, with respect to tropospheric OH calculated relative to 6.1 years for CH_3CCl_3, assuming an average temperature of 272 K.

for 2005 emissions. The estimated uncertainties in indirect GWPs are high, and are caused by a 70% uncertainty in determining the radiative forcing from ozone depletion. Also, direct and indirect GWPs cannot be simply added, because these are global averages.

NMHCs can have an indirect radiative forcing through tropospheric chemistry interactions. As described in Section 2.5.3.2, methane concentrations increase when the hydroxyl radicals are consumed by the more reactive organic compounds emitted during anthropogenic activities. The photochemical production of ozone, also a greenhouse gas, is also enhanced by an increased burden of NMHCs. Indirect GWPs for some alkanes and alkenes have been estimated from global averages by Collins *et al.* (2002) and are given in Table 2.8. This study also considered the increase in CO_2 as a result of the oxidation of these compounds. The GWPs attributed to the indirect chemical effects are much more important than their direct contribution to radiative forcing. However, the indirect GWPs of the alkanes and alkenes considered as replacement refrigerants are relatively small compared with other non-ozone-depleting halocarbons, and are highly uncertain. This uncertainty arises because the indirect radiative impacts of alkanes and alkenes depend strongly on the location and season of the emission, so it is difficult to give a single number that covers all circumstances and eventualities. The indirect GWPs in Table 2.8 are meant to be averages over the year for hydrocarbons emitted in polluted environments in the major Northern Hemisphere continents.

2.5.5 *Future radiative forcing*

This section presents future direct radiative forcings of ODSs and their replacements based on the emissions scenarios from IPCC SRES (IPCC, 2000) and on emission scenarios from Chapter 11. The direct radiative forcings were calculated from the simplified expression $\Delta F = \alpha(X - X_0)$ (IPCC, 2001, Chapter 6), where X is the projected concentration of a substance in ppb, X_0 is its pre-industrial global concentration, and α is its radiative efficiency (from Table 2.6). In this section the term radiative forcing refers to the future change in direct radiative forcing

relative to pre-industrial conditions (in 1750) unless otherwise specified. Past radiative forcings (to the year 2000) are given in Table 1.1 in Chapter 1.

For the time frame up to 2100, the emissions from the IPCC SRES scenarios were used in IPCC (2001) to estimate a range of radiative forcings possible for HFCs and PFCs. The individual radiative forcings of the HFCs are plotted in Figure 2.8 using the B1 and A1B scenarios, which represent the low and a high future estimates in the SRES scenarios. The total radiative forcings of the HFCs and PFCs are plotted in Figure 2.9 for several SRES scenarios. Alternative scenarios up to 2015 for HFCs were derived by Ashford *et al.* (2004) (see also Chapter 11) based on future demands for the refrigeration and foams sectors.

The contributions of HFCs and PFCs to the total radiative forcing of long-lived greenhouse gases around 2003 (relative to 1750) are about 0.0083 W m⁻² (0.31%) and 0.0038 W m⁻² (0.15%), respectively. A natural background concentration of 40 ppt is assumed for CF_4 (IPCC, 2001). Radiative forcings of CFCs, HCFCs and HFCs up to 2003 are shown in Figure 2.10.

The emissions used to calculate the future radiative forcing of CFCs and HCFCs are based on WMO (2003, Chapter 1) and of HFCs on the emissions scenarios from Ashford *et al.* (2004) (see also Chapter 11). The estimated radiative forcing of HFCs in 2015 is in the range of 0.022–0.025 W m⁻² based on the SRES projections and in the range of 0.019–0.030 W m⁻² based on scenarios from Ashford *et al.* (2004). The radiative forcing of PFCs in 2015 is about 0.006 W m⁻² based on SRES projections. The HFC and PFC radiative forcing corresponds to about 6–10% and 2%, respectively, of the total estimated radiative forcing due to CFCs and HCFCs in 2015 (estimated at 0.297 W m⁻² for the baseline scenario). Alternatively, the HFC and PFC radiative forcing corresponds to about 0.8% and 0.2%, respectively, of the estimated radiative forcing of all well-mixed greenhouse gases, with a contribution of the ODSs of about 10%.

Projections over longer time scales become more uncertain because of the growing influences of uncertainties in technological practices and policies. Based on the SRES emission scenarios (Figure 2.9), by 2050 the upper limit of the range of the

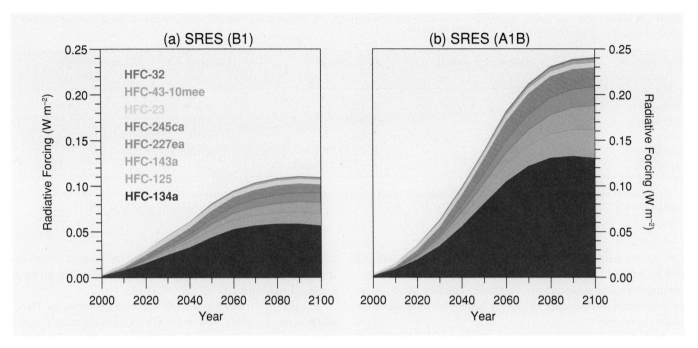

Figure 2.8. Radiative forcings of the individual HFCs for the B1 and A1B SRES scenario (Appendix II of IPCC, 2001).

radiative forcing from HFCs is 0.14 W m^{-2} and from PFCs it is 0.015 W m^{-2}. By 2100 the upper limit of the range of the radiative forcing from HFCs is 0.24 W m^{-2} and from PFCs it is 0.035 W m^{-2}. In comparison, the SRES forcing from CO_2 by 2100 ranges from about 4.0 to 6.7 W m^{-2}. The SRES emission scenarios suggest that HFC-134a contributes 50–55% to the total radiative forcing from HFCs (Figure 2.8), and that CF_4 contributes about

80% to the total radiative forcing from PFCs. Uncertainties in SRES emissions should be recognized – for example, the long-term nearly linear growth in HFC emissions up to 2050 as envisaged in SRES scenarios is highly unlikely for the longer time frame. The contribution of the long-lived ODSs (CFCs, HCFCs and CCl_4) decreases gradually from a maximum of 0.32 W m^{-2} around 2005 to about 0.10 W m^{-2} in 2100 (IPCC, 2001).

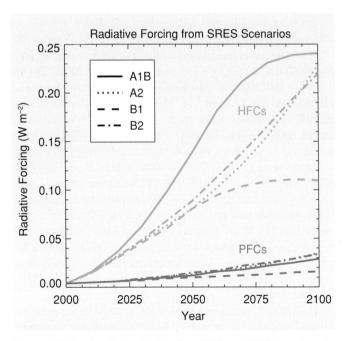

Figure 2.9. Total radiative forcing from HFCs and PFCs based on the A1B, A2, B1 and B2 SRES scenarios (Appendix II of IPCC, 2001).

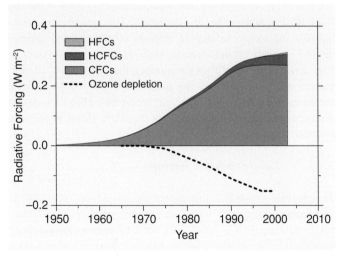

Figure 2.10. Changes in radiative forcing due to halocarbons. The radiative forcing shown is based on observed concentrations and the WMO Ab scenario. Radiative forcing from observed ozone depletion is taken from Table 6.13 of IPCC (2001, Chapter 6); see also Section 1.5 in this report.

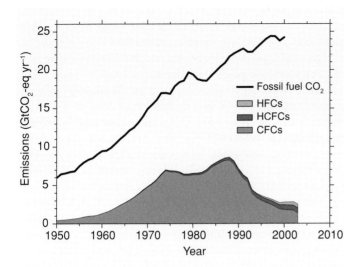

Figure 2.11. GWP-weighted emissions (for a 100-year time horizon) from CFCs and HCFCs (from WMO, 2003), and from HFCs (from Ashford *et al.*, 2004). Total CO_2 emissions shown are those from fossil-fuel combustion and cement production as estimated by Marland *et al.* (2003).

Actions taken under the Montreal Protocol and its Adjustments and Amendments have led to replacement of CFCs with HCFCs, HFCs and other substances, and to changing of industrial processes. These actions have begun to reduce atmospheric chlorine loading, the radiative forcings of the CFCs (Figure 2.10), and have also reduced the total GWP-weighted annual emissions from halocarbons (Figure 2.11). Observations of the annual concentrations of the major halocarbons at multiple sites accurately quantify past changes in direct radiative forcing and emissions (see also Section 2.3.2). The combined CO_2-equivalent emissions, calculated by multiplying the emissions of a compound by its GWP, of CFCs, HCFCs and HFCs have decreased from a peak of about 7.5 ± 0.4 GtCO$_2$-eq yr^{-1} around 1990 to about 2.5 ± 0.2 GtCO$_2$-eq yr^{-1} (100-year time horizon) in 2000, or about 10% of the CO_2 emissions due to global fossil-fuel burning in that year.

Continued observations of CFCs and other ODSs in the atmosphere enable improved validation of estimates of the lag between production and emission to the atmosphere, and of the associated banks of these gases. Because the production of ODSs has been greatly reduced, annual changes in concentrations of those gases widely used in applications – such as refrigeration, air conditioning, and foam blowing – are increasingly dominated by releases from existing banks. For example, CFC-11 is observed to be decreasing at a rate about 60% slower than would occur in the absence of emissions, and CFC-12 is still increasing slightly rather than declining. The emission rates of these two gases for 2001–2003 were estimated to be about 25–35% of their 1990 emission rates. In contrast, CFC-113 and CH_3CCl_3, which are used mainly as solvents with no significant banking, are currently decreasing at a rate consistent with a de-

crease of more than 95% in their emission rates compared with their 1990 emission rates. Thus, continued monitoring of a suite of gases is placing tighter constraints on the derived emissions and their relationship to banks. This information provides new insights into the significance of banks and end-of-life options for applications using HCFCs and HFCs as well.

Current banked halocarbons will make a substantial contribution to future radiative forcing of climate for many decades unless a large proportion of these banks is destroyed. Large portions of the global inventories of CFCs, HCFCs and HFCs currently reside in banks. For example, it is estimated that about 50% of the total global inventory of HFC-134a currently resides in the atmosphere, whereas 50% resides in banks. Figure 2.12 shows the evolution of the GWP-weighted bank size (for a 100-year time horizon) for halocarbons, based on a top-down approach using past reported production and detailed comparison with atmospheric concentrations (WMO, 2003). This approach suggests that release of the current banks of CFCs, HCFCs and HFCs to the atmosphere would correspond conservatively to about 2.8, 3.4 and 1.0 GtCO$_2$-eq, respectively, for a 100-year time horizon. Although observations of concentrations constrain emissions well, estimates of banks depend on the cumulative difference between production and emission and so are subject to larger uncertainties. The bottom-up analysis by sectors as presented in this report (Chapter 11) suggests possible banks of CFCs, HCFCs and HFCs of about 15.6, 3.8, and 1.1 GtCO$_2$-eq, for a 100-year time horizon. These approaches both show that current banks of halocarbons represent a substantial possible contribution to future radiative forcing.

If all of the ODSs in banks in 2004 were not released to the atmosphere, their direct positive radiative forcing could be reduced by about 0.018–0.025 W m^{-2} by 2015. Over the next two decades this positive radiative forcing change is expected to be about 4–5% of that caused by CO_2 emissions over the same pe-

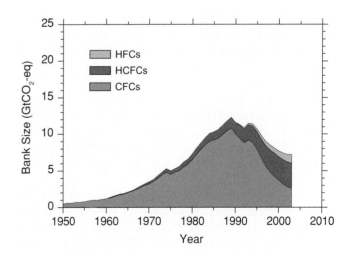

Figure 2.12. Evolution of GWP-weighted bank sizes (for a 100-year time horizon) for halocarbons. The data for CFCs and HCFCs are based on WMO (2003), and the data for HFCs are based on Ashford et al. (2004).

riod. The lower limit is based on the banks from WMO (2003, Chapter 1) and the upper limit on the banks from Ashford *et al.* (2004).

As mentioned before, observations of concentrations in the atmosphere constrain the past emissions, but the banks are the difference between cumulative emissions and production and have much larger uncertainties. The values of the emissions and banks of CFCs and HCFCs used for the Figures 2.11 and 2.12 are based on WMO (2003). Ashford *et al.* (2004) reported banks of CFCs and HCFCS that were considerably larger than those of WMO (2003), especially for CFC-11 and CFC-12. The CFC-11 bank in 2002 is 0.59 Mt according to WMO (2003) compared with 1.68 Mt according to Ashford *et al.* (2004); the CFC-12 bank in 2003 is virtually zero according to WMO (2003) compared with 0.65 Mt according to Ashford *et al.* (2004). The larger banks in Ashford *et al.* (2004) are accompanied by significantly lower emissions of CFC-11 and CFC-12 over the past 10 years, compared with the emissions reported in WMO (2003). The lower emissions cannot support the observations in the atmosphere – that is, they lead to global atmospheric concentrations about 40 ppt lower than those observed in 2002 for both CFC-11 and CFC-12.

The uncertainty in the accumulated top-down emissions is estimated at about 3%, based on an estimated 2–3% uncertainty in observed concentrations by different measurement networks (see Table 2.1). Based on an inverse calculation, the accumulated emissions for 1990–2001 needed to support the observations of CFC-11 and CFC-12 have an uncertainty of about 4%, assuming an uncertainty in the observed concentration trends of 1%. The uncertainty in cumulative production is harder to estimate. The total 1990–2001 production of CFC-11 and CFC-12 in Ashford *et al.* (2004) is about 30–35% lower than in WMO (2003). This difference can be attributed largely to the neglect of production data not reported to AFEAS (2004) but which is included in UNEP production data. See Chapter 11 for a discussion of uncertainties in bottom-up emission estimates. But because the additional UNEP production is most likely used in rapid-release applications (Chapter 11), it does not change the size of the bank. Uncertainties in accumulated emission and production add up to the uncertainty in the change in the bank from 1990 to 2002.

The virtually zero bank of CFC-12 reported in WMO (2003) is not in agreement with Chapter 6 of this report because CFC-12 is still present in current air conditioners. The banks of CFC-11 and CFC-12 from WMO (2003) could therefore be a lower limit of the real banks, whereas those of Ashford *et al.* (2004) could be an upper limit.

As discussed in Section 2.5.3.2, the concept of ESSC can be used to estimate the indirect GWP for the ODSs. The best estimate of the year when EESC is projected to return to 1980 values for the baseline scenario Ab of WMO (2003, Chapter 1) is 2043.9, if the banks of all ODSs are as estimated in WMO (2003, Chapter 1); if the banks of CFCs and HCFCs are as estimated in this report (see Ashford *et al.*, 2004, and Chapters 4 to 11), this date would be estimated to occur in the year 2046.4.

Some emissions of banked CFCs (such as the slow emissions of CFC-11 from foams) could occur after ozone recovery. Such emissions would reduce the effect of the banks on ozone recovery but would still contribute to the positive direct radiative forcing as greenhouse gases. Thus the larger banks of some ODSs estimated in this report could lead to a maximum delay in ozone recovery of the order of two to three years compared with the baseline scenario. If no emissions occurred from the ODS banks as estimated in WMO (2003, Chapter 1), the date when EESC is estimated to return to its 1980 value is 2038.9, or an acceleration of about five years. Finally, if emissions of all ODSs were stopped in 2003, the date when EESC is estimated to return to its 1980 value was given as 2033.8 in WMO (2003, Chapter 1), an acceleration of about ten years. Thus, these changes in future ODS emissions, within ranges compatible with the present uncertainties in the banks, have relatively small effects on the time at which EESC recovers to 1980 levels and therefore on stratospheric ozone and the indirect radiative forcing of these gases. The uncertainties in the physical and chemical processes involved in the time of ozone recovery (as shown in the Figure in Box 1.7 in Chapter 1) are larger than the uncertainties in the scenarios associated with these alternative emissions.

2.6　Impacts on air quality

2.6.1　Scope of impacts

Halocarbons and their replacements may exert an impact on the composition of the troposphere and hence may influence air quality on global, regional and local scales. Air quality impacts of replacement compounds will depend on (1) their incremental emissions relative to the current and future emissions from other sources of these and other compounds that affect air quality, (2) their chemical properties, particularly the atmospheric reactivity, which is inversely related to the atmospheric lifetime, and (3) for the more reactive compounds, the geographic and temporal distribution of their incremental emissions.

This section provides a framework for evaluating the impacts of halocarbons and their replacements on the global self-cleaning capacity (Section 2.6.2) and on urban and regional air quality (Section 2.6.3). For the urban and regional scales in particular, such impacts are difficult to assess because of many factors, including uncertainties in the relative spatial and temporal distributions of emission of different reactive compounds, non-linear chemical interactions (especially in polluted regimes), and complex patterns of atmospheric mixing and transport. Therefore only the methodology for assessing these impacts is presented here. Although some numerical examples are provided, they are not intended as a prediction of future impacts, but only as an illustration of the methodology required to perform a full assessment of the effects of replacement compounds (e.g., through the use of reactivity indices, such as the Photochemical Ozone-Creation Potential, POCP, and the Maximum Incremental Reactivity, MIR) in the context of known local and regional emissions of all the other compounds that contribute to poor air quality. Actual impacts will depend not only on the specific replacement compound being considered, but also on the current air quality and the prevailing meteorological conditions at any given location, as well as on a detailed knowledge of all other pollution sources.

2.6.2   *Global-scale air quality impacts*

Globally, organic compounds added to the troposphere can lead simultaneously to increased concentrations of tropospheric O_3 (a greenhouse gas) and to decreased concentrations of hydroxyl (OH) radicals. The resulting impacts on the global distributions of the greenhouse gases CH_4 and O_3 and their radiative forcing have already been discussed in Section 2.5, where they were quantified using the indirect GWP concept (see Table 2.8 for the GWP of short-lived hydrocarbons). Here we address more generally the potential impacts of halocarbons and their replacements on the global distribution of OH concentrations. These concentrations are nearly synonymous with the tropospheric 'self-cleaning' capacity because OH-initiated oxidation reactions are a major mechanism for the removal of numerous trace gases, including CH_4, NMHCs, nitrogen and sulphur oxides (NO_x and SO_x), as well as HFCs and HCFCs.

A direct way to evaluate these impacts is to compare the consumption rate of OH radicals by halocarbons and their replacements with that by the current burden of trace gases such as CH_4 and CO. For long-lived species, this comparison is straightforward because their concentrations are relatively uniform on the global scale. Table 2.9 lists the OH consumption rates for selected long-lived species. The rates are computed as $k_{(OH+X)}[X]$, where $k_{(OH+X)}$ is the rate constant for the reaction of substance X with OH, and [X] is its observed atmospheric concentration. For the purpose of obtaining an upper limit, the rate constants and concentrations are evaluated for surface conditions (1 atm, 287 K). For the CFCs, halons, HCFCs and HFCs shown in the table, the OH radical consumption rates are orders of magnitudes smaller than those of CH_4 and CO. Similar estimates for many other CFCs, halons, HCFCs and HFCs, as well as the PFCs and HFEs, are not possible because of lack of measurements of their atmospheric concentrations or of their rate constant for reaction with OH. However, their OH consumption rates are expected to be small. CFCs, halons and PFCs are expected to react very slowly with OH because no H atom is available on these molecules for OH to abstract (see Section 2.4.1). For HFCs, the known OH reaction rate constants (0.02–3.2×10^{-14} cm^3 molecule^{-1} s^{-1}) are generally smaller than those of HCFCs (0.02–9.9×10^{-13} cm^3 molecule^{-1} s^{-1}), and their global mean concentrations are expected to be lower than those of HCFCs because of their shorter emission histories, so that the OH radical consumption rate of HFCs should be smaller than that of HCFCs. Thus, the OH radical consumption rates and consequently the impacts on global air quality of CFCs, HCFCs, halons, HFCs and PFCs, are not significant compared to those of CH_4 and CO.

For short-lived compounds, such as volatile organic com-

Table 2.9. Comparison of the OH radical consumption rates of selected CFCs, halons, HCFCs, HFCs, methane (CH_4) and carbon monoxide (CO).

Species	OH Consumption Rate (s^{-1})[a]
Substances controlled by the Montreal Protocol[b]	
CFC-11	2.0×10^{-8}
CFC-12	6.0×10^{-8}
CFC-113	6.3×10^{-7}
CFC-114	2.2×10^{-7}
Halon-1301	6.6×10^{-9}
Halon-1211	1.0×10^{-8}
HCFC-22	1.4×10^{-5}
HCFC-123	2.5×10^{-8}
HCFC-124	2.8×10^{-7}
HCFC-141b	1.6×10^{-6}
HCFC-142b	7.5×10^{-7}
HFCs[c]	
HFC-23	8.0×10^{-8}
HFC-125	5.5×10^{-8}
HFC-134a	1.3×10^{-6}
HFC-152a	1.4×10^{-6}
Others	
CH_4	0.23
CO	0.61
Sum of CH_4 and CO	**0.84**

[a] The consumption rate is calculated as $k_{(OH+X)}[X]$, where $k_{(OH+X)}$ is the rate constants of the reactions of a substance with OH radicals, and [X] is the observed atmospheric concentrations of that substance. The concentrations and rate constants are evaluated for surface conditions (1 atm, 287 K) to provide upper limits to consumption rates. Rate-constant data are from Atkinson *et al.* (2002), except for CFC-113 and CFC-114, which are from Atkinson (1985).
[b] Based on concentrations (abundances) for the year 2000 from Table 2.1.
[c] Based on concentrations reported by WMO (2003, Chapter 1).

pounds (VOCs), the impacts on global OH are more difficult to estimate because of the large spatial and temporal variations in their concentrations, known chemical nonlinearities, and the complexity of transport and mixing of clean and polluted air. Although concentration measurements of many of these compounds have been made in both polluted and clean environments, they are not sufficiently representative to estimate globally averaged values. A crude estimate can be made by comparing the globally integrated emissions of anthropogenic VOCs with those of other gases that react with OH, especially CO, CH_4 and natural (biogenic) VOCs. This comparison is premised on the approximation that all emitted VOCs are eventually removed by reaction with OH, so that the rate of consumption of OH is related stoichiometrically to the emission rate of any VOC. The globally integrated emissions of CO, CH_4 and the VOCs are given in Table 2.10. Anthropogenic VOCs are a small fraction of the global reactive-carbon emissions, with total contributions of 2.1% on a per-mole basis and 6.1% on a per-carbon basis. The impact of increased emissions of any of the compounds listed in Table 2.10 can be estimated from the table. For ex-

Table 2.10. Global emissions of volatile organic compounds (VOCs), HFCs, CH_4 and CO.

Species	Emissions (Tg yr^{-1})[a]	Average molecular weight, (g mol^{-1})	% of total on a molar basis	Number of carbons[b]	% of total on a carbon basis
Anthropogenic VOCs					
Alkanols	10.7	46.2	0.2	2.3	0.3
Ethane	8.2	30	0.2	2	0.3
Propane	7.6	44	0.1	3	0.3
Butanes (n-butane and isobutane)	14.1	57.8	0.2	4	0.5
Pentanes (n-pentane and isopentane)	12.4	72	0.1	5	0.5
Hexanes and higher alkanes	23.3	106.8	0.2	7.6	0.9
Ethene	10.3	28	0.3	2	0.4
Propene	4.8	42	0.1	3	0.2
Ethyne	4.0	26	0.1	2	0.2
Other olefins	6.8	67	0.1	4.8	0.3
Benzene	5.8	78	0.1	6	0.3
Toluene	6.7	92	0.1	7	0.3
Xylenes	4.5	106	0.0	8	0.2
Trimethylbenzene	0.8	120	0.0	9	0.0
Other aromatics	3.9	126.8	0.0	9.3	0.2
Esters	2.6	104.7	0.0	5.2	0.1
Ethers	4.3	81.5	0.0	4.8	0.1
Chlorinated HCs	2.4	138.8	0.0	2.6	0.0
Formaldehyde	2.0	30	0.0	1	0.0
Other alkanals	4.8	68.6	0.0	3.7	0.1
Ketones	3.0	75.3	0.0	4.4	0.1
Alkanoic acids	18.6	59.1	0.2	1.9	0.3
Other NMHCs	12.4	86.9	0.1	4.9	0.4
Subtotal	*174.2*		*2.1*		*6.1*
Natural VOCs					
Isoprene	250	68.1	2.5	5	10.3
Terpenes	144	136.3	0.7	10	5.9
Acetone	48	58.1	0.6	3	1.4
HFCs[c]					
HFC-125	0.002	120.0	0.0	2	0.0
HFC-134a	0.043	102.0	0.0	2	0.0
HFC-143a	0.001	84.0	0.0	2	0.0
HFC-152a	0.007	66.1	0.0	2	0.0
HFC-227ea	0.007	170.0	0.0	3	0.0
HFC-23	0.008	70.0	0.0	1	0.0
Other					
CH_4	598	16	25.7	1	20.9
CO	2789	28	68.5	1	55.6
Total	**4003**		**100.0**		**100.1**

[a] Emissions of anthropogenic VOCs and HFCs are from GEIA (www.geiacenter.org), and emissions for natural VOCs, CH_4 and CO are from IPCC (2001).
[b] Weighted averages of industrial and biomass values given in IPCC (2001).
[c] Emitted in 1997.

ample, a doubling of the current emissions of butanes would increase the emissions of reactive carbon by about 0.5% (on the more conservative per-carbon basis), and is therefore unlikely to have a significant impact on global OH concentrations. A similar argument can also be made for HFCs (see Table 2.10 for emissions), which currently account for a negligible fraction of the global reactive-carbon emissions.

The total molar emission rate of compounds that remove OH (summed from Table 2.10) is about 4×10^{13} mol yr^{-1} and is dominated by CO and CH_4. By comparison, the molar emission rate of the major CFCs (CFC-11, CFC-12 and CFC-113) and of CCl_4 at their maximum values in the late 1980s is estimated as about 8×10^9 mol yr^{-1}. Substitution of these CFC emission rates by an equal number of moles of compounds that react with OH (e.g., VOCs) would increase the OH loss rate by only about 0.02%.

This simple comparison neglects the nonlinear interactions that result from chemical and mixing processes involving different air masses, any non-OH removal processes (such as reactions of some VOCs with O_3), and the secondary reactions of OH with intermediates of VOC oxidation. The reactive intermediates may be particularly important for the larger VOCs, making the comparisons of the relative OH consumption more conservative on a per-carbon rather than on a per-mole basis, but many of these intermediates are also believed to be scavenged by aerosols (Seinfeld and Pankow, 2003) rather than by OH. Regeneration of OH – for example, by the photolysis of H_2O_2 produced during VOC oxidation, or by NO_x chemistry – is not considered, but would, in any case, reduce the impacts. In view of these uncertainties, the relative reactive-carbon emissions should not be equated to the relative OH consumption rates, but may be useful in establishing whether incremental emissions could or could not be significant.

2.6.3   Urban and regional air quality

2.6.3.1   Volatile organic compounds (VOCs)

Organic compounds, in the presence of sunlight and NO_x, take part in ground-level ozone formation and thereby contribute to the deterioration in regional and urban air quality, with adverse effects on human health and biomes. Each organic compound exhibits a different propensity to form ozone, which can be indexed in a reactivity scale. In North America, ground-level ozone formation is seen as an urban-scale issue and reactivity scales, such as the Maximum Incremental Reactivity (MIR) scale, describe the contributions by different organic compounds to the intense photochemical ozone formation in urban plumes (Carter, 1994). In Europe, ground-level ozone formation occurs on the regional scale in multi-day episodes, and a different reactivity scale, the Photochemical Ozone Creation Potential (POCP), has been developed to address long-range transboundary formation and transport of ozone. These indices of reactivity are useful only if fully speciated organic emissions inventories are known. Such information is frequently available for European and North American locations, but is not generally available for cities and regions of less developed nations.

Table 2.11 presents the MIR scale for conditions appropriate to North America and the POCP scale for conditions appropriate to Europe for a range of chemical species relevant to this report. MIR values generally fall in the range from 0, unreactive, to about 15, highly reactive. POCP values cover the range from 0, unreactive, to about 140, highly reactive. In general terms, the ODSs and HFCs in common use have very low MIR and POCP values and take no part in ground-level ozone formation. In contrast, the alkanes (ethane, propane and isobutane) and alkenes (ethylene and propylene) exhibit steadily increasing reactivities from ethane, which is unreactive, to propylene, which is highly reactive. Alkanes have low reactivity on the urban scale and so have low MIR values. However, alkanes generate ozone efficiently on the multi-day scale and so appear much more reactive under European conditions on the POCP scale.

Table 2.11. Propensity of VOCs to form tropospheric ozone according to two reactivity scales: the Maximum Incremental Reactivity (MIR) for single-day urban plumes appropriate to North American conditions, and the Photochemical Ozone Creation Potential (POCP) for multi-day regional ozone formation appropriate to European conditions.

Species	Reactivity	
	North America **MIR scale** **(g-O_3/g-substance)**	**Europe** **POCP scale** **(relative units)**
ODSs[a]		
Methyl chloroform	<0.1	0.2
HCFC-22	<0.1	0.1
HCFC-123	<0.1	0.3
HCFC-124	<0.1	0.1
HCFC-141b	<0.1	0.1
HCFC-142b	<0.1	0.1
HCFC-225ca	<0.1	0.2
HCFC-225cb	<0.1	0.1
HFCs[a]		
HFC-23	<0.1	0
HFC-32	<0.1	0.2
HFC-125	<0.1	0
HFC-134a	<0.1	0.1
HFC-143a	<0.1	0
HFC-152a	<0.1	1
HFC-227ea	<0.1	0
Hydrocarbons and other compounds[b]		
Ethane	0.31	12
Propane	0.56	18
n-Butane	1.32	35
Isobutane	1.34	31
Isopentane	1.67	41
Ethylene	9.07	100
Propylene	11.57	112
Dimethylether	0.93	17
Methylene chloride	0.07	7
Methyl chloride	0.03	0.5
Methyl formate	0.06	3
Isopropanol	0.71	14
n-Pentane	1.53	40
Trichloroethylene	0.60	33

[a] Values from Hayman and Derwent (1997). The values for North America were estimated from the POCP values given.

[b] Values for North America from Carter *et al.* (1998, update of 5 February 2003, http://pah.cert.ucr.edu/~carter/reactdat.htm); values for Europe from Derwent *et al.* (1998).

On the basis of reactivity scales, it therefore appears that among all the proposed substances relevant to the phase-out of ODSs used in refrigeration, air conditioning and foam blowing sectors, only alkenes, some alkanes and some oxygenated organics have the potential to significantly influence ozone formation on the urban and regional scales. Substitutes for smaller sectors, such as aromatics and terpenes as solvents, are potentially more reactive but are not within the scope of this report.

Table 2.12. The change in POCP-weighted emissions in the UK following the hypothetical substitution of a 1 kt yr^{-1} emission of HFC-134a by the emission of 1 kt yr^{-1} of each species. Also shown are the POCP-weighted emissions in kt yr^{-1} from all sources in the UK.

Species	Chemical Formula	POCP-Weighted Emissions (kt yr^{-1})	
		Substitution of 1 kt yr^{-1}	Current Emissions[a]
Ethane (R170)	C_2H_6	12	559
Propane (R290)	C_3H_8	18	926
n-Butane (R600)	$CH_3(CH2)_2CH_3$	35	5319
Isobutane (R600a)	$(CH_3)_2CHCH_3$	31	1256
Isopentane (R601a)	$(CH_3)_2CHCH_2CH_3$	41	2842
Ethylene (R1150)	CH_2CH_2	100	4640
Propylene (R1270)	CH_3CHCH_2	112	2916
Dimethylether	CH_3OCH_3	17	29
Methylene chloride	CH_2Cl_2	7	6
Methyl chloride	CH_3Cl	0.4	0.8
Methyl formate	$C_2H_4O_2$	3	12
Isopropanol	$CH_3CHOHCH_3$	14	118
n-Pentane	$CH_3(CH_2)_3CH_3$	40	2549
Trichloroethylene	CCl_2CHCl	33	691

[a] Emissions in kt yr^{-1} multiplied by POCP.

When it comes to assessing the possible impacts of the substitution of an ODS by an alkane or alkene, what needs to be considered is the reactivity-weighted mass emission involved with the substitution. Generally speaking such substitutions show relatively small impacts because there are already huge anthropogenic urban sources of these particular alkanes and alkenes.

Speciated VOCs inventories are required to calculate POCP values, and among European countries only the United Kingdom has such data. Table 2.12 shows the current POCP-weighted emissions of selected VOCs from all sources in the UK. It also shows the incremental POCP-weighted emission from the replacement of 1 kt yr^{-1} mass emission of these compounds with an equal mass of HFC-134a, which illustrates the use of POCP values for estimating impacts on regional ozone: Consider a hypothetical replacement of HFC-134a by any of the compounds

listed in Table 2.12. The POCP for HFC-134a is 0.1 (see Table 2.11), so that substitution of 1 kt yr^{-1} of HFC-134a by an equal mass of VOCs will lead to an increase in POCP-weighted emissions as shown in the third column of Table 2.12, and therefore to some deterioration in regional ozone-related air quality. For a few organic compounds, such as ethylene and propylene, such substitutions of HFC-134a would increase POCP-weighted emissions by three orders of magnitude. However, there are already large sources of these compounds in UK emissions, so the increase in POCP-weighted emissions from the substitution is small compared with the current POCP-weighted emissions from all other sources, as is shown in the comparison between the third and fourth columns of Table 2.12. There are a few species for which the substitution of 1 kt yr^{-1} of HFC-134a would lead to an increase in POCP-weighted emissions that are large

Table 2.13. The change in MIR-weighted emissions in Mexico City following the hypothetical substitution of a 1 kt yr^{-1} emission of HFC-134a by the emission of 1 kt yr^{-1} of each species. Also shown are the MIR-weighted emissions in kt yr^{-1} from all sources.

Species	Chemical Formula	MIR-weighted emissions (kt yr^{-1})	
		Substitution of 1 kt yr^{-1}	Current Emissions[a]
Ethane (R170)	C_2H_6	0.31	0.3
Propane (R290)	C_3H_8	0.56	16
n-Butane (R600)	$CH_3(CH2)_2CH_3$	1.34	16
Isobutane (R600a)	$(CH_3)_2CHCH_3$	1.34	8
Isopentane (R601a)	$(CH_3)_2CHCH_2CH_3$	1.67	39
Ethylene (R1150)	CH_2CH_2	9.07	35
n-Pentane	$CH_3(CH_2)_3CH_3$	1.53	9
All other explicit VOCs			1700
Total			**1823**

[a] Emissions in kt yr^{-1} multiplied by MIR.

Table 2.14. The change in MIR-weighted emissions in the Los Angeles area following the hypothetical substitution of a 1 kt yr^{-1} emission of HFC-134a by the emission of 1 kt yr^{-1} of each species. Also shown are the 1987 MIR-weighted emissions in kt yr^{-1} from all sources.

Species	Chemical Formula	MIR-weighted emissions (kt yr^{-1})	
		Substitution of 1 kt yr^{-1}	Current Emissions[a]
Ethane (R170)	C_2H_6	0.31	7.8
Propane (R290)	C_3H_8	0.56	6.0
n-Butane (R600)	$CH_3(CH2)_2CH_3$	1.34	44.1
Isobutane (R600a)	$(CH_3)_2CHCH_3$	1.34	15.5
Isopentane (R601a)	$(CH_3)_2CHCH_2CH_3$	1.67	48.6
Ethylene (R1150)	CH_2CH_2	9.07	363
Propylene	CH_3CHCH_2	11.57	195
n-Pentane	$CH_3(CH_2)_3CH_3$	1.53	26.2
All other explicit VOCs			1474
Total			**2180**

[a] Emissions from Harley and Cass (1995), multiplied by MIR.

in comparison with current emissions; these species include dimethyl ether, methylene chloride, methyl chloride and methyl formate. However, these species tend to have relatively low ozone production capacities, so the impact on air quality of the increase in POCP-weighted emissions from the substitution of HFC-134a is likely to be insignificant.

Table 2.13 illustrates the use of the MIR scale for the Mexico City Metropolitan Area, which represents one of the worst air pollution cases in the world and is the only city in the developing world that has the speciated inventory required for MIR calculations. These estimates were obtained from a hydrocarbon source-apportionment study (Vega *et al.*, 2000) using a chemical mass-balance receptor model applied to measurements at three sites, and using the current emission estimate reported for total VOCs as 433,400 t yr^{-1} (SMA-GDF, 2003). As can be seen from the table, the introduction of 1 kt yr^{-1} of a particular hydrocarbon to substitute for, say, 1 kt yr^{-1} of HFC-134a, is negligible in comparison with the current MIR-weighted emission rates of all VOCs.

Table 2.14 presents a similar calculation for the Los Angeles area (South Coast Air Basin). Speciated emissions were reported by Harley and Cass (1995) for the summer of 1987. More recent speciated emissions are not available in the scientific literature, but it should be noted that total emissions of reactive organic carbon (ROG) have decreased by a factor of about 2.2 from 1985 to 2000 (Alexis *et al.*, 2003).

Non-speciated VOC emission inventories are becoming available for other regions of the world. Although MIR- and POCP-based reactivity estimates are not possible without detailed speciation, the total VOC emissions are generally much larger than those anticipated from ODS replacements. For example, Streets *et al.* (2003) have estimated that the total non-methane VOC emissions from all major anthropogenic sources, including biomass burning, in 64 regions in Asia was 52.2 Tg in 2000, with 30% of these emissions from China. In Europe, urban, industrial and agricultural emissions are a major concern,

and the national VOC emission ceiling in the EU is expected to be 6.5 Mt in 2010 (EC, 2001). However, the 1990 annual total VOC emissions for Europe were estimated to be 22 Mt, according to the recent European 'CityDelta' study. The study was conceived in support of the CAFE (Clean Air For Europe) programme on EU environmental legislation related to NO_x, VOCs and particulate matter, organized by the Joint Centre of the European Commission – Institute for Environment and Sustainability (JRC-IES). CityDelta is focused on exploring changes in air quality in eight cities (London, Paris, Prague, Berlin, Copenhagen, Katowice, Milan and Marseille) in seven countries caused by changes in emissions as predicted by atmospheric models (Thunis and Cuvelier, 2004). Table 2.15 shows the total anthropogenic VOC emissions for CityDelta Project countries observed in 1990 and 2000, and projected for 2010.

Global estimates of incremental emissions of VOCs used as replacement compounds cannot be used directly to estimate regional and local air quality impacts. These impacts depend also on the reactivity of the selected replacement compounds and on how their incremental emissions are distributed geographically and in time (e.g., seasonally). Such detailed information is

Table 2.15. Anthropogenic emissions of VOCs (in kt yr^{-1}) in the CityDelta Project countries (UN-ECE, 2003).

Country	Year		
	1990	2000	2010
Czech Republic	394	227	209
Denmark	162	129	83
France	2473	1726	1050
Germany	3220	1605	1192
Italy	2041	1557	1440
Poland	831	599	804
UK	2425	1418	1200

not normally available and therefore several additional assumptions need to be made. For illustration, consider (as in Section 2.6.2) the global replacement of the major CFCs emissions at their maximum values in the late 1980s (of about 8×10^9 mol yr^{-1}) by an equal number of moles of a single compound, isobutane. The corresponding global isobutane emission increment is about 460 kt yr^{-1}, representing a 3.5% increase over its current global emissions (see Table 2.10). Regional and local increments can, however, be substantially larger, depending on how these emissions are distributed. If additional assumptions are made that the incremental emissions have no seasonal variation and are distributed spatially according to population, the increments would be about 4.6 kt yr^{-1} for the UK (approximate population 60 million), 1.5 kt yr^{-1} for Mexico City (approximate population 20 million) and 1.15 kt yr^{-1} for the Los Angeles area (approximate population 15 million). The POCP-weighted isobutane emission increment for the UK would then be 143 kt yr^{-1}, which is 11% of the current isobutane emissions in the UK but only 0.6% of the total POCP-weighted VOC emissions shown in Table 2.12. Similarly, for other European countries the current and future VOC emissions (see Table 2.15) are much larger than this hypothetical population-weighted increment in isobutane emissions. For Mexico City, the MIR-weighted isobutane increment would be 2.0 kt yr^{-1}, which is 25% of the current isobutane emission but 0.1% of the total MIR-weighted VOC emissions shown in Table 2.13. For Los Angeles, the MIR-weighted isobutane increment would be 1.5 kt yr^{-1}, which is 10% of the 1987 isobutane emission but 0.07% of the total MIR-weighted VOC emissions shown in Table 2.14. Even considering the substantial reductions in VOC emissions in recent years (Alexis *et al.*, 2003), it is clear that such isobutane increments would increase total MIR-weighted VOC emissions by less than 1%. However, even such small increases in reactivity may be of some concern in urban areas that currently fail to meet air quality standards.

2.6.3.2 *Ammonia (NH₃)*

Gaseous ammonia (NH_3) is an important contributor to the mass of atmospheric aerosols, where it exists primarily as ammonium sulphate and ammonium nitrate (Seinfeld and Pandis, 1998). Recent evidence also suggests that NH_3 may promote the formation of new sulphate particles (Weber *et al.*, 1999). NH_3 is removed relatively rapidly (within a few days) from the atmosphere mostly by dry deposition and rain-out, and to a lesser extent by gas-phase reaction with OH (see Section 2.4.4). Incremental emissions of NH_3 from its use as a replacement for ODSs must be considered in the context of the uncertain current local, regional and global NH_3 emissions.

Most of the anthropogenic emissions of NH_3 result from agricultural activities (livestock, fertilizer use, crops and crop decomposition), biomass burning, human waste and fossil-fuel combustion. Together with contributions from natural sources, such as the oceans, natural soils and natural vegetation, the total (1990) global emission of NH_3 was 54 MtN yr^{-1}, of which about 80% was anthropogenic (FAO, 2001). Spatially distributed (1°

Table 2.16. Anthropogenic emissions of NH_3 (in kt yr^{-1}) in the City-Delta Project countries (UN-ECE, 2003).

Country	Year		
	1990	2000	2010
Czech Republic	156	74	62
Denmark	133	104	83
France	779	784	780
Germany	736	596	579
Italy	466	437	449
Poland	508	322	468
UK	341	297	297

$\times 1°$) emissions inventories have been developed by Bouwman *et al.* (1997) and Van Aardenne *et al.* (2001), and are available from the Global Emissions Inventory Activity (GEIA, www.geiacenter.org) and EDGAR2.0 (Olivier *et al.*, 1996) databases.

For Europe, the 1990 annual total anthropogenic NH_3 emissions were estimated to be 5.7 Mt, with an average emission per capita of 12 kg (11 kg per capita in the EU) (CORINAIR, 1996). European NH_3 emissions have decreased by 14% between 1990 and 1998 (Erisman *et al.*, 2003). Anthropogenic emissions of NH_3 for 1990 and 2000 in the CityDelta Project countries are summarized in Table 2.16, together with the 2010 projection. According to the table, total ammonia emissions have decreased by 6–52% between 1990 and 2000, with the largest reductions (140 kt yr^{-1}) occurring in Germany, primarily because of the fall in livestock numbers and the decreasing use of mineral nitrogen fertilizers.

Incremental NH_3 emissions from substitution of ODSs are expected to be small on a global basis. For example, the equimolar substitution of the major CFCs (CFC-11, CFC-12 and CFC-113) and of CCl_4 at their maximum emission rates in the late 1980s would require about 140 kt yr^{-1} of NH_3, or a 0.2% increase in global emissions. If this global increase in NH_3 emissions is distributed geographically according to population, the additional national and regional emissions would be negligible compared with current emissions (see for example Table 2.16). However, larger local impacts could result from highly localized NH_3 release episodes.

References

AFEAS, 2004: Production, sales and atmospheric releases of fluorocarbons through 2001. Alternative Fluorocarbons Environmental Acceptability Study, Arlington, VAI, USA. (Available at http://www.afeas.org)

Alexis, A., J. Auyeung, V. Bhargava, P. Cox, M. Johnson, M. Kavan, C. Nguyen, and Y. Yajima, 2003: *The 2003 California Almanac of Emissions and Air Quality.* California Air Resources Board. (Available at http://www.arb.ca.gov)

Andersen, M. P. S., M. D. Hurley, T. J. Wallington, J. C. Ball, J. W. Martin, D. A. Ellis, S. A. Mabury, and O. J. Nielsen, 2003: Atmospheric chemistry of C_2F_5CHO: Reaction with Cl atoms and OH radicals, IR spectrum of $C_2F_5C(O)O_2NO_2$. *Chemical Physics Letters,* **379**(1–2), 28–36.

Arnold, F., V. Burger, B. DrosteFanke, F. Grimm, A. Krieger, J. Schneider, and T. Stilp, 1997: Acetone in the upper troposphere and lower stratosphere: Impact on trace gases and aerosols. *Geophysical Research Letters,* **24**(23), 3017–3020.

Ashford, P., D. Clodic, A. McCulloch, L. Kuijpers, 2004: Emission profiles from the foam and refrigeration sectors compared with atmospheric concentrations, part 2 – Results and discussion. *International Journal of Refrigeration,* **27**(7), 701–716.

Atkinson, R., 1985: Kinetics and mechanisms of the gas-phase reactions of the hydroxyl radical with organic compounds under atmospheric conditions. *Chemical Reviews,* **85**, 69–201.

Atkinson, R., D. L. Baulch, R. A. Cox, L. N. Crowley, R. F. Hampson Jr., J. A. Kerr, M. J. Rossi, and J. Troe, 2002: *Summary of evaluated kinetic and photochemical data for atmospheric chemistry.* IUPAC Subcommittee on Gas Kinetic Data Evaluation for Atmospheric Chemistry, web version, December 2002.

Bakwin, P. S., D. F. Hurst, P. P. Tans, and J. W. Elkins, 1997: Anthropogenic sources of halocarbons, sulfur hexafluoride, carbon monoxide, and methane in the southeastern United States. *Journal of Geophysical Research – Atmospheres,* **102**(D13), 15,915–15,925.

Ballard, J., R. J. Knight, D. A. Newnham, J. Vander Auwera, M. Herman, G. Di Lonardo, G. Masciarelli, F. M. Nicolaisen, J. A. Beukes, L. K. Christensen, R. McPheat, G. Duxbury, R. Freckleton, and K. P. Shine, 2000: An intercomparison of laboratory measurements of absorption cross-sections and integrated absorption intensities for HCFC-22. *Journal of Quantitative Spectroscopy & Radiative Transfer,* **66**(2), 109–128.

Barnes, D. H., S. C. Wofsy, B. P. Fehlau, E. W. Gottlieb, J. W. Elkins, G. S. Dutton, and S. A. Montzka, 2003a: Urban/industrial pollution for the New York City–Washington, DC, corridor, 1996–1998: 1. Providing independent verification of CO and PCE emissions inventories. *Journal of Geophysical Research – Atmospheres,* **108**(D6), 4185, doi:10.1029/2001JD001116.

Barnes, D. H., S. C. Wofsy, B. P. Fehlau, E. W. Gottlieb, J. W. Elkins, G. S. Dutton, and S. A. Montzka, 2003b: Urban/industrial pollution for the New York City–Washington, DC, corridor, 1996–1998: 2. A study of the efficacy of the Montreal Protocol and other regulatory measures. *Journal of Geophysical Research – Atmospheres,* **108**(D6), 4186, doi:10.1029/2001JD001117.

Barry, J., G. Locke, D. Scollard, H. Sidebottom, J. Treacy, C. Clerbaux, R. Colin, and J. Franklin, 1997: 1,1,1,3,3,-pentafluorobutane (HFC-365mfc): Atmospheric degradation and contribution to radiative forcing. *International Journal of Chemical Kinetics,* **29**(8), 607–617.

Berends, A. G., J. C. Boutonnet, C. G. de Rooij, and R. S. Thompson, 1999: Toxicity of trifluoroacetate to aquatic organisms. *Environmental Toxicology and Chemistry,* **18**(5), 1053–1059.

Berntsen, T. K., I. S. A. Isaksen, G. Myhre, J. S. Fuglestvedt, F. Stordal, T. A. Larsen, R. S. Freckleton, and K. P. Shine, 1997: Effects of anthropogenic emissions on tropospheric ozone and its radiative forcing. *Journal of Geophysical Research – Atmospheres,* **102**(D23), 28,101–28,126.

Biraud, S., P. Ciais, M. Ramonet, P. Simmonds, V. Kazan, P. Monfray, S. O'Doherty, T. G. Spain, and S. G. Jennings, 2000: European greenhouse gas emissions estimated from continuous atmospheric measurements and radon 222 at Mace Head, Ireland. *Journal of Geophysical Research – Atmospheres,* **105**(D1), 1351–1366.

Boutonnet, J. C., P. Bingham, D. Calamari, C. de Rooij, J. Franklin, T. Kawano, J. M. Libre, A. McCulloch, G. Malinverno, J. M. Odom, G. M. Rusch, K. Smythe, I. Sobolev, R. Thompson, and J. M. Tiedje, 1999: Environmental risk assessment of trifluoroacetic acid. *Human and Ecological Risk Assessment,* **5**(1), 59–124.

Bouwman, A. F., D. S. Lee, W. A. H. Asman, F. J. Dentener, K. W. VanderHoek, and J. G. J. Olivier, 1997: A global high-resolution emission inventory for ammonia. *Global Biogeochemical Cycles,* **11**(4), 561–587.

Bowden, D. J., S. L. Clegg, and P. Brimblecombe, 1996: The Henry's law constant of trifluoroacetic acid and its partitioning into liquid water in the atmosphere. *Chemosphere,* **32**(2), 405–420.

Butler, J. H., M. Battle, M. L. Bender, S. A. Montzka, A. D. Clarke, E. S. Saltzman, C. M. Sucher, J. P. Severinghaus, and J. W. Elkins, 1999: A record of atmospheric halocarbons during the twentieth century from polar firn air. *Nature,* **399**(6738), 749–755.

Cahill, T. M., C. M. Thomas, S. E. Schwarzbach, and J. N. Seiber, 2001: Accumulation of trifluoroacetate in seasonal wetlands in California. *Environmental Science & Technology,* **35**(5), 820–825.

Carter, W. P. L., 1994: Development of ozone reactivity scales for volatile organic compounds. *Journal of Air and Waste Management Association,* **44**, 881–899.

Carter, W. P. L., 1998: *Updated maximum incremental reactivity scale for regulatory applications.* Air Pollution Research Center and College of Engineering, Center for Environmental Research and Technology University of California, Riverside, California. (Available at ftp://ftp.cert.ucr.edu/pub/carter/pubs/r98tab.pdf)

Christidis, N., M. D. Hurley, S. Pinnock, K. P. Shine, and T. J. Wallington, 1997: Radiative forcing of climate change by CFC-11 and possible CFC-replacements. *Journal of Geophysical Research – Atmospheres,* **102**(D16), 19,597–19,609.

Cicerone, R. J., 1979: Atmospheric carbon tetrafluoride: A nearly inert gas. *Science,* **206**, 59–61.

Clerbaux, C. and R. Colin, 1994: Determination of the infrared cross-sections and global warming potentials of 1,1,2-trifluoroethane (HFC-143). *Geophysical Research Letters,* **21**(22), 2377–2380.

Clerbaux, C., R. Colin, P. C. Simon, and C. Granier, 1993: Infrared cross-sections and global warming potentials of 10 alternative hydrohalocarbons. *Journal of Geophysical Research – Atmospheres,* **98**(D6), 10,491–10,497.

Collins, W. J., R. G. Derwent, C. E. Johnson, and D. S. Stevenson, 2002: The oxidation of organic compounds in the troposphere and their global warming potentials. *Climatic Change,* **52**(4), 453–479.

CORINAIR, 1996: CORINAIR 1990: Summary Report 1. European Environment Agency (EEA), Copenhagen.

Culbertson, J. A., J. M. Prins, E. P. Grimsrud, R. A. Rasmussen, M. A. K. Khalil, and M. J. Shearer, 2004: Observed trends for CF_3-containing compounds in background air at Cape Meares, Oregon, Point Barrow, Alaska, and Palmer Station, Antarctica. *Chemosphere,* **55**, 1109–1119.

Crutzen, P. J. and M. G. Lawrence, 2000: The impact of precipitation scavenging on the transport of trace gases: A 3-dimensional model sensitivity study. *Journal of Atmospheric Chemistry,* **37**(1), 81–112.

Cunnold, D. M., P. J. Fraser, R. F. Weiss, R. G. Prinn, P. G. Simmonds, B. R. Miller, F. N. Alyea, and A. J. Crawford, 1994: Global trends and annual releases of $CFCl_3$ and CF_2Cl_2 estimated from ALE/GAGE measurements from July 1978 to June 1991. *Journal of Geophysical Research,* **99**, 1107–1126.

Daniel, J. S. and S. Solomon, 1998: On the climate forcing of carbon monoxide. *Journal of Geophysical Research – Atmospheres,* **103**(D11), 13,249–13,260.

Daniel, J. S., S. Solomon, and D. L. Albritton, 1995: On the evaluation of halocarbon radiative forcing and global warming potentials. *Journal of Geophysical Research,* **100**, 1271–1285.

Debruyn, W. J., S. X. Duan, X. Q. Shi, P. Davidovits, D. R. Worsnop, M. S. Zahniser, and C. E. Kolb, 1992: Tropospheric heterogeneous chemistry of haloacetyl and carbonyl halides. *Geophysical Research Letters,* **19**(19), 1939–1942.

Dentener, F. J. and P. J. Crutzen, 1994: A three-dimensional model of the global ammonia cycle. *Journal of Atmospheric Chemistry,* **19**, 331–369.

Dentener, F., W. Peters, M. Krol, M. van Weele, P. Bergamaschi, and J. Lelieveld, 2003a: Interannual variability and trend of CH_4 lifetime as a measure for OH changes in the 1979–1993 time period. *Journal of Geophysical Research – Atmospheres,* **108**(D15), 4442, doi:10.1029/2002JD002916.

Dentener, F., M. van Weele, M. Krol, S. Houweling, and P. van Velthoven, 2003b: Trends and inter-annual variability of methane emissions derived from 1979–1993 global CTM simulations. *Atmospheric Chemistry and Physics,* **3**, 73–88.

Derwent, R. G., M. E. Jenkin, S. M. Saunders, and M. J. Pilling, 1998: Photochemical ozone creation potentials for organic compounds in northwest Europe calculated with a master chemical mechanism. *Atmospheric Environment,* **32**(14–15), 2429–2441.

Derwent, R. G., W. J. Collins, C. E. Johnson, and D. S. Stevenson, 2001: Transient behaviour of tropospheric ozone precursors in a global 3-D CTM and their indirect greenhouse effects. *Climatic Change,* **49**(4), 463–487.

EC, 2001: Directive 2001/81/EC of the European Parliament and of the Council on national emission ceilings for certain atmospheric pollutants. European Commission.

Ehhalt, D. H., F. Rohrer, A. Wahner, M. J. Prather, and D. R. Blake, 1998: On the use of hydrocarbons for the determination of tropospheric OH concentrations. *Journal of Geophysical Research,* **103**, 18,981–18,997.

Ellis, D. A., M. L. Hanson, P. K. Sibley, T. Shahid, N. A. Fineberg, K. R. Solomon, D. C. G. Muir, and S. A. Mabury, 2001: The fate and persistence of trifluoroacetic and chloroacetic acids in pond waters. *Chemosphere,* **42**(3), 309–318.

Ellis, D. A., J. W. Martin, A. O. De Silva, S. A. Mabury, M. D. Hurley, M. P. Sulbaek Andersen, and T. J. Wallington, 2004: Degradation of fluorotelomer alcohols: A likely atmospheric source of perfluorinated carboxylic acids. *Environmental Science and Technology,* **38**, 3316–3321.

Erisman, J. W., P. Grennfelt, and M. Sutton, 2003: The European perspective on nitrogen emission and deposition. *Environment International,* **29**(2–3), 311–325.

FAO, 2001: Global estimates of gaseous emissions of NH_3, NO and N_2O from agricultural land. Food and Agriculture Organization (FAO) of the United Nations, Rome, Italy. (Available at http://www.gm-unccd.org/FIELD/Multi/FAO/FAO3.pdf)

Finlayson-Pitts, B. J. and J. N. Pitts, 2000: *Chemistry of the Upper and Lower Atmosphere: Theory, Experiments, and Applications.* Academic Press, San Diego, CA, USA.

Fisher, D. A., C. H. Hales, W. C. Wang, M. K. W. Ko, and N. D. Sze, 1990: Model calculations of the relative effects of CFCs and their replacements on global warming. *Nature,* **344**(6266), 513–516.

Folkins, I., P. O. Wennberg, T. F. Hanisco, J. G. Anderson, and R. J. Salawitch, 1997: OH, HO_2, and NO in two biomass burning plumes: Sources of HO_x and implications for ozone production. *Geophysical Research Letters,* **24**(24), 3185–3188.

Forster, P. M. de F., J. B. Burkholder, C. Clerbaux, P. F. Coheur, M. Dutta, L. K. Gohar, M. D. Hurley, G. Myhre, R. W. Portmann, K. P. Shine, T. J. Wallington, and D. J. Wuebbles, 2005: Resolving uncertainties in the radiative forcing of HFC-134a. *Journal of Quantitative Spectroscopy and Radiative Transfer,* **93**(4), 447–460.

Frank, H., E. H. Christoph, O. Holm-Hansen, and J. L. Bullister, 2002: Trifluoroacetate in ocean waters. *Environmental Science & Technology,* **36**(1), 12–15.

Freckleton, R. S., E. J. Highwood, K. P. Shine, O. Wild, K. S. Law, and M. G. Sanderson, 1998: Greenhouse gas radiative forcing: Effects of averaging and inhomogeneities in trace gas distribution. *Quarterly Journal of the Royal Meteorological Society,* **124**(550), 2099–2127.

Fuglestvedt, J. S., T. K. Berntsen, I. S. A. Isaksen, H. T. Mao, X. Z. Liang, and W. C. Wang, 1999: Climatic forcing of nitrogen oxides through changes in tropospheric ozone and methane; global 3D model studies. *Atmospheric Environment,* **33**(6), 961–977.

Gohar, L. K., G. Myhre, and K. P. Shine, 2004: Updated radiative forcing estimates of four halocarbons. *Journal of Geophysical Research – Atmospheres,* **109**(D1), D01107, doi:10.1029/2003JD004320.

Gregory, J. M., W. J. Ingram, M. A. Palmer, G. S. Jones, P. A. Stott, R. B. Thorpe, J. A. Lowe, T. C. Johns, and K. D. Williams, 2004: A new method for diagnosing radiative forcing and climate sensitivity. *Geophysical Research Letters,* **31**(3), L03205, doi:10.1029/2003GL018747.

Hansen, J., M. Sato, and R. Ruedy, 1997: Radiative forcing and climate response. *Journal of Geophysical Research – Atmospheres,* **102**(D6), 6831–6864.

Hansen, J., M. Sato, L. Nazarenko, R. Ruedy, A. Lacis, D. Koch, I. Tegen, T. Hall, D. Shindell, B. Santer, P. Stone, T. Novakov, L. Thomason, R. Wang, Y. Wang, D. Jacob, S. Hollandsworth, L. Bishop, J. Logan, A. Thompson, R. Stolarski, J. Lean, R. Willson, S. Levitus, J. Antonov, N. Rayner, D. Parker, and J. Christy, 2002: Climate forcings in Goddard Institute for Space Studies SI2000 simulations. *Journal of Geophysical Research – Atmospheres,* **107**(D18), 4347, doi:10.1029/2001JD001143.

Harley, R. A., and G. R. Cass, 1995: Modeling the atmospheric concentrations of individual volatile organic compounds. *Atmospheric Environment,* **29**(8), 905–922.

Harnisch, J., R. Borchers, P. Fabian, H. W. Gaeggeler, and U. Schotterer, 1996: Effect of natural tetrafluoromethane. *Nature,* **384**, 32.

Harnisch, J. R., R. Borchers, P. Fabian, and M. Maiss, 1999: CF_4 and the age of mesospheric and polar vortex air. *Geophysical Research Letters,* **26**, 295–298.

Hauglustaine, D. A. and G. P. Brasseur, 2001: Evolution of tropospheric ozone under anthropogenic activities and associated radiative forcing of climate. *Journal of Geophysical Research – Atmospheres,* **106**(D23), 32,337–32,360.

Hayman, G. D. and R. G. Derwent, 1997: Atmospheric chemical reactivity and ozone-forming potentials of potential CFC replacements. *Environmental Science & Technology,* **31**(2), 327–336.

Highwood, E. J. and K. P. Shine, 2000: Radiative forcing and global warming potentials of 11 halogenated compounds. *Journal of Quantitative Spectroscopy & Radiative Transfer,* **66**(2), 169–183.

Highwood, E. J., K. P. Shine, M. D. Hurley, and T. J. Wallington, 1999: Estimation of direct radiative forcing due to non-methane hydrocarbons. *Atmospheric Environment,* **33**(5), 759–767.

Höhne, N., and J. Harnisch, 2002: Comparison of emission estimates derived from atmospheric measurements with national estimates of HFCs, PFCs and SF_6. In Proceedings of Third International Symposium on Non-CO_2 Greenhouse Gases (NCGG-3), Maastricht, the Netherlands, 21–23 January 2002.

Huang, J. and R. Prinn, 2002: Critical evaluation of emissions of potential new gases for OH estimation. *Journal of Geophysical Research,* **107**, 4748, doi:10.1029/2002JD002394.

Hurley, M., T. J. Wallington, G. Buchanan, L. Gohar, G. Marston, and K. Shine, 2005: IR spectrum and radiative forcing of CF_4 revisited. *Journal of Geophysical Research,* **110**, D02102, doi:10.1029/2004JD005201.

Hurst, D. F., P. S. Bakwin, and J. W. Elkins, 1998: Recent trends in the variability of halogenated trace gases over the United States. *Journal of Geophysical Research – Atmospheres,* **103**(D19), 25,299–25,306.

IPCC, 1990: *Scientific Assessment of Climate change – Report of Working Group I* [Houghton, J. T., G. J. Jenkins, and J. J. Ephraums (eds.)]. Cambridge University Press, Cambridge, United Kingdom, and New York, NY, USA, 365 pp.

IPCC, 1994: *Radiative Forcing of Climate Change and an Evaluation of the IPCC IS92 Emissions Scenarios* [Houghton, J. T., L. G. Meira Filho, J. Bruce, H. Lee, B. A. Callander, E. Haites, N. Harris, and K. Maskell (eds.)]. Cambridge University Press, Cambridge, United Kingdom, and New York, NY, USA, 339 pp.

IPCC, 1996: *Climate Change 1995: The Science of Climate Change. Contribution of Working Group I to the Second Assessment Report of the Intergovernmental Panel on Climate Change* [Houghton, J. T., L. G. Meira Filho, B. A. Callander, N. Harris, A. Kattenberg, and K. Maskell (eds.)]. Cambridge University Press, Cambridge, United Kingdom, and New York, NY, USA, 572 pp.

IPCC, 2000: *Emissions Scenarios.* Special Report of the Intergovernmental Panel on Climate Change [Nakicenovic, N. and R. Swart (eds.)]. Cambridge University Press, Cambridge, United Kingdom, and New York, NY, USA, 570 pp.

IPCC, 2001: *Climate Change 2001: The Scientific Basis. Contribution of Working Group I to the Third Assessment Report of the Intergovernmental Panel on Climate Change* [Houghton, J. T., Y. Ding, D. J. Griggs, M. Noguer, P. J. van der Linden, X. Dai, K. Maskell, and C. A. Johnson (eds.)]. Cambridge University Press, Cambridge, United Kingdom, and New York, NY, USA, 944 pp.

Isaksen, I. S. A. and Ö. Hov, 1987: Calculation of trends in the tropospheric concentration of O_3, OH, CH_4, and NO_x. *Tellus,* **39B**, 271–285.

Jacquinet-Husson, N., E. Arie, J. Ballard, A. Barbe, G. Bjoraker, B. Bonnet, L. R. Brown, C. Camy-Peyret, J. P. Champion, A. Chedin, A. Chursin, C. Clerbaux, G. Duxbury, J. M. Flaud, N. Fourrie, A. Fayt, G. Graner, R. Gamache, A. Goldman, V. Golovko, G. Guelachvili, J. M. Hartmann, J. C. Hilico, J. Hillman, G. Lefevre, E. Lellouch, S. N. Mikhailenko, O. V. Naumenko, V. Nemtchinov, D. A. Newnham, A. Nikitin, A. Orphal, A. Perrin, D. C. Reuter, C. P. Rinsland, L. Rosenmann, L. S. Rothman, N. A. Scott, J. Selby, L. N. Sinitsa, J. M. Sirota, A. M. Smith, K. M. Smith, V. G. Tyuterev, R. H. Tipping, S. Urban, P. Varanasi, and M. Weber, 1999: The 1997 spectroscopic GEISA databank. *Journal of Quantitative Spectroscopy & Radiative Transfer,* **62**(2), 205–254.

Jain, A. K., B. P. Briegleb, K. Minschwaner, and D. J. Wuebbles, 2000: Radiative forcings and global warming potentials of 39 greenhouse gases. *Journal of Geophysical Research – Atmospheres,* **105**(D16), 20,773–20,790.

Jöckel, P., C. A. M. Brenninkmeijer, and P. J. Crutzen, 2003: A discussion on the determination of atmospheric OH and its trends. *Atmospheric Chemistry and Physics,* **3**, 107–118.

Johnson, C. E. and R. G. Derwent, 1996: Relative radiative forcing consequences of global emissions of hydrocarbons, carbon monoxide and NO_x from human activities estimated with a zonally-averaged two-dimensional model. *Climatic Change,* **34**(3–4), 439–462.

Jordan, A. and H. Frank, 1999: Trifluoroacetate in the environment. Evidence for sources other than HFC/HCFCs. *Environmental Science & Technology,* **33**(4), 522–527.

Karlsdottir, S. and I. S. A. Isaksen, 2000: Changing methane lifetime: Possible cause for reduced growth. *Geophysical Research Letters,* **27**(1), 93–96.

Kim, B. R., M. T. Suidan, T. J. Wallington, and X. Du, 2000: Biodegradability of trifluoroacetic acid. *Environmental Engineering Science,* **17**(6), 337–342.

Ko, M. K. W., R. L. Shia, N. D. Sze, H. Magid, and R. G. Bray, 1999: Atmospheric lifetime and global warming potential of HFC-245fa. *Journal of Geophysical Research – Atmospheres,* **104**(D7), 8173–8181.

Kotamarthi, V. R., J. M. Rodriguez, M. K. W. Ko, T. K. Tromp, N. D. Sze, and M. J. Prather, 1998: Trifluoroacetic acid from degradation of HCFCs and HFCs: A three-dimensional modeling study. *Journal of Geophysical Research – Atmospheres,* **103**(D5), 5747–5758.

Krol, M. and J. Lelieveld, 2003: Can the variability in tropospheric OH be deduced from measurements of 1,1,1-trichloroethane (methyl chloroform)? *Journal of Geophysical Research – Atmospheres,* **108**(D3), 4125, doi:10.1029/2001JD002040.

Krol, M., P. J. van Leeuwen, and J. Lelieveld, 1998: Global OH trend inferred from methylchloroform measurements. *Journal of Geophysical Research – Atmospheres,* **103**(D9), 10,697–10,711.

Krol, M., P. J. van Leeuwen, and J. Lelieveld, 2001: Comment on 'Global OH trend inferred from methylchloroform measurements' by Maarten Krol *et al.* – Reply. *Journal of Geophysical Research – Atmospheres,* **106**(D19), 23,159–23,164.

Krol, M. C., J. Lelieveld, D. E. Oram, G. A. Sturrock, S. A. Penkett, C. A. M. Brenninkmeijer, V. Gros, J. Williams, and H. A. Scheeren, 2003: Continuing emissions of methyl chloroform from Europe. *Nature,* **421**, 131–135.

Lawrence, M. G., P. Jockel, and R. von Kuhlmann, 2001: What does the global mean OH concentration tell us? *Atmospheric Chemistry and Physics,* **1**, 37–49.

Lelieveld, J., P. J. Crutzen, and F. J. Dentener, 1998: Changing concentration, lifetime and climate forcing of atmospheric methane. *Tellus Series B – Chemical and Physical Meteorology,* **50**(2), 128–150.

Lelieveld, J., W. Peters, F. J. Dentener, and M. C. Krol, 2002: Stability of tropospheric hydroxyl chemistry. *Journal of Geophysical Research – Atmospheres,* **107**(D23), 4715, doi:10.1029/2002JD002272.

Maiss, M. and C. A. M. Benninkmeijer, 1998: Atmospheric SF$_6$ trends, sources and prospects. *Environmental Science and Technology,* **32**, 3077–3086.

Manning, A. J., D. B. Ryall, R. G. Derwent, P. G. Simmonds, and S. O'Doherty, 2003: Estimating European emissions of ozone-depleting and greenhouse gases using observations and a modeling back-attribution technique. *Journal of Geophysical Research – Atmospheres,* **108**(D14), 4405, doi:10.1029/2002JD002312.

Marland, G., T. A. Boden, and R. J. Andres, 2003: Global, regional, and national fossil fuel CO$_2$ emissions. In Trends: A Compendium of Data on Global Change, Carbon Dioxide Information Analysis Center, Oak Ridge National Laboratory, US Department of Energy, Oak Ridge, TN, USA (Available at http://cdiac.esd.ornl.gov/trends/emis/em_cont.htm)

Martin, J. W., J. Franklin, M. L. Hanson, K. R. Solomon, S. A. Mabury, D. A. Ellis, B. F. Scott, and D. C. G. Muir, 2000: Detection of chlorodifluoroacetic acid in precipitation: A possible product of fluorocarbon degradation. *Environmental Science & Technology,* **34**(2), 274–281.

Martin, J. W., S. A. Mabury, C. S. Wong, F. Noventa, K. R. Solomon, M. Alaee, and D. C. G. Muir, 2003: Airborne haloacetic acids. *Environmental Science & Technology,* **37**(13), 2889–2897.

Martin, J. W., M. M. Smithwick, B. M. Braune, P. F. Hoekstra, D. C. G. Muir, and S. A. Mabury, 2004: Identification of long-chain perfluorinated acids in biota from the Canadian Arctic. *Environmental Science & Technology,* **38**(2), 373–380.

McCulloch, A., P. Ashford, and P. M. Midgley, 2001: Historic emissions of fluorotrichloromethane (CFC-11) based on a market survey. *Atmospheric Environment,* **35**(26), 4387–4397.

McCulloch, A, P. M. Midgley, and P. Ashford, 2003: Releases of refrigerant gases (CFC-12, HCFC-22 and HFC-134a) to the atmosphere. *Atmospheric Environment,* **37**(7), 889–902.

McDaniel, A. H., C. A. Cantrell, J. A. Davidson, R. E. Shetter, and J. G. Calvert, 1991: The temperature-dependent, infrared absorption cross sections for the chlorofluorocarbons – CFC-11, CFC-12, CFC-13, CFC-14, CFC-22, CFC-113, CFC-114, and CFC-115. *Journal of Atmospheric Chemistry,* **12**(3), 211–227.

Mickley, L. J., P. P. Murti, D. J. Jacob, J. A. Logan, D. M. Koch, and D. Rind, 1999: Radiative forcing from tropospheric ozone calculated with a unified chemistry-climate model. *Journal of Geophysical Research – Atmospheres,* **104**(D23), 30,153–30,172.

Møgelberg, T. E., O. J. Nielsen, J. Sehested, T. J. Wallington, and M. D. Hurley, 1994: Atmospheric chemistry of CF$_3$COOH – Kinetics of the reaction with OH radicals. *Chemical Physics Letters,* **226**(1–2), 171–177.

Montzka, S. A., C. M. Spivakovsky, J. H. Butler, J. W. Elkins, L. T. Lock, and D. J. Mondeel, 2000: New observational constraints for atmospheric hydroxyl on global and hemispheric scales. *Science,* **288**(5465), 500–503.

Moody, C. A. and J. A. Field, 1999: Determination of perfluorocarboxylates in groundwater impacted by fire-fighting activity. *Environmental Science & Technology,* **33**(16), 2800–2806.

Moody, C. A., W. C. Kwan, J. W. Martin, D. C. G. Muir, and S. A. Mabury, 2001: Determination of perfluorinated surfactants in surface water samples by two independent analytical techniques: Liquid chromatography/tandem mass spectrometry and F-19 NMR. *Analytical Chemistry,* **73**(10), 2200–2206.

Moody, C. A., J. W. Martin, W. C. Kwan, D. C. G. Muir, and S. C. Mabury, 2002: Monitoring perfluorinated surfactants in biota and surface water samples following an accidental release of fire-fighting foam into Etohicoke Creek. *Environmental Science & Technology,* **36**(4), 545–551.

Morris, R. A., T. M. Miller, A. A. Viggiano, J. F. Paulson, S. Solomon, and G. Reid, 1995: Effects of electron and ion reactions on atmospheric lifetimes of fully fluorinated compounds. *Journal of Geophysical Research – Atmospheres,* **100**(D1), 1287–1294.

Müller, J. F. and G. Brasseur, 1999: Sources of upper tropospheric HO_x: A three-dimensional study. *Journal of Geophysical Research – Atmospheres,* **104**(D1), 1705–1715.

Myhre, G. and F. Stordal, 1997: Role of spatial and temporal variations in the computation of radiative forcing and GWP. *Journal of Geophysical Research – Atmospheres,* **102**(D10), 11,181–11,200.

Naik, V., A. K. Jain, K. O. Patten, and D. J. Wuebbles, 2000: Consistent sets of atmospheric lifetimes and radiative forcings on climate for CFC replacements: HCFCs and HFCs. *Journal of Geophysical Research – Atmospheres,* **105**(D5), 6903–6914.

Nemtchinov, V. and P. Varanasi, 2004: Absorption cross-sections of HFC-134a in the spectral region between 7 and 12 μm. *Journal of Quantitative Spectroscopy & Radiative Transfer,* **83**(3–4), 285–294.

Nielsen, O. J., B. F. Scott, C. Spencer, T. J. Wallington, and J. C. Ball, 2001: Trifluoroacetic acid in ancient freshwater. *Atmospheric Environment,* **35**(16), 2799–2801.

Nölle, A., H. Heydtmann, R. Meller, W. Schneider, and G. K. Moortgat, 1992: UV absorption-spectrum and absorption cross-sections of COF_2 at 296 K in the range 200–230 nm. *Geophysical Research Letters,* **19**(3), 281–284.

O'Doherty, S., D. M. Cunnold, A. Manning, B. R. Miller, R. H. J. Wang, P. B. Krummel, P. J. Fraser, P. G. Simmonds, A. McCulloch, R. F. Weiss, P. Salameh, L. W. Porter, R. G. Prinn, J. Huang, G. Sturrock, D. Ryall, R. G. Derwent, and S. A. Montzka, 2004: Rapid growth of hydrofluorocarbon 134a and hydrochlorofluorocarbons 141b, 142b, and 22 from Advanced Global Atmospheric Experiment (AGAGE) observations at Cape Grim, Tasmania, and Mace head, Ireland. *Journal of Geophysical Research – Atmospheres,* **109**, D06310, doi:10.1029/2003JD004277.

Olivier, J. G. J., 2002: Part III: Greenhouse gas emissions: 1. Shares and trends in greenhouse gas emissions; 2. Sources and methods; Greenhouse gas emissions for 1990 and 1995. In: *CO_2 emissions from fuel combustion 1971–2000,* 2002 Edition, pp. III.1–III.31. International Energy Agency (IEA), Paris.

Olivier, J. G. J. and J. J. M. Berdowski, 2001: Global emissions sources and sinks. In: *The Climate System* [Berdowski, J., R. Guicherit, and B. J. Heij (eds.)]. A. A. Balkema Publishers/Swets & Zeitlinger Publishers, Lisse, The Netherlands, pp. 33–78.

Olivier, J. G. J., A. F. Bouwman, C. W. M. v. d. Maas, J. J. M. Berdowski, C. Veldt, J. P. J. Bloos, A. J. H. Visschedijk, P. Y. J. Zandveld, and J. L. Haverlag, 1996: Description of EDGAR Version 2.0: A set of global emission inventories of greenhouse gases and ozone-depleting substances for all anthropogenic and most natural sources on a per country basis and on 1 degree x 1 degree grid. RIVM Report 771060002, National Institute of Public Health and the Environment (RIVM), Bilthoven, The Netherlands, 171 pp. (Available at http://www.rivm.nl/bibliotheek/rapporten/771060002.pdf)

Oram, D. E., W. T. Sturges, S. A. Penkett, A. McCulloch, and P. J. Fraser, 1998: Growth of fluoroform (CHF_3, HFC-23) in the background atmosphere. *Geophysical Research Letters,* **25**, 35–38.

Orkin, V. L., A. G. Guschin, I. K. Larin, R. E. Huie, and M. J. Kurylo, 2003: Measurements of the infrared absorption cross-sections of haloalkanes and their use in a simplified calculational approach for estimating direct global warming potentials. *Journal of Photochemistry and Photobiology A – Chemistry,* **157**(2–3), 211–222.

Pinnock, S., M. D. Hurley, K. P. Shine, T. J. Wallington, and T. J. Smyth, 1995: Radiative forcing of climate by hydrochlorofluorocarbons and hydrofluorocarbons. *Journal of Geophysical Research – Atmospheres,* **100**(D11), 23,227–23,238.

Prather, M. J., 1985: Continental sources of halocarbons and nitrous oxide. *Nature,* **317**(6034), 221–225.

Prather, M. J., 1996: Time scales in atmospheric chemistry: Theory, GWPs for CH_4 and CO, and runaway growth. *Geophysical Research Letters,* **23**(19), 2597–2600.

Prather, M. J., 2002: Lifetimes of atmospheric species: Integrating environmental impacts. *Geophysical Research Letters,* **29**(22), 2063, doi:10.1029/2002GL016299.

Prather, M. J. and D. J. Jacob, 1997: A persistent imbalance in HO_x and NO_x photochemistry of the upper troposphere driven by deep tropical convection. *Geophysical Research Letters,* **24**(24), 3189–3192.

Prather, M. and R. T. Watson, 1990: Stratospheric ozone depletion and future levels of atmospheric chlorine and bromine. *Nature,* **344**, 729–734.

Prinn, R. G., R. F. Weiss, B. R. Miller, J. Huang, F. N. Alyea, D. M. Cunnold, P. J. Fraser, D. E. Hartley, and P. G. Simmonds, 1995: Atmospheric trends and lifetime of CH_3CCl_3 and global OH concentrations. *Science,* **269**, 187–192.

Prinn, R. G., R. F. Weiss, P. J. Fraser, P. G. Simmonds, D. M. Cunnold, F. N. Alyea, S. O'Doherty, P. Salameh, B. R. Miller, J. Huang, R. H. J. Wang, D. E. Hartley, C. Harth, L. P. Steele, G. Sturrock, P. M. Midgley, and A. McCulloch, 2000: A history of chemically and radiatively important gases in air deduced from ALE/GAGE/AGAGE. *Journal of Geophysical Research – Atmospheres,* **105**(D14), 17,751–17,792.

Prinn, R. G., J. Huang, R. F. Weiss, D. M. Cunnold, P. J. Fraser, P. G. Simmonds, A. McCulloch, C. Harth, P. Salameh, S. O'Doherty, R. H. J. Wang, L. Porter, and B. R. Miller, 2001: Evidence for substantial variations of atmospheric hydroxyl radicals in the past two decades. *Science,* **292**(5523), 1882–1888.

Prinn, R. G., and J. Huang, 2001: Comment on 'Global OH trend inferred from methylchloroform' by M. Krol *et al. Journal of Geophysical Research,* **106**, 23,151–23,157.

Ravishankara, A. R., S. Solomon, A. A. Turnipseed, and R. F. Warren, 1993: Atmospheric lifetimes of long-lived halogenated species. *Science,* **259**(5092), 194–199.

Ravishankara, A. R., A. A. Turnipseed, N. R. Jensen, S. Barone, M. Mills, C. J. Howard, and S. Solomon, 1994: Do hydrofluorocarbons destroy stratospheric ozone? *Science,* **263**(5143), 71–75.

Reimann, S., D. Schaub, K. Stemmler, D. Folini, M. Hill, P. Hofer, B. Buchmann, P. G. Simmonds, B. Greally, and S. O'Doherty, 2004: Halogenated greenhouse gases at the Swiss High Alpine Site of Jungfraujoch (3580 m asl): Continuous measurements and their use for regional European source allocation. *Journal of Geophysical Research,* **109,** D05307, doi:10.1029/2003JD003923.

Rodriguez, J. M., M. K. W. Ko, N. D. Sze, and C. W. Heisey, 1993: Two-dimensional assessment of the degradation of HFC-134a: Tropospheric accumulations and deposition of trifluoroacetic acid. In: *Kinetics and Mechanisms for the Reactions of Halogenated Organic Compounds in the Troposphere,* AFEAS, Univ. College, Dublin.

Rothman, L. S., A. Barbe, D. C. Benner, L. R. Brown, C. Camy-Peyret, M. R. Carleer, K. Chance, C. Clerbaux, V. Dana, V. M. Devi, A. Fayt, J. M. Flaud, R. R. Gamache, A. Goldman, D. Jacquemart, K. W. Jucks, W. J. Lafferty, J. Y. Mandin, S. T. Massie, V. Nemtchinov, D. A. Newnham, A. Perrin, C. P. Rinsland, J. Schroeder, K. M. Smith, M. A. H. Smith, K. Tang, R. A. Toth, J. Vander Auwera, P. Varanasi, and K. Yoshino, 2003: The HITRAN molecular spectroscopic database: Edition of 2000 including updates through 2001. *Journal of Quantitative Spectroscopy & Radiative Transfer,* **82**(1–4), 5–44.

Ryall, D. B., R. G. Derwent, A. J. Manning, P. G. Simmonds, and S. O'Doherty, 2001: Estimating source regions of European emissions of trace gases from observations at Mace Head. *Atmospheric Environment,* **35**(14), 2507–2523.

Sander, S. P., R. R. Friedl, W. B. DeMore, D. M. Golden, M. J. Kurylo, R. F. Hampson, R. E. Huie, G. K. Moortgat, A. R. Ravishankara, C. E. Kolb, and M. J. Molina, 2000: Chemical kinetics and photochemical data for use in stratospheric modeling. Supplement to Evaluation 12: Update of key reactions. Evaluation Number 13. JPL Publication No. 00-3, National Aeronautics and Space Administration, Jet Propulsion Laboratory, California Institute of Technology, Pasadena, CA. (Available at http://jpldataeval.jpl.nasa.gov/pdf/JPL_00-03.pdf)

Sander, S. P., R. R. Friedl, A. R. Ravishankara, D. M. Golden, C. E. Kolb, M. J. Kurylo, R. E. Huie, V. L. Orkin, M. J. Molina, G. K. Moortgat, and B. J. Finlayson-Pitts, 2003: Chemical kinetics and photochemical data for use in atmospheric studies. Evaluation Number 14. JPL Publication No. 02-25, National Aeronautics and Space Administration, Jet Propulsion Laboratory, California Institute of Technology, Pasadena, CA. (Available at http://jpldataeval.jpl.nasa.gov/pdf/JPL_02-25_rev02.pdf)

Schmidt, M., H. Glatzel-Mattheier, H. Sartorius, D. E. Worthy, and I. Levin, 2001: Western European N_2O emissions: A top-down approach based on atmospheric observations. *Journal of Geophysical Research – Atmospheres,* **106**(D6), 5507–5516.

Scott, B. F., D. MacTavish, C. Spencer, W. M. J. Strachan, and D. C. G. Muir, 2000: Haloacetic acids in Canadian lake waters and precipitation. *Environmental Science & Technology,* **34**(20), 4266–4272.

Seinfeld, J. H. and S. N. Pandis, 1998: *Atmospheric Chemistry and Physics: From Air Pollution to Climate Change.* John Wiley, New York, 1326 pp.

Seinfeld, J. H. and J. Pankow, 2003: Organic atmospheric particulate matter. *Annual Review of Physical Chemistry,* **54,** 121–140.

Shindell, D. T., J. L. Grenfell, D. Rind, V. Grewe, and C. Price, 2001: Chemistry-climate interactions in the Goddard Institute for Space Studies general circulation model 1. Tropospheric chemistry model description and evaluation. *Journal of Geophysical Research – Atmospheres,* **106**(D8), 8047–8075.

Shine, K. P., J. Cook, E. J. Highwood, and M. M. Joshi, 2003: An alternative to radiative forcing for estimating the relative importance of climate change mechanisms. *Geophysical Research Letters,* **30**(20), 2047, doi:10.1029/2003GL018141.

Shine, K. P., J. S. Fuglestvedt, K. Hailemariam, and N. Stuber, 2005: Alternatives to the global warming potential for comparing climate impacts of emissions of greenhouse gases. *Climatic Change,* **68**(3), 281–302.

Sihra, K., M. D. Hurley, K. P. Shine, and T. J. Wallington, 2001: Updated radiative forcing estimates of 65 halocarbons and nonmethane hydrocarbons. *Journal of Geophysical Research – Atmospheres,* **106**(D17), 20,493–20,505.

Simmonds, P. G., R. G. Derwent, A. McCulloch, S. O'Doherty, and A. Gaudry, 1996: Long-term trends in concentrations of halocarbons and radiatively active trace gases in Atlantic and European air masses monitored at Mace Head, Ireland from 1987–1994. *Atmospheric Environment,* **30**(23), 4041–4063.

Simmonds, P. G., S. O'Doherty, J. Huang, R. Prinn, R. G. Derwent, D. Ryall, G. Nickless, and D. Cunnold, 1998: Calculated trends and the atmospheric abundance of 1,1,1,2-tetrafluoroethane, 1,1-dichloro-1-fluoroethane, and 1-chloro-1,1-difluoroethane using automated in-situ gas chromatography mass spectrometry measurements recorded at Mace Head, Ireland, from October 1994 to March 1997. *Journal of Geophysical Research – Atmospheres,* **103**(D13), 16,029–16,037.

Singh, H. B., M. Kanakidou, P. J. Crutzen, and D. J. Jacob, 1995: High concentrations and photochemical fate of oxygenated hydrocarbons in the global troposphere. *Nature,* **378**(6552), 50–54.

SMA-GDF, 2003: Inventario de Emisiones a la Atmósfera, Zona Metropolitana del Valle de México 2000. Secretaría del Medio Ambiente, Gobierno del Distrito Federal. Mexico.

Solomon, K. R., X. Y. Tang, S. R. Wilson, P. Zanis, and A. F. Bais, 2003: Changes in tropospheric composition and air quality due to stratospheric ozone depletion. *Photochemical & Photobiological Sciences,* **2**(1), 62–67.

Spivakovsky, C. M., J. A. Logan, S. A. Montzka, Y. J. Balkanski, M. Foreman-Fowler, D. B. A. Jones, L. W. Horowitz, A. C. Fusco, C. A. M. Brenninkmeijer, M. J. Prather, S. C. Wofsy, and M. B. McElroy, 2000: Three-dimensional climatological distribution of tropospheric OH: Update and evaluation. *Journal of Geophysical Research – Atmospheres,* **105**(D7), 8931–8980.

Stordal, F., N. Schmidbauer, P. Simmonds, B. Greally, A. McCulloch, S. Reimann, M. Maione, E. Mahieu, J. Notholt, I. Isaksen, and R. G. Derwent, 2002: System for Observation of halogenated Greenhouse gases in Europe (SOGE): Monitoring and modelling yielding verification and impacts of emissions. In Proceedings of the Third Symposium on Non-CO_2 Greenhouse Gases (NCGG-3), Maastricht, the Netherlands, 21–23 January 2002.

Streets, D. G., T. C. Bond, G. R. Carmichael, S. D. Fernandes, Q. Fu, D. He, Z. Kimont, S. M. Nelson, N. Y. Tsai, M. Q. Wang, J.-H. Woo, and K. F. Tarber, 2003: An inventory of gaseous and primary aerosol emissions in Asia in the year 2000. *Journal of Geophysical Research – Atmospheres,* **108**(D21), 8809, doi:10.1029/2002JD003093.

Sturges, W. T., H. P. McIntyre, S. A. Penkett, J. Chappellaz, J. M. Barnola, R. Mulvaney, E. Atlas, and V. Stroud, 2001: Methyl bromide, other brominated methanes, and methyl iodide in polar firn air. *Journal of Geophysical Research – Atmospheres,* **106**(D2), 1595–1606.

Sturrock, G. A., D. M. Etheridge, C. M. Trudinger, P. J. Fraser, and A. M. Smith, 2002: Atmospheric histories of halocarbons from analysis of Antarctic firn air: Major Montreal Protocol species. *Journal of Geophysical Research – Atmospheres,* **107**(D24), 4765, doi:10.1029/2002JD002548.

Takahashi, K. T. Nakayama, Y. Matsumi, S. Solomon, T. Gejo, E. Shigemasa, and T. J. Wallington, 2002: Atmospheric lifetime of SF$_5$CF$_3$. *Geophysical Research Letters,* **29**(15), 1712, doi:10.1029/2002GL015356.

Taniguchi, N., T. J. Wallington, M. D. Hurley, A. G. Guschin, L. T. Molina and M. J. Molina, 2003: Atmospheric chemistry of C$_2$F$_5$C(O)CF(CF$_3$)$_2$: Photolysis and reaction with Cl atoms, OH radicals, and ozone. *Journal of Physical Chemistry A,* **107**, 2674–2679.

Tang, X., S. Madronich, T. Wallington, and D. Calamari, 1998: Changes in tropospheric composition and air quality. *Journal of Photochemistry and Photobiology B – Biology,* **46**(1–3), 83–95.

Thom, M., R. Bosinger, M. Schmidt, and I. Levin, 1993: The regional budget of atmospheric methane of a highly populated area. *Chemosphere,* **26**(1–4), 143–160.

Tromp, T. K., M. K. W. Ko, J. M. Rodriguez, and N. D. Sze, 1995: Potential accumulation of a CFC-replacement degradation product in seasonal wetlands. *Nature,* **376**, 327–330.

Thunis, P. and C. Cuvelier, 2004: CityDelta: A European modelling intercomparison to predict air quality in 2010. *Proceedings of the 26th NATO/CCMS international technical meeting on air pollution modelling and its application XVI.* C. Borrego and S. Incecik, (eds.), 26–30 May 2003 Istanbul, Kluwer Academic/Plenum Publishers, 205–214.

Tokuhashi, K., A. Takahashi, M. Kaise, S. Kondo, Λ. Sekiya, S. Yamashita, and H. Ito, 2000: Rate constants for the reactions of OH radicals with CH$_3$OCF$_2$CHF$_2$, CHF$_2$OCH$_2$CF$_2$CHF$_2$, CHF$_2$OCH$_2$CF$_2$CF$_3$, and CF$_3$CH$_2$OCF$_2$CHF$_2$ over the temperature range 250–430 K. *Journal of Physical Chemistry A,* **104**(6), 1165–1170.

Trudinger, C. M., D. M. Etheridge, P. J. Rayner, I. G. Enting, G. A. Sturrock, and R. L. Langenfelds, 2002: Reconstructing atmospheric histories from measurements of air composition in firn. *Journal of Geophysical Research – Atmospheres,* **107**(D24), 4780, doi:10.1029/2002JD002545.

US EPA, 2003: Preliminary risk assessment of the developmental toxicity associated with exposure to perflourooctanoic acids and its salts. Office of Pollution Prevention and Toxics, Risk Assessment Division, United States Environmental Protection Agency. (Available at http://www.epa.gov/opptintr/pfoa/pfoara.pdf)

UN-ECE, 2003: Present state of emission data. United Nations Economic Commission for Europe, EB.AIR/GE.1/2003/6.

Van Aardenne, J. A., F. J. Dentener, J. G. J. Olivier, C. Goldewijk, and J. Lelieveld, 2001: A 1° × 1° resolution data set of historical anthropogenic trace gas emissions for the period 1890–1990. *Global Biogeoehemical Cycles,* **15**(4), 909–928.

Varanasi, P., Z. Li, V. Nemtchinov, and A. Cherukuri, 1994: Spectral absorption-coefficient data on HCFC-22 and SF$_6$ for remote-sensing applications. *Journal of Quantitative Spectroscopy & Radiative Transfer,* **52**(3–4), 323–332.

Vega, E., V. Mugica, R. Carmona, and E. Valencia, 2000: Hydrocarbon source apportionment in Mexico City using the chemical mass balance receptor model. *Atmospheric Environment,* **34**(24), 4121–4129.

Wallington, T. J. and M. D. Hurley, 1993: Atmospheric chemistry of HC(O)F – Reaction with OH radicals. *Environmental Science & Technology,* **27**(7), 1448–1452.

Wallington, T. J., W. F. Schneider, D. R. Worsnop, O. J. Nielsen, J. Sehested, W. J. Debruyn, and J. A. Shorter, 1994: The environmental impact of CFC replacements – HFCs and HCFCs. *Environmental Science & Technology,* **28**(7), A320–A326.

Wallington, T. J., W. F. Schneider, J. Sehested, and O. J. Nielsen, 1995: Hydrofluorocarbons and stratospheric ozone. *Faraday Discussions,* **100**, 55–64.

Wallington, T. J., M. D. Hurley, J. M. Fracheboud, J. J. Orlando, G. S. Tyndall, J. Sehested, T. E. Mogelberg, and O. J. Nielsen, 1996: Role of excited CF$_3$CFHO radicals in the atmospheric chemistry of HFC-134a. *Journal of Physical Chemistry,* **100**(46), 18,116–18,122.

Wang, Y. H. and D. J. Jacob, 1998: Anthropogenic forcing on tropospheric ozone and OH since preindustrial times. *Journal of Geophysical Research – Atmospheres,* **103**(D23), 31,123–31,135.

Wang, C., and R. Prinn, 1999: Impact of emissions, chemistry and climate on atmospheric carbon monoxide: 100 year predictions from a global chemistry model. *Chemosphere (Global Change),* **1**, 73–81.

Wang, J. S., J. A. Logan, M. B. McElroy, B. N. Duncan, I. A. Megretskaia, and R. M. Yantosca, 2004: A 3-D model analysis of the slowdown and interannual variability in the methane growth rate from 1988 to 1997. *Global Biogeochemical Cycles,* **18**, GB3011, doi:10.1029/3003GB002180.

Warneck, P., 1999: *Chemistry of the Natural Atmosphere.* Academic Press, San Diego, CA, USA.

Warwick, N. J., S. Bekki, K. S. Law, E. G. Nisbet, and J. A. Pyle, 2002: The impact of meteorology on the interannual growth rate of atmospheric methane. *Geophysical Research Letters,* **29**(20), 1947, doi:10.1029/2002GL015282.

Weber, R. J., P. H. McMurry, R. L. Mauldin, D. J. Tanner, F. L. Eisele, A. D. Clarke, and V. N. Kapustin, 1999: New particle formation in the remote troposphere: A comparison of observations at various sites. *Geophysical Research Letters,* **26**(3), 307–310.

Wennberg, P. O., T. F. Hanisco, L. Jaegle, D. J. Jacob, E. J. Hintsa, E. J. Lanzendorf, J. G. Anderson, R. S. Gao, E. R. Keim, S. G. Donnelly, L. A. Del Negro, D. W. Fahey, S. A. McKeen, R. J. Salawitch, C. R. Webster, R. D. May, R. L. Herman, M. H. Proffitt, J. J. Margitan, E. L. Atlas, S. M. Schauffler, F. Flocke, C. T. McElroy, and T. P. Bui, 1998: Hydrogen radicals, nitrogen radicals, and the production of O_3 in the upper troposphere. *Science,* **279**(5347), 49–53.

Wild, O., M. J. Prather, and H. Akimoto, 2001: Indirect long-term global radiative cooling from NO_x emissions. *Geophysical Research Letters,* **28**(9), 1719–1722.

WMO, 1995: *Scientific Assessment of Ozone Depletion: 1994.* Global Ozone Research and Monitoring Project – Report No. 37, World Meteorological Organization (WMO), Geneva.

WMO, 1999: *Scientific Assessment of Ozone Depletion: 1998.* Global Ozone Research and Monitoring Project – Report No. 44, World Meteorological Organization (WMO), Geneva.

WMO, 2003: *Scientific Assessment of Ozone Depletion: 2002.* Global Ozone Research and Monitoring Project – Report No. 47, World Meteorological Organization, Geneva, 498 pp.

Yvon-Lewis, S. A. and J. H. Butler, 2002: Effect of oceanic uptake on atmospheric lifetimes of selected trace gases. *Journal of Geophysical Research – Atmospheres,* **107**(D20), 4414, doi:10.1029/2001JD001267.

3

Methodologies

Coordinating Lead Authors
Jochen Harnisch (Germany), José Roberto Moreira (Brazil)

Lead Authors
Paul Atkins (United Kingdom), Daniel Colbourne (United Kingdom), Martin Dieryckx (Belgium),
H.S. Kaprwan (India), Fred Keller (USA), Archie McCulloch (United Kingdom),
Stephan Sicars (Germany), Bishnu Tulsie (St. Lucia), Jinhuang Wu (China)

Contributing Authors
Paul Ashford (United Kingdom), Kirsten Halsnæs (Denmark), Atsushi Inaba (Japan)

Review Editors
Pauline Midgley (United Kingdom), Mahi Sideridou (Greece)

Contents

EXECUTIVE SUMMARY

The phase out of CFCs and HCFCs under the Montreal Protocol requires the selection of replacement technologies, and in many cases are these alternative fluids. These technologies have differing impacts on global climate change, health, safety and other environmental endpoints, and different private and social costs. Analyses that focus on one or more of these types of impacts can help decision-makers to make choices about competing replacement technologies. However the outcomes of such analyses are influenced by many factors not intrinsic to the technologies. Examples of these are the analytical approach (e.g., top-down compared with bottom-up), the degree of product or process optimization, service and disposal practices, regional circumstances and a wealth of other inputs and assumptions. Therefore in order to make informed choices, decision-makers need to be aware of the sensitivities, uncertainties and limitations inherent in each type of analysis, and must be able to evaluate whether the approach and assumptions used in an analysis are reasonable for the regions and time periods in which the competing technologies are to be applied.

The purpose of this chapter is to provide an overview of the different types of analyses as well as concise guidance on how to evaluate and apply these. For each type of analysis, the most important analytical approaches and variables are discussed, along with the associated sensitivities, uncertainties and limitations. The requirements and limitations of each method are explained. This provides a point of reference for the selection of assessment methods in the technical chapters of this report and gives a framework which helps to harmonize the reporting of results. Further the chapter provides an introduction to the technical chapters and to their subsequent application, non-application and specific-default macrodata. A description of the key methods used in these chapters is also given.

An overview is given of the key approaches used to compare the lifetime, and the direct and indirect emissions of different types of systems. These range from the modelling of partial or complete systems to measured values of individual systems or of representative equipment populations. There can be significant differences between the results from different approaches and therefore relevant policy comparisons can only be made if there is maximum transparency and a harmonization of assumptions. Reference values for emissions from energy consumption in the production of fluids and products are given along with values for CO_2 emissions from electricity generation for the national electricity grids. These values differ significantly over time and between regions, suggesting that great care should be taken if results of system comparisons based on total equivalent warming impact (TEWI) or Life Cycle Assessment (LCA) are transferred to related applications or other regions.

In any economic analysis, the cost of mitigation is calculated as the difference in costs (defined in monetary units) between a reference situation and a new one characterized by lower emissions. Both situations should, as far as possible, be defined such that the assessment can include all major economic and social impacts of the policies and the resulting impact on greenhouse-gas emissions. At the project level, the simplest cost assessment using cost-benefit analysis considers that a new technology requires capital investment to cover costs accounted in the project. Major categories of costs accounted in a project are labour, land, material, energy, investments, environmental services and foreign exchange. These costs are known as the direct engineering and financial costs. A more complete cost evaluation requires the inclusion of externalities, the costs not paid directly by the private commercial entity developing the project. External costs are paid by society and where available, these should be considered for wider policy assessments. In principle, the total cost to society (*social cost*) consists of both the *external cost* and the *private cost*. For the assessment of engineering-type measures relevant to this report, the focus is best placed on private costs, expressed in terms of their net present value (NPV) or as levelized costs to account for the time distribution of costs and investments during the project lifetime.

Health and safety issues are an integral aspect of decisions concerning the choice of fluids. These decisions can have far-reaching consequences for the workforce, the population, industry, the environment and the economy. The prevention of negative health and safety impacts requires methodologies such as risk assessment, risk management, and policy and regulatory controls. Health and safety issues are considered under the following criteria: Flammability, acute toxicity, chronic toxicity, carcinogenicity, acute ecotoxicity, chronic ecotoxicity, and persistence. Information on the substances covered by this report was drawn from several sources. The technical chapters provide detailed information on the substances and the products these are used in is provided in the subsequent technical chapters. The information is divided into general characteristics of a group of fluids followed by typical characteristics of certain fluids within the group. There are significant differences between the various fluid groups, and in some cases within the fluid groups, with respect to flammability, acute toxicity, ecotoxicity and persistence. The design of systems and processes should reflect the specific weaknesses of the fluids used.

A fairly wide range of assessment methods is described so that the impacts of different technologies on the environment and climate can be compared and understood. One significant problem identified is that methodologies like LCA and TEWI are installation-specific and therefore do not provide meaningful results for entire sectors. It is also demonstrated that the available assessment methodologies are not generic but have been developed for a specific purpose. Each of the methodologies can play a role in technology choices if used appropriately.

The use of fluorocarbons is specific to certain technical sectors. In these sectors, technology selection and product development are influenced by the customers and a number of other factors, such as the enforcement of legislation, of a local, national or regional nature. An overview of the regional factors that should be reflected in the inputs for analyses is given. These factors include climate, labour costs, the availability of capital, skilled labour and spare parts, the replacement rates of systems, and disposal pathways. In this section a regional partitioning of developed and developing countries is used according to the Montreal Protocol.

For applications involving banked amounts of fluids, an overview is given of the available approaches and associated uncertainties for the modelling of process emissions compared to emissions in the usage phase and upon disposal. For both areas, the future usage pattern and emissions can be estimated using bottom-up and top-down approaches. The former ideally involves the modelling of the emissions of individual substances or substance classes from populations of equipment. The properties of these pieces of equipment will often be modelled differently for different years and in different regions so as to reflect anticipated technological changes. In contrast to these technology-rich approaches, top-down models rely on historically established relations between sales into certain sectors and economic growth. This approach is typically weakest in capturing long-term technological changes but is good for the appropriate capture of mid- to long-term growth and wealth effects. However for technologies involving the use of fluorinated greenhouse gases, uncertainties associated with both bottom-up

and top-down models become so significant that projections beyond the year 2020 are unreliable.

In the past, too little attention was paid to ensuring the comparability and transferability of results from different technology assessments. The treatment of uncertainties is often incomplete and therefore the resulting recommendations are often not robust enough to be transferred across a sector. To address these concerns, analysts and decision-makers should ensure, wherever possible, that the assumptions and methods used to compare competing technologies are consistent and that uncertainties and sensitivities are identified and quantified. The development of simple and pragmatic standard methodologies and the respective quality criteria should be continued. It is recommended that future efforts should focus on increasing the involvement of relevant stakeholders and introducing additional measures to increase transparency for outside users, for example by providing more extensive documentation. For certain regions and policy questions the amount of resources needed for some of the assessment methods is prohibitive. There is a need for simple and pragmatic assessment methods that can also be used in regions with very limited resources.

3.1   Introduction

This chapter provides public and corporate decision-makers with an overview of assessment methodologies that can be used to support informed decisions on technologies to replace the use of ozone-depleting substances (ODSs). Such decisions are taken in the context of the phase out of ozone-depleting substances under the Montreal Protocol, and national and international policies aimed at reducing emissions of anthropogenic greenhouse gases. The latter include not only direct emissions of fluids during production, use and disposal but also energy-related emissions over the lifetime of products and equipment. Decisions on technology choices and the definition of appropriate practices will often take place at the level of individual projects but could also be important in designing policies. This chapter has therefore been designed as a toolbox for decision-makers. The intention is to give an overview of the most common assessment methodologies for evaluating competing technologies, including a description of how these are applied and their practical limitations. It also provides information used by several subsequent chapters in this report.

Fluorinated substances such as hydrofluorocarbons (HFCs) and perfluorocarbons (PFCs) are being widely used or are being considered for future use as refrigerants, blowing agents in foam production, propellants for aerosol applications, solvents, surfactants, fire-fighting agents and anaesthetics. These substances are used in many technical applications over a wide range of conditions. They are replacing the use of CFCs and HCFCs, which were and are being banned due to their significant impact as ODSs. In quantitative terms, stationary refrigeration and air conditioning, mobile air conditioning and foam blowing dominate usage and emissions in most countries.

As these substances are potent greenhouse gases, considerable efforts are underway to reduce the emissions of these from products or processes to well below the levels that can be obtained on the basis of current technical and economic drivers. The principal available emission reduction options – over the lifetime of the product equipment – for the aforementioned applications fall into five main groups:

a) Improved containment of fluorinated gases during the life cycle of a product or system (manufacture, use and decommissioning/disposal);
b) Use of technologies with a lower fluid charge;
c) Use of alternative fluids with a zero/low global-warming potential (GWP);
d) Use of not-in-kind (NIK) technologies;
e) Process modifications to avoid byproduct formation or emission.

The benefits of reducing emissions of fluorinated gases clearly need to be offset against potential changes in terms of energy efficiency, safety and costs or the impact on environmental categories such as air quality and the continuation of damage to the ozone layer (see Figure 3.1). The specific details of a sector – or even the application concerned – need to be taken into account when making decisions about technological options.

As a result of international concerted action, many of the sectors and applications covered in this report have undergone a rapid transition away from the use of ODSs. This mandated transition has also led to a significant increase in knowledge about technological alternatives to the use of ODSs and has resulted in increased innovation. This high rate of innovation has made it more difficult to appropriately characterize technology options and then to assess them in terms of their performance or costs. Therefore it is now more important than ever to apply consistent methodologies, as outlined in this chapter, for the purpose of producing valid comparisons upon which robust technology choices can be based.

The subsequent sections of this chapter cover the following aspects of technology assessment relevant to the sectors using fluorinated gases: Key performance characteristics such as direct and indirect emissions (3.2), categories of costs (3.3), consideration of health and safety issues (3.4), assessment of climate and environmental impacts (3.5), regional dimensions of technology choices (3.6), basics of emission projections (3.7) and future methodological developments (3.8).

3.2   Direct and indirect emissions

This section considers aspects related to emissions from products and equipment using HFCs or PFCs. Direct and indirect emissions need to be identified to account for the full inventory of such emissions. Indirect emissions are usually associated with the amount of energy consumed for the operation of equipment loaded with the fluid. Table 3.1 gives an overview of the relative contribution of direct HFC emissions to the total greenhouse-gas emissions associated with systems, for example, a domestic refrigerator, a refrigerated truck, a supermarket cooling system or the energy losses through an insulated area of building surface. The table shows that for several important technical systems (e.g. mobile air conditioners or supermarket refrigeration) direct and indirect emissions are of the same order of magnitude but that for several other applications, energy-related indirect emissions outweigh direct emissions by one or two orders of magnitude. The methodologies stated in Table 3.1 are described in greater detail in Section 3.5. Specific values are highly dependent on emission factors, end-of-life treatment of equipment, the GWP of the fluids used and the carbon intensity of the energy supply system.

3.1.1   Direct emissions

Emissions are possible throughout the lifetime of the fluid, from the initiation of fluid manufacture through use within the intended equipment, to its destruction. The following paragraphs describe key stages within the lifetime of the fluid where direct emissions occur. The identification of emissions from these various stages is necessary for both the environmental impact and safety assessments of applications that use any type of fluid.

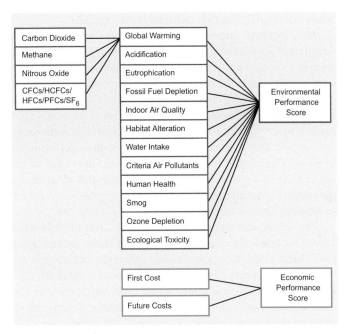

Figure 3.1. Example of impacts to be accounted when taking decisions about the introduction of new technologies or products (Lippiatt, 2002 adapted by authors).

3.2.1.1   Emissions by fluid life stage

Fluid manufacture [1]

The manufacture of fluids requires feedstock materials, which are sourced, produced and used worldwide. Emissions from these feedstock materials, their intermediates and the end substance can occur in chemical processing plants. For the production of many HFCs, these direct emissions can be significant in terms of CO_2 equivalent (on a GWP(100) basis – see Section 3.5.1). For example, the substances emitted during the production of HFC-227ea were estimated to amount to 120 kg CO_2-eq kg^{-1} of material manufactured (Banks *et al.*, 1998), whereas for HFC-134a this figure was relatively low with estimates ranging from 2–40 kg CO_2-eq kg^{-1}. (McCulloch and Lindley, 2003 and Banks and Sharratt, 1996, respectively). These values are strongly dependent on the fluid manufacturing process and the integrity of the chemical plant. The good design and operation of the plant can lead to lower emissions. It should be noted that some of the intermediate substances might be ODSs (although these are converted into the end product without being emitted). For example CFC-114a, HCFC-133 and HFCF-124 are used in particular routes for the manufacture of HFC-134a (Frischknecht, 1999). The emissions of GHGs are significantly less for alternative fluids. HC-290, HC-600a and HC-1270 are generally extracted from natural gas mixtures and leakage from the process plant will generally comprise a variety

---

[1] Only direct emissions from the production process are considered here and not indirect emissions from electrical energy requirements.

of hydrocarbons. Gover *et al.* (1996) estimate 0.14 kg CO_2, 0.5 g HC and 0.7 g methane emissions per kg of propane/butane, which equates to 0.2 kg CO_2-eq kg^{-1}, and data from Frischknecht (1999) are lower than this. Ammonia is normally produced from natural gas and emissions of GHGs will only comprise methane and CO_2. Frischknecht (1999) estimates between 1.5 and 2.3 kg CO_2 kg^{-1} ammonia produced, consistent with the value in Campbell and McCulloch (1998). CO_2 is a slightly different case, as commodity CO_2 is generally a recovered byproduct from numerous other chemical manufacturing processes and it is therefore difficult to specifically attribute emissions to CO_2. Frischknecht (1999) reports emissions of CO_2 and methane, equivalent to 0.2 kg CO_2 kg^{-1} CO_2 produced.

Distribution of fluids

Once manufactured, fluids can be shipped nationally or internationally, often as bulk shipment or in individual cylinders. Typically, national or regionally organized distribution chains deliver bulk quantities to product manufacturers or transfer the substances into smaller containers for use by manufacturers and service companies, in the case of refrigerants and solvents. The distribution chain stops when the substance enters into the management or control of the 'user', such as a manufacturer or a service company. Losses normally occur during the transfer of fluids, connection and disconnection of hoses, and leakage from containers and pressure relief devices. These losses tend to be relatively small in relation to the large quantities of materials handled (Banks *et al.*, 1998), as transfer operations take place under carefully-controlled conditions and are subject to international regulations for the prevention of releases (e.g. UN, 2002; IMO, 2000).

Manufacture and distribution of products

HFCs and alternative fluids are generally used in the manufacture of refrigeration systems, foams, aerosols and fire protection equipment, whereas solvent applications (e.g. cleaning) tend to apply fluids directly on-site during the in-use stage. The manufacture of refrigeration products requires transfer of the fluid. Losses normally occur due to the connection and disconnection of hoses and valves. Emissions also originate from storage vessels and the associated piping, but these are less frequent.

Some sources of emissions are specific to certain applications and products. For example, refrigerants are often used to rinse out air, moisture and other contaminants from refrigeration equipment to ensure internal cleanliness prior to charging. Moreover, the refrigerant may also be employed as a tracer for the detection of leaks. Equipment leaks may be found after charging of the refrigerant, in which case further emissions will occur during the recovery and evacuation process prior to repairing the leaks. During the manufacture of foams, fluids are used as blowing agents. The blowing-agent emissions during the preparation of foam formulations, when the blowing agent is mixed with other raw materials such as polyols, are low due to the use of closed systems. Depending on the type of foam

Table 3.1. Percentage contribution of direct (HFC) emissions to total lifetime greenhouse-gas emissions in various applications (emissions associated to functional unit) – selected indicative examples.

Application	Method applied	HFC emissions as percentage of lifetime systemgreenhouse-gas emissions (using GWP-100)	Characterization of system and key assumptions	Publication
Mobile Air Conditioning	TEWI	**40–60%** – Current systems (gasoline engine) **50–70%** – Current systems (diesel engine)	Passenger vehicle; HFC-134a Sevilla (Spain)	Barrault *et al.* (2003)
Commercial Refrigeration	LCCP	**20–50%** – for a wide range of sensitivity tests on leakage rate, energy efficiency and energy supply	Direct Expansion Refrigeration Unit; Supermarket (1000m²); HFC-404A; Germany	Harnisch *et al.* (2003)
Domestic Refrigeration	TEWI	**2–3%** – No recovery at end-of-life	European standard domestic refrigerator; HFC-134a; World average electricity mix	Chapter 4 of this report
Insulation Foam of Domestic Refrigerators	LCCP	**6%** – with 90% blowing agent recovered at disposal **17%** – with 50% blowing agent recovered at disposal	HFC-245 fa; Europe	Johnson (2004)
PU Insulation Foam in Refrigerated Truck	LCCP	**2%** – with full recovery of HFC at disposal **13%** – without recovery of HFC at disposal	Refrigerated Diesel truck; Germany	Harnisch *et al.* (2003)
PU Spray Foam Industrial Flat Warm Roof	LCA	**13%** – with full recovery of HFC at disposal **20%** – without recovery of HFC at disposal	4 cm thickness; HFC-365 mfc; Germany	Solvay (2000)
PU Boardstock in Private Building Cavity Wall	LCA	**4%** – with full recovery of HFC at disposal **17%** – without recovery of HFC at disposal	5 cm thickness; HFC-365 mfc; Germany	Solvay (2000)
PU Boardstock in Private Building Pitched Warm Roof	LCA	**10%** – with full recovery of HFC at disposal **33%** – without recovery of HFC at disposal	10 cm thickness; HFC-365 mfc; Germany	Solvay (2000)

and process, either a small fraction (rigid closed cell foam) or the entire blowing agent (flexible open cell foam) is emitted during the manufacturing process. The foaming usually takes place under ambient atmospheric conditions. The blowing agent emitted from the foaming process mixes with the air and is then vented or incinerated. Aerosols are generally charged in factory premises and, as for refrigerating systems, the disconnection of charging heads results in small releases. Cleaning machinery is occasionally filled with solvent, although emissions tend to be minimal at this stage because fluids with a low vapour pressure are generally used.

Use

The in-use stage of the product lifetime tends to result in the greatest emissions for most applications, particularly for refrigeration, aerosols, solvents and fire protection systems. For refrigeration and air-conditioning systems, for example, in-use emissions can vary widely depending on the service and disposal practices as well as other operational and environmental factors not intrinsic to the technology. Emissions from domestic, commercial and industrial refrigerating equipment generally originate from failures in system components or from the handling of refrigerant during servicing, with the size and frequency of leaks depending upon several factors. External factors include usage patterns, frequency of equipment relocation, weathering and aggressive environments, repair quality and frequency of preventative maintenance. Rapid fluctuations of temperature or pressure and excessive vibration of piping, joints and so forth, due to fan motors and compressors or other external sources are major causes of in-use leakage. Other integral aspects are associated with the design and construction of the equipment. The design quality of the equipment and the use of suitable components have a significant impact on leakage rates. Control systems are also responsible for emissions, such as pressure-relief devices. Most industry guidance on emissions focuses on leak prevention for the in-use stage (e.g., see Institute of Refrigeration, 1995; Butler, 1994; ETSU, 1997). The multiple factors contributing to emissions of refrigerant often lead to a fairly broad distribution of net leakage rates for individual pieces of equipment. Figure 3.2 shows an example

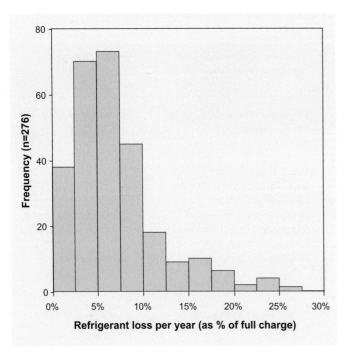

Figure 3.2. Distribution of annual leakage rates of mobile air conditioning systems in a fleet of 276 European passenger vehicles (after Schwarz & Harnisch, 2003.

from a survey carried out on mobile air-conditioning systems in Europe (Schwarz and Harnisch, 2003). A similar distribution is seen with supermarket systems (Radford, 1998). It is clear that the mean leakage rate for an equipment population can result from a broad distribution that includes many very tight, a large number of fairly tight and a small number of very leaky systems.

Decision-makers should realise that, in general, a leakage rate is not necessarily intrinsic to a certain technology but can also strongly depend on a number of environmental and operational factors. If properly understood, these factors can be systematically influenced to reduce emission levels. Whilst the quality of maintenance and repairs to equipment influence the leakage during operation, the actual handling of refrigerant may also lead to significant emissions, for example the removal of fluid from a system and the subsequent recharge. The degree of leakage is strongly dependent on the behaviour of the personnel and the tools they use, and will range from the minimum (that associated with residual gas within hoses and the system) up to quantities exceeding the system charge. Despite the legislation against venting in a number of countries, practices frequently prevail in which the whole system is vented, leak-tested with the fluid (possibly several times until all leaks are repaired) and flushed with the fluid before final recharging. Poorly connected hoses and fittings, and poor-quality recovery equipment can significantly contribute to high emissions.

The majority of emissions from aerosols, fire protection systems and solvents also occur during the in-use stage, although these are intentional since a release is part of the functional-

ity of the equipment. Emissions from aerosols are simply a function of the amount of propellant used. Fire protection systems produce emissions upon demand, as is the intention, although undesired emissions occur from false alarms or faulty signals, or from mandatory system qualification tests that require full system discharge. Sometimes training also involves the operation of the fire protection system. Within the use phase of certain applications, such as fire protection systems, equipment servicing and maintenance may require the removal of fluid from a system and subsequent recharge, and emissions associated with this. Fire protection systems also exhibit slow passive emissions throughout the in-use stage, particularly in the case of high-pressure cylinder systems. Similarly systems that are being maintained can produce significant emissions if the system is accidentally initiated, or when servicing of the systems requires checking and confirmation of operation. Processes that use solvents generally lead to emissions as a result of residual fluid evaporation of treated items, and to a lesser extent whenever equipment is opened as part of its use.

Annual emissions from foams are generally minor during the in-use stage. The blowing agent in closed cell foams is released over time and typically only part of the blowing agent/insulant will be released during the useful life of the product. Caleb (2000) has calculated emission factors for different rigid foams based on a survey and the collection of data from emission factors reported in the literature. Lee and Mutton (2004) recently did the same for extruded polystyrene or XPS foam.

Decommissioning
Fluids may be vented or recovered during the decommissioning of refrigeration equipment, unexpelled fire protection systems and equipment using solvents. Over the past decade the price of recovery equipment has fallen significantly and the sales of such equipment have expanded, indicating that recovery has become much more widespread than was the case during the 1980s and first half of the 1990s. Blowing agents from foams and propellant from aerosols can also be recovered at end-of-life, but current methods have a low effectiveness. Emission levels of refrigeration and air-conditioning systems, and of foam applications are very sensitive to disposal practices.

For refrigerants, fluids from fire protection systems and solvents, the sources of emissions associated with removal at end-of-life are the same as those associated with normal handling, as discussed previously for the servicing aspect during the in-use stage. Fluids used to be vented when equipment was decommissioned but where legislation has been introduced, the expected practice is to recover the fluid. If the fluid is recovered, it may be re-used (in its recovered form), recycled (using on-site machinery) or taken back to the supplier or recycling centre for cleaning (and re-use) or disposal. Uncontaminated fluids can normally be reused and whether a fluid can be cleaned on-site or has to be returned to the supplier, depends on the type and degree of contamination and the process needed to return it to the purity of virgin fluid. However, in many situations, the resources required to clean up contaminated refrigerant rarely outweigh the risks and benefits to the service company. Therefore in most countries recovery rates have remained low if the chemical to be recovered is relatively inexpensive. Some suppliers or governments (e.g. Australia) have introduced cash incentives for the return of 'minimally-contaminated' refrigerant, but the success rate of such schemes is still unknown.

In the case of rigid insulating foams, the majority of blowing agents remain in the foam until end-of-life. Some rigid insulating foams such as 'board stock' and 'sandwich panels' can be recovered or re-used if they are not adhered to substrates or can be easily separated from these. A similar approach can be used with aerosol cans that contain residual propellant, although this procedure is not normally used.

Disposal
Following the recovery, and potentially recycling and reclamation, of refrigeration equipment, fluids meant for disposal are stored ready for destruction. The handling and storage of refrigerants and fire protection system fluids, solvents and recovered aerosol propellants prior to destruction can lead to emissions in the same way as in the distribution of fluids stage.

In the case of rigid insulating foams, the disposal is complicated because the foams are only a small part of overall systems such as a refrigerator or a building. Refrigerator foams can be shredded and incinerated to destroy all blowing agents. Alternatively they can be sent to landfills, where the blowing agents will slowly be emitted and/or decompose (Kjeldsen and Scheutz, 2003).

In general, destruction by incineration produces a small amount of emissions (of the original material). Destruction (or destruction and recovery) efficiencies are typically between 99% and 99.99%, resulting in 0.1–10 g released per kg of material (UNEP-TEAP, 2002). Gases such as CO_2, resulting from the combustion of fuel and fluid, are also emitted.

3.2.1.2 Types of emission
Table 3.2 identifies the types of emission common to the various applications in the subsequent technical chapters. These can be categorized into four general groups. Mitigation of emissions for each category requires attention for a particular process within the life of the equipment: Material selection (passive), mechanical design/construction (rupture), technician training (handling) and usage patterns (intentional/functional).

Passive
Passive emissions are generally small, 'seeping' leaks that occur constantly and are normally the result of permeation through construction materials and metal fatigue. This applies in particular to refrigerants, fire protection system fluids and aerosols when materials gasket, plastics and elastomers for seals and hosing are used. Emissions are minimized by selecting the correct materials for the fluids used. In making this choice, the influence of other fluids within the mixture such as refrigeration oils and aerosol fragrances, which can affect permeability, should also be considered. In rigid insulating foams, the passive emission of blowing agents occurs through diffusion. The diffusion coefficients of HFCs through the rigid insulating foams vary considerably, dependent on the type of blowing agent. The use of non-permeable facer materials such as aluminium foils and metal skins can significantly reduce the diffusion of blowing agents. Passive emissions typically occur throughout all stages of the fluid and foam life, but as they are relatively constant, they predominate during the in-use stage.

Table 3.2. Sources of emissions by type and application.

Application	Production				Use				Decommissioning				Disposal			
	Passive	Rupture	Handling	International	Passive	Rupture	Handling	International	Passive	Rupture	Handling	International	Passive	Rupture	Handling	International
Refrigeration		✓	✓	✓	✓	✓	✓	✓			✓	✓			✓	✓
Air conditioning		✓	✓	✓	✓	✓	✓	✓			✓	✓			✓	✓
Mobile air conditioning		✓	✓	✓	✓	✓	✓	✓			✓	✓			✓	✓
Foams		✓	✓		✓				✓		✓				✓	✓
Aerosols		✓	✓		✓	✓		✓			✓	✓	✓		✓	✓
Fire protection		✓	✓		✓	✓	✓	✓			✓	✓			✓	✓
Solvents		✓	✓		✓	✓		✓			✓	✓	✓		✓	✓

Note: Fluid manufacture and distribution are excluded from this table since they do not pertain to any one specific application.

Rupture

Ruptures are accidental breaks in pressure systems, such as fractures in pipework, vessels and components. These are normally associated with fluids under high pressure and therefore tend to occur throughout most stages of the fluid life. Ruptures can result from external forces applied to components, inherent material weaknesses, the influence of pressure and temperature changes, vibration, the ageing of materials and corrosion. Large ruptures are generally the most notable type of leaks and tend to cause rapid and often complete release of the fluid. Rupture leaks or emissions do not apply to foam because the pressure differential within and outside the foams is very small. Ruptures can be minimized by appropriately designing piping and components to account for anticipated stresses, protecting against external impact and avoiding a chemically aggressive environment.

Handling

Handling emissions are unintentional releases that occur with human intervention, for example, where a fluid is being transferred into or out of equipment and complete recovery is impractical. Residual amounts of fluid occur in transfer hoses, within components and systems following recovery, and absorbed in certain materials such as oil. The fluid will subsequently migrate to the atmosphere when hoses are disconnected or systems are opened following recovery. Such releases occur throughout the life of the fluid regardless of the application, but a lack of training and insufficient awareness of the environmental impact of fluids means that personnel are less likely to mitigate emissions when handling them.

Intentional/Functional

Intentional releases are determined by human activity and may be unnecessary or required because of the function of the application (e.g. aerosols, fire protection). Unnecessary intentional releases occur where the operator chooses to directly release the fluid to the atmosphere. Examples include venting where the fluid is not recovered following removal from an application, the fluid being released directly through opening valves or cutting into pipe work, or the operator employing the fluid for ulterior uses such as blowing dust from pipes, and so forth.

Functional emissions occur specifically with fire protection, aerosols and some solvent uses. In situations where the fluid is within storage facilities, or within pressure systems during the in-use stage, intentional releases can occur in response to uncontrolled circumstances, for example, when pressure relief devices vent in the event of fire. Similarly, fire protection systems may release fluid in response to false signals from heat, smoke or light detectors, thus discharging the whole system.

3.2.1.3 Measurement and estimation of emissions

Quantifying emissions from a specific application is useful for several purposes, including the retrospective environmental impact assessment or the evaluation of operating costs.

Unfortunately for most applications it is difficult to measure the field leakage and even where this is possible, the measurement is imprecise. Laboratory studies are largely impractical for most applications, for example large supermarket refrigeration systems that are built and maintained by a number of different companies. However, emissions from foams are usually less sensitive to external conditions and so laboratory measurements are normally highly appropriate. Releases during fluid manufacture, fluid distribution and the manufacture/distribution of equipment would normally be measured by monitoring the mass flow of material into and out of facilities. The same applies at end-of-life, when recovered fluid is returned to suppliers and delivered to recycling or incineration facilities.

Since releases at the in-use stage tend to predominate, these are more frequently monitored. The problem with existing methods such as gas detection is that they do not measure the mass of fluid released; concentrations of leaked gas detected indicate the presence of a leak, but do not permit this to be quantified (Van Gerwen and Van der Wekken, 1995). Recent developments in leak detection for refrigerating systems include intrinsic detectors, where a reduction in refrigerant inventory is measured within the system (as opposed to measuring the presence of refrigerant outside it) (Peall, 2003, www.nesta.org.uk/ourawaardees/profiles/3763/index.html). Such an approach is particularly useful for accurately measuring the loss of charge but does not provide information on losses from handling activities. The most accurate method is to record the mass usage of the substance. For refrigerating equipment, fire protection systems and solvent use, this involves tracking the quantities of chemicals that are acquired, distributed and used to fill a net increase in the total mass (charge) of the equipment. Quantities not accounted for are assumed to have been emitted. Entities that contract equipment maintenance to service companies can estimate their emissions by requesting the service company to track the quantities of refrigerant recovered from and added to systems.

Different approaches can be used to prediction emissions associated with particular equipment or systems. The least accurate but most simplistic approach is to apply annual leak rates (% yr^{-1}) for the appropriate sector or equipment types to the mass of fluid used. However, these are generally approximated using bottom-up methods, combined with limited measured data and industry interviews (e.g. March, 1996). At best, leak rates may be found for equipment manufacture, aggregate in-use stage and end-of-life recovery. More detailed analyses may be conducted by calculating releases from each element throughout the equipment life. This also depends on good information about the flow of material throughout its life, data on component dimensions and knowledge about the behaviour of technician handling (US EPA, 1995). For example, Colbourne and Suen (1999) provide empirically-derived emission indexes for the leakage of different components and different servicing frequencies in order to estimate leakage for a whole system. With this more detailed approach, the actual design of systems and equipment can be more accurately assessed and 'emission optimization' can be applied to the design and operation.

Nevertheless, it is known that leakage rates are highly erratic and for two similar systems these may vary from 0% to over 100% yr[1].

The emission of foam-blowing agents through diffusion can be derived from model calculations (Vo and Paquet, 2004; Albouy *et al.*, 1998). Alternatively, the residual quantity of blowing agents can be determined analytically, although caution must be exercised when using these measurements to estimate in-use emissions. Although the emission of the blowing agent out of a well-defined foam system (e.g., a refrigerator or a piece of rigid foam) can be estimated, obtaining accurate emissions from the foam sector remains difficult due to the variety of foams, blowing-agent initial concentrations, diffusion rates of specific blowing agents, product thicknesses, densities, usage conditions and cell structures.

Direct measurement of greenhouse gas and other fluid emissions is only possible in a few cases and so the values described as 'real' data have to be calculated from secondary data. These calculations should conform to the standard methodologies already developed for emissions trading, which cover refrigeration and air-conditioning systems and chemical processes (DEFRA, 2003), and the standards and guidance for greenhouse-gas emissions inventories (IPCC, 1997; IPCC, 2000a).

The IPCC Good Practice Guidance on National Greenhouse Gas Inventories published in 2000 (IPCC, 2000a) includes three methods for estimating emissions of ODS substitutes. The tier 1 method that equates emissions to consumption (potential emissions), the tier 2a bottom-up method that applies country-specific emission factors to estimates of equipment stock at different life-cycle stages, and the tier 2b top-down method that uses a country-level, mass-balance approach. For sectors where the chemical is banked into equipment, the tier 1 method is significantly less accurate than the other two approaches. This is particularly the case where the equipment bank is being built up and this is precisely the current situation for air conditioning and refrigeration equipment during the transition from CFCs and HCFCs to HFCs. In this situation, most of the chemical consumed is used to fill new equipment volume (charge) rather than to replace emitted gas, and therefore the tier 1 method greatly overestimates emissions. To a lesser extent, the tier 2b country-level, mass-balance approach is also inaccurate during the period of bank building, but in this case, the error is an underestimate (see the discussion in Section 3.2.1.3 for an explanation of this underestimate). That is why the IPCC recommends supplementing the tier 2b approach with the tier 2a emission-factor based approach. The disadvantage of the tier 2a approach is that for the first few years of equipment life, the emission factors will necessarily be based on engineering estimates rather than empirical experience. However, when the first cohort of equipment is serviced, the emission factors can be corrected as necessary. Ultimately, when the HFC-using equipment starts to be retired, the tier 2b method can be used on its own with a high level of accuracy.

More detailed requirements for calculations are stated in IPCC (1997). IPCC (2000a) should be consulted for additional guidelines on emission estimation methods, particularly for the different sectors.

3.2.2   Indirect emission

This section examines aspects related to emissions arising from the energy consumption associated with the manufacturing of products and their components, the use of these products during their useful life and their disposal. The use phase is outlined for cooling applications, heat pumps and foams, as this phase usually dominates the energy consumption of these systems. More application-specific energy aspects are covered in the specific sections of Chapters 4 to 10 of this report.

3.2.2.1   Use phase
ODS substitutes are typically used in cooling applications and thermal insulation foams. For aerosols, solvents and fire protection, the energy consumption during the use phase is not relevant.

Cooling applications and heat pumps
The refrigeration, heat pump and air-conditioning sectors use different approaches to establish and compare energy efficiencies for various technologies:
- Modelled efficiencies based on modelled coefficients of performance (COP);
- Measured efficiencies of products in the research and development phase (established under standard test conditions)[2];
- Measured efficiencies of products in mass productions (established under standard test conditions);
- Measured consumption under representative real conditions.

The coefficient of performance (COP[3]) of a refrigerating plant, product or system is a key parameter for characterizing the energy efficiency of a process. It is the ratio of refrigerating capacity to the input power required to operate the compressor, pumps, fans and other ancillary components.

$$COP = Q_0 / P \qquad (3.1)$$

Where:

Q_0 is the refrigerating capacity (cooling mode) or heating capacity (heating mode) including an allowance for losses in any secondary circuit (kW), and

---

[2] It is important that the standard conditions represent the situation in which the equipment will be used; for example it would not be appropriate to use standard test results gained at an ambient temperature of 15°C when the real ambient temperature of operation of the equipment is 35°C.

[3] In European standards and some ISO standards the term EER (Energy Efficiency Ratio) is used for the cooling mode of the cycle and COP (Coefficient Of Performance) is used for the heating mode. For the purpose of this report only one term is used, COP. Where this term is used, it should be specified whether it relates to cooling or heating.

P is the total power consumption (kW) of the compressors, controls, fans and pumps required to deliver that capacity to the place where it is used.

The COP primarily depends on the working cycle and the temperature levels (evaporating/condensing temperature) but is also affected by the properties of the refrigerant and the design of the system. Energy consumption estimated from known models does not usually take into account the effects due to different 'real-world' handling practices, imperfect design, assembly, capacity control and maintenance. Small fluid leakage, for example, not only impacts direct emission but also indirect emission due to a decrease in the COP of equipment. Figure 3.3 gives an example of the real world spread of measured energy consumptions for a large group of homogenous refrigeration systems. Another typical example is the use of capacity control technologies to optimize the energy consumption. Good capacity control results in smaller pressure differences (evaporating/ condensing temperature) leading to improvements of the overall energy efficiency of the system.

Insulation foams

There are two fundamental approaches to assess the effects on energy consumption of using insulation materials with differing insulation properties as a result of a changed blowing agent (Table 3.3).

Further the specific usage pattern of the system should be noted and addressed if it is material to the calculation. One example might be the effect of day to night temperature changes on a cold store that has traffic through it 24 hours per day as opposed to one that is shut up all night. Both approaches need to be documented with references to standard calculation methods and sources of information.

Approach B in Table 3.3 depends on more assumptions than approach A and is conceptually more challenging. However circumstances can arise in which thickness compensation is

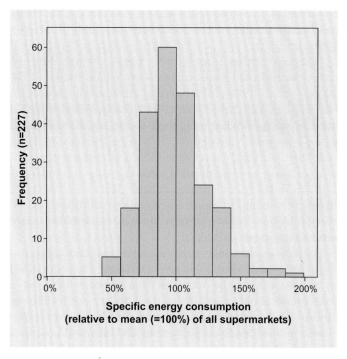

Figure 3.3. Distribution of measured specific energy consumptions (originally expressed as kWh d^{-1} m^{-1} of cooled display cases) expressed as percentage of the mean for 227 standard stores using standard HFC technology of a German supermarket chain – after Harnisch *et al.* (2003).

technically not possible due to space constraints and then approach B is the only feasible method. Even if partial thickness compensation is possible, the calculation of energy impacts will still require approach B.

In general these comparisons are limited to identical systems where only one element has been changed. Comparisons between completely different systems are much more difficult

Table 3.3. Comparison of two approaches for comparing the impact of blowing agent choice on energy consumption.

Approach A) Thickness compensation:

The thickness is increased for any loss of insulation value so that unchanged energy consumption is assumed to be achieved by a material penalty. This may have an environmental impact due to increased direct emissions and the energy used for production. However, for the energy consumption during the usage phase, the key requirements are relatively simple and involve characterization of:
- The functional unit and its desired service (e.g. required common additional heat resistance R value m^{-2} of application and time);
- The insulant types and their reference and replacement thicknesses, and the extent to which additional foam thickness is possible, (e.g. respective thickness and density or mass);
- The insulation properties of the foam(s) using a test appropriate to the duty (e.g. respective thermal conductivity values).

Approach B) Comparative energy modelling:

Where it is not possible to compensate for a change in insulation value, the energy consumption of the system in which the foam is used is calculated for both the reference case and alternative case. This is a more complex process and, in addition to the parameters outlined above, it will require definition of:
- The type and efficiency of the process supplying heat or cold (to accommodate change in efficiency with temperature);
- The appropriate ambient conditions and the internal temperature profile require or alternatively a description of an appropriate proxy (for example heating or cooling degree days);
- Internal sources of heat or cold and how the demand will change with temperature.

to interpret. Examples of comparisons of different foam types, where the model assumptions are clear and do not compromise the results, are given in Caleb (1998, 1999), Enviros March (1999), ADL (1999), Krähling and Krömer (2000), Harnisch and Hendriks (2000) and Harnisch *et al.* (2003).

3.2.2.2 Production energy of fluids and fugitive emissions

Table 3.4 details the small number of estimates of production energy (also known as embodied energy) and fugitive emissions associated with the production of the materials used in systems covered by this report. These estimates are also summarized and applied in Frischknecht (1999a and 1999b) and Harnisch *et al.* (2003). It is worth noting that the newer study shows lower values for HFC-134a due to more widespread application of vent gas treatment to destroy 'fugitive' HFC streams (in previous studies these were emitted into the atmosphere). In-use emissions usually dominate environmental impact and production energy, whereas fugitive emissions only make small contributions.

3.2.2.3 Indirect emissions from energy use

Indirect emissions (of carbon dioxide) from the energy used to operate a system comprise those arising from the generation of the electricity consumed (for example by the compressor, controls, pumps and fans of a building air-conditioning system) and any fuel used directly by the system (e.g. gas used for gas-driven compressors, fuel used by absorption systems or the additional gasoline usage associated with automobile air conditioning). The total lifetime emission of carbon dioxide from the energy used to operate the system (E_I) is:

$$E_I = Q_E \times I_E + \sum (Q_{Fi} \times I_{Fi}) \qquad (3.2)$$

Where:
Q_E is the total lifetime use of electricity;

I_E is the carbon dioxide intensity of electricity production (from Table 3.5);
Q_{Fi} is the total lifetime use of fuel i, and
I_{Fi} is the carbon dioxide intensity of that fuel (from Table 3.5)

Carbon dioxide emissions associated with the generation of electricity vary greatly between countries depending on the specific mix of generation technologies and fuels (e.g. coal, natural gas, combined cycle systems, hydroelectricity, etc.) used. Increasingly complex calculations could be performed to define the minute details of energy-related emissions of carbon dioxide and other greenhouse gases. However for the generic approach described here, it is assumed that the most important emission is carbon dioxide and that the methods employed by the International Energy Agency (IEA) will give internationally consistent estimates. Accordingly, national and regional carbon dioxide intensities of electricity are shown in Table 3.5. These intensities are calculated as the ratio of national carbon dioxide emission from electricity generation, taken from IEA (2002b), to the quantity of electricity used nationally, obtained from IEA (2002a) or IEA (2003). The IEA statistics take into account electricity trading between countries, in the form of an annual average.

For those countries or regions not shown in Table 3.5, the national carbon dioxide intensity of their electrical power (I_E) may be calculated as the sum of the total of each fuel used in electricity generation, multiplied by its carbon dioxide intensity (also quoted in Table 3.5), divided by the total national quantity of electricity *delivered* to customers.

$$I_E = \sum (Q_{Fi} \times I_{Fi}) / D_E \qquad (3.3)$$

Where:
Q_{Fi} is the total annual quantity of fuel i used in electricity generation and I_{Fi} is the carbon dioxide intensity of that fuel (from Table 3.5), and

Table 3.4. Overview of production energy requirements and associated CO_2 emissions.

Material	Production Energy Requirement MJ kg⁻¹	Equivalent Production CO_2 Emissions kg CO_2-eq kg⁻¹	Reference
Aluminium	170	-	Lawson (1996)
		7.64	Ingots : SAEFL (1998)
		2.06–6.56	Pira (2001)
Steel/iron		2.95	Sheet: SAEFL (1998)
		1.60–2.78	Pira (2001)
Stainless Steel	38	-	Lawson (1996)
Copper	100	-	Lawson (1996)
Brass		2.97	Plate: SAEFL (1998)
		11.4–16.1	Norgate and Rankin (2000)
Glass	13	-	Lawson (1996)
HFC-134a (I)	64–105	6–9	Campbell and McCulloch (1998)
HFC-134a (II)	-	4.5	McCulloch and Lindley (2003)
Cyclopentane	24	1	Campbell and McCulloch (1998)
Ammonia	37	2	Campbell and McCulloch (1998)

Table 3.5. Carbon dioxide intensities of fuels and electricity for regions and countries.

Region	Carbon Dioxide Intensity Of Electricity kg CO_2 kWh^{-1}	Note a	Country	Carbon Dioxide Intensity Of Electricity kg CO_2 kWh^{-1}	Note
Africa	0.705	b	Argentina	0.319	b
Asia	0.772	b	Australia	0.885	c
EU	0.362	c	Austria	0.187	c
Europe (OECD)	0.391	c	Belgium	0.310	c
Europe (non-OECD)	0.584	b	Brazil	0.087	b
Latin America	0.189	b	Canada	0.225	c
Middle East	0.672	b	China	1.049	b
N America	0.567	c	Denmark	0.385	c
Pacific	0.465	c	Finland	0.222	c
Former USSR	0.367	c	France	0.078	c
			Germany	0.512	c
Carbon Dioxide Intensities Of Fuels Used In The Calculations	**g CO_2 MJ^{-1}**		Greece	0.876	c
			India	1.003	b
			Indonesia	0.715	b
Fuel			Ireland	0.722	c
Natural gas	56.1	d	Italy	0.527	c
Gasoline	69.3	d	Japan	0.389	c
Kerosene	71.5	d	Malaysia	0.465	b
Diesel Oil	74.1	d	Mexico	0.689	b
Liquefied Petroleum Gas	63.1	d	Netherlands	0.487	c
Residual Fuel Oil	77.4	d	New Zealand	0.167	c
Anthracite	98.3	d	Norway	0.003	c
Bituminous Coal	94.6	d	Pakistan	0.524	b
Sub-bituminous coal	96.1	d	Philippines	0.534	b
Lignite	101.2	d	Portugal	0.508	c
Oil Shale	106.7	d	Russia	0.347	b
Peat	106.0	d	S Africa	0.941	b
			Saudi Arabia	0.545	b
			Singapore	0.816	b
			Spain	0.455	c
			Sweden	0.041	c
			Switzerland	0.007	c
			UK	0.507	c
			USA	0.610	c

Notes:

a. Regions as defined in IEA (2002a) and IEA (2003).

b. Carbon dioxide from "Public Electricity and Heat Production"5 (units Mtonnes CO_2) in summary tables of IEA (2002b), divided by Total Final Consumption electricity and heat6 given as ktonne Oil Equivalent in IEA (2002a), further divided by 11.63 to convert to kg CO_2 kWh^{-1}.

c. Carbon dioxide as in 2 above, divided by Total Final Consumption4 given as GWh in IEA (2003), multiplied by 1000 to convert to kg CO2 kWh^{-1}.

d. Values from Table 3 of IEA (2002b) multiplied by 44/12 to convert to mass of CO_2.

e. Using this category has the effect that all energy inputs to systems that generate electricity and heat are counted against both the electricity and heat generated.

f. Total Final Consumption is electricity or heat available at the consumer net of transmission and distribution losses.

D_E is the total annual national delivery of electricity. Table 3.6 shows examples of the application of this method.

Where there is a nationally agreed energy plan for the future, figures from this may be used for assumptions about future indirect emissions from energy use.

3.3   Categories of costs

The cost of climate change mitigation is an important input to decision-making about climate policy goals and measures. This section provides an oerview of key concepts and assumptions that can be applied to the assessment of policy options related

Table 3.6. Carbon dioxide intensity calculation for representative countries.

	Fuel mix (for electricity generation only)									Total CO$_2$ emissions	Electricity generated less distribution losses	Carbon intensity
	Coal			Fuel oil		LPG		Natural gas				
Type	Usage	Calorific value	Emission factor	Usage	Emission factor	Usage	Emission factor	Usage	Emission factor			
	kt	TJ/kt	tCO$_2$/TJ	kt	tCO$_2$/TJ	kt	tCO$_2$/TJ	kt	tCO$_2$/TJ	ktCO$_2$	GWh$_e$	kgCO$_2$/kWh$_e$
A	B	C	D	E	F	G	H	I	J	K	L	M
First Country Example												
Sub bituminous	112,775	18	96.1	118,712	77.4	0	63.1	127,574	56.1	21,0558	234,000	0.900
Second Country Example												
Bituminous	11,430	27	94.6	57,173	77.4	304	63.1	47,558	56.1	36,288	62,059	0.585
Third Country Example												
None				3,892	77.4	0	63.1	3,160	56.1	48	1,750	0.273

Notes:

Values in italics are constants available from standard tables (see Table 3.5 here)

A.   Both the calorific value and the carbon content (and thus emission factor) vary with the quality of coal.

B.   More than one quality of coal may be used (and separate rows should be used, then added together).

C.   Calorific value varies between sources and should be determined by testing.

D, F, H, J.   CO$_2$ emission factors are shown in Table 3.5

K.   The value in K is equal to (B x C x D + E x F + G x H + I x J) divided by 1000.

L.   This is the total electrical production minus only the amount lost in distribution.

M.   The carbon intensity is equal to K divided by L.

to technologies and production processes using fluorinated gases. The use of consistent and well-defined cost concepts is recommended for the assessment of the various technologies and options described in detail in various parts of this Special Report, and for reporting the assumptions and concepts applied in mitigation studies in a thorough and transparent manner.

3.2.1   Introduction

Actions to abate emissions of fluorinated gases generally divert resources from other alternative uses, and the aim of a cost assessment is to measure the total value that society places on the goods and services foregone due to resources being diverted to climate protection. Where possible the assessment should include all resource components and implementation costs and should therefore take into account both the costs and benefits of mitigation measures.

A key question in cost analysis is whether all relevant dimensions (e.g. technical, environmental, social) can be measured in the same units as the costs (i.e. monetary). It is generally accepted that some impacts, such as avoided climate change, cannot be fully represented by monetary estimates and it is imperative that the cost methodology is supplemented by a broader assessment of impacts measured in quantitative and if needs be qualitative terms.

In any economic analysis of climate change mitigation, the cost of mitigation is calculated as a difference in costs and benefits between a baseline case and a policy case that implies lower emissions. Where possible, the definitions of the baseline and policy cases should include all major social, economic and environmental impacts (at minimum from GHG emissions and ODP emissions). In other words, the system boundary of the cost analysis should facilitate the inclusion of all major impacts. The system boundary can be a specific project, one or more sectors, or the entire economy.

The project, sector and macroeconomic levels can be defined as follows:

1. *Project.* A project level analysis considers a 'stand-alone' investment that is assumed not to have significant impacts on markets beyond the activity itself. The activity can be the implementation of specific technical facilities, demand-side regulations, technical standards, information efforts, and so forth. Methodological frameworks to assess the project level impacts include cost-benefit analysis, cost-effectiveness analysis and Life Cycle Assessment.

2. *Sector.* Sector level analysis considers sectoral policies in a 'partial equilibrium' context, for which all other sectors and the macroeconomic variables are assumed to be as given.

3. *Macroeconomic.* A macroeconomic analysis considers the impacts of policies across all sectors and markets.

Costs and benefits can be reported in present values or as levelized values (alternative ways to generate time-consistent values for flows of costs and benefits that occur at different points in time). Further details about these approaches can be found in Box 3.1.

This report focuses on project level cost analysis in particular because a); the scale of the mitigation policies analyzed can be considered small enough to exclude significant sectoral and economy-wide impacts; b) the basis for conducting a sectoral level cost analysis is weak since the literature does not include sectoral modelling studies for activities which involve the production and use of ODSs and their substitutes; c) and finally the current section is a first attempt to define consistent cost concepts applied to the assessment of climate change and ODS mitigation policies.

3.3.2   Direct engineering and financial cost approach

At the project level, the simplest cost assessment considers the financial costs of introducing a new technology or a production process that has lower emissions than the baseline case. Such practices can imply capital costs of new investments and changed operation and maintenance costs. When the system boundary is defined to include only the financial costs associated with the project implementation, some studies been termed this the direct engineering or financial cost approach.

Policy implementation can require upfront capital costs and changes (decreases or increases) in operation and maintenance costs compared with the baseline case over the lifetime of the project. Major categories of costs accounted in a financial cost assessment are capital, labour, land, materials, maintenance and administrative costs. The various costing elements need to be transformed into values that are comparable over the time frame and as such the cost assessment depends on assumptions about discount rates. The time dimension of costs can be dealt with using various policy evaluation approaches and an overview of some of those applied to the cost-effectiveness analysis of projects is given in Box 3.1.

3.3.2.1   Discounting
There are two approaches to discounting (IPCC, 1996b). One approach (known as ethical) gives special attention to the wealth of future generations and uses a social discounting rate. Another approach (known as descriptive) is based on the discount rates savers and investors actually apply on their day-to-day decisions and uses a higher, private cost discount rate. The former leads to relatively low rates of discount (around 2–3% in real terms) and the latter to relatively higher rates (at least 6% and in some case very much higher rates) (IPCC, 2001b, pp. 466).

For climate change, a distinction needs to be drawn between the assessment of mitigation programmes and the analysis of impacts caused by climate change. The discount rate applied in cost assessment depends on whether the social or private perspective is taken. The issues involved in applying discount rates in this context are addressed below. For mitigation effects, the

---

BOX 3.1 – The NPV and Levelized Cost Concepts

Guidelines for project assessment use a number of different concepts to compare the cost-effectiveness of projects. The most frequently used concepts are net present value (NPV) and levelized cost. These concepts basically provide similar project ranking.

The NPV concept
The NPV concept can be used to determine the present value of net costs, *NPVC*, incurred in a time period T, by discounting the stream of costs (C_t) back to the beginning of the base year (t = 0) at a discount rate I:

$$NPVC = \sum_{t=0}^{T} C_t / (1+i)^t$$

The levelized cost concept
The levelized cost represent a transformation of the *NPCV* into constant annual cost values, C_0, over the lifetime of the project. The levelized costs are calculated as a transformation of the *NPVC* using the formula:

$$C_0 = NPVC \left(i / \left(1 - (1+i)^{-n} \right) \right)$$

where n is the time horizon over which the investment is evaluated.

The levelized costs can directly be compared with annual emission reductions if these are constant over the project lifetime.

The use of NPV and levelized costs as project ranking criteria is valid, given a number of assumptions:

NPV
An investment I_1 is more favourable than another investment I_2 if *NPVC₁/GHG reduction* < *NPVC₂/GHG reduction*. It should here be noticed that the use of NPVCs to compare the cost-efficiency of projects requires that some discounting criteria be applied to the annual greenhouse-gas emission reductions. The NPVC/GHG ratio can be used to rank investments with different time horizons.

Levelized cost
An investment I_1 is more favourable than another investment I_2 if the levelized cost of I_1 per unit of annual emission reduction is less than the levelized cost of I_2 per unit of annual emission reduction. The levelized costs should be calculated for similar investment lifetimes. The lifetimes of the investments if necessary can be made uniform by adding terminal values to investments with relatively long life time, or by replicating investments that have a relatively short lifetime.

---

country must base its decisions, at least in part, on discount rates that reflect the opportunity cost of capital. In developed countries, rates of around 4%–6% are probably justified (Watts, 1999). In developing countries the rate could be as high as 10–12%. These rates do not reflect private rates of return, which typically need to be considerably higher to justify the project, potentially between 10% and 25%.

For climate change impacts, the long-term nature of the problem is the key issue. The specific benefits of a GHG emission reduction depend on several factors such as the timing of the reduction, the atmospheric GHG concentration at the time of reduction and afterwards.. These are difficult to estimate. Any 'realistic' discount rate used to discount the impacts of increased climate change impacts would render the damages, which occur over long periods of time, very small.

3.3.3 Investment cycle and sector inertia

In the area of technologies associated with the manufacture and use of fluorinated gases, observations about replacement rates of old products have shown that it has been possible to undertake investments with a payback period as low as 1 to 5 years. This evidence suggests that from an economic and technical point of view[4], it is possible to rely on mitigation policies that are fully implemented over less than a decade. However, social structures and personal values also interact with society's phys-

---

[4] There are some concerns about the capacity to maintain such fast technical transitions for future technical evolution in these sectors. Transition from CFCs to HCFCs was relatively easy because the chemistry was already known.

ical infrastructure, technology applications and institutions, and these combined systems have in many cases evolved relatively slowly. An example of such system inertia is seen in relation to the energy consumption for heating, cooling and transport, and the impacts of urban design and infrastructure. Markets sometimes tend to 'lock in' to specific technologies and practices that are economically and environmentally suboptimal, because the existing infrastructure makes it difficult to introduce alternatives. Similarly the diffusion of many innovations can be in conflict with people's traditional preferences and other social and cultural values (IPCC, 2001a, pp. 92-93).

At the same time, it should also be recognized that social and economic time scales are not fixed. They are sensitive to social and economic forces, and are influenced by policies as well as the choices made by individuals and institutions. In some cases behavioural and technological changes have occurred rapidly under severe economic conditions, for example during the oil crises of the 1970s (IPCC, 2001a, pp. 93). Apparently, the converse can also be true: In situations where the pressure to change is small, inertia is large.

Both of these issues should be considered when building scenarios about future GHG emissions from a sector. These lessons suggest that new policy approaches are needed. Instead of looking solely for least-cost policies given current preferences and social norms, policies could also aim at reshaping human behaviour and norms. This could support fast technology penetration and from a longer time perspective in particular, could imply cost reductions through learning and market development.

3.3.4   *Wider costing methodologies – concepts*

Up until now we have only considered the direct engineering and financial costs of specific technical measures. However, the implementation of policy options that mitigate climate change and ODSs will often imply a wider range of social and environmental impacts that need to be considered in a cost analysis.

3.3.4.1   *Social and financial costs*
In all work on costs, a basic distinction can be drawn between the social cost of any activity and the financial cost. Social cost is the full value of the scarce resources used in the activity measured from the point of view of society. Financial cost measures the costs from the perspective of a private company or an individual and bases its values on the costs that actually face these agents. A difference between social and financial costs arises when private agents do not take full account of the costs that they impose on other agents through their activities – such a cost is termed an external cost. Positive impacts which are not accounted for in the actions of the agent responsible, are referred to as external benefits.

External costs and benefits are distinct from the costs and benefits that companies or other private agents take into account when determining their outputs such as the prices of fuel, labour, transportation and energy, known as conventional com-

pany costs, and also from environmental costs usually accounted in more complete evaluations of company costs (see Table 3.7). Categories of costs that influence an individual's decision-making are referred to as private costs. The total cost to society is the sum of the external and private costs, which together are defined as social cost:

$$Social\ Cost = External\ Cost + Private\ Cost \qquad (3.4)$$

The scope of the social and private costs is illustrated in Figure 3.4.

External costs typically arise when markets fail to provide a link between the person who creates the 'externality' and the person who is affected by it, or more generally when property rights for the relevant resources are not well defined. Externalities do not necessarily arise when there are effects on third parties. In some cases, these effects may already be recognized, or 'internalized' and included in the price of goods and services. Figure 3.4 illustrates different subcategories of environmental costs, including external costs and private costs as faced by a private company. The centre box represents company costs that are typically considered in conventional decision-making. The next box (private costs) includes the typical costs plus other internal environmental costs that are potentially overlooked in decision-making, including regulatory, voluntary, up-front, operational, back-end, overhead, future, contingent and image/relationship costs. These 'private costs' include internal intangible costs (e.g., costs that could be experienced by a company related to delays in permitting, and so forth, and due to disputation with regulators and others). The box labelled societal includes environmental costs that are external to a company. These are costs incurred as a result of a company affecting the environment or human health, but for which the company is not currently held legally or fiscally responsible. These 'externalities' include environmental degradation and adverse effects on humans,

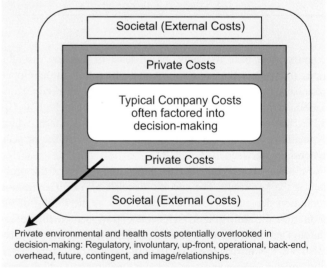

Private environmental and health costs potentially overlooked in decision-making: Regulatory, involuntary, up-front, operational, back-end, overhead, future, contingent, and image/relationships.

Figure 3.4. Scope of full costs (Adapted from US EPA, 1995).

Table 3.7. Examples of environmental costs incurred by firms.

Potential Hidden Costs		
Regulatory	**Upfront**	**Voluntary (Beyond Compliance)**
Notification	Site studies	Community relations/outreach
Reporting	Site preparation	Monitoring/testing
Monitoring/testing	Permitting	Training
Studies/Modelling	R&D	Audits
Remediation	Engineering at procurement	Qualifying suppliers
Record keeping	Installation	Reports (e.g., annual environmental reports)
Plans		Insurance
Training		Planning
Inspections	Conventional Costs	Feasibility Studies
Manifesting	Capital equipment	Remediation
Labelling	Materials	Recycling
Preparedness	Labour	Environmental studies
Protective equipment	Supplies	R&D
Medical surveillance	Utilities	Habitat and wetland protection
Environmental insurance	Structures	Landscaping
Financial assurance	Salvage Value	Other environmental projects
Pollution control researchers		Financial support to environmental groups and/or
Spill response	Back-End Costs	
Storm water management	Closure/decommissioning	
Waste management	Disposal of inventory	
Taxes/fees	Post-closure care	
	Site survey	
Contingent Costs		
Future compliance costs	Remediation	Legal expenses
Penalties/fines	Property damage	Natural resource damage
Response to future releases	Personal injury damage	Economic loss damages
Image and Relationship Costs		
Corporate image	Relationship with professional staff and workers	Relationship with lenders
Relationship with customers	Relationship with insurers	Relationship with communities
Relationship with investors	Relationship with suppliers	Relationship with regulators

Note. In upfront cost category, the centred box surrounded by dashed lines represents conventional costs, which are usually accounted.
Source: US EPA (1995).

property and welfare associated with emissions/activities that are performed in compliance with regulatory requirements. The figure does not directly portray the benefits that may be associated with alternative decisions.

3.3.4.2 Welfare basis of costs

The external effects described above cannot be valued directly from market data, because there are no 'prices' for the resources associated with the external effects (such as clean air or clean water). Indirect methods must therefore be used. Values have to be inferred from decisions of individuals in related markets, or by using questionnaires to directly determine the individuals' willingness to pay (WTP) to receive the resource or their willingness to accept payment (WTA) for the environmental good.

3.3.4.3 Ancillary costs and benefits

Projects or policies designed for GHG and ODS mitigation frequently have significant impacts on resource use efficiency, reductions in local and regional air pollution, and on other issues such as employment (IPCC, 2001b, pp. 462). When estimating the social costs of using technologies that impact climate change and/or ODS, all changes in cost arising from this activity have to be taken into account. If some of them imply a reduction (increase) in external costs, they are sometimes referred to as secondary, indirect benefits (costs) or ancillary benefits (costs).

3.3.5   Wider costing methodologies – cost categories

3.3.5.1   Project Costs

This item has already been discussed in the introduction of Section 3.3.2. However, the cost categories listed there may need to be adjusted when carrying out the wider cost methodology. Adjustments in land costs, labour, investments, materials, energy costs, environmental services and foreign exchange may be needed for private costs and external costs, and a detailed list is provided by Markandya and K. Halsnæs (2000).

3.3.5.2   Implementation cost

In addition to the above, the costs of implementation deserve special attention. Many aspects of implementation are not fully covered in conventional cost analyses (see Table 3.7). A lot of work needs to be done to quantify the institutional and other costs of programmes, so that the reported cost figures represent the full costs of policy implementation. As shown in Table 3.7, implementation costs depend on institutional and human capacities, information requirements, market size and the learning potential, as well as on market prices and regulations in the form of taxes and subsidies.

3.3.6   Key economic drivers and uncertainty

For various reasons cost estimates are shrouded by uncertainty and therefore any presentation of cost estimates should include transparent information about various keys to uncertainty that relate to both the baseline case and the new project case. Uncertainty in baseline cases is best dealt with by reporting cost estimates for multiple as opposed to single baselines. With this costs will not be given as single values, but as ranges based on the full set of plausible baselines (see for example IPCC, 2001b, pp. 30-37).

Uncertainties in cost estimates are related to both private and external cost components. Private cost figures tend to be less uncertain than external cost components, since the private costs primarily relate to market-based economic transactions. However, there is a particular uncertainty related to projections of future efficiency, and the costs and penetration rates of new technologies. One way to handle this uncertainty is to undertake a sensitivity analysis based on scenarios for high, low and medium case values (Markandya and Halsnæs, 2000). Another way of accounting is to consider some kind of 'learning curve', that is an expected cost reduction as a function of the increasing amount of products using the technology.

3.3.7   Other issues

3.3.7.1   Baseline Scenarios

Quite often the costs of a programme are evaluated against a situation where the programme is not implemented. This situation is defined by a baseline scenario, which tries to infer future conditions without the implementation of the programme. There are assumptions embedded in the baseline to forecast the future, for example, inefficient baseline, or 'business-as-usual' baseline. It is important to note that a programme's cost and benefit will vary according to this baseline scenario definition. For a mitigation programme, the cost will be larger if an economically efficient baseline is set rather than an inefficient one.

3.3.7.2   Macroeconomic costs

The cost of a programme can be measured using a macroeconomic analysis based on dynamic models of the economy. These models examine the impacts of a programme at an integrated level and allow for intersectoral effects. This means that they are more suitable for programmes large enough to produce impacts on other sectors of the economy.

On the other hand macroeconomic cost estimates generally provide less detail about technological options and externalities than project or sectoral cost estimates.

3.3.7.3   The equity issue

Equity considerations are concerned with the issues of how the costs and benefits of a programme are distributed and the climate change impacts avoided, as input to a more general discussion about the fairness of climate change policies. Equity concerns can be integrated in cost analysis by reporting the distribution of costs and benefits to individuals and society as a supplement to total cost estimates. Some authors also suggest applying income distribution weightings to the costs and benefits to reflect the prosperity of beneficiaries and losers (Ray, 1984; Banuri *et al.*, 1996).

3.3.8   Conclusions

For most of the mitigation measures discussed in this Special Report, the specific measures (e.g. technical facilities, infrastructure, demand-side regulation, supply-side regulation, information efforts, technical standards) can be considered to have relatively small economic impacts outside of a narrow project border and can therefore be regarded as 'stand-alone' investments that are assessed using a project assessment approach. However, this does not imply that the cost assessment should solely limit itself to a consideration of the financial cost elements. A project system boundary allows a fairly detailed assessment of GHG emissions and the economic and social impacts generated by a specific project or policy. Accordingly various direct and indirect social costs and benefits of the GHG reduction policies under consideration should be included in the analysis.

Furthermore, it should be realized that as industrial competition increases, an increasing number of companies might become interested in using the most advanced production paradigms. For example, this was the case for lean production, an approach which evolved in Japan during the post-war period and implied greater flexibility in production and working partners. Many typical company features have included environmental concerns as well as broader issues of sustainable development as an evolving feature of the lean production paradigm.

In other words the companies have expanded the view about the boundaries of their own production.

Companies set boundaries around the activities they manage directly as well as those they do not control or manage. A distinctive feature of the lean production system has been an increasing transparency across firms that are dealing with different elements of the production chain. There has also been a tendency towards integrating management functions along the supply chain in order to examine the entire production chain for added value sources, irrespective of the current legal company boundaries along the chain. The application of information technology to business processes has facilitated the introduction of these new management systems. Without this the application of quantitative methods would have proved too complex (Wallace, 1996).

This new integrated management approach is illustrated in Figure 3.5. The figure shows a typical production chain, where the dotted line represents the boundary within the production process. Within that boundary, the management of the process may be integrated, irrespective of the number of companies involved or their exact legal relationship. This boundary might also include the extraction of raw materials, various product end-uses and even the disposal of the materials after use.

Every stage of production and consumption implies important environmental impacts. Companies are increasingly being required to explicitly manage these environmental impacts in response to formal regulations and pollution charges. Alternatively they might voluntarily adopt cleaner production technologies and tools, such as eco-auditing, in response to increasing expectations from society. Another driving force can be the increasing legislative liabilities of companies with respect to pollution. It is therefore useful for analysts to consider a system-wide company boundary that includes all stages of production in life cycle assessment, as shown in Figure 3.5.

3.4   Consideration of health and safety issues

Health and safety issues are an integral aspect of deciding the choice of fluids when alternatives are available, and the decisions can have far-reaching consequences for the workforce, domestic consumers, industry, the environment and the economy. Assessment methods for health and safety should first of all focus on minimizing negative health and safety impacts, and then consider risk management, policy and regulatory controls. This approach should be used for each step of the life cycle of the product including production, distribution, use, maintenance, repair and the end-of-life treatment such as destruction, re-use or recycling. Sometimes it can be wise to accept an increase in one life-cycle stage so as to arrive at an overall improvement in the impact accumulated over the life cycle. During the switch from ozone-depleting substances (CFCs and HCFCs) to HFCs, the health and safety risks of both groups of chemicals were similar. Here the main concern was energy efficiency and reliability. During the switch from higher GWP fluids to lower GWP fluids, health and safety often become a

key issue. Some of the key considerations are examined in the following sections.

3.4.1   *Prevention of negative health and safety impacts*

Chemical exposure can cause or contribute to many serious health effects such as heart ailments, damage to the central nervous system, kidney and lungs, sterility, cancer, burns and rashes (US DoL, 1998). The impact of these effects includes lower productivity, absenteeism, increased health-care costs, litigation, and economic downturn at both the enterprise and national levels. Most countries have passed occupational health and safety laws in response to this, but the enforcement of such legislation is difficult, particularly in developing countries. This is borne out by the fact that only 5−10 % of workers in developing countries and 20–50 % of workers in industrialized countries (with a few exceptions) are estimated to have access to adequate occupational health services (Chemical Hazard communication: US Department of Labour (1998 revised). However, given the potential negative impacts in the absence

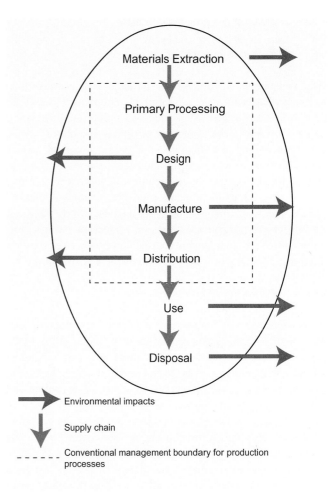

Figure 3.5. Environmental impacts along the supply chain (Wallace, 1996)

of adequate health and safety precautions, it is in the interest of both businesses and government to minimize these. Periods in which businesses are undergoing technological changes provide good opportunities to institute measures for preventing negative health and safety impacts on their workers and operations. They also provide governments with opportunities to implement measures to ensure that these matters are considered during the period of change.

3.4.2   Risk assessment of chemicals

Risk assessment is central to safety. It provides the scientifically sound basis for actions, including policy and regulatory actions to manage potential risks from the production, storage, transportation, use and disposal of chemicals. A number of parameters must be considered when undertaking risk assessments, such as chemical composition, stability and reactivity, hazards identification and classification, transportation, storage and handling, ecological impacts, physical and chemical properties, routes of exposure, effects of exposure, exposure limits, and toxicological information. As well as guiding the decision on the choice of a chemical to be used for a particular application, the assessment will also inform decisions on risk management as well as policy and regulatory controls.

3.4.3   Risk management of systems

Risk management is a broad term for the process that uses the outcomes of a scientific risk assessment to implement best practices, which are usually supported by appropriate policy and regulatory frameworks. A number of options are usually available for managing risk. These depend on the nature of the risk and the technological, economic and policy options available to address this. Effective risk management includes a wide range of measures such as information provision, training and/or retraining, risk assessment training, redesignated work practices, use of personal protective equipment, evaluation and monitoring of both the immediate and wider environments, redefinition of exposure limits and standards, and medical examinations.

3.4.4   Policy and regulatory controls

Most countries have occupational health and safety laws that require employers (as far as it is reasonable and practicable) to provide safe working environments and to develop and implement policies and measures to educate and protect their employees. In general, laws are developed on the principles of precaution and reasonableness, and these need to be adjusted when new processes, technologies or inputs are introduced into the economy that have health and safety implications not covered under the existing framework. Governments are responsible for ensuring that an appropriate regulatory and policy framework exists to protect human and ecological health and safety as well as property, and to ensure compliance. When such a framework and the associated legal requirements are compiled, the health and safety of the user, worker and those in the locality must be the main

priorities. Product liability laws have been established in several areas of the world to protect users, workers and members of the public with respect to health and safety or any other damage. The liability legislation in a country is an important factor in the choice of the system and fluid chosen for the application, irrespective of what the safety standard specifies. When drawing up regulations, the combination of several regulatory requirements, including product liability, needs to be taken into account.

3.4.5   Health and safety criteria

For the purposes of this report, health and safety issues are considered under the following criteria:

Flammability: Ability to support combustion; a high capacity for combustion; burning velocity and expansion ratio.

Acute toxicity: Adverse effects are observed within a short time after exposure to a chemical. This exposure may be a single dose, a short period of continuous exposure, or multiple doses over a period of 24 hours or less.

Chronic toxicity: Adverse effects observed following repeated exposure to a chemical during a substantial fraction of an organism's lifespan. For human chronic toxicity typically means exposure over several decades; for experimental animals it is typically more than 3 months.

Carcinogenicity: The ability of a substance or agent to produce or provoke cancer.

Acute ecotoxicity: Adverse effect on ecosystems and/or the organisms within the ecosystem within a short period of time after exposure to a chemical.

Chronic ecotoxicity: Adverse effects to an ecosystem and/or the organisms within the ecosystem following exposure to a chemical during a substantial fraction of the ecosystem's or organism's lifespan.

Accumulation: The action or process of accumulating within biological tissues.

Persistence: Continued presence of a chemical or its effects in the environment after source or cause has stopped.

3.4.6   Health and safety data for relevant substances

The data for health and safety are extensive and this report only includes references to the databases. Most data can be found on the site of the International Programme on Chemical Safety (IPCS) (www.inchem.org), a collaborative venture of the World Health organization (WHO), the United Nations Environment Programme (UNEP) and the International Labour Organization

(ILO). The IPCS site refers to the ICSCs, CICADs, and EHCs. The International Chemical Safety Cards (ICSCs) (www.inchem. org/pages/icsc.html) provide a structured overview of the data for most of the substances under consideration. The Concise International Chemical Assessment Documents (CICADs) (www.inchem.org/pages/cicads.html) provide extensive data for a very limited number of substances. They are similar to the Environmental Health Criteria Monographs (EHC) (www. inchem.org/pages/ehc.html) which provide internationally accepted reviews on the effects of chemicals or combinations of chemicals on human health and the environment.

Additional data and substances can be found in the databases of the IPCS INTOX Programme, the US EPA, the US National Institute for Occupational Safety and Health (NIOSH), the University of Oxford Physical and Theoretical Chemistry Laboratory, the Programme for Alternative Fluorocarbon Toxicity Testing (www.afeas.org/paft/) and documents from ISO Technical Committees TC 86 "Refrigeration and air conditioning" and ISO TC 21 "Equipment for fire protection and fire fighting". If the data are not available from these sources, then national standards or the material safety data sheet from the supplier can be used as the source of information. Care is needed when using material safety data sheets from suppliers, as these data are not always peer reviewed. The most recent peer-reviewed data agreed at an international level (IPCS, PAFT or ISO) should be used in preference to other data.

The information required for health and safety considerations depend on the subsector and application involved. For example, the data required for refrigeration and air conditioning are different from that for fire protection and medical aerosols. Even within a sector, regional differences exist for detailed data. Each sector shall use the appropriate data valid for it to perform the risk assessment and management with respect to health and safety. For refrigeration and air conditioning an ISO work item has been approved to unify the data and resolve the regional differences (ISO TC 86/SC8/WG5). For fire protection this is handled by ISO TC 21.

3.5 Assessing climate and environmental impacts

This chapter describes approaches in which the environmental comparisons are made systematically using standardized procedures and factors. They are most suitably used for making comparisons between individual installations or items of equipment and do not provide 'generic' information. There is a hierarchy among the system-based approaches, which depends on the scope of treatment, but they all seek to apply data in the same rigorous manner. In every case, care should be taken to examine and clearly define the scope of the analysis, taking into account the requirements of those who commissioned the study.

3.5.1 *Environmental impact categories and respective indicators including approaches for their ranking*

A rational choice of systems, such as heating and cooling, to provide for societal needs should include an assessment of their environmental impact so that excessive demands on the environment can be identified and avoided. Environmental impact depends as much on the quantity of the material emitted as it does on the material's properties. Climate change and ozone depletion are clearly prioritized in this report. Within another framework, other impact categories such as energy-related acidification or resource depletion could be emphasized. An exhaustive list of potential impact parameters or a definition of the process of life cycle assessment fall outside of the scope of this report. However, the principal environmental impacts that may be considered for systems using HFCs, PFCs and other replacements for ozone-depleting substances are:

Climate Change The radiative effects of CFCs and their alternatives on climate is discussed in detail in Chapter 2 and, for the purposes of comparisons between climate impacts, the most important parameter is *global-warming potential.* This is a conversion factor that relates the climatic impact from an emission of particular greenhouse gas to that of an equivalent amount of emissions of carbon dioxide. It is calculated by integrating the radiative forcing from an emission of one kilogram of the greenhouse gas over a fixed time period (the *integration time horizon, ITH*) and comparing it to the same integral for a kilogram of carbon dioxide; units are $(kg\ CO_2\ equivalent)(kg\ emission)^{-1}$. Commonly quoted integration time horizons are 20, 100 and 500 years with impacts beyond each ITH being ignored (see Table 2.1). The calculation has to be performed in this way because the reference gas, carbon dioxide, has a very long environmental lifetime; for example its impact up to 20 years is only 9% of that up to 500 years (IPCC, 1996a). The standard values for the emissions accounting required by the Kyoto Protocol are those in the Second Assessment Report of the IPCC (IPCC, 1996a) at the 100-year time horizon. The 20-year time scale does not meet the time criterion for judging sustainability; focusing on 20 years would ignore most of the effect on future generations (WCED, 1987). GWPs from the Second Assessment Report at the 100-year time horizon represent the standard for judging national performance. For the purpose of system comparisons the most recent IPCC GWPs could be used, for example, as presented earlier in this report. However, it should be noted that GWPs are parameters constructed to enable the ranking of emissions of greenhouse gases and do not reflect absolute environmental impact in the same way as, for example, the calculated future radiative forcing described in Chapter 2.

Ozone depletion gases that contain reactive halogens (chlorine, bromine and iodine) and are sufficiently unreactive to be transported to the stratosphere, can cause the halogen concentration in the ozone layer to rise. They are therefore *ozone-depleting substances.* For any given gas the efficiency of ozone depletion depends on the extent to which material released at ground level is transported into the stratosphere, how much halogen each molecule carries and the potency of that halogen for ozone depletion, and how the gas decomposes in the stratosphere and hence how much of its halogen content can affect the ozone layer. These factors are combined in mathematical models of the atmosphere to give relative *ozone depletion potentials (ODPs)* based on a scale where the ODP of CFC-11 (CCl_3F) is unity (Daniel *et al.*, 1995; Albritton *et al.*, 1999); values important for Life Cycle Assessments are shown in Table 1.1.

Acidification: The two groups principally involved in acidification are sulphur and nitrogen compounds and, with the exception of ammonia, neither the ODS nor their substitutes have a direct effect in this category. However, energy-related emissions can exhibit significant acidification potential, and degradation products of substances such as HF or HCl could have considerable acidification potential. Indicators for potential acid deposition onto the soil and in water have been developed with hydrogen ions as the reference substance. These factors permit computation of a single index for potential acidification (in grams of hydrogen ions[5] per functional unit of product), which represents the quantity of hydrogen ion emissions with the same potential acidifying effect:

$$\text{acidification index} = \sum i \; mi \times APi \qquad (3.5)$$

Where:
mi is the mass (in grams) of flow *i*, and
APi are the millimoles of hydrogen ions with the same potential acidifying effect as one gram of flow i, as listed in Table 3.8.

However, the acidification index may not be representative of the actual environmental impact, as this will depend on the susceptibility of the receiving systems (soil and water, in this case).

Photo-oxidant formation: The relative potencies of compounds in atmospheric oxidation reactions are characterized by their photochemical ozone creation potentials (POCP), on a scale where ethene is 100 (Derwent *et al.*, 1998). The hydrocarbon substitutes for ODSs have POCPs ranging from 30 to 60 but HFCs and PFCs are not implicated in any significant photo-oxidant formation (Albritton *et al.*, 1989) and are among the lowest priority category for volatile organic compound regulation (UN-ECE, 1991).

Resource depletion: The production of all of the chemicals considered in this report will deplete resources and the extent of this should become apparent in a Life Cycle Assessment. For example, an important consideration for fluorinated gases is the extraction of fluorspar mineral, as most of this is destined for the manufacture of fluorochemicals (Miller, 1999).

Eutrophication is the addition of mineral nutrients to soil or water. In both media, the addition of large quantities of mineral nutrients (such as ammonium, nitrate and phosphate ions) results in generally undesirable shifts in the number of species in

Table 3.8. Acidification-potential characterization factors (Alternatively, in the literature sulphuric oxides are often used as reference).

Flow (i)	AP_i (hydrogen-ion equivalents)
Ammonia (NH_3)	95.49
Hydrogen chloride (HCl)	44.70
Hydrogen cyanide (HCN)	60.40
Hydrogen fluoride (HF)	81.26
Hydrogen sulphide (H_2S)	95.90
Nitrogen oxides (NO_x as NO_2)	40.04
Sulphur oxides (SO_x as SO_2)	50.79
Sulphuric acid (H_2SO_4)	33.30

Source: Lippiatt, 2002

ecosystems and a reduction in ecological diversity. In water it tends to increase algal growth, which can cause a depletion in oxygen and therefore the death of species such as fish.

Characterization factors for potential eutrophication have been developed, in a similar vein to those for the global-warming potential, with nitrogen as the reference substance. These factors permit the computation of separate indices for the potential eutrophication of soil and water (in grams of nitrogen per functional unit of product), which represent the quantity of nitrogen with the same potential nutrifying effect:

$$\text{eutrophication index (to water)} = \sum i \; mi \times EPi \qquad (3.6)$$

Where:
mi is the mass (in grams) of inventory flow *i*, to water, and
EPi are the grams of nitrogen with the same potential nutrifying effect as one gram of inventory flow i, as listed in Table 3.9.

The calculation for soil eutrophication is similar but, for both soil and water the actual impact will vary, dependent on the ability of the local environment to cope with an additional stress of this sort, as was the case for acidification.

Ecotoxicity is the introduction of a compound that is persistent, toxic and can accumulate in the biosphere (commonly shortened to PTB). All three attributes are required for environmental releases to accumulate to the point at which there is a toxic response. No such compounds are known to be directly associated with the production and use of any of the fluorocarbons considered in this report. An in-depth discussion of ecotoxicity issues can be found in Hauschild and Wenzel (1998), Heijungs (1992) and Goedkoop (1995).

3.5.2   System-based approaches

In these approaches the environmental comparisons are made systematically using standardized procedures and factors. They

---

[5] The hydrogen release potentials are criticized by some authors. They have proposed alternative factors based on UN-ECE-LRTAP models. See www.scientificjournals.com/sj/lca/pdf/aId/6924.

Table 3.9. Eutrophication Potential Characterization Factors. (Alternatively, in the literature PO4+ is often used as a reference).

Flow (i)	EPi (nitrogen-equivalents)
Ammonia (NH_3)	0.12
Nitrogen Oxides (NOx as NO_2)	0.04
Nitrous Oxide (N_2O)	0.09
Phosphorus to air (P)	1.12
Ammonia (NH_4^+, NH_3 as N)	0.99
BOD5 (Biochemical Oxygen Demand)	0.05
COD (Chemical Oxygen Demand)	0.05
Nitrate (NO_3^-)	0.24
Nitrite (NO_2^-)	0.32
Nitrogenous Matter (unspecified, as N)	0.99
Phosphates (PO_4^{3-}, HPO_4^{2-}, $H_2PO_4^-$, H_3PO_4, as P)	7.29
Phosphorus to water (P)	7.29

Source: Lippiatt, 2002

are best used for making comparisons between individual installations or items of equipment and do not provide 'generic' information. There is a hierarchy among the system-based approaches, which depends on the scope of treatment, but they all seek to apply data in the same rigorous manner. In every case, the scope of the analysis should be clearly examined and defined, taking into account the requirements of those who commissioned the study.

Life Cycle Assessment (LCA) is clearly the most comprehensive and formal approach to assessing and comparing the environmental impacts of technologies. The methodology for LCAs has been developed and formalized in the ISO 14040 series of international standards. On the other hand, TEWI (Total Equivalent Warming Impact) has the most limited scope, but has been applied most widely for the technologies within the remit of this report. It addresses the climatic impact of equipment operation and the disposal of operating fluids at end-of-life but, although it may be appropriate for most of the common systems, it does not consider the energy embodied in the fluid or equipment. This energy may be important in some cases and this consideration has led to the concept of LCCP (Life Cycle Climate Performance).

In LCCP a more complete climatic impact of the fluid is calculated and includes the impacts from its manufacture, the impacts from operating and servicing the system and finally those associated with disposal of the fluid at the system's end of useful life. However, both TEWI and LCCP consider just the climatic impact; this is reasonable for cases where the predominant environmental impact is on climate. Life cycle assessment (LCA) is the broadest-based approach and this includes the environmental impacts of other inputs and outputs to the system, in addition to those associated with energy.

LCCP can be seen as a submethod of LCA and TEWI as a submethod of LCCP. To a large extent the approach chosen

will depend on the context. If the information required is the relative climate impacts of a number of alternative approaches for achieving a societal good, then TEWI or LCCP are likely to provide adequate information. However this will ignore all other environmental impacts that are addressed in LCA, assuming that these will be similar for the alternative technologies. Although the three approaches differ in their scope, all of data should be derived and all of the analyses performed with the same rigour.

3.5.2.1 Total equivalent warming impact (TEWI)

Arguably the largest environmental impact from many refrigeration and air-conditioning applications arises from their energy consumption and emissions during their operation. Similarly, the energy saved by thermal insulating foam is the principal offset for any effect due to fluid emissions. In order to help quantify these effects, TEWI sets out to standardize the calculation of climate-change impact in terms of emissions over the service life of the equipment, including emissions arising from the disposal of the fluids it contains. The units of TEWI are mass of CO_2 equivalent.

TEWI using generic or default data.

The analysis is performed by calculating the direct emissions of the fluids contained in each system from leakage during operation over its entire service lifetime. This includes servicing and the system's eventual decommissioning and disposal. In this context, the system does not cover the full life cycle (ISO, 1997) but includes the operation, decommissioning and disposal of the application.

The total mass emission of each greenhouse-gas component is converted to CO_2-equivalent emissions using GWP (see discussion in Section 3.5.1) as the conversion factor (see Table 2.1). These figures are then added to the emissions of actual carbon dioxide arising from the energy used during operation (see 3.2.2.3) to give a TEWI value for the lifetime of the equipment. Examples of 'equipment' are a refrigeration or air-conditioning system, or a building (particularly if it is insulated). There is often a combination of energy-consuming and energy-conserving parts, and different direct releases of greenhouse gases. Typically:

$$TEWI_S = \sum OR_i \times GWP_i + \sum DR_i \times GWP_i + E_I \qquad (3.7)$$

Where:

$TEWI_S$ is the total equivalent warming impact from system S (for example, a particular refrigeration system or building installation) the units of which are mass of CO_2 equivalent;

OR_i is the operational release of each greenhouse gas i (the mass total of the releases of each gas during the system's operating lifetime);

GWP_i is the global-warming potential of greenhouse gas i (at the 100-year integration time horizon, as discussed below);

DR_i is the total mass of each greenhouse gas i released when the system is decommissioned, and

E_I is the indirect emission of carbon dioxide resulting from the energy used to operate the system (for its whole lifetime), already discussed in Section 3.2.1.3 and calculated according to Equation 3.2, above.

While this is apparently an absolute value, it can carry a high uncertainty associated with the assumptions and factors used in the calculation. TEWI is most effectively used to compare alternative ways of performing a service, where the same assumptions apply to all of the alternatives and the effect of these on relative ranking is minimized. A TEWI value calculated for one system using one methodology (i.e. set of assumptions, equations, procedures and source data) is not comparable with a TEWI value calculated for another system using another methodology. Then a comparison of TEWI is meaningless.

Depending on the quantity of information available and the needs of the study, there are several levels of complexity in the application of TEWI. At the simplest level, a default emission function could be used for the fluid release together with calculated energy requirements and regional carbon dioxide intensity. Default emission functions have been developed by AFEAS to calculate global emissions from refrigeration and closed-cell (insulating) foams (AFEAS, 2003). One feature of these functions is that all of the substance used is eventually released (in some cases after many years service) and this can have a profound effect on the application's impact.

In this case, the quantity released is equal to the amount originally charged into the system, plus any amount added during the system's period of service:

$$OR_i + DR_i = C_i + QA_i \qquad (3.8)$$

Where OR_i and DR_i are the operational and decommissioning releases of substance i as described above;
C_i is the mass of i originally charged into the system, and
QA_i is the mass of i added into the system during its service life.

For hermetic refrigeration systems (such as domestic refrigerators and window air conditioners), units are rarely, if ever, serviced and therefore QA_i is set at zero because of its insignificance. Yet for systems which require frequent servicing, such as mobile air conditioning, that default condition is not appropriate and a value for QA_i could be derived by analogy from the operation of similar systems.

Energy (either as power required to operate the system or the energy saved by thermal insulation) can be calculated using standard engineering methods. In many cases, electricity is used to power the equipment and this will have been produced by technologies that vary between countries and regions, with large differences in the fossil fuels used as primary sources (for more information see Section 3.2.1.3). Table 3.5 lists some regional and national carbon dioxide intensities for electricity.

Such a calculation is only suitable for showing major differences (say within a factor of two) due to the extensive use of default factors. Nevertheless, it is useful for identifying the more important areas of the calculation that would repay further refinement (Fischer *et al.*, 1991 and 1992; McCulloch, 1992 and 1994a). Uncertainties can be significantly reduced by using appropriate specific data.

GWP and integration time horizon
For the conversion of other greenhouse-gas emissions into their CO_2-equivalents, GWPs at the 100-year integration time horizon are usually used and the source of the GWP values must be clearly stated. For example, TEWI analyses are now usually performed using the most recent GWP values published by the IPCC, even though this is not the normative standard. To ensure that the results are as portable as possible and to facilitate intercomparisons, the standard values from the Second Assessment Report of the IPCC (IPCC, 1996a) as used in the emissions accounting reported under the Kyoto Protocol and UNFCCC, have frequently been used in existing TEWI analyses.

TEWI using specific data
The next level of complexity goes beyond the use of generic data. It requires real emission patterns obtained from field trials and operating experience and, preferably, the range of values obtained from such studies should be indicated and used in a sensitivity analysis (Fischer *et al.*, 1994; ADL, 1994 and 2002; Sand *et al.*, 1997; IPCC/TEAP, 1999). As the disposal of the fluid can have a significant impact, it is important to incorporate the real emissions on disposal. If the systems under consideration do not yet exist, the methods for calculating emissions patterns should have been verified against real operating systems.

Similarly, the actual energy consumption based on trials should be used in the more thorough analysis, together with the carbon dioxide intensity of the energy that would actually be used in the system. Many systems are powered electrically and therefore the procedures already discussed in Section 3.2.1.3 should be applied so as to facilitate comparisons between similar systems operated in different countries.

As in many cases electrical energy is the most important energy carrier, the sum of the other fuel usages and intensities can be neglected so that Equation 3.2 becomes:

$$E_I = Q_E \times I_E \qquad (3.9)$$

Where
E_I is the total indirect lifetime emission of carbon dioxide from the energy used to operate the system;
Q_E is the total lifetime use of electricity, and
I_E is the average carbon dioxide intensity of national electricity production (from Table 3.5)[6].

Uncertainty
When most of the impact arises from fluid emissions, the crite-

rion for significance is set by the uncertainty of the GWP values and this is typically 35% (IPCC, 1996a). Where the impact is a combination of fluids and CO_2 from energy, the more common case, the total uncertainty should be assessed. A rigorous uncertainty analysis may not be meaningful in all cases or might not possible due to the poorly quantified uncertainties of emission factors, emissions from the energy supply systems, specific energy consumption and the like. However, the sensitivity of uncertainty in the data is valuable because the effort required to gather the information needed for increasingly detailed calculations, will only be repaid if these show significant differences.

Uses
TEWI is particularly valuable in making choices about alternative ways of performing a function in a 'new' situation but it also can be used to minimize climate impact in existing operations by providing information on the relative importance of sources, so that remedial actions can be prioritized. This is, however, methodologically restricted to those cases in which the original and alternative technologies remain reasonably similar throughout their life cycles.

A standard method of calculation which includes the concepts and arithmetic described here has been developed for refrigeration and air-conditioning systems, and the principles of this may be applied to other systems (BRA, 1996, consistent with EN378, 2000). TEWI can also be used to optimize the climate performance of existing installations and methods of working (McCulloch, 1995a; DETR, 1999). A particularly valuable application is in the construction or refurbishment of buildings, where TEWI can be used to facilitate the choice between different forms of insulation, heating and cooling. The affect of the design on both the TEWI and cost can be investigated, and significant greenhouse-gas emissions abated (DETR, 1999). The interaction between TEWI and cost is particularly useful when additional equipment is required to achieve an acceptable level of protection. For example, the cost of that safety equipment could have been invested in efficiency improvements (ADL, 2002; Hwang *et al.*, 2004).

If sufficient information is available, TEWI can also be used to examine the climate and cost incentives of targeting operations at particular periods of the day or year when the carbon dioxide intensity of electrical power is lower (Beggs, 1996).

3.5.2.2   Life cycle climate performance (LCCP)
Like TEWI, this form of analysis concentrates on the greenhouse-gas emissions from direct emissions of operating fluids together with the energy-related CO_2 but, it also considers the fugitive emissions arising during the manufacture of the operating fluids and the CO_2 associated with their embodied energy. Like TEWI, LCCP is most effective when applied to individual installations, and a 'generic' LCCP will only be representative only if the data used to calculate it are representative of the types of installation being examined.

A comprehensive study has been made of representative LCCPs for alternative technologies in the areas of domestic refrigeration, automobile air conditioning, unitary air conditioning, large chillers, commercial refrigeration, foam building insulation, solvents, aerosols (including medical aerosols) and fire protection (ADL, 2002). The results were very similar to those of the TEWI analyses (see particularly Sand *et al.* (1997)). For example, the LCCP of domestic refrigerators was dominated by their energy use and there was no clear difference between the refrigerant fluids or blowing agents used in insulating foams. However the end-of-life disposal method has a significant impact on the LCCP (Johnson, 2003). As long as the disposal of all of the systems was treated in the same way, automobile air conditioning was most heavily influenced by the conditions under which it was operated (the climatic and social conditions of different geographical areas). Thus the highest LCCP values arose for vehicles in the southern USA, and this value was higher still if the fluid chosen did not allow for efficient operation (ADL, 2002).

LCCP is a useful addition to the TEWI methodology, even though in many cases the influence of fugitive emissions and embodied energy (which account for most of the difference between TEWI and LCCP results) can be small compared to the lifetime impact of using the system.

$$LCCP_S = TEWI_S + \sum OR_i \times (EE_i + FE_i) + \sum DR_i \times (EE_i + FE_i) \quad (3.10)$$

Where:
$LCCP_S$ is the system Life Cycle Climate Performance;
$TEWI_S$ is the system TEWI, as defined by Equation 3.5.1 above;
OR_i and DR_i are, respectively, the quantities of fluid i released from the system during operation and at decommissioning;
EE_i is the embodied energy of material i (the specific energy used during the manufacture of unit mass, expressed as CO_2 equivalent), and
FE_i is the sum of fugitive emissions of other greenhouse gases emitted during the manufacture of unit mass of i (expressed as their equivalent CO_2 mass), so that:

$$EE_i = \sum (EE_j \times I_{Fj}) \quad (3.11)$$

for j sources of energy used during the production of material i, each with a carbon dioxide intensity of I_{Fj} (see also Equation 3.2), and:

---

[6] The differing practice of using the carbon intensity of the most expensive fuel in an attempt to show the situation for an additional demand in a deregulated energy market, could be misleading. This carries unwarranted assumptions: the demand may not be additional, even if the system represents a new load. It is most probable that the effect of the new system on the energy balance would be, at least in part, to replace demand from elsewhere. And even if demand is additional, it may not result in the most expensive energy being used. That would depend on the daily, weekly and seasonal demand pattern, which is beyond the scope of this level of TEWI analysis.

$$FE_i = \sum (FE_j \times GWP_j) \qquad (3.12)$$

for j greenhouse gases emitted during the production of material i, each with a global-warming potential at 100 years of GWP_j.

A comparison of Equations 3.7, 3.10, 3.11 and 3.12 shows that the difference between LCCP and TEWI is that the GWP of each greenhouse-gas component is augmented by the embodied energy and fugitive emissions of that component. However, the effect of these is now generally quite small as in much of the world the practice is to minimize or destroy process emissions (compare Section 3.2.1.1).

A relatively straightforward application for TEWI and LCCP analyses is the study of emissions attributable to a household refrigerator. In this case, either the refrigerant choice or the blowing agent choice may be studied, or both. Table 3.10 gives a summary of the items that would typically be considered in such studies.

3.5.2.3 Life Cycle Assessment (LCA)

LCA is a technique for assessing the environmental aspects of a means of accomplishing a function required by society (a 'product or service system' in LCA terminology) and their impacts. A life cycle assessment involves compiling an inventory of relevant inputs and outputs of the system itself and of the systems that are involved in those inputs and outputs (Life Cycle Inventory Analysis). The potential environmental impacts of these inputs and outputs are then evaluated. At each stage it is important to interpret the results in relation to the objectives of the assessment (ISO, 1997). Only an assessment which covers the full life cycle of the product system can be described as an 'LCA'. The methodology is also applicable to other forms of assessment, such as TEWI described above, provided that the scope of the assessment and its result are clearly defined. So whereas LCA studies describe the environmental impacts of product systems from raw material acquisition to final disposal, studies that are conducted to the same rigorous standards for

Table 3.10. Items considered in TEWI and LCCP studies for a refrigerator (Johnson, 2003).

	Considered in	
	TEWI	**LCCP**
Refrigerant		
Emissions of refrigerant:		
• During manufacturing of the refrigerant		X
• Fugitive emissions at the refrigerator factory	X	X
• Emissions during the life of the product (leaks and servicing)	X	X
• Emissions at the time of disposal of the product	X	X
Emissions of CO_2 due to energy consumption:		
• During manufacturing of the refrigerant		X
• During transportation of the refrigerant		X
Blowing agent		
Emissions of blowing agent:		
• During manufacturing of the blowing agent		X
• At the refrigerator factory	X	X
Emissions of blowing agent from the foam:		
• During the life of the product	X	X
• At the time of disposal of the product	X	X
• After disposal of the product	X	X
Emissions of CO_2 due to energy consumption:		
• During manufacturing of the blowing agent		X
• During transportation of the blowing agent		X
The refrigerator		
Emissions of CO_2 due to energy consumption related to the product:		
• During manufacturing of components		X*
• During assembly of the refrigerator		X*
• During transportation of the refrigerator		X*
• By the refrigerator during its useful life	X	X
• During transportation of the refrigerator for disposal		X*
• During disposal of the refrigerator (usually shredding)		X*

* These items are relatively small in comparison with emissions related to the power consumed by the refrigerator and those related to emission of the blowing agent, and are independent of the refrigerant and blowing agent. They may therefore be neglected in some LCCP studies.

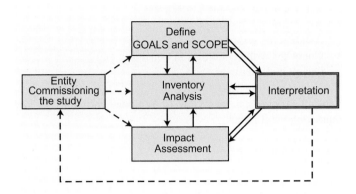

Figure 3.6. Outline phases of a Life Cycle Assessment and interactions with the commissioning entity.

subdivisions of the product chain (for example up to the sale of a unit to a customer) or subdivisions of the environmental impact (for example LCCP and TEWI) are just as valid, and may better meet the requirements of those commissioning the study. Indeed as the use phase usually dominates the environmental impact of the CFC substitutes, the LCA will be application-specific. However, attempts to provide 'generic' LCA results can help to identify the most relevant life-cycle stages and impact categories. These attempts will also tend to be tests of the extent to which the assumptions made about this use phase actually match the real performance of the application.

Figure 3.6 shows the steps in the general methodology.

For any given analysis it is essential that the objective is clearly defined. This should include the application, reasons for the study and the intended recipients of the results. To meet this objective, the scope of the study needs to specify the performance characteristics of the product and to define a 'functional unit' that will be used to quantify these characteristics (ISO, 1997 and 1998). This is a particularly important step and is best illustrated with examples:

Once a functional unit that is practical and meets the needs of those commissioning the assessment has been defined, a reference flow can be established from which all of the ancillary inputs and outputs may be calculated. For example, in a cold store this could be the annual throughput of foodstuff from which the number of receipts and deliveries, the energy load, and so forth can be estimated. In any LCA it is essential that the system boundaries are clearly and unambiguously defined. This includes not only the boundaries of the primary physical system under assessment (in this example the cold store) but also the extent to which inputs and outputs will be traced back to elementary flows (material drawn from or placed into the environment). The documentation for the decisions on system boundaries should be sufficient to judge whether or not more detailed examination is desirable and to permit subsequent changes to the assessment if new information becomes available.

Once the system has been defined, it should then be described in terms of its unit processes and their interrelation-

ships. Each of these unit processes has its own set of inputs and outputs, enabling a matrix of flows, based on the reference flow, to be constructed. This constitutes the Life Cycle Inventory. In the subsequent stages of LCA, the environmental impacts of the inputs and outputs identified in the inventory are assessed. For the materials addressed in this report, the most obvious impact categories are climate change and ozone depletion but some or all of the additional categories described in 3.5.1 above may be important (ISO, 2000a and 2000b). The end product is a description of the environmental impacts of a defined product system in terms of the effects on the most appropriate individual categories, together with an indication of how significant that impact is for each category. There is no scientific basis for reducing LCA results to a single overall score or number (ISO, 1997) and, similarly, there is little justification for closely ranking impacts (although it is worth noting when one impact, such as that on climate change, clearly outweighs the rest). A data documentation format for Life Cycle Assessments has been developed (ISO, 2002) to facilitate common input. Data collection formats have also been developed for specific applications, such as motor vehicle manufacture (Finkbeiner *et al.*, 2003) and plastics (O'Neill, 2003) and, similarly, large consistent databases are now sold (for example at www.ecoinvent.ch/).

It is difficult to characterize the uncertainty in LCA; there are a large number of variables with varying degrees of autocorrelation and for which formal uncertainty analyses may not exist. However, it should be possible to perform sensitivity analyses with comparatively little effort so as to provide a commentary on the significance of the impacts determined in the assessment (Ross *et al.*, 2002). The requirements for data quality assessment are described in ISO (1997, 1998, 2000b).

In order to facilitate complete LCA studies involving refrigerants and foam blowing agents, several studies have been performed to characterize the environmental impacts of fluid manufacture. The general conclusion was that the impact from producing the fluids was small compared to that arising from their use during service and their eventual disposal (Banks *et al.*, 1998; Campbell and McCulloch, 1998; McCulloch and Campbell, 1998; McCulloch and Lindley, 2003). The most significant contribution to the impact for producing fluids comes from material, such as other fluorocarbons, released during the manufacturing process and there is a wide variation in the values used. The highest values were calculated by Banks *et al.* (1998) who used maximum permitted emissions rather than real values. The other studies used actual process records so that the amount of material released was not only lower but also decreased substantially in recent years as the treatment of vent gases to avoid their release to the atmosphere became more commonplace, particularly in the new plants to manufacture HFCs (Campbell and McCulloch, 1998; McCulloch and Campbell, 1998; McCulloch and Lindley, 2003).

As for studies of complete systems, Yanagitani and Kawahari (2000) confirmed that for air-conditioning systems the largest source of environmental impact arose from energy use, but that proper waste management at end-of-life could significantly reduce the impact. The predominant influence of energy production on the impact on global warming, acidification, aquatic ecotoxicity, photochemical ozone creation, terrestrial toxicity and the proliferation of radionuclides was demonstrated by Frischknecht (1999) in LCA studies of generic heat pumps, building air conditioning, and industrial and commercial refrigeration.

3.5.2.4   Other system-based approaches

Environmental burden
Environmental burden is a method for assessing the environmental impact of a production facility. Mass emissions of individual compounds released from the site are multiplied by a 'potency' that characterizes the impact of the compound on a particular environmental end-point (for example ozone depletion or global warming). The sum of these values in each impact category is the environmental burden of the facility (Allen *et al.*, 1996). The resulting site-specific review of environmental impact can be used in environmental management (as described in ISO 14001).

Eco-efficiency
This combines Life Cycle Assessment of similar products or processes with a total cost determination of each alternative. Economic and ecological data are plotted on an x/y graph, with costs shown on the horizontal axis and the environmental impact on the vertical axis. The graph reveals the eco-efficiency of a product or process compared to other products or processes, with alternatives that have high cost and high impact occupying the upper right-hand quadrant. Similarly, those with low impact and low-cost occupy the lower left-hand quadrant, close to the origin. (BASF, 2003). However, such analyses demand a great deal of accurate data.

3.6   Regional dimensions

The use of fluorocarbons is specific to certain technical sectors. The technology selection in these sectors, their customers and product developments are influenced by a number of factors, which are of a local, national or regional nature (e.g. EU regulation (COM(2003)0492), under preparation). In addition to technical requirements, those factors can also include cost, environmental considerations, legal requirements, health and safety issues, energy inputs and costs and market characteristics. Therefore prescriptions on how to arrive at these decisions are not possible, as each country, and each enterprise within it, must make its own decision. Against this background, this section presents some of the more general characteristics and considerations that will influence the choice of technology at both the national and enterprise levels.

  This section also highlights some of the regional differences that influence technology choice. For these purposes, countries are considered in groupings recognized under the Montreal Protocol, namely:

- Latin America and the Caribbean;
- Africa and the Indian Ocean;
- Asia/Pacific region;
- Countries with economies in transition; and
- Developed countries.

Table 3.11 gives an overview of regional variations in key methodological issues.

3.6.1   Sector characteristics

3.6.1.1   Refrigeration and air conditioning
Growth in the demand for refrigeration has paralleled the demand for food preservation, processing, freezing, storage, transport and display, as well as final storage in homes. The more centralized food production becomes, and thus further removed from the consumer, the greater the amount of refrigeration. Consequently, societies with a more complex food supply structure and countries with a higher urban population will have a higher demand for refrigeration than countries supplying food from more local sources.

Large air-conditioning systems with capacities of about 1 MW cooling capacity upwards, are used in most of the large commercial buildings, hospitals and hotels around the globe, irrespective of the local climate (UNEP, 2003c). The occurrence of such systems roughly matches the occurrence of the type of buildings described. Smaller air-conditioning systems are largely desired in countries with warm climates (UNEP, 2003c), but there is an increasing market for these in areas with a more moderate climate, for example Central and Northern Europe. Therefore, the influencing factors for the spread of such systems are the occurrence of high ambient temperatures, and high humidity, as well as available income.

  There is an almost universal preference for mobile air conditioning, even in colder climates. The only limiting factor is the cost of the system, which typically has to be covered when purchasing the vehicle. Certain types of systems – refrigerators, unitary air conditioning products and water chillers – have universal usage characteristics and can therefore be manufactured in centralized facilities. This simplifies quality control and reduces the likelihood of leaks, and thus the need for service. Nevertheless, since high ambient temperatures create an increased demand for servicing due to higher mechanical stress on the systems and longer periods of operation, and considering that most repairs currently lead to emissions of the refrigerant, hot climates tend to have higher levels of refrigerant emissions than cooler climates. In other sectors, for example most commercial refrigeration systems, the installations are too custom- or location-specific to be manufactured in a centralized facility, although research is underway to change this (UNEP, 2003c).

  Maintenance philosophies which encourage preventive maintenance of refrigeration equipment, have lower emissions and maintain a stable energy-efficiency performance. The decision to have preventive maintenance or to request service

Table 3.11. Overview of regions and specific methodological dimensions.

Region Dimension	Latin America and Caribbean	Africa and Indian Ocean	Asia-Pacific	Countries with economies in transition	Developed countries
3.2 Key Technical Performance Indicators					
3.2.1 lifetime perspectives	No specific differences with the exception that the more expensive equipment generally has a longer lifetime. Such expensive equipment is sold more in developed countries, where standards are higher, and enforced.				
3.2.2 Fluid emission rates	Some care during fluid production. Frequent maintenance requirement due to high ambient temperature, yielding more fluid emissions. Poor care for fluid emissions at service and disposal, but this is being addressed under initiatives funded by the Multilateral Fund of the Montreal Protocol.	Frequent maintenance requirement due to high ambient temperature, yielding more fluid emissions. Poor care for fluid emissions at service and disposal, but this is being addressed under initiatives funded by the Multilateral Fund of the Montreal Protocol.	Some care during fluid production. Poor care for fluid emissions at service and disposal, but this is being addressed under initiatives funded by the Multilateral Fund of the Montreal Protocol.		Significant care during fluid production. Significant care during servicing and some during disposal of equipment.
3.2.3 Energy aspects	The energy aspect is not the driving factor when buying new equipment and material. Main factor is initial cost.				High concern with use of highly-efficient equipment.
	Significant share (72%)[1] of renewable electricity. Some concern about energy efficiency.	Moderate share (19%)[1] of renewable electricity.	Significant use of fossil fuel (79%)[1] for electricity. Some concern about energy efficiency.	Average use (63%)[1] of fossil fuel for electricity	Average use (60%)[1] of fossil fuels for electricity.
3.3 Categories Of Cost					
3.3.2 Direct engineering and financial cost	Always considered. Focus on manufacturing, mostly assembly.		Always considered. Some R&D and component manufacturing.	Always considered Some component manufacturing.	Includes liability provision. Significant R&D and component manufacturing.
3.3.2.1 The time dimension in cost 3.3.2.2 Discounting	High interest rate.	High interest rate.	Average interest rate.	Average interest rate.	Low interest rate.
3.3.3 Investment cycle and Sector Inertia	Shortage of capital. Large inertia due to unavailability of resources for transition away from HCFC.		High economic development/ modest inertia. Little emphasis on transition away from HCFC.	Shortage of capital/large inertia. Transition away from HCFC according to Montreal Protocol schedule.	Strong pressure from legislation/low inertia.

Table 3.11. Continued

Region Dimension	Latin America and Caribbean	Africa and Indian Ocean	Asia-Pacific	Countries with economies in transition	Developed countries
3.3.4 Wider costing Methodologies-Concepts	Not accounted.				Modest consideration.
3.3.5 Wider costing methodologies- cost categories	Life Cycle Cost (LCC) generally not considered.				LCC used as a marketing tool.
3.3.6 Economic Key drivers and technology uncertainty	Montreal Protocol Fund; Growing domestic and export markets; Low uncertainty since technology is generally imported; High uncertainty on HCFC future price. Large fluid producer.		Montreal Protocol Fund; Growing equipment market (domestic and export) Medium uncertainty due to fast transition. Large fluid producer.	GEF support for transition based on Montreal Protocol schedule, but some difficulties to achieve targets. Large fluid producer.	Market Leadership; Medium uncertainty due to fast transition.
3.3.7 Other issues	Increasing legislative framework to control trade in ODSs and trade in related technologies under the Montreal Protocol.				National and regional legislation more restrictive than Montreal Protocol.
3.4 Health And Safety Issues					
3.4.1 Health and Safety considerations	Modest concern due to modest liability. Main influence is from the USA.	Modest concern due to modest liability. Mainly influenced by Europe.	Growing concerns in production facilities. Mainly influenced by Europe.	Growing concerns in production facilities. Mainly influenced by Europe.	These two issues are the driving factors in USA and Europe policy design.
3.5 Climate And Environmental Impacts	Modest contribution; Main driver is ozone layer protection.		Significant contribution; Main driver is ozone layer protection.		Large contribution. Europe and Japan are taking a leading role in mitigation.

[1] Assessments based on data from International Energy Agency database, and considers electricity production from fossil and non-fossil (including nuclear) fuels.

only in the case of system failures, is not only dependent on the labour costs but also the business culture and the use-specific importance of uninterrupted delivery of refrigeration capacity. Labour costs, business culture and the value of reliable services are country-specific.

The widespread use of refrigeration and air-conditioning technology, and the accompanying high demand for service and repairs, makes the diffusion of improved techniques important. However, the large number of servicing companies makes it difficult to introduce new maintenance practices and to ensure that these are adhered to (UNEP, 2003c). Through the Montreal Protocol, networks of service technicians have been established in several countries for the diffusion of information within the service sector enterprises, and in some cases there

are also mechanisms in place to facilitate a certain maintenance quality. For low-cost, factory-manufactured equipment, such as refrigerators or small- and medium-sized, air-conditioning systems, high labour costs reduce the demand for servicing and instead favour early replacement. Although this results in lower emissions during maintenance, there is the potential of repair-worthy systems being dumped in countries with a lower level of income. This further complicates the situation in poorer countries as ageing refrigeration systems tend to have higher emissions and energy consumptions, and require more frequent repairs (UNEP, 2003c).

Methods of disposal at the end-of-life of the equipment also have implications for the life-cycle GHG emissions of the equipment. (IPCC, 2001b, Chapter 3 Appendix). Given the widespread ownership of refrigeration systems and the high costs associated with recycling, appropriate disposal at end-of-life is at present (2004) more the exception than the rule in most regions. However, several countries have established legislation requiring certain disposal practices, although enforcement is still generally a challenge. For example in Japan, a recent system of CFC coupons compulsorily acquired by car owners when the car reaches end-of-life and are transferred to car dismantlers as they recover the fluid, is not performing well (www.yomiuri.co.jp/newse/20030511wo12.htm). The need for enforcement is further amplified by the fact that users have purchased the equipment many years, if not decades, before the disposal takes place, making data availability and the link between the manufacturer and the user fairly weak.

Customs tariffs have, according to experience gained under the Montreal Protocol, not significantly hindered the spread of new technologies. A further complication is the adherence to different national or regional regulations and standards, which are often mandatory in nature. These might not be compatible with the use of alternative technologies and their characteristics, or might hinder supplies of refrigerants and spare parts. These difficulties can be both substantial and long-lasting, thereby delaying the introduction of new technologies by several years.

In the case of mass-produced refrigeration systems, in particular refrigerators and air-conditioning systems, high labour costs in some regions have made the migration of industrial production an issue. With decreasing freight tariffs, refrigerators are also being increasingly transported over long distances, although differences in local requirements mean that the product is less standardized than air conditioners.

Investment capacities and interest to invest are also significant drivers for technology diffusion, for both manufacturers and consumers. Typically, manufacturers invest in new technology in response to consumers' demands and/or legal requirements[7]. For small manufacturers and technicians in the informal sector, investments are very complicated, especially as there are often few options for obtaining loans in many less-developed countries.

3.6.1.2   Foam sector

During the implementation of the Montreal Protocol, the consumption of CFCs in foam manufacture was largely phased out, and these were replaced by different technologies. In closed-cell rigid insulating foams, hydrocarbons including cyclopentane, n-pentane, isopentane and blends have been widely used in foam subsectors, where energy efficiency, safety and product performance criteria can be met. In flexible foams, CO_2 (water) technology has been successfully introduced. Currently, the most significant uses of HCFCs (developing countries) and HFCs (mostly developed countries) are in rigid insulating foam subsectors, where safety, cost, product performance and energy conservation are important (UNEP-TEAP, 2003).

For closed-cell rigid insulating foams, a large portion of the blowing agent remains in the foam until the end of its useful life (UNEP, 2003b). Consequently, the disposal practice (land-fill compared with incineration) has a large influence on the direct emissions from a system (i.e., a refrigerator or a building) insulated with foams blown with fluorocarbons. Foam disposal requires collection from a large number of individual users or retrieval from a large quantity of mixed solid waste such as demolished building rubble. This is further complicated by the fact that, in some cases, the foams are integrated with other materials, for example, when used as building material it is adhered to substrates. These factors will make collection a major logistical and legal undertaking, which has not been mastered except in certain subsectors like domestic refrigeration, and even there, only a few countries are implementing such measures (UNEP, 2003b). As most foam products are lightweight compared to their volume, transportation costs prohibit their transportation over long distances, unless the foam is only a small fraction of the final product or system (i.e., a refrigerator). Addition, the movement of foam products, in particular building insulating foams, is further hindered by building construction traditions and building code requirements, which differ significantly between countries.

3.6.1.3   Solvents

The solvent sector is characterized on the one hand by the small scale, open use of solvents and on the other hand by use in industrial processes or closed machines. Both industrial uses and closed machine uses have undergone significant improvements as part of the efforts to reduce the use of ODSs under the Montreal Protocol. A completely different issue is the open use of solvents. For some open uses in medium-size consuming operations, investments might lead to a transition towards closed uses with internal recycling of the solvents. In other uses, in particular cleaning in smaller workshops, the solvent will evaporate into the atmosphere. Low labour costs and, thus, less automated production tend to support the open use of solvents. As in the case of the refrigeration sector, the large number of users in non-homogenous solvent applications makes the spreading of know-how a complex and labour-intensive undertaking (UNEP, 2003d). There are no substantial technical barriers to phasing out ODSs in the solvent sector. Alternatives are available that will meet the needs of all solvent users with very few exceptions. The main barrier to overcoming the obstacles in developing countries is communication and education about suitable alternatives. (UNEP, 2003d).

3.6.1.4   Aerosols/MDIs

Since the beginning of the Montreal Protocol, most aerosol uses of fluorocarbons have been converted to other motive agents, particularly hydrocarbons (UNEP, 2003e). Even so in Japan alone, 1850 tonnes of HFCs were distributed in about 4.5 million cans in 2003. This is a considerable increase compared to the 1050 tonnes distributed in 1995. It is estimated that 80% of these cans are used to blow away dust. The fluid used (HFC-

---

[7] See for example the EU regulation on Fluorinated Gases under discussion (COM(2003) 0492).

134a) has a high GWP and measures adopted by the government to replace it with a fluid with a lower GWP (HFC-152a) are only slowly having an effect (www.asahi.com/english/business/TKY200405270126.html). Furthermore, certain critical technical and/or laboratory uses of CFCs remain that are not controlled under the Protocol. For the users of such specialised aerosols the associated costs are less important. In these specific sectors, the introduction of alternatives to fluorocarbons is very knowledge intensive. However, due to the limited number of manufacturers, the number of specialists needed for technology transfer is limited (UNEP, 2003e). National legislation with respect to imports and standards is important because several products, in particular pharmaceuticals, need to adhere to national or regional standards. Most countries require intensive testing of new pharmaceuticals before the lengthy approval process is initiated and this can delay the introduction of new technologies. For all specialized products, manufacturers will often face very significant investments for research and development, testing, licenses and approval. The cost of converting production facilities to utilize alternative technologies could also be high.

3.6.1.5   *Fire protection*

Fire protection is a knowledge-intensive sector, which only needs a few specialists to service the limited amount of facilities. The diffusion of new technologies can therefore be undertaken with a limited amount of effort. Appropriate servicing of fire protection equipment and, where applicable, the subsequent appropriate disposal and destruction are key elements in the overall climate impact of these applications. The specific nature of this sector provides good opportunities for implementing containment measures for remaining applications, for example the banking of halons under the Montreal Protocol. For the introduction of new technologies, fire protection has a similar characteristic requirement to pharmaceuticals. The safe and efficient use of the agents has to be proven before new technologies can be accepted. This can cause significant delays in technology transfer. The costs of the systems are, within certain limits, secondary for the user because fire protection systems are required and/or essential and form an integral part of the purchase of buildings, military equipment or aircraft.

3.7   Emission projections

HFC and PFC emissions arise from two distinct sources (process emissions and releases when the product is in use (including disposal)) that require different methodologies for accounting historic and current mass emissions or for projecting mass emissions in the future.

Process emissions that occur during chemical production are subject to pollution-control regulations in many countries. These originate from a relatively small number of large facilities and are potentially simple to monitor. For example, there are some thirteen companies throughout the world that produce HCFC-22 and hence could be sources of the HFC-23 byproduct

(AFEAS, 2003; EU, 2003). Together with their subsidiaries and associates, and the other independent facilities in a small number of developing countries, these constitute a set of 50 potential emission sites, which are point sources. A standard methodology exists to monitor the release of HFC-23 from these facilities (DEFRA, 2003; IPCC, 2000a) and future emissions will depend on production activity and the extent to which byproducts are abated at the sources.

Emissions arising during use of a fluorocarbon, or on disposal of the system containing it, occur over a much wider geographical area than the point source emissions described above. Furthermore, the losses are spread out over the service lifetime of the system with system-dependent rates of release; for most applications this will result in an emissions pattern that covers several years after the system is charged. Future releases will therefore depend on the release pattern from the current deployment, future changes in the number of systems, how widespread the use of fluids is and the extent to which the fluids are contained during usage and disposal. Methodological guidance is available for monitoring current releases of HFCs and PFCs from refrigeration and air-conditioning systems, foam blowing, aerosols, solvents and fire-fighting applications (IPCC, 2000a) and there is a standard methodological protocol for calculating releases from refrigeration systems (DEFRA, 2003).

Almost all predictions of future emissions are extrapolations of current quantities and trends and, the primary difference in methodology is the extent to which this is based on either:
a) An appreciation of the details of the market for the systems and the way those, and the emission rates of fluids, will change in the future (bottom-up approach)[8], or b) A view of the economy as a whole and the emissions arising from the niches filled by HFCs and PFCs, so that trends are governed by overall economic parameters (top-down approach).

3.7.1   *Process emissions*

This category includes emissions of HFC-23 from the production of HCFC-22 which, in recent years, has been the largest fluorocarbon contribution to potential climate change. This release of HFC-23 is used to exemplify the requirements for forecasting emissions.

In general, process emissions can be related to process activity:

$$E_i = A_i \times F_i \qquad (3.12)$$

Where:
E_i is the annual emission in year i;
A_i is the activity in that year, and
F_i is a factor relating activity to emissions.

---

8 This terminology is widely applied but has a variety of meanings. In this part of the report, all predictions based on study of HFC and PFC markets will be called bottom-up. Top-down will be used only for predictions based on macroeconomic parameters.

In the case of HFC-23, A_i is the annual rate of production of HCFC-22 for all uses, whether or not the HCFC-22 is released into the atmosphere or used as feedstock for fluoropolymer manufacture; so that the calculation of future activity will be a projection of both dispersive and feedstock end-uses. The global estimates of the production of HCFC-22 for dispersive use that were made for comparison with atmospheric measurements (McCulloch *et al.*, 2003) can be extrapolated in accordance with the provisions of the Montreal Protocol as outlined in Montzka and Fraser (2003). It is important that such estimates include the significant changes in production in the developing world that are evident from UNEP (2003a). The fluoropolymers are products in their own right and have different markets and growth rates from that of HCFC-22, which, in this case, is simply a raw material. These growth parameters will need to be extrapolated separately and explicitly for developed and developing economies in order to calculate a credible total activity.

HFC-23 (trifluoromethane) is formed at the reactor stage of the manufacture of HCFC-22 (chlorodifluoromethane) as a result of over-fluorination. Its formation is dependent upon the conditions used in the manufacturing process and amounts to between 1.5–4.0 % of the quantity of HCFC-22 produced. Its production can be minimized by optimizing process conditions but the most effective means of elimination is destruction by thermal oxidation (Irving and Branscombe, 2002). Thus, the emission factor F_i for HFC-23 lies between zero and 4%. Use of a single value (3%) as the default emission rate (Irving and Branscombe, 2002), although allowed in the methodology for calculating national greenhouse-gas emissions (IPCC, 2000a), is not likely to give a credible forecast. In many cases, actual HFC-23 emission rates are recorded in national greenhouse-gas inventories (UNFCCC, 2003) and these can be used as information on the trends in emission rate (either the absolute rate or the rate relative to HCFC-22 production). It should also be possible to take into consideration national regulations that will affect such emissions in order to generate more robust predictions.

3.7.2 Calculating releases of fluorocarbons during use and disposal from sales data

Models used for extrapolation of emissions need to match historical data, including trends and, at the simplest level this means that the extrapolated data must start from the recorded baseline. Furthermore, the projections need to match the shape of the historical growth (or decline) in the sales from which emissions are calculated.

There is a long record of historic data for audited production for all of the major CFCs, HCFCs and HFCs (AFEAS, 2003). These data are consistent with the aggregated values for CFC and HCFC production and consumption reported under the Montreal Protocol (UNEP, 2002). The annual releases of CFC-11 and CFC-12, HCFC-22 and HFC-134a have been calculated from the audited production and sales (which are reported in categories having similar emission functions), and been shown to be consistent with the atmospheric concentrations observed

for these species (McCulloch *et al.*, 2001; 2003). This indicates that the emission functions for these compounds from use in refrigeration, air conditioning, foam blowing, solvent applications and aerosol propulsion are robust. Comparisons between atmospheric concentrations and production or sales can also be used to refine emission functions (Ashford *et al.*, 2004).

The primary variables that respond to economic parameters are the activities for the product, which in this case is the use (or sales) of that product in the categories listed above. Emissions are then secondary variables calculated from the deployment in these categories (commonly called the banks) according to models of the time pattern of the extent of emissions. These time patterns may change in response to factors such as legislation (McCulloch *et al.*, 2001).

Consideration of the long-term production databases for a wide range of industrial halocarbons, including CFCs and HCFCs, has shown that the compound growth model can be seriously misleading. It fails to replicate the shape of the demand curve over time for any of the materials examined (McCulloch, 2000). This appears to be because it does not address the changes which occur over the product lifetime. Therefore whereas growth during the early stages of product life may be compound, and hence directly related to an economic parameter such as GDP, it assumes a linear relationship with time when the product becomes more mature and then starts to assume an asymptote at full maturity. As S-curve has been shown to best represent the actual shape of the demand curve over time, the curve used to describe the growth and decline of biological populations (Norton *et al.*, 1976; McCulloch, 1994b, 1995b and 1999). There is however no fixed time cycle and some products reach maturity far sooner than others. In the short term (say ten to fifteen years) it may be permissible to forecast future demand on the basis of a relatively simple function based on the historical demand. A product in its early life could be forecast to grow at a rate governed by the growth in GDP (the compound growth model); similarly, one that has reached more mature status may have a linear growth rate, increasing by the same mass rate each year. The completely mature product will have reached a constant demand (and may, in fact, be subject to falling demand if there is replacement technology).

3.7.3 Modelling future sales and emissions from bottom-up methodologies

The first step is to construct a history of the demand for the material in its individual end-uses, both those where it is currently used and those where it has the potential to be used. These demands may then be extrapolated from starting points that reflect the current status. Although it is possible, given sufficient resources and access to much information that may be considered confidential, to construct separate demand models for each new compound (Enviros March, 1999; Haydock *et al.*, 2003), the most common methodology involves constructing models of the overall demand in a particular sector, for example hermetic refrigeration. Extrapolation of that demand into the future can be based on a mathematical analysis of the prior changes in functions with time (as outlined above in McCulloch, 1994b, 1995b and 1999) or on the application of an external economic function, such as a compound growth rate (for example the growth in GDP as detailed in McFarland, 1999). Although over the relatively short time-period considered in this report (up to the year 2015), the difference between linear and compound growth for future demand may be small, any robust model should show the sensitivity of the forecasts to assumptions about future growth rates and should justify those rates by reference to the historic growth or models of comparable systems.

Once a forecast for the overall demand for a function has been established, the extent to which the HFC or PFC is deployed in that function can be estimated using a substitution fraction and a view of how that fraction might change in the future. There is now a body of data that describes recent substitution fractions and the changes expected in the coming decades (Enviros March, 1999; Forte, 1999; McFarland, 1999; Harnisch and Hendriks, 2000; Harnisch and Gluckman, 2001; Haydock *et al.*, 2003). The most accurate substitution data will be found by examining current technical data for each compound in each application. In almost all cases, the potential for substitution is greater than the actual extent of substitution. Enforced changes in technology have provided the opportunity to switch to completely different materials and techniques (the not-in-kind solutions), to improve recovery and recycling (McFarland, 1999) and to significantly reduce charge in each installation (Baker, 1999). The requirements for continuing economic and environmental improvements will serve to drive these in the direction of further reductions in the substitution fractions.

Emission functions, factors applied to the quantities of material in use at each stage of the equipment life cycle, can also be predicted. In many studies the functions are based on AFEAS methodology, or variants of it. This allows for an initial loss, a loss during use and a final loss on disposal (AFEAS, 2003; McCulloch *et al.*, 2001, 2003; Haydock *et al.*, 2003; Enviros March, 1999; Harnisch and Hendriks, 2000). Default values for the emission functions for each category of end-use are provided in the IPCC Guidance on Good Practice in Emissions Inventories (IPCC, 2000a). However these emission functions have been shown to change in response to changes in technology and regulations (McCulloch *et al.*, 2001) and predictions should take this into account, either explicitly or as a sensitivity case.

Finally the future evolution of emissions for each compound may be calculated by combining the temporally developed emission functions with the forecast demands. This approach of building a quantified description of emissions from databases that can be separately verified, requires a large body of information to be gathered. However, this can be reduced by making assumptions and by estimating quantities and parameters by analogy. During the course of this process it will be relatively easy to identify the parts of the analysis that rely on such assumptions and to calculate the sensitivities of the results to changes in them. This is much more difficult if 'top-down' methods are used because the assumptions are unlikely to be explicit.

3.7.4   Modelling future sales and emissions using top-down parameterization

This form of methodology uses forecast changes in major econometric parameters, such as GDP, to predict future emissions. A typical example is the series of scenarios for future emissions of HFCs and PFCs contained in the IPCC Special Report on Emissions Scenarios (IPCC, 2000b). Such long-term forecasts of emissions are desirable for predicting future climate change but the real implications of these forecasts need to be established and the predictions then need to be modified accordingly.

One significant advantage of top-down forecasting is that it depends on a parameter that should be common to all other forecasts of greenhouse-gas emissions – GDP or similar. It is therefore readily scaleableas expectations for the economic parameter change. However, the assumptions must be completely clear and the sensitivity of the result to changes in the assumptions must be an integral part of the analysis.

In the ideal case, the historical connection should be established between economic parameters and the demand for refrigeration, insulation and other categories in which HFCs and PFCs can be used, so that the parameter with the best fit can be chosen. Preferably, the analysis should be done on a regional or even a national basis, but it is unlikely that sufficient data would be available for this. Then the same methodology as used in the bottom-up models should be applied to translate this demand into emissions of individual compounds. In the form of this model it is unlikely that the timing of emissions can be rigorously estimated and so the sensitivity of the result to changes in the assumptions about the timing of emissions is essential. This allows the major failing of top-down models (that deviations from reality are perpetuated throughout the modelled period) to be addressed, with the possibility of making changes in the light of technological developments.

Technology change, diffusion and transfer
In all forms of modelling it is essential to establish the drivers of technological change and to assess their effects on demand. In the case of refrigeration and air conditioning as a whole, continuing improvements to system engineering have resulted in significant reductions in the absolute rates of leakage. In turn, this has allowed similar reductions in charge size, so that the inventory of refrigerants has been reduced together with the quantities required by original equipment manufacturers (OEMs) and for servicing. The scope for such reductions was reported in IPCC/TEAP (1999). Changes of this nature may originate in developed economies but is by no means confined to them. Furthermore, the adoption of new technology in the rest of the world could be expected to accelerate as the manufacturing base shifts from North America, Europe and Japan towards developing economies.

Predictions of future emissions need to take into account the probability of technical change, in terms of both the primary innovation and its diffusion and transfer into the global manufacturing sector. At the very least analyses of the sensitivities to such changes should be incorporated, which address both the magnitude and rate of change, and take into account both geographical and economic factors. Ideally a system should be developed to simply calculate their effects on the predictions.

Uncertainty

As the results are predictions, formal statistical analyses have relatively little meaning. The key to the exercise is how well these predictions will match the future reality and this cannot be tested now. However, the models can be subjected to certain tests and the simplest is replicating the current situation; models that cannot match this from historic data are likely to give meaningless predictions. If the model output does match with reality, then the sensitivity of the result to changes in the historic parameters will provide useful information for predictions of the future. Sensitivity analyses are likely to be the most that will be necessary. Given sufficient resources it may be possible to apply Monte Carlo methods to the predictions in order to derive a more statistically rigorous uncertainty for the result. However, the value of going to such lengths is questionable bearing in mind that the models are based on assumptions and not observations.

3.8 Outlook: Future methodological developments

Some of the assessment methodologies described in this chapter are very comprehensive. For example Life Cycle Assessment was developed for high volume products in mass markets and not for the customized systems found, for example, in commercial refrigeration or fire-protection. for certain areas of application or regions, the amount of resources required for some of these assessment methods will probably be considered inappropriate. There is an evident need for simple and pragmatic assessment methods in many world regions. TEWI and LCCP analyses – using standard assumptions and boundaries – are likely to have a strong role in fulfilling this need.

An important future task in using assessment methods such as TEWI and LCCP lies in achieving consistency and comparability among different studies from different authors and years. One example for this is the choice of weighting factors for the climate impact of emissions of different substances, for example, the choice of sets of GWP values which can come from the second or third IPCC assessment report or from other more recent sources, or the application of other metrics for the radiative impact of substance emissions (see Fuglestvedt *et al.* (2003) for an overview). It would therefore seem advisable for authors to publish enough interim results to allow the recalculation of modified parameters, such as more recent emission factors or values for indirect emissions from a country's electricity production.

In the past little attention was paid to ensuring the comparability and transferability of results from different technology assessments. The treatment of uncertainties was often incomplete and the resulting recommendations were often not robust

enough to be transferred across a sector. Researchers and the users of their results should therefore pay more attention to determining the circumstances under which clear and robust conclusions on the relative performance of different technologies can be drawn, and on where uncertainties preclude such rankings. Carefully designed and performed sensitivity tests of the results for variations of key parameters are crucial for obtaining these insights.

In view of the many assumptions and different methodologies, an important role has been identified for comparisons between technologies using a common set of methods and assumptions as well as for the development of simple and pragmatic standard methodologies and the respective quality criteria. A future international standardization of simple as well as more complex or comprehensive evaluation methodologies will be important. An advanced level of international consistency has already been achieved in the field of health and safety. However, international standardization processes consume considerable amounts of time and resources (from scoping, via drafting and review to finalization). More timely and flexible processes will therefore be required as well. Whereas current standards are mainly based on low-toxicity, non-flammable fluids, standardization committees must also prepare proper standards which consider the limits and conditions for the safe use of fluids that are flammable or show higher toxicity. An essential input for this is the global standardization of international levels of requirements in respect to health, safety and environmental performance as well as respective test methods and classifications.

Policymakers need to have such information that is valid for entire sectors and this warrants additional methodology development. Future work will need to bridge the gap between application-specific comparisons and results which are robust enough to be used for policy design in entire subsectors. These analyses will have to be based on extensive databases on equipment populations, which comprise empirical data on fluid emissions and energy consumption. These databases should ideally be consistent and compatible with national greenhouse-gas emission inventories. Information on fluid sales to the different parties involved in the subsector will need to be made available. Significant resources will be required for these fairly comprehensive data requirements to support robust sectoral policies, and a number of resulting confidentiality issues will need to be addressed cautiously. In their efforts to achieve acceptability across subsectors, decision-makers could focus on increasing the involvement of relevant stakeholders and introducing additional measures to increase the transparency for outside users by means of more extensive documentation.

It is important to bear in mind that the methodologies and policies discussed above may be subject to misuse and neglect. For example, although industry uses such assessments, they are rarely determining factors in selecting a particular alternative. In fact, environmental assessment methods that are sensitive to inputs are often employed to justify the suitability of a technology that has already been selected for other reasons.

Policymakers should therefore recognize other parameters industry uses to choose technologies, so that they are aware of the factors affecting the outcomes of an analysis and of the market forces which may counter the spirit of environmental policies.

Cost is clearly one of the most important factors driving decisions for or against certain technologies. Private decision-makers usually take a life-cycle cost perspective based on their enterprises' rules for depreciation times and capital costs for their investment. Policymakers commonly use different rules and parameters to judge the cost-effectiveness of different measures. As this Special Report has shown, there is still little public cost information available which policymakers can use to reach a judgement about the cost-effectiveness of measures. Many firms give considerably less weight to social costs than to private costs in making their decisions. Initial exploratory studies would seem to be a worthwhile means of filling this gap. In the future it might be useful to apply uniform costing methodologies with common standards for transparency and data quality.

In summary the following points can be highlighted as key results:

A systems perspective is usually used to select a technology. This takes into account the system's life-cycle costs, its energy consumption and associated emissions, health and safety impacts, and other environmental impacts. The available assessment methods for each of these attributes have been described in this chapter. These will often need to be adapted to the specific application region concerned. A decision-maker can avoid inconsistencies by initiating concerted technology comparisons of competing technologies under common rules. In any case decision-makers need to make their decisions in the light of the remaining uncertainties and limitations of the available assessment methods, such as Total Equivalent Warming Impact (TEWI), the Life Cycle Climate Performance (LCCP) or a Life Cycle Assessment (LCA).

Ensuring the faithful application of existing assessment methodologies by all players in order to provide information relevant for decisions, is an ongoing challenge. A decision-maker may want to ensure that the full life cycle of the application has been considered, that all relevant stakeholders have been involved in the scoping and execution of the analysis and in the review of its results, that accepted emissions monitoring protocols have been applied for direct and indirect emissions, that all costs are properly accounted based in the best available figures, and that the uncertainties, sensitivities and limitations of the analysis have all been clearly identified.

References

ADL (A.D. Little, Inc.), 1994: Update on Comparison of Global Warming Implications of Cleaning Technologies Using a Systems Approach, AFEAS, Arlington, USA.

ADL (A.D. Little, Inc), 1999: Global Comparative Analysis of HFC and Alternative Technologies for Refrigeration, Air Conditioning, Foam, Solvent, Aerosol Propellant, and Fire Protection Applications. Final Report to the Alliance for Responsible Atmospheric Policy, August 23, 1999, Prepared by J. Dieckmann and H. Magid (available online at www.arap.org/adlittle-1999/toc.html), Acorn Park, Cambridge, Massachusetts, USA, 142pp.

ADL (A.D. Little, Inc.), 2002: Global Comparative Analysis of HFC and Alternative Technologies for Refrigeration, Air Conditioning, Foam, Solvent, Aerosol Propellant, and Fire Protection Applications. Final Report to the Alliance for Responsible Atmospheric Policy, March 21, 2002 (available online at www.arap.org/adlittle/toc.html), Acorn Park, Cambridge, Massachusetts, USA, 150pp.

AFEAS (Alternative Fluorocarbons Environmental Acceptability Study), 2003: *Production, Sales and Atmospheric Releases of Fluorocarbons through 2001* (available online at www.afeas.org), Arlington, Va., USA.

Albouy, A., Roux, J.D., Mouton, D. and J. Wu, 1998: Development of HFC Blowing Agents. Part II: Expanded Polystyrene Insulation Board. *Cellular Polymers*, **17**, 163.

Albritton, D.L. and R.T. Watson (eds.), 1989: *Scientific Assessment of Stratospheric Ozone: 1989, Vol II The AFEAS Report*. World Meteorological Organization Global Ozone Research and Monitoring Project report No 20, WMO, Geneva.

Albritton, D.L., P.J. Aucamp, G. Megie and R.T. Watson (eds), 1999: *Scientific Assessment of Ozone Depletion: 1998*. World Meteorological Organization Global Ozone Research and Monitoring Project, Report No 44, WMO, Geneva.

Allen, D., R. Clift and H. Sas, 1996: Environmental Burden: The ICI Approach. ICI plc, London, United Kingdom.

Ashford, P., D. Clodic, A. McCulloch and L. Kuijpers, 2004: Emission profiles from the foam and refrigeration sectors, Comparison with atmospheric concentrations. Part 2: results and discussion, *Int. J. Refrigeration*, **27** (2004), 701-716.

Baker, J.A., 1999: Mobile Air Conditioning: HFC-134a Emissions and Emission Reduction Strategies. Proceedings of the Joint IPCC/TEAP Expert Meeting on Options for the Limitation of Emissions of HFCs and PFCs, L. Kuijpers, R. Ybema (eds.), 26-28 May 1999, Energy Research Foundation (ECN), Petten, The Netherlands (available online at www.ipcc-wg3.org/docs/IPCC-TEAP99).

Banks, R.E. and P.N. Sharrat, 1996: Environmental Impacts of the Manufacture of HFC-134a, Department of Chemical Engineering, UMIST, Manchester, United Kingdom.

Banks, R.E., E.K. Clarke, E.P. Johnson and P.N. Sharratt, 1998: Environmental aspects of fluorinated materials: part 3, Comparative life-cycle assessment of the impacts associated with fire extinguishants HFC-227ea and IG-541, *Trans IChemE*, **76B** (1998), 229-238.

Banuri, T.K., K. Göran-Mäler, M. Grubb, H.K. Jacobson and F. Yamin, 1996: Equity and Social Considerations. In *Climate Change 1995 – Economic and Social Dimensions of Climate Change*, J. Bruce, H. Lee and E.F. Haites (eds.), Contribution of Working Group III to the Second Assessment Report of the Intergovernmental Panel on Climate Change, Cambridge University Press, Cambridge, United Kingdom, Chapter 3, pp. 79-124.

Barrault, S., J. Benouali and D. Clodic, 2003: Analysis of the Economic and Environmental Consequences of a Phase Out or Considerable Reduction Leakage of Mobile Air Conditioners. Report for the European Commission, Armines, Paris, France.

BASF, 2003: Eco-Efficiency (available online at www.basf.de/en/corporate/sustainability/oekoeffizienz).

Beggs, C.B., 1996: A Method for Estimating the Time-of-Day Carbon Dioxide Emissions per kWh of Delivered Electrical Energy in England and Wales. Proc. CIBSEA, *Building Service Engineering Research Technology*, **17**(3), 127-134.

BRA (British Refrigeration Association)**,** 1996: Guideline Method for Calculating TEWI: BRA Specification. Issued under UK DTI Best Practice Programme, ISBN 1 870623 12 6, Federation of Environmental Trade Associations, Bourne End, United Kingdom.

Butler, D.J.G., 1994: Minimising Refrigerant Emissions from Air Conditioning Systems in Buildings. Information Paper 1/94, IP1/94, Building Research Establishment (BRE), Garston, Watford, United Kingdom.

Caleb, 1998: Assessment of Potential for the Saving of Carbon Dioxide Emissions in European Building Stock, Caleb Management Services, Bristol, United Kingdom.

Caleb, 1999: The Cost Implications of Energy Efficiency Measures in the Reduction of Carbon Dioxide Emissions from European Building Stock, Caleb Management Services, Bristol, United Kingdom.

Caleb, 2000: Final Report prepared for AFEAS on the Development of a Global Emission Function for Blowing Agents used in Closed Cell Foam, Caleb Management Services, Bristol, United Kingdom.

Campbell, N.J. and A. McCulloch, 1998: The climate change implications of manufacturing refrigerants: a calculation of 'production' energy contents of some common refrigerants. *Trans IChemE.,* **76B** (1998), 239-244.

Colbourne, D. and K.O. Suen, 1999: Expanding the Domain of TEWI Calculations. Proceedings of the 20th International Congress of Refrigeration, Sydney, Australia, 19-24 September 1999. D.J. Cleland and C. Dixon (eds.), International Institute of Refrigeration, (IIR/IIF), Paris, France.

Daniel, J.S., S. Solomon and D.L. Albritton, 1995: On the evaluation of halocarbon radiative forcing and global warming potentials. *J. Geophys. Res.,* **100**(D1), 1271-1285.

DEFRA, (UK Department for Environment, Food and Rural Affairs), 2003: Guidelines for the Measurement and Reporting of Emissions by Direct Participants in the UK Emissions Trading Scheme. Report No. UK ETS(01)05rev1, DEFRA, London, United Kingdom, 120pp.

Derwent, R.G., M.E. Jenkin, S.M. Saunders and M.J. Pilling, 1998: Photochemical ozone creation potentials for organic compounds in northwest Europe calculated with a master chemical mechanism, *Atmos. Environ.,* **32** (14-15), 2429-2441.

DETR (UK Department of the Environment, Transport and the Regions), 1999: Environmental Reporting: Guidelines for Company reporting on Greenhouse Gas Emissions. DETR, London, United Kingdom, 46 pp.

EN378, 2000**:** Refrigerating systems and heat pumps – Safety and environmental requirements. Part 1: Basic requirements, definitions, classification and selection criteria, Part 2: Design, Construction, testing, marking and documentation, Part 3: Installation, site and personal protection, Part 4: Operation, maintenance, repair and recovery. British Standards Institution, London, United Kingdom.

Enviros March, 1999: UK Emissions of HFCs, PFCs and SF6 and Potential Emission Reduction Options. Report for UK Department of Environment, Transport and the Regions, Enviros March, London, United Kingdom, 169 pp.

ETSU (Energy Technology Support Unit), 1997: Cutting the cost of refrigerant leakage. Good Practice Guide 178. Prepared for the Department of the Environment, Transport and the Regions by ETSU and March Consulting Group, London, United Kingdom, 22 pp.

EU (European Union)**,** 2003: Risk Assessment Report Chlorodifluoromethane (Draft). Office for Official Publications of the European Communities, Luxembourg.

Finkbeiner, M., S. Krinke, D. Oschmann, T. Saeglitz, S. Schäper, W.P. Schmidt and R. Schnell, 2003: Data collection format for Life Cycle Assessment of the German Association of the Automotive Industry, *Int. J. Life Cycle Assess.,* **8**(6), 379-381.

Fischer, S.K., P.J. Hughes, P.D. Fairchild, C.L. Kusik, J.T. Dieckmann, E.M. McMahon and N. Hobday, 1991: Energy and Global Warming Impacts of CFC Alternative Technologies, U.S. Department of Energy and AFEAS, Arlington, Va., USA.

Fischer, S.K., P.D. Fairchild and P.J. Hughes, 1992: Global warming implications of replacing CFCs, *ASHRAE Journal,* **34**(4), 14-19.

Fischer, S.K., J.J. Tomlinson and P.J. Hughes, 1994: Energy and Global Warming Impacts of Not in Kind and Next Generation CFC and HCFC Alternatives, U.S. Department of Energy and AFEAS, 1200 South Hayes Street, Arlington, Va., USA.

Forte, R., 1999: The Impact of Regulation on the Use of Halocarbons in the United States. Proceedings of Fluorspar '99, San Luis Potosi, Mexico, Industrial Minerals Information Limited, London, United Kingdom.

Frischknecht, R., 1999: Environmental Impact of Natural Refrigerants – Eco-balances of heat Pumps and Cooling Equipment (in German: Umweltrelevanz natürlicher Kältemittel – Ökobilanzen von Wärmepumpen und Kälteanlagen). Forschungsprogramm Umgebungs- und Abwärme, Wärme-Kraft-Kopplung (UAW), Report prepared for the Bundesamt für Energie by ESU-Services, Bern, 199 pp.

Fuglestvedt, J.S., T.J. Berntsen, O. Godal, R. Sausen, K.P. Shine and T. Skodvin, 2003: Metrics of climate change: Assessing radiative forcing and emission indices. *Climatic Change,* **58**(2003), 267–331.

Goedkoop, M.J., 1995: The Eco-Indicator 95. PRé consultants, Amersfoort, The Netherlands, 85 pp.

Gover, M.P., S.A. Collings, G.S. Hitchcock, D.P. Moon and G.T. Wilkins, 1996: Alternative Road Transport Fuels – A Preliminary Life-Cycle Study for the UK, Volume 2. ETSU (Energy Technology Support Unit), Harwell, United Kingdom.

Harnisch, J. and C. Hendriks, 2000: Economic Evaluation of Emission Reductions of HFCs, PFCs and SF_6 in Europe. Report prepared for the European Commission DG Environment, Ecofys, Cologne/Utrecht, Germany/Netherlands, 70 pp.

Harnisch, J. and R. Gluckman, 2001: Final Report on the European Climate Change Programme Working Group Industry, Work Item Fluorinated Gases. Report prepared for the European Commission (DG Environment and DG Enterprise), Ecofys and Enviros, Cologne, Germany, 58 pp.

Harnisch, J., N. Höhne, M. Koch, S. Wartmann, W. Schwarz, W. Jenseit, U. Rheinberger, P. Fabian and A. Jordan, 2003: Risks and Benefits of Fluorinated Greenhouse Gases in Practices and Products under Consideration of Substance. Report prepared for the German Federal Environmental Protection Agency (Umweltbundesamt) by Ecofys GmbH (Köln/Nürnberg), Öko-Recherche GmbH (Frankfurt), Öko-Institut e.V. (Darmstadt/Berlin), TU München (München), Max-Planck-Institut für Biogeochemie (Jena), Berlin, 128 pp.

Hauschild, M.Z. and H. Wenzel (eds.), 1998: *Environmental Assessment of Products. Volume 2: Scientific Background.* Chapman and Hall, London, United Kingdom, 565 pp.

Haydock, H., M. Adams, J. Bates, N. Passant, S. Pye, G. Salway and A. Smith, 2003: Emissions and Projections of HFCs, PFCs and SF_6 for the UK and Constituent Countries. Report No AEAT/ED50090/R01 prepared for the UK Department for Environment, Food and Rural Affairs, London, United Kingdom.

Heijungs, R., (ed.), 1992: *Environmental Assessment of Products. Guide and Background* (in Dutch: Milieugerichte levenscyclusanalyses van producten. I. Handleiding, II Achtergronden). Centrum voor Milieukunde (CML), Leiden University, Leiden, Netherlands.

Hwang, Y., D.H. Jin and R. Radermacher, 2004: Comparison of Hydrocarbon R-290 and Two HFC Blends R-404a and R-410a for Medium Temperature Refrigeration Applications. Final Interim Report to ARI, Global Refrigerant Environmental Evaluation Network (GREEN) Program, CEEE Department of Engineering, University of Maryland, USA.

IEA (International Energy Agency), 2002a: *Energy Statistics of Non-OECD Countries: 1999-2000,* IEA/OECD, Paris, France.

IEA (International Energy Agency), 2002b: *CO_2 Emissions from Fuel Combustion: 1971-2000,* IEA/OECD, Paris, France.

IEA (International Energy Agency), 2003: *Energy Statistics of OECD Countries: 2000-2001,* IEA/OECD, Paris, France.

Institute of Refrigeration, 1995: Code of Practice for the Minimisation of Refrigerant Emissions from Refrigerating Systems, London, United Kingdom.

IMO (International Maritime Organization), 2000: *International Maritime Dangerous Goods Codes (IMDG) 2000.* IMO, London, United Kingdom.

IPCC (Intergovernmental Panel on Climate Change), 1996a: *Climate Change 1995: The Science of Climate Change. Contribution of Working Group I to the Second Assessment Report of the Intergovernmental Panel on Climate Change* [Houghton, J.T., L.G. Meira Filho, B.A. Callander, N. Harris, A. Kattenberg and K. Maskell (eds.)]. Cambridge University Press, Cambridge, United Kingdom, and New York, NY, USA, 572 pp.

IPCC, 1996b: *Climate Change 1995: Impacts, Adaptation and Mitigation of Climate Change: Scientific Technical Analyses. Contribution of Working Group II to the Second Assessment Report of the Intergovernmental Panel on Climate Change* [Watson, R.T., M.C. Zinyowera and R.H. Moss (eds.)]. Cambridge University Press, Cambridge, United Kingdom, and New York, NY, USA, 878 pp.

IPCC, 1997: *Revised 1996 Guidelines for National Greenhouse Gas Inventories – Reference Manual* [Houghton, J.T., L.G. Meira Filho, B. Kim, K. Treanton, I. Mamaty, Y. Bonduki, D.J. Griggs and B.A. Callender (eds.)]. Published by UK Meteorological Office for the IPCC/OECD/IEA, Bracknell, United Kingdom.

IPCC, 2000a: *Good Practice Guidance and Uncertainty Management in National Greenhouse Gas Inventories* [Penman, J., M. Gytarsky, T. Hiraishi, T. Krug, D. Kruger, R. Pipatti, L. Buendia, K. Miwa, T. Ngara, K. Tanabe and F. Wagner (eds.)]. Published by the Institute for Global Environmental Strategies (IGES) for the IPCC, Hayama, Kanagawa, Japan.

IPCC, 2000b: *Emissions Scenarios.* Special Report on Emissions Scenarios to the Intergovernmental Panel on Climate Change [Nakicenovic, N. and R. Swart (eds.)]. Cambridge University Press, Cambridge, United Kingdom, and New York, NY, USA, 570 pp.

IPCC, 2001a: *Climate Change 2001 – Synthesis Report. A Contribution of Working Groups I, II, and III to the Third Assessment Report of the Intergovernmental Panel on Climate Change* [Watson, R.T. and the Core Writing Team (eds.)]. Cambridge University Press, Cambridge, United Kingdom, and New York, NY, USA, 398 pp.

IPCC, 2001b *Climate Change 2001 – Mitigation. Contribution of Working Group III to the Third Assessment Report of the Intergovernmental Panel on Climate Change* [Metz, B., O. Davidson, R. Swart and J. Pan (eds.)] Cambridge University Press, Cambridge, United Kingdom, and New York, NY, USA, pp 752.

IPCC/TEAP, 1999: Options for the Limitation of Emissions of HFCs and PFCs. Proceedings of the Joint IPCC/TEAP Expert Meeting on Options for the Limitation of Emissions of HFCs and PFCs, L. Kuijpers, R. Ybema (eds.), 26-28 May 1999, Energy Research Foundation (ECN), Petten, The Netherlands (available online at www.ipcc-wg3.org/docs/IPCC-TEAP99).

Irving, W.N. and M. Branscombe, 2002: HFC-23 Emissions from HCFC-22 Production. In *Background Papers – IPCC Expert Meetings on Good Practice Guidance and Uncertainty Management in National Greenhouse Gas Inventories,* IPCC/OECD/IEA Programme on National Greenhouse Gas Inventories, published by the Institute for Global Environmental Strategies (IGES), Hayama, Kanagawa, Japan, pp. 271-283 (available online at www.ipcc-nggip.iges.or.jp/public/gp/gpg-bgp.htm).

ISO (International Organization for Standardization), 1997: ISO 14040:1997 Environmental Management – Life Cycle Assessment – Principles and Framework. International Organization for Standardization, Geneva, Switzerland.

ISO, 1998: ISO 14041:1998 Environmental Management – Life Cycle Assessment – Goal and Scope Definition and Inventory Analysis. International Organization for Standardization, Geneva, Switzerland.

ISO, 2000a: ISO 14042:2000 Environmental Management – Life Cycle Assessment – Life Cycle Impact Assessment. International Organization for Standardization, Geneva, Switzerland, 2000a.

ISO, 2000b: ISO 14043:2000 Environmental Management - Life Cycle Assessment – Life Cycle Interpretation. International Organization for Standardization, Geneva, Switzerland.

ISO, 2002: ISO 14048:2002 Environmental Management – Life Cycle Assessment – Data Documentation Format. International Organization for Standardization, Geneva, Switzerland.

Johnson, R.W., 2003, The Effect of Blowing Agent Choice on Energy and Environmental Impact of a Refrigerator in Europe. Proceedings of the Polyurethanes Expo 2003 International Technical Conference & Exposition, 1-3 October 2003, Orlando, Florida, USA, Alliance for the Polyurethanes Industry (API), Arlington, VA, USA, pp. 513.

Kjeldsen, P. and Scheutz, C., 2003, Attenuation of Alternative Blowing Agents in Landfills. Proceedings of the Polyurethanes Expo 2003 International Technical Conference & Exposition, 1-3 October 2003, Orlando, Florida, USA, Alliance for the Polyurethanes Industry (API), Arlington, VA, USA, pp. 473.

Krähling, H. and S. Krömer, 2000: HFC-365mfc as Blowing and Insulation Agent in Polyureathane Rigid Foams for Thermal Insulation, Solvay Management Support, Hannover, Germany.

Lawson, B., 1996: Building Materials, Energy and the Environment – Towards Ecologically Sustainable Development, Royal Australian Institute of Architects (RAIA), Canberra, Australia.

Lee, S. and Mutton, J., 2004: Global Extruded Polystyrene Foams Emissions Scenarios. Proceedings of the 15th Annual Earth Technologies Forum, April 13-15, 2004, Washington, D.C., USA.

Lippiatt, B.C., 2002: BEES 3 - Building for Environmental and Economic Sustainability Technical Manual and User Guide, NISTIR 6916. National Institute of Standards and Technology, Technology Administration, US Department of Commerce, Washington, DC, USA.

March Consulting Group, 1996: UK Use and Emissions of Selected Hydrofluorocarbons – A Study for the Department of the Environment carried out by March Consulting Group. HMSO, London, United Kingdom.

Markandya, A. and K. Halsnæs, 2000: Costing Methodologies. In *Guidance Papers on the Cross-Cutting Issues of The Third Assessment Report of the IPCC*, R. Pachauri, T. Taniguchi, K. Tanaka (eds.), published by the Global Industrial and Social Progress Research Institute (GISPRI) for the IPCC, Tokyo, Japan, pp. 15-31. (available online at www.ipcc.ch/pub/support.htm).

McCulloch, A., 1992: Depletion of the Ozone Layer and Development of Zero ODP Compounds. Proceedings of the Second World Renewable Energy Congress 'Renewable Energy: Technology and the Environment', A.A.M. Sayigh (ed.), Reading, September 1992, pp. 77-82.

McCulloch, A., 1994a: Life cycle analysis to minimise global warming impact. *Renewable Energy,* **5**(II), 1262-1269.

McCulloch, A., 1994b: Sources of Hydrochlorofluorocarbons, Hydrofluorocarbons and Fluorocarbons and their potential emissions during the next twenty five years. *Env. Mon. and Assessment,* **31**, 167-174.

McCulloch, A., 1995a: Total Equivalent Warming Impact (TEWI) of a Company's Activities. Proceedings of the International Conference on Climate Change, Washington DC, May 23, 1995, Alliance for Responsible Atmospheric Policy, Arlington USA.

McCulloch, A., 1995b: Future consumption and emissions of hydrofluorocarbon (HFC) alternatives to CFCs: Comparison of estimates using top-down and bottom-up approaches. *Environ. Internat.,* **21**(4), 353-362.

McCulloch, A., 1999: Halocarbon Greenhouse Gas Emissions During the Next Century. Proceedings of the Joint IPCC/TEAP Expert Meeting on Options for the Limitation of Emissions of HFCs and PFCs, L. Kuijpers, R. Ybema (eds.), 26-28 May 1999, Energy Research Foundation (ECN), Petten, The Netherlands (available online at www.ipcc-wg3.org/docs/IPCC-TEAP99).

McCulloch, A., 2000: Halocarbon Greenhouse Gas Emissions During The Next Century. In: *Non-CO₂ Greenhouse Gases: Scientific Understanding, Control and Implementation. Proceedings of the Second International Symposium, Noordwijkerhout, The Netherlands, September 8-10, 1999*, J. van Ham, A.P.M. Baede, L.A. Meyer and R. Ybema (eds.), Kluwer Academic Publishers, Dordrecht, The Netherlands, pp. 223-230.

McCulloch, A. and N.J. Campbell, 1998: The Climate Change Implications of Producing Refrigerants. In: *Natural Working Fluids '98. Proceedings of the IIR-Gustav Lorentzen Conference, June 2-5, 1998; Oslo, Norway*, International Institute of Refrigeration, Paris, France, p.191-199.

McCulloch, A. and A.A. Lindley, 2003: From mine to refrigeration: a life cycle inventory analysis of the production of HFC-134a. *International Journal of Refrigeration,* **26**, 865–872.

McCulloch, A., P. Ashford and P.M. Midgley, 2001: Historic emissions of Fluorotrichloromethane (CFC-11) based on a market survey. *Atmos. Environ.,* **35**(26), 4387-4397.

McCulloch, A., P.M. Midgley and P. Ashford, 2003: Releases of refrigerant gases (CFC-12, HCFC-22 and HFC-134a) to the atmosphere. *Atmos. Environ.,* **37**(7), 889-902.

McFarland, M., 1999: Applications and Emissions of Fluorocarbon Gases: Past, Present and Prospects for the Future. Proceedings of the Joint IPCC/TEAP Expert Meeting on Options for the Limitation of Emissions of HFCs and PFCs, L. Kuijpers, R. Ybema (eds.), 26-28 May 1999, Energy Research Foundation (ECN), Petten, The Netherlands (available online at www.ipcc-wg3.org/docs/IPCC-TEAP99).

Miller, M.M., 1999: Fluorspar - 1999, In *U.S. Geological Survey Minerals Yearbook 1999,* U.S. Geological Survey, USA, pp. 28.1-28.7.

Montzka, S.A., P.J. Fraser (lead authors), J.H. Butler, P.S. Connell, D.M. Cunnold, J.S. Daniel, R.G. Derwent, S. Lal, A. McCulloch, D.E. Oram, C.E. Reeves, E. Sanhueza, L.P. Steele, G.J.M. Velders, R.F. Weiss, R.J. Zander, 2003: Controlled Substances and Other Source Gases. In *Scientific Assessment of Ozone Depletion: 2002,* Global Ozone research and Monitoring project - Report No. 47, World Meteorological Organization, Geneva, Chapter 1.

Norgate, T.E. and W.J. Rankin, 2000: Life Cycle Assessment of Copper and Nickel. In: *Minprex 2000. Proceedings of the International Congress on Mineral Processing and Extractive Metallurgy, 11-13 September 2000, Melbourne, Australia,* Australian Institute of Mining and Metallurgy, Carlton, Vic, Australia, pp. 133-138

Norton, L., R. Simon, H.D. Brereton and A.E. Bogden, 1976: Predicting the course of Gompertzian growth. *Nature, 264,* 542-544.

O'Neill, T.J., 2003: Life cycle assessment and environmental impact of plastic products. *Rapra Review Reports, 13*(12), 156.

Peall, D., 2003: Refrigeration Leakage – A Global Problem. Healthcare Engineering Seminar, London.

Pira Ltd, 2001: Ecopackager. Streamlined Life Cycle Assessment Model. Pira Ltd, United Kingdom.

Radford, P., 1998: The Benefits of Refrigerant Inventory Control. In: *R22 Phase Out - Impact on End Users Today. Proceedings of the 1998 Conference of the Institute of Refrigeration,* London, United Kingdom.

Ray, A., 1984: Cost Benefit Analysis: Issues and Methodologies, Johns Hopkins Press, Baltimore, USA.

Ross, S., D. Evans and M. Webber, 2002: How LCA studies deal with uncertainty. *Int. J. Life Cycle Anal., 7*(1), 47-52.

SAEFL (Swiss Agency for the Environment, Forests and Landscape), 1998: Life Cycle Inventories for Packagings. Environmental Series No. 250/1 Waste (Volume 1). Bundesamt für Umwelt, Wald und Landschaft (BUWAL), Berne, Switzerland.

Sand, J.R., S.K. Fischer and V.D. Baxter, 1997: Energy and Global Warming Impacts of HFC Refrigerants and Emerging Technologies, U.S. Department of Energy and AFEAS, Arlington, Va, USA.

Schwarz, W. and J. Harnisch, 2003: Establishing the Leakage Rates of Mobile Air Conditoners. Report prepared for DG Environment of the European Commission, Ecofys, Öko-Recherche and Ecofys, Frankfurt, Germany.

UN (United Nations), 2002: UN Recommendations on the Transport of Dangerous Goods. Model Regulations (11th edition), Geneva, Switzerland.

UN-ECE (United Nations Economic Commission for Europe), 1991: Protocol to the 1979 Convention on Long-Range Transboundary Air Pollution Concerning the Control of Emissions of Volatile Organic Compounds or their Transboundary Fluxes, 18 November 1991, Geneva, Switzerland.

UNEP (United Nations Environment Programme), 2002: Production and Consumption of Ozone Depleting Substances under the Montreal Protocol, 1986-2000. UNEP Ozone Secretariat (Secretariat to the Vienna Convention for the Protection of the Ozone Layer and the Montreal Protocol on Substances that Deplete the Ozone Layer), Nairobi, Kenya.

UNEP, 2003a: Handbook for the International Treaties for the Protection of the Ozone Layer – The Vienna Convention (1985), The Montreal Protocol (1987). Sixth edition (2003). [M. Gonzalez, G.M. Bankobeza (eds.)]. UNEP Ozone Secretariat, Nairobi, Kenya.

UNEP, 2003b: 2002 Report of the Rigid and Flexible Foams Technical Options Committee – 2002 Assessment. UNEP Ozone Secretariat, Nairobi, Kenya.

UNEP, 2003c: 2002 Report of the Refrigeration, Air Conditioning and Heat Pumps Technical Options Committee – 2002 Assessment. [L. Kuijpers (ed.)]. UNEP Ozone Secretariat, Nairobi, Kenya.

UNEP, 2003d: 2002 Report of the Solvents, Coatings and Adhesives Technical Options Committee – 2002 Assessment. [B. Ellis (ed.)]. UNEP Ozone Secretariat, Nairobi, Kenya.

UNEP, 2003e: 2002 Report of the Aerosols, Sterilants, Miscellaneous Uses and Carbon Tetrachloride Technical Options Committee – 2002 Assessment. [H. Tope (ed.)]. UNEP Ozone Secretariat, Nairobi, Kenya.

UNEP-TEAP, 2002: April 2002 Report of the Technology and Economic Assessment Panel, Volume 3b, Report of the Task Force on Destruction Technologies. [S. Devotta, A. Finkelstein and L. Kuijpers (eds.)]. UNEP Ozone Secretariat, Nairobi, Kenya.

UNEP-TEAP, 2003: May 2003 Report of the Technology and Economic Assessment Panel, HCFC Task Force Report. [L. Kuijpers (ed.)]. UNEP Ozone Secretariat, Nairobi, Kenya.

UNFCCC (United Nations Framework Convention on Climate Change), 2003: National Submissions of Greenhouse Gas Emissions on the Common Reporting Format. UNFCCC Secretariat, Bonn, Germany (available online at http://unfccc.int).

US DoL (Department of Labor), 1998: Chemical Hazard Communication, OSHA 3084, 1998 revised, Washington, D.C., USA.

US EPA (Environmental Protection Agency), 1995: An Introduction to Environmental Accounting as a Business Management Tool: Key Concepts and Terms. US EPA 742-R-95-001, Washington, D.C., USA.

Van Gerwen, R.J.M. and B.J.C. van der Wekken, 1995: Reduction of Refrigerant Emissions. Proceedings of the XIXth International Congress of Refrigeration – New Challenges in Refrigeration, The Hague, The Netherlands, International Instiute of Refrigeration, Paris, France.

Vo, C.V. and A.N. Paquet, 2004: An evaluation of the thermal conductivity for extruded polystyrene foam blown with HFC-134a or HCFC-142b. *Journal of Cellular Plastics, 40*(3), 205-228.

Wallace, D., 1996: *Sustainable Industrialization, Energy and Environmental Programme.* The Royal Institute of International Affairs, Earthscan Publication Ltd, London, 87 pp.

Watts, W., 1999: Discounting and Sustainability. The European Commission, DGII, Brussels, Belgium.

WCED (World Commission on Environment and Development, Chair: G.H. Brundtland), 1987: *Our Common Future*. Oxford University Press, Oxford, United Kingdom, 383 pp.

Yanagitani, K. and K. Kawahari, 2000: LCA study of air conditioners with an alternative refrigerant. *Int. J. Life Cycle Anal.,* **5**(5), 287-290.

4

Refrigeration

Coordinating Lead Authors
Sukumar Devotta (India), Stephan Sicars (Germany)

Lead Authors
Radhey Agarwal (India), Jason Anderson (USA), Donald Bivens (USA), Daniel Colbourne (United Kingdom), Guy Hundy (United Kingdom), Holger König (Germany), Per Lundqvist (Sweden), Edward McInerney (USA), Petter Nekså (Norway)

Contributing Authors
Ayman El-Talouny (Egypt)

Review Editors
Eduardo Calvo (Peru), Ismail Elgizouli (Sudan)

Contents

EXECUTIVE SUMMARY

Domestic refrigeration

Domestic refrigerators and freezers are used throughout the world for food storage in dwelling units and in non-commercial areas such as offices. More than 80,000,000 units are produced annually with internal storage capacities ranging from 20 litre to greater than 850 litre. With an estimated average unit lifespan of 20 years, this means there is an installed inventory of approximately 1500 million units. As a result of the Montreal Protocol, manufacturers initiated transition from CFC refrigerant applications during the early 1990s. This transition has been completed in developed countries and significant progress has been made in developing countries. The typical lifespan for domestic refrigerators means that products manufactured using CFC-12 refrigerant still comprise approximately one-half of units in the installed base. This has significantly slowed down the rate of reduction in the demand for CFC-12 refrigerant in the servicing sector.

Isobutane (HC-600a) and HFC-134a are the dominant alternative refrigerants for replacing CFC-12 in new domestic refrigeration appliances. Each of these has demonstrated mass production capability for safe, efficient, reliable and economic use. Both refrigerants give rise to similar product efficiencies. Independent studies have concluded that application design parameters introduce more efficiency variation than that attributable to the refrigerant choice. Comprehensive refrigerant selection criteria include safety, environmental, functional, cost and performance requirements. The choice of refrigerant can be strongly influenced by local regulatory and litigation environments. Each refrigerator typically contains 50–250 grams of refrigerant contained in a factory-sealed hermetic system. A simplified summary of the relative technical considerations for these two refrigerants is:

- HC-600a uses historically familiar mineral oil lubricants. Manufacturing processes and designs must fully take into account the flammable nature of the refrigerant. For example, the need for proper factory ventilation and appropriate electrical equipment, preventing leaking refrigerant from gaining access to electrical components, using sealed or non-sparking electrical components, and the use of proper brazing techniques or preferably the avoidance of brazing operations on charged systems. Service procedures must similarly include appropriate precautions for working with flammable refrigerants;
- HFC-134a uses moisture-sensitive polyolester oils. Manufacturing processes should ensure that low moisture levels are maintained. Long-term reliability requires a more stringent avoidance of contaminants during production or servicing compared to either CFC-12 or HC-600a practices.

The use of the hydrocarbon blend propane (HC-290)/isobutane (HC-600a) allows CFC-12 volumetric capacity to be matched and avoids the capital expense of retooling compressors. These blends introduce manufacturing complexities and require the use of charging techniques suitable for refrigerant blends which have components with different boiling points. The application of these blends in Europe during the 1990s was an interim step towards the transition to HC-600a using retooled compressors. The same safety considerations apply to hydrocarbon blends as to HC-600a.

Alternative refrigeration technologies such as the Stirling cycle, absorption cycle, thermoelectrics, thermionics and thermoacoustics continue to be pursued for special applications or for situations with primary drivers that differ from conventional domestic refrigerators. These technology options are not expected to significantly alter the position of vapour compression technology as the choice for domestic refrigeration.

Vapour compression technology is mature and readily available worldwide. The availability of capital resources is dictating the timing of conversion to HC-600a and HFC-134a. Current technology designs typically use less than half the electrical energy required by the units they replace. This reliable performance is provided without resorting to higher cost or more complex designs. Continued incremental improvements in unit performance and/or energy efficiency are anticipated. Government regulations and voluntary agreements on energy efficiency and labelling programmes have demonstrated their effectiveness in driving improved efficiency product offerings in several countries.

Good design and the implementation of good manufacturing and service practices will minimize refrigerant emissions; however, special attention must be given to the retirement of the large number of units containing CFC-12. With a typical 20-year lifespan, refrigerator end-of-life retirement and disposal happens to about 5% of the installed base each year. This means approximately 75 million refrigerators containing 100 grams per unit, or 7500 total tonnes of refrigerant are disposed of annually. This refrigerant will be predominantly CFC-12 for at least another 10 years. The small refrigerant charge means that refrigerant recovery is not economically justifiable. Regulatory agencies around the world have therefore provided incentives or non-compliance penalties to promote recovery of this ODS.

In 2002, the total amount of refrigerants banked in domestic refrigeration amounted to 160,000 tonnes, with annual refrigerant emissions of 5.3% of banked system charge. The annualized HFC emissions rate from this sector was 1.0% in 2002. HFC emissions mostly occur during useful life. Production transition to HFCs started during 1995; consequently in 2002 the installed product age was 7 years or less compared to a typical lifespan of 20 years. Further, recovery during service and disposal is required in most early conversion countries.

Commercial refrigeration

Commercial refrigeration makes fresh and frozen food available to customers at the appropriate temperature levels: chilled food in the range of 1°C–14 °C and frozen food in the range of –12°C to –20°C.

On a global basis, commercial refrigeration is the refrigeration subsector with the largest refrigerant emissions calculated as CO_2-equivalents. This amounts to 40% of the total annual refrigerant emissions, see Table 11.5. In 2002 worldwide commercial refrigeration emission rates were reported to be 30% yr^{-1} of the installed commercial refrigeration banked inventory of 605,000 tonnes refrigerant. This means that in an environment with an average energy mix, the refrigerant emissions represent about 60% of the total emissions of GHG resulting from system operation, the rest being indirect emissions caused by power production.

Refrigeration equipment types vary considerably in terms of size and application. Stand-alone equipment consists of systems where the components are integrated, such as beverage vending machines, ice cream freezers and stand-alone display cases. Refrigerant charge sizes are small (0.2–1 kg), and the CFCs CFC-12 and R-502 are being replaced by HFCs HFC-134a, R-404A and R-507A. HCFC-22 is also used, but is subject to phase-out requirements. Refrigerant emissions are low in these mainly hermetic systems and are similar to domestic refrigerator emissions. However, end-of-life recovery is almost non-existent on a global basis and this results in an average annual leakage of 7–12% of the refrigerant charge, dependent on the equipment lifetime. Some stand-alone equipment using hydrocarbons as refrigerants have been developed and are available in European countries, with refrigerant charge sizes in accordance with the limitations imposed by European and national safety standards.

Condensing units are small commercial systems with compressors and condensers located external to the sales area, and the evaporators located in display cases in the sales area, or in a cold room for food storage. These units are installed in shops such as bakeries, butchers and convenience stores as well as in larger food retailer stores. Similar refrigerants are used in these applications as in stand-alone equipment; however, with the larger refrigerant charges in these systems (1–5 kg), hydrocarbon refrigerant applications may be limited by national safety standards. Refrigerant emissions depend on the robustness of the system design, installation, monitoring and refrigerant recovery at end of equipment lifetime.

Full supermarket systems can be categorized by whether refrigerant evaporation occurs in the display cabinets and cold stores, or whether a low-temperature, secondary heat transfer fluid that is cooled centrally, is circulated to the display cabinets and cold stores. The first type is termed a direct system and the second type an indirect system.

Supermarket centralized direct systems consist of a series of compressors and condensers located in a remote machinery room, providing a cooling medium to display cabinets and cold storage rooms in other parts of the building. The size of systems can vary from cooling capacities of 20 kW to more than 1 MW, as used in larger supermarkets. Refrigerant charge sizes can range from 100–2000 kg. The most common form of centralized system is direct expansion. Specific units can be dedicated to low-temperature or medium-temperature evaporators. HCFC-22 continues to be extensively used in these systems, with R-502 for low-temperature applications being replaced by

R-404A and R-507A. Due to European regulations on HCFCs that have been in force since January 2001, R-404A and R-507A are the most commonly used refrigerants for large capacity low- and medium-temperature systems in Europe.

The 'distributed' system is a variation of the direct system. In this the compressors are located in sound-proof boxes near the display cases, permitting the shortening of refrigerant circuit length and a corresponding 75% reduction of refrigerant charge. Condensing units can be air-cooled or water-cooled. When compressor systems are installed as small packs with roof-mounted, air-cooled condensers, or as small packs adjacent to the sales area in conjunction with remote air-cooled condensers, they are sometimes referred to as close coupled systems. The refrigerants used are mainly HCFC-22 and the low-temperature refrigerants R-404A and R-507A. Other refrigerants such as R-410A are also being considered. With the close-coupled system design, refrigerant emissions are estimated to be 5–7% of charge on an annual basis. Compared to the centralized systems, the absolute reduction in refrigerant emissions is much greater due to the considerable reduction in refrigerant charge size.

The design of indirect systems for supermarkets permits refrigerant charge size reduction of 75–85%. Fluorocarbon-based refrigerants are generally used in these systems. However, if the centralized refrigeration system can be located in a controlled-access room away from the customer area, indirect systems may also use flammable and/or toxic refrigerants, dependent on system safety measures and national safety regulations. Refrigerant emissions are reduced to about 5% of charge yr^{-1} due to the reductions in the reduced piping lengths and the number of connecting joints.

Systems that use ammonia and hydrocarbons as primary refrigerants in indirect systems operate in several European countries. Published results show that ammonia and hydrocarbon indirect systems have a 10–30% higher initial cost than direct expansion systems and an energy consumption 0–20% higher than that of direct expansion systems, due to the additional system requirements (heat exchanger and circulating pumps with their costs and energy penalties). Development work on indirect systems design is continuing with the goals of reducing the cost and energy penalties in these systems.

Carbon dioxide is being evaluated in direct systems for both low- and medium-temperature applications, and in cascade systems with carbon dioxide at the low-temperature stage and ammonia or R-404A at the medium-temperature stage. Thirty cascade systems have been installed in supermarkets and the initial costs and energy consumption are reported to be similar to R-404A direct expansion systems.

Important considerations in the selection of designs for supermarket refrigeration systems and refrigerants are safety, initial cost, operating cost and climate change impact (refrigerant emissions and carbon dioxide from electricity produced to operate the refrigeration systems). In the 1980s the centralized direct systems had annual refrigerant emissions up to 35% of charge. Recent annualized emission rates of 3–22% (average

18%) were reported for 1700 supermarket systems in several European countries and the USA. The reduced emission rates were due to a combination of factors aimed at improving refrigerant containment, such as system design for tightness, maintenance procedures for early detection and repairs of leakage, personnel training, system leakage record keeping, end-of-life recovery of refrigerant and in some countries, increasing the use of indirect cooling systems.

In 2002, worldwide commercial refrigeration emission rates were reported to be 30% yr^{-1} of the installed commercial refrigeration banked inventory of 605,000 tonnes of refrigerant. The higher worldwide emission rates indicate less attention was paid to refrigerant containment and end-of-life recovery than in the limited survey data reported above.

Traditional supermarket centralized direct systems must be designed for lower refrigerant emissions and higher energy efficiency in order to reduce climate change impact. From an overall perspective, significant research and development is underway on several designs of supermarket refrigeration systems to reduce refrigerant emissions, use lower global-warming refrigerants and reduce energy consumption. Life cycle climate performance calculations indicate that direct systems using alternative refrigerants, distributed systems, indirect systems and cascade systems employing carbon dioxide will have significantly lower CO_2-equivalent emissions than centralized direct systems that have the above-stated, historically-high refrigerant emission rates.

Food processing, cold storage and industrial refrigeration

Food processing and cold storage is one of the important applications of refrigeration for preserving and distributing food whilst keeping food nutrients intact. This application of refrigeration is very significant in terms of size and economic importance in both developed and developing countries. The annual consumption of frozen food worldwide is about 30 Mtonnes yr^{-1}. Over the past decade, consumption has increased by 50% and is still growing. The amount of chilled food is about 10–12 times greater than the supply of frozen products. Frozen food in long-term storage is generally kept at –15°C to –30°C, while –30°C to –35°C is typical for freezing. Chilled products are cooled and stored at temperatures from –1°C–10°C.

The majority of refrigeration systems for food processing and cold storage are based on reciprocating and screw compressors. Ammonia, HCFC-22, R-502 and CFC-12 are the refrigerants historically used, with other refrigerant options being HFCs, CO_2 and hydrocarbons. HFC refrigerants are being used instead of CFC-12, R-502 and HCFC-22 in certain regions. The preferred HFCs for food processing and cold storage applications are HFC-134a and HFC blends with an insignificant temperature glide such as R-404A, R-507A and R-410A. Ammonia/CO_2 cascade systems are being introduced in food processing and cold storage.

Some not-in-kind (non-vapour compression) technologies, such as vapour absorption technology and compression-absorption technology, can be used for food processing and cold storage applications. Vapour absorption technology is well established, whereas compression-absorption technology is still under development.

For this category, limited data are available on TEWI/LCCP. A recent study of system performance and LCCP calculations for a 11 kW refrigeration system operating with R-404A, R-410A and HC-290 showed negligible differences in LCCP, based on the assumptions used in the calculations.

Industrial refrigeration includes a wide range of cooling and freezing applications in the chemical, oil and gas industries as well as in industrial ice-making, air liquefaction and other related industry applications. Most systems are vapour compression cycles, with evaporator temperatures ranging from 15°C down to –70°C. Cryogenic applications operate at even lower temperatures. Capacities of units vary from 25 kW to 30 MW, with systems often being custom made and erected on-site. Refrigerant charge size varies from 20–60,000 kg. The refrigerants used are preferably single component or azeotropes, as many of the systems use flooded evaporators to achieve high efficiency. Some designs use indirect systems (with heat transfer fluids) to reduce refrigerant charge size and to the risk of direct contact with the refrigerant.

These refrigeration systems are normally located in industrial areas with limited public access, and ammonia is the main refrigerant. The second refrigerant in terms of volume use is HCFC-22, although the use of HCFC-22 in new systems is forbidden for all types of refrigerating equipment by European regulations since January 2001. Smaller volume CFC refrigerants CFC-12 and R-502 are being replaced by HFC-134a and R-404A, and R-507A and R-410A. CFC-13 and R-503 are being replaced by HFC-23 and R-508A or R-508B. HCFC-22 is being replaced by R-410A, as the energy efficiency of R-410A is slightly higher than that of HCFC-22. The energy efficiency of R-410A can be similar to that of ammonia for evaporation temperatures down to –40°C, dependent on the compressor efficiency. Hydrocarbon refrigerants have historically been used in large refrigeration plants within the oil and gas industry.

Carbon dioxide is another non-HFC refrigerant which is starting to be used in industrial applications, as the energy efficiency of carbon dioxide systems can be similar to that of HCFC-22, ammonia and R-410A in the evaporator temperature range of –40°C to –50°C for condensing temperatures below the 31°C critical temperature of carbon dioxide. Cascade systems with ammonia in the high stage and carbon dioxide in the low stage show favourable cost and energy efficiency. Carbon dioxide is also being used as a heat-transfer fluid in indirect systems.

Attempts are being made to reduce refrigerant emissions in industrial refrigeration, food processing and cold storage by improving the system design, minimizing charge quantities, ensuring proper installation, improving the training of service personnel with respect to the detection of potential refrigerant leakage, and improving procedures for recovery and re-use of refrigerant. The total amount of refrigerants banked in the combined sectors of industrial refrigeration, food processing and cold storage was 298,000 tonnes in 2002, with ammonia at 35% and HCFC-22 at 43% of the total banked inventory. Annual refrigerant emissions were 17% of banked system charge.

Transport refrigeration

Transport refrigeration consists of refrigeration systems for transporting chilled or frozen goods. Transport takes place by road, rail, air and sea; further, containers as refrigerated systems are used with moving carriers. All transport refrigeration systems must be sturdily built to withstand movements, vibrations and accelerations during transportation, and be able to operate in a wide range of ambient temperatures and weather conditions. Despite these efforts, refrigerant leakage continues to be a common issue. It is imperative that refrigerant and spare system parts are available on-board and along the transport routes. Ensuring safe operation with all working fluids is essential, particularly in the case of ships where there are limited options for evacuation.

Ships with cargo-related, on-board refrigeration systems have either refrigerated storage spaces or provide chilled air supply. There are about 1100 such ships, with HCFC-22 being the main refrigerant. In addition, there are approximately 30,000 merchant ships which have refrigerated systems for crew food supply, again mainly using HCFC-22. Alternative refrigerants being implemented are R-404A/R-507A, R-410A, ammonia and ammonia/CO_2.

Refrigerated containers allow storage during transport on rail, road and seaways. There are more than 500,000 such containers that have individual refrigeration units of about 5 kW cooling capacity. Refrigerants in this sector are transitioning from CFC-12 to HFC-134a and R-404A/R-507A.

Refrigerated railway transport is used in North America, Europe, Asia and Australia. The transport is carried out with either refrigerated railcars, or, alternatively, refrigerated containers (combined sea-land transport; see Section 4.6.2) or swap bodies (combined road-land transport; see Section 4.6.4).

Road transport refrigeration systems (with the exception of containers) are truck-mounted systems. The refrigerants historically used were CFC-12, R-502 and HCFC-22. New systems are using HFC-134a, R-407C, R-404A, R-410A and decreasing amounts of HCFC-22. There are about 1 million vehicles in operation, and annual refrigerant use for service is reported to be 20–25% of the refrigerant charge. These high leakage rates call for additional design changes to reduce leakage, which could possibly follow the lead of newer mobile air-conditioning systems. Another option would be for systems to use refrigerants with a lower global-warming potential (GWP).

The non-HFC refrigerant hydrocarbons, ammonia and carbon dioxide are under evaluation, and in some cases these are being used for transport applications in the various sectors, with due consideration for regulatory, safety and cost issues. Fishing trawlers in the North Pacific Ocean already use ammonia for refrigeration, with a smaller number of trawlers using R-404A or R-507A. Carbon dioxide is a candidate refrigerant for low-temperature refrigeration, but the specific application conditions must be carefully considered. Carbon dioxide systems tend towards increased energy consumption during high-temperature ambient conditions, which may be significant when containers are closely stacked on-board ships, leading to high condensation/gas cooler temperatures because of lack of ventilation. In the case of reefer ships and fishing vessels, a promising alternative technology is equipment with ammonia/carbon dioxide systems. These systems have similar energy efficiency to existing refrigeration systems but higher initial costs.

Low GWP refrigerant options will technically be available for transport refrigeration uses where fluorocarbon refrigerants are presently used. In several cases, these low GWP options may increase the costs of the refrigeration system, which is an important consideration for owners of transport equipment. A technology change from an HFC, such as R-404A, to a low-GWP fluid will usually lead to a reduction of TEWI, if the energy consumption is not substantially higher than in existing systems.

The total amount of refrigerants banked in transport refrigeration was 16,000 tonnes in 2002, with annual refrigerant emissions of 38% of banked system charge consisting of CFCs, HCFCs and HFCs.

4.1   Introduction

The availability and application of refrigeration technology is critical to a society's standard of living. Preservation throughout the food chain and medical applications are examples of key contributors to quality of life. Integrated energy consumption information is not available, but this largest demand sector for refrigerants is estimated to use about 9% of world power generation capacity (Bertoldi, 2003; EC, 2003; ECCJ, 2004; EIA, 2004; ERI, 2003; UN-ESCAP, 2002, Table 1.1.9). This consumption of global power-generation capacity means that the relative energy efficiency of alternatives can have a significant impact on indirect greenhouse-gas (GHG) emissions.

Refrigeration applications vary widely in size and temperature level. Sizes range from domestic refrigerators requiring 60–140 W of electrical power and containing 40–180 g of refrigerant, to industrial and cold storage refrigeration systems with power requirements up to several megawatts and containing thousands of kilograms of refrigerant. Refrigeration temperature levels range from –70°C to 15°C. Nearly all current applications use compression-compression refrigeration technology. The potential market size for this equipment may approach US\$ 100,000 million annually. This diversity has resulted in unique optimization efforts over the decades, which has resulted in solutions optimized for different applications. For discussion purposes, the refrigeration sector is divided into the five subsectors:

- Domestic Refrigeration: the refrigerators and freezers used for food storage primarily in dwelling units;
- Commercial Refrigeration: the equipment used by retail outlets for holding and displaying frozen and fresh food for customer purchase;
- Food Processing and Cold Storage: the equipment to preserve, process and store food from its source to the wholesale distribution point;
- Industrial Refrigeration: the large equipment, typically 25 kW to 30 MW, used for chemical processing, cold storage, food processing and district heating and cooling;
- Transport Refrigeration: the equipment to preserve and store goods, primarily foodstuffs, during transport by road, rail, air and sea.

Data in Table 4.1 indicate that the annualized refrigerant emission rate from the refrigeration sector was 23% in 2002. This includes end-of-life losses. There is a wide range of annualized emissions from the five subsectors, from 5% for domestic refrigeration to 30% for commercial refrigeration to 38% for transport refrigeration. For commercial refrigeration, the 30% annual refrigerant emissions represent typically 60% of the total emissions of GHGs resulting from system operation, the rest being indirect emissions from power production. This indicates the importance of reducing refrigerant emissions from this sector, in addition to the importance of the energy efficiency of systems stated above.

4.2   Domestic refrigeration

4.2.1   Background

Domestic refrigerators and freezers are used for food storage in dwelling units and in non-commercial areas such as offices throughout the world. More than 80,000,000 units are produced annually with internal storage capacities ranging from 20 litre to greater than 850 litre. With an estimated typical unit life of 20 years (Weston, 1997), the installed inventory is approximately 1500 million units. Life style and food supply infrastructures strongly influence consumer selection criteria, resulting in widely differing product configurations between different global regions. Products are unitary factory assemblies employing hermetically-sealed, compression refrigeration systems. These typically contain 50–250 g of refrigerant.

4.2.2   Refrigerant options

Conversion of the historic application of CFC-12 refrigerant in these units to ozone-safe alternatives was initiated in response to the Montreal Protocol. Comprehensive refrigerant selection criteria include safety, environmental, functional, performance and cost requirements. A draft refrigerant selection-decision map and a detailed discussion of requirements were included in the 1998 report of the Refrigeration, Air Conditioning, and Heat Pumps Technical Options Committee (UNEP, 1998). The integration of these requirements with other potential drivers such as global-warming emissions reduction, capital resource availability and energy conservation results in a comprehensive analysis of refrigerant options for strategic consideration. Two different application areas must be addressed: (1) new equipment manufacture, and (2) service of the installed base. New equipment manufacture can be addressed more effectively, since the ability to redesign avoids constraints and allows optimization.

4.2.2.1   New Equipment Refrigerant options
Most new refrigerators or freezers employ either HC-600a or HFC-134a refrigerant. Each of these refrigerants has demonstrated mass production capability for safe, efficient, reliable and economic use. There are no known systemic problems with properly manufactured refrigerator-freezers applying either of these primary options. The key variables influencing selection between these two refrigerants are refrigerator construction details, energy efficiency, building codes, environmental considerations and the economics of complying with standards. Other selected alternative refrigerants or selected refrigerant blends have had limited regional appeal, driven by either niche application requirements or by availability of suitable compressors or refrigerants. Some brief comments about selected refrigerant use are now given.

Isobutane (HC-600a) refrigerant
HC-600a applications use naphthenic mineral oil, the historic

Table 4.1. Refrigerant bank and direct emissions of CFCs, HCFCs, HFCs and other substances (hydrocarbons, ammonia and carbon dioxide) in 2002, the 2015 business-as-usual scenario and the 2015 mitigation scenario, for the refrigeration sector, the residential and commercial air-conditioning and heating sector ('stationary air conditioning') and the mobile air-conditioning sector.

	Banks (kt)					Emissions (kt yr⁻¹)					Emissions (MtCO₂-eq yr⁻¹) SAR/TAR[2]	Emissions (MtCO₂-eq yr⁻¹) This Report[3]
	CFCs	HCFCs	HFCs	Other	Total	CFCs	HCFCs	HFCs	Other	Total		
2002												
Refrigeration	330	461	180	108	1079	71	132	29	18	250	848	1060
- Domestic refrigeration	107	-	50	3	160	8	-	0.5	0.04	9	69	91
- Commercial refrigeration	187	316	104	-	606	55	107	23	-	185	669	837
- Industrial refrigeration[1]	34	142	16	105	298	7	24	2	18	50	92	110
- Transport refrigeration	2	4	10	-	16	1	1	3	-	6	19	22
Stationary Air Conditioning	84	1028	81	1	1194	13	96	6	0.2	115	222	271
Mobile Air Conditioning	149	20	249	-	418	60	8	66	-	134	583	749
Total 2002	**563**	**1509**	**509**	**109**	**2691**	**144**	**236**	**100**	**18**	**499**	**1653**	**2080**
2015 BAU												
Refrigeration	64	891	720	136	1811	13	321	115	21	471	919	1097
- Domestic refrigeration	37	-	189	13	239	5	-	8	1	13	51	65
- Commercial refrigeration	6	762	425	-	1193	5	299	89	-	393	758	902
- Industrial refrigeration[1]	21	126	85	123	356	4	21	11	21	56	88	104
- Transport refrigeration	0.1	2.8	20.3	-	23.2	0.1	1.3	7.4	-	9	22	26
Stationary Air Conditioning	27	878	951	2	1858	7	124	68	0	199	314	370
Mobile Air Conditioning	13	23	635	4	676	5	11	175	1	191	281	315
Total 2015-BAU	**104**	**1792**	**2306**	**143**	**4345**	**25**	**455**	**359**	**23**	**861**	**1514**	**1782**
2015 Mitigation												
Refrigeration	62	825	568	186	1641	8	202	52	15	278	508	607
- Domestic refrigeration	35	-	105	60	200	3	-	3	1	6	27	35
- Commercial refrigeration	6	703	378	-	1087	3	188	40	-	230	414	494
- Industrial refrigeration[1]	21	120	65	126	331	3	13	5	14	36	53	63
- Transport refrigeration	0.1	2.8	20.3	-	23.2	0.0	0.9	4.3	-	5	13	15
Stationary Air Conditioning	27	644	1018	2	1691	3	50	38	0	91	145	170
Mobile Air Conditioning	13	23	505	70	611	3	7	65	7	82	119	136
Total 2015 Mitigation	**102**	**1493**	**2090**	**259**	**3943**	**14**	**259**	**155**	**22**	**451**	**772**	**914**

[1] Including food processing/cold storage

[2] Greenhouse gas CO₂-equivalent (GWP-weighted) emissions, using direct GWPs, taken from IPCC (1996 and 2001) (SAR/TAR)

[3] Greenhouse gas CO₂-equivalent (GWP-weighted) emissions, using direct GWPs, taken from Chapter 2 in this report

choice for CFC-12 refrigerant, as the lubricant in the hermetic system. Competent manufacturing processes are required for reliable application but cleanliness control beyond historic CFC-12 practices is not required. HC-600a has a 1.8% lower flammability limit in air, increasing the need for proper factory ventilation and appropriate electrical equipment. This flammable behaviour also introduces incremental product design and servicing considerations. These include preventing leaking refrigerant access to electrical components or using sealed or non-sparking electrical components, using proper brazing methods or preferably avoiding brazing operations on charged systems, and ensuring a more robust protection of refrigerant system components from mechanical damage to help avoid leaks.

HFC-134a refrigerant: HFC-134a applications require synthetic polyolester oil as the lubricant in the hermetic system. This oil is moisture sensitive and requires enhanced manufacturing process control to ensure low system moisture level. HFC-134a is chemically incompatible with some of the electrical insulation grades historically used with CFC-12. Conversion to the electrical insulation materials typically used for HCFC-22 applications may be necessary. HFC-134a is not miscible with silicone oils, phthalate oils, paraffin oils or waxes. Their use should be avoided in fabrication processes for components in contact with the refrigerant. Common items for concern are motor winding lubricants, cutting fluids in machining operations and drawing lubricants. Careful attention to system cleanliness is required to avoid incompatible contaminants. Trace contaminants can promote long-term chemical degradation within the system, which can reduce cooling capacity or cause system breakdown. Necessary process controls are not technically complex but do require competent manufacturing practices and attention to detail (Swatkowski, 1996).

Isobutane (HC-600a)/propane (HC-290) refrigerant blends
The use of these hydrocarbon blends allows matching CFC-12 volumetric capacity and avoids capital expense for retooling compressors. These blends introduce design and manufacturing complexities. For example, they require charging techniques suitable for use with blends having multiple boiling points. The use of HC-600a/HC-290 blends in Europe during the 1990s was an interim step towards a final transition to HC-600a using retooled compressors. Unique application considerations are consistent with those discussed above for HC-600a.

Other refrigerants and refrigerant blends
Example applications of additional refrigerants in new equipment include HC-600a/HFC-152a blends in Russia, HCFC-22/HFC-152a blends in China and HCFC-22 replacing R-502 in Japan. These all are low volume applications supplementing high-volume primary conversions to HC-600a or HFC-134a refrigerant. Demand for <u>all</u> refrigerants other than HC-600a and HFC-134a totals less than 2% of all Original Equipment Manufacturer (OEM) refrigerant demand (UNEP, 2003). These special circumstance applications will not be further discussed in this report.

4.2.2.2   *Service of existing equipment*
Service options range from *service* with original refrigerant, to *drop-in*, where only the refrigerant is changed, to *retrofit*, which changes the refrigerant and other product components to accommodate the specific refrigerant being used. Several binary and ternary blends of various HFC, HCFC, PFC and hydrocarbon refrigerants have been developed to address continuing service demand for CFC-12. These blends are tailored to have physical and thermodynamic properties compatible with the requirements of the original CFC-12 refrigerant charge. Their application has been successful and is growing. Some of these are near-azeotrope blends; others have disparate boiling points or glide. If refrigerants and lubricants other than original design specification are proposed for use, their compatibility with the specific refrigerator-freezer product configuration and its component materials must be specifically reviewed. An extended discussion of domestic refrigerator service options was included in the 1998 Report of the Refrigeration, Air Conditioning and Heat Pumps Technical Options Committee (UNEP, 2003).

4.3.3   *Not-in-kind alternatives*

Alternative refrigeration technologies such as the Stirling cycle, absorption cycle, thermoelectrics, thermoacoustics and magnetic continue to be pursued for special applications or situations with primary drivers that differ from conventional domestic refrigeration. Two examples of unique drivers are portability or absence of dependence on electrical energy supply. The 1994 Report of the Refrigeration, Air Conditioning and Heat Pumps Technical Options Committee concluded that no identified technology for domestic refrigerator-freezers was competitive with conventional compression-compression technology in terms of cost or energy efficiency (UNEP, 1994). The 1998 and 2002 reports of this committee reaffirmed this conclusion (UNEP, 1998; UNEP, 2003). No significant near-term developments are expected to significantly alter this conclusion.

4.2.4   *Energy efficiency and energy standards*

Relative refrigerator energy efficiency is a critical parameter in the assessment of alternatives. In practice, similar refrigeration system efficiency results from the use of either HFC-134a or HC-600a refrigerant. Independent studies have concluded that the relative energy efficiencies of these two primary alternatives are comparable. Efficiency differences from normal manufacturing variation exceed the differences introduced by the refrigerant choice (Sand *et al.*, 1997; Fischer *et al.*, 1994; D&T, 1996; Wenning, 1996). Energy efficiency of a product is strongly influenced by configuration, component hardware selection,"thermal insulation, heat exchange surfaces and control algorithms. Effective options are readily available from multiple commercial sources. The improved energy efficiency of domestic refrigeration products is a national initiative in several countries. Energy labelling and energy standards are both being effectively used to facilitate these initiatives. The

Table 4.2. Global refrigerant demand for domestic refrigeration (tonnes) (UNEP, 2003; Euromonitor, 2001).

Refrigerant	Refrigerant demand (tonnes)		
	1992	1998	2000
New equipment			
CFC-12	10,130	4460	3330
HFC-134a	-	5520	7150
HC-600a	-	430	1380
Other [1]	80	200	230
Sub-total New equipment	**10,210**	**10,610**	**12,090**
Field service			
CFC-12	4458	5002	4484
HFC-134a	-	349	391
HC-600a	-	4	146
Other [2]	15	40 [3]	- [3]
Sub-total Field service	**4473**	**5395**	**5021**
Total Global demand	**14,683**	**16,005**	**17,111**

[1] HCFC-22, HFC-152a and HC-190 refrigerants
[2] Three refrigerants above, plus numerous HFC and/or HCFC and/or HC blends
[3] Reliable demand data not available due to disperse nature of demand

Collaborative Labelling and Appliance Standards Program (CLASP) maintains a website with substantive information including links to various national programmes (URL: http://www.claxponline.org).

Energy standards and tests procedures

Energy test procedures provide the basis for energy regulations and labelling initiatives. These test procedures must be reproducible and repeatable and should ideally provide an indication of energy consumption under consumer use conditions. They should also provide an effective amendment protocol to accommodate evolving product technologies. Test procedures have been developed by several global standards organizations. The tests are different, and the results from one should never be directly compared with the results from another. Each can provide a relative energy consumption value for the test conditions specified. The interested reader is referred to instructive discussions and comparisons of energy test procedures, their limitations and their future needs (Bansal and Kruger, 1995; Meier and Hill, 1997; Meier, 1998).

4.2.5 Consumption and consumption trends

Table 4.2 presents consumption of the three most used domestic refrigerator refrigerants during 1992, 1996 and 2000 (UNEP, 2003). Table 4.3 presents consumption details by global region for the year 2000 (UNEP, 2003). New equipment conversions from CFC-12 to ozone-safe alternatives are occurring in advance of the Montreal Protocol requirements. By the year 2000, 76% of new unit production had been converted: 53% to HFC-134a, 21% to HC-600a and 2% to all other (UNEP, 2003). Subsequent developments have maintained this trend with an apparent increase in the percentage converting to hydrocarbon refrigerants. Two large market examples are the production in India converting to either HFC-134a or an HC-600a/HC-290

Table 4.3. Global refrigerant demand in 2000 for domestic refrigeration by global region, in tonnes (UNEP, 2003; Euromonitor, 2001).

Region	Segment[1]	Global refrigerant demand in 2000 (tonnes)				
		CFC-12	HFC-134a	HC-600a	Other [2]	Total
Western Europe	OEM		900	770		**1670**
	Service	34	14	12	n.a.	**60**
Eastern Europe	OEM	230	420	40	30	**720**
	Service	180	17	4	n.a.	**201**
North America	OEM		2460			**2460**
	Service	60	60		n.a.	**120**
Central and South America	OEM	200	1200			**1400**
	Service	990	20		n.a.	**1010**
Asia and Oceania	OEM	2030	1900	570	200	**4700**
	Service	2420	230	130	n.a.	**2780**
Africa and Mid-east	OEM	870	270			**1140**
	Service	800	50		n.a.	**850**
World	OEM	3330	7150	1380	230	**12,090**
	Service	4484	391	146	n.a.	**5021**
	Total	**7814**	**7541**	**1526**	**230**	**17,111**

[1] OEM: Original Equipment Manufacturer
[2] n.a.: data not available

blend (UNEP, 2000) and some units converting from HFC-134a to HC-600a in Japan.

Conversion of the service demand has been less successful. Field service procedures typically use the originally specified refrigerants. The long useful product life (up to 30+ years for some units with an average around 20 years), large installed base (approximately 1500 million units) and uncertainties with field conversion to alternative refrigerants have resulted in a strong continuing service demand for CFC-12. The estimated percentage of refrigerators requiring post-warranty service of the hermetically sealed system and replacement of the original refrigerant charge at sometime during their service life is 1% in industrialized countries and 7% in developing countries. In industrialized countries this demand is typically satisfied with reclaimed or stockpiled CFC-12 when not contrary to local regulations. (Note: 'Reclaimed' refrigerant refers to recovered refrigerant that has been purified to original specifications. Unpurified recovered refrigerant should never be used in long-life domestic refrigerators. Probable impurities are likely to catalyze systems degradation and cause premature failures.) CFC-12 is normally the lowest cost refrigerant in developing countries. Regardless of location, use of the CFC-free refrigerant service blends mentioned above only becomes significant when CFC-12 availability becomes limited. Since post-warranty service is typically provided by small, independent businesses, reliable service demand data are not available. Further, the limited capital resources in developing countries promotes the labour-intensive refurbishing of units compared to retirement and replacement with new units. This not only prolongs the phase-out of CFC-12, but also results in increased failure rates from the highly variable quality of workmanship.

4.2.6   *Factors affecting emissions*

A text-box example included in Chapter 3 of this report (see Table 3.5) tabulates emission factors which must be considered for comprehensive Life Cycle Climate Performance (LCCP) or Total Equivalent Warming Impact (TEWI) of domestic refrigeration design options. There are two general types of emissions: *direct*, which for discussion of the refrigerant choice are limited to the refrigerant itself; and *indirect*, which depends on the refrigerator design and the infrastructures of supporting services in the use environment (Sand *et al.*, 1997). The emission of insulating foam blowing-agents is not addressed in this chapter. Insulating foam is addressed in Chapter 7

Direct emissions

Efficient factory operations and effective process and product designs will minimize emissions at the start of the product life cycle. Significant process variables include refrigerant transfer and storage, charge station operations, maintenance protocols and factory process efficiencies. Key refrigerator design variables include hermetic-system internal volume, number of joints, mechanical fatigue and abuse tolerance and, of course, the choice of refrigerant. Domestic refrigerators contain refrig-

erant in factory-sealed, hermetic systems. Once sealed, refrigerant emission can only occur if there is a product defect or quality issue. Typical examples are brazing defects or containment component fatigue. In all cases, excluding life-ending failures, the defect will require the product hermetic system to be repaired. Variables influencing emissions during service include refrigerant recovery procedure usage and efficiency, charge technique employed, technician training and technician work standards. Refrigerant recovery during service is practiced in many countries. The small refrigerant charge quantity, typically 50–250 g per unit, make this recovery economically unattractive. Regulations and non-compliance penalties are usually required to provide incentives for recovery. Audits and refrigerant charge logs can provide useful metrics for the quality of refrigerant recovery practices.

Clodic and Palandre (2004) have detailed worldwide refrigerant bank and emissions data for domestic refrigerators (see Table 4.1). CFC emissions for domestic refrigeration in 2002 were estimated to be 8000 tonnes. HFC emissions for domestic refrigeration in 2002 were estimated to be 0.3% of the domestic refrigerant bank, or 500 tonnes. Domestic HFC refrigerant emission estimates will have increased in 2015 to 3000–8000 tonnes, dependent on the extent of refrigerant containment and recovery assumed.

With a typical 20-year lifespan, refrigerator end-of-life retirement and disposal happens to about 5% of the installed base every year. In quantified terms this means approximately 75 million refrigerators containing approximately 100 g per unit or 7500 total tonnes of refrigerant are retired and disposed of annually. For the next few years, or possibly even decades, CFC-12-containing product will continue to be a significant fraction of the waste stream. As is the case for service recovery, the small refrigerant charge makes end-of-life recovery uneconomical. Equipment and procedures commonly used for refrigerant recovery during service can be used but recovery is more typically accomplished in central disposal locations. This allows the use of faster, less labour-intensive procedures to moderate recovery costs. Nevertheless, regulations and non-compliance penalties normally provide incentives for this recovery. Regulating agencies in various global regions administer these requirements and are an appropriate source for further information.

Indirect emissions

Product design affects indirect emissions through refrigerator operating efficiency, and ease of refrigerator disassembly and separation for recycling. Higher efficiency units consume less electricity which, in turn, proportionately reduces the emissions derived from electrical power generation and distribution. Parameters influencing energy efficiency are fundamental design considerations such as heat exchangers, control efficacy, refrigerant systems, heat losses, parasitic power demands such as fans and anti-sweat heaters and product safety. The design approaches taken and options selected are directly related to the desired product features, performance and regulatory environment. A comprehensive discussion of detailed design parameters is beyond the scope of this report. Information is commercially available from multiple sources. Example references listing areas of opportunity are the UNEP Refrigerants Technical Options Committee assessment reports (UNEP, 1994, 1998, 2003), the International Energy Agency energy efficiency policy profiles report (IEA, 2003) and the Arthur D. Little global comparative analysis of HFC and alternative technologies (ADL, 2002). Objective discussions of many options are contained in the Technical Support Documents of the US Department of Energy rulemakings for domestic refrigerators and freezers (US DOE, 1995).

4.2.7 Comparison of emissions from alternative technologies

HFC-134a and HC-600a are clearly the significant alternative refrigerants for domestic refrigeration. Consequently, the significant global-warming emissions comparison for this application sector is HFC-134a compared to HC-600a. An accurate comparison of these is very complex. The multiple and widely diverse product configurations available globally are the consequence of consumer needs and choices. Comparative analysis results will be influenced by the example scenario selected and its assumed details. The available degrees of freedom are too high to achieve a comprehensive perspective within a manageable number of scenarios. Any single technical solution will not provide an optimized solution.

Harnisch and Hendriks (2000) and March (March, 1998) estimated the conversion cost from HFC-134a to HC-600a, expressed as unit emissions avoidance cost. Harnisch and Hendriks assumed no product cost or performance impact and a 1 million per manufacturing site conversion cost which yielded an avoidance cost of 3.4 per tonne CO_2-eq. March (1998) assumed higher product and development costs, also with no performance impact, which resulted in an avoidance cost of 400 per tonne of CO_2. These two estimates differ by more than two orders of magnitude in direct emissions abatement costs with assumed equivalent indirect emissions. Table 4.4 summarizes emission abatement opportunities with increased application of HC-600a refrigerant in the three most common domestic refrigerator configurations. Estimates for manufacturing cost premiums, development costs and required implementation investments are also included. Emission abatement opportunities are based on Clodic and Palandre (2004).

The objective is to assess the total emissions from *direct* and *indirect* sources. HC-600a clearly has the advantage of minimizing *direct* GHG emissions. *Indirect* emissions can dominate overall results using some scenarios or assumptions. The energy consumption of basic HFC-134a and HC-600a refrigeration systems is similar. At issue is what product modifications are required or allowed when converting to an alternative refrigerant and what effect these modifications have on product efficiency and performance. This uncertainty is particularly applicable to larger, auto-defrost refrigerators where a trade-off between system efficiency and other product attributes necessary to maintain product safety is not obvious. The consequences of trends in consumer purchase choices and their influence on the rate of emissions reduction are also difficult to predict. LCCP and TEWI are powerful, complementary tools, but results are sensitive to input assumptions. Assumptions should be carefully validated to ensure they are representative of the specific sce-

Table 4.4. Domestic refrigeration, current status and abatement options.

Product Configuration		Cold Wall	Open Evaporator Roll Bond	No-Frost
Cooling capacity	From	60 W	60 W	120 W
	To	140 W	140 W	250 W
Refrigerant charge (HFC)	From	40 g	40 g	120 g
	To	170 g	170 g	180 g
Approximate percentage of sector refrigerant bank (160 kt) in configuration		20 units * 100 g average 18% of 160 kt	15 units * 100 g average 14% of 160 kt	50 units * 150 g average 68% of 160 kt
Approximate percentage of sector refrigerant emissions (8950 tonnes) in subsector		18% of 8950 tonnes	14% of 8950 tonnes	68% of 8950 tonnes
Predominant technology		HC-600a	HFC-134a	HFC-134a
Other commercialized technologies		HFC-134a, CFC-12	HC-600a, CFC-12	HC-600a, CFC-12
Low GWP technologies with fair or better than fair potential for replacement of HCFC/HFC in the markets		R-600a	HC-600a	HC-600a
Status of alternatives		Fully developed and in production	Fully developed and in production	Fully developed and in production
R-600a Mfg. Cost Premium		No Premium	3–5 US$	8–30 US$
Capital Investment		0	45–75 million US$	400–1500 million US$
Emission reduction		1432 tonnes	1253 tonnes	6086 tonnes

narios of interest.

Several investigators have analyzed total emission scenarios comparing HFC-134a and HC-600a for domestic refrigeration:
- An Arthur D. Little, Inc. LCCP study (ADL, 1999) estimated that approximately 14 grams (10% of initial charge) of HFC-134a would be the total net lifetime emissions from a domestic refrigerator in the USA regulatory environment. Using US power generation emission data, this equates to a 0.3% energy consumption increase over a typical 20-year product life;
- Ozone Operations Resource Group of the World Bank Report No. 5: 'The Status of Hydrocarbon and Other Flammable Alternatives Use in Domestic Refrigeration' (World Bank, 1993) cited TEWI assessments presented at the 1993 German National Refrigeration Congress in Nurnberg. Regarding the relative refrigerant selection effects, this TEWI analysis concluded that 'The *direct* contribution of HFC-134a to global warming ... should not be given serious consideration within this rough estimate because it does not amount to more than a few percent of the *indirect* contribution caused by the energy consumption of the appliance' (Lotz, 1993).

4.2.8   *Emission abatement opportunities*

The following emission abatement opportunities are available for domestic refrigerators:
- Conversion to alternatives having reduced GWP: The refrigerant direct emission contribution ranges from less than 2% up to 100% of total emissions. Direct emissions of 100% reflect the condition where the power generation and distribution infrastructure has zero dependence on fossil fuel energy sources. Direct emissions favour HC-600a over HFC-134a. Regional regulatory and product liability considerations can hamper the viability of HC-600a application. Indirect emissions depend upon relative product energy efficiency. Thermodynamic cycle efficiencies of the alternatives are comparable. Product efficiency is dependent upon design attributes required to accommodate the flammability of HC-600a. There is no penalty with the cold-wall evaporator configurations common in Europe. Information concerning configurations commonly used for forced-convection, automatic-defrost products is limited or proprietary;
- Reduction of refrigerant leakage during service life: Annual leakage rates for the factory-sealed, hermetic systems in domestic units are typically less than 1%...This leakage typically drives service demand;
- Recovery of refrigerant during end-of-life disposal or during field repair: Approximately 5% of the installed base are retired each year. The annual service call rate is significantly less than that. Recovery efficiency is a critical variable;
- Reduction of indirect emissions through improved product energy efficiency: The indirect emission contribution for domestic refrigeration ranges from zero to more than 98% of total emissions. Current production refrigerators consume less than half the energy of the typically 20-year-old unit they replace. With a 5% yr^{-1} retirement rate, this translates to a 2.5% yr^{-1} improvement in indirect emissions from the installed base;
- Opportunities for reduced indirect emissions exist via improved product energy efficiency. The IEA energy efficiency policy profiles report (IEA, 2003) estimated the potential improvements to be 16–26%, dependent upon product configuration. Average cost inflation was estimated to be €23 (US$ 31) for manufacturing and €66 (US$ 88) for purchase. The report presents comparative Least Life Cycle Cost analyses for alternatives. Arthur D. Little conducted Life Cycle Climate Performance studies of HFC and other refrigerant alternatives (ADL, 2002). Their report gives heavy domestic refrigeration emphasis on the relative energy efficiency and Total Equivalent Warming Impact assessment of various blowing-agent alternatives.

4.3   Commercial refrigeration

Commercial Refrigeration is the part of the cold chain comprising equipment used by retail outlets for preparing, holding and displaying frozen and fresh food and beverages for customer purchase.

For commercial systems, two levels of temperature (medium temperature for preservation of fresh food and low temperature for frozen products) may imply the use of different refrigerants. Chilled food is maintained in the range 1°C–14°C but the evaporating temperature for the equipment varies between –15 °C and 5 °C dependent upon several factors: the type of product, the type of display case (closed or open) and the type of system (direct or indirect). Frozen products are kept at different temperatures (from –12 °C to –18 °C) depending on the country. Ice cream is kept at –18 to –20 °C. Usual evaporating temperatures are in the range of –30 to –40 °C.

On a global basis, commercial refrigeration is the refrigeration subsector with the largest refrigerant emissions calculated as CO_2 equivalents. These represent 40% of the total annual refrigerant emissions, see Table 11.5. Annual leakage rates higher than 30% of the system refrigerant charge are found when performing a top-down estimate (Clodic and Palandre, 2004; Palandre *et al.*, 2004). This means that in an environment with an average energy mix, the refrigerant emissions might represent 60% of the total emissions of GHG resulting from system operation, the rest being indirect emissions caused by power production. This indicates how important emission reductions from this sector are.

There are five main practices in order to reduce direct GHG emissions:
1. A more widespread use of non-HFC refrigerants;
2. Leak-tight systems;
3. Lower refrigerant charge per unit of cooling capacity;
4. Recovery of refrigerant during service and end-of-life;
5. Reduced refrigeration capacity demand.

4.3.1   Sector background

Commercial refrigeration comprises three main types of equipment: stand-alone equipment, condensing units and full supermarket systems.

Stand-alone equipment consists of systems where all the components are integrated: wine coolers, beer coolers, ice cream freezers, beverage vending machines and all kinds of stand-alone display cases. This equipment is installed in small shops, train stations, schools, supermarkets and corporate buildings. Annual growth is significant. All types of stand-alone equipment are used intensively in industrialized countries and are the main form of commercial refrigeration in many developing countries. These systems tend to be less energy efficient per kW cooling power than the full supermarket systems described below. A main drawback to stand-alone units is the heat rejected to ambient air when placed indoors. Therefore, the heat must be removed by the building air conditioning system when there is no heating requirement.

Condensing units are used with small commercial equipment. They comprise one or two compressors, a condenser and a receiver which are normally located external to the sales area. The cooling equipment includes one or more display cases in the sales area and/or a small cold room for food storage. Condensing units are installed in specialized shops such as bakeries, butchers and convenience stores in industrialized countries, whilst in developing countries a typical application is the larger food retailers.

Full supermarket systems can be categorized by whether refrigerant evaporation occurs in the coolers, or whether a low-temperature secondary heat transfer fluid (HTF) that is cooled centrally is circulated in a closed loop to the display cabinets and cold stores. The first type is termed 'direct expansion' or direct system and the second type is termed indirect system. Direct systems have one less thermal resistance and no separate fluid pumping equipment, which gives them an inherent efficiency and cost advantage. The HTF circulated in an indirect system normally gains sensible heat, but may gain latent heat in the case of ice slurry or a volatile fluid like CO_2.

Many different designs of full supermarket systems can be found. *Centralized systems* consist of a central plant in the form of a series of compressors and condenser(s) located in a machinery room or an outside location. This provides refrigerant liquid or an HTF at the correct temperature levels to cabinets and cold stores in other parts of the building. Each rack of multiple compressors is usually associated with a single air-cooled condenser. Specific racks are dedicated to low-temperature or medium-temperature evaporators. The quantity of refrigerant is related to the system design, refrigerating capacity and refrigerant choice varies. The centralized systems can be either direct or indirect systems. Centralized direct systems constitute by far the largest category in use in supermarkets today. The size can vary from refrigerating capacities of about 20 kW to more than 1 MW. The centralized concept is flexible in order to utilize heat recovery when needed (Arias, 2002).

Distributed Systems are characterized by having smaller compressors and condensers close to or within the coolers, so that many sets of compressor/condenser units are distributed around the store. The compressors can be installed within the sales area with remote condensers. When they are installed as small packs with roof-mounted, air-cooled condensers, or as small packs adjacent to the sales area in conjunction with remote air-cooled condensers they are sometimes referred to as *Close Coupled Systems*. The quantity of such units could range from just a few to upwards of 50 for a large supermarket. They are direct systems, but when installed inside the building that may employ a HTF, usually water, for collecting heat from the different units.

Hybrid systems cover a range of possibilities where there is a combination of types. An example is a variation of the distributed system approach, where low-temperature cabinets and cold stores comprise individual water-cooled condensing units, which are supplied by the medium-temperature HTF. Thus, in the indirect medium temperature section, the refrigerant charge is isolated mainly to the machinery room, whilst an HTF is circulated throughout the sales and storage areas at this temperature level.

In some countries, indirect, close-coupled, distributed and hybrid systems have been employed in increasing number in recent years because they offer the opportunity of a significant reduction in refrigerant charge. Additionally, with indirect systems the refrigerant charge is normally located in a controlled area, enabling the use of low-GWP refrigerants that are flammable and/or have higher toxicity. This approach has been adopted in certain European countries due to regulatory constraints on HCFCs and HFCs (Lundqvist, 2000). A review of possible system solutions is provided by Arias and Lundqvist (1999 and 2001). The close-coupled systems offer the advantages of low charge, multiple compressors and circuits for part load efficiency and redundancy, as well as the efficiency advantage of a direct system (Hundy, 1998).

4.3.2   Population/production

There is a lot of variation in the geographical distribution of commercial refrigeration systems, even in neighbouring countries, due to differing consumption habits, regulation of opening hours, leadership of brand names, state of the economy and governmental regulations.

A number of leading US and European manufacturers are expanding worldwide, especially into Eastern European countries and other countries with fast growing economies, such as: Argentina, Brazil, China, Indonesia, Mexico, Thailand and Tunisia. The growth of all types of commercial refrigerating systems in China is one of the most significant of the past 4 years. For example, the number of small supermarkets (average total sales area of about 380 m²) has increased by a factor of six in the past 4 years.

Table 4.5 shows the average total sales area of supermarkets, which differs significantly per country. The 'hypermarket' concept of selling food, clothing and all types of household

Table 4.5. Typical sales areas of supermarkets in selected countries (UNEP, 2003).

	Brazil	China	France	Japan	USA
Average surface of supermarkets (m²)	680	510	1500	1120	4000
Average surface of hypermarkets (m²)	3500	6800	6000	8250	11,500

goods, is expanding worldwide.

Table 4.6 shows an estimate of supermarket and hypermarket populations and Table 4.7 an evaluation of the population of smaller commercial units.

It is only possible to evaluate the refrigerant quantities based on the number of supermarkets if additional data are used concerning the total sales area of fresh and frozen food and the type of refrigerating system. Nevertheless in terms of the number of supermarkets, China represents more than 30% of the total global population of supermarkets (UNEP, 2003).

For small commercial supermarkets, China represents about 40% of the total global population, with the exception of vending machines. The growth of vending machines is still very significant, especially in Europe (UNEP, 2003).

4.3.3 HFC and HCFC technologies, current usage and emissions

4.3.3.1 Refrigerant choices

Refrigerant choices for new equipment vary according to national regulations and preferences.

Europe: Following CFC phase out for new equipment and servicing in Europe, commercial refrigeration tended towards the use of HCFC-22 and HCFC-22 blends. However, in the Nordic countries, the period with HCFC-22 was very short, and HFCs such as R-404A became the preferred solution from 1996 onwards. Since 2000, European Regulation 2037/2000 (Official Journal, 2000) has prohibited HCFCs in all type of new refrig-

Table 4.6. Number of supermarkets and hypermarkets (UNEP, 2003).

	Number of Supermarkets	Number of Hypermarkets
EU	58,134	5410
Other Europe	8954	492
USA	40,203	2470
Other America	75,441	7287
China	101,200	100
Japan	14,663	1603
Other Asia	18,826	620
Africa, Oceania	4538	39
Total	**321,959**	**18,021**

Table 4.7. Evaluation of the number of items of commercial equipment (UNEP, 2003).

	Condensing Units	Hermetic groups in stand alone equipment	Vending Machines
EU	6,330,500	6,400,700	1,189,000
Other Europe	862,000	754,700	113,900
USA	247,500	217,400	8,807,900
Other America	3,321,300	2,430,600	411,800
China	13,000,000	12,316,600	385,000
Japan	2,216,000	2,470,600	2,954,500
Other Asia	5,750,400	5,750,600	758,200
Africa, Oceania	843,700	831,400	87,000
Total	**32,571,400**	**31,172,600**	**14,707,300**

erating equipment. HFC-404A and HFC-507A are now the most commonly used refrigerants for larger capacity low- and medium-temperature systems, such as condensing units and all types of centralized systems. For stand-alone systems, HFC-134a is used for medium-temperature applications, while both HFC-134a and HFC-404A are used for low-temperature applications.

Japan: In Japan where HCFCs are still permitted, a voluntary policy is followed by OEMs and more than one-third of new equipment employs HFCs, with the remainder using HCFC-22. Typically, HFC-407C is used for medium temperature and HFC-404A for low temperature in all categories of commercial systems.

USA and Russia: In the USA, HCFC-22 and HCFC-22 blends are commonly used in existing systems, primarily for medium-temperature applications. HCFC-22 continues to dominate new supermarket systems, but HFC-404A and HFC-507A are becoming more widely used. HFC-404A and HFC-134a are used in new stand-alone equipment. These trends are also seen in Russia, where alternatives include HFC-134a, HCFC-22 and HCFC-22 blends as well as HFC-404A in a broad range of commercial equipment.

Developing countries: Stand-alone equipment is the main form of commercial refrigeration in developing countries, with condensing units being used by larger food retailers. CFCs, HCFCs and HFCs are all being used, with trends towards HFCs HFC-134a and HFC-404A in the future.

In China, HCFC-22 and HFC-134a are the major refrigerants for commercial refrigeration, with R-404A showing rapid growth. Only limited amounts of CFC-12 and R-502 are in use, as most of the systems designed for these refrigerants have converted to HFC-134a and R-404A. HCFC blends have very little application, as Chinese regulatory groups prefer to switch directly to HFCs instead of using any transitional HCFC blends.

4.3.3.2 Emissions

Emission rates derived with a bottom-up approach suggest a

global annual emissions rate from the commercial refrigeration sector of 30% of the refrigerant charge (leakage and non-recovery) (Palandre *et al.*, 2004). Expressed in CO_2 equivalents, commercial refrigeration represents 40% of total annual refrigerant emissions, see Table 4.1. The emission levels (including fugitive emissions, ruptures, emissions during servicing and at end-of-life) are generally very high, especially for supermarkets and hypermarkets. The larger the charge, the larger the average emission rate. This is due to the due to the very long pipes, the large numbers of fittings and valves, and the huge emissions when ruptures occur.

In the 1980s, the reported average commercial refrigeration emission rates for developed countries were in the range of 20–35% of refrigerant charge per year (Fischer *et al.*, 1991; AEAT, 2003; Pedersen, 2003). The high emission rates were due to design, construction, installation and service practices being followed without an awareness of potential environmental impact. In some countries emissions from these systems have been decreasing due to industry efforts and governmental regulations with respect to refrigerant containment, recovery and usage record keeping, increased personnel training and certification, and improved service procedures, as well as increased attention for many system mechanical details including the reduction or elimination of threaded joints and a reduction in the number of joints in refrigerating systems.

Recent annualized emission rates in the range of 3–22% (average of 18%) were reported for 1700 supermarket systems in several European countries and the USA. The country-specific data and references are listed in Table 4.8. It may be concluded that if the emission estimates of Palandre *et al.* (2003 and 2004) are correct, the above-reported values of 3–22% must represent selected company data within countries that have a strong emphasis on emission reductions.

Emission rates vary considerably between equipment categories. Annual emission rates for the several categories are listed in Table 4.9. Individual system leak rates, however, can

Table 4.8. Leakage rates of supermarket refrigeration systems.

Country	Year(s)	Annual Refrig. Loss	References
The Netherlands	1999	3.2	Hoogen *et al.*, 2002
Germany	2000–2002	5-10%	Birndt *et al.*, 2000; Haaf and Heinbokel, 2002
Denmark	2003	10%	Pedersen, 2003
Norway	2002–2003	14%	Bivens and Gage, 2004
Sweden	1993	14%	Bivens and Gage, 2004
	1998	12.5%	
	2001	10.4%	
United Kingdom	1998	14.4%	Radford, 1998
USA	2000–2002	13%, 18%, 19%, 22%	Bivens and Gage, 2004

Table 4.9. Indicative leakage rates from commercial refrigeration equipment categories found in the literature.

Category	Annual Refrigerant Loss	References
Stand-alone hermetic	≤1%	March, 1999; ADL, 2002
Small condensing unit	8–10%	March, 1999; AEAT, 2003
Centralized direct (DX)	3–22%	Several; see main text
Distributed	4%	ADL, 2002
Indirect (secondary loop)	2–4%	ADL, 2002

range from zero to over 100% yr^{-1}. It should also be noted that end-of-life recovery data are mostly not included, and therefore the annual average leakage rates may be 5–10% higher than the values given in the tables.

It is important to note that in certain cases, data collection should be considered in context. Some of the base data used in emission and emission projection studies has been collated from telephone interviews, and other similar techniques, from historical reports. This reliance on anecdotal data may suggest underestimated emissions, since both the end-users and refrigeration contractors have an interest in reporting low values because of exposure of poor practices and the threat of restrictive legislation.

As well as measures designed to decrease emissions, there are also drivers – typically at field level – that inhibit emission reduction and must be addressed at a policy level. These include partial success in finding system leaks, end-users employing contractors on an 'as-the-need-arises' basis rather than a preventative basis, additional attendance time for refrigerant recovery and leak testing, and a financial incentive for contractors to sell more refrigerant to the end-user.

There have been important observations on system emission characteristics and how emission reductions have been accomplished. Some are listed below.

In the Netherlands, emissions have been significantly reduced through national mandatory regulations established in 1992 for CFCs, HCFCs and HFCs. These measures have been assisted by an industry supported certification model (STEK, which is the abbreviation for the institution for certification of practices for installation companies in the refrigeration business). Elements of the regulation are detailed by Gerwen and Verwoerd (1998), and include the technical requirements to improve tightness, system commissioning to include pressure and leakage tests, refrigerant record keeping, periodic system-leak tightness inspections, and maintenance and installation work by certified companies and servicing personnel (Gerwen and Verwoerd, 1998). The STEK organization was founded in 1991 to promote competency in the handling of refrigerants and to reduce refrigerant emissions. STEK is responsible for company certification, personnel certification and the setting-up of training courses.

The success of the Dutch regulations and the STEK organization in reducing refrigerant emissions was demonstrated by the results from a detailed study in 1999 of emission data from the refrigeration and air conditioning sectors. For commercial refrigeration, annual refrigerant emissions (emissions during leakage plus disposal) as low as 3.2% of the total bank of refrigerant contained in this sector were reported (Hoogen and Ree, 2002).

In Germany, the report by Birndt *et al.* (2000) found that no leaks were identified in 40.3% of the systems, 14.4% of the leaks contributed to 85% of the refrigerant loss and 83% of the leaks occurred in the assembly joints. The report by Haaf and Heinbokel (2002) was on R-404A systems in medium- and low-temperature supermarket refrigeration. Data was taken on systems installed after 1995 with improved technologies for leak tightness, plus a reduction of refrigerant fill quantities by 15%. Annual leakage rates were determined to be 5% of charge, which represented a 10% reduction on the level reported in previous years.

The data from Sweden showed annual refrigerant losses decreasing from 14% in 1993 to 10.4% in 2002 (Bivens and Gage, 2004), with the lower emissions being attributed, in part, to an increased application of indirect cooling systems with reduced refrigerant charges in supermarkets.

A set of 2000–2001 USA emissions data were available for 223 supermarkets in the California South Coast Air Quality Management District (Los Angeles area). The data were reported in system charge sizes from 23 kg up to 1285 kg. Over the two-year period, 77% of the smaller charge size systems (23–137 kg) required no refrigerant additions, 65% of the medium charge size systems (138–455 kg) required no refrigerant additions and 44% of the larger charge size systems (456–1285 kg) required no refrigerant additions. These are the outcomes expected, based on larger charge size systems having longer piping runs, more assembly joints, more valves and more opportunities for refrigerant leakage. For the 223 supermarkets, total averaged refrigerant emission rates were 13% of charge in 2000, and 19% in 2001 (Bivens and Gage, 2004).

The data from Germany and the USA indicate that, since the average emission rates include systems with no emissions, the leaking systems have higher loss rates than the averages. This amplifies the importance of monitoring refrigerant charge using sight glasses and liquid levels, and of periodic checking with leak detectors. These both represent a significant opportunity for identifying and repairing high leakage rate sources. Procedures for emission reduction have been developed by ANSI/ASHRAE, 2002.

The trend away from the ozone-depleting CFCs and HCFCs, and towards an increased use of HFCs means, that despite lower leakage rates, HFC leakage from refrigeration is set to increase considerably. For example in Europe, a 50% cut in leakage rates due to the initiation of STEK-like programmes in every member state would result in emissions rising from 2.5–4.3 Mtonnes CO_2-eq in 1995 to around 30 Mtonnes CO_2-eq in 2010, in-

stead of 45 Mtonnes CO_2-eq under a business-as-usual scenario (Harnisch and Hendriks, 2000; Enviros, 2003).

The continuing collection of reliable emissions data is an important factor in getting a clear picture of the leakage situation and thereby establishing progress in the reduction of refrigerant emissions. Palandre *et al.* (2003 and 2004) and US EPA (2004) report two global programmes for the collection of such data. Data from the Clodic and Palandre (2004) permits the calculation of the worldwide commercial refrigeration emission rate for the year 2002 and this amounted to 30% of the commercial refrigeration systems inventory. The US EPA model information also permits the calculation of a potential 20–30% reduction business-as-usual HFC emissions from commercial refrigeration in the year 2015, by applying abatement options that require a more aggressive leak detection and repair and the increased use of distributed and indirect systems (Bivens and Gage, 2004).

4.3.4 Non-HFC technologies (vapour compression)

A number of HCs, ammonia and CO_2 systems of different refrigerating capacities have been installed in various European countries during the past 5 years. A few examples of these are now given.

4.3.4.1 Stand-alone equipment and condensing units
Some well-established beverage companies and ice cream manufacturers have recently stated (2000–2001) that by 2004 they will no longer purchase new equipment that uses HFCs in their refrigerant systems, provided that alternative refrigerants or technologies become available at an acceptable cost. HCs, CO_2 and Stirling technology are being evaluated by one of the companies (Coca Cola, 2002). The HFC-free strategy of the companies were confirmed during June 2004 (RefNat, 2004).

4.3.4.1.1 Hydrocarbons
Various companies in several countries have developed vending machines and small commercial equipment using HCs. The equipment uses HC-600a, HC-290 and HC-based blends. Limitations on charge sizes are specified by safety standards (e.g. EN 378, IEC 60335-2-89), where maximum amounts per circuit are 2.5 kg, 1.5 kg and 150 g, dependent on the application. Nevertheless, HC charges tend to be about 50% less than equivalent HFCs and HCFCs due to lower densities which minimize the impact of such limits. Recent developments with charge-reduction techniques (Hoehne and Hrnjak, 2004) suggest that charges for future systems will become even less. Christensen (2004) reports on the experience with stand-alone equipment installed in a restaurant in Denmark. Results from a detailed quantitative risk assessment model that examined the safety of hydrocarbons in commercial refrigeration systems are reported in Colbourne and Suen (2004).

4.3.4.1.2 Carbon dioxide (CO_2)
CO_2 is being evaluated by a European company interested in

developing stand-alone equipment with a direct expansion CO_2 system (Christensen, 1999) and is also one of the refrigerants being evaluated by Coca Cola (2002). The company confirmed their HFC-free strategy in June 2004 and announced that CO_2-based refrigeration is their current choice for future equipment (Coca Cola, 2004). R&D activities for CO_2-based solutions have also been announced by another company (McDonalds, 2004).

4.3.4.2   Full supermarket systems

4.3.4.2.1   Direct systems

CO_2 direct systems
CO_2 is non-flammable, non-toxic and has a GWP value of only 1. It is therefore highly suited for use in direct refrigeration systems, as long as acceptable energy efficiency can be achieved at a reasonable cost. There are two basic types of CO_2 direct systems using only CO_2 as a refrigerant and cascade systems.

Direct systems using only CO_2 as a refrigerant have been developed with a transcritical/subcritical cycle, depending on ambient temperature, for both low- and medium-temperature refrigeration are developed. In addition to giving a totally non-HFC solution, reduced pipe diameters due to higher pressures, good heat transfer characteristics of CO_2 and the possibility to obtain energy efficient heat recovery can be mentioned. Five medium-sized supermarkets have been installed with this concept by the beginning of June 2004, in addition to some smaller field test systems (Girotto and Nekså, 2002; Girotto *et al.* 2003; Girotto *et al.* 2004).

Cascade systems are being developed with CO_2 at the low-temperature stage associated with ammonia or other refrigerants (R-404A for example) at the medium-temperature stage. Several of these systems have been installed in the field and are currently being evaluated in different European countries. Haaf and Heinbokel (2003) have described 33 such CO_2 cascade systems from one manufacturer that were in service in 2003. It is emphasized that this technology could receive widespread interest because it has also been developed for the food industry (Rolfsmann, 1999; Christensen, 1999).

In addition to these two options, a third distributed system concept was described by Nekså *et al.* (1998). Self-contained display cabinets, each with CO_2 refrigeration units, are connected to a hydronic heat-recovery circuit that heats service water and buildings. A large temperature glide in the hydronic circuit, typically 50–60 K, and a correspondingly low volume flow rate and small pipe dimensions can be achieved by using the transcritical CO_2 process. Waste heat with a high temperature (70–75°C) is available for tap water and/or space heating. Excess heat is ejected to the ambient air.

The Institute of Refrigeration in London has released a 'Safety Code for Refrigerating Systems Utilizing Carbon Dioxide'. This contains a lot of relevant information despite much of the focus being on larger capacity industrial sized systems.

4.3.4.2.2   Indirect systems

Ammonia and hydrocarbons (HCs)
The quantity of ammonia can be 10% of the usual HFC refrigerant charge, due to indirect system design and the thermodynamic properties such as latent heat vaporization and liquid density (Presotto and Süffert, 2001). For HCs, the refrigerant charge is typically 10% of the direct system HFC reference charge (Baxter, 2003a,b).

In Northern Europe, ammonia or HCs (including HC-1270, HC-290 and HC-290/170 blends) have been used as refrigerants for the same type of indirect systems. For safety purposes, the refrigerant circuits are either separated in a number of independent circuits to limit the charge of each system or a number of independent chiller circuits are used (Powell *et al.*, 2000).

Heat transfer fluids (HTF)
The HTFs used in indirect systems require special attention, especially at low temperatures where pumping power may be excessive. The choice of the correct HTF to obtain the desired energy efficiency is critical and a handbook on fluid property data is available from IIR/IIF (Melinder, 1997).

CO_2 as a heat transfer fluid
For indirect systems, CO_2 can be used as either a standard HTF without phase change or as a two-phase HTF that partially evaporates in the display case evaporators and condenses in the primary heat exchanger.

At low temperatures, phase-changing CO_2 HTF shows promising results. Due to the viscosity constraint of other alternatives at low temperatures and the good heat transfer properties of CO_2, the use of CO_2 as a low-temperature HTF has received more consideration than the alternatives. When CO_2 is used with phase change, the diameter of the tubes can be significantly reduced, and the heat transfer in the display case heat-exchanger is far more effective. If the temperature can be maintained below −12°C, traditional technologies in which the tubes and heat exchanger are designed for a maximum operating pressure of 25 bar, are possible. About 50 such systems are in operation in Europe. Expansion vessels, cold finger concepts or simply using the cold stored in the goods, are possible alternatives for keeping the pressures within acceptable limits.

Ice slurry as a heat transfer fluid
An interesting new technology for medium temperature, which offers the possibility of energy storage and high-energy efficiency, is indirect systems that use ice slurry as the HTF. Research has been carried out in some pilot installations. A handbook is currently being developed by the International Institute of Refrigeration and several recent papers on various aspects of the technology are described in Egolf and Kauffeld (2005).

4.3.5 Not in-kind technologies (non-vapour compression)

There are very few examples of the successful implementation of 'not-in-kind' technologies in this sector. One possible example is the Stirling cycle. For low capacity, high-temperature lift applications in particular (>60 K), the Stirling cycle may reach competitive COP values. Although Stirling systems have been developed, cost is still an issue (Lundqvist 1993; Kagawa, 2000). This technology is also being evaluated for display cabinets (Coca Cola, 2002). Another interesting recent technology is thermoacoustic refrigeration (Poese, 2004).

Heat-driven cycles have not found their way into commercial refrigeration. The use of heat-driven cycles such as absorption and adsorption in supermarkets have been discussed in literature (Maidment and Tozer, 2002). Some attempts with solar-driven refrigeration for fresh food handling have also been developed and tested (Pridasawas and Lundqvist, 2003). The use of sorption technologies for dehumidification, thus lowering the cooling load on display cabinets, is an interesting option.

4.3.6 Relevant practices to reduce refrigerant emissions

As stated at the beginning of this section, several abatement strategies can be used to reduce refrigerant greenhouse gas emissions. New design ideas have been mentioned throughout the chapter and these may be summarized as a general trend towards lower refrigerant charge, using direct or indirect systems, and the use of non-HFC refrigerants such as ammonia, CO_2 or HCs. These options should be considered on the basis of a balanced evaluation of refrigerant emission reductions, initial investment costs, safety, operating costs and energy consumption.

The European Commission has proposed a new regulation to reduce the emissions of fluorinated greenhouse gases, including HFCs from refrigeration equipment. In addition to a general obligation to avoid leakage, installations with over 3 kg of charge will require at least annual inspections, and a refrigerant detector will be required for systems over 300 kg. Reports will also be required for the import and export of refrigerants, and end-of-life recovery, recycling or destruction of the refrigerant. Additional information on country initiatives for refrigerant conservation is described in the 2002 UNEP TOC report, Sections 4.7 and 10.1 to 10.9 (UNEP, 2003).

Several programmes, for example in the Netherlands and Sweden, have shown good results with respect to leakage mitigation for existing plants. A common denominator has been a combination of regulation, education and accreditation of service personnel (STEK, 2001). In Denmark and Norway a tax on refrigerants in proportion to their GWP value has proven successful in curbing emissions and promoting systems that use non-HFC refrigerants.

A reduction of the refrigeration capacity demand, for example by using better insulation and closed rather than open cabinets, might indirectly reduce refrigerant emissions, but also the power consumption of a supermarket. Design integration with air-conditioning and heating systems are also important measures in this respect.

An interesting development in new design tools for supermarkets, opens up new possibilities for improving the design of systems using an LCC perspective, which favours more energy efficient systems with lower operating costs (Baxter, 2003a,b).

4.3.7 Comparison of HFC and non-HFC technologies

The current rapid developments in the subsector are moving the targets for energy efficiency, charge reduction and cost. This makes a comparison between technologies difficult. Furthermore, the relatively complex links between energy efficiency, emissions and the costs of systems and their maintenance means that it sis difficult to make fair comparisons.

4.3.7.1 Energy consumption of supermarkets

Depending on the size of the supermarket, the refrigeration equipment energy consumption represents between 35–50% of the total energy consumption of the store (Lundqvist, 2000). This ratio depends on a number of factors such as lighting, air conditioning, and so forth. For typical smaller supermarkets of around 2000 m^2, refrigeration represents between 40–50% of the total energy consumption and for even smaller stores it could be up to 65%.

New, high-efficiency commercial supermarkets have been designed in some European countries and the USA by using a number of efficient technologies. These references can be seen as prototypes and one example from UK presents energy consumption figures which are a factor of two lower compared to usual stores (Baxter, 2003a,b). Most examples however show reductions of between 10–20%.

The high annual growth rate of stand-alone equipment, which tends to be less energy efficient per kW cooling power than centralized systems, should be addressed. Integrating heat rejection from the individual cabinets in a water circuit may be one way of obtaining improved energy efficiency. Excessive heat rejection within the store might also lead to an increased demand for air conditioning, further increasing the energy demand.

4.3.7.2 Energy efficiency of direct systems

The energy efficiency of refrigeration systems first of all depends on the temperature levels for which refrigeration is provided and on the global design of the system. Measurements of system efficiency can be found in the literature, for example in UNEP (2003). However, comparisons of different systems are often difficult because the boundary conditions are rarely comparable.

Potential energy-saving measures may be divided into four different groups: advanced system solutions, utilization of natural cold (free cooling), energy-efficient equipment (display cases, efficient illumination, night curtains, etc.) and indoor climate/building-related measures. Energy-efficient illumination has the double effect of reducing loads on display cases as well as direct electric consumption. Heat recovery from condensers is sometimes preferred in cold climates but internal heat generation from plug-in units and illumination is often enough to heat the premises (Lundqvist, 2000).

The refrigeration system efficiency also depends on a number of parameters: pressure losses related to the circuit length, system control and the seasonal variation of the outside temperature. For a number of global companies energy consumption, and with this the energy efficiency of refrigerating systems, has become an important issue, especially in countries where electricity prices are high. One approach to energy savings is to utilize 'floating condensing temperature' in which the condensing temperature follows ambient temperature. The issue of climate change and the desire to reduce GHG emissions has also heightened the interest in increasing the energy efficiency.

4.3.7.3 Energy efficiency and cost of indirect systems

The evaluation of the additional energy consumption related to indirect systems is an ongoing process. Direct field comparisons between direct and indirect systems are difficult (Lundqvist, 2000; Baxter, 2003a). Moreover, the main driver for centralized systems is initial cost. Due to the design of heat exchangers in display cases (especially medium-temperature, open-type) the performances of some indirect systems can be equal or even slightly better than direct systems (Baxter, 2003a). For low temperatures, the energy penalty can be substantial depending on the design.

On the other hand, the relative energy consumption of indirect systems – compared to conventional direct expansion systems – can show an increased energy consumption of up to 15%. Conclusions can only be drawn if reference lines for the energy consumption of centralized systems are plotted in which the origin of energy inefficiencies are apparent. Due to the extra temperature difference required, inherently indirect systems should give higher energy consumptions compared to direct systems. Recent practical experiences and experimental studies (Mao *et al.*, 1998; Mao and Hrnjak, 1999; Lindborg, 2000; Baxter, 2003a,b), however, indicate that well-designed indirect systems may have energy efficiencies approaching those of good direct systems. Further research is clearly needed to clarify the reasons for this. More efficient defrost, better part-load characteristics, better expansion device performance and more reliable systems are believed to contribute to indirect system energy efficiency. The costs might be 10–30% higher, but these can potentially be reduced (Yang, 2001; Christensen and Bertilsen, 2003).

4.3.7.4 Energy efficiency and cost of ammonia systems

There are several indirect systems in operation that are successfully using ammonia as the primary refrigerant (Haaf and Heinbokel, 2002). As ammonia is toxic and may create panic due to the strong smell at low concentrations appropriate safety precautions are required. Excellent energy efficiency can be achieved with properly-designed systems. The drawbacks for ammonia systems are limited service competence (Lindborg, 2002) and higher initial costs, typically 20–30%. A life-cycle cost evaluation is therefore required.

4.3.7.5 Energy efficiency and cost of hydrocarbon systems

Full supermarket systems using hydrocarbons in an indirect design have been installed in several European countries. A dedicated ventilation system (if installed in a machine room), gas detectors, gas-tight electric equipment and so forth have been installed for safety reasons. The use of hydrocarbons has increased the R&D effort to significantly minimize refrigerant charge. A small prototype system of approximately 4 kW cooling capacity using 150 g of HC-290 and micro-channel heat exchanger technology has been demonstrated by Fernando *et al.* (2003). Cost is still an issue, typically up to 30% higher, but further development is expected to reduce costs. HC-290 and HC-1270 are excellent refrigerants from a thermodynamic point of view and equipment design is relatively straightforward. The availability of some standard components is still limited, but to a certain extent the hydrocarbon systems can use the same type of system components as HFC systems. Cascade systems with HC-290 and CO_2 for full supermarket systems were reported to have an energy efficiency equal to conventional direct system design (Baxter, 2003a,b).

For stand-alone equipment and condensing units, several references report a higher efficiency of HC refrigerants systems compared to equivalent systems with HFCs, for example Elefsen *et al.* (2002). Others claim that higher efficiency can be achieved with HFC systems, if the extra costs used for the safety precautions of HC systems are used to improve system efficiency of the HFC system (Hwang *et al.*, 2004).

4.3.7.6 Energy efficiency and cost of CO_2 systems

Centralized CO_2 direct systems for both medium and low temperature, operating in either transcritical or subcritical cycle dependent on the ambient temperature, are reported to require about 10% higher energy consumption than a state-of-the-art R-404A direct system (Girotto *et al.*, 2003 and 2004). Several measures for improvements have however been identified. The cost is reported to be about 10–20% higher than for direct expansion R-404A systems and this difference is mainly due to components produced in small series.

Haaf and Heinbokel (2003), report energy consumption and investment costs for R-404A/CO_2 cascade systems that are similar to R-404A direct systems. This is due to the fact that components for CO_2 cascade systems are more similar to R-404A components (maximum pressure 40 bar), allowing more standard components to be used. Girotto *et al.* (2004) report higher costs for cascade systems (see also comment about HC-290/CO_2 above).

4.3.7.7   Energy efficiency and cost of HFC systems with reduced emissions

Ongoing R&D efforts to minimize refrigerant charge without compromising energy efficiency are applicable to HFC refrigerants as well. The standard approach to charge minimization is to use no receiver or hermetic compressors and to keep piping as short as possible. Tight systems require brazed joints and these are most reliably made when systems are factory assembled. The potential for charge reduction is illustrated by Fernando *et al.* (2003). They present HC systems using as little as 150 g refrigerant for a 5 kW domestic heat pump. The density of HFC refrigerant is approximately twice that of HC and therefore a comparable system using 300 g of HFC refrigerant is within reach, if further heat exchanger development is undertaken. More complex cycles are another way of improving systems. Beeton and Pham (2003), report a 41% capacity increase and a 20% efficiency increase for low-temperature economizer systems using R-404A and R-410A.

4.3.8   Comparison of LCCP and mitigation costs

The number of publications that give TEWI or LCCP data for commercial refrigeration systems is limited but is growing rapidly. Harnisch *et. al.* (2003) calculated the LCCP for several different types of full supermarket refrigeration systems in Germany. They used a straightforward model which took production, emissions and energy usage into account. CO_2 emissions from power production were calculated using an average emission factor of 0.58 kg CO_2 kWh^{-1}. The transparent method to evaluate the various systems allows sensitivity analyses to be performed using other literature references referred to in this report. Table 4.10 presents characteristic figures from Harnisch *et. al.* (2003) and compares these to calculated results based on representative data from other literature references.

The data used for the table are selected as follows: The 30% emission is based on Palandre *et al.*, 2004 and the 11.5% and 6.5% emission scenarios are based on Harnisch *et al.* (2003),

Bivens and Gage (2004) Baxter (2003a) and ADL (2002). The 11.5% and 6.5% emissions represent 10% and 5% emissions yr^{-1}, with a 15% end-of-life recovery loss apportioned over a 10-year lifetime. Energy consumption figures are extracted from Harnisch *et al.* (2003), Haaf and Heinbokel (2002), Girotto *et al.* (2003) and Baxter (2003b). It is clear that several different alternatives result in reductions in CO_2 equivalent emissions of the same order of magnitude. The same applies for an HFC alternative, if the annual emission rate can be as low as 5%. If a 5% leakage is possible, the dominating contribution from most systems is an indirect effect due to power production.

Supermarket system and mitigation cost estimates are scarce. Harnich *et al.* (2003) give data for German supermarkets with costs ranging from 20–280 US$ per tCO_2-eq mitigated. The lowest values are given for a system using direct expansion CO_2 for low temperature and direct expansion with R-404A for high temperature. Mitigation costs are estimated using a 10% leakage rate and a 1.5 % recovery loss, with a 10-year lifetime and a discount rate of 10% to reflect commercial decision-making. An average cost of 100 US$ m^{-2} of supermarket area is used as a baseline for cost estimates. This figure is confirmed by Sherwood (1999) who reports on cost figures for a 3200 m^2 supermarket in the USA.

Using this data with a broader range of leakage rates and estimated costs, significantly reduces the typical mitigation costs per tonne of CO_2 suggested by Harnisch *et al.* (2003) but also expands the total range to values of 10–300 US$ per tonne CO_2-eq mitigated.

Additional mitigation costs for the various systems suggested in the chapter have not been calculated. Cost estimates for various technologies given in the literature suggest a cost increase between 0 and 30% for alterative technologies compared to a baseline, full supermarket, direct system using R-404A as refrigerant. Some detailed figures are already given under each section and a general summary is given for each technology in Table 4.11.

Table 4.10. LCCP values of full supermarket systems.

Configuration	Refrigerant Emissions % charge yr^{-1}	Energy Consumption	LCCP, in tCO_2-eq yr^{-1}		
			Indirect	Direct	Total
Direct Expansion (DX)	30%	baseline	122	183	305
DX (Harnisch *et al.*, 2003, data)	11.5%	baseline	122	70	192
DX distributed 75% charge reduction	6.5%	baseline	122	10	132
Sec. Loop R-404A 80% charge reduction	6.5%	baseline + 15%	140	8	148
Sec. Loop propane 80% charge reduction	6.5%	baseline + 10%	134	0	134
Sec. Loop ammonia 80% charge reduction	6.5%	baseline + 15%	140	0	140
DX R-404A and DX CO_2 50% charge reduction	6.5%	baseline	122	20	142
DX CO_2/CO_2	11.5%	baseline + 10%	134	0	134

Table 4.11. Sector summary for commercial refrigeration – current status and abatement options.

Subsector	Stand-alone Equipment	Condensing Units	Full supermarket system			
			Direct Centralized	Indirect Centralized	Distributed	Hybrids
Cooling capacity From	0.2 kW	2 kW	20 kW			
To	3 kW	30 kW	>1000 kW			
Refrigerant charge From	0.5 kg	1 kg	100 kg	20	*	*
To	~2 kg	15 kg	2000 kg	500 kg	*	*
Approximate percentage of sector refrigerant bank in subsector	11% of 606 kt	46% of 606 kt	43% of 606 kt			
Approximate percentage of sector refrigerant emissions in subsector	3% of 185 kt	50% of 185 kt	47% of 185 kt			
2002 Refrigerant bank, percentage by weight	CFCs 33%, HCFCs 53%, HFCs 14%					
Typical annual average charge emission rate	30%					

Subsector	Stand-alone Equipment	Condensing Units	Full supermarket system			
			Direct Centralized	Indirect Centralized	Distributed	Hybrids
Technologies with reduced LCCP EmR – Direct Emission Reduction (compared to installed systems) ChEU – Change in Energy Usage (+/-) (compared to state of the art) ChCst – Change in Cost (+/-) (compared to state of the art)	**Improved HFC** SDNA **HC** SDNA **CO$_2$** SDNA	**Improved HFC** SDNA **R-410A** SDNA **HC** SDNA **CO$_2$** SDNA	**Improved HFC** EmR 30% ChEU 0% ChCst 0 ±10% **CO$_2$ (all-CO$_2$)** EmR 100% ChEU 0 ±10% ChCst 0±10%	**Ammonia** EmR 100% ChEU 0–20% ChCst 20–30% **HC** EmR 100% ChEU 0–20 % ChCst 20–30% **HFC** EmR 50–90% ChEU 0–20% ChCst 10–25%	**HFC** EmR 75% ChEU 0–10% ChCst 0–10% **Economized- HFC-404A** SDNA **Economized- HFC-410A** SDNA **CO$_2$** SDNA	**Cascade- HFC/CO$_2$** EmR 50–90% ChEU 0% **Cascade- Ammonia/CO$_2$** SDNA **Cascade- HC/CO$_2$** SDNA
SDNA – Sufficient data on emission reduction, energy usage and change in cost not available from literature						
LCCP reduction potential (world avg. emission factor for power production)	SDNA		35–60%			
Abatement cost estimates (10 yr lifetime, 10% interest rate)	SDNA		20-280 US$ per tonne CO$_2$ mitigated			

* Alternatives in these categories have been commercialized, but since the current number of systems are limited, they are only referenced as options below

In this report energy efficiency has been treated as relative changes in energy usage for several recent types of systems, the main purpose of which is to mitigate emissions, see also Table 4.11. However, the systems investigated are based on current technological standards for components such as heat exchangers, compressors and so forth. No attempts have been made to predict the future energy-saving potential in commercial refrigeration applications if future possible improvements are achieved. Several possibilities for reducing the energy con- sumption of refrigeration systems exists, but in principle most of these may be applied irrespective of the refrigerant used in the system. These options may also lead to negative mitigation costs, as for instance reported in Godwin (2004) and March (1998).

The authors firmly believe that the ongoing technical de- velopment of components and systems together with various energy-saving measures (such as heat recovery, more efficient compressors and display cases, larger heat exchangers, floating condensation, energy efficient buildings and so forth) may lower supermarket energy consumption considerably.

4.4 Food processing and cold storage

4.4.1 Introduction

Food processing and cold storage is one of the important applications of refrigeration, and is aimed at preserving and distributing food whilst keeping its nutrients intact. This application of refrigeration is very significant in terms of size and economic importance, and this also applies to developing countries. Food processing includes many subsectors such as dairy products, ice cream, meat processing, poultry processing, fish processing, abattoirs, fruit & vegetable processing, coffee, cocoa, chocolate & sugar confectionery, grain, bread & flour confectionery & biscuits, vegetable, animal oils & fats, miscellaneous foods, breweries and soft drinks (March, 1996).

The annual global consumption of frozen foods is about 30 Mtonnes yr^{-1}. Over the past decade, consumption has increased by 50% and it is still growing. The USA accounts for more than half of the consumption, with more than 63 kg per capita. The average figure for the European Union (EU) is 25 kg and for Japan 16 kg. The amount of chilled food is about 10–12 times greater than the supply of frozen products, giving a total volume of refrigerated food of around 350 Mtonnes yr^{-1} (1995) with an estimated annual growth of 5% (IIR, 1996; UNEP, 2003). Like chilling and freezing, food processing is also of growing importance in developing countries. This is partly due to the treatment of high-value food products for export. Even in 1984, about half of the fish landed in developing countries (more than 15 Mtonnes) was refrigerated at certain stages of processing, storage or transport (UNEP, 1998). The estimated annual growth rate in food processing between 1996 and 2002 was 4% in developed countries and 7% in developing countries (UNEP, 2003).

Frozen food in long term-storage is generally kept at −15°C to −30°C, while −30°C to −35 °C is typical for the freezing process. In so-called 'super-freezers', the product is kept at −50°C. Chilled products are cooled and stored at temperatures from −1°C to 10°C.

The majority of refrigerating systems for food processing and cold storage are based on reciprocating and screw compressors. System size may vary from cold stores of 3 kW cooling demand to large processing plants requiring several MW of cooling. Reciprocating compressors are most frequently used in the lower capacity range, whereas screw compressors are common in larger systems (UNEP, 2003).

4.4.2 Technical options

Most of the refrigerating systems used for food processing and cold storage are based on vapour compression systems of the direct type, with the refrigerant distributed to heat exchangers in the space or apparatus to be refrigerated. Such systems are generally custom-made and erected on site. Indirect systems with liquid chiller or ice banks are also commonly used in the food processing industry for fruit and vegetable packing, meat processing and so forth. Ammonia, HCFC-22, R-502 and CFC-12 are the refrigerants historically used. The current technical options are HFCs, and non-fluorocarbons refrigerants such as ammonia, CO_2 and hydrocarbons (UNEP, 2003). Table 4.12 gives the main refrigerant technical options along with percentage annual emissions for food processing, cold storage and industrial refrigeration applications (Clodic and Palandre, 2004).

4.4.3 HFC technologies

HFC refrigerants are being replaced by place of CFC-12, R-502 and HCFC-22 in certain regions. The preferred HFCs for food processing and cold storage applications are HFC-134a, HFC blends with insignificant temperature glide such as R-404A, R-410A and azeotropic blends like R-507A. The HFC blend R-407C is also finding application as a replacement for HCFC-22.

HFC-134a has completely replaced CFC-12 in various applications of refrigeration. However, CFC-12 was not widely used in food processing and cold storage because it requires considerably greater compressor swept volume than HCFC-22 or ammonia to produce the same refrigerating effect. There is limited use of HFC-134a in this subsector.

R-404A, R-407C and R-507A are currently the most used HFCs for cold storage and food processing. These blends are preferred to HFC-134a due to the higher volumetric capacity

Table 4.12. Food processing, cold storage and industrial refrigeration (2002).

	CFCs (CFC-12 and R-502)[1]	HCFC-22	NH$_3$	HFCs (HFC-134a, R-404A, R-507A, R-410A)[1]
Cooling Capacity	25 kW–1000 kW	25 kW–30 MW	25 kW–30 MW	25 kW–1000 kW
Emissions, t yr^{-1}	9500	23,500	17,700	1900
Refrigerant in bank, tonnes	48,500	127,500	105,300	16,200
Emissions % yr^{-1}	20%	16%	17%	12%

[1] See Annex V for an overview of refrigerant designations for blends of compounds.

and lower system cost (UNEP, 1998). In spite of minor temperature glides, R-404A has proven to be applicable even in flooded systems (Barreau *et al.*, 1996). R-404A and R-507A are the primary replacements for R-502. The coefficients of performance are comparable to R-502 but significantly lower compared to those of NH_3 and HCFC-22, especially at high condensing temperatures. Air-cooled condensers should be avoided as far as possible. The liquid should be subcooled to achieve optimal efficiencies and high cost-effectiveness for systems with R-404A and R-507A. In chill applications it may also be necessary to add significant superheat to the suction gas in order to avoid refrigerant condensation in the oil separator (UNEP, 2003).

R-410A is also one of the HFC blends which is expected to gain a market share in food processing and cold storage applications due to the lower compressor swept volume requirements in comparison to other refrigerants (except to CO_2). The compressor efficiencies, pressure drop in suction lines and heat transfer efficiency will benefit from high system pressure (UNEP, 1998). Due to the high volumetric capacity (40% above that of HCFC-22), R-410A compressor efficiency has been reported to be higher than with HCFC-22 (Meurer and König, 1999). R-410A can have system energy efficiency similar to that of ammonia and HCFC-22 and significantly higher than that of R-404A and R-507A for evaporation temperatures down to –40°C.

4.4.4   Non-HFC technologies

Ammonia

Ammonia is one of the leading refrigerants for food processing and cold storage applications. The current market share in several European countries, especially in the north, is estimated to be up to 80% (UNEP, 1998). In the USA ammonia has approximately 90% market share in systems of 100 kW cooling capacity and above in custom-engineered process use (IIR, 1996).

Recently designed ammonia-based systems have improved quality with respect to design, use of low-temperature materials and better welding procedures. Low charge is another positive development. However, more important is that these factory made units or systems represent a new level of quality improvement. These systems are not likely to break or release their charge in another way unless there is a human error or direct physical damage. Charge reduction has been achieved by using plate-type heat exchangers or direct expansion tube and shell evaporators (UNEP, 2003).

HCFC-22

The use of HCFC-22 is declining in food processing and cold storage applications in most developed countries. In Europe some of the end-users prefer ammonia and CO_2 wherever possible, whereas HCFC-22 has become the most common refrigerant to replace CFCs in food processing and cold storage in the USA (UNEP, 2003).

In developing countries, HCFC-22 is still an important replacement refrigerant for CFCs in new systems, as from a technical point of view HCFC-22 could replace CFC-12 and CFC-502 in new systems. Another important consideration in developing countries is that HCFC-22 will be available for service for the full system lifetime.

Hydrocarbons (HCs)

A growing market for low charge hydrocarbon systems has been observed in some European countries. So far market shares are small, which may be due to the flammability of these refrigerants. Nevertheless, several manufacturers have developed a wide range of products.

Commercialized refrigerants used in food processing and cold storage applications include HC-290, HC-1270 and HC-290/600a blends, although pure substances will be preferred in flooded systems. All of these refrigerants possess vapour pressures very similar to those of HCFC-22 and R-502. System performance with regard to system efficiency is comparable to, and in some cases even superior to, that of the halocarbons. Hydrocarbons are soluble with all lubricants, and compatible with materials such as metals and elastomers that are traditionally used in refrigeration equipment. As long as safety aspects are duly considered, standard refrigeration practice for HCFCs and CFCs can be used without major system detriment to system integrity (UNEP, 1998, 2003).

Given the flammability concerns, design considerations as detailed in the relevant safety standards should be adhered to. Additional safety measures should be considered for repairing and servicing. Several national and European standards permit the use of HCs in industrial applications and lay down specific safety requirements (ACRIB, 2001; UNEP, 2003).

Carbon dioxide

Carbon dioxide technology for low temperatures such as food freezing is an attractive alternative, especially in cascade systems with CO_2 in lower stage and ammonia in the upper stage, due to its excellent thermophysical properties along with zero ODP and negligible GWP.

Further, the volumetric refrigerating capacity of CO_2 is five times higher than HFC-410A and eight times higher than for ammonia and other refrigerants. This means that the size of most of the components in the system can be reduced (Roth and König, 2001). However, application of CO_2 places a limitation on evaporating temperatures due to the triple point (the temperature and pressure at which liquid, solid and gaseous CO_2 are in equilibrium) of –56.6°C at 0.52 MPa.

CO_2 technology has been applied to food processing and cold storage, both as a conventional and as secondary refrigerant. It is expected that CO_2 market share will increase in this subsector, especially for freezing and frozen food storage.

Not-in-kind technologies

There are some not-in-kind (non-vapour compression) technologies like air cycle, vapour absorption technology and compression-absorption technology which can be used for food processing and cold storage applications. Vapour absorption technology is well-established whereas compression-absorption technology is still under development.

Vapour absorption technology

Vapour absorption is a tried and tested technology. Absorption technologies are a viable alternative to vapour compression technology wherever low cost residual thermal energy is available. The most commonly used working fluid in food processing and cold storage applications is ammonia with water as the absorbent. The use of absorption technology is often limited to sites that can utilize waste heat, such as co-generation systems.

Compression-absorption technology

Compression-absorption technology has been developed by combining features of the vapour-compression and vapour-absorption cycles. About 20 compression-absorption systems on both a laboratory and full-scale have been developed and tested successfully so far. Various analytical and experimental studies have shown that the COP of compression-absorption systems is comparable to that of vapour compression systems. However, this system suffers from the inherent disadvantage of being capital intensive in nature. The technology is still at developmental stage (Pratihar *et al.*, 2001, Ferreira and Zaytsev, 2002).

4.4.5 Factors affecting emission reduction

Design aspects

The refrigeration system design plays a vital role in minimizing the refrigerant emissions. A proper system design including heat exchangers, evaporators and condensers can minimize the charge quantity and hence reduce the potential amount of emissions. Every effort should be made to design tight systems, which will not leak during the system's lifespan. The potential for leakage is first affected by the design of the system; therefore designs must also minimize the service requirements that lead to opening the system. Further, a good design and the proper manufacturing of a refrigerating system determine the containment of the refrigerant over the equipment's intended life. The use of leak tight valves is recommended to permit the removal of replaceable components from the cooling system. The design must also provide for future recovery, for instance by locating valves at the low point of the installation and at each vessel for efficient liquid refrigerant recovery (UNEP, 2003).

Minimizing charge

The goal of minimal refrigerant charge is common for all systems due to system and refrigerant costs. Normally the designer calculates the amount of charge. In large systems such as food processing and cold stores, very little attention was generally given to determining the full quantity of refrigerant charge for the equipment. Its quantity is not often known (except for small factory built units). Charging the refrigerant into the system is done on site to ensure stable running conditions.

Improved servicing practices

Servicing practices in refrigeration systems must be improved in order to reduce emissions. Topping-off cooling systems with refrigerants is a very common practice, especially for the large systems normally used in food processing and cold storage industry, which causes greater emissions of refrigerant. However in general, proper servicing has proven to be more expensive than topping-off refrigeration systems. It is therefore necessary to make end-users understand that their practice of topping-off the systems without fixing leaks must cease because of the increased emissions to the environment. The good service practices are preventive maintenance, tightness control and recovery during service and at disposal.

Installation

After proper designing of the system, installation is the main factor that leads to proper operation and containment during the useful life of the equipment. Tight joints and proper piping materials are required for this purpose. Proper cleaning of joints and evacuation to remove air and non-condensable gases will minimize the service requirements later on and results in reduced emissions. Careful system performance monitoring and leak checks should also be carried out during the first days of operation and on an ongoing basis. The initial checks also give the installer the opportunity to find manufacturing defects before the system becomes fully operational. The proper installation is critical for maximum containment over the life of the equipment (UNEP, 2003). The refrigeration system should be designed and erected according to refrigeration standards (e.g. EN378 (CEN, 2000/2001)) and current codes of good practice.

Recovery and recycling

The recovery and recycling of refrigerants is another important process that results in significant reductions in emissions. The purpose of recovery is to remove as much refrigerant as possible from a system in the shortest possible amount of time. For applications where maintenance operations require opening the circuit, the difference between deep recovery and 'normal recovery' can represent 3–5% of the initial charge (Clodic, 1997). However many countries have adopted final recovery vacuum requirements of 0.3 or 0.6 atm absolute depending on the size of the cooling system and saturation pressure of the refrigerant. This provides a recovery rate of 92–97% of the refrigerant (UNEP, 1998).

The recovered fluorocarbon refrigerants can be recycled and then reused. The process of recycling is expected to remove oil, acid, particulate, moisture and non-condensable contaminants from the used refrigerant. The quality of recycled refrigerant can be measured on contaminated refrigerant samples according to standardized test methods (ARI 700). However, recycling is not common practice in the case of large food processing and cold storage units, where the preference is to recover and re-use the refrigerant.

4.4.6   *Trends in consumption*

The lifetime emissions of a refrigerant are dependent on the installation losses, leakage rate during operation, irregular events such as tube break and servicing losses including recovery loss and end-of-life loss during reinstallation/reconstruction. In some European countries, HCFC systems with more than 10 kg charge (including all application areas) showed an annual emission rate of 15% of the charge in the early 1990s (Naturvardsverket, 1996). This figure dropped to 9% in 1995 (UNEP 1998). Emissions from HFC systems are reported to be less than this, and this is probably due to more leak-proof designs. On a global scale, current CFC and HCFC annual emission rates are likely to be in the range of 10–12% of the charge (UNEP, 1998). A recent study (Clodic and Palandre, 2004) has provided estimates of emissions of CFCs, HCFCs, HFCs and ammonia from the combined sector of Food Processing, Cold Storage, and Industrial Refrigeration. Table 4.12 gives the percentage annual emissions of these refrigerants.

The consumption and banks of HFCs and other fluorocarbons have been estimated for the industrial refrigeration sector as a whole. Both a top-down (UNEP, 2003) and a bottom-up approach (Clodic and Palandre, 2004) are presented here (Table 4.13). The data for 2002 and the 2015 business-as-usual projections of Clodic and Palandre (2004) are used as a basis for the refrigeration subsectors in this report, so as to ensure consistency with other refrigeration and air conditioning subsectors (see Table 4.1). However, Table 4.13 clearly illustrates the differences between both approaches, which are clearly significant for CFCs.

Food processing and cold storage are assumed to account for 75% of the combined emissions and industrial refrigeration for the remaining 25% (UNEP, 2003).

The business-as-usual projections for 2015 show a significant increase in HFC consumption.

4.4.7   *Comparison of HFC and non-HFC technologies*

Energy efficiency and performance
As stated above R-404A and R-507A are the proven replacements for R-502 and this also includes the application in flooded evaporators used for food processing and cold storage systems. The cycle efficiencies are comparable to R-502 but significantly lower compared to those of ammonia (non-HFC technology) especially at high condensing temperatures (UNEP, 2003).

R-410A is another important HFC refrigerant in this sector. The energy efficiency of R-410A systems can be similar to ammonia for evaporation temperatures down to –40 °C, depending on compressor efficiency and condensing temperature. The efficiency below –40 °C until its normal boiling point of –51.6 °C is slightly higher for R-410A than that of ammonia and other refrigerants. The compressor efficiencies are also reported to be higher compared to HCFC-22, due to the high volumetric capacity (40% above that of HCFC-22) of R-410A (Meurer and König, 1999).

CO_2 technology is another non-HFC technology which is gaining momentum. CO_2 as a refrigerant is being used in food processing and cold storage units in cascade systems with ammonia in higher cascade. It has been reported that the volumetric refrigerating capacity of CO_2 is five times higher than HFC 410A and eight times higher than that of ammonia and other refrigerants. Therefore the size of most components in the system can be reduced (Roth and König, 2001). The efficiency of CO_2/ammonia cascade system in the temperature range of –40 °C to –55°C is comparable to a two-stage system with R-410A. CO_2 also shows a strong cost benefit in large systems (Axima, 2002).

Life cycle climate performance LCCP
Very limited data are available for TEWI/LCCP for this refrigeration sector. A recent publication (Hwang *et al.*, 2004) reports a comprehensive experimental study of system performance and LCCP for an 11 kW refrigeration system operating with R-404A, R-410A and propane (HC-290) at evaporator temperatures of –20°C to 0°C. For a comparison on an equal first cost basis, the increased cost of safety features for HC-290 was used for a larger condenser for the HFC systems. The LCCP of the R-410A system was 4% lower and the LCCP of R-404A was 2% higher than that of the HC-290 system at an annual refrigerant emission rate of 2%. The LCCP values for R-410A and HC-290 were equal at an annual refrigerant emission rate of 5%.

4.5   Industrial refrigeration

4.5.1   *Introduction*

One characteristic of industrial refrigeration is the temperature range it embraces. While evaporating temperatures may be as high as 15°C, the range extends down to about –70°C.

Table 4.13. Estimated consumption and banks of halocarbons refrigerants for industrial refrigeration, including food processing and cold storage for 2002. (UNEP, 2003 and Clodic and Palandre, 2004).

		Consumption (kt yr⁻¹)				Refrigerant Banks (kt)			
		CFCs	**HCFCs**	**HFCs**	**NH₃**	**CFCs**	**HCFCs**	**HFCs**	**NH₃**
2002	UNEP (2002)	12	28	5	-	109	165	9	-
2002	Clodic and Palandre (2004)	7	27	6	22	34	142	16	105
2015	Clodic and Palandre (2004)	4	24	18	27	21	126	85	123

Table 4.14. Major applications and refrigerants used in industrial refrigeration.

Application	Refrigerant	Other Refrigerants	CFC, HCFC Replacements
freeze drying	NH_3, HCFC-22	R-502	CO_2, R-410A
separation of gases	CFC-12, CFC-13, HCFC-22, R-503	-	PFC14, PFC-116, R-404A, R-507A, CO_2
solidification of substances	HCs, CFC-13, HCFC-22, CFC-12	-	HCs, CO_2, PFC-14, R-404A, R-507A
reaction process	Various		
humidity control of chemicals	CFC-12, CFC-13, HCFC-22, R-503		PFC-14, PFC-116, R-404A, R-507A, CO_2, NH_3, Air
industrial process air conditioning	NH_3, HCFC-22	R-502	NH_3, R-404A, HFC-134a, Water
refrigeration in manufacturing plants	Various		
refrigeration in construction	NH_3, R-502	HCFC-22	NH_3, CO_2, R-410A, R-404A, R-507A
ice rinks	NH_3, HCFC-22		NH_3, CO_2, R-404A, R-507A
wind tunnel	NH_3, R-502, HCFC-22, CFC-12	-	NH_3, R-404A, R-507A, HFC-134a
laboratories	Various		

At temperatures much lower than about −70°C the so-called 'cryogenics' technology comes into play. This produces and uses liquefied natural gas, liquid nitrogen, liquid oxygen and other low-temperature substances. Industrial refrigeration in this section covers refrigeration in chemical plants (separation of gases, solidification of substances, removal of reaction heat, humidity control of chemicals), process technology (industrial process air conditioning, refrigeration in manufacturing plants, refrigeration in construction), ice rinks and winter sports facilities, and laboratories where special conditions such as low temperatures, must be maintained. Some definitions of industrial refrigeration include food processing and cold storage; these are described in Section 4.4 of this chapter.

Industrial systems are generally custom-made and erected on site. A detailed description of industrial refrigeration and cold storage systems can be found in the 'Industrial Refrigeration Handbook' (Stoecker, 1998). Industrial refrigeration often consists of systems for special and/or large refrigerating purposes. The cooling/heating capacity of such units vary from 25 kW to 30 MW or even higher. These refrigeration systems are based on reciprocating, screw and centrifugal compressors, depending on the capacity and application.

Industrial refrigeration has mainly operated with two refrigerants: ammonia (60–70%) and HCFC-22 (15–20%). To a lesser extent CFC-502 (5–7%) has been used and other minor refrigerants complete the rest of industrial applications. Replacement refrigerants for CFCs and HCFCs, plus other non-fluorocarbon fluids are included in Table 4.14.

The refrigerants used are preferably single compound or azeotropic mixture refrigerants, as most of the systems concerned use flooded evaporators to achieve high thermodynamic efficiencies. Industrial refrigeration systems are normally located in industrial areas with very limited public access. For this reason ammonia is commonly used in many applications where the hazards of toxicity and flammability are clearly evident, well-defined, well understood and easily handled by competent personnel. Hydrocarbons may be used as an alternative to ammonia within sectors handling flammable fluids, such as chemical processing.

There are clear differences in how countries have developed the technology for industrial refrigeration since the starting of the CFC phase-out. In Europe, the use of HCFC-22 and HCFC-22 blends in new systems has been forbidden for all types of refrigerating equipment by European regulation 2037/00 (Official Journal, 2000) since 1 January 2001. The use of CFCs is also forbidden, that is no additional CFC shall be added for servicing. HFCs are occasionally used where ammonia or hydrocarbons are not acceptable, although they are not often preferred in Europe, as European users are expecting regulations limiting the use of GHGs in stationary refrigeration (see proposals in EU, 2004).

4.5.2   *Technical options*

Most of the industrial refrigerating systems use the vapour compression cycle. The refrigerant is often distributed with pumps to heat exchangers in the space or apparatus to be refrigerated. Indirect systems with heat transfer fluids are used to reduce the risk of direct contact with the refrigerant. Ammonia is the main refrigerant in this sector. HCFC-22, R-502 and CFC-12 are the historically used refrigerants from the group of CFCs and HCFCs. Beside the increasing share of ammonia for new systems, the current technical options to replace CFCs are HFCs, HCFC-22 and non-fluorocarbon technologies such as CO_2 and HCs.

4.5.3   *Factors affecting emission reduction*

The refrigerant charge in industrial systems varies from about 20 kg up to 10,000 kg or even more. Large ammonia refrigeration systems contain up to 60,000 kg of refrigerant. The high costs of the refrigerants, with the exception of ammonia and

CO_2, and the large refrigerant charge required for the proper operation of the plant have led to low emissions in industrial systems. In these systems annual average leakage rates of 7–10% are reported (UNEP, 2003); smallest leakage rates are observed in ammonia systems because of the pungent smell. Clodic and Palandre (2004) estimate somewhat higher annual leakage rates (17%) for the category of industrial refrigeration (which also includes food processing and cold storage). The abatement costs of refrigerant emissions from industrial refrigeration was determined to be in the range of 27–37 US$ (2002) per tonne CO_2-eq (March, 1998). Cost data were calculated with a discount rate of 8%.

Design aspects

Industrial refrigeration systems are custom made and designs vary greatly from case to case. Due to the increasing requirements concerning safety and the quantity of refrigerant used in refrigeration systems, a design trend towards indirect systems has been observed over the past 10 years. Yet whenever possible, the majority of systems are still direct. Refrigerant piping fabrication and installation has changed from direct erection on site towards pre-assembled groups and welded connections (see Design aspects in Section 4.4.5 for additional information on design and installation practices to minimize refrigerant emissions).

Minimizing charge

There are limits on design optimization in terms of balancing low charge on the one hand against achieving high COPs or even stable conditions for liquid temperatures to be delivered to heat exchangers on the other. For example, flooded type evaporators represent the best technology available for a low temperature difference between the liquid to be cooled and distributed and the evaporating refrigerant, yet this requires large quantities of refrigerant. The increased use of plate heat exchangers, plate and shell heat exchangers, and printed circuit heat exchangers over the past 15 years has enabled the design of lower charge systems with flooded evaporators. Efforts to minimize the amount of refrigerant charge will continue with improvements in system technology.

Improved servicing practices

Trained service personnel are, according to safety standards (CEN-378, 2000/2001; ISO-5149, 1993), required for the maintenance and operation of industrial systems. Service and maintenance practices on industrial systems are updated periodically according to safety standards and service contracts, which are negotiated with the plant owner in the majority of the cases.

Recovery and recycling

Recovery of refrigerants from industrial plants is common in many countries and is sometimes also a requirement (CEN-378, 2000 and 2001). The recovery of small quantities of ammonia (less than 3 kg) through absorption in water is common practice. Larger quantities are recovered by special large recovery units and pressure vessels.

The recovery rate from industrial systems is high due to the high costs and quantity of the refrigerants, especially CFCs, HCFCs and HFCs. The recovery rate is estimated to be 92–97% of the refrigerant charge (UNEP, 2003). The recovered refrigerants are dried in the recovery systems on site and re-used.

Lifetime refrigerant emissions

As most industrial refrigeration systems are designed for specific manufacturing processes, information on lifetime emissions of refrigerants are not readily available. Even data on cooling capacities are often not official, as the production capacity of the manufactured product could be estimated from this data. However, the estimated annual leakage rates referred to in Section 4.5.3 should be noted.

4.5.4   HFC technologies

HFC refrigerants are options to replace CFC-114, CFC-12, CFC-13, R-502 and HCFC-22. The preferred HFCs are HFC-134a, HFC-23 and HFC-blends with insignificant temperature glide such as R-404A, R-410A and azeotropic blends like R-507A.

HFC-134a has completely replaced CFC-12 because of its comparable thermodynamic properties in various applications of industrial refrigeration. CFC-12 and HFC-134a are used in large systems for higher temperatures with evaporator temperatures from 15°C down to –10°C.

HFC-23 has replaced CFC-13 and to lesser extent CFC-503 for the same reasons as mentioned for CFC-12. The evaporator temperature range varies from –80°C to –55°C in the low-temperature applications of these refrigerants.

HFC-245fa, HFC-365mfc and HFC-236fa are possible replacements for CFC-114 in high-temperature heat pumps. No single fluid is ideal, especially for large industrial heat pumps, because the dew line of the fluids requires large superheat to avoid compression in the vapour region. The temperature range for condensing temperatures varies from 75°C to 100°C.

HFC-410A is not comparable to refrigerants HCFC-22 or CFC-12 in terms of thermodynamic properties because of its considerably higher vapour pressure. HFC-410A compressor efficiencies, the pressure drop in suction lines and the heat transfer efficiency will benefit from high system pressure. This HFC is used mainly in new industrial systems designed for the refrigerant, especially in terms of low condensing temperatures of 35°C to avoid pressures higher than 25 bar. In industrial refrigeration, the evaporator temperature range for HFC-410A varies from –60°C to –35°C. Due to the high volumetric capacity (40% above that of HCFC-22), compressor efficiency has been reported to be higher than with HCFC-22 (Meurer and König, 1999). R-410A has a COP similar to NH_3 and HCFC-22 and slightly higher than that of R-404A, R-507A. Further, at temperatures below –40°C and up to the to HFC-410A normal boiling point at –51.6°C, COPs are slightly higher for R-410A compared to other refrigerants (Roth *et al., 2002*).

R-404A and R-507A are the main refrigerants to replace R-502 in the temperature range from –50°C to –30°C and these have comparable cycle efficiencies (COPs) and slightly lower GWP values than R-502. For systems with R-404A and R-507A, the liquid should be subcooled to achieve optimum efficiencies and high cost-effectiveness . In chiller applications with screw compressors it may also be necessary to add significant superheat to the suction gas in order to avoid refrigerant condensation in the oil separator.

4.5.5 Non-HFC technologies

In large systems CFCs and recently introduced HFCs have a lower average share in industrial refrigeration than NH_3 and HCFC-22. For new industrial systems designers and plant owners will mainly need to decide between NH_3, HFCs and HCFC-22 (except in the EU where HCFC-22 is forbidden in new systems) (Stoecker, 1998). Non-HFC technologies described in the following subsection are sorted by refrigerant.

Ammonia
Ammonia is one of the leading refrigerants for industrial refrigeration, based on performance and safety, and is used in large quantities in locations physically separated from general public access. The current market share in several European countries, especially in Northern Europe, is estimated to be up to 80% (UNEP, 1998). In the USA, ammonia has approximately 90% market share in systems of 100 kW cooling capacity and above that use custom-engineered processes (IIR, 1996).

New ammonia systems have an improved design, use low-temperature materials and standardized welding procedures, and the systems operation and maintenance are under continuous monitoring. A human error or direct physical damage is often the reason for failure (Lindborg, 2003).

Charge reduction has been achieved through dry or direct expansion in plate type heat exchangers and shell and tube evaporators. With soluble oils, it has been possible to reduce charge by 10%. New developments showed charges of 28g ammonia per kW cooling capacity for low overall capacity down to 100 kW (Behnert and König, 2003). With these low charges, new opportunities for applications not previously considered for ammonia have been realized, such as water chillers for air conditioning (Stoecker, 1998). This new ammonia technology with high COP was regarded as being fully practical, but strong market penetration has not been achieved due to price competition with HFC-based units (UNEP, 2003).

HCFC-22
The use of HCFC-22 is declining in industrial refrigeration in Europe, as the use of HCFC-22 in new systems is forbidden by European regulations (Official Journal, 2000).

In developing countries, HCFC-22 is still an important replacement refrigerant for CFCs in new systems, as from a technical point of view HCFC-22 could replace CFC-12 and CFC-502 in new systems. Another important consideration in developing countries is that under the Montreal Protocol the production of HCFC-22 is allowed until 2040, or the full lifetimes of equipment installed in the next 15 years or so.

Hydrocarbons (HCs)
HCs can fit into any temperature range for evaporating temperatures down to –170°C. Historically, their use as working fluids has been restricted to large refrigeration plants within the oil and gas industry. A certain registered increase in hydrocarbon consumption has mainly appeared in these sectors (Stoecker, 1998).

Commercialized products used in industrial refrigeration equipment include HC-290 and HC-1270. System performance with regard to system efficiency is comparable to and, in some cases even superior to, that of the halocarbons. Hydrocarbons are soluble with mineral oils and compatible with materials such as metals and elastomers that are traditionally used in refrigeration equipment. The use of hydrocarbons in screw compressors may be problematic due to the strong dilution of mineral oil. Other less soluble lubricants such as PAG or PAO may be required. As long as safety aspects are taken into consideration, standard refrigeration practices used for HCFCs and CFCs can be used for hydrocarbon fluids without major system detriment.

Given the flammability concerns, design considerations as detailed in the relevant safety standards should be adhered to. Additional safety measures are required for repairing and servicing. Several national and European standards permit the use of HCs in industrial applications and lay down specific safety requirements. Industry guidelines for the safe use of hydrocarbon refrigerants are available (ACRIB, 2001).

Carbon dioxide (CO_2)
As well as being non-ODP and having a GWP of 1, carbon dioxide (CO_2) offers a number of other advantages:
- Excellent thermophysical properties, leading to high heat transfer;
- Efficient compression and compact system design due to high volumetric capacity;
- Non-flammable and low toxicity;
- Low system costs at evaporation temperatures below 45°C (depending on system design);
- Widely available at low cost.

CO_2 systems can be used for industrial refrigeration applications with evaporation temperatures down to –52°C and condensing temperatures up to 5°C. CO_2 is also increasingly being used in the low stage of cascade systems for industrial refrigeration. CO_2 is also commonly used as a secondary refrigerant. The design requires the same pressure of 25 bar for the secondary refrigerant systems and for the CO_2 used as the refrigerant, except for ice rinks and some other limited systems which are designed for 40 bar. Defrosting was an open issue, but the most recent developments show that several different techniques such as electrical heating, hot gas defrosting, high-pressure liquid evaporation and the distribution of hot gas have been realized in plants (Siegel and Metger, 2003).

A comparative study for low temperatures has been carried out using a typical system design with cooling capacities of 600kW at −54°C for R-410A, R-507 and ammonia used as a single fluid in two-stage systems and for NH_3/CO_2 and HFC-134a/R-410A as refrigerants in high/low stage cascade systems. The volumetric refrigerating capacity of CO_2 is five times higher than HFC-410A and eight times higher than for ammonia and other refrigerants. Therefore the size of most components in the system can be reduced. The study found the lowest cost for NH_3/CO_2 (Roth *et al.*, 2002). If CO_2 is used as the refrigerant, the cost break-even point for industrial refrigeration compared to NH_3 and R-410A is approximately at an evaporating temperature of −40°C to −45°C. Below this temperature, lower costs for CO_2/NH_3 cascade systems have been achieved. It is expected that costs for screw or reciprocating units, including compressors and oil separation circuit, will be further reduced (Roth *et al.*, 2002). The efficiency of CO_2 systems in this low temperature range is similar to other refrigerants such as R-410A or ammonia.

CO_2 shows strong cost benefits if the system size is increased, especially in cases where evaporators or heat exchangers are distributed and long piping systems are required. In industrial refrigeration applications, cost benefits have been achieved with total pipe runs of more than 2500 m (Siegel and Metzger, 2003).

In food processing, a trend can be observed towards CO_2 as a refrigerant at temperatures lower than −45°C and as an HTF for cooling temperatures lower than −5°C (Pirard, 2002).

Some examples are given to illustrate the use of CO_2 as refrigerant in low-temperature applications:

- In the USA, the first large CO_2 system was being erected in 2003 with cooling capacities of 6 MW (Stellar, 2003);
- In Japan a standard low-temperature cascade system has been developed with NH_3/CO_2 as refrigerants. The systems are designed for evaporating temperatures of −40 to −55°C with cooling capacities of 80−4450 kW;
- In Europe more than 30 large systems with CO_2 as the heat transfer fluid and refrigerant have been installed since 1998 and are operating with total cooling capacities of more than 25 MW (Pearson, 2004a,b);
- At least two large systems in Europe have been retrofitted from HCFC-22 (1.5 MW at −45 to −55°C) and from CFC-13B1 (2.4 MW at −35°C) to NH_3/CO_2 cascade systems (Gebhardt, 2001; König, 2002).

4.5.6   Trends in consumption

The trends in the consumption of refrigerants for industrial refrigeration as well as the food processing and cold storage subsector are discussed in Section 4.4.6.

4.5.7   Comparison of HFC and non-HFC technologies

4.6.7.1   Energy efficiency and performance
On a worldwide basis, only two refrigerants have significant market share in industrial refrigeration: ammonia and HCFC-

22. Stoecker (1998) provides a comparison of both refrigerants. Compared to ammonia and HCFC-22, the market share of HFCs and non-HFC technologies is small. Nevertheless, one point of comparison is the cost of the refrigerant to be used in the system, and the lowest costs are found for ammonia (Stoecker, 1998).

The energy efficiency comparisons for HFCs 404A, 507A, and 410A with ammonia and HCFC-22 are described in Section 4.5.4. HFC410A has an energy efficiency similar to ammonia and HCFC-22, and slightly higher than R-404A and R-507A (Roth *et al.*, 2002).

CO_2 technology is a non-HFC technology which is gaining momentum. The energy efficiency of CO_2 systems in the temperature range of −40°C to −45°C is similar to HCFC-22 and HFC refrigerants such as HFC-410A. CO_2 also shows strong cost benefit if the system size is large (Siegel and Metzger, 2003).

4.5.7.2   TEWI/LCCP/LCA
For various reasons, only limited TEWI/LCCP/LCA data are available for industrial refrigeration systems. Such systems are normally custom-designed for special requirements and are erected on site. The design differs not only in terms of cooling capacities and temperatures, but also in terms of temperature control requirements (air blast cooling systems), size of piping, distance to consumers and charge of refrigerant. There are therefore only a few references which compare TEWI and costs for the same application (Pearson , 2004a; Roth *et al.*, 2002). Roth *et al.* (2002) give a comparative example for a manufacturing plant with a cooling capacity of 600 kW at −54°C (see Figures 4.1 and 4.2). In this investigation the combination of CO_2 and ammonia was more competitive than other solutions.

In addition to the above references, examples of LCCP calculations for supermarket refrigeration systems provide general guidance for selecting systems and refrigerants with a lower LCCP (see Section 4.3). Lower LCCP results from systems with low energy consumption, and in the case of fluorocarbon refrigerants, low refrigerant charge size and low refrigerant emissions. LCCP calculations should be used to optimize the choice of refrigerant and system design for the lowest environmental impact.

4.6   Transport refrigeration

4.6.1   Introduction

The transport refrigeration subsector consists of refrigeration systems for transporting chilled or frozen goods. Typically the task of a transport refrigeration system is to keep the temperature constant during transport. The technical requirements for transport refrigeration units are more severe than for many other applications of refrigeration. The equipment has to operate in a wide range of ambient temperatures and under extremely variable weather conditions (sun radiation, rain, etc.); it also has to be able to carry any one of a wide range of cargoes with

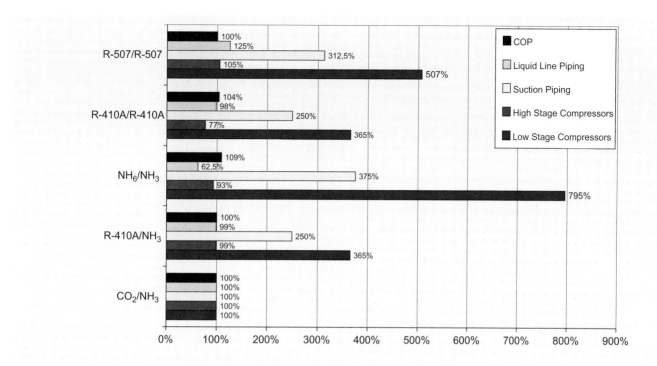

Figure 4.1. COP and size comparison of different components for different refrigerant combinations (Q_0 = 600 kW, t_C = 35°C; t_0 = −54°C; CO_2/ NH_3-cascade system is equal to 100 %) (Roth and König, 2002).

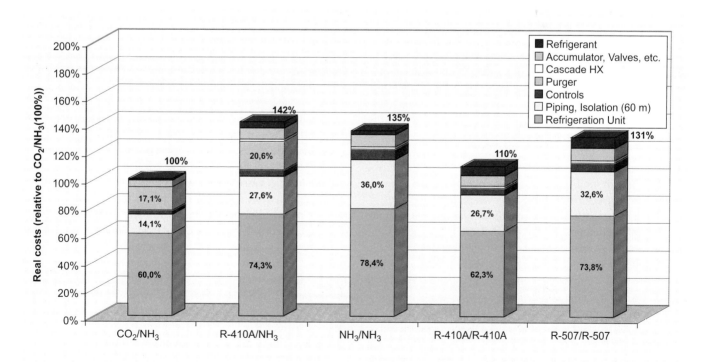

Figure 4.2. Cost comparison for different refrigerant combinations (Q_0 = 600 kW, t_C = 35°C; t_0 = −54°C; CO_2/NH_3-cascade system is equal to 100 % (Roth and König, 2002).

differing temperature requirements, and it must be robust and reliable in the often severe transport environment (IIR, 2003). Typical modes of transport are road, rail, air and sea. In addition, systems which are independent of a moving carrier are also used; such systems are generally called 'intermodal' and can be found as containers (combined sea-land transport) as well as swap bodies (combined road and rail transport). This section covers also the use of refrigeration in fishing vessels where the refrigeration systems are used for both food processing and storage.

The technology used in transport refrigeration is mainly the mechanically- or electrically-driven vapour compression cycle using refrigerants such as CFC, HCFC, HFC, ammonia or carbon dioxide. Due to the complete and worldwide phase-out of CFC consumption by the end of 2009, CFCs are not addressed in this chapter. In addition, a number of refrigeration systems are based on using substances in discontinuous uses. This type of equipment can be found as open uses with solid or liquid CO_2, ice, or liquid nitrogen and in these cases the refrigerant is being completely emitted and lost after removing the heat (Viegas, 2003). Closed systems such as eutectic plates (Cube *et al.*, 1997) or flow-ice, reuse the same substance (Paul, 1999). Such systems used to be very commonplace in transport refrigeration, and are still used on a significant scale. Some propose that their use should be increased in the future.

All transport refrigeration systems need to be compact and lightweight, as well as highly robust and sturdy so that they can withstand movements and accelerations during transportation. Despite these efforts, leaks within the refrigeration system occur due to vibrations, sudden shocks and so forth. The likelihood of leaks or ruptures is also greater than with stationary systems, due to a higher risk of collisions with other objects. Ensuring safe operation with all working fluids is essential, particularly in the case of ships where there are no options to evacuate a larger area (SCANVAC, 2001). The safety is either inherent in the fluids or is ensured through a number of technical measures (Stera, 1999).

4.6.2   Container transport

Refrigerated containers allow uninterrupted storage during transport on different types of mobile platforms, for example railways, road trucks and ships. The two main types of refrigerated containers are porthole containers and integral containers. Porthole containers are the older of the two concepts and are insulated containers with two front apertures and no built-in refrigeration systems. Some predict that by 2006, transport will have been completely converted to integral containers (Hochhaus, 2003; Wild, 2003).

Integral refrigerated containers are systems which have their own small refrigeration unit of about 5 kW refrigeration capacity on board. There were more than 550,000 of these in 2000, representing the transport capacity of 715,000 20-foot containers, and there numbers are set to strongly increase (UNEP, 2003; Sinclair, 1999; Stera, 1999). The electrical power needed

to drive the system is supplied from an external power supply via an electrical connection. These systems typically use HFC-134a, R-404A and HCFC-22, and in some cases R-407C (Wild, 2003). Newer systems generally have a more leak-resistant design (Crombie, 1999; Stera, 1999; Yoshida *et al.*, 2003; Wild, 2003). In 1998, when older design systems were prevalent, an average annual leakage rate of 20% of the charge of about 5 kg was assumed for a lifetime of 15 years (Kauffeld and Christensen, 1998).

4.6.3   Sea transport and fishing vessels

Virtually all of the 35,000 plus merchant ships worldwide larger than 500 gross tonnes (Hochhaus, 1998) have some on-board refrigeration system. The majority of systems use HCFC-22. These refrigeration systems and options for emission abatement are referred to in Sections 4.2 and 4.3 of this report. In terms of technology and performance, chillers for air conditioning or, in case of naval vessels for electronics and weapon system cooling, are similar to stationary systems (see section 5.1 on air conditioning). The following remarks and information relate to ship-bound refrigeration systems essential to the main purpose of non-naval vessels, namely the transportation of perishable products, the chilling of fish and the like.

Refrigerated transport vessels, also called reefers, provide transportation for perishable foodstuff at temperatures between –30°C and 16°C (Cube *et al.*, 1997). It is estimated that there are around 1300 to 1400 reefer vessels in operation (Hochhaus, 2002; Hochhaus, 1998), a number which has been constant for quite some time and is expected to decrease. In 2001, it was reported that more than 95% of the refrigeration installations on these vessels use HCFC-22 as a refrigerant (SCANVAC, 2001), although various HFCs such as HFC-134a, R-404A, R-507 and R-407C as well as ammonia are being used. About two-thirds of the systems are direct systems with up to 5 tonnes of refrigerant per system and the remaining are indirect systems with a charge below 1 tonne of refrigerant (UNEP, 2003). Estimates of current annual leakage rates based on known refrigerant consumption are 15–20% of the system charge (SCANVAC, 2001).

Worldwide there about 1.3 million decked and about 1.0 million undecked, powered fishing vessels. In 2001, more than 21,500 fishing vessels over 100 gross tonnes were recorded (FAO, 2002), with a slightly decreasing trend. Vessels of that size are assumed to operate internationally and to be equipped with significant refrigeration equipment. Within a wide range, the average larger fishing vessel has a refrigerant charge in the order of 2000 kg with 15–20% annual leakage rate. In 2001 more than 95% of such vessels in Europe used HCFC-22 as the refrigerant (SCANVAC, 2001). It is assumed that 15% of the fleet have full size refrigeration systems, while the remaining fleet is assumed to be equipped with small refrigeration systems that have a filling mass of approximately 100 kg.

Specialized tankers are used to transport liquefied gases, in particular liquefied petrol gas (LPG) and liquefied natural gas (LNG). Medium and large LNG tankers transport LNG at normal pressure. The refrigeration effect needed for this type of transport is provided by evaporating the LNG, which is recondensed using specialized refrigeration units. Since the number of such ships is limited (about 150 ships of above 50,000 tonnes were registered in 1996) (Cube *et al.*, 1997) and the refrigeration equipment typically uses the transported, low GWP hydrocarbon gases as the refrigerant, refrigeration use in gas tankers is not further considered in this report.

4.6.4 Road transport

Road transport refrigeration units, with the exception of refrigeration containers, are van, truck or trailer mounted systems. Some trailers are equipped to be mounted or have their main bodies mounted on railroad systems; these are so-called swap-bodies. In a number of uses those systems are of the discontinuous type, using eutectic plates in closed systems (Cube *et al.*, 1997) or liquid nitrogen, liquid carbon dioxide or solid carbon dioxide in open systems (UNEP, 2003). These systems are frequently used in local frozen food distribution, for example, delivery directly to the customer (Cube *et al.*, 1997). Liquid nitrogen for cooling purposes is used by more than 1000 vehicles in the UK. Liquid carbon dioxide is reported to be used in 50 trucks in Sweden (UNEP, 2003). In general, the necessity for storage and filling logistics, the hazardous handling of very cold liquids and solids and the energetically unfavourable low-temperature storage reduce the widespread application of these historically frequently used technologies.

The predominant technology in road transport, covering virtually all of the remaining refrigerated road transport equipment, is the mechanical vapour compression cycle. Trailers usually have unitary equipment that consists of a diesel engine, compressor, condenser, engine radiator and evaporator with fans as well as refrigerant and controls. These systems are also used for swap bodies. Larger trucks often have similar equipment as trailers. However, as the truck size decreases an increasing proportion of systems have the compressor being driven by the drive engine (ASHRAE, 2002). Alternatively, some truck systems use a generator coupled with the truck engine to generate electricity, which is then used to drive the compressor (Cube *et al.*, 1997).

In 1999, it was estimated that in North America alone 300,000 refrigerated trailers were in use (Lang, 1999). For the 15 countries of the European Union in 2000, 120,000 small trucks, vans and eutectic systems with 2 kg refrigerant charge were estimated to be in use, with 70,000 mid-size trucks of 5 kg refrigerant filling and 90,000 trailers with 7.5 kg refrigerant filling (Valentin, 1999). The worldwide numbers in 2002 were estimated to total 1,200,000 units, with 30% trailer units, 40% independent truck units and 30% smaller units. The annual amount of refrigerant needed for service is reported to be 20–25% of the refrigerant charge (UNEP, 2003). The refrigerant typically chosen is HFC-134a for applications where only cooling is needed, and predominantly R-404A and R-410A for freezing applications and general-purpose refrigeration units (UNEP, 2003).

4.6.5 Railway transport

Refrigerated railway transport is used in North America, Europe, Asia and Australia. The transport is carried out by using either refrigerated railcars, or refrigerated containers (combined sea-land transport; see Section 4.6.2) or swap bodies (combined road-land transport; see Section 4.6.4). This section concentrates on transport in refrigerated railcars.

Different technologies have been used in the past: Solid CO_2 as well as ice have been used in discontinuous emissive systems to date (CTI, 2004). Mechanically-driven refrigeration systems have also been used and are now the prime choice because of the typically long duration of trips, which makes refilling of the emitted refrigerant in discontinuous emissive systems a challenge for both logistical and cost reasons.

Mechanically driven systems are almost completely equipped with diesel engines to supply the necessary energy to the refrigeration unit. The existing fleets of railcars in Asia still seem to mostly operate on one-stage (cooling) and two-stage (freezing /combined use) CFC-12 systems (UNEP, 2003). The European railcars have been converted to HFC-134a (Cube *et al.*, 1997), and this has been facilitated by European regulations phasing out the use of CFCs (EC No 2037/2000 (Official Journal, 2000)). In North America, existing older systems have been converted to HFC-134a, while newer systems utilize HFC-134a and R-404A (DuPont, 2004).

The lifetime of newer rail refrigeration systems, which are often easily replaceable units originally developed for road transport and only adapted for rail use, is believed to be 8 to 10 years with a running time of 1000 to 1200 hours per annum (refrigeratedtrans.com, 2004). Older units specifically designed for rail use have a lifetime of typically 40 years and a refrigerant filling of approximately 15 kg (UNEP, 2003). The annual leakage rate may be assumed to be at least similar to the leakage rate experienced in road transport, that is 20–25% of the refrigerant charge (UNEP, 2003).

4.6.6 Air transport

In order to provide constant low temperature during the flight, containers to be loaded upon aircraft are provided with refrigeration systems. There are some battery powered mechanical refrigeration systems (Stera, 1999), but the total number of these is believed to be small. Other, more commonly used systems are discontinuous with solid carbon dioxide (Sinclair, 1999; ASHRAE, 2002), or ice (ASHRAE, 2002). As the amount of ODS replacement during use is apparently very small, air transport will not be detailed further in this report.

4.6.7 Abatement options

4.6.7.1 General
Based on the study of Clodic and Palandre (2004), the total amount of refrigerant contained in transport refrigeration systems is estimated to be 16,000 tonnes; 6000 tonnes of this are

emitted annually. It should be noted that the widespread use of R-404A as a non-ODS alternative with a relatively high GWP of 3800 kg CO_2 kg[-1] leads to very high CO_2-equivalent emissions. Using alternatives in systems with a more moderate GWP than R-404A, such as the HFC mixture R-410A, would cut the CO_2-equivalent emissions substantially.

Current system requirements lead to a refrigerant selection which is largely limited to HFC-134a and refrigerant mixtures with a relatively high global-warming impact such as R-404A. Since the emission rates in operation are significant, improvements in energy consumption, alternative substances and not-in-kind technologies are the main options for emission abatement. R-404A is the main refrigerant in current use (IIR, 2002) and is popular because of its flexibility (medium- and low-temperature applications) and safety. Only a limited number of TEWI calculations are available in the literature; the only investigation comparing different technologies such as CFC-12, HFC-134a, R-404A, HC-600a/HC-290, ammonia and CO_2, states that R-404A systems are at least sustainable from the different options investigated for reefer ships (Meffert and Ferreira, 2003).

4.6.7.2   Containment

As there are already considerable incentives to optimize design and to minimize leakage, further containment in most uses would require a new approach not yet seen. One example might be to use fully hermetic systems for road transport (Chopko and Stumpf, 2003a, b), although the effect of this on energy consumption has yet to be determined. The development of hermetic scroll compressors for container systems with acceptable energy efficiency for both cooling and freezing applications (Yoshida *et al.*, 2003; DeVore, 1998) allows their widespread use, and leads to less service requirements and therefore less related refrigerant losses. In addition, these compressors are hermetic, which further decreases leaks (Wild, 2003). This technology has already been introduced and is penetrating the market as existing equipment is gradually replaced.

Recovery and recycling is a statutory requirement in many countries and is probably adhered to since the equipment contains a considerable, but still easy-to-handle, amount of refrigerant and due to its mobility it can easily be transported to a recovery facility (except seagoing). On the other hand due to the large emission rates in operation, the improvements through recovery and recycling, which encompass only the refrigerant losses during service and disposal, are likely to be limited.

An alternative approach to improving systems to reduce leaks might be to improve operating conditions to reduce wear, likeliness of ruptures and refrigerant losses during service. The potential of such measures compared to system-related improvements has yet to be assessed, but might be considerable.

4.6.7.3   Improvement in energy efficiency

Most refrigeration systems operate under partial load conditions for a large proportion of their useful life (Meffert and Ferreira, 1999). Different methods for partial load control have been investigated for both electrically driven compressors and open compressors (e.g. Crombie, 1999). Potential energy savings for electrical systems using frequency converters are said to reach up to 25.8% per voyage (Han and Gan, 2003). Other sources compare a range of control possibilities (Meffert and Ferreira, 1999 and 2003). These sources concluded that energy efficiency gains of more than 70% can be achieved under part-load conditions.

4.6.7.4   Discontinuous processes

The use of ice as well as solid CO_2 are both established alternatives to vapour compression systems. Besides the logistical necessities of such systems, there are also temperature limitations for the use of ice as well as handling and energy issues when using solid CO_2 (as heat absorption is energetically unfavourable at $-78.4°C$). These issues are even more valid for the use of liquid nitrogen, producing an unnecessarily low temperature of $-195.8°C$, or liquid air with $-194.3°C$. Nevertheless, refrigerated systems using ice and solid CO_2 systems remain abatement option for HFC in suitable cases.

The commercialization of a fully self-powered liquid CO_2 system with a moderate evaporation temperature of $-51°C$ for the delivery of frozen product to customers was reported by Viegas (2003), and this addressed handling as well as energy efficiency issues. The system, which needs a service infrastructure, has been commercialized in Sweden (Viegas, 2003) and the UK (UNEP, 2003) and is therefore available as an abatement option, especially for local and short-haul transport.

The use of a pumpable suspension of ice crystals in water ('binary ice', 'flow ice'), has been developed for certain transport uses. The suspension is pumped into the hollow walls, floors, ceilings or trays of a containment to be refrigerated. While equipment for service trolleys for passenger trains is already commercially available, the same principle is being suggested for containers (Paul, 1999). Although the remaining technical issues seem to be standard engineering tasks, the technology is not yet commercially available for cooling of full containers, trucks or vans.

4.6.7.5   Sorption processes

Sorption processes are well known, heat-driven processes using water, methanol or ammonia as a refrigerant, and solids such as activated coal, zeolite or silica gel (adsorption) as well as liquids such as lithium bromide (LiBr) and water (absorption) as sorbents in a closed circuit. The heat to drive such processes can come from a variety of sources; in the case of transport refrigeration, the waste heat from the transporter's engine could be used. Such a use has been proposed for several years, especially for ship-bound systems (Cube *et al.*, 1997).

LiBr-water systems, are frequently used in stationary applications and for the capacity range of 200 kW – 600 kW, these have been reported to operate successfully and produce chilled water in certain specialized ships (Han and Zheng, 1999). Below zero refrigeration is not feasible with LiBr-water systems. As such systems have already been successfully employed on ships, their utilization might be increased at a relatively short notice.

The applicability to modes of transportation other than ships might be limited because of downscaling problems as well as design restrictions on those systems.

For truck-mounted refrigeration systems, the use of waste heat from the truck engine has been suggested to drive a water-ammonia sorption cycle (Garrabrant, 2003). For medium and small fishing vessels, adsorption ice-makers with carbon-methanol are being proposed, which utilize the exhaust heat of the ship's engine as an energy source (Wang *et al.*, 2003).

4.6.7.6 Hydrocarbons

Hydrocarbon cooling systems for the recondensation of transported hydrocarbons have successfully been installed in gas tankers. International activities are underway to develop hydrocarbon systems for reefer ships (Jakobsen, 1998). In road transport refrigeration, commercially available systems have been developed in Australia, Germany and other European countries using HC-290 (propane). The systems require a leak detector in the trailer and special driver training to fulfil safety-related legal requirements (UNEP, 2003; Frigoblock, 2004).

Technically this solution could be adopted worldwide in certain road and railroad systems, especially in compact systems. Nevertheless, either certain existing regulations or present system use patterns would have to be adapted. The flammability of hydrocarbons will require additional safety measures, thus increasing the costs of the system, and in the beginning at least probably insurance rates as well. Containers might also require changes in the transporting ships.

4.6.7.7 Ammonia

Ammonia as refrigerant is being increasingly used in marine refrigeration equipment. Applications include its use in reefers (Stera, 1999), as a proposed refrigerant for sorption ice machines (Garrabrant, 2003), and the use in fishing vessels both as a single refrigerant (UNEP, 2003; Berends, 2002) and in combination with CO_2 (Nielsen and Lund, 2003). The applicability has been sufficiently proven. Ammonia as a refrigerant requires certain design considerations as well as the presence of additional safety equipment on board (SCANVAC, 2001).

4.6.7.8 Carbon Dioxide

Carbon dioxide as a refrigerant in mechanically-driven vapour compression systems, might be used as a subcritical refrigerant (critical point at 31°C) with a condensing temperature well below the critical point in cascade systems or in applications where low-temperature cooling options means are available. Alternatively, it can be used as a near-critical or, more likely, a super-critical working fluid. If the condensing temperature of CO_2 is below 15°C (border of subcritical region), this refrigerant typically offers, but not always, significant advantages in terms of efficiency and costs in comparison to other refrigerants. This advantage can only be utilized in cascade systems with other refrigerants or where low-temperature heat sinks are available. Near- or super-critical uses require a much higher pressure resistance of the equipment than is currently usual for

other refrigerants, and such uses are often energetically less favourable than other refrigerants in the same temperature range.

For low-temperature uses, combinations of ammonia and CO_2 have been developed and built into ships. A comparison shows that the efficiency for a –40°C evaporation and 25°C condensing temperature is 17% higher than for a 2-stage HCFC-22 system (25% improvement at –50°C/25°C) (Nielson and Lund, 2003). The advantage of using CO_2 in such applications is that the necessary components (in particular the compressor) are commercially available or require only minor modifications, while consuming less space than other solutions.

CO_2 has also been proposed for container systems, where it would typically be used in a super-critical manner. A prototype system has yet to be reported as until recently no suitable compressor was available. A prototype CO_2 system for trucks has been developed, laboratory tested and optimized (Sonnekalb, 2000). The calculated TEWI shows a 20% decrease compared to a R-404A system.

4.6.7.9 Air

The air cycle for transport refrigeration purposes has been investigated for a number of years (e.g. Halm, 2000). A prototype system has been developed and tested, but has never been commercialized. Presently air cycle equipment for transport refrigeration does not seem to represent a suitable short- or medium-term abatement option due to the lack of suitable and reliable components.

4.6.8 Comparison of alternatives

Emissions of halocarbons in the transport refrigeration sector are related to four subsectors: Sea transport and fishing, road transport, rail transport and intermodal transport, that is containers and swap bodies. An overview can be found in Table 4.15.

There are a number of possibilities to improve those transport refrigeration systems built today to achieve a lowering of direct or energy-consumption related emissions without changing the working fluid or technology. A number of measures have been proposed and these have in part already been implemented to improve the energy efficiency, for example, the use of efficient compressors, frequency control for part load conditions, water-cooled condensers for containers on board ships, regular preventive maintenance and so forth. Measures to control direct emissions have mainly been proposed for mass-produced systems (e.g. container units) in terms of design improvements.

An alternative to improving the currently predominant halocarbon technologies is the replacement of those refrigerants by fluids or technologies with a lower GWP. Technically there are or will be low GWP replacement options available for all transport refrigeration uses where CFCs, HCFCs or HFCs are currently used. However in several cases these might increase the costs of the refrigeration system.

In case of reefer ships and fishing vessels, the most promising and already implemented non-halocarbon abatement tech-

Table 4.15. Subsectors of transport refrigeration, characteristics and alternatives.

Subsector		Sea Transport & Fishing	Road Transport	Rail Transport	Intermodal Transport
Cooling capacity	From	5 kW	2 kW	10 kW	Approx. 5 kW
	To	1400 kW	30 kW	30 kW	
Refrigerant charge	From	1 kg	1 kg	10 kg	Approx. 5 kg
	To	Several tonnes	20 kg	20 kg	
Approximate percentage of sector refrigerant bank in subsector		52% of 15,900 tonnes	27% of 15,900 tonnes	5% of 15,900 tonnes	16% of 15,900 tonnes
Approximate percentage of sector refrigerant emissions in subsector		46% of 6000 tonnes	30% of 6000 tonnes	6% of 6000 tonnes	18% of 6000 tonnes
Predominant technology		HCFC-22	HFC-134a, HFC-404A, HFC-410A	HFC-134a, HFC-404A, HFC-410A	HFC-404A
Other commercialized technologies		Various HFCs, ammonia, ammonia, CO_2/ammonia for low temperatures; hydrocarbon systems for gas tankers; sorption systems for part of the cooling load	Hydrocarbon, liquid CO_2; with unknown systems for liquefaction/freezing: liquid CO_2, ice slurry; with on-board HCFC/HFC refrigeration systems: Eutectic plates	Solid CO_2 (with unknown systems for freezing)	HFC-134a, HCFC-22
Low GWP technologies with fair or better than fair potential for replacement of HCFC/HFC in the markets		Ammonia, CO_2/ammonia for low temperatures	Hydrocarbon, CO_2 compression systems; for short haul combination of stationary hydrocarbon or ammonia with liquid CO_2, ice slurry or eutectic plates	Hydrocarbon, CO_2 compression systems; for specific transports (certain fruits, ...) combination of stationary hydrocarbon or ammonia with liquid CO_2, ice slurry or eutectic plates	CO_2 compression system
Status of alternatives		Fully developed. Some cost issues related to additional safety for ammonia plants on ships. Hydrocarbon practical mainly for ships which are built according to explosion-proof standards (gas carriers, ...)	Hydrocarbon mini-series successfully field tested, lack of demand/add. requirements on utilization (driver training, parking, ...). Liquid CO_2 systems commercialized. CO_2 compression tested in proto-types, but open compressor needed for most systems in combination with leaks remains an issue	Solid CO_2 standard use, but not very energy efficient, difficult handling, high infrastructure requirements, therefore presently being phased out. Increasingly use of systems designed for trailer use with optimization for rail requirements (shock resistance, ...)	Under development – prototype testing; might be available in the near future if demanded

nology is equipment with ammonia or ammonia/CO_2 systems. These systems are likely to operate at least as energy efficiently as existing systems. One source (SCANVAC, 2001) estimates the additional costs for a ship-bound ammonia system to be 20–30% higher if retrofitted into an existing vessel and potentially lower if included in the ships planning from the start. Another source (Nielsen and Lund, 2003) assumes that small industrial ammonia/CO_2 systems might be more expensive than conventional systems, but large systems might have a more or less equivalent price for the same capacity.

In the case of container systems, CO_2 in a vapour compression cycle could develop into a promising alternative. The costs for the refrigeration system might be higher than for current conventional systems. The energy consumption will probably be higher if the containers are only air-cooled but if additional water-cooling is installed, as has already implemented on some vessels, the systems could be energetically as good as or even better than existing equipment.

Options using CO_2 and hydrocarbons exist for road transport. The hydrocarbon technology is technically implementable within a short time frame. For larger systems, CO_2 systems or hydrocarbon refrigerants are potential options, depending on the safety issues. The same alternatives could be used for new railway systems. Certain types of refrigerated road transport, such as short-range distribution trucks, might use discontinuous systems with evaporating CO_2 or nitrogen as alternative.

As transport refrigeration systems have very significant emissions and a limited runtime, which is typically far below

100%, direct emissions play a very important role in the calculation of the TEWI. The replacement options currently being considered by manufacturers do not significantly increase transport weight or volume. The data is sufficient to state that in several applications, a substantial reduction in TEWI could be achieved by introducing a low GWP technology.

References

ACRIB, 2001: Guidelines for the Use of Hydrocarbon Refrigerant in Static Refrigeration and Air-conditioning Systems. Air Conditioning and Refrigeration Industry Board, (ACRIB), Carshalton, UK.

ADL (A.D. Little, Inc), 1999: Global Comparative Analysis of HFC and Alternative Technologies for Refrigeration, Air Conditioning, Foam, Solvent, Aerosol Propellant, and Fire Protection Applications. Final Report to the Alliance for Responsible Atmospheric Policy, August 23, 1999, Prepared by J. Dieckmann and H. Magid (available online at www.arap.org/adlittle-1999/toc.html), Acorn Park, Cambridge, Massachusetts, USA, 142pp.

ADL (A.D. Little, Inc.)**,** 2002: Global Comparative Analysis of HFC and Alternative Technologies for Refrigeration, Air Conditioning, Foam, Solvent, Aerosol Propellant, and Fire Protection Applications. Final Report to the Alliance for Responsible Atmospheric Policy, March 21, 2002 (available online at www.arap.org/adlittle/toc.html), Acorn Park, Cambridge, Massachusetts, USA, 150pp.

AEAT (AEA Technology), 2003: Emissions and Projections of HFCs, PFCs and SF_6 for the UK and Constituent Countries. Report prepared for the Global Atmosphere Division of the UK Department for Environment, Food, and Rural Affairs, July 2003.

ANSI/ASHRAE, 2002: Standard 147-2002 Reducing the Release of Halogenated Refrigerants from Refrigeration and Air-Conditioning Equipment and Systems. American Society of Heating, Refrigerating, and Air-Conditioning Engineers, Inc. (ASHRAE), Atlanta, GA, USA.

Arias, J. and P. Lundqvist, 1999: Innovative System Design in Supermarkets for the 21st Century. Proceedings of the 20th International Congress of Refrigeration, Sydney, Australia, 19-24 September 1999. D.J. Cleland and C. Dixon (eds.). International Institute of Refrigeration, (IIR/IIF), Paris, France.

Arias, J. and P. Lundqvist, 2001: Comparison of Recent Refrigeration Systems in Supermarkets. Proceedings of the 6th Ibero-American Congress of Air-Conditioning and Refrigeration, Buenos Aires, Argentina, 14-17 August, 2001.

Arias, J. and P. Lundqvist, 2002: Heat Recovery in Recent Refrigeration Systems in Supermarkets. Proceedings of the 7th International Energy Agency Heat Pump Conference 2002: Heat Pumps – Better by Nature, Beijing, China, 19-22 May 2002. IEA Heat Pump Centre, Sittard, The Netherlands.

ASHRAE, 2002: *2002 ASHRAE Handbook – Refrigeration*, (SI edition). American Society of Heating, Refrigeration and Air-Conditioning Engineers Inc., Atlanta, GA, USA.

Axima, 2002: Fish processing factory with CO_2 as brine and NH_3 as refrigerant. Axima Refrigeration GmbH, Company Information, Lindau, Germany.

Bansal, P., and R. Kruger, 1995: Test standards for household refrigerators and freezers I: Preliminary comparisons. *International Journal of Refrigeration*, **18**(1), 4-17.

Barreau, M., S. Macaudiere, P. Weiss, and M. Joubert, 1996: R-404A in Industrial Refrigeration Application for R-502 and HCFC-

22 Replacement. System with Recirculation-type Evaporator. Proceedings of the International Conference on Ozone Protection Technologies, October 21-23, 1996, Washington DC, USA.

Baxter, V.D. (ed.), 2003a: Advanced Supermarket Refrigeration / Heat Recovery Systems, Vol 1 – Executive Summary. IEA Heat Pump Centre, Sittard, The Netherlands, 73 pp (ISBN: 90-73741-48-3).

Baxter, V.D. (ed.), 2003b: Advanced Supermarket Refrigeration / Heat Recovery Systems, Vol 2 - Country Reports. IEA Heat Pump Centre, Sittard, The Netherlands (CD-ROM, ISBN: 90-73741-49-1).

Beeton, W.L. and H.M. Pham, 2003: Vapor injected scroll compressors. *ASHRAE Journal*, **45**(4), 22-27.

Behnert, T. and H. König, 2003: Entwicklung eines NH_3-Standard-Flüssigkeitskühlsatzes mit minimaler Füllmenge. *Die Kälte- und Klimatechnik*, 6/2003, 32-37 (in German).

Berends, E., 2002: Ammoniak (her)ontdekt voor scheepskoelinstallaties? (The rediscovery of ammonia as a refrigerant on new refrigerated cargo vessels?). *Koude & Luchtbehandeling*, **95**(9), 24-33 (in Dutch).

Bertoldi, P., 2003: The European End-Use Energy Efficiency Potential and the Needed Policies. European Commission, DG JRC, Presentation at Fondazione Eni Enrico Mattei, Milan, Italy, December 8, 2003 (http://www.feem.it/NR/Feem/resources/pdf/cop9/20031208/Bertoldi.pdf).

Birndt, R., R. Riedel and J. Schenk, 2000: Dichtheit von Gewerbekälteanlagen (Tightness of Commercial Refrigeration Systems. *Die Kälte- und Klimatechnik*, 9/2000, 56-63 (in German).

Bivens, D. and C. Gage, 2004: Commercial Refrigeration Systems Emissions. Proceedings of the 15th Annual Earth Technologies Forum, April 13-15, 2004, Washington, D.C., USA.

CEN, 2000/2001: Refrigerating systems and heat pumps – Safety and environmental requirements. Part 1: Basic requirements, definitions, classification and selection criteria (2001), Part 2: Design, Construction, testing, marking and documentation (2000), Part 3: Installation, site and personal protection (2000), Part 4: Operation, maintenance, repair and recovery (2000). European Standards, EN CEN 378.

Chopko, R.A. and A. Stumpf, 2003a: Advantages of All-electric Transport Refrigeration Systems. Proceedings of the 21st International Congress of Refrigeration, Washington, DC, USA, 17-22 August 2003. International Institute of Refrigeration (IIR/IIF), Paris, France.

Chopko, R.A. and A. Stumpf, A., 2003b: Survey of Multi-Temperature Transport unit refrigeration design. Proceedings of the 21st International Congress of Refrigeration, Washington, DC, USA, 17-22 August 2003. International Institute of Refrigeration (IIR/IIF), Paris, France.

Christensen, K.G., 1999: Use of CO_2 as Primary and Secondary Refrigerant in Supermarket Applications. Proceedings of the 20th International Congress of Refrigeration, Sydney, Australia, 19-24 September 1999. D.J. Cleland and C. Dixon (eds.). International Institute of Refrigeration (IIR/IIF), Paris, France.

Christensen, K.G., 2004: The World's First McDonalds Restaurant Using Natural Refrigerants. Proceedings of 6th IIR Gustav Lorentzen Conference – Natural Working Fluids 2004, Glasgow, Scotland, August 29 – September 1, 2004. International Institute of Refrigeration (IIR/IIF), Paris, France.

Christensen, K.G. and P. Bertilsen, 2003: Refrigeration Systems in Supermarkets With Propane and CO_2 – Energy Consumption and Economy. Proceedings of the 21st International Congress of Refrigeration, Washington, DC, USA, 17-22 August 2003. International Institute of Refrigeration (IIR/IIF), Paris, France.

Clodic, D., 1997: *Zero Leaks*. American Society of Heating, Refrigerating, and Air-Conditioning Engineers, Inc. (ASHRAE), Atlanta, GA, USA, 189 pp.

Clodic, D. and L. Palandre, 2004: Determination of Comparative HCFC and HFC Emission Profiles for the Foam and Refrigeration Sectors Until 2015. Part 1: Refrigerant Emission Profiles. Centre d'Energetique (Ecole des Mines de Paris/Armines), Report for US EPA and ADEME, 132 pp.

Coca Cola, 2002: Our Environmental Values – The Coca Cola Company Environmental Report 2002, The Coca Cola Company, Atlanta, GA, USA.

Coca Cola, 2004: The Coca Cola Company – Alternative Refrigeration Backgrounder. Proceedings of the Conference: Refrigerants, Naturally, 22 June 2004, Brussels, Belgium (available at http://www.refrigerantsnaturally.com).

Colbourne, D. and Suen, K. O., 2004, Appraising the flammability hazards of hydrocarbon refrigerants using quantitative risk assessment model. Part II: Model evaluation and analysis. *Int. J. Refrigeration*, **27**(2004), pp. 784-793.

Crombie, D., 1999: New Technologies Allow Radical Energy Savings in Seagoing Container Refrigeration Systems. Proceedings of the 20th International Congress of Refrigeration, Sydney, Australia, 19-24 September 1999. D.J. Cleland and C. Dixon (eds.). International Institute of Refrigeration (IIR/IIF), Paris, France.

CTI (Cryo-Trans, Inc.), 2004: Cryo-Trans, Inc. webpage: www.cryo-trans.com/history2.htm.

Cube, H.L. von, F. Steimle, H. Lotz and J. Kunis (eds.), 1997: *Lehrbuch der Kältetechnik – Vol. 2*, C.F. Müller Verlag, Heidelberg, Germany, 4th Edition, 850 pp.

D&T (Deloitte & Touche Consulting Group), 1996: Assessment of the Prospects for Hydrocarbon Technology in the Global Domestic Refrigeration Market, London, United Kingdom.

DeVore, T.A., 1998: Development of an Open Drive Scroll Compressor for Transport Refrigeration. Proceedings of the Purdue Compressor Conference, 1998. Purdue Printing Services, West Lafayette, IN, USA, pp. 231-236.

Dupont, 2004: Dupont website: www.dupont.com/suva/na/usa/about/success/transport1.html

EC, 2003: European Energy and Transport Trends to 2030. European Commission, Directorate-General for Energy and Transport, brussels, Belgium. http://europa.eu.int/comm/dgs/energy_transport/figures/trends_2030/index_en.htm

ECCJ (Energy Conservation Center Japan), 2004: Website of the Energy Conservation Center Japan, Tokyo, Japan. http://www.eccj.or.jp/index_e.html, http://www.eccj.or.jp/summary/local0303/eng/03-01.html, http://www.eccj.or.jp/databook/2002-2003e/03_04.html, http://www.eccj.or.jp/databook/2002-2003e/03_05.html

EIA, 2004: Annual Energy Outlook 2004 with Projections to 2025

– Market Trends – Energy Demand. Energy Information Administration, US Department of Energy, Washington, DC, USA http://www.eia.doe.gov/oiaf/aeo/

Elefsen, F., J. Nyvad, A. Gerrard and R. van Gerwen, 2002: Field Test of 75 R404A and R290 Ice Cream Freezers in Australia. 5th IIR Gustav Lorentzen Conference – Natural Working Fluids 2002, September 17-20, Guangzhou, China. International Institute of Refrigeration (IIR/IIF), Paris, France.

Egolf, P. and M. Kauffeld, 2005: From physical properties to industrial ice slurry applications. *Int. J. of Refrigeration*, **20**(2005), pp. 4-12.

Enviros, 2003: Assessment of the Costs & Implication on Emissions of Potential Regulatory Frameworks for Reducing Emissions of HFCs, PFCs & SF_6. Report prepared for the European Commission (reference number EC002 5008), London, United Kingdom, pp. 39.

ERI (Energy Research Institute), 2003: China's Sustainable Energy Future – Scenarios of Energy and Carbon Emissions. Energy Research Institute of the National Devleopment and reform commission, People's Republic of China, with Lawrence Berkeley National Laboratory, USA [Sinton, J.E., J.I. lewis, M.D. levine, Z. Yuezhong (eds.)] http://china.lbl.gov/pubs/china_scenarios_summary_final.pdf

EU (European Union), 2004: Regulation of the European Parliament and of the Council on certain fluorinated greenhouse gases, Draft Regulation 2003/0189 (COD), 9 July 2004, Brussels, Belgium.

Euromonitor International Inc., 2001: Global Appliance Information System, http://www.euromonitor.com (August 2001)

FAO (Food and Agriculture Organization of the United Nations), 2002: The State of the World Fisheries and Aquaculture, Rome, Italy, pp. 23.

Fernando, W.P.D. B. Palm, E. Granryd and K. Andersson, 2003: Mini-Channel Aluminium Heat Exchangers with Small Inside Volumes. Proceedings of the 21st International Congress of Refrigeration, Washington, DC, USA, 17-22 August 2003. International Institute of Refrigeration (IIR/IIF), Paris, France.

Ferreira, C.A.I. and D. Zaytsev, 2002: Experimental Compression – Resorption Heat Pump for Industrial Applications. Proceedings of the Purdue Compressor Conference, 2002. Purdue Printing Services, West Lafayette, IN, USA.

Fischer, S.K., P.J. Hughes, P.D. Fairchild, C.L. Kusik, J.T. Dieckmann, E.M. McMahon and N. Hobday, 1991: Energy and Global Warming Impacts of CFC Alternative Technologies, U.S. Department of Energy and AFEAS, Arlington, Va., USA.

Fischer, S.K., J.J. Tomlinson and P.J. Hughes, 1994: Energy and Global Warming Impacts of Not in Kind and Next Generation CFC and HCFC Alternatives, U.S. Department of Energy and AFEAS, 1200 South Hayes Street, Arlington, Va., USA.

Frigoblock, 2004: Frigoblock GmbH webpage: http://www.frigoblock.de

Garrabrant, M.A., 2003: Proof-of-concept design and experimental validation of a waste heat driven absorption transport refrigerator. *ASHRAE Transactions*, **109**(2003), pp. 1-11.

Gebhardt, H., 2001: HFCKW-Ausstieg – Umrüstung einer Industriellen Tieftemperaturkälteanlage von R22 auf CO_2/NH_3-Kaskadensystem (HCFC Phase Out – Retrofit of Industrial Low Temperature Refrigeration System from HCFC-22 to CO_2/NH_3 Cascade System). Proceedings of the Deutsche Kälte-Klima-Tagung, Ulm, 22-23 November 2001, Deutscher Kälte- und Klimatechnischer Verein DKV, Stuttgart, Germany (in German).

Gerwen, R.J.M. van, M. Verwoerd, 1998: Dutch Regulations for reduction of Refrigerant Emissions: Experiences with a Unique Approach over the period 1993-1998. ASERCOM Symposium: Refrigeration/Air conditioning and Regulations for Environment Protection – A Ten Years Outlook for Europe, 7 October 1998, Nürnberg, Germany.

Girotto, S. and P. Nekså, 2002: Commercial Refrigeration Systems with CO_2 as Refrigerant, Theoretical Considerations and Experimental Results. Proceedings of the IIR conference New Technologies in Commercial Refrigeration, July 22-23, University of Illinois, USA. International Institute of Refrigeration (IIR/IIF), Paris, France.

Girotto, S., S. Minetto and P. Nekså, 2003: Commercial Refrigeration with CO_2 as Refrigerant, Experimental Results. Proceedings of the 21st International Congress of Refrigeration, Washington, DC, USA, 17-22 August 2003. International Institute of Refrigeration (IIR/IIF), Paris, France.

Girotto, S., S. Minetto and P. Nekså, 2004: Commercial Refrigeration System Using CO_2 as the Refrigerant. *Int. J. of Refrigeration*, **27**(7), 717-723.

Godwin, D., 2004: Analysis of Costs to Abate International Ozone Depleting Substance Substitute Emissions. US EPA Report 430-R-04-006, US Environmental Protection Agency, Washington, DC, USA.

GTZ, 2002: India Servicing Sector Project Proposal, submitted September 2002 to the Multilateral Fund Secretariat, GTZ.

Haaf, S. and B. Heinbokel, 2002: Alternative Kaltemittel fur Supermarkt-Kalteanlagen (Alternative Refrigerants for Supermarket Refrigeration Installations). Proceedings of the Deutsche Kälte-Klima-Tagung, Magdeburg, 21-22 November 2002, Deutscher Kälte- und Klimatechnischer Verein DKV, Stuttgart, Germany, pp 29-42 (in German).

Haaf, S. and B. Heinbokel, 2003: Supermarkte mit alternativen Kaltemitteln. *KI Luft und Kaltetechnik* **39**(11), 508-512 (in German).

Halm, N.P., 2000: Air-Cycle Technology used for Truck Air-Conditioning. Proceedings of 4th IIR Gustav Lorentzen Conference – Natural Working Fluids 2000, West Lafayette, USA, July 25-28, 2000. International Institute of Refrigeration (IIR/IIF), Paris, France.

Han, H.D and Q.R. Zheng, 1999: Research on the Replacement for R12 and R22 used in Marine Rrefrigerating Units for Coming 21st century. Proceedings of the 20th International Congress of Refrigeration, Sydney, Australia, 19-24 September 1999. D.J. Cleland and C. Dixon (eds.). International Institute of Refrigeration (IIR/IIF), Paris, France.

Han, H., and W. Gan, 2003: Energy Conservation of Ocean-going refrigerated container transportation. Proceedings of the 21st International Congress of Refrigeration, Washington, DC, USA, 17-22 August 2003. International Institute of Refrigeration (IIR/IIF), Paris, France.

Harnisch, J. and C. Hendriks, 2000: Economic Evaluation of Emission Reductions of HFCs, PFCs and SF_6 in Europe. Report prepared for the European Commission DG Environment, Ecofys, Cologne/Utrecht, Germany/Netherlands, 70 pp.

Harnisch, J., N. Höhne, M. Koch, S. Wartmann, W. Schwarz, W. Jenseit, U. Rheinberger, P. Fabian and A. Jordan, 2003: Risks and Benefits of Fluorinated Greenhouse Gases in Practices and Products under Consideration of Substance. Report prepared for the German Federal Environmental Protection Agency (Umweltbundesamt) by Ecofys GmbH (Köln/Nürnberg), Öko-Recherche GmbH (Frankfurt), Öko-Institut e.V. (Darmstadt/Berlin), TU München (München), Max-Planck-Institut für Biogeochemie (Jena), Berlin, 128 pp.

Hochhaus, K.H., 1998: Reefer container ships and reefer containers. Sea Transportation. *Fruit World International*, **3**(1998), 198-204.

Hochhaus, K.H., 2002: Reefer deliveries in the doldrums. Sea Transportation. *Fruit World International*, **3**(2002), 210-216.

Hochhaus, K.H., 2003: Australia service, change in refrigerated container system. Sea Transportation, *Fruit World International*, **1**(2003), 39-46.

Hoehne, M. and P.S. Hrnjak, 2004: Charge minimisation in hydrocarbon systems. Proceedings of 6th IIR Gustav Lorentzen Conference – Natural Working Fluids 2004, Glasgow, Scotland, August 29 – September 1, 2004. International Institute of Refrigeration (IIR/IIF), Paris, France.

Hoogen, B. van den, and H. van der Ree, 2002: The Dutch Approach to Reduce Emissions of Fluorinated Greenhouse Gases. Proceedings of the IIR Conference on "Zero Leakage - Minimum Charge. Efficient Systems for Refrigeration, Air Conditioning and Heat Pumps", Stockholm, Seden, August 26-28, 2002. International Institute of Refrigeration (IIR/IIF), Paris, France.

Hundy, G.F.: Application of Scroll Compressors for Supermarket Refrigeration. Proceedings of the IMechE Seminar "Design, Selection and Operation of Refrigerator and Heat Pump Compressors", London, November 1998.

Hwang, Y., D.H. Jin and R. Radermacher, 2004: Comparison of Hydrocarbon R-290 and Two HFC Blends R-404a and R-410a for Medium Temperature Refrigeration Applications. Final Interim Report to ARI, Global Refrigerant Environmental Evaluation Network (GREEN) Program, CEEE Department of Engineering, University of Maryland, USA.

IEA, 2003: *Cool Appliances, Policy Strategies for Energy Efficient Homes.* International Energy Agency (IEA), Paris, France.

IIR (International Institute of Refrigeration), 1996: The Role of Refrigeration in Worldwide Nutrition. Informatory Note on Refrigeration and Food (November 1996). International Institute of Refrigeration (IIR/IIF), Paris, France.

IIR, 2002: Industry as a Partner for Sustainable Development – Refrigeration. International Institute of Refrigeration (IIR/IIF), Paris, France, 84 pp.

IIR, 2003: 16th Informatory Note on Refrigerating Technologies, Refrigerated Transport: Progress Achieved and Challenges to Be Met. International Institute of Refrigeration (IIR/IIF), Paris, France.

IPCC (Intergovernmental Panel on Climate Change), 1996a: *Climate Change 1995: The Science of Climate Change. Contribution of Working Group I to the Second Assessment Report of the Intergovernmental Panel on Climate Change* [Houghton, J.T., L.G. Meira Filho, B.A. Callander, N. Harris, A. Kattenberg and K. Maskell (eds.)]. Cambridge University Press, Cambridge, United Kingdom, and New York, NY, USA, 572 pp.

IPCC, 2001: *Climate Change 2001: The Scientific Basis. Contribution of Working Group I to the Third Assessment Report of the Intergovernmental Panel on Climate Change* [Houghton, J. T., Y. Ding, D. J. Griggs, M. Noguer, P. J. van der Linden, X. Dai, K. Maskell, and C. A. Johnson (eds.)]. Cambridge University Press, Cambridge, United Kingdom, and New York, NY, USA, 944 pp.

ISO (International Organization for Standardization), 1993: ISO 5149:1993 Mechanical Refrigerating Systems Used for Cooling and Heating – Safety Requirements. International Organization for Standardization, Geneva, Switzerland.

Jakobsen, A., 1998: Improving Efficiency of Trans-critical CO_2 Refrigeration Systems for Reefers. Proceedings of the IIR Conference, Commission D2/3, with D1, Cambridge, UK. International Institute of Refrigeration (IIR/IIF), Paris, France, pp. 130-138.

Kagawa, N., 2000: *Regenerative Thermal Machines (Stirling and Vuilleumier Cycle Machines) for Heating and Cooling.* International Institute of Refrigeration (IIR/IIF), Paris, France, ISBN 2-913149-05-7, 214 pp.

Kauffeld, M. and Christensen, K.G., 1998: A New Energy-efficient Reefer Container Concept Using Carbon Dioxide as Refrigerant. Proceedings of the IIR Gustav Lorentzen Conference – Natural Working Fluids, 2-5 June 1998, Oslo, Norway.

König, H., 2002: *Overview of applications with CO_2 as heat transfer fluid and as refrigerant,* Güntner Symposium Mai 2002, *Die Kälte- und Klimatechnik,* 7/2002 (in German)

Lang, D., 1999: Customer Demands Unfreeze Potential of Refrigerated Trailers. Transport Topics News from website: www.ttnews.com/members/printEdition/0002015.html.

Lindborg, A., 2000: Rätt utförda Indirekta Kylsystem förbrukar mindre Energi än Direkta, *Scandinavian Refrigeration,* **2**(2000), pp. 6-8 (in Swedish)

Lindborg, A., 2002: Mangelndes Wissen ist das grösste Hindernis für eine verbreiterte Verwendung von Ammoniak (Lack of knowledge is the main barrier for further use of ammonia). *Die Kälte- und Klimatechnik,* 1/2002, 34-39 (in German).

Lindborg, A., 2003: Sicherheit von NH_3-Anlagen, Schadensanalyse (Security of NH_3 systems – risk analysis). Proceedings of the Annual Conference of the Deutsche Kälte und Klimatechnischer Verein e.V. (DKV), Bonn, November 2003.

Lotz, H., 1993: Ermittlung des TEWI-Beitrags am Beispiel von Haushaltskuhlgeraten werksmontierter sowie feldmontierter Kalteanlagen. Proceedings of the Deutsche Kälte-Klima-Tagung, Nurnberg, 18-19 November 1993, Deutscher Kälte- und Klimatechnischer Verein DKV, Stuttgart, Germany (in German).

Lundqvist, P., 1993: *Stirling Cycle Heat Pumps and Refrigerators.* Diss. TRITA Refr. Report No 93/9, Stockholm, Sweden (ISSN 1102-0245).

Lundqvist, P., 2000: Recent refrigeration equipment trends in supermarkets: energy efficiency as leading edge. *Bulletin of the International Institute of Refrigeration*, 5/2000, 2-29.

Maidment, G.G. and R.M. Tozer, 2002: Combined cooling heat and power in supermarkets, *J. of Applied Thermal Engineeering*, **22**(2002), 653-665.

Mao, Y., W.J. Terrell and P.S. Hrnjak, 1998: Heat Exchanger Frosting Patterns with Evaporating R404A and Single Phase Secondary Refrigerants. Proceedings of the Purdue Compressor Conference, 1998. Purdue Printing Services, West Lafayette, IN, USA, pp. 277-282.

Mao, Y. and P.S. Hrnjak, 1999: Defrost Issues of Display Cabinets with DX Evaporators and Heat Exchangers with Secondary Refrigerants. Proceedings of the 20th International Congress of Refrigeration, Sydney, Australia, 19-24 September 1999. D.J. Cleland and C. Dixon (eds.). International Institute of Refrigeration (IIR/IIF), Paris, France.

March, 1996: UK Use and Emissions of Selected Hydrofluorocarbons – A Study for the Department of the Environment carried out by March Consulting Group. HMSO, London, United Kingdom.

March, 1998: Opportunities to Minimise Emissions of Hydrofluorocarbons from the European Union. March Consulting Group, London, United Kingdom.

March, 1999: UK Emissions of HFCs, PFCs and SF_6 and Potential Emission Reduction Options. March Consulting Group, London, United Kingdom

Meurer, C. and H. König, 1999: Effects of Accelerated HCFC Phase-Out Scenarios on Options to Replace R-22 in Heat Pumps. Proceedings of the 6th International Energy Agency Heat Pump Conference 1999: Heat Pumps – A Benefit fot the Environment, Berlin, Germany, 31 May – 2 June 1999. IEA Heat Pump Centre, Sittard, The Netherlands.

McDonalds, 2004: McDonald's Alternative Refrigeration Backgrounder, http://www.refrigerantsnaturally.com/doc/McDonalds%20backgrounder.pdf

Meffert, H., and A. Ferreira, 1999: Energy Efficiency of Transport Refrigeration Units. Proceedings of the 20th International Congress of Refrigeration, Sydney, Australia, 19-24 September 1999. D.J. Cleland and C. Dixon (eds.). International Institute of Refrigeration (IIR/IIF), Paris, France.

Meffert, H., and A. Ferreira, 2003: Part-Load Control Systems, Refrigerant Selection and Sustainability. Proceedings of the 21st International Congress of Refrigeration, Washington, DC, USA, 17-22 August 2003. International Institute of Refrigeration (IIR/IIF), Paris, France.

Meier, A., and J.E. Hill, 1997: Energy test procedures for appliances. *Energy and Buildings* 26(1), 22-33.

Meier, A., 1998: Energy Test Procedures for the Twenty-First Century. Proceedings of the 1998 Appliance Manufacturer Conference & Expo, 13-14 October 1998, Nashville, TN, USA.

Melinder, Å., 1997: *Thermophysical Properties of Liquid Secondary Refrigerants*. International Institute of Refrigeration (IIR/IIF), Paris, France.

Naturvardsverket, 1996: Års rapporteringen for användningen av köldmedier CFC/HCFC (1995) – Sammanstallning med kommentarer (1995 Annual report regarding the use of the refrigerants CFC/HCFC – Summary and comments). Naturvardsverket (Swedish Environmental Protection Agency), Stockholm, Sweden.

Nekså, P., S. Girotto and P.A. Schiefloe, 1998: Commercial Refrigeration Using CO_2 as Refrigerant – System Design and Experimental Results. Proceedings of the IIR Gustav Lorentzen Conference – Natural Working Fluids, 2-5 June 1998, Oslo, Norway.

Nielson, P. S. and T. Lund, 2003: Introducing a New Ammonia/CO_2 Cascade Concept for Large Fishing Vessels. Proceedings of the International Institute of Ammonia Refrigeration (IIAR), 2003, Technical Paper #11.

Official Journal, 2000: Regulation (EC) No 2037/2000 of the European Parliament and of the Council of 29 June 2000 on substances that deplete the ozone layer. *Official Journal of the European Communities*, OJ L244, 29 September 2000.

Palandre, L., A. Zoughaib, D. Clodic and L. Kuijpers, 2003: Estimation of the World-wide Fleets of Refrigerating and Air-conditioning Equipment in Order to Determine Forecasts of Refrigerant Emissions. Proceedings of the 14th Annual Earth Technologies Forum, April 22-24, 2003, Washington, D.C., USA.

Palandre, L., D. Clodic, and L. Kuijpers, 2004: HCFCs and HFCs emissions from the refrigerating systems for the period 2004-2015. Proceedings of the 15th Annual Earth Technologies Forum, April 13-15, 2004, Washington, D.C., USA.

Paul, J., 1999: Novel Transport Cooling System Utilizing Liquid, Pumpable Ice. Proceedings of the 20th International Congress of Refrigeration, Sydney, Australia, 19-24 September 1999. D.J. Cleland and C. Dixon (eds.). International Institute of Refrigeration (IIR/IIF), Paris, France.

Pearson, A, 2004a: Carbon Dioxide in the Spotlight. c-dig meeting September 2003, Glasgow, Scotland. Carbon Dioxide Interest Group (c-dig), United Kingdom (http://www.c-dig.org/).

Pearson, A., 2004b: Case Studies with CO_2. c-dig meeting 18-19 March 2004, Dresden, Germany. Carbon Dioxide Interest Group (c-dig), United Kingdom (http://www.c-dig.org/).

Pedersen, P.H., 2003: Evaluation of the Possibilities of Substituting Potent Greenhouse Gases (HFCs, PFCs and SF_6). Environmental Project No. 771, Danish Environmental Protection Agency, Copenhagen, Denmark.

Pirard, M., 2002: CO_2/NH_3 Refrigeration in Food Processing. Güntner Symposium Mai 2002, *Die Kälte- und Klimatechnik*, 7/2002 (in German)

Poese, M., 2004: Thermoacoustic Refrigeration for Ice Cream Sales. Proceedings of 6th IIR Gustav Lorentzen Conference – Natural Working Fluids 2004, Glasgow, Scotland, August 29 – September 1, 2004. International Institute of Refrigeration (IIR/IIF), Paris, France.

Powell, L., P. Blacklock, C. Smith and D. Colbourne, 2000: The Use of Hydrocarbon Refrigerants in Relation to Draft UK Government Policy on Climate Change. Proceedings of 4th IIR Gustav Lorentzen Conference – Natural Working Fluids 2000, West Lafayette, USA, July 25-28, 2000. International Institute of Refrigeration (IIR/IIF), Paris, France.

Pratihar, A.K., S.C. Kausik and R.S. Agarwal, 2001: Thermodynamic

Modelling and Feasibility Analysis of Compression-Absorption Refrigeration System. Proceedings of the International Conference on Emerging Technologies in Air-conditioning and Refrigeration, 26-28 September 2001, New Delhi, India, pp. 207-215.

Presotto, A. and C.G. Süffert, 2001: Ammonia Refrigeration In Supermarkets. *ASHRAE Journal*, **43**(10), 25-30.

Pridasawas, W. and P. Lundqvist, 2003: Feasibility and Efficiency of Solar-driven Refrigeration Systems. Proceedings of the 21st International Congress of Refrigeration, Washington, DC, USA, 17-22 August 2003. International Institute of Refrigeration (IIR/IIF), Paris, France.

Radford, P., 1998: The Benefits of Refrigerant Inventory Control, Proceedings of the Institute of Refrigeration Conference: R22 Phase Out – Impact on End Users Today. Institute of Refrigeration, London, United Kingdom.

RefNat, 2004: Proceedings of the Conference: Refrigerants, Naturally, 22 June 2004, Brussels, Belgium (available at http://www.refrigerantsnaturally.com).

Refrigeratedtrans.com, 2004: Refrigerated Transporter website: http://refrigeratedtrans.com/mag/transportation_bnsf_buys_first/

Rolfsman, L., 1999: Plant Design Considerations for Cascade Systems Using CO_2. Proceedings of the 20th International Congress of Refrigeration, Sydney, Australia, 19-24 September 1999. D.J. Cleland and C. Dixon (eds.). International Institute of Refrigeration (IIR/IIF), Paris, France.

Roth, R. and H. König, 2001: Experimental Investigation and Experiences with CO_2 as Refrigerant in Cascade System. Proceedings of the Deutsche Kälte-Klima-Tagung, Ulm, 22-23 November 2001, Deutscher Kälte- und Klimatechnischer Verein DKV, Stuttgart, Germany (in German).

Roth, R. and H. König, 2002: Wirtschaftlichkeitsanalyse für Industrie-Kälteanlagen mit CO_2 als Tieftemperaturkältemittel, *KI Klima, Kälte, Heizung*, 3/2002. C.F. Müller, Karlsruhe, Germany (in German).

Sand, J.R., S.K. Fischer and V.D. Baxter, 1997: Energy and Global Warming Impacts of HFC Refrigerants and Emerging Technologies, U.S. Department of Energy and AFEAS, Arlington, Va, USA.

SCANVAC, (Nordic Council), 2001: Alternative to HCFC as refrigerant in shipping vessels. *SCANVAC*, 1/2001, 4-5.

Siegel, A. and A. Metzger, 2003: Modernstes Fischverarbeitungszentrum Europas in Sassnitz, Part I and II. *Die Kälte- und Klimatechnik*, Nov./Dec. 2003, (in German).

Sinclair, J., 1999: *Refrigerated Transportation*. Witherby Publishers, London, United Kingdom, 151 pp.

Sonnekalb, M., 2000: Einsatz von Kohlendioxid als Kältemittel in Busklimaanlagen und Transportkälteanlagen, Messung und Simulation. Forschungsberichte des Deutsche Kälte-und Klimatechnischen Vereins (DKV), No 67, 331 pp.

Stellar (The Stellar group), 2003: Stellar Report Summer 2003, Food Processing Plant at Jonesboro, Arkansas with CO_2 as refrigerant, Company Report, www.thestellargroup.com, 2003

Stera, A.C., 1999: Long Distance Refrigerated Transport into the Third Millennium. Proceedings of the 20th International Congress of Refrigeration, Sydney, Australia, 19-24 September 1999. D.J.

Cleland and C. Dixon (eds.). International Institute of Refrigeration (IIR/IIF), Paris, France.

STEK, 2001: Consumption of Refrigerants in the Netherlands, Report based on the National Investigation on Refrigerant Flows, Zero measurement for 1999. Stichting Erkenningsregeling voor de Uitoefening van het Koeltechnisch Installatiebedrijf (STEK), Baarn, the Netherlands.

Stoecker, W.F., 1998: *Industrial Refrigeration Handbook*, McGraw-Hill, New York, NY, USA, ISBN: 007061623X, 782 pp.

Swatkowski, L., 1996: US Experience in Phasing Out CFC's. Proceedings of the International Conference on Ozone Protection Technologies, October 21-23, 1996, Washington DC, USA.

UNEP (United Nations Environment Programme), 1994: Report of the Refrigeration, Air Conditioning and Heat Pumps Technical Options Committee (1995 Assessment). UNEP Ozone Secretariat (Secretariat to the Vienna Convention for the Protection of the Ozone Layer and the Montreal Protocol on Substances that Deplete the Ozone Layer), Nairobi, Kenya.

UNEP, 1998: 1998 Report of the Refrigeration, Air Conditioning and Heat Pumps Technical Options Committee – 1998 Assessment. [L. Kuijpers (ed.)]. UNEP Ozone Secretariat, Nairobi, Kenya.

UNEP, 2000: Report of the Thirtieth Meeting of the Executive Committee of the Multilateral Fund for the Implementation of the Montreal Protocol (UNEP/OzL.Pro/ExCom/30/41), 29-31 March 2000, UNEP, Executive Committee of the Multilateral Fund for the Implementation of the Montreal Protocol, Montreal, Canada.

UNEP, 2003: 2002 Report of the Refrigeration, Air Conditioning and Heat Pumps Technical Options Committee – 2002 Assessment. [L. Kuijpers (ed.)]. UNEP Ozone Secretariat, Nairobi, Kenya.

UN-ESCAP (United Nations Economic and Social Commission for Asia and the Pacific), 2002: *Guidebook on Promotion of Sustainable Energy Consumption: Consumer Organizations and Efficient Energy Use in the Residential Sector*. United Nations Economic and Social Commission for Asia and the Pacific, Bangkok, Thailand, 158 pp (available at http://www.unescap.org/esd/energy/publications/psec/, http://www.unescap.org/esd/energy/publications/psec/tables/guidelines-table-119.jpg

US DOE, 1995: Technical Support Document: Energy Efficiency Standards for Consumer Products: Refrigerators, Refrigrator-Freezers, & Freezers Including Draft Environmental Assessment Regulatory Impact Analysis. US Department of Energy, Report # DOE/EE-0064, Washington, DC, USA (available at http://www.papyrus.lb.gov/index.nsf8/0/a35eec2400d2e3c88256ced006dd732?OpenDocument&Highlight=2,011521)

US EPA, 2004: International ODS Substitute Emissions 1990-2020: Inventories, Projections, and Opportunities for Reductions. Office of Air and Radiation, US Environmental Protection Agency, Washington, DC, USA.

Valentin, B., 1999: Limitation des emissions dans les transports frigorifiques (Emission reduction in transport refrigeration). *Revue Général du Froid*, **992**, 42-45 (in French).

Viegas, H., 2003: Liquid Carbon Dioxide Transport Refrigeration System. Proceedings of the 21st International Congress of Refrigeration, Washington, DC, USA, 17-22 August 2003. International Institute of Refrigeration (IIR/IIF), Paris, France.

Wang, S., J. Wu, R. Wang and Y. Xu, 2003: Adsorption Ice-maker Driven by Exhaust Gas Medium and Small Sized Fishing Vessels. Proceedings of the 21st International Congress of Refrigeration, Washington, DC, USA, 17-22 August 2003. International Institute of Refrigeration (IIR/IIF), Paris, France.

Wenning, U., 1996: Three Years Experience with Hydrocarbon Technology in Domestic Refrigeration. Proceedings of the International Conference on Ozone Protection Technologies, October 21-23, 1996, Washington DC, USA.

Weston, 1997: Recycling Rate Determinant Study – Phase I Report. Roy W. Weston, Inc, Norcross, Georgia, USA.

Wild, Y., 2003: Der Einsatz von Kühlcontainern im Seetransport. Proceedings of the Annual Conference of the Deutsche Kälte und Klimatechnischer Verein e.V. (DKV), Bonn, November 2003 (in German).

World Bank, 1993: The Status of Hydrocarbon and Other Flammable Alternatives use in Domestic Refrigeration, Ozone Operations Resource Group, OORG Report No. 5.

Yang, Y., 2001: Investigation of Deep-freeze Refrigeration Systems in Supermarket Application. Diss. Royal Institute of Technology, Stockholm, Sweden.

Yoshida, Y., K. Yoshimura, R. Kato and S. Hiodoshi, 2003: Development of Scroll Compressors with the High Performance Used for Marine Container Refrigeration Unit. Proceedings of the 21st International Congress of Refrigeration, Washington, DC, USA, 17-22 August 2003. International Institute of Refrigeration (IIR/IIF), Paris, France.

5

Residential and Commercial Air Conditioning and Heating

Coordinating Lead Authors
Roberto de Aguiar Peixoto (Brazil)

Lead Authors
Dariusz Butrymowicz (Poland), James Crawford (USA), David Godwin (USA), Kenneth Hickman (USA), Fred Keller (USA), Haruo Onishi (Japan)

Review Editors
Makoto Kaibara (Japan), Ari D. Pasek (Indonesia)

Contents

EXECUTIVE SUMMARY

The various applications, equipment and products included in residential and commercial air-conditioning and heating sector can be classified in three groups: stationary air conditioners (including both equipment that cools air and heat pumps that directly heat air), chillers and water-heating heat pumps.

Stationary Air Conditioners (Heat Pumps for Cooling and Heating)

Air conditioners and air-heating heat pumps generally fall into four distinct categories:
*   window-mounted, portable, and through-the-wall;
*   non-ducted split residential and commercial;
*   ducted residential split and single packaged;
*   ducted commercial split and packaged.

The vast majority of stationary air conditioners (and air-heating heat pumps) use vapour-compression cycle technology with HCFC-22 refrigerant. This refrigerant is already being phased out in some countries ahead of the schedule dictated by the Montreal Protocol. In Europe HCFC-22 had been phased out of new equipment by 31 December 2003. In the USA, production of HCFC-22 for use in new equipment will end on 1 January 2010. In Japan, HCFC-22 is to be phased out of new equipment on 1 January 2010; however, almost all new equipment has already been converted to HFCs.

The refrigerant options being considered as replacements for HCFC-22 are the same for all of the stationary air conditioner categories: HFC-134a, HFC blends, hydrocarbons, and CO_2. At present, two of these are being used: HFC blends in the vast majority of systems and hydrocarbons in a very small number of smaller systems.

It is estimated that more than 90% of the installed base of stationary air conditioners currently use HCFC-22, and an estimated 368 million air-cooled air conditioners and heat pumps are installed worldwide. This represents an installed bank of approximately 548,000 tonnes of HCFC-22 (UNEP, 2003).

Water Chillers

Water chillers combined with air handling and distribution systems frequently provide comfort air conditioning in large commercial buildings (e.g., hotels, offices, hospitals and universities) and to a lesser extent in large multi-family residential buildings. Water chillers using the vapour-compression cycle are manufactured in capacities ranging from approximately 7 kW to over 30,000 kW. Two generic types of compressors are used: positive displacement and centrifugal. Heat-activated absorption chillers are available as alternatives to electrical vapour-compression chillers. However, in general these are only used where waste heat is available or the price of electricity, including demand charges, is high.

HFCs (particularly HFC-134a) and HFC blends (particularly R-407C and R-410A) are beginning to replace HCFC-22 in new positive-displacement chillers. Ammonia is used in some positive-displacement chillers in Europe. The high discharge temperatures associated with ammonia permit a greater use of heat recovery than is the case for other refrigerants. Some chillers which use hydrocarbon refrigerants (as substitute for HCFC-22), are also produced in Europe each year.

Centrifugal compressors are generally the most efficient technology in units exceeding 1700 kW capacity. HCFC-123 and HFC-134a have replaced CFC-11 and CFC-12, respectively, in new centrifugal chillers produced since 1993.

Water-Heating Heat Pumps

Water-heating heat pumps using vapour-compression technology are manufactured in sizes ranging from 1 kW heating capacity for single room units, to 50–1000 kW for commercial/institutional applications, and tens of MW for district heating plants.

Various heat sources exist: air, water from ponds and rivers, and the ground. Integrated heat pumps that simultaneously heat water and cool air are also available.

In developed countries, HCFC-22 is still the most commonly used refrigerant but HFC alternatives are being introduced. In developing countries, CFC-12 is also used to a limited extent. HFC refrigerants are used in Europe in equipment produced after 2003 (EU, 2000).

In the area of non-HFC refrigerants, carbon dioxide is being introduced in domestic, hot-water heat pumps in Japan and Norway, ammonia is being used in medium-size and large-capacity heat pumps in some European countries, and several northern-European manufacturers are using propane (HC-290) or propylene (HC-1270) as refrigerants in small residential and commercial water-to-water and air-to-water heat pumps.

Reduction in HFC emissions

Options for reducing HFC emissions in residential and commercial air-conditioning and heating equipment involve containment in HFC vapour-compression systems (applicable worldwide and for all equipment) and the use of non-HFC systems (applicable in certain cases but not all due to economic, safety and energy efficiency considerations). Non-HFC systems include vapour-compression cycles with refrigerants other than HFCs, and alternative cycles and methods to produce cooling and heating.

Containment can be achieved through:
*   the improved design, installation and maintenance of systems to reduce leakage;
*   designs that minimize refrigerant charge quantities in systems
*   the recovery, recycling and reclaiming of refrigerant during servicing, and at equipment disposal.

A trained labour force using special equipment is needed to minimize installation, service and disposal emissions. However, implementing best practices for the responsible use of HFCs requires an infrastructure of education, institutions and equipment that is not widely available in much of the developing world. There is also a role for standards, guidelines, and regulations on HFC emission reduction that are appropriate for regional or local conditions.

A number of other non-traditional technologies have been examined for their potential to reduce the consumption and emission of HFCs. With only a few exceptions, these all suffer such large efficiency penalties that the resultant indirect effects would overwhelm any direct emission reduction benefit.

Global warming effects
Several factors influence the direct and indirect emission of greenhouse gases associated with residential and commercial air-conditioning and heating equipment. In those warm climate regions where electricity is predominantly generated using fossil fuels, the generation of energy to power air conditioners can cause greenhouse-gas emissions that are greater than the direct refrigerant emissions by an order of magnitude or more. Therefore, improving the integrity of the building envelope (re-duced heat gain or loss) and other actions to reduce building energy consumption can have a very significant impact on indirect emissions. In cooler climates where air conditioning is used less often, or in locations where power generation emits little or no carbon dioxide, the direct emissions can exceed the indirect greenhouse-gas emissions.

Residential and commercial air-conditioning and heating units are designed to use a given charge of a refrigerant, and not to emit that refrigerant to the atmosphere; however, emissions can occur due to numerous causes. The effects of refrigerant gas emissions are quantified by multiplying the emissions of a refrigerant in kg by its global warming potential (GWP). The emissions calculated are on a $kgCO_2$-equivalent basis. If more than a few specific systems are analyzed then it is appropriate to use average annual emission rates for each type of system to calculate the comparative direct greenhouse-gas emissions.

5.1 Stationary air conditioners (heat pumps for cooling and heating)

The several applications, equipment and products that are included in the sector of residential and commercial air conditioning and heating can be classified in three groups: stationary air conditioners (this section), chillers (section 5.2), and water heating heat pumps (section 5.3).

Air-cooled air conditioners and heat pumps, ranging in size from 2.0–700 kW, account for the vast majority of the residential and light-commercial air-conditioning market. In fact, over 90% of the air-conditioning units produced in the world are smaller than 15 kW. In the rest of this chapter the term *air conditioners* will be used for air conditioners and heat pumps that directly cool or heat air.

5.1.1 *Technologies and applications*

The vast majority of air conditioners use the vapour-compression cycle technology, and generally fall into four distinct categories:
- window-mounted, portable and through-the-wall air conditioners;
- non-ducted or duct-free split residential and commercial air conditioners;
- ducted residential split and single package air conditioners;
- ducted commercial split and packaged air conditioners.

5.1.1.1 *Window-mounted, through-the-wall, and portable air conditioners*

Due to their small size and relatively low cost, window-mounted, through-the-wall, and portable air conditioners[1] are used in small shops and offices as well as private residences. They range in capacity from less than 2.0 kW to 10.5 kW. These types of air conditioners have factory-sealed refrigerant cycles that do not require field-installed connections between the indoor and outdoor sections. Therefore refrigerant leaks resulting from imperfect installation practices do not occur in these systems unless the unit is damaged during installation and service and a leak results. Representative refrigerant leakage rates are in the order of 2–2.5% of the factory charge per year (UNEP, 2003).

5.1.1.2 *Non-ducted (or duct-free) split air conditioners*

In many parts of the world, non-ducted split air conditioners are used for residential and light-commercial air-conditioning. Non-ducted split air conditioners include a compressor/heat exchanger unit installed outside the space to be cooled or heated.

The outdoor unit is connected via refrigerant piping to one ('single-split') or more ('multi-split') indoor units (fan coils) located inside the conditioned space. Capacities range from 2.2–28 kW for a single split, and from 4.5–135 kW for a multi-split. Representative leakage rates for single split are in the order of 4–5% of the nominal charge per year (UNEP, 2003). As multi-split air conditioners have more connections the probability of leaks is higher.

5.1.1.3 *Ducted split residential air conditioners*

Ducted split residential air conditioners have a duct system that supplies cooled or heated air to each room of a residence or individual zones within commercial or institutional buildings. A compressor/heat exchanger unit outside the conditioned space supplies refrigerant to a single indoor coil (heat exchanger) installed within the duct system or air handler. Capacities range from 5–17.5 kW. Representative leakage rates are in the order of 4–5% of the nominal charge per year (UNEP, 2003).

5.1.1.4 *Ducted, commercial, split and packaged air conditioners*

Ducted, commercial, split-system units must be matched with an indoor air handler and heat exchanger. Packaged units contain an integral blower and heat exchanger section that is connected to the air distribution system. The majority of ducted, commercial split and single package air conditioners are mounted on the roof of office, retail or restaurant buildings or on the ground adjacent to the building. The typical range of capacities for these products is 10-700 kW.

Representative leakage rates are in the order of 4–5% of the factory charge per year (UNEP, 2003).

5.1.2 *Refrigerant use and equipment population*

There are no global statistics on the percentage of air-cooled air conditioners that have been manufactured with non ozone depleting refrigerants. However, it is estimated that more than 90% of the installed base of stationary air conditioners currently uses HCFC-22 (UNEP, 2003).

Estimates of the installed base (number of units) and refrigerant inventory were made using a computer model which predicts the number of units and refrigerant in the installed population on the basis of production data and product longevity models (UNEP, 2003).

An estimated 358 million air-cooled air conditioners (cooling and heating) are installed worldwide with a total capacity of 2.2×10^9 kW cooling. Refrigerant charge quantities vary in relation to the capacity. Assuming an average charge of 0.25 kg per kW of capacity, those 358 million units represent an installed bank of approximately 550,000 tonnes of HCFC-22 (Table 5.1).

HCFC-22 is already being phased out in some countries, which elected to phase out ahead of the schedule dictated by the Montreal Protocol. In Europe HCFC-22 had been phased out of new equipment by 31 December 2003. In the USA HCFC-22

---

[1] Portable air conditioners are a special class of room air conditioners designed to be rolled from room to room. They draw condenser air from the conditioned space or from outdoors and exhaust it outdoors. The air flows from and to outdoors through small flexible ducts which typically go through a window. In some models condenser cooling is further augmented by the evaporation of condensate and water from a reservoir in the unit.

Table 5.1. Units manufactured in 1998 and 2001, unit population and refrigerant inventory.

Product Category	Units Manufactured 2001 (millions)	Units Manufactured 1998 (millions)	Estimated Unit Population (2001) (millions)	Estimated HCFC-22 Inventory (ktonnes)	Estimated Refrigerant Bank HFC (ktonnes)[1]
Window-mounted and Through-the-Wall (Packaged Terminal) Air Conditioners	13.6	12.1	131	84	4
Non-ducted or duct-free Split Residential and Commercial Air Conditioners	24.2	16.3	158	199	10
Ducted Split and single Packaged Residential Air conditioner	5.9	5.7	60	164	9
Ducted commercial split and packaged air conditioners	1.7	1.7	19	101	5
TOTAL	**45.4**	**35.8**	**368**	**548**	**28**

[1] These values were calculated assuming that HCFC-22 bank is 95% of the total, for each category

Source: ARI, 2002; JARN, 2002b; DRI, 2001

will be phased out of new equipment on 1 January 2010. In Japan HCFC-22 is due to be phased out of new equipment on 1 January 2010, but almost all new equipment has already been converted to HFCs.

The refrigerant options being considered as replacements for HCFC-22 are the same for all of the stationary air conditioner categories: HFC-134a, HFC blends, hydrocarbons, and CO_2. At present, two of these are being used: HFC blends, and hydrocarbons (propane, a propane/ethane blend, and propylene).

5.1.2.1   HFC blends

To date, the vast majority of air conditioners using non ozone depleting refrigerants have used HFC blends. Two HFC blends currently dominate the replacement of HCFC-22 in new air-cooled air conditioners. These are R-407C and R-410A. A few other HFC blends have been investigated and/or commercialized as refrigerants; however, none have been widely used in new or existing (retrofit) air conditioners. There is a limited use of R-419A and R-417A as 'drop-in' refrigerants in some CEIT countries.

R-407C

Systems that use R-407C can be designed to match the performance of HCFC-22 systems if appropriate adjustments are made, such as changing the size of the heat exchangers. This is demonstrated by the availability of R-407C systems in Europe and Japan at capacities and efficiencies equal to the HCFC-22 units which they replace. In Europe, R-407C has been predominantly used as the replacement for HCFC-22 in air-to-air air-conditioning applications. In Japan, R-407C has primarily been used in the larger capacity duct-free and multi-split products.

R-410A

R-410A is being used to replace HCFC-22 in new products in

some markets. R-410A air conditioners (up to 140 kW) are currently available on a commercial basis in the USA, Asia and Europe. A significant proportion of the duct-free products sold in Japan use R-410A. In 2002, approximately 5% of the equipment sold into the US ducted residential market used R-410A. It is likely that the US ducted residential market will mainly use R-410A as the HCFC-22 replacement.

5.1.2.2   Hydrocarbons and CO_2

The use of hydrocarbons in air-conditioning applications has been limited due to the safety concerns inherent in the application of flammable refrigerants.

Propane (HC-290) has mainly been used in portable (factory sealed) air conditioners. Approximately 90,000 HC-290 portable air conditioners are reported to have been sold in Europe in 2003. The typical charge quantity used in these units is approximately 0.10 kg kW^{-1}.

To date, CO_2 units have been essentially limited to custom built applications or demonstration units. A component supply base from which to manufacture CO_2 systems does not currently exist.

5.1.3   Options for reducing HFC emissions

Options for reducing HFC emissions include refrigerant conservation in HFC vapour-compression systems and the use of non-HFC systems. These options are discussed below.

5.1.3.1   HFC vapour-compression systems

Residential and commercial air-conditioning and heating units are designed to use a specified charge of a refrigerant, and not to emit that refrigerant to the atmosphere during normal operation. However, refrigerant emissions due to losses can occur as a result of several factors:

• Refrigerant leaks associated with poor design or manufacturing quality, such as leaks from valves, joints, piping and heat exchangers represent on average 2–5% of the factory refrigerant charge per year;

- Leaks in poorly installed field-interconnecting tubing, which can emit 5–100% of factory charge within the first year of installation;
- Accidental releases due to mechanical failure or damage of equipment components can result in up to 100% loss of the system charge;
- Intentional venting of refrigerant during servicing (e.g., air purging) or disposing of equipment (in many countries this practice is still legal). This type of emission can represent anywhere from a small percentage to the total system charge;
- Losses of refrigerant during equipment disposal (up to 100% of the system charge).

For air conditioners working on the vapour-compression cycle and using any refrigerant, there are several practical ways to promote refrigerant conservation, and to reduce refrigerant emissions. The most significant are:

- Improved design and installation of systems to reduce leakage and consequently increase refrigerant containment;
- Design to minimize refrigerant charge quantities in systems;
- Adoption of best practices for installation, maintenance and repairing of equipment, including leak detection and repair;
- Refrigerant recovery during servicing;
- Recycling and reclaiming of recovered refrigerant;
- Refrigerant recovery at equipment decommissioning;
- Appropriate government policies to motivate the use of good practices and to promote refrigerant conservation.

Standards and good practice guidelines, like ANSI/ASHRAE[2] Standard 147-2002, outline practices and procedures to reduce the inadvertent release of halogenated refrigerants from stationary refrigeration, air conditioning, and heat pump equipment during manufacture, installation, testing, operation, maintenance, repair, and disposal.

5.1.3.1.1 Developing country aspects

Developing countries face specific issues with respect to the containment and conservation of refrigerants. Since the manufacturing process is approaching a global standard, and most of the developing countries are importers and not manufacturers of air-conditioning equipment, the specific issues faced by these countries are mostly related to servicing, training of technicians, legislation and regulations. Important points, in addition to those mentioned above, are:

- Technician training and awareness are essential to the success of refrigerant conservation, especially where preventive maintenance procedures have not been routine in the past;
- Developing countries could devote resources to developing a reclamation infrastructure, with the necessary refrigerant recovery and reclaiming network, or emphasize on-site refrigerant recycling. The Multi-Lateral Fund of the Montreal Protocol supports this practice;
- In many developing countries, preventive maintenance of air-conditioning and refrigeration equipment has been rare. Conservation approaches, which rely heavily on regular maintenance, could be successfully implemented if countries were to provide incentives to encourage routine scheduled maintenance (UNEP, 2003).

5.1.3.2 Non-HFC systems

Non-HFC systems include vapour-compression cycles with refrigerants other than HFCs, and alternative cycles and methods to produce refrigeration and heating. The four stationary air conditioner categories described in Section 5.1.1 have the non-HFC system options described below.

5.1.3.2.1 Vapour-compression cycle with non-HFC refrigerants

Many factors need to be taken into consideration when designing an air-conditioning product with a new refrigerant, for example, environmental impact, safety, performance, reliability, and market acceptance. Non-HFC refrigerants that are currently being investigated and used are now detailed.

Hydrocarbon refrigerants

An extensive literature review on the performance of hydrocarbon refrigerants was performed in 2001 (ARTI, 2001). Many articles reported that refrigerants such as propane offer similar or slightly superior efficiency to HCFC-22 in air-conditioning systems. Few rigorous comparisons of fluorocarbon and hydrocarbon systems have been reported. However, the available data suggest that efficiency increases of about 2–5% were common in drop-in 'soft-optimized' system tests. In a system specifically optimized for hydrocarbons, it might be possible to achieve efficiency increases somewhat greater than 5% by using propane rather than HCFC-22, assuming no other fire safety measures need to be taken which would reduce efficiency. In certain countries safety regulations require the use of a secondary loop and this significantly reduces the efficiency of the hydrocarbon system and increases its cost compared to the HCFC-22 system. In order to offer equipment which meets the market requirements for the lowest cost, manufacturers will need to determine how the costs of safety improvements required for hydrocarbon systems compare with the costs required to raise the efficiency of competing systems. Safety standards are likely to vary around the world and this may lead to different choices. Vigorous debates among advocates of hydrocarbon and competing refrigerants are likely to continue, due to the differences in perceived

---

[2] ANSI is the American National Standards Institute, Inc. ASHRAE is the American Society of Heating, Refrigerating, and Air-Conditioning Engineers, Inc.

and acceptable risk in different countries. However under the appropriate conditions, for example limited charge and sealed circuits hydrocarbons can be used safely (ARTI, 2001).

Carbon dioxide
Carbon dioxide (CO_2) offers a number of desirable characteristics as a refrigerant: availability, low-toxicity, low direct GWP and low cost. CO_2 systems are also likely to be smaller than systems using other common refrigerants but will not necessarily be cheaper (Nekså, 2001). There is a significant amount of conflicting data concerning the efficiency of CO_2 in air- conditioning applications. Some of the data indicate very low efficiencies compared to HCFC-22 systems while other references indicate parity to better performance. Additional research and development will be needed to arrive at a definitive determination of the efficiency of CO_2 in comfort air-conditioning applications.

5.1.3.2.2 Alternative technologies to vapour-compression cycle

The absorption cycle offers a commercially-available alternative to the vapour-compression cycle. At least two Japanese manufacturers have had commercially-available, split-type absorption air conditioners available for about 5 years. One Italian manufacturer has also been selling small-scale absorption units for some commercial installations. It is reported that over 360,000 gas-fired absorption units with capacities below 7.5 kW have been produced in Europe and North America using the ammonia-water cycle (Robur, 2004). The performance of a direct-fired absorption system will generally result in a higher total-equivalent-warming-impact (TEWI) value than for a vapour-compression system, unless the regional electrical power generation has a high CO_2 emission factor.

A number of other non-traditional technologies have been examined for their potential to reduce consumption and emission of HFCs. These include desiccant cooling systems, Stirling cycle systems, thermoelectrics, thermoacoustics and magnetic refrigeration. With the exception of the Stirling cycle and desiccants, all of these alternatives suffer such large efficiency penalties that the consequent indirect effects would overwhelm any direct benefit in emission reduction. In the USA, the Stirling cycle has remained limited to niche applications, despite the research interest and very substantial funding by the US Department of Energy, and has never been commercialized for air conditioning. In high latent-load applications, desiccant systems have been used to supplement the performance of conventional mechanical air conditioning.

5.1.4 Global warming effects

Several factors influence the emission of greenhouse gases associated with residential and commercial air-conditioning and heating equipment. These include direct emissions during equipment life and at the end of life, refrigerant properties, system capacity (size), system efficiency, carbon intensity of the electrical energy source, and climate. Some of the sensitivities

are illustrated by the examples in Figures 5.1 to 5.9. In regions with cooler climates, where air conditioning is used less often or where the electricity generation energy source is not carbon-intensive, direct emissions can outweigh the indirect effects.

Including the life-cycle climate performance (LCCP) as a design criterion is one aspect that can minimize the GWP of residential and commercial air-conditioning and heating equipment. Factoring LCCP into the design methodology will result in an optimum design that is different from one just optimized for lowest cost. By optimizing for the best LCCP, the designer will also improve on a number of other parameters, for example, the design for energy efficiency, the type and amount of refrigerant used in the unit (determined by refrigerant cycle design), reduced leakage (service valve design, joining technologies, manufacturing screening methods, sensor technologies for the early detection of refrigerant leaks), and reduced installation and service losses (factory sealed refrigerant circuits, robust field connection technologies, service valves that reduce losses during routine service). The investment required to achieve a given reduction in LCCP will differ per factor.

5.1.4.1 LCCP examples for air conditioners

Several examples of LCCP calculation are now given for technologies typical of those described above (i.e. vapour compression cycle with HCFC and HFC refrigerants), as well as technologies that have been studied for their potential to reduce greenhouse-gas emissions from air-conditioning applications. As stated previously, the results obtained by these studies are dependent on the assumptions made (leakage rate, recovery rate, use of secondary loop, etc.). Changing these assumptions can lead to different results.

Figure 5.1 compares LCCP values for 3 tonne (10.5 kW) air-conditioning and heat pump units operating in Atlanta, Georgia, USA. LCCP values are calculated for three efficiency levels – seasonal energy efficiency ratio (SEER) levels of 10, 12, and 14 Btu Wh^{-1}. By 2010 when HCFC-22 has been phased out for new equipment and higher energy efficiency standards (13 SEER in the US) are in place, an HFC blend refrigerant is likely to represent a large part of the market for new equipment. The results generally show that direct warming impacts due to life-cycle refrigerant emissions are less than 5% of the LCCP. The difference in the indirect warming component of LCCP at different efficiency levels is much greater. Propane and CO_2 emissions have a negligible warming impact. However, the possible additional cost for using propane safely or for achieving a given efficiency level with CO_2, exceeds the difference in cost between the 12 and 14 SEER performance levels, which have a greater impact on LCCP than the direct warming from refrigerant emissions. (Figure 5.1 is based upon annual make-up losses of 2% of charge and an end-of-life loss of 15% of charge, electrical generation with emissions of 0.65 kg CO_2 kWh^{-1}, annual cooling load of 33.8 million Btu and heating load of 34.8 million Btu, and a 15-year equipment lifetime) (ADL, 2002).

Figure 5.2 provides a comparison of LCCP values for a small, commercial, rooftop air conditioner in Atlanta, Georgia

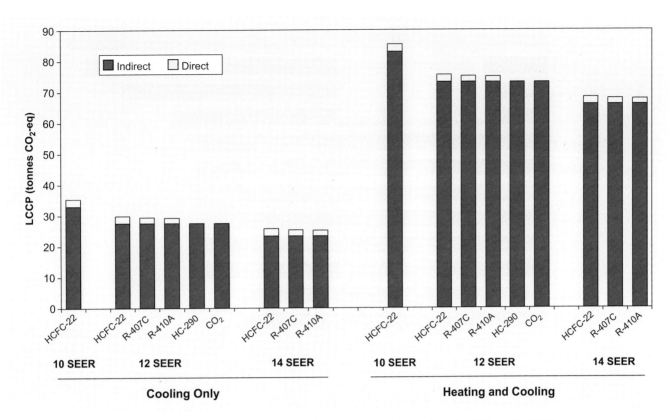

Figure 5.1. LCCP values for 3 tonne (10.5 kW) air conditioner units operating in Atlanta, Georgia, USA (ADL, 2002).

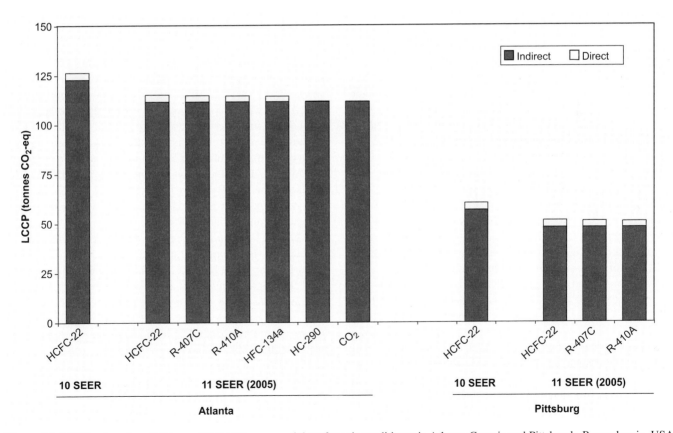

Figure 5.2. LCCP values for a 7.5 tonne (26.3 kW) commercial rooftop air conditioner in Atlanta, Georgia and Pittsburgh, Pennsylvania, USA (ADL, 2002; Sand *et al.*, 1997).

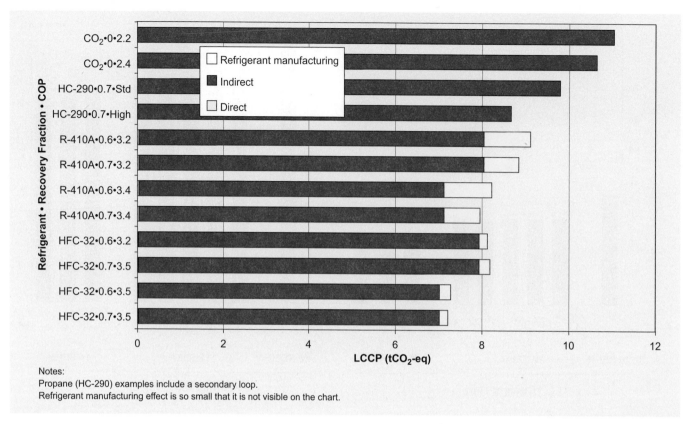

Figure 5.3. LCCP values for 4 kW mini-split air conditioner units in Japan with COPs varying from 2.2–3.5 and with end-of-life refrigerant recovery rates of 60% or 70% (Onishi *et al.*, 2004).

and Pittsburgh, Pennsylvania, USA. The results are similar to those shown in Figure 5.1. Differences in efficiency have a much greater effect on LCCP than the direct effect of refrigerant emissions. (Figure 5.2 is based upon annual make-up losses of 1% of charge and an end-of-life loss of 15% of charge, electrical generation with emissions of 0.65 kg CO_2 kWh^{-1}, and equivalent full load cooling hours of 1400 in Atlanta and 600 in Pittsburgh, and a 15-year equipment lifetime) (ADL, 2002).

Figure 5.3 presents LCCP values for 4 kW mini-split heat pump units in Japan. The chart compares units with 4 different refrigerants: CO_2, propane (HC-290), R-410A, and HFC-32. Other parameters varied in this chart are the assumptions about the amount of refrigerant recovered at the end of the equipment life (recovery is assumed to be consistent with normal practice in Japan; 60% and 70% are analyzed) and the coefficient of performance (COP) level of the equipment (standard models compared with high COP models – values shown on the chart). Equipment life is taken to be 12 years with no refrigerant charge added during life, and power generation emissions of 0.378 kg CO_2 kWh^{-1} are assumed. The units are assumed to run for 3.6 months for cooling and 5.5 months for heating according to Japanese Standard JRA4046-1999 (JRAIA, 1999). The figure shows that the LCCP for these mini-splits is dominated by the COP, which is why CO_2 has such a high LCCP. The source for the data assumed that a secondary heat transfer loop would be

required for propane, reducing COP and adding a 10–20% cost penalty. The two end-of-life refrigerant recovery rates examined have only a secondary effect on LCCP. This is only apparent for R-410A, which has the highest GWP of those compared. In Japan and many other countries, it is unclear whether HFC-32 (a flammable refrigerant) in mini-splits would be permitted for use in direct expansion or whether it would require a secondary loop (work to determine this is still underway including an IEC[3] standard). The LCCP penalty for a secondary loop in the HFC-32 system is not shown here. This penalty would make HFC-32 systems less attractive than R-410A.

Figure 5.4 shows LCCP values for 56 kW multi-split air conditioners for commercial applications in Japan. The refrigerants compared are propane, R-407C, and R-410A with two rates of refrigerant recovery at the end of the equipment life, 50% and 70%. Each system has a COP level shown on the chart, which has been obtained by using the variable compressor speed (inverter drive). For comparative purposes, a propane system without inverter has been added. The source for the data assumed that the propane system would have a secondary heat transfer loop. Multi-split air conditioners for commercial application units are assumed to operate 1941 h yr^{-1} for cooling

---

[3] IEC is the International Eletrotechnical Commission

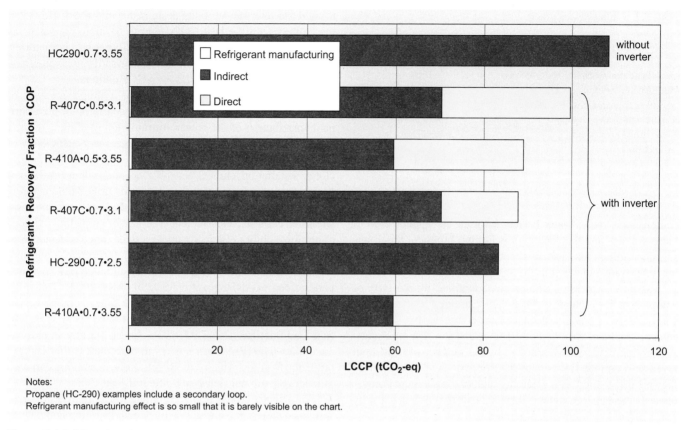

Notes:
Propane (HC-290) examples include a secondary loop.
Refrigerant manufacturing effect is so small that it is barely visible on the chart.

Figure 5.4. LCCP values for 56 kW multi-split air conditioners in Japan with COPs varying from 2.5–3.55 and end-of-life refrigerant recovery rates of 50% or 70% (Onishi *et al.*, 2004).

and 888 h yr^{-1} for heating (JRAIA, 2003). Equipment life is assumed to be 15 years with no additional charge required during the operating life. Power generation emissions are assumed to be 0.378 kg CO_2 kWh^{-1} (Onishi *et al.*, 2004). The figure shows that the combination of COP improvements obtained with inverter drive plus the lower emissions rate for power generation in Japan, mean that the indirect component of LCCP is less important than in the previous US cases. Also the higher the COP, the less important differences in COP are to the overall LCCP. This figure clearly shows the value of achieving a high recovery rate of refrigerant at the end of service life.

5.1.4.2   *Global refrigerant bank*

Table 5.1 estimates the refrigerant banks in 2001 as 548,000 tonnes of HCFC-22 and 28,000 tonnes of HFCs. Another source (Palandre *et al.*, 2004) estimates that in 2002, the stationary AC bank consisted of over 1,000,000 tonnes of HCFCs and nearly 81,000 tonnes of HFCs. Although stationary AC includes more than just air-cooled air conditioners and heat pumps, it is clear that this type of equipment constitutes a large part of the HCFC bank. Emission estimates for stationary air conditioners are given in Section 5.4.

5.2   Chillers

Comfort air conditioning in large commercial buildings (including hotels, offices, hospitals, universities) is often provided by water chillers connected to an air handling and distribution system. Chillers cool water or a water/antifreeze mixture which is then pumped through a heat exchanger in an air handler or fan-coil unit to cool and dehumidify the air.

5.2.1   *Technologies and applications*

Two types of water chillers are available, vapour-compression chillers and absorption chillers.

The principal components of a vapour-compression chiller are a compressor driven by an electric motor, a liquid cooler (evaporator), a condenser, a refrigerant, a refrigerant expansion device, and a control unit. The refrigerating circuit in water chillers is usually factory sealed and tested; the installer docs not need to connect refrigerant-containing parts on site. Therefore leaks during installation and use are minimal.

The energy source for absorption chillers is the heat provided by steam, hot water, or a fuel burner. In absorption chillers, two heat exchangers (a generator and an absorber) and a solution pump replace the compressor and motor of the vapour-

Table 5.2. Chiller capacity ranges.

Chiller Type	Capacity Range (kW)
Scroll and reciprocating water-cooled	7–1600
Screw water-cooled	140–2275
Positive displacement air-cooled	35–1500
Centrifugal water-cooled	350–30,000
Centrifugal air-cooled	630–1150
Absorption	Less than 90, and 140–17,500

Source: UNEP, 2003

compression cycle. Water is frequently the refrigerant used in these systems and the absorbent is lithium bromide. Small absorption chillers may use an alternative fluid pair: ammonia as the refrigerant and water as the absorbent.

Vapour-compression chillers are identified by the type of compressor they employ. These are classified as centrifugal compressors or positive displacement compressors. The latter category includes reciprocating, screw, and scroll compressors. Absorption chillers are identified by the number of heat input levels they employ (i.e., single-stage or two-stage), and whether they are direct-fired with a burning fuel, or use steam or hot water as the heat energy source. Table 5.2 lists the cooling capacity range offered by each type of chiller.

For many years, centrifugal chillers were the most common type of chillers above 700 kW capacity. Reciprocating compressors were used in smaller chillers. From the mid-1980s onwards, screw compressors became available as alternatives to reciprocating compressors in the 140–700 kW range and as alternatives to centrifugal compressors in the range up to about 2275 kW. Scroll compressors were introduced at about the same time and have been used as alternatives to reciprocating compressors in the 7 to over 90 kW range.

The Japan Air-Conditioning, Heating, and Refrigeration News (JARN, 2001) estimates that:

- The market for centrifugal and large screw chillers is divided between 40% in the USA and Canada, 25–30% in Asia, and smaller percentages in other regions in the world;
- The market for large absorption chillers is highly concentrated in Japan, China, and Korea with the USA and Europe as the remaining significant markets;
- The world market for smaller, positive displacement chillers (with hermetic reciprocating, scroll, and screw compressors) is much larger in absolute terms than for the other chiller types.

The coefficient of performance (COP) is one of the key criteria used to describe chillers. Other efficiency parameters are kW tonne[-1] (electrical power consumption in relation to cooling capacity) and energy efficiency ratio (EER) or Btu Wh[-1] (cooling capacity related to power consumption).

Each type of chiller and refrigerant combination has a best-in-class COP level that can be purchased. This COP level tends

to increase over the years as designs are improved. However, the chillers with the best COPs tend to be more expensive as they employ larger heat exchangers and other special features. In the absence of minimum efficiency standards, many purchasers choose to buy lower-cost, lower-COP chillers.

Full-load COP is commonly used as a simple measure of chiller efficiency. With the increasing recognition of the dominant contribution of the power consumption of chillers to their GWP, more attention has been paid to the energy efficiency of chillers at their more common operating conditions. In a single-chiller installation, chillers generally operate at their full-load or design point conditions less than 1% of the time. Manufacturers developed techniques such as variable-speed compressor drives, advanced controls, and efficient compressor unloading methods to optimize chiller efficiency under a wide range of conditions. In the US, ARI developed an additional performance measure for chillers called the Integrated Part Load Value (IPLV) which is described in ARI Standard 550/590 (ARI, 2003). The IPLV metric is based on weighting the COP at four operating conditions by the percentage of time assumed to be spent at each of four load fractions (25%, 50%, 75%, and 100%) by an individual chiller. The IPLV metric takes into account chiller energy-reducing features which are increasingly becoming common practice, but are not reflected in the full-load COP.

For a single chiller it is appropriate to use IPLV as the performance parameter, multiplied by actual operating hours when calculating the LCCP. For multiple chiller installations, which constitute about 80% of all installations, the calculation of LCCP includes full load COP and the IPLV based on the actual operating hours estimated for each load condition.

The ARI IPLV calculation details are based on single chiller installations and an average of 29 distinct US climate patterns. A modified version is being considered for Europe (Adnot, 2002).

Most installations have two or more chillers, so ARI recommends use of a comprehensive analysis that reflects the actual weather data, building load characteristics, number of chillers, operating hours, economizing capabilities, and energy for auxiliaries such as pumps and cooling towers to determine the overall chiller-plant system performance (ARI, 1998).

5.2.2 *Refrigerant use and equipment population*

Estimates and data about refrigerant use and equipment population, for the different types of chillers are presented below.

5.2.2.1 *Centrifugal chillers*

Centrifugal chillers are manufactured in the United States, Asia, and Europe. Prior to 1993, these chillers were offered with CFC-11, CFC-12, R-500, and HCFC-22 refrigerants. Of these, CFC-11 was the most common. With the implementation of the Montreal Protocol, production of chillers using CFCs or refrigerants containing CFCs (such as R-500) essentially ended in 1993. Centrifugal chillers using HCFC-22 rarely were produced after the late 1990s.

Table 5.3. Centrifugal chiller refrigerants.

Refrigerant	Capacity Range (kW)
CFC-11	350–3500
CFC-12	700–4700
R-500	3500–5000
HCFC-22	2500–30,000
HCFC-123	600–13,000
HFC-134a	350–14,000

The refrigerant alternatives for CFC-11 and CFC-12 or R-500 are HCFC-123 and HFC-134a, respectively. These refrigerants began to be used in centrifugal chillers in 1993 and continue to be used in 2004 in new production chillers.

Chillers employing HCFC-123 are available with maximum COPs of 7.45 (0.472 kW tonne^{-1}). With additional features such as variable-speed drives, HCFC-123 chillers can attain IPLV values of up to 11.7. Chillers employing HFC-134a are available with COPs of 6.79 (0.518 kW tonne^{-1}). With additional features such as variable-speed drives, HFC-134a chillers can attain IPLV values of up to 11.2.

Table 5.3 shows the range of cooling capacities offered for centrifugal chillers with several refrigerants. Table 5.4 shows the equipment population in a number of countries. This table provides estimates of the refrigerant bank in these chillers, assuming an average cooling capacity of 1400 kW in most cases and approximate values for the refrigerant charge for each refrigerant. The refrigerant charge for a given cooling capacity may vary with the efficiency level of the chiller. For any given refrigerant, higher efficiency levels often are associated with larger heat exchangers and, therefore, larger amounts of charge.

Production of a new refrigerant, HFC-245fa, as a foam-blowing agent commenced in 2003, and it has been considered as a candidate for use in new chiller designs. It has operating pressures higher than those for HCFC-123 but lower than for HFC-134a. Its use requires compressors to be redesigned to match its properties, a common requirement for this type of compressor. Unlike those for HCFC-123, heat exchangers for HFC-245fa must be designed to meet pressure vessel codes. Chillers employing HFC-245fa are not available yet. No chiller manufacturer has announced plans to use it at this time.

Centrifugal chillers are used in naval submarines and surface vessels. These chillers originally employed CFC-114 as the refrigerant in units with a capacity of 440–2800 kW. A number of CFC-114 chillers were converted to use HFC-236fa as a transitional refrigerant. New naval chillers use HFC-134a.

5.2.2.2 Positive displacement chillers

Chillers employing screw, scroll, and reciprocating compressors are manufactured in many countries around the world. Water-cooled chillers are generally associated with cooling towers for heat rejection from the system. Air-cooled chillers are equipped with refrigerant-to-air finned-tube condenser coils and fans to reject heat from the system. The selection of water-cooled as opposed to air-cooled chillers for a particular application varies

Table 5.4 Centrifugal chiller population and refrigerant inventory.

Country or Region	Refrigerant	Avg. Capacity (kW)	Avg. Charge Level (kg kW^{-1})	No. Units	Refrigerant Bank (tonnes)	Source of Unit Nos.
USA	CFC-11	1400	0.28	36,755	14,400	Dooley, 2001
USA	HCFC-123	1400	0.23	21,622	7000	Dooley, 2001
	HFC-134a	1400	0.36	21,622	10,900	with 50% split
Canada	CFC-11	1400	0.28	4212	1650	HRAI, 2003
Canada	HCFC-123	1400	0.23	637	205	HRAI, 2003
	HFC-134a	1400	0.36	637	320	with 50% split
Japan	CFC-11	1100	0.40	7000	3080	JARN, 2002c
	HCFC-123 and HFC-134a	1600	0.40	4500	2880	JRAIA, 2004
India	CFC-11	1450	0.28	1100	447	UNEP, 2004
China	CFC-11	65% of total are	0.28	3700	2540	UNEP, 2004
	CFC-12	1400–2450, rest	0.36	338	300	Digmanese, 2004
	HCFC-22	are 2800–3500:	0.36	550	485	
	HCFC-123	2450 avg.	0.23	3200	1800	
	HFC-134a		0.36	3250	2870	
Brazil	CFC-11	1350	0.28	420	160	UNEP, 2004
	CFC-12	1450	0.36	280	145	
17 Developing Countries	CFC-11		*Avg. unit charge*	11,700	4000	UNEP, 2004
	CFC-12		*of 364 kg*	280	145	

Source for charge levels: Sand *et al.*, 1997; for HFC-134a, ADL, 2002.

Table 5.5. Positive displacement chiller refrigerants and average charge levels.

Refrigerant and Chiller Type	Evaporator Type	kg kW^{-1}
HCFC-22 and HFC-134a screw and scroll chillers	DX	0.27
R-410A and R-407C scroll chillers	DX	0.27
HCFC-22 and HFC-134a screw chillers	flooded	0.35
HCFC-22 reciprocating chillers	DX	0.26
Ammonia (R-717) screw or reciprocating chillers[1]	DX	0.04–0.20
Ammonia (R-717) screw or reciprocating chillers[1]	flooded	0.20–0.25
Hydrocarbons	DX	0.14

Source: UNEP, 2003
[1] Charge levels for R-717 chillers tend to decrease with capacity and are lowest for plate-type heat exchangers rather than with tube-in-shell (UNEP, 1998)

with regional conditions and owner preferences.

When they were first produced in the mid-1980s, **screw chillers** generally employed HCFC-22 as the refrigerant. HFC-134a chillers have recently been introduced by a number of manufacturers and in some cases these have replaced their HCFC-22 products.

Screw chillers using a higher pressure refrigerant, R-410A, have recently been introduced. Screw chillers using ammonia as the refrigerant are available from some manufacturers and these are mainly found in northern-European countries. The numbers produced are small compared to chillers employing HCFC-22 or HFCs.

Air-cooled and water-cooled screw chillers below 700 kW often employ evaporators with refrigerant flowing inside the tubes and chilled water on the shell side. These are called direct-expansion (DX) evaporators. Chillers with capacities above 700 kW generally employ flooded evaporators with the refrigerant on the shell side. Flooded evaporators require higher charges than DX evaporators (see Table 5.5), but permit closer approach temperatures and higher efficiencies.

Scroll chillers are produced in both water-cooled and air-cooled versions using DX evaporators. Refrigerants offered include HCFC-22, HFC-134a, R-410A, and R-407C. For capacities below 150 kW, brazed-plate heat exchangers are often used for evaporators instead of the shell-and-tube heat exchangers employed in larger chillers. Brazed-plate heat exchangers reduce system volume and refrigerant charge.

Air-cooled chiller systems are generally less expensive than the equivalent-capacity water-cooled chiller systems that include a cooling tower and water pump. However, under many conditions water-cooled systems can be more efficient due to the lower condensing temperatures.

Reciprocating chillers are produced in both water-cooled and air-cooled versions using DX evaporators. Air-cooled versions have increased their market share in recent years. Prior to the advent of the Montreal Protocol, some of the smaller reciprocating chillers (under 100 kW) were offered with CFC-12 as the refrigerant. Most of the smaller chillers, and nearly all the larger chillers, employed HCFC-22 as the refrigerant. Since the Montreal Protocol, new reciprocating chillers have employed HCFC-22, R-407C, and to a small extent, HFC-134a and propane or propylene. Some water-cooled reciprocating chillers were manufactured with ammonia as the refrigerant but the number of these units is very small compared to the number of chillers employing fluorocarbon refrigerants. As with scroll chillers, the use of brazed-plate heat exchangers reduces the system volume and system charge.

Table 5.5 shows approximate charge levels for each type of positive displacement chiller with several refrigerants.

The refrigerant blend R-407C is being used as a transitional replacement for HCFC-22 in direct expansion (DX) systems because it has a similar cooling capacity and pressure levels. However, R-407C necessitates larger and more expensive heat exchangers to maintain its performance. For R-407C DX evaporators, some of this difficulty is offset in new equipment by taking advantage of the refrigerant's 'glide' characteristic ('glide' of about 5°C temperature variation during constant-pressure evaporation) in counter-flow heat exchange. The glide also can be accommodated in the conventional condensers of air-cooled chillers. In time, the higher-pressure blend, R-410A, is expected to replace the use of R-407C, particularly in smaller chillers (UNEP, 2003).

5.2.2.3 Absorption chillers

Absorption chillers are mainly manufactured in Japan, China, and South Korea. A few absorption chillers are manufactured in North America. Absorption chiller energy use can be compared to electrical chiller energy by using calculations based on primary energy. Absorption systems have higher primary energy requirements and higher initial costs than vapour-compression chillers. They can be cost-effective in applications where waste heat is available in the form of steam or hot water, where electricity is not readily available for summer cooling loads, or where high electricity cost structures (including demand charges) make gas-fired absorption a lower-cost alternative. In Japan, government policy encourages absorption systems so as to facilitative a more balanced gas import throughout the year and to reduce summer electrical loads.

Single-stage absorption applications are typically limited to sites that can use waste heat in the form of hot water or steam as the energy source. Such sites include cogeneration systems where waste engine heat or steam is available. Two-stage absorption chillers, driven by steam or hot water or directly fired by fossil fuels, were first produced in large numbers in Asia (primarily in Japan) for the regional market during the 1980s. Two-stage chillers were produced in North America shortly afterwards, often through licensing from the Asian manufacturers. Small single-stage gas-fired absorption chillers with capacities below 90 kW are produced in Europe and North America using ammonia as the refrigerant and water as the absorbent.

5.2.2.4   World market characteristics

Table 5.6 summarizes the market for chillers in 2001. It shows that air-cooled positive displacement chillers represented nearly 75% of the number of units in the positive displacement category. Chillers larger than 100 kW are dominant in the Americas, the Middle East, and southern Asia while smaller air-cooled chillers and chiller heat pumps for residential and light commercial use are more common in East Asia and Europe.

In a number of countries the commercial air-conditioning market appears to be moving away from small chillers toward ductless single-package air conditioners or ducted unitary systems, due to the lower installation cost (JARN, 2002a).

Market conditions in China are particularly interesting due to the recent rapid development of its internal market, chiller manufacturing capabilities, and export potential. The centrifugal chiller population in China is included in Table 5.4. Significant growth began in the 1990s. Before 1995, most centrifugals were imported. After 1995, increasing numbers of chillers were produced in China by factories using US designs (primarily HCFC-123 (30%) and HFC-134a (70%)) (ICF, 2003). For chillers of all types, China is now the largest market in the world with sales of 34,000 units in 2001 and a growth of over 8.5% yr^{-1}. The main market is East China where there is a growing replacement market. Over half of all chiller sales are now reversible heat pumps that can provide cooling and heating. Screw and scroll chiller sales, mostly using HCFC-22, are rising as their technology becomes more familiar to the major design institutes. Demand for absorption chillers has been slowing since 1999 when national energy policy changed to relax controls on electricity for commercial businesses. China has a major residential market for chillers with fan coil units (BSRIA, 2001).

5.2.3   Options for HFC emissions reduction

As with stationary air conditioners, options for reducing HFC emissions in chillers include refrigerant conservation as described in Section 5.1.3.1 and the use of non-HFC systems. These options are now detailed.

5.2.3.1   HFC vapour-compression systems

Over the past 30 years, the life-cycle refrigerant needs of chillers have been reduced more than tenfold (Calm, 1999) due to design improvements and, in particular, the improved care of equipment in the field. The approaches that have been used to reduce CFC emissions over the last 30 years can also be applied to HCFCs and HFCs.

The starting points for reducing HFC emissions from chillers were designing the chiller and its components to use a reduced amount of refrigerant charge, employing a minimum number of fittings that are potential leakage sources, avoiding the use of flare fittings on tubing, and including features that minimize emissions while servicing components such as shut-off valves for oil filters and sensors. Many manufacturers have already implemented such changes.

Service technicians can be trained and certified to perform their tasks while minimizing refrigerant emissions during installation and refrigerant charging, servicing, and ultimately taking equipment out of service. Charging and storing the refrigerant in the chiller at the factory prior to delivery can reduce emissions at installation. Refrigerant should be recovered at the end of equipment life. Appropriate government policies can be effective in accomplishing these objectives. Some countries require annual inspections of equipment or monitoring of refrigerant use to determine whether emissions are becoming excessive and require action if this is the case.

Remote monitoring is becoming an established method for monitoring the performance of chillers. It is also being used to detect leakage either directly through leak detectors or indirectly through changes in system characteristics (e.g., pressures). Remote monitoring can provide alerts to maintenance engineers and system managers so as to ensure that early action is taken to repair leaks and maintain performance.

Table 5.6. The world chiller sales in 2001 (number of units).

Chiller Type	North and South America	Middle East, S. Asia, Africa	East Asia and Oceania	Europe	World Total
Positive Displacement	16,728	11,707	66,166	77,599	172,200
Air cooled	12,700	7749	43,714	61,933	126,096
Water cooled	4028	3958	22,542	15,666	46,104
<100 kW	2721	1678	48,444	58,624	111,467
>100 kW	14,007	10,029	17,722	18,975	60,733
Centrifugal	5153	413	2679	664	8908
Absorption >350 kW	261	289	5461	528	6539
Total chillers	**22,142**	**12,409**	**74,306**	**78,791**	**187,648**

Source: JARN, 2002b

5.2.3.2 Non-HFC systems

5.2.3.2.1 Vapour-compression cycle with non-HFC refrigerants

5.2.3.2.1.1 Positive displacement chillers

The non-HFC refrigerants that have been used in positive displacement compressor chillers are presented below.

Ammonia

Chillers using ammonia as the refrigerant are available in the capacity range 100–2000 kW and a few are larger than this. The use of ammonia is more complex than that of many other refrigerants because ammonia is a strong irritant gas that is slightly toxic, corrosive to skin and other membranes, and flammable. Recommended practice (ASHRAE, 2001a; ISO, 1993; CEN, 2000/2001) limits the use of large ammonia systems in public buildings to those systems, which use a secondary heat transfer fluid (which is intrinsic in chillers), so that the ammonia is confined to the machine room where alarms, venting devices, and perhaps scrubber systems can enhance safety. Guidelines are available for the safe design and application of ammonia systems (IEA, 1998, Chapter 4; ASHRAE, 2001a). Modern, compact factory-built units contain the ammonia far more effectively than old ammonia plants.

The high discharge temperatures associated with ammonia permit a far greater degree of heat recovery than with other refrigerants.

The wider acceptance of ammonia requires public officials being satisfied that ammonia systems are safe under emergency conditions such as building fires or earthquakes, either of which might rupture refrigerant piping and pressure vessels. The most important factor is the establishment of building codes that are acceptable to safety officials (e.g., fire officers).

Hydrocarbons

Hydrocarbon refrigerants have a long history of application in industrial chillers in petrochemical plants. Before 1997 they were not used in comfort air-conditioning chiller applications due to reservations about the system safety. European manufacturers now offer a range of hydrocarbon chillers. About 100 to 150 hydrocarbon chiller units are sold each year, mainly in northern Europe (UNEP, 2003). This is a small number compared to the market for more than 78,000 HCFC and HFC chillers in Europe (Table 5.6). The major markets have been office buildings, process cooling, and supermarkets.

In a system optimized for hydrocarbons, one might be able to achieve efficiency increases of more than 5% by using propane instead of HCFC-22. In the literature, efficiency comparisons for HCFC, HFC, and HC systems sometimes show substantial differences but do not represent rigorous comparisons. This issue was discussed in Section 5.1.3.2.1. The cost of HC chillers is higher than that of HCFC or HFC equivalents, partly due to the fact that hydrocarbon chillers still are a niche market.

A major disadvantage of hydrocarbons is their flammability, which deters their consideration for use in many applications. Refrigeration safety standards have been developed for hydrocarbon systems, for example, IEC 60355-2-40, AMD. 2 ED. 4, 'Safety of Household and Similar Electrical Appliances, Part 2'. Typical safety measures include proper placement and/ or gas tight enclosure of the chiller, application of low-charge system design, fail-safe ventilation systems, and gas detector alarm activating systems. An alternative is outdoor installation (ARTI, 2001). Comprehensive guidelines for safe design, installation, and handling of hydrocarbon refrigerants have been produced (ACRIB, 2001). These guidelines limit the charge for domestic/public applications to <1.5 kg for a sealed system or <5 kg in a special machinery room or outdoors. For commercial and private applications the limits are <2.5 kg and <10 kg respectively.

Carbon dioxide

Carbon dioxide is being investigated for a wide range of potential applications. However, CO_2 does not match the cycle energy efficiencies of fluorocarbon refrigerants for typical water chilling applications (ASHRAE, 2001b). Therefore, there is usually no environmental incentive to use CO_2 in chillers instead of HFCs. In Japan, CO_2 has not been used in a chiller on a commercial basis, but one demonstration unit has been built.

5.2.3.2.1.2 Centrifugal chillers

The non-HFC refrigerants that have been used in centrifugal compressor chillers are discussed below.

Hydrocarbons

Hydrocarbon refrigerants are used in centrifugal chillers in petrochemical plants where a variety of hazardous materials are routinely used and where the staff are highly trained in safety measures and emergency responses. Hydrocarbon refrigerants have not been used elsewhere due to concerns about system safety due to the large charges of flammable refrigerants.

Ammonia

Ammonia is not a suitable refrigerant for centrifugal chillers due to the large number of compressor stages required to produce the necessary pressure rise ('head') for the ammonia chiller cycle.

Water

Water is a very low-pressure refrigerant, with a condensing pressure of 4.2 kPa (0.042 bar) at 30°C and a suction pressure of 1.6 kPa (0.016 bar) at 9°C. Traditionally, water has been used in specialized applications with steam aspirators, and rarely with vapour compressors. The low pressures and very high volumetric flow rates required in water vapour-compression systems necessitate compressor designs that are uncommon in the air-conditioning field.

The few applications that use water as a refrigerant, use it to chill water or produce an ice slurry by direct evaporation from a pool of water. These systems carry a cost premium of more than

50% above conventional systems. The higher costs are inherent and are associated with the large physical size of water vapour chillers and the complexity of their compressor technology.

Recent studies indicate that there are no known compressor dsigns or cycle configurations of any cost that will enable water vapour-compression cycles to reach efficiencies comparable to existing technology (ARTI, 2000; ARTI, 2004).

5.2.3.2.2 Alternative technologies to vapour-compression cycle

Absorption Chillers

Absorption chillers are inherently larger and more expensive than vapour-compression chillers. They have been successful in specific markets as described in Section 5.2.2.3.

Some countries have implemented the use of water-LiBr absorption chillers in trigeneration systems. Trigeneration is the concept of deriving three different forms of energy from the primary energy source, namely, heating, cooling and power generation. This is also referred to as CHCP (combined heating, cooling, and power generation). This option is particularly relevant in tropical countries where buildings need to be air-conditioned and many industries require process cooling and heating. Although cooling can be provided by conventional vapour-compression chillers driven by electricity, heat exhausted from the cogeneration plant can drive the absorption chillers so that the overall primary energy consumption is reduced.

5.2.4   Global warming effects

5.2.4.1   LCCP examples for chillers

Figure 5.5 presents LCCP values for chiller technology alternatives at 350 tonnes rated capacity (1230 kW) applied to a typical office building in Atlanta, Georgia, USA. LCCP values for centrifugal and screw chillers fall within a ±8% range and refrigerant emissions account for less than 3% of the LCCP of any of these technology options. Ammonia has been included as a technical option, but local codes may affect its use. The data source did not calculate LCCP for a hydrocarbon system. However, hydrocarbon refrigerants have not been used in centrifugal chillers in office buildings due to concerns about safety with large charges of flammable refrigerants (UNEP, 2003).

The major portion of LCCP is the indirect warming associated with energy consumption. Direct warming due to refrigerant emissions only amounts to between 0.2 and 3.0% of the total LCCP. The LCCP values of the vapour-compression alternatives fall within a reasonably narrow range and show the clear superiority of vapour compression over absorption in terms of LCCP.

The LCCP of a typical direct-fired, two-stage water-LiBr absorption chiller is about 65% higher than the average LCCP for vapour-compression cycle chillers.

The basic assumptions used to create Figure 5.5 include 2125 annual operating hours, 30-year equipment life, 0.65 kg CO_2 kWh^{-1} power plant emissions, and inclusion of cooling

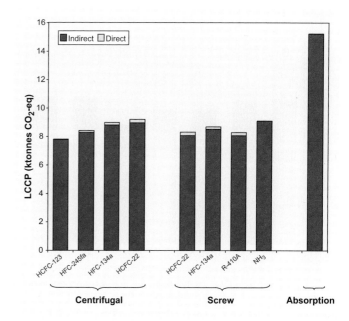

Figure 5.5. LCCP values for 1230 kW chiller-technology alternatives in an office building in Atlanta, Georgia, USA with a 1% refrigerant annual make-up rate (ADL, 2002).

tower fan and pump power. The annual charge loss rates are assumed to be 1% yr^{-1} for the vapour-compression chillers to account for some end-of-life losses and accidental losses in the field (ADL, 2002).

Figure 5.6 compares TEWI values for 1000 tonne (3500 kW chillers) with a 1% refrigerant annual make-up rate. CFC-11 and CFC-12 chiller data for equipment with 1993 vintage efficien-

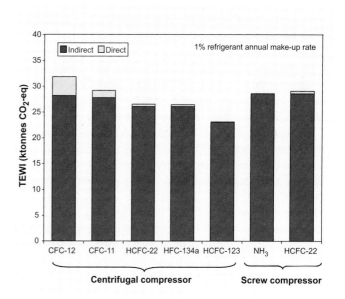

Figure 5.6. LCCP TEWI values for 1000 tonne (3500 kW) chillers with a 1% refrigerant annual make-up rate in an Atlanta office application (Sand *et al.*, 1997).

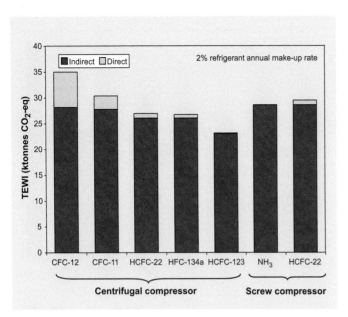

Figure 5.7. LCCP TEWI values for 1000 tonne (3500 kW) chillers with a 2% refrigerant annual make-up rate in an Atlanta office application (Sand *et al.*, 1997).

cies are shown because many chillers are still operating with these refrigerants. The figure shows the environmental benefits obtained by replacing CFC chillers with chillers employing non-CFC refrigerants that have higher COPs and a lower direct warming impact. In practice, the environmental benefits from replacement are greater because older CFC chillers are likely to have refrigerant leak rates of 4% or more, which is higher than the 1% rate assumed in this figure.

Figure 5.7 shows the effect on TEWI for chillers in Figure 5.6, if the annual refrigerant make-up rate is doubled to 2% for the chillers and the end-of-life refrigerant loss is 5%. The impact of the increased loss rate on TEWI is small, especially for the non-CFC chillers.

The leakage rates of 1% and 2% used in Figures 5.6 and 5.7 are lower than the historical average for chillers, but 2 to 4 times the best-practice value of 0.5% per year available today in the leading centrifugal and screw chillers.

The basic assumptions for Figures 5.6 and 5.7 are the same as for Figure 5.5 with the exception of the increased cooling capacity, the CFC chiller characteristics mentioned above, and the additional assumption of a 5% loss of charge when the chiller is scrapped.

Figure 5.8 compares LCCP values for air-cooled 25 kW scroll chiller/heat pumps in Japan. Two refrigerants, propane and R-407C, are compared for these chiller/heat pumps with two levels of end-of-life refrigerant recovery, 50% and 70%. The units are assumed to operate 700 h yr^{-1} for cooling and 400 h yr^{-1} for heating No additional charge has been added during the 15-year life of the chiller, and the emissions from power generation are taken to be 0.378 kg CO_2 kWh^{-1} for Japan (Onishi *et al.*, 2004). The figure shows that the indirect component of

LCCP is dominant for this application but the effect of only 50% end-of-life recovery is not negligible. In this comparison, the propane system is not equipped with a secondary heat transfer loop with its added COP penalties. This is because chillers and chiller/heat pumps inherently contain secondary loops in their water-to-water or air-to-water systems. However, propane chiller/heat pumps will have a 10–20% cost increase for safety features compared to a system with a non-flammable refrigerant and the same COP. For the same 10–20% cost increase, an increase in COP from 2.65–3.05 should be achievable for the R-407C system. This makes the LCCP with R-407C lower than that of an equivalent-cost propane system (Onishi *et al.*, 2004).

Figure 5.9 shows LCCP values for 355 kW air-cooled screw chillers in Japan using propane, HFC-134a, or R-407C as refrigerants in systems with several levels of COP. The chiller life is assumed to be 25 years with end-of-life refrigerant recovery assumed to be either 70% or 80%. Also, during the life of the equipment it is assumed that a 10% additional charge is needed to compensate for emissions. The units are assumed to operate 700 h yr^{-1} for cooling and 400 hours yr^{-1} for heating. Emissions from power generation are assumed to be 0.378 kg CO_2 kWh^{-1} for Japan (Onishi *et al.*, 2004). The figure shows that the indirect component of LCCP is dominant for this application and end-of-life refrigerant recovery rate. A comparison of the LCCP values for the propane and HFC-134a air-cooled screw chiller/heat pumps reveals that only a modest increase in COP is required for the HFC-134a system to have a better LCCP than propane. This COP increase with HFC-134a could be achieved by investing the cost of safety features for flammable refrigerant systems in performance improvements to the HFC systems instead (Onishi *et al.*, 2004).

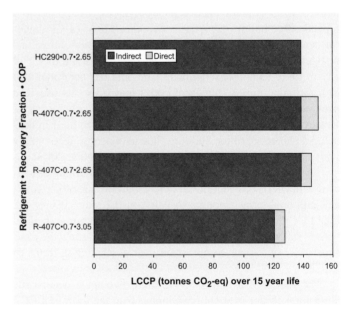

Figure 5.8. LCCP values for air-cooled 25 kW scroll chiller/heat pumps in Japan for R-290 and R-407C, with end-of-life recovery of 70% or 50% of the refrigerant charge, and with a system COP of 2.65 or 3.05 (Onishi *et al.*, 2004).

Table 5.7. HFC consumption and emission estimates for chillers.

	Year	
	2000	**2010**
HFC consumption, kt yr[-1]	2.5	3.5–4.5
HFC consumption, MtC-eq yr[-1]	1	2.3–3.0
HFC emissions, kt yr[-1]	0.2	0.5–0.7
HFC emissions, MtC-eq yr[-1]	0.1	0.3–0.5

Source: IPCC, 2001

5.2.4.2   Global refrigerant bank and emissions

Table 5.4 provides a sample of the large and varying refrigerant bank in chillers. One source (Palandre *et al.*, 2004) estimates the Stationary AC banks in 2002 to be nearly 84,000 tonnes CFCs and nearly 81,000 tonnes of HFCs. Although Stationary AC includes more than just chillers, it is clear that the CFC-11 and CFC-12 banks in chillers make up nearly the entire CFC bank estimated. The growing use of HFC-134a in chillers contributes substantially to the HFC bank as well.

Table 5.7 shows estimates of global HFC consumption and emission for chillers in 2000 and 2010. These estimates are based on information from IPCC (IPCC, 2001). Additional information on emission estimates is provided in Section 5.4.

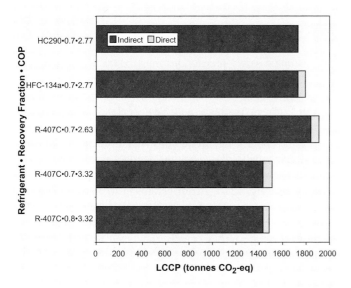

Figure 5.9. LCCP values for 355 kW air-cooled screw chillers in Japan for HFC-134a, R-290, and R-407C, with end-of-life refrigerant recovery of 70% or 80%, and COPs of 2.63, 2.77, or 3.32 (Onishi *et al.*, 2004).

5.3   Water-heating heat pumps

This section describes equipment and refrigerants for heating water with heat pumps[4].

5.3.1   Technologies and applications

5.3.1.1   Vapour-compression cycle, heat-pump water heaters

Almost all heat pumps work on the principle of the vapour-compression cycle. Heating-only, space-heating heat pumps are manufactured in a variety of sizes ranging from 1 kW heating capacity for single room units, to 50–1000 kW for commercial/institutional applications, and tens of MW for district heating plants. Most small to medium-sized capacity heat pumps in buildings are standardized factory-made units. Large heat pump installations usually are custom-made and are assembled at the site.

In several countries water heating for swimming pools is provided by heat pumps. This is a growing market for heat pumps.

Heat sources include outdoor, exhaust and ventilation air, sea and lake water, sewage water, ground water, earth, industrial wastewater and process waste heat. Air-source and ground-coupled heat pumps dominate the market. For environmental reasons, many countries discourage the use of ground water from wells as a heat pump source (ground subsidence, higher-value uses for well water). In countries with cold climates such as in northern Europe, some heat pumps are used for heating only. In countries with warmer climates, heat pumps serve hydronic systems with fan coils provide heat in the winter and cooling in the summer. Heat pumps with dual functions, such as heating water and cooling air simultaneously, are also available.

In mature markets, such as Sweden, heat pumps have a significant market share as heating systems for new buildings and are entering into retrofit markets as well. In Europe, comfort heating dominates heat pump markets – mostly with hydronic systems using outside air or the ground. There is increasing use of heat pumps that recover a portion of exhaust heat in ventilation air to heat incoming air in balanced systems. This reduces the thermal load compared to having to heat the incoming air with primary fuel or electricity. Heat pumps in Germany and Sweden provide up to 85% of the annual heating in some buildings. For these buildings, supplementary heat is required only on the coldest days.

Heat pumps have up to a 95% share of heating systems in new buildings in Sweden. This is due to the initial development support and subsidies from the government that made the units reliable and popular, high electricity and gas prices, widespread use of hydronic heating systems, and rating as a 'green' heating system by consumers (IEA, 2003a).

Heat pumps for combined comfort heating and domestic

---

[4] Heat pumps that heat air are included in section 5.2 on Stationary Air Conditioners.

hot-water heating are used in some European countries. Most of the combined systems on the market alternate between space and water heating, but units simultaneously serving both uses are being introduced (IEA, 2004).

The heat pumps for comfort heating have capacities up to 25 kW. Supply temperatures are 35–45°C for comfort heat in new constructions and 55°–65°C for retrofits. Regulations in a number of European countries require domestic water heaters to produce supply temperatures of 60–65°C.

Small capacity (10–30 kW) air-to-water heat pump chillers for residential and light commercial use in combination with fan-coil units are popular in China as well as Italy, Spain, and other southern-European countries. Hot water delivery temperatures are in the 45–55°C range. In the future, the market growth of small air-to-water heat pumps may be slowed in some markets by the growing popularity of variable-refrigerant-flow systems combined with multiple, indoor fan coil units connected to a refrigerant loop for direct refrigerant-to-air heat transfer.

In Japan, heat pump chillers are mainly for commercial applications above 70 kW. Commercial size heat pump chillers of up to 700 or 1000 kW capacity are used for retrofit, replacing old chillers and boilers to vacate machine room space and eliminate cooling towers (JARN, 2002b).

Night-time electricity rates in Japan are only 25% of daytime rates. As a consequence, domestic hot-water heat pumps are a rapidly-growing market. They are operated only at night and the hot water is stored for daytime use. Germany and Austria have been installing dedicated domestic hot water (DHW) heat pumps for a number of years (IEA, 2004).

5.3.1.2   Absorption heat pumps

Absorption heat pumps for space heating are mostly gas-fired and commonly provide cooling simultaneously with heating. Most of the systems use water and lithium bromide as the working pair, and can achieve about 100°C output temperature. Absorption heat pumps for the heating of residential buildings are rare. In industry, absorption heat pumps are only employed on a minor scale.

5.3.2   *Refrigerant use and equipment population*

In the past, the most common refrigerants for vapour-compression heat pumps were CFC-12, R-502, HCFC-22, and R-500. In developed countries, HCFC-22 still is used as one of the main refrigerants in heat pumps, but manufacturers have begun to introduce models using HFC alternatives (HFC-134a, R-407C, R-404A) or hydrocarbons to replace their HCFC-22 models.

Data on the installed base of water-heating heat pumps are not readily available for most countries. In particular, the data needed to estimate the bank of various refrigerants in use in these heat pumps do not seem to exist. Global sales of these heat pumps were small until 1995, but have increased steadily since. The installed base of ground-source heat pumps was estimated to be about 110,000 units in 1998 (IEA, 1999).

Total sales of water-heating heat pumps in Sweden were close to 40,000 units in 2002. Of these, 60% used water or brine as a source, and the rest used air or exhaust air. Fifty percent of these heat pump sales are for retrofits. The Swedish Energy Agency estimates that over 300,000 heat pumps are in operation there, a small portion of which are 'air-to-air' (IEA, 2003b).

In the rest of Europe, heat pumps are primarily used in new construction and provide combined operation – comfort heating and DHW heating.

Switzerland had heat pump sales of 7500 units in 2002, of which 50% were air-to-water and 43% were brine-to-water. Heat pump sales in Germany were 12,500 units in 2002, of which 43% were ground-source combined heat pumps and 33% were for DHW heating only (JARN, 2004a).

In China, the use of heat pumps is rapidly increasing and had reached 35,000 units in 2002. Sales have increased as a result of nationwide housing development projects where the preference is for hydronic systems. More than half of the sales volume is for units with a capacity of less than 30 kW (JARN, 2003).

5.3.3   *Options for reducing HFC emissions*

5.3.3.1   *HFC vapour-compression systems*
The actions described in Section 5.1.3.1 can also be used to reduce emissions in heat pumps.

5.3.3.2   *Vapour-compression cycle with non-HFC refrigerants*

Hydrocarbons
In most applications HC-290 will yield an energy efficiency comparable to or slightly higher (e.g., 5–10% higher) than that of HCFC-22. The performance difference increases in heat pumps at lower ambient temperatures. When designing new heat pump systems with propane or other flammable refrigerants, adequate safety precautions must be taken to ensure safe operation and maintenance. Several standards that regulate the use of hydrocarbons in heat pumps exist or are being developed in Europe, Australia, and New Zealand. An example is European Standard EN 378 (CEN, 2000/2001).

In some countries hydrocarbons are considered to be a viable option in small, low-charge residential heat pumps. Several northern-European manufacturers are using propane (HC-290) or propylene (HC-1270) as refrigerants in small residential and commercial water-to-water and air-to-water heat pumps. The hydrocarbon circuit is located outdoors using ambient air, earth, or ground water sources, and is connected to hydronic floor heating systems (IEA, 2002).

Carbon Dioxide
The transcritical CO_2 cycle exhibits a significant temperature glide on the high temperature side. Such a glide can be advantageous in a counter-flow heat exchanger. Heat pumps generating water temperatures of 90°C have been developed in Japan

for home use. Typical heating capacities are 4.5 kW. The COP achieved by CO_2 water-heating heat pumps is 4.0 and is slightly higher for 'mild climates'. This COP also is attained by R-410A heat pumps, but the highest water temperature available is about 80°C (JARN, 2004b).

Carbon dioxide is being introduced as a refrigerant for heat pumps, particularly those with a DHW function. Japan and Norway have been leaders in the development of CO_2 water-heating heat pumps. Because there is government support in Japan for the introduction of high-efficiency water heaters, 37,000 heat pump water heaters were sold in Japan in 2002 that used CO_2 or R-410A as refrigerants. The sales are estimated to have increased to 75,000–78,000 units in FY 2003 (JARN, 2004b).

Ammonia

Ammonia has been used in medium-sized and large capacity heat pumps, mainly in Scandinavia, Germany, Switzerland, and the Netherlands (IEA, 1993, 1994, 1998 (Chapter 4); Kruse, 1993). System safety requirements for ammonia heat pumps are similar to those for ammonia chillers, which were discussed in Section 5.2.

5.3.4   *Global warming effects*

There are no known published data on the global warming effects of water heating heat pumps.

5.4   Estimates for refrigerant emissions and costs for emission reductions

There are many data sources that can be used to estimate discrete equipment inventories and refrigerant banks (e.g., ICF, 2003; JARN, 2002b, and JARN, 2002c). Several studies have used these data along with 'bottom-up' methodologies to estimate refrigerant banks and/or refrigerant emissions, for past, current and/or future years, and for various countries, regions or the world.

These studies point to the dynamic and competitive nature of the air-conditioning market, especially as the transitions from CFCs and HCFCs to HFCs and other refrigerants, as described earlier in this chapter, occur. Therefore due consideration must be given to the data used, the assumptions made, and the methodologies employed in estimating refrigerant banks and emissions. The differences that arise for current estimates of banks and emissions are large, and are often further exacerbated when projecting future banks and emissions. Some of the aspects that may vary from study to study are:

- Equipment Inventories. What type of equipment is included? How is it disaggregated?
- Refrigerant Charge. What is the average refrigerant charge? Are different charges used for different types or different vintages of equipment?
- Emission Sources. Are various emissions sources (e.g., installation, operating, servicing, end-of-life disposal) evaluated separately, or is an average emission rate used?

- Equipment Lifetimes. How long is equipment assumed to exist? Are emission rates assumed to be constant over the lifetime?
- Emissions Scope. Are all refrigerants included, or just those reported in national inventories under the UNFCCC (i.e., HFCs and PFCs)? Does the source also estimate indirect emissions from power generation?
- Geographical Extrapolation. If data are only available for a particular region, how are the data extrapolated to other countries or regions or disaggregated into individual countries within the region?
- Temporal Extrapolation. How are data extrapolated into the future? Do emission rates or refrigerant charges change in the future? If so, by how much and on what basis?
- Global Warming Potentials. What source is used for GWPs? If CFCs and HCFCs are included in estimates, do GWPs represent the direct effect or include the indirect effect as well?

Table 5.8 compiles several estimates for recent (1996–2005) emission rates. The data shown are direct emissions only. Estimates for residential and commercial air conditioning and heating (also called 'stationary air conditioning and heating') are sometimes divided into subcategories. For instance, some studies report separate estimates for air conditioners (for cooling and/or heating) and chillers, as described in Sections 5.1 and 5.2, respectively. No studies were found that contained separate emissions of water-heating heat pumps as described in Section 5.3.

Table 5.8 mostly shows estimates for the entire world, with two examples for industrialized Europe to further highlight the differences in the literature. Estimates for the entire air conditioning and refrigeration sector are included to provide a perspective; see Chapter 4 and Chapter 6 for more information on Refrigeration and Mobile Air Conditioning.

It is clear that different sources provide vastly different emission estimates. Similar differences are seen for refrigerant banks. Some of the differences shown above can be explained by the transition from ODS to non-ODS refrigerants (e.g., in 1996 relatively few HFC units existed, whereas by 2005 substantially more HFC units had been installed). However, the major difference in the estimates is due to the data and methodologies used.

Table 5.9 provides some example estimates of future emissions under 'baseline' or 'business-as-usual' conditions, in 2010 and 2015. Again, the data is for direct emissions only. The data for 2010 are mainly included to show that any given source is not always consistently higher or lower than another source (e.g., compare estimates from sources Harnisch *et al.*, 2001, and US EPA, 2004).

Some authors also examine various options for reducing the predicted emissions and the costs associated with this. As with the emission estimates, there is a lot of variation between the sources and the results are heavily influenced by the assumptions made. The economic factors used, such as the monetary

Table 5.8. Refrigeration and air conditioning emission estimates (MtCO$_2$-eq yr^{-1}) for past and current years.

Region	Substance	Year	Application(s) Refrigeration and Air Conditioning — Stationary AC and Heating — Chillers	Commercial AC&H	Residential AC&H	Refrigeration	MAC	Source/Notes
EU-15	HFCs	1995	4.3					March Consulting Group, 1998
West. Europe		1996	16.1					Harnisch *et al.*, 2001
World	HFCs	1996	0.2			X	X	Harnisch *et al.*, 2001
		2001	>19.9	2.0–2.4				See note[1]
		2002	8.4			X	X	Palandre *et al.*, 2004
			222.7					
		2005	7.6	1.7	1.4	X	X	US EPA, 2004
			10.7			X	X	
			219.7					
	HFCs, HCFCs, CFCs	1996	20.0			X	X	Harnisch *et al.*, 2001
			638.0					
		2002	222.8			X	X	Palandre *et al.*, 2004
			1676.8					

X = applications not included in emission estimate(s) shown

[1] Air-conditioner emissions calculated using Table 5.1 for bank, and averages from Section 5.1.1 for annual emission rates. Range assumes 0% to 100% R-407C with the remainder R-410A. GWPs of blends calculated using GWPs from Table 2.6. Minimum chiller emissions calculated as total centrifugal chiller HFC-134a bank for USA, Canada and China as shown in Table 5.4 (note Table 5.4 does not represent the complete world inventory) multiplied by emission rate of 1% yr^{-1} as used in Figures 5.5 and 5.6, and the same GWP source as above.

Table 5.9. Unmitigated refrigeration and air conditioning emission estimates (MtCO2-eq yr-1) for future years.

Region	Substance	Year	Scenario	Application(s) Refrigeration and Air Conditioning — Stationary AC and Heating — Chillers	Commercial AC&H	Residential AC&H	Refrigeration	MAC	Source
EU-15	HFCs	2010	BAU	28.2					March Consulting Group, 1998
			Base	36.6					US EPA, 2004
West. Europe			Base	68.8					Harnisch *et al.*, 2001
World	HFCs	2015	Base	9.2	31.7	49.4	X	X	US EPA, 2004
				90.3			X	X	
				472.0					
			Sc1	100.1			X	X	Palandre *et al.*, 2004
				667.0					
			Base	14.8			X	X	Harnisch *et al.*, 2001
	HFCs, HCFCs, CFCs	2015	Sc1	322.8			X	X	Palandre *et al.*, 2004
				1527.2					
			Base	23.5			X	X	Harnisch *et al.*, 2001
				293.5					

X = applications not included in emission estimate(s) shown
Base = Baseline scenario
BAU = Business-as-usual scenario
Sc1 = Scenario 1 (business-as-usual) in Palandre *et al.*, 2004.

Table 5.10. Abatement options applicable for residential and commercial air conditioning and heating.

Application	Option	Region	Cost per tCO$_2$–eq	Monetary Unit	Discount Rate	Source
AC	Alternative Fluids and Leak Reduction	EU-15	23 to 26	ECU (year not stated)	8%	March Consulting Group, 1998
AC	Energy Efficiency Improvements	EU-15	−79 to −70[1]	ECU (year not stated)	8%	March Consulting Group, 1998
AC	HC Refrigerant	EU-15	114[1]	1999 Euro	4%	Harnisch, 2000
AC	Leak Reduction	EU-15	44	1999 Euro	4%	Harnisch, 2000
Chillers	HC and Ammonia Refrigerant	EU-15	49	1999 Euro	4%	Harnisch, 2000
Chillers	Leak Reduction	EU-15	173	1999 Euro	4%	Harnisch, 2000
Stationary AC	Leak Reduction and Recovery	World	38	1999 USD	5%	Harnisch et al., 2001
Stationary AC and others	STEK-like Programme	EU-15	18.3	Euro (year not stated)	Not stated	Enviros, 2003
AC and others	Recovery	World	0.13	2000 USD	4%	US EPA, 2004
AC and others	Recovery	World	0.13	2000 USD	20%	US EPA, 2004
Chillers and others	Leak Repair	World	−3.20	2000 USD	4%	US EPA, 2004
Chillers and others	Leak Repair	World	−1.03	2000 USD	20%	US EPA, 2004
AC and others	Recovery	World	1.47	2000 USD	4%	Schaefer et al., 2005
Chillers and others	Leak Repair	World	1.20	2000 USD	4%	Schaefer et al., 2005

[1] These costs incorporate savings or additional costs due to assumed changes in energy efficiency; see the referenced source for more details.

Table 5.11 Mitigated refrigeration and air-conditioning emission estimates (MtCO$_2$-eq yr^{-1}) for future years and mitigation costs (USD per tCO$_2$-eq abated).

Region	Substance	Year	Scenario	Refrigeration and Air Conditioning — Stationary AC and Heating — Chillers	Commercial AC&H	Residential AC&H	Refrigeration	MAC	Source
World	HFCs	2015	Mit	8.9 @ -3.20 to -1.03	29.2 @ 0.13	45.5 @ 0.13	X	X	US EPA, 2004
				83.6 @ −3.20 to 0.13			X	X	
				364.9 @ −75 to 49					
			Sc2	67.9			X	X	Palandre et al., 2004
				452.8					
			Sc3	43.0			X	X	
				286.8					
			Mit	7.6 @ 38.26			X	X	Harnisch et al., 2001
	HFCs, HCFCs, CFCs	2015	Mit	9.4 @ 8.37–41.14			X	X	Harnisch et al., 2001
				109.6 @ 1.05–85.14					
			Sc2	225.2			X	X	Palandre et al., 2004
				1114.2					
			Sc3	149.6			X	X	
				783.5					

X = applications not included in emission estimate(s) shown
Sc2 = Scenario 2 (some mitigation of emissions) in Palandre et al., 2004.
Sc3 = Scenario 3 (partial HFC phase-out) in Palandre et al., 2004.

unit (USD, EUR, etc.) and discount rates, also vary. Table 5.10 tabulates several examples of abatement options. Only one source was found to estimate the effectiveness of energy efficiency improvements. This indicates the confidential nature of any such data. When these savings were included in the calculations this option proved to be by far the most cost-effective. The remaining options concentrate on other items highlighted earlier in this chapter (e.g., recovery, alternative refrigerants). Many of these options are assumed to partially exist in the baseline (e.g., recovery occurs to some extent) and are assumed to increase if the option is applied. Note that some costs are negative, indicating that energy-efficiency improvements or lower refrigerant costs render the option cost effective under the assumptions applied.

A few of the studies assume certain market penetration, beyond that assumed in the baseline, of the aforementioned abatement options and predict mitigated emissions under various scenarios. These estimates (again, only direct refrigerant emissions) along with the cost-effectiveness of the mitigation option are shown in Table 5.11. Note that the cost-effectiveness of the mitigation option is shown as '@ ###' per tonne CO_2-eq, where ### is the cost using the monetary unit and discount rate shown in Table 5.10.

References

ACRIB, 2001: Guidelines for the Use of Hydrocarbon Refrigerant in Static Refrigeration and Air-conditioning Systems. Air Conditioning and Refrigeration Industry Board, (ACRIB), Carshalton, UK.

ADL (A.D. Little, Inc.), 2002: Global Comparative Analysis of HFC and Alternative Technologies for Refrigeration, Air Conditioning, Foam, Solvent, Aerosol Propellant, and Fire Protection Applications. Final Report to the Alliance for Responsible Atmospheric Policy, March 21, 2002 (available online at www.arap.org/adlittle/toc.html), Acorn Park, Cambridge, Massachusetts, USA, 150pp.

Adnot, J., 2002: Energy Efficiency and Certification of Central Air Conditioners (EECAC). Interim Report for the European Commission DG-TREN, Contract DGXVII-4.1031/P/00-009, Armines, France, September 2002, 86 pp..

ARI, 1998: ARI White Paper, ARI Standard 550/590-98, Standard for Water Chilling Packages Using the Vapor Compression Cycle. Air-Conditioning and Refrigeration Institute (ARI), Arlington, VA, USA, 5 pp. (available online at http://www.ari.org/wp/550.590-98wp.pdf)

ARI, 2002: Statistical Release: Industry Shipment Statistics for Small and Large Unitary Products. Air Conditioning and Refrigeration Institute (ARI), Arlington, VA, USA June 2002, 2 pp.

ARI, 2003: Standard 550/590, Standard for Water Chilling Packages Using the Vapor Compression Cycle. Air-Conditioning and Refrigeration Institute (ARI), Arlington, VA, USA, 36 pp.

ARTI, 2000: The Efficiency Limits of Water Vapor Compressors Suitable for Air-Conditioning Applications, Phase I, Report ARTI-21CR/605-10010-01. Air-Conditioning and Refrigeration Technology Institute (ARTI), Arlington, VA, USA, 260 pp.

ARTI, 2001: Assessment of the Commercial Implications of ASHRAE A3 Flammable Refrigerants in Air Conditioning and Refrigeration, Final Report, ARTI 21-CR/610-50025-01. Air-Conditioning and Refrigeration Technology Institute (ARTI), Arlington, VA, USA, 2001, 116 pp.

ARTI, 2004: Use of Water Vapor as a Refrigerant; Phase II - Cycle Modifications and System Impacts on Commercial Feasibility, Report ARTI-21CR/611-10080-01. Air-Conditioning and Refrigeration Technology Institute (ARTI), Arlington, VA, USA, 257 pp.

ASHRAE, 2001a: Safety Standard for Refrigeration Systems. American Society of Heating, Refrigerating, and Air-Conditioning Engineers (ASHRAE) Standard 15, ASHRAE, Atlanta, GA 30329, USA, 2001, 34 pp.

ASHRAE, 2001b: Fundamentals Handbook. American Society of Heating, Refrigerating, and Air-Conditioning Engineers (ASHRAE), ASHRAE, Atlanta, GA 30329, USA, 2001, 897 pp.

BSRIA, 2001: The Chinese Air Conditioning Market: Poised to Become World Number 1 in 2005. BSRIA Ltd., Press Release No. 30/01, Bracknell, Berkshire, UK.

Calm, J.M., 1999: Emissions and Environmental Impacts from Air-Conditioning and Refrigerating Systems. Proceedings of the Joint

IPCC/TEAP Expert Meeting on Options for the Limitation of Emissions of HFCs and PFCs, L. Kuijpers, R. Ybema (eds.), 26-28 May 1999, Energy Research Foundation (ECN), Petten, The Netherlands (available online at www.ipcc-wg3.org/docs/IPCC-TEAP99).

CEN, 2000/2001: Refrigerating systems and heat pumps – Safety and environmental requirements. Part 1: Basic requirements, definitions, classification and selection criteria (2001), Part 2: Design, Construction, testing, marking and documentation (2000), Part 3: Installation, site and personal protection (2000), Part 4: Operation, maintenance, repair and recovery (2000). European Committee for Standardization. Standard EN 378, 2000, Brussels, Belgium, 123 pp.

Digmanese, T., 2004: Information on Centrifugal Chillers in China. Contribution to the TEAP Report of the Chiller Task Force (UNEP-TEAP, 2004), January 2004, pp. 37-40.

Dooley, E., 2001: Survey of chiller manufacturers. *Koldfax*, 2001(5), Newsletter of the Air Conditioning and Refrigeration Institute (ARI), Arlington, VA, USA.

DRI, 2001: HVAC Industry Statistics. Data Resource International, 2001.

Enviros Consulting Ltd., 2003: Assessment of the Costs & Implication on Emissions of Potential Regulatory Frameworks for Reducing Emissions of HFCs, PFCs & SF_6. Report prepared for the European Commission (reference number EC002 5008), London, United Kingdom, pp. 39.

EU, 2000: Regulation (EC) No. 2037/2000 of the European Parliament and of the Council of 29 June 2000 on substances that deplete the ozone layer, Official Journal of the European Communities, No. 29.9.2000, 24 pp.

Harnisch, J. and C. Hendriks, 2000: Economic Evaluation of Emission Reductions of HFCs, PFCs and SF_6 in Europe. Report prepared for the European Commission DG Environment, Ecofys, Cologne/Utrecht, Germany/Netherlands, 70 pp.

Harnisch, J., O. Stobbe and D. de Jager, 2001: Abatement of Emissions of Other

Greenhouse Gases: 'Engineered Chemicals', Report for IEA Greenhouse Gas R&D Programme, M754, Ecofys, Utrecht, The Netherlands, 85 pp.

HRAI, 2003: Canadian CFC Chiller Stock Decreases More Rapidly in 2002. News Release from the Heating, Refrigeration, & Air Conditioning Institute of Canada, Mississaugua, ON, Canada, 11 June 2003.

ICF, 2003: International Chiller Sector Energy Efficiency and CFC Phaseout. ICF Consulting, Draft Revised Report prepared for the World Bank, Washington, DC, USA, May 2003, 78 pp.

IEA, 1993: Trends in heat pump technology and applications, *IEA Heat Pump Centre Newsletter*, **11**(4), December 1993, p. 7.

IEA, 1994: Heat pump working fluids, *IEA Heat Pump Centre, Newsletter*, **12**(1), March 1994, p.8.

IEA, 1998: Guidelines for Design and Operation of Compression Heat Pump, Air Conditioning and Refrigerating Systems with Natural Working Fluids - Final Report. [J. Stene (ed.)], December 1998, Report No. HPP-AN22-4, IEA Heat Pump Centre, Sittard, The Netherlands.

IEA, 1999: Ground-source heat pumps, *IEA Heat Pump Centre, Newsletter*, **17**(1), pp. 28.

IEA, 2002: Hydrocarbons as Refrigerant in Residential Heat Pumps and Air Conditioners – IEA Heat Pump Centre Informative Fact Sheet HPC-IFS1. IEA Heat Pump Centre, Sittard, The Netherlands, January 2002.

IEA, 2003a: IEA National Presentation for Sweden by Peter Rohlin, Swedish Energy Agency, 2003.

IEA, 2003b: Heat pump systems in cold climates, *IEA Heat Pump Centre, Newsletter*, **21**(3).

IEA, 2004: Test Procedure and Seasonal Performance Calculation for Residential Heat Pumps with Combined Space and Domestic Hot Water Heating. Interim Report IEA HPP Annex 28, [Wemhöner,C. and Th. Afjei (eds.)], University of Applied Sciences Basel, Institute of Energy, Basel Switzerland, February 2004.

IPCC, 2001: *Climate Change 2001 – Mitigation. Contribution of Working Group III to the Third Assessment Report of the Intergovernmental Panel on Climate Change* [Metz, B., O. Davidson, R. Swart and J. Pan (eds.)] Cambridge University Press, Cambridge, United Kingdom, and New York, NY, USA, pp 752.

ISO, 1993: ISO 5149:1993 Mechanical Refrigerating Systems Used for Cooling and Heating – Safety Requirements. International Organization for Standardization, Geneva, Switzerland, 34 pp.

JARN, 2001: *Japan Air Conditioning, Heating & Refrigeration News*, Serial No. 394-S, November 2001.

JARN, 2002a: *Japan Air Conditioning, Heating & Refrigeration News*, Serial No. 406-S, November 25, 2002.

JARN, 2002b: World Air Conditioning Market, 2001. *Japan Air Conditioning, Heating & Refrigeration News*. Serial No. 403-S25, August 2002.

JARN, 2002c: *Japan Air Conditioning, Heating & Refrigeration News*, Serial No. 397-S, February 2002.

JARN, 2003: *Japan Air Conditioning, Heating & Refrigeration News*, Serial No.418-S, November 25, 2003.

JARN, 2004a: *Japan Air Conditioning, Heating & Refrigeration News*, Serial 427-S, August, 2004

JARN, 2004b: *Japan Air Conditioning, Heating & Refrigeration News*. Serial No. 424-36, May 2004

JRAIA, 1999: Calculating Method of Annual Power Consumption for Room Air Conditioners, Standard JRA4046. Japan Refrigeration and Air conditioning Industry Association (JRAIA), Tokyo, Japan, 21 pp (in Japanese).

JRAIA, 2003: Calculating Method of Annual Power Consumption for Multi Split Package Air Conditioners, Standard JRA4055. Japan Refrigeration and Air conditioning Industries Association (JRAIA), Tokyo, Japan, 2003, 58 pp (in Japanese).

JRAIA, 2004: Extract from JRAIA inventory database. Japan Refrigeration and Air conditioning Industries Association (JRAIA), Tokyo, Japan Data provided by H. Sagawa, 1 June 2004.

Kruse, H., 1993: European Research and Development Concerning CFC and HCFC Substitutes, Refrigerants. Conference on R-22/R-502 Alternatives, Gaithersburg, MD, USA, August 19-20, 1993 ASHRAE, Atlanta, GA, 30329, USA, pp. 41-57.

March Consulting Group, 1998: Opportunities to Minimise Emissions of Hydrofluorocarbons (HFCs) from the European Union, Final Report, Prepared by March Consulting Group, UK, ENVIROS Group, Cambourne, Cambridge, 123 pp.

Nekså, P. and J. Pettersen, 2001: Prospective for the Use of CO_2 in Refrigeration and Heat Pump Systems. 37 Annual Meeting of the Norwegian Society of Refrigeration, Trondheim, 23-25 March 2001.

Onishi, H., R. Yajima and S. Ito, 2004: LCCP of Some HVAC & R Applications in Japan. Proceedings of the 15th Annual Earth Technologies Forum, April 13-15, 2004, Washington, D.C., USA, 18 pp.

Palandre, L, D. Clodic and L. Kuijpers, 2004. HCFCs and HFCs emissions from the refrigerating systems for the period 2004-2015. Proceedings of the 15th Annual Earth Technologies Forum, April 13-15, 2004, Washington, D.C., USA, 13 pp.

Robur, 2004: Webpage Robur gas-fired absorption chillers and chillers/heaters. http://www.gasforce.com/gascool/robur.html (1 November 2004).

Sand, J.R., S.K. Fischer and V.D. Baxter, 1997: Energy and Global Warming Impacts of HFC Refrigerants and Emerging Technologies. Report prepared by Oak Ridge National Laboratory for the Alternative Fluorocarbons Environmental Acceptability Study (AFEAS) and the US Department of Energy, Arlington, Va, USA, 215 pp.

Schaefer, D. O., D. Godwin and J. Harnisch, 2005: Estimating future emissions and potential reductions of HFCs, PFCs and SF_6. *Energy Policy*.

UNEP (United Nations Environment Programme), 1998: 1998 Report of the Refrigeration, Air Conditioning and Heat Pumps Technical Options Committee – 1998 Assessment. [L. Kuijpers (ed.)]. UNEP Ozone Secretariat, Nairobi, Kenya, 285 pp.

UNEP, 2003: 2002 Report of the Refrigeration, Air Conditioning and Heat Pumps Technical Options Committee – 2002 Assessment. [L. Kuijpers (ed.)]. UNEP Ozone Secretariat, Nairobi, Kenya, 197 pp.

UNEP-TEAP, 2004: Report of the TEAP Chiller Task Force. [L. Kuijpers (ed.)]. UNEP Ozone Secretariat, Nairobi, Kenya, 73 pp.

US EPA, 2004: Analysis of Costs to Abate International Ozone-Depleting Substance Substitute Emissions. US Environmental Protection Agency report 430-R-04-006, D.S. Godwin (ed.), Washington, D.C. 20460, USA, 309 pp..

6

Mobile Air Conditioning

Coordinating Lead Author
Denis Clodic (France)

Lead Authors
James Baker (USA), Jiangping Chen (China), Toshio Hirata (Japan), Roland Hwang (USA), Jürgen Köhler (Germany), Christophe Petitjean (France), Aryadi Suwono (Indonesia)

Contributing Authors
Ward Atkinson (USA), Mahmoud Ghodbane (USA), Frank Wolf (Austria), Robert Mager (Germany), Greg Picker (Australia), Frank Rinne (Germany), Jürgen Wertenbach (Germany)

Review Editors
Stephen O. Andersen (USA), Tim Cowell (United Kingdom)

Contents

EXECUTIVE SUMMARY

Mobile air-conditioning (MAC) systems have been mass-produced in the USA since the early 1960s and in Japan since the 1970s. However, the numbers of air-conditioned cars in both Europe and developing countries only started to significantly increase much later, in about 1995.

The rapid switch from CFC-12 to HFC-134a was a global decision taken by all car manufacturers in developed countries. The first HFC-134a system was installed in 1990 and by 1994 almost all vehicles including cars, light commercial vehicles, and truck cabins, sold in developed countries used this refrigerant. In developing countries the transition from CFC-12 to HFC-134a will be completed in 2008. Since 1990, significant progress has been made in limiting refrigerant emissions and the implementation of the Montreal Protocol has contributed to this. For example, recovery equipment has been developed, some countries have encouraged the intensive training of service technicians to improve servicing practices, component manufacturers have designed low-emission components (fittings, hoses, etc.) and car manufacturers have usually lowered the original refrigerant charge. These efforts led to a significant decrease in emission levels from about 400 g yr^{-1} in the late 1980s to about 200 g yr^{-1} in the mid-1990s (UNEP, 1998) and efforts are still underway to further reduce these emission levels to 100 g yr^{-1} or less (Schwarz and Harnisch, 2003).

If all types of refrigerants are considered, emissions from MAC systems range from approximately 105,000 tonnes of CFC-12 in 1990 to 137,000 tonnes of CFC-12 and HFC-134a in 2003, with CFC-12 still representing 46% of those emissions. A business-as-usual scenario forecasts emissions in the range of 180,000 tonnes in 2015 (see Figure 6a), most of which will be HFC-134a. The CO_2-eq emissions fell from about 850 MtCO$_2$-eq in 1990 to 609 Mtonnes CO_2-eq in 2003 (see Figure 6b) and a business-as-usual scenario forecasts CO_2-eq emissions in the range of 270 Mtonnes CO_2-eq in 2015. The progressive phase out of CFC-12 in the fleet and its replacement by HFC-134a has led to a significant decrease in CO_2-eq. emissions, despite the continued growth of the MAC system fleet.

CFC-12 was not widely used within the EU due to the very low number of air-conditioned cars before 1995. However, there have been increasing concerns about the significant increase of HFC-134a use in Europe. In its Fluorinated Gas regulation project, the European Commission has proposed the phase-out of HFC-134a, thereby creating a totally new context for refrigerant choices.

Evaluating the additional fuel consumption of MAC system operation is a complex task. Nevertheless, initial evaluations indicate that in 2003, CO_2 emission due to global MAC operation was in the order of 100 Mtonnes.

Options for greenhouse gas emission mitigation
There are three main options for reducing greenhouse gas (GHG) emissions: Improving the current HFC-134a system or switching to other low global warming potential (GWP) refrigerants such as HFC-152a or CO_2. Other options such as hydrocarbons (HCs) have been investigated but car manufacturers did not support these as a possible alternative technology. Other refrigeration techniques such as sorption systems and air cycle have been tested, but are not energy efficient enough to replace the vapour compression technology.

Current regulatory policies ignore the additional fuel consumption of air-conditioning (AC) system operation. It might be technically possible to reduce the fuel consumption of the AC system by up to 30–40%, depending on the option chosen and the climate. This could lead to a global reduction in CO_2 emissions of 40 Mtonnes yr^{-1}.

Improved HFC-134a systems have a low leakage rate and a higher energy efficiency. Moreover, further emission reductions might be obtained with a full professional service that includes recovery, recycling and systematic end-of-life (EOL) recovery. This would require refrigerant management policies governed by national regulations and incentives.

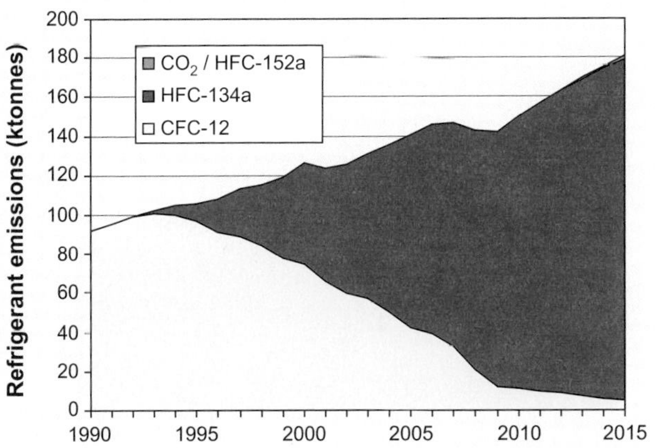

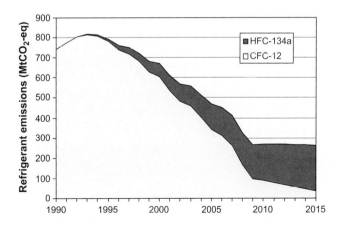

Figure 6a. MAC refrigerant emissions from 1990 to 2015 (Clodic *et al.,* 2004).

Figure 6b. MAC refrigerant emissions in CO_2-eq from 1990 to 2015 (Clodic *et al.,* 2004).

Alternate low GWP refrigerant systems

AC systems that use CO_2 have an entirely new design unique to this refrigerant as well as a safety system. A new infrastructure for servicing will also need to be implemented. The cooling performances of CO_2 systems have been demonstrated in passenger vehicles from most major car manufacturers as well as some commercial buses and an electrical hermetic system in prototype fuel cell vehicles. Direct emissions have a very low impact due to CO_2 GWP. For indirect emissions, CO_2 systems have shown energy efficiency comparable to an improved HFC-134a system. There are several barriers to the commercialization of this technology. These are the risk of suffocation, outstanding technical and cost issues and the need for an entirely new servicing infrastructure.

Systems using HFC-152a contain the same components as the HFC-134a system but have an added safety system. Cooling performances of HFC-152a systems have been demonstrated in passenger vehicles. Direct emissions (CO_2-eq) are very low (more than 90% reduction compared to HFC-134a base line) and HFC-152a systems have an energy efficiency comparable to the improved HFC-134a system. Barriers to the commercialization of this technology are the risk of flammability and guaranteeing a commercial supply of HFC-152a.

System comparison and design considerations

Safety concerns can be mitigated by using identified, existing technologies. Risk assessments for the CO_2 and HFC-152a systems are currently being undertaken.

The reference cost of a typical European HFC-134a system with an internally-controlled compressor is estimated to be about 215 US\$. Assuming mass production, the additional cost of a CO_2 system (including the safety system) is estimated to be between 48 and 180 US\$. The additional cost of an HFC-152a system is estimated to be about 48 US\$ for the safety devices used for a direct system, or for a secondary loop for the indirect system.

If development, validation and production lead times are considered, the improved HFC-134a system is likely to be ready first, followed by the HFC-152a system and then the CO_2 system.

The servicing infrastructure will need to be prepared for the transition to a new refrigerant. This is a significant issue for each alternative refrigerant system and the burden this will place on the service industry needs to be considered, especially in developing countries.

6.1   Introduction

Mobile air-conditioning (MAC) systems have been mass-produced in the USA since the early 1960s and in Japan since the 1970s. However, the numbers of air conditioned cars in Europe and in developing countries only started to significantly increase much later in about 1995.

The rapid switch from CFC-12 to HFC-134a was a global decision taken by all car manufacturers in developed countries. The first HFC-134a system was installed in 1990 and by 1994 almost all vehicles sold in developed countries used this refrigerant. In developing countries the transition from CFC-12 to HFC-134a will be completed in 2008.

This chapter addresses an evaluation of the additional fuel consumption due to MAC operation. In 2003, this was estimated to be in the order of 100 Mt CO_2-eq emissions. For the same year, CO_2 equivalent emissions from HFC-134a were about 96 Mtonnes and those for the remaining CFC-12 in cars were 514 Mtonnes.

There are three main options for reducing GHG emissions: Improving the current HFC-134a system or switching to other low GWP refrigerants such as HFC-152a or CO_2.

6.2   Current refrigerant banks and emission forecast

Emissions from mobile air-conditioning systems of cars, light commercial vehicles (LCVs)and truck cabins are detailed by Palandre *et al.*, (2002), Clodic *et al.*, (2004) and Palandre *et al.*, (2004) (see also UNEP, 2003). Even though the car industry switched from CFC-12 to HFC-134a with effect from 1992, the number of vehicles and CFC-12 bank in this fleet is still significant (see Table 6.1).

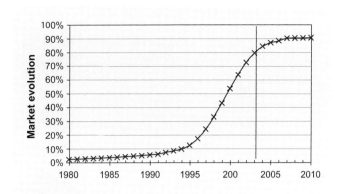

Figure 6.1. AC rate typical evolution: Western Europe example (Palandre *et al.*, 2000).

The average lifetime of cars depends on the country. It is typically 12 years in Europe and Japan and 16 years in North America.

For a number of countries, the last 10 years have seen a rapid penetration of air-conditioning systems in cars, as shown in Figure 6.1 for Europe. This S curve is typical for the introduction of a major technology and a corresponding change of consumer habits.

In 2000, the worldwide automotive fleet numbered some 720 million of vehicles. Nearly half of the fleet is equipped with AC (361 million).

6.2.1   AC system characteristics

The average refrigerant charge depends on the type of car and the type of refrigerant. Usually HFC-134a systems have about a 15% lower charge than CFC-12 charge due to the difference in liquid densities of those two refrigerants. Current HFC-134a charges vary from 500–900 g. The lifetime of the MAC system (9–12 years) is usually considered to be shorter than the car lifetime (12–16 years) due to the relatively high cost of repairing AC in older vehicles.

A number of studies (Palandre *et al.*, 2002; UNEP, 2003; Schwarz and Harnisch, 2003) have shown that annual emission rates, including accidents and servicing losses, are between 10 and 15% of the original charge. Servicing is assumed to take place for recharging when the residual charge is between 50 and 75%. Dependent on servicing habits and regulations, the recovery during servicing varies from 0–40% of recoverable refrigerant.

6.2.2   CFC-12, HFC-134a emissions from 1990 to 2003 and emission forecasts for the period 2004 to 2015

The emission inventory method described in Ashford *et al.* (2004) is based on IPCC inventory guidelines (IPCC, 1997). Using the results of Clodic *et al.* (2004), Figure 6.2 and Table 6.2 show the evolution of the refrigerant banks of HFC-134a

Table 6.1. Evolution of the CFC-12 and HFC-134a fleet (Clodic *et al.*, 2004).

Year	AC vehicle fleet (million)	
	CFC-12	**HFC-134a**
1990	212	-
1991	220	-
1992	229	0.7
1993	229	10
1994	222	27
1995	215	49
1996	206	74
1997	197	100
1998	186	128
1999	175	161
2000	163	198
2001	149	238
2002	134	285
2003	119	338

Table 6.2 MAC refrigerant bank (Clodic *et al.*, 2004).

Year	CFC-12 (tonne)	HFC-134a (tonne)	CO_2 / HFC-152a (tonne)
1990	254,273	-	-
2003	131,365	287,185	-
2015	12,863	633,893	3177

Table 6.3 MAC refrigerant emissions (Clodic *et al.*, 2004).

Year	CFC-12 (tonne)	HFC-134a (tonne)	CO_2 / HFC-152a (tonne)
1990	104,707	-	-
2003	63,431	73,956	-
2015	5192	175,174	934

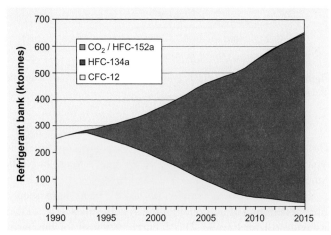

Figure 6.2. MAC refrigerant bank evolution from 1990 to 2015 (Clodic *et al.*, 2004).

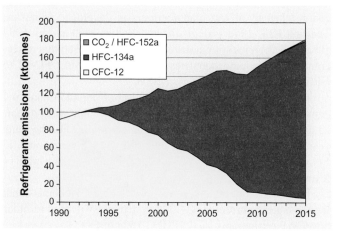

Figure 6.3. MAC refrigerant emissions from 1990 to 2015 (Clodic *et al.*, 2004).

and CFC-12 in the MAC fleet in the past (from 1990 to 2003) and for the future (up to 2015) according to a business-as-usual scenario[1].

Figure 6.3 and Table 6.3 present refrigerant emissions for the two refrigerant banks. Figure 6.4 and Table 6.4 transform CFC-12 and HFC-134a emissions into CO_2-eq emissions based on their GWP (values from IPCC Second Assessment Report (IPCC, 1996)).

Due to the huge difference of the GWP of these two refrigerants, Figure 6.4 underlines the significant decrease from approximately 900 MtCO_2-eq in 1993 to 610 MtCO_2-eq in 2003. The switch from CFC-12 to HFC-134a had a clear and positive effect on lowering global warming due to refrigerant emissions from MACSs.

6.3 Fuel consumption due to MAC operation

Nowadays the primary energy source for operating MACs is supplied by the car engine via two mechanisms: Firstly by the direct transfer of mechanical power to the compressor through a belt connected to the engine and secondly as the electrical power used by the fans and the control system.

Vehicle ACs consume more energy (including its weight) than any other auxiliaries currently present in vehicles. Yet

Table 6.4. MAC refrigerant emissions in CO_2-eq (Clodic *et al.*, 2004).

Year	CFC-12 (MtCO_2-eq)	HFC-134a (MtCO_2-eq)	CO_2 / HFC-152a (MtCO_2-eq)
1990	848	-	
2003	514	96	
2015	42	228	0.11

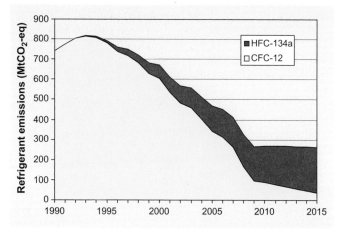

Figure 6.4. MAC refrigerant emissions in CO2-eq from 1990 to 2015 (Clodic et al., 2004).

---

[1] The Business as Usual Scenario integrates several assumptions detailed in Ashford *et al.*, (2004) and Clodic *et al.*, (2004): the Montreal Protocol schedule for CFC phase-out in developed and developing countries is taken into account, and economic growth rates are defined for 10 different regions as well as for the penetration of AC systems for the annual production of new cars.

this additional energy consumption is not currently taken into account in fuel economy regulatory test cycles. The fuel consumption due to MAC operation can only be measured if the usage profile and system energy consumption characteristics are known. Up until now, such data has been sparse and incomplete.

6.3.1   Hours of operation by climate/region

In order to estimate the annual fuel consumption of MACs, both the climatic conditions (including temperature, humidity and solar load) and the usage profile (including average mileage, time of day when driving occurs and urban or extra-urban driving conditions) must be known. As indicated in Figure 6.5, there are significant differences in the cooling energy required due to climatic conditions. Temperate climates such as that in Frankfurt, differ from hot and humid climates, such as in Orlando or Miami, by nearly a factor 5 in terms of heat load and therefore in cooling energy.

The usage profile first of all includes the cool-down operation with a higher energy consumption and the stabilized operating conditions with a lower energy consumption. To assess the global energy consumption impact of MAC, the repartition of urban and extra-urban driving durations needs to be known

and these differ per country.

At present there are no widely accepted standards for these parameters per country or region, even though there is a clear need for these. However, initial estimates can be made on the basis of several published publications (Benouali *et al.*, 2003b; Rugh *et al.*, 2004).

6.3.2   Fuel consumption and equivalent CO$_2$ emissions

Several factors need to be known in order to calculate CO$_2$ emissions from MAC operation systems. Firstly the mechanical and electrical energies used by MACs must be measured. Secondly the type of engine and the management of the energy for the MACs required from the engine must be considered. Then from the additional fuel consumption the CO$_2$ emission of MAC can be calculated.

The additional fuel consumption due to MAC operation mainly depends on climatic conditions. Table 6.5 shows the calculated mechanical power required by the compressor. This varies from 0.4–3.4 kW depending on the ambient temperature and the engine speed (Barrault *et al.*, 2003; Benouali *et al.*, 2003b).

MAC systems operate for about 24% of the vehicle running time in northern Europe, 60% in southern Spain and up

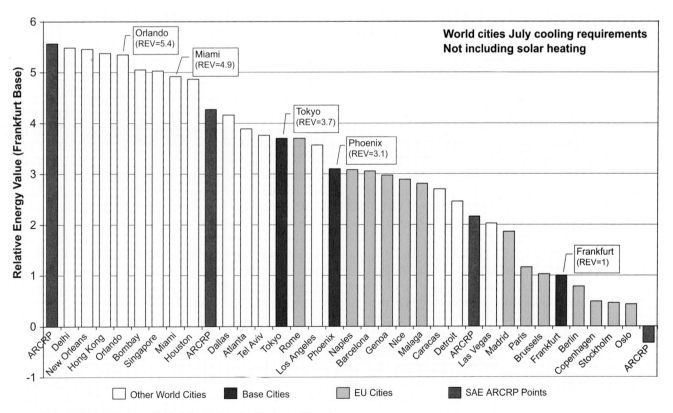

Figure 6.5. MAC cooling energy required depending on climatic conditions (based on Atkinson *et al.*, 2003).

Table 6.5. Mechanical power for HFC-134a system (kW) (Barrault *et al.*, 2003).

Rotation speed (RPM)	Mechanical power (kW)						
	Ambient air temperature (°C)						
	15°C	20°C	25°C	30°C	35°C	40°C	45°C
900	0.4	0.7	0.9	1.1	1.25	1.32	1.39
1500	0.5	1.2	1.8	1.8	1.9	2.3	2.7
2500	0.7	1.4	2.1	2.4	2.7	2.9	3.1
3500	0.7	1.5	2.2	2.6	3.0	3.2	3.4

Table 6.6. Fuel consumption and CO_2 emissions in Seville and Frankfurt (Barrault *et al.*, 2003).

City	Engine type	Total (kWh)	Energy consumption (L yr^{-1})	System weight (L yr^{-1})	Total (L yr^{-1})	% of the annual fuel consumption	Emissions (kg CO_2 yr^{-1})
Seville	Gasoline (HFC-134a)	473.00	55.54	7.61	63.15	6.8	148.0
	Diesel (HFC-134a)	769.25	75.79	7.92	83.71	6.4	221.5
Frankfurt	Gasoline (HFC-134a)	134.07	15.74	7.61	23.35	2.5	54.7
	Diesel (HFC-134a)	218.33	21.51	7.92	29.43	2.3	77.9

to 70% in Phoenix, Arizona. For European diesel engines the additional consumption ranges from 23 l yr^{-1} (Paris) to about 80 l yr^{-1} (Sevilla) (see Table 6.6).

If climatic conditions, engine type (diesel or gasoline) and user profile are taken into account, the annual additional fuel consumption is between 2.5 and 7.5%. This corresponds to an additional CO_2 emission due to MAC operation of between 54.7 and 221.5 kg CO_2 yr^{-1} per vehicle.

6.3.3   *Estimates of fleet-wide fuel consumption*

For the USA, an NREL study has estimated that 26 billion litres of gasoline per year (62 Mtonnes CO_2) are currently used for cooling (not including defrost). This is equivalent to 5.5% of US light-duty vehicle gasoline consumption (Rugh *et al.*, 2004). The study assumed that compressor power varied with rotations per minute (rpm) up to 3.7 kW at 3500 rpm. The assumed AC load penalized the fuel economy by 18% (22 mpg[2] to 18 mpg) for US cars and 14% (18.8 mpg to 16.2 mpg) for US trucks. Using the same methodology NREL (Rugh *et al.*, 2004) estimates the EU MAC fuel consumption to be 6.9 billion litres (equivalent to 16 MtCO$_2$) and the Japanese to be 1.7 billion litres (equivalent to 4 MtCO$_2$). These figures represent 3.2 and 3.5% of vehicle fuel use respectively. Worldwide, the total emissions of air-conditioned cars is estimated to be about 100 Mtonnes of CO_2 Tables 6.7 and 6.8 show the growth of air conditioned vehicles in China and India respectively. In 2002 Hu (Hu *et al.*, 2004) estimates the air-conditioned fleet to be 6.4

million vehicles. Based on data by Kanwar (2004), the Indian AC fleet was estimated to be 4.2 million cars in 2003.

6.3.4   *Future trends and possible improvements in energy efficiency*

Most of the MACs currently used in the total fleet are still basic technology (especially in the USA) with a fixed displacement compressor and manual control. These systems were not designed with a view to fuel savings. They produce an excess cooling capacity and the desired comfortable temperature is obtained by mixing heated air from the heater core to compensate for the excess cooling capacity. An on/off control is used to manage the system with respect to driving conditions (outside temperature and engine rpm). The Improved Mobile Air-conditioning (I-MAC) partnership is a current global effort to reduce direct refrigerant emissions by 50% and indirect emissions by 30%.

Since the mid-1990s, variable capacity technology (internal control) has been introduced, mainly in Europe and Japan, to limit the energy consumption of MACs. Comparisons of the three types of compressor control: On/off, internal control and external control (introduced in 2000) show that significant energy savings can be achieved with the appropriate control (Karl *et al.*, 2003).

However, a reduction in the requested cooling capacity is another option for reducing fuel consumption. Insulation of doors and roof, limitation of window size and special glass for solar load limitation are all parameters which influence the final energy consumption of MACs. In practice these solutions have a limited usefulness due to the additional cost, extra weight, consumer demand for larger windows or safety issues related to driver visibility.

---

[2] 'mpg' stands for 'miles per gallon'. 1 mile per US gallon equals about 0.425 km per litre.

Table 6.7. Evolution of air-conditioned vehicle fleets in China (Hu, Li and Yi, 2004).

	1995	1996	1997	1998	1999	2000	2001	2002
Air-conditioning automobile[1] population[2] (thousand units)	1650	2090	2650	3210	3900	4700	5530	6430
CFC-12 air-conditioning automobile population (thousand units)	1590	1930	2330	2590	2770	2910	2920	2760
HFC-134a automobile population (thousand units)	60	160	320	620	1130	1790	2610	3670

[1] automobiles = vehicles including buses, trucks and cars.
[2] population means fleet

Table 6.8. Evolution of the annual sales of air-conditioned cars in India (Kanwar, 2004).

Year	1996	1997	1998	1999	2000	2001	2002	2003
Car[1] AC sales (thousand units)	202.5	259	278.2	369.6	464.9	459.8	488.6	616.2

[1] car means cars only

Low consumption MAC systems can only be developed if there are accepted standards for measuring the energy consumption of the MAC system and the additional fuel consumption of the vehicle equipped with a given MAC system.

6.3.5 *Future vehicles and MACs*

Environmental concerns are also driving technology changes in powertrains, which in turn is affecting MAC design.

The first trend is improvements in the energy efficiency of internal combustion engines (diesel or gasoline). This trend affects both the heating and the cooling modes of the MAC system. In cooling mode, a higher engine energy efficiency results in a higher relative energy penalty due to MAC. This requires significant efforts to improve MAC energy efficiency. High-efficiency engines result in a heating cabin deficit which necessitates the installation of an additional heating function. This is why efficient heat pump systems are still being developed in competition with current existing solutions based on electrical heaters or burners.

A second trend is to introduce new functions such as stop-start (engine is stopped at idle and then restarted), stop & go (engine is stopped at idle, then restarted so that a short electrical driving period is possible) and electrical driving mode (as with a hybrid powertrain). In order to improve fuel consumption, especially in city traffic conditions, all of these solutions are based on a more or less deep electrification of the car. However, with current MACs, stopping the engine means stopping the AC and vehicle occupants do not accept such a loss of comfort. Usually the only strategy for sustaining comfortable conditions is to restart the engine whilst still in idle. To solve this problem, different solutions have been investigated with respect to the availability of electrical energy: Cooling power storage, hybrid compressors and electrically-driven compressors. Some of these have being introduced onto the market. For example, the two-way driven compressor has been developed for mass-produced, mild-hybrid vehicles which do not generate enough electrical power (Sakai *et al.*, 2004). This compressor can be driven by the engine belt whilst the engine is running and by the electric motor during engine stop. An electrically-driven compressor has also been developed for mass-produced, strong-hybrid vehicles (Petitjean *et al.*, 2000; Takahashi *et al.*, 2004). Even if the engine stops, the air-conditioning can still be operated efficiently.

A third trend is to change the energy source of the powertrain by replacing the internal combustion engine with a fuel cell. MACs adapted to this type of powertrain will benefit from the high level of electrical power available and will be able to compensate long stop & go as well as heating or cooling deficits of the cabin. One example exists in Japan, where there is a fuel cell vehicle equipped with electrical hermetic CO_2 heat pump systems.

Adapted efficient MACs will be a key issue for the marketability of these new powertrains. Methods to qualify efficiency or fuel consumption will have to be adapted to take into account both the special energy management of this new type of powertrain and the adapted driving cycle, especially in city traffic.

6.4 Technical options for reducing direct and indirect emissions

Concerns about global warming associated with HFC-134a emissions mean that the impact of the MAC industry on anthropogenic greenhouse gas emissions needs to be assessed.

This involves assessing both industry emissions and proposed emission-reduction initiatives for their environmental impact and cost-effectiveness. Such initiatives include improving the current system as well as investigating possible alternative refrigerant systems with a lower climate impact.

6.4.1 Improved HFC-134a systems

HFC-134a systems can be improved in two ways. Firstly by improving the containment of the refrigerant and secondly by increasing the energy efficiency. Improvements to the HFC-134a system are concerned with optimizing current systems and not in developing a completely new design.

6.4.1.1 Improved leak tightness

The future European Union F-Gas regulation will limit maximum annual emissions. This will lead to significant new efforts for improving the leak tightness of MAC systems.

At present there is not an agreed leakage test method to guarantee the leakage rate of an MAC system installed in a car. Each supplier of compressors, condensers, evaporators, hoses, etcetera, has their own methodology, using trace gases such as helium, for checking the leak tightness of their components. SAE standards (SAE, 1989, 1993) define a test method for hose permeation but none of these permit certification of the emission level of AC systems installed in a car. Several projects are underway to define a certification method that can ensure the certified level of refrigerant emissions. Various publications (Clodic and Fayolle, 2001; Clodic and Ben Yahia, 1997) indicate a large range of leakage rates depending on the technology used for hoses, compressor shaft seals and service valves. A certification method will certainly need to be developed for each component: The refrigerant lines, the two heat exchangers and the compressor. Complementary calculations from test results are necessary to predict annual emission. Moreover, the leak rates depend on the refrigerant pressure and so in turn on the climatic conditions under which the car operates throughout the year.

6.4.1.2 Servicing issues and vehicle end of life

In developed countries car servicing is performed by at least three categories of garage: After-sales service from car manufacturers, private garages and specialized networks servicing only certain parts of vehicles (batteries, exhaust pipes, brakes, tyres). Europe now has specialized networks for MAC servicing. In the USA the MACS Association (Mobile Air-Conditioning Society) has developed a training programme which has generalized recovery systems in nearly all garages in line with the US regulation which has made the recovery of CFC-12 mandatory since 1992 and HFC-134a mandatory since 1996. In order to significantly reduce service-related emissions, technicians should be trained to use a recovery and recharge system with near zero emissions as a standard procedure. Leak detection with high sensitivity leak detectors (in the range of 1 g yr^{-1}) and sufficient knowledge to fix the leaks when found, are comple-

mentary steps towards realizing low refrigerant emissions during servicing. Nevertheless many of these steps forward can be compromized by end-users recharging MACs themselves if disposable cans of HFC-134a are available. The global emissions related to these practices are high, and disposable cans of HFC-134a will be forbidden by new EU regulation.

In many European countries, recovery during servicing is not yet mandatory and this might result in high emissions. However, good practices have been adopted in many after-sales service garages and specialized networks, which systematically use standard recovery equipment when the circuit needs to be opened.

The increased use of air-conditioned cars in developing countries requires the training of servicing personnel and the availability of recovery/recycling equipment. The current phase out of CFC-12 has permitted the implementation of good practices in some developing countries, but a number of efforts are still needed to avoid large emission rates due to poor servicing practices.

Looking to the future, the aim is to produce very low-emission HFC-134a systems by decreasing the level of AC system leak tightness to the extent that no recharge is required during the system's lifetime.

The recovery of refrigerant at vehicle end of life could be integrated into the global management of the recycling process of cars. Such a global recovery and recycling process would significantly reduce the additional cost associated with refrigerant recovery. At the EU level, refrigerant recovery will be integrated in the End-of-Life Vehicle Directive (OJEC, 2000).

6.4.1.3 High efficiency systems

As indicated in section 6.3.4, a number of programmes are underway to achieve better control of compressor power. Despite this, see (Benouali *et al.*, 2003c; Benouali *et al.*, 2003a), the coefficient of performance (COP) measured under realistic conditions (typical driving conditions) is low, 1.3–1.4 for an outdoor temperature of 35°C. The use of an external control compressor yields significant gains in the range of 25–35% on the COP, even if the mechanical efficiencies of external control compressors are in the same range as internal control compressors. Up until now systems have been designed for low cost, reliability and cooling capacity but not for energy efficiency. A number of improvements can still be realized in the development of high efficiency compressors. Improvements in the evaporator and condenser design as well as the control system will result in better airflow management and therefore energy efficiency.

With respect to energy efficiency improvements, progress could be made in standardizing the measurement of energy consumption due to MAC operation in order to facilitate the comparison of technical proposals (see section 6.3.4).

6.4.1.4 LCCP/TEWI

To calculate Life Cycle Climate Performance (LCCP), all types of emissions over the entire cycle should be taken into account, including energy consumption related to refrigerant production,

Table 6.9. Example of TEWI calculations for a MAC system operating in the Frankfurt area (Barrault *et al.*, 2003).

	Current system		Improved system	
	Diesel	**Gasoline**	**Diesel**	**Gasoline**
Indirect emissions (kg CO_2 yr^{-1})	77.9	54.7	54.5	38.3
Direct emissions (kg CO_2-eq yr^{-1})	138–345	138–345	30.4	30.4
TEWI (kg CO_2 yr^{-1})	215.9–422.9	192.7–399.7	84.9	68.7

direct emissions throughout the lifetime and indirect emissions due to energy consumption by the AC system. For HFC-134a AC systems, a number of publications (UNEP, 2003; Barrault *et al.*, 2003; Schwarz and Harnisch, 2003) indicate that the most significant parameter is the direct emission level. Depending on the recovery policy, the annual refrigerant emissions for current systems are estimated to be 0.106–0.266 kg yr^{-1} leading to 138–345 kg CO_2-eq yr^{-1}.

If the previous results of CO_2 emissions related to MAC operation and those different emission levels are taken into account (see section 6.2.2), the total equivalent warming impact (TEWI) of a given MAC system can be calculated. Table 6.9 presents an example of TEWI calculations for gasoline and diesel engines for a MAC system operating in the Frankfurt area.

Table 6.9 summarizes possible gains on indirect emissions due to energy consumption and direct emissions of refrigerant. Considering the extreme values, it can be seen that refrigerant emission reductions can be a highly effective method for reducing emission impacts. However, at this stage it has yet to be demonstrated that MAC systems can be designed with servicing limited to ruptures and accidents.

Based on current knowledge of possible energy efficiency improvements as well as reduction of emissions, Table 6.10 compares the old CFC-12 and the current HFC-134a systems to the improved HFC-134a, CO_2 and HFC-152a systems.

6.4.2 CO_2 systems

6.4.2.1 Description of system

Carbon dioxide has zero Ozone Depletion Potential (ODP). Its GWP of 1 is negligible compared to the GWP of HFC-134a. Moreover, there is no need to specifically produce carbon dioxide, since it can be recovered from industrial waste gas. Carbon dioxide (when used as refrigerant it is called R-744) is cheap, readily available in sufficient quantities throughout the world and its properties are well known and documented. If used as refrigerant there is no need for recovery, which would necessitate servicing, and the recycling cost would be totally eliminated. The refrigerating equipment safety standard (ASHRAE 34, EN-378) classifies CO_2 as an A1 refrigerant, a non-toxic and non-flammable refrigerant. But if accidentally lost in large quantities, CO_2 has a certain asphyxiation risk that needs to be mitigated by engineering systems. A refrigeration cycle with carbon dioxide as refrigerant operates at comparatively high pressures (5–10 times higher than HFC-134a systems).

However the safety risk of these high pressures is comparable to current HFC-134a systems due to the small volume used and the low refrigerant mass of the system.

As the critical temperature is approached or even exceeded, the ideal cycle efficiency decreases with increasing high-side pressure. On an ideal basis, the efficiency of a transcritical CO_2 cycle is unfavourable compared with that of a HFC-134a cycle. However, for real cycles this is compensated by the favourable transport properties of CO_2 (high heat transfer and low pressure loss), the inclusion of an internal heat exchanger and the improved compressor efficiency due to low pressure ratios. Under normal operating conditions for vehicle air-conditioning systems the carbon dioxide cycle often operates in the 'transcritical' mode due to the low critical temperature (31°C). For example, Pettersen (1994) showed that the high-side pressure influences the COP of the system and there is an optimum pressure for this. He suggested that during normal operation a high-side pressure close to the optimum pressure should be selected in order to maintain a satisfactory COP. This means the pressure should be kept at a certain level dependent on the heat rejection temperature (ambient air temperature).

A special flow circuit enabling efficient operation and high-side pressure control has been designed and patented by Lorentzen *et al.* (1993) (see also Lorentzen and Pettersen, 1993). This particular refrigeration cycle is shown in Figure 6.6 and serves as the worldwide basis for the majority of car AC systems which use carbon dioxide as a refrigerant. The p-h diagram in Figure 6.6 depicts two different high-side pressures for different control conditions for which the compression process has been calculated with the same isentropic efficiency.

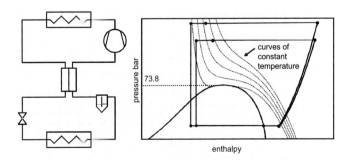

Figure 6.6. Refrigeration cycle and pressure-enthalpy diagram of the transcritical cycle commonly used for car applications (Lorentzen and Pettersen, 1993).

Table 6.10. Comparison of MAC options.

	HFC-134a (reference)	Improved HFC-134a	CFC-12 (old type) development)	CO_2 (under development)	HFC-152a (direct system, under devolpment)
Substance characteristics					
Radiative efficiency (W m^{-2} ppb^{-1})	0.16	0.16	0.32	See ch. 2	0.09
Atmospheric lifetime (yr)	14	14	100	See ch. 2	1.4
Direct GWP (100-yr time horizon)	1410	1410	10,720	1	122
Technical data					
Stage of development	Commercial	Near commercial	Commercial	Demonstration	Demonstration
System lifetime	12–16	12–16	12–16	12–16	12–16
Cooling capacity (kW)	6	5	6	6	6
Charge (kg/system)					
- range	0.7–0.9	0.6–0.75	1–1.2	0.5–0.7	0.45–0.55
- relative figures (%)	100	80	125	70	70
# Charges over lifetime	2–3	1–2	4	2–4	1–2
Coefficient of Performance (COP)	0.9–1.6	1.2–2.5	0.9–1.2	0.9–2.0	1.2–2.0
Seasonal Performance Factor (SPF)	NA	NA	NA	NA	NA
Energy consumption (relative figures)	100	80	130	70	70
Emissions per functional unit					
Direct emissions					
- in % of charge yr^{-1}	15	7	20	15	7
- in kg CO_2-eq yr^{-1}	166	64	1782	0.09	4.9
- relative figures (%)	100	40	1043	0.05	2.9
Indirect CO_2-emissions (kg CO_2 yr^{-1})					
- Sevilla	184	147	239	129	129
- Tokyo	126	101	163	88	88
- Phoenix	369	295	480	258	258
End-of-life emissions recovery efficiency[2]	0	50	0	0	50
TEWI (kg CO_2-eq 14 years)[1]					
- Sevilla	4900	2954	28,294	1807	1875
- Tokyo	4088	2310	27,230	1233	1301
- Phoenix	7490	5026	31,668	3613	3681
(without recovery)					
Costs per functional unit					
Investment costs (US$)	(215 US$)	+24–36 US$	NA	+48–180 US$	+48 US$

[1] The GWP values used for the calculations are taken from the Second Assessment Report (IPCC, 1996).

[2] Recovery has not been taken into account for the TEWI calculations, due to large uncertainties of the recovery effectiveness and so the average direct emission per year for 'improved HFC-134a systems' is 100 g yr^{-1}.

With this refrigeration cycle the high-side pressure can be controlled, for example, by an electronic expansion valve. Each high-side pressure corresponds to a particular refrigerant charge circulating in the cycle. The receiver located between the evaporator and the internal heat exchanger on the low-pressure side stores the refrigerant charge which is not needed for the particular operating condition. The presence of the internal heat exchanger provides protection for the compressor from liquid slugging and may yield to an increased cooling capacity at high ambient air temperature conditions.

A CO_2 refrigeration cycle different from the Lorentzen/ Pettersen cycle approach was used by Sonnekalb (2002) (see also Sonnekalb and Köhler, 2000) for bus AC systems which use carbon dioxide as refrigerant. Based on the fact that the COP compared with high-side pressure curve is relatively flat, Sonnekalb designed a low-cost CO_2 system for bus air-conditioning. This system (which can be seen in Figure 6.7) has only four major components with a fixed refrigerant charge.

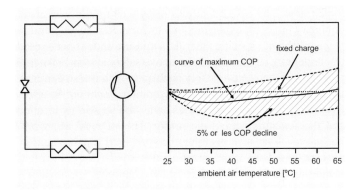

Figure 6.7. Refrigeration cycle and charge-ambient air temperature diagram of a low-cost transcritical cycle used for bus applications (Sonnekalb, 2002).

Sonnekalb showed that it is possible to operate this particular cycle over a wide range of ambient air temperatures close to the maximum COP without a high-side pressure control. Figure 6.7 presents on the right-hand side a charge-ambient air temperature diagram with the optimal COP curve. The range of a 5% or less decline of this optimal COP can also be seen as a shaded area. The dashed horizontal line in the diagram shows that it is possible to cover a wide range of ambient air temperature conditions with a fixed charge whilst incurring only minor COP losses (5% or less).

In the years 1996 and 1997 two CO_2 prototypes based on this cycle were installed in standard city buses in Bad Hersfeld, Germany. By the end of 2003 the AC systems of these two buses had together accumulated more than 4000 running hours on the road, indicating that this CO_2 system is technically and economically feasible for bus AC (for more details see also Köhler *et al.*, 1998 and Foersterling *et al.*, 2002).

Since 1997 prototype CO_2 systems have been demonstrated by many vehicle manufacturers and their suppliers in a wide range of vehicles (Wolf 2004). For example, BMW successfully ran a small test fleet of 7 vehicles with prototype AC systems between 2000 and 2002 (Mager 2002). Toyota have also run a first fleet test in Japan and the USA using approximately 15 fuel cell hybrid vehicles .

If used in a heat pump mode, CO_2 has a superior refrigerant performance (see for example Heyl and Fröhling, 2001 or Hesse *et al.*, 2002). Modern fuel-efficient direct injection gasoline or diesel engines may not have enough waste heat to sufficiently heat the passenger compartment. Additional heaters (for example positive temperature coefficient (PTC) heaters, fuel burners, or other technologies) increase fuel consumption and CO_2 emissions. The CO_2 heat pump might be an attractive future alternative with respect to performance, energy efficiency and total cost (AC plus heater). More research is needed to determine which is the most suitable for car application.

In summary, CO_2 systems need a new design for all components (compared to usual HFC-134a technology). This involves an ongoing learning process.

6.4.2.2 *Energy efficiency*

Carbon dioxide is a typical high-pressure refrigerant with a correspondingly high volumetric cooling capacity. Therefore, compared to low-pressure HFCs, higher cooling capacities can be achieved in a given volume. This results in excellent cool down curves (or heat up curves for the heat pumps) as demonstrated, for example, by Vetter and Memory (2003). However, comparing the energy efficiency and fuel consumption of a carbon dioxide and corresponding HFC system is far more difficult. From the results in a number of publications it can be surmised that compared to an improved HFC-134a system, carbon dioxide has a comparable or even better COP and a correspondingly lower fuel consumption, especially for lower ambient air temperatures. At high ambient air temperatures, improved HFC-134a systems have advantages at lower compressor speeds. Based on measurements made by Hrnjak (2003), Wertenbach (2004) calculated the AC energy demand for typical European and US driving cycles. His results show different energy demands for Europe and the USA are due to the selected running time in the drive cycles and the different climatic conditions. Nevertheless the energy efficiency benefits of the systems are about the same for Europe and the USA. Compared to a baseline HFC-134a system, an improved HFC-134a system showed 23% less energy demand whereas an efficiency-optimized CO_2 system showed about 30% less energy demand. Hafner *et al.* (2004) obtained comparable results in their energy analysis.

Compared to HFC-134a, CO_2 has a disadvantage of low efficiency at idling and high load conditions. Compact vehicles are generally at a disadvantage as far as the refrigeration cycle efficiency is concerned compared to medium or large vehicles. This is due to space constraints which limit the condenser and evaporator sizes. Actual vehicle tests have shown that in compact vehicles, the increase in compressor power consumption with CO_2 has an adverse effect on vehicle driving characteristics such as acceleration and fuel economy, even though equivalent or better cooling performance can be maintained (Kobayashi *et al.*, 1999). Further technical developments are required for CO_2 compact vehicles to counteract this. An ejector, which can recover the expansion energy and reduce the compressor power might be a potential solution (Takeuchi *et al.*, 2004).

6.4.2.3 *Cost*

The real cost of a CO_2 AC system (or the additional cost compared to current AC systems) is hard to estimate as the CO_2 system has not yet entered mass production. The additional costs compared to a current system are expected to be between 48 and 180 US$ (Mager *et al.*, 2003; Barrault *et al.*, 2003; NESCCAF, 2004) for the first generation of mass-produced systems. No cost data are available regarding the conversion of the service system.

6.4.2.4 *LCCP/TEWI*

Aspects concerning the certainty of energy efficiency also apply to calculations of the Life Cycle Climate Performance (LCCP) or the Total Equivalent Warming Impact (TEWI). The results from different publications are not always comparable, as the specific results depend strongly on the assumptions made. In an LCCP study, Hafner *et al.* (2004) compared an improved HFC-134a system with a corresponding CO_2 system (based on measured data published by Hrnjak, 2003). The authors assumed NEDC and US FTP 75 driving cycles. The total hours with AC on were based on regional climate data and usage profiles provided by Sand *et al.* (1997) and Duthie *et al.* (2002). The direct HFC-134a emissions were calculated assuming total system losses of 60 g yr-[1]. The end-of-life recovery was assumed to be 80%. The authors also took into account the 77 kg CO_2-eq emissions caused by the production of 1 kg of HFC-134a (see Campbell and McCulloch, 1998). The authors analyzed the LCCP of the HFC-134a and the CO_2 systems for different climatic regions in the USA and Europe. Their results show a 17–49% reduction in LCCP for the CO_2 system compared to the improved HFC-134a system. Table 6.10 shows a reduction of nearly 40% in the LCCP for the CO_2 system.

6.4.2.5   *Leakage rate of complete system*
Froehling *et al.*, (2003) assessed the upper acceptable limits of refrigerant leakage. He assumes that the operating time of CO_2 AC systems without recharging should be about 6 years. This leads to an annual emission rate of 25g or less, half of which is related to the shaft seal of the compressor. Additional research and development on low leakage hoses and fittings, compressor shaft seals, durability and accurate methods of leak rate measurements is currently being undertaken to realize this goal.

6.4.2.6   *Reliability*
Air-conditioning with CO_2 is a new system and virtually all of the components need to be redesigned. In terms of reliability the main tasks to be completed are:
- Development of economical, leak-tight CO_2 hoses and compressor shaft seals;
- Demonstration of reliability under real-world conditions;
- Development of service and maintenance best practices.

In order to ensure uniform quality and reliability, a workgroup of German OEMs has proposed a component specification Standard CO_2 AC System (Mager *et al.*, 2004). This specification defines the main and auxiliary components of the system and sets their operating conditions, performance and limits.

6.4.2.7   *Safety*
Due to the high pressure of the CO_2 systems risks of rupture which could lead to burst hazards must be mitigated. The basic design concept of car air-conditioners is described in SAE J639 (SAE, 1999). CO_2 car air-conditioners are currently being standardized and the revisions of the set values for safety devices such as the design pressure, proof pressure and pressure relief valve took into account the significant pressure levels that can arise in CO_2-refrigerant car air-conditioners (Kim *et al.*, 2003; Wertenbach, 2004).
Leakage of CO_2 into the passenger cabin creates risks of

impairment. The US EPA allows 6% v/v and 4% v/v as upper boundary limits for exposure of up to 2 minutes and 60 minutes respectively and specifies that the exposure should not exceed 6% under any circumstance. An active safety system, such as a refrigerant-based sensor safety concept (which is also currently under scrutiny for flammable refrigerants), might not be effective enough due to uncertainties about the sensor, its location and the actuators behind the sensor. The VDA ad hoc working group of car manufacturers has proposed a passive safety concept based on supplementing the standard specifications mentioned in Section 6.4.2.6. These additional specifications would cover all AC systems from manual to automatic control and would consider all vehicle operating conditions (Rebinger, 2004). This passive safety concept is based on the so-called 'safe evaporator', which means a negligible risk of leaking or bursting under any operating condition of the car.

As well as the aforementioned issues, the safety of the 'service' offered in the marketplace – how to handle CO_2 safely – is another important safety issue. Devices such as a refrigerant charging machine, which can read the amount of refrigerant from bar-coded information in the service manual and automatically charge CO_2, have already been developed (Lorenz, 2002).

6.4.3   **HFC-152a systems**

6.4.3.1   *Description of a direct expansion system*
HFC-152a systems are identical to the current HFC-134a system, but may have added safety features, such as solenoid valves or discharge devices.

A direct expansion HFC-152a system was demonstrated at the 2002 SAE Alternate Refrigerants Symposium (SAE, 2002a). This was a joint venture between the US government, a major car manufacturer and a major MAC system supplier (Andersen *et al.*, 2002). The large vehicle system contained just 0.45 kg of refrigerant and compared to the equivalent current HFC-134a system it showed a better cooling performance for high loads and a comparable cooling performance for low loads. In a direct refrigerant replacement with no component changes, HFC-152a showed a 3–17% lower energy usage than HFC-134a and was on average about 10% lower (Ghodbane *et al.*, 2003).

Leakage of HFC-152a into the passenger cabin creates risks of fire. The US EPA allows 3.7% v/v for up to 15 s as an upper boundary limit in any part of the free space of the passenger cabin.

The safety system (Figure 6.8) consists of an HFC-152a sensor, mounted in the evaporator case, capable of detecting the presence of HFC-152a. If detected above a threshold level, a signal is sent to a controller, which activates a directed relief mechanism. The relief mechanism consists of two squib-initiated discharge devices, one on the high-pressure side and one on the low-pressure side of the system. The discharged refrigerant is directed away from the vehicle by hoses attached to the discharge devices. Therefore, the system is designed to safely discharge the refrigerant away from the vehicle when leakage

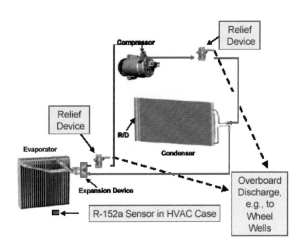

Figure 6.8. Direct HFC-152a System with Safety System.

is detected or a vehicle crash situation occurs. In the case of a crash, the discharge would be activated by the airbag crash-sensing system. Such a discharge quickly reduces the pressure in the system. This rapidly eliminates the driving force (pressure) for unwanted leakage elsewhere, thereby reducing any risk. The sensing system is capable of continuous monitoring when the vehicle is being driven and intermittent monitoring when the vehicle is parked.

In the event of a leak into the passenger cabin the safety system is reported to be able to discharge the refrigerant so fast that no significant amount of HFC-152a enters the passenger cabin.

6.4.3.2 Description of a HFC-152a indirect system

At the 2003 SAE Alternate Refrigerants Symposium (SAE, 2003), a direct expansion HFC-152a system and a secondary loop HFC-152a system were demonstrated (Ghodbane and Fernqvist, 2003).

The secondary loop system (Figure 6.9) was jointly demonstrated by a major European car manufacturer and a major MAC system supplier (Ghodbane and Fernqvist, 2003). The system was demonstrated on a Sport Utility Vehicle (SUV) with front and rear evaporators. A secondary loop system (Figure 6.9) would overcome the concern of leakage into the passenger cabin by allowing the refrigerant to be contained under the hood where it is completely separated from the passenger compartment. The refrigerant cools a chiller (liquid-liquid heat exchanger), which, in turn, chills a water-glycol mixture that is then pumped into the passenger cabin heat exchanger(s) for cooling. This system would use the same components as the current HFC-134a system, with the addition of a chiller and liquid pump. It would therefore require a limited capital investment. For the dual evaporator SUV, the HFC-134a charge was 1.3 kg compared to just 0.59 kg of HFC-152a for the secondary loop system – a significant reduction. This vehicle had one of the highest cooling performances of all of the vehicles demonstrated at the symposium.

From an energy perspective, a secondary loop requires about 10% more energy than a direct system. However, the greater energy efficiency of HFC-152a compared to HFC-134a (about 10%) would result in comparable cooling performance with comparable energy efficiency (Ghodbane and Fernqvist, 2003).

From a general safety perspective, the small amount of HFC-152a refrigerant needed, combined with its relatively low flammability and an effective safety system such as those described above, would result in a vehicle that would be both safe to drive and safe to park in a garage. Of course, safety during vehicle servicing and repair must also be considered.

6.4.3.3 Energy efficiency

The HFC-152a system described in 6.3.3.1 was an HFC-134a system that was first tested with HFC-134a and then simply charged with HFC-152a to provide a direct comparison with no component changes. HFC-152a provided better cooling (2–3°C) and used 7–20% less energy than HFC-134a, reinforcing the results of the 2002 study. This vehicle had one of the highest cooling performances of all of the vehicles demonstrated at the symposium.

6.4.3.4 System costs and availability of HFC-152a

At present, only relatively small amounts of HFC-152a are produced globally; the major manufacturer is in the USA. Three manufacturers producing small quantities exist in China and one in Japan. Production capacity would have to be developed for HFC-152a to be used in MAC systems and with the necessary capital investment this would take several years to achieve.

Due to the simpler processing, HFC-152a ought to cost less than HFC-134a. However, the MAC system would cost an additional 25–48 US$ to cover the cost of the safety system for a

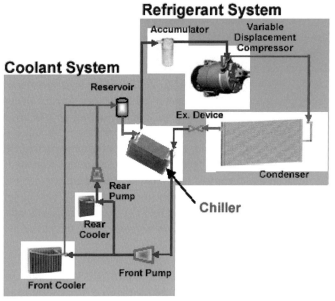

Figure 6.9. Secondary Loop HFC-152a System.

direct system or the additional cost of the secondary loop for an indirect system.

6.4.3.5   LCCP, TEWI

HFC-152a has been proposed as a replacement for HFC-134a because its global warming potential is about 10 times lower and it can be used with current HFC-134a components without any major modification (Baker *et al.*, 2003; Ghodbane, 2003; Scherer *et al.*, 2003).

HFC-152a has been assigned a GWP of 122 as opposed to 1410 for HFC-134a leading to a reduction in global warming impact of more than 90%. Moreover, the lower liquid density of HFC-152a compared to HFC-134a implies a refrigerant charge that is only 65% of the reference HFC-134a charge. For example, an HFC-134a system using 0.5 kg would be replaced by an HFC-152a system using 0.325 kg. This reduction in refrigerant charge will lead to lower emissions in the event of rupture and at vehicle end of life. The greater energy efficiency of the HFC-152a over HFC-134a would result in enough fuel savings to offset the GWP of HFC-152a. The net result of replacing HFC-134a with HFC-152a would be the elimination of the direct climate impact associated with current HFC-134a MAC systems. This has also been shown by Hafner *et al.* (2004). Table 6.10 shows a reduction of almost 40% in LCCP for the HFC-152a system compared to an energy-optimized HFC-134a system.

6.4.3.6   Safety aspects of future vehicle air-conditioning systems

HFC-152a is a moderately flammable gas and is listed by the American Society of Heating, Refrigerating and Air-Conditioning Engineers (ASHRAE) Standard 34 (ASHRAE, 2001) as a Class A2 refrigerant, defined as being of lower flammability than Class A3 (hydrocarbons such as propane). As a reference, HFC-134a is Class A1, low toxicity and non-flammable. The flammability of HFC-152a will probably require added safety features such as those described below.

HFC-134a systems and the CFC-12 systems they replaced, are considered safe for the intended use. Replacement refrigerant systems must also provide a comparable level of safety for both vehicle occupants and service technicians. SAE J639 (SAE, 1999), SAE J1739 (SAE, 2002b) and refrigerant manufacturer's safety data information all serve as reference documents for designing safe systems. The latest draft of SAE J639 (SAE, 1999) provides an example of the conditions for safe design that must be met. For example, 'For refrigerants contained in components that are located within the passenger compartment, or in direct contact with passenger compartment airflow (e.g., evaporators and associated lines and fittings), system design shall be such that neither a harmful nor a flammable condition will result from a release of refrigerant into the passenger compartment.'.

SAE J1739 (SAE, 2002b) requires a safety assessment of proposed systems (by Failure Mode and Effects Analysis using the Risk Priority Number) for the purpose of identifying all anticipated events that result in safety concerns and then taking appropriate preventative, or mitigating action.

In addition, all regional, national and local regulations and codes for the safe handling of new refrigerants and new refrigerant systems must be satisfied (e.g., flammability and pressure vessel requirements). Managing costs and risks will be an important part of implementing alternatives.

6.4.4   HC-blend systems

In Australia and the USA, hydrocarbon blends, mostly HC-290/600a (propoane/isobutane) blends, have been introduced as drop-in refrigerants to replace CFC-12 and to a lesser extent for HFC-134a. The real number of cars that have been retrofitted with such HC refrigerant blends is unknown. Maclaine-Cross (2004) has estimated that it is about 330,000 vehicles in Australia, although no data exist to confirm this. No accidents have been reported to date. These retrofits with HCs are legal in some Australian states and illegal in others and in the USA. US EPA has forbid the uses of HCs for retrofit but has considered the possible use of HCs for new systems, providing safety issues are mitigated.

HCs or HC-blends, when correctly chosen, present suitable thermodynamic properties for the vapour compression cycle and permit high energy efficiency to be achieved with well-designed systems. They have zero ODP and low GWPs, but their high flammability (lower flammability limit in the range of 1.8–2.1% v/v) means that they have not been considered by car manufacturers.

Some studies have been carried out using hydrocarbons in indirect systems, in which a water-glycol circuit provides cooling capacity in the cabin. This water-glycol coolant is in turn refrigerated by the refrigerating system using hydrocarbons (same system as the secondary loop system presented above for HFC-152a). Nevertheless, even with indirect systems, HCs are not seen by vehicle manufacturers as replacement fluids for mass-produced AC systems.

6.4.5   Alternative technologies to vapour compression

There are two types of alternative technologies: The Brayton-Joule air cycle and a number of heat-generated cooling systems.

6.4.5.1   Air cycle

The air cycle has been thoroughly investigated by Bhatti (1998). Even if air is used as a refrigerant, with a zero GWP, the energy efficiency of the system is very low. Calculations of air cycle efficiency compared to the current HFC-134a system show an energy penalty of about 35%. Nevertheless, from the TEWI analysis the author concludes that the amount of CO_2 emitted per year is in the same range (303 kg CO_2 for the air cycle compared with 282 kg CO_2 for current HFC-134a systems).

6.4.5.2   Metal hydrides

A review by Johnson (2002) evaluates the opportunities for different cooling cycles using the available heat from the exhaust gas. A first approach is the use of a metal hydride heat pump which makes use of the adsorption and desorption of hydrogen by metals. Desorption is an endothermic reaction and so absorbs heat, whereas adsorption is exothermic and releases heat. Metal hydride systems are interesting due to the small number of moving parts. No refrigerant is needed but the COP is significantly lower, in the range of 0.4–1.5. The possible heat recovery from the exhaust gas is also interesting. A first prototype has been developed in the USA, with a 5 kW AC system. The system weight was 22 kg and the COP was 0.33.

6.4.5.3   Absorption systems

Absorption systems that use either ammonia water or water-LiBr as working solution are well known. These systems use mainly heat but they need electrical energy for the fans and pump. The main drawback of absorption systems for cars is the heating-up time needed before cooling capacity can be produced to create the temperature level necessary in the boiler to produce refrigerant vapour. This system therefore needs a refrigerant storage as well as a minimum operating time of about half an hour before efficient cooling operation.

6.4.5.4   Adsorption zeolite systems

Zeolite systems are similar to metal hydride systems but use zeolite and water instead of metal hydride and hydrogen. The adsorption / desorption cycles usually need several separate reaction chambers for successive cycles in order to reach pseudo-continuous operation. Studies have been actively carried out for electrical vehicles using a burner to operate the AC system. The COP is considered to be relatively low (0.3–1.2). No prototype has been built in real cars in order to verify the performances.

6.4.5.5   Thermo-acoustics

Sound waves create small temperature oscillations. If those oscillations release or absorb heat within a regenerator, a significant temperature difference can be realized between the hot and the cold ends. A resonant cavity is used to enhance the efficiency of thermo-acoustics systems. A COP up to 2 has been achieved on a laboratory test bench. However, the integration of the system into cars could be difficult due to the possible size of system needed for a capacity of about 3 kW.

6.5   Heat-load reduction and new design of AC systems

Independent of the vapour compression system, research is underway to improve comfort conditions in car cabins. If energy consumption is not integrated as a constraint in those developments then it might be compromized by them. For example, panoramic roofs may increase heat loads in the cabin, but development of efficient reflecting glasses can mitigate the extra heat loads while improving the vision of car passengers. New designs such as ventilated seats associated with heating and cooling systems may lower the needed cooling or heating capacity, due to a far more effective diffusion of cooling or heat close to the body. Insulation of the car itself may have contradictory effects for summer and winter conditions and compromises will have to be found in order to improve the fast heat-up of the vehicle in the winter and fast cool-down in the summer. The AC system and car cabin must be jointly designed to achieve thermal comfort and low energy consumption.

References

Andersen, S.O., J.A., Baker, M., Ghodbane and W.R., Hill, 2002: R152a Mobile Air Conditioning System. Proceedings of the 2002 SAE Alternate Refrigerant Forum, 9-11 July, Scottsdale, AZ, USA, Society of Automotive Engineers (SAE), Warrendale, PA, USA.

Ashford, P., D. Clodic, A. McCulloch and L. Kuijpers, 2004: Emission profiles from the foam and refrigeration sectors, Comparison with atmospheric concentrations. Part 1: Methodology and data. *Int. J. Refrigeration*, **27**(7), 687-700.

ASHRAE, 2001: Standard 34-2001 Safety Classification of Refrigerants. American Society of Heating Refrigerating and Air Conditioning Engineers, Inc. (ASHRAE), Atlanta, GA, USA.

Atkinson, W., J.A. Baker and W. Hill, 2003: Mobile Air Conditioning Industry Overview, SAE Interior Climate Control Standards Committee, Automotive Industry Executive Summit on Vehicle Climate Control, Society of Automotive Engineers (SAE), Troy, MI, USA, February 2003, 24 pp.

Baker, J.A., M. Ghodbane, L.P. Scherer, P.S. Kadle, W.R. Hill and S.O. Andersen, 2003: HFC-152a Refrigerant System for Mobile Air Conditioning. SAE Technical Paper 2003-01-0731, Society of Automotive Engineers (SAE), Warrendale, PA, USA, 7 pp.

Barrault, S., J. Benouali and D. Clodic, 2003: Analysis of the Economic and Environmental Consequences of a Phase Out or Considerable Reduction Leakage of Mobile Air Conditioners. Report by Ecole des Mines de Paris/Armines for the European Commission, Paris, France, 53 pp.

Benouali, J., D. Clodic, S. Mola, G. Lo Presti, M. Magini and C. Malvicino, 2003a: Fuel Consumption of Mobile Air Conditioning, Method of Testing and Results. Proceedings of the 14th Annual Earth Technologies Forum, April 22-24, 2003, Washington, D.C., USA.

Benouali, J., S. Mola, C. Malvicino and D. Clodic, 2003b: Méthode de Mesure et Mesures des Surconsommations de Climatisations Automobiles. Report by Ecole des Mines de Paris for the French Agency for Environments and Energy Management (ADEME), Agreement 01 66 067, Paris, France, 194 pp.

Benouali, J., C. Malvicino and D. Clodic, 2003c: Possible Energy Consumption Gains for MAC Systems Using External Control Compressors. SAE Technical Paper 2003-01-0732, Society of Automotive Engineers (SAE), Warrendale, PA, USA.

Bhatti, M.S., 1998: Open Air Cycle Air Conditioning System for Motor Vehicles. SAE Technical Paper 980289, Society of Automotive Engineers (SAE), Warrendale, PA, USA, pp. 579-599.

Campbell, N.J. and A. McCulloch, 1998: The climate change implications of manufacturing refrigerants – A calculation of production energy contents of some common refrigerants. *Transactions of The Institute of Chemical Engineers*, **76**, Part B, August 1998, 239-244.

Clodic, D. and M. Ben Yahia, 1997: New Test Bench for Measuring Leak Flow Rate of Mobile Air Conditioning Hoses and Fittings. Proceedings of the International Conference on Ozone Protection Technologies, November 12-13, 1997, Baltimore, MA, USA, pp. 385-391.

Clodic, D. and F. Fayolle, 2001: Test-Bench for Measurement of Leak Flow Rate of MAC Compressors. SAE Technical Paper 2001-01-0794, Society of Automotive Engineers (SAE), Warrendale, PA, USA.

Clodic, D. and L. Palandre, 2004: Determination of Comparative HCFC and HFC Emission Profiles for the Foam and Refrigeration Sectors until 2015. Part 1: Refrigerant Emission Profiles. Report prepared by Armines for ADEME and US EPA, Paris, France, 132 pp.

Duthie, G.S., S. Harte, V. Jajasheela and D. Tegart, 2002: Average Mobile A/C Customer Usage Model – Development and Recommendations. Proceedings of the 2002 SAE Alternate Refrigerant Forum, 9-11 July, Scottsdale, AZ, USA, Society of Automotive Engineers (SAE), Warrendale, PA, USA.

Fösterling, S., W. Tegethoff and J. Köhler, 2002: Theoretical and Experimental Investigations on Carbon Dioxide Compressors for Mobile Air-Conditioning Systems and Transport Refrigeration. Proceedings of the Sixteenth International Compressor Engineering Conference, 14-19 July 2002, Purdue University, West Lafayette, IL, USA, Paper R11-9.

Fröhling, J., M. Lorenz-Börnert, F. Schröder, V. Khetarpal and S. Pitla, 2003: Component Development for CO_2. Proceedings of the 2003 SAE Alternative Refrigerant Symposium, 15-17 July, Scottsdale, AZ, USA, Society of Automotive Engineers (SAE), Warrendale, PA, USA.

Ghodbane, M. and H. Fernqvist, 2003: HFC-152a Secondary Loop Mobile A/C System. Proceedings of the 2003 SAE Alternative Refrigerant Symposium, 15-17 July, Scottsdale, AZ, USA, Society of Automotive Engineers (SAE), Warrendale, PA, USA.

Ghodbane, M., 2003: Potential Applications of HFC-152a Refrigerant in Vehicle Climate Control. Proceedings of the 6th Vehicle Thermal Management Systems (VTMS-6), 18-21 May, 2003, Brighton, UK, Paper C599/083/03, John Wiley & Sons, USA.

Ghodbane, M., J.A., Baker, W.R., Hill and S.O. Andersen, 2003: R-152a Mobile A/C with Directed Relief Safety System. Proceedings of the 2003 SAE Alternative Refrigerant Symposium, 15-17 July, Scottsdale, AZ, USA, Society of Automotive Engineers (SAE), Warrendale, PA, USA.

Hafner, A., P. Nekså and J. Pettersen, 2004: Life Cycle Climate Performance (LCCP) of Mobile Air-Conditioning Systems with HFC-134a, HFC-152a and R-744. Proceedings of the Mobile Air Conditioning Summit 2004, 15 April 2004, Washington, DC, USA. Earth Technologies Forum, Arlington, VA, USA.

Hesse, U., M. Arnemann and T. Hartmann, 2002: Ergebnisse von R-744 Wärmepumpen-Applikationen im Fahrzeug. Proceedings of the Deutsche Kälte-Klima-Tagung, Magdeburg, 21-22 November 2002, Deutscher Kälte- und Klimatechnischer Verein DKV, Stuttgart, Germany, **29**, pp. III.63-III.74.

Heyl, P. and J. Fröhling, 2001: Proceedings of the Deutsche Kälte-Klima-Tagung, Ulm, 22-23 November 2001, Deutscher Kälte- und Klimatechnischer Verein DKV, Stuttgart, Germany, **28**, pp. III.103-III.116.

Hrnjak, P., 2003: Design and Performance of Improved R-744 System Based on 2002 Technology. Proceedings of the 2003 SAE Alternative Refrigerant Symposium, 15-17 July, Scottsdale, AZ, USA, Society of Automotive Engineers (SAE), Warrendale, PA, USA.

Hu, J., C. Li and X. Yi, 2004: Growing Markets of MAC with HFC-134a. Proceedings of the Mobile Air Conditioning Summit 2004, 15 April 2004, Washington, DC, USA. Earth Technologies Forum, Arlington, VA, USA.

IPCC (Intergovernmental Panel on Climate Change), 1996: *Climate Change 1995: The Science of Climate Change. Contribution of Working Group 1 to the Second Assessment Report of the Intergovernmental Panel on Climate Change* [Houghton, J.T., L.G. Meira Filho, B.A. Callander, N. Harris, A. Kattenberg and K. Maskell (eds.)]. Cambridge University Press, Cambridge, United Kingdom, and New York, NY, USA, 572 pp.

IPCC (Intergovernmental Panel on Climate Change), 1997: *Revised 1996 Guidelines for National Greenhouse Gas Inventories – Reference Manual* [Houghton, J.T., L.G. Meira Filho, B. Kim, K. Treanton, I. Mamaty, Y. Bonduki, D.J. Griggs and B.A. Callender (eds.)]. Published by UK Meteorological Office for the IPCC/OECD/IEA, Bracknell, United Kingdom.

Johnson, V.H., 2002: Heat-Generated Cooling Opportunities in Vehicles. Proceedings of 2002 SAE Future Car Congress, June 2002, Arlington, VA, USA, SAE Technical Paper 2002-01-1969, Society of Automotive Engineers (SAE), Warrendale, PA, USA.

Kanwar, V., 2004: Indian Scenario: Refrigerants in MACs and the importance of Fuel Efficiency. Proceedings of the Mobile Air Conditioning Summit 2004, 15 April 2004, Washington, DC, USA. Earth Technologies Forum, Arlington, VA, USA.

Karl, S., C. Petitjean, E. Mace, J.M. Liu and M. Ben Yahia, 2003: Reduction of the Power Consumption of an A/C System. Proceedings of the 6th Vehicle Thermal Management Systems (VTMS-6), 18-21 May, 2003, Brighton, UK, Paper L07/C599-15, John Wiley & Sons, USA.

Kim, M.H., J. Pettersen and C.W. Bullard, 2004: Fundamental process and system design issues in CO_2 vapor compression systems. *Progress in Energy and Combustion Science*, **30**(2004), 119-174.

Kobayashi, N., 1999: Concerns of CO_2 A/C System for Compact Vehicles. Proceedings of the Second Annual Earth Technologies Forum, September 27-29, 1999, Washington, DC, USA.

Köhler, J., M. Sonnekalb and H. Kaiser, 1998: A Transcritical Refrigeration Cycle with CO_2 for Bus Air Conditioning and Transport Refrigeration and Heat Pumps. Proceedings of the 1998 International Refrigeration Conference, July 14-17, 1998, Purdue University, Purdue Printing Services, West Lafayette, IN, USA.

Lorentzen, G. and J. Pettersen, 1993: A new, efficient and environmentally benign system for car air-conditioning. *Int. J. Refrigeration*, **16**(1), 4-12.

Lorentzen, G., J. Pettersen and R.R. Bang, 1993: Method and Device for High-Side Pressure Regulation in Transcritical Vapor Compression Cycle. U.S. Patent 5,245,836.

Lorenz, M., R.Knorr, H.Mittelstrass, D.Schroeder, J.Schug and C.Walter, 2002: Safety when Handling Carbon Dioxide (CO_2) Systems. Proceedings of the 2002 SAE Alternate Refrigerant Forum, 9-11 July, Scottsdale, AZ, USA, Society of Automotive Engineers (SAE), Warrendale, PA, USA.

Maclaine-Cross, I.L., 2004: Usage and risk of hydrocarbon refrigerants in motor cars for Australia and the United States. *Int. J. Refrigeration*, **27**(4), 339-345.

Mager, J., 2003: New Technology: CO_2 (R-744) as an Alternative Refrigerant. Proceedings of the Mobile Air Conditioning Summit, February 10-11, 2003, Brussels, Belgium.

Mager, R., 2002: Experience of a R744 Fleet Test. Proceedings of the 2002 SAE Alternate Refrigerant Forum, 9-11 July, Scottsdale, AZ, USA, Society of Automotive Engineers (SAE), Warrendale, PA, USA.

Mager, R., J. Wertenbach, P. Hellmann and C. Rebinger, 2004: Standard R744-Kälteanlage: Ein Vorschlag einer Systemspezifikation. In *PKW-Klimatisierung III Klimakonzepte: Regelungsstrategien und Entwicklungsmethoden*, D. Schlenz (ed.), Haus der Technik Fachbuch Band 27, Expert Verlag, Renningen, Germany.

NESCCAF, 2004: Reducing Greenhouse Gas Emissions from Light-Duty Motor Vehicles. Northeast States Center for a Clean Air Future (NESCCAF), Boston, MA, USA, Interim Report, March 2004, 108 pp..

OJEC (Official Journal of the European Community), 2000: Directive 2000/53/EC of the European Parliament and of the Council of 18 September 2000 on end-of-life vehicles, *Official Journal of the European Communities*, L 269/34, 21 October 2000.

Palandre, L., D. Clodic and L. Kuijpers, 2002: Global Inventories and Emission Previsions of Refrigerants: A Case-Study, mobile air-conditioning systems. Proceedings of the Earth Technologies Forum, 25-27 March 2002, Washington, DC, USA.

Palandre, L., D. Clodic and L. Kuijpers, 2004: HCFCs and HFCs Emissions From the Refrigerating Systems for the Period 2004 – 2015. Proceedings of the 15th Annual Earth Technologies Forum, April 13-15, 2004, Washington, D.C., USA.

Petitjean, C., G. Guyonvarch, M. Ben Yahia and R. Bauvis, 2000: TEWI Analysis for Different Automotive Air Conditioning Systems. Proceedings of 2000 SAE Future Car Congress, April 2-6, 2000, Arlington, VA, USA. Society of Automotive Engineers (SAE), Warrendale, PA, USA, paper 00VCC-39.

Pettersen, J., 1994: An efficient new automobile air-conditioning system based on CO_2 vapor compression. *ASHRAE Transactions*, **100**(2), OR-94-5-3.

Rebinger, C., 2004: Safety Concept Proposal for R744-A/C-Systems in Passenger Cars. VDA Alternate Refrigerant Winter Meeting, Saalfelden, February 18-19, 2004, Austria, Verband der Automobilindustrie, Austria.

Rugh, J., V. Hovland and S.O. Andersen, 2004: Significant Fuel Savings and Emission Reductions by Improving Vehicle Air Conditioning. Proceedings of the Mobile Air Conditioning Summit 2004, 15 April 2004, Washington, DC, USA. Earth Technologies Forum, Arlington, VA, USA.

SAE (Society of Automotive Engineers), 1989: SAE Standard J51, Automotive Air Conditioning Hose (revised May 1989). Society of Automotive Engineers (SAE), Warrendale, PA, USA.

SAE, 1993: SAE Standard J2064, HFC-134a Refrigerant Automotive Air Conditioning Use. Society of Automotive Engineers (SAE), Warrendale, PA, USA.

SAE, 1999: SAE Standard J639, Safety and Containment of Refrigerant for Mechanical Vapor Compression Systems used for Mobile Air Conditioning Systems. Society of Automotive Engineers (SAE), Warrendale, PA, USA.

SAE, 2002a: Proceedings of the 2002 SAE Alternate Refrigerant Forum, 9-11 July, Scottsdale, AZ, USA, Society of Automotive Engineers (SAE), Warrendale, PA, USA.

SAE, 2002b: SAE Standard J1739, Potential Failure Mode and Effects Analysis in Design (Design FMEA) and Potential Failure Mode and Effects Analysis in Manufacturing and Assembly Processes (Process FMEA) and Effects Analysis for Machinery (Machinery FMEA). Society of Automotive Engineers (SAE), Warrendale, PA, USA.

SAE, 2003: Proceedings of the 2003 SAE Alternative Refrigerant Symposium, 15-17 July, Scottsdale, AZ, USA, Society of Automotive Engineers (SAE), Warrendale, PA, USA.

Sakai, T., M. Ueda, M. Iguchi, T. Adaniya and A. Kanai, 2004: 2-way Driven Compressor for Hybrid Vehicle Climate Control System. SAE Technical Paper 2004-01-0906 (presented at the SAE 2004 World Congress & Exhibition, March 2004, Detroit, MI, USA), Society of Automotive Engineers (SAE), Warrendale, PA, USA.

Sand, J.R., S.K. Fischer and V.D. Baxter, 1997: Energy and Global Warming Impacts of HFC Refrigerants and Emerging Technologies, U.S. Department of Energy and AFEAS, Arlington, Va, USA, 215 pp.

Scherer, L., M. Ghodbane, J.A. Baker and P.S. Kadle, 2003: On Vehicle Performance Comparison of HFC-152a and HFC-134a Heat Pump. SAE Technical Paper 2003-01-0733 (presented at the SAE 2003 World Congress & Exhibition, March 2003, Detroit, MI, USA), Society of Automotive Engineers (SAE), Warrendale, PA, USA.

Schwarz, W. and J. Harnisch, 2003: Establishing the Leakage Rates of Mobile Air Conditioners. Report prepared for DG Environment of the European Commission, Ecofys, Öko-Recherche and Ecofys, Frankfurt, Germany.

Sonnekalb, M., 2002: Einsatz von Kohlendioxid als Kältemittel in Busklimaanlagen und Transportkälteanlagen, Messung und Simulation (Use of Carbon Dioxide as Refrigerant in Bus Air Conditioning and Transport refrigeration, Measurement and Simulation). Ph.D. Thesis, DKV Research Report No 67, DKV, Stuttgart, Germany (in German).

Sonnekalb, M. and J. Köhler, 2000: Compression Refrigeration Unit, U.S. Patent 6,085,544.

Takahashi, K., 2004: Product Development of Air Conditioner System with Electrically-Driven Compressor for Hybrid Vehicle, JSAE Paper 20045168 (presented at the 2004 JSAE Annual Congress, May 19-21, 2004, Yokohama, Japan), Society of Automotive Engineers of Japan (JSAE), Tokyo, Japan.

Takeuchi, H., T. Ikemoto and H. Nishijima, 2004: World's First High Efficiency Refrigeration Cycle with Two-Phase Ejector: "Ejector Cycle". SAE Technical Paper 2004-01-0916 (presented at the SAE 2004 World Congress & Exhibition, March 2004, Detroit, MI, USA), Society of Automotive Engineers (SAE), Warrendale, PA, USA.

UNEP (United Nations Environment Programme), 1998: 1998 Report of the Refrigeration, Air Conditioning and Heat Pumps Technical Options Committee – 1998 Assessment. [L. Kuijpers (ed.)]. UNEP Ozone Secretariat, Nairobi, Kenya, 285 pp.

UNEP, 2003: 2002 Report of the Refrigeration, Air Conditioning and Heat Pumps Technical Options Committee – 2002 Assessment. [L. Kuijpers (ed.)]. UNEP Ozone Secretariat, Nairobi, Kenya, 209 pp.

Vetter, F. and S. Memory, 2003: Automotive AC/HP systems using R-744 (CO_2). Proceedings of the 6th Vehicle Thermal Management Systems (VTMS-6), 18-21 May, 2003, Brighton, UK, Paper C599/098/03, John Wiley & Sons, USA.

Wertenbach, J., 2004: Energy Analysis of Refrigerant Circuits. VDA Alternate Refrigerant Winter Meeting, Saalfelden, February 18-19, 2004, Austria, Verband der Automobilindustrie, Austria.

Wolf, F., 2004: Development of Alternate Refrigerant Technology. VDA Alternate Refrigerant Winter Meeting, Saalfelden, February 18-19, 2004, Austria, Verband der Automobilindustrie, Austria.

7

Foams

Coordinating Lead Author
Paul Ashford (United Kingdom)

Lead Authors
Andrew Ambrose (Australia), Mike Jeffs (United Kingdom), Bob Johnson (USA), Suzie Kocchi (USA),
Simon Lee (United Kingdom), Daniel Nott (USA), Paulo Vodianitskaia (Brazil), Jinhuang Wu (China)

Contributing Authors
Theresa Maine (USA), John Mutton (Canada), Bert Veenendaal (The Netherlands)

Review Editors
Jorge Leiva Valenzuela (Chile), Lalitha Singh (India)

Contents

EXECUTIVE SUMMARY

Following the mainstream introduction of HFC use in 2002, the phase-out of the use of ozone-depleting substances in the foam sector in the majority of developed countries is only now progressing towards completion. Consequently, predicting HFC usage patterns has been notoriously difficult and the downsizing of HFC demand estimates has been a feature of the last 3–5 years, being driven significantly by the costs of HFC-based systems when compared with other options.

Nevertheless, HFCs are being used in those applications where investment cost, product liability, process safety, and thermal efficiency are particularly important elements in the decision process. Where HFCs are used, careful consideration has been given to the use of blowing-agent blends for both cost and environmental reasons. Co-blowing with CO_2 (water)[1] has emerged as an important way of limiting HFC consumption in key applications such as the US appliance industry and the global spray-foam market.

It is now estimated that global HFC consumption in the foam sector is unlikely to exceed 75,000 tonnes annually in the period to 2015. This represents around 20% of the consumption that would have been associated with CFCs in the absence of the Montreal Protocol and around 50% of the uptake anticipated when the situation was first reviewed at the IPCC/TEAP Petten Conference in 1999. Not-in-kind insulation materials, such as mineral fibre, have continued to be the predominant choice in most global markets, primarily because of cost. Accordingly, foams are only used where their properties add value.

Hydrocarbon technologies have been widely adopted in several foam sub-sectors, including domestic appliances, water heaters, polyurethane (PU) sandwich panels, PU boardstock and some PU integral skin applications, and are expected to represent >55% of overall blowing-agent usage globally in the period after 2005. Progress into other sub-sectors has been thwarted by specific investment cost, product and process safety and, to a lesser extent, thermal performance.

HCFCs will continue to be used in foam applications in developing countries throughout the period and consumption has been predicted to grow to just under 50,000 tonnes by 2015. This assessment concurs broadly with the estimate in the 2003 TEAP HCFC Task Force Report (UNEP-TEAP, 2003), despite the different method used. Table 7A summarizes the projected consumption and emission pattern as at 2015, together with an assessment of the remaining blowing agent bank. The fact that annual emissions are not equivalent to annual consumption figures reflects the fact that most applications for which HCFCs, HFCs and hydrocarbons are considered as closed cell foams with time-delayed emission profiles.

As shown in Table 7A, emissions of HCFCs are expected to plateau in the 20,000–25,000 tonnes per annum range in the period after 2005, with decreases in production emissions from phase-out in developed countries being offset by emissions from domestic refrigerators at end of life. In contrast, the trend for HFC emissions will continue to be gradually upwards through the period of assessment as the bank builds. However, the most significant bank remains that related to CFCs. This is partly associated with the long period of historic use, but the impact of emissions on climate change is accentuated by the high average global warming potential (GWP) of the CFCs emitted. On the basis of current projections, and if there is no effort to instigate further end-of-life emission reduction options, the impact of CFC emissions will remain predominant, at least until 2050.

There is significant potential for managing emissions of CFCs, HCFCs and HFCs through further substitution, reduction in emissions during foam production/installation and improvements in end-of-life management. Some of these options involve innovation in building construction, allowing for disassembly as part of end-of-life management (e.g. the wider use of pre-fabricated building elements), while others involve the improved engineering of foam processes. As highlighted by Table 7A, the delayed release of blowing agents from both <u>new</u> and <u>existing</u> foamed products generates significant banks of blowing agent and there is a particular opportunity to focus on end-of-life issues both for recovery of existing materials containing ozone-depleting substances and also for future HFCs. The technical and economic potential for recovery still has to be fully quantified, although a lot of work is on-going now to

Table 7A. Projected global consumption and emissions of blowing agent by type as at 2015 (based on 2001 consumption data cited in the 2002 UNEP Foams Technical Options Committee Report (UNEP-FTOC, 2003)).

Blowing Agent	CFCs	HCFCs	HFCs	Hydrocarbons
Consumption (metric tonnes) – 2002	11,300	128,000	11,200	79,250
Consumption (metric tonnes) – 2015	Nil	50,000	73,000	177,250
Emissions (metric tonnes)	16,100	20,650	18,050	33,600
Bank as at 2002 (metric tonnes)	1,860,000	1,125,000	11,650	316,800
Remaining bank as at 2015 (metric tonnes)	**1,305,000**	**1,502,000**	**566,100**	**1,232,000**

---

[1] The CO_2 (water) option refers to carbon dioxide generated by the reaction of water with excess isocyanate in the formulation for polisocyanurate and polyurethane processes.

Table 7B. Cumulative emission savings resulting from the emission reduction scenarios outlined.

Measure	Year	CFCs	HCFCs	HFCs	CO2-eq	Est. Costs
		(tonnes)	*(tonnes)*	*(tonnes)*	*(Mtonnes)*	*US$/tCO2-eq*
HFC consumption reduction	2015	0	0	31,775	36	
(2010–2015)	2050	0	0	225,950	259	15-100
	2100	0	0	352,350	411	
Production/installation	2015	78	14,450	16,700	36	
improvements	2050	58	31,700	32,700	68	Varying
	2100	47	24,350	26,500	55	
End-of-life management						
options	2015	8,545	16,375	105	52	
	2050	64,150	144,650	88,540	540	10–50[1]
	2100	137,700	358,300	194,800	1200	

[1] Cost range for recovery of blowing agents from appliances only.

do so. However, this potential is likely to differ widely between sectors and applications. Although some emission management measures can be implemented relatively quickly, implementation in most cases is expected to be after 2007. Recognizing that the precise mix of measures cannot be fully defined at this stage, this chapter assesses the comparative impact of three scenarios:

- a linear decrease in the use of HFCs between 2010 and 2015, leading to a 50% reduction by 2015;
- the adoption of strategies for the reduction of production emissions in the block foam sub-sector from 2005 and from 2008 onwards in other foam sub-sectors;
- the extension of existing end-of-life measures to all appliances and steel-faced panels by 2010, together with a 20% recovery rate from other building-based foams from 2010 onwards.

Table 7B indicates the cumulative emission savings resulting from these three scenarios and gives an early indication of the likely costs based on initial assessments documented in the literature.

It can be seen that focusing on the reduction of HFC consumption provides the most significant saving in the period to 2015 and that, if any such reduction is extrapolated to use patterns after 2015, it offers the greatest 'HFC-specific' benefit up to 2100 as well. In contrast, end-of-life measures deliver lower savings in the period to 2015, but have the potential to deliver more overall savings in the period to 2100 if all blowing-agent types are considered. The value is particularly significant for CFCs, where GWPs are substantive and there is an incremental effect on ozone depletion.

The future consumption patterns for HFCs in the foam sector beyond 2005 are heavily reliant on the wider role of foams in energy conservation and, in particular, building renovation. It is expected that spray foam techniques may have a particular role to play. The development of well-grounded responsible-use guidelines will be vital for both the foam industry and climate change policymakers in order to ensure the most appropriate use of HFCs and the control of emissions, where applicable. However, in all cases, it will be necessary to assess the overall incremental energy benefit, together with the non-preventable direct emissions, in a policy-level comparison.

To illustrate the significance of energy-related emission savings, Table 7C provides typical ratios of CO_2 emissions saved

Table 7C. Ratios of indirect savings to direct emissions in various building applications

Application Sector	Method applied	Ratio of indirect (energy-related) savings to direct (HFC) greenhouse emissions	Key assumptions
Polyurethane spray foam, industrial flat warm roof	Life- cycle assessment (LCA)	15 – with full recovery of HFC at disposal 8 – without recovery of HFC at disposal	4 cm thickness; HFC-365mfc Germany
Polyurethane boardstock in private building cavity wall	LCA	140 – with full recovery of HFC at disposal 21 – without recovery of HFC at disposal	5 cm thickness; HFC-365mfc; Germany
Polyurethane boardstock in private building, pitched warm roof	LCA	92 – with full recovery of HFC at disposal 14 – without recovery of HFC at disposal	10 cm thickness; HFC-365mfc; Germany

by HFC-based foam to direct emissions of the HFC blowing agent during the lifecycle. In this instance, three typical applications were evaluated (Solvay, 2000).

These ratios do not imply that it is always essential to use HFC-blown foams to achieve the best overall thermal performance, but do illustrate that small incremental changes in thermal efficiency from the use of HFCs can have substantial overall emission benefits, primarily as a result of the long lifetime of most buildings. The benefits can be particularly significant if HFCs are recovered at disposal.

HCFC consumption in developing countries will be frozen after 2015 under the Montreal Protocol and will be phased out by 2040. In the business-as-usual analysis in this chapter, it is assumed that the decline is linear between 2030 and 2040. There is less certainty about HFC use beyond 2015, but the degree of innovation that has marked the last 20 years would suggest that the foam industry is unlikely to be reliant on HFCs beyond 2030. Accordingly, the business-as-usual analysis proposes a freeze in consumption at 2015 levels and a linear decline in use between 2020 and 2030.

7.1   Foam markets

7.1.1   *Foams by broad grouping and application*

Foamed (or cellular) polymers have been used historically in a variety of applications, utilizing the potential for creating both flexible and rigid structures. Flexible foams continue to be used effectively for furniture cushioning, packaging and impact management (safety) foams. Rigid foams are used primarily for thermal insulation applications of the types required for appliances, transport and in buildings. Rigid foams are also used to provide structural integrity and buoyancy.

For thermal insulation applications (the majority of rigid foam use), mineral-fibre alternatives (e.g. glass fibre and rock wool) have been, and continue to be, major not-in-kind alternatives. Table 7.1 illustrates the major benefits and limitations of both approaches.

The implications of these relative benefits and limitations vary substantially, both between products, within a category and between applications. This makes a generic conclusion about preferences impossible. The current thermal insulation market supports a variety of solutions (at least 15 major product types) and this reflects the range of properties demanded for the applications served. Unfortunately, only limited data are available covering the thermal insulation market at the global and regional levels. One of the complexities of global-market analysis is that building practice around the world varies, often responding to material availability and climatic conditions.

In reviewing the not-in-kind options, it is important to acknowledge continuing development. It looks likely that the use of vacuum panels in domestic refrigerators and freezers will increase and already most Japanese units contain at least one such panel in strategic design positions. However, the price implications of such technology choices make them relevant only when they are required to meet legislation relating to energy efficiency or where the market is clearly willing to pay the necessary premium.

There are at least twenty sub-sectors to be considered in any assessment of the market for foams. These are broken down in the Tables 7.2 and 7.3 according to whether they relate primarily to thermal insulation applications or not. This means that there is a mix of flexible and rigid foams in the Table 7.3.

Table 7.2 gives an indication of how this segmentation works for insulating foams. There are also a number of applications for non-insulation foams. These are shown in Table 7.3.

7.1.2   *Energy savings*

One of the prime reasons for using thermally insulating materials is to save energy. This objective is particularly important when the materials are used in the fabric of buildings and their related services, and also when they are used in appliances and equipment in our homes, offices and factories. As indicated in Tables 7.2 and 7.3, various materials meet the requirements for both these application areas and the following sections illustrate the importance of their role in the two categories.

7.2.2.1   *Buildings (including building services, renovations)*

Global assessments of carbon dioxide emissions have consistently revealed that emissions from the use of buildings (including appliances used in buildings) represent 30–40% of the global total. (Price *et al.*, 1998). In the industrialized countries of the northern hemisphere, this can rise as high as 40–50% (Ashford, 1998). The reduction of these emissions depends on two parallel strategies. The first consists of reducing the carbon intensity of the energy used by fuel by switching to alternatives with less carbon (e.g. natural gas) or by using more renewable energy. The second consists of improving the energy efficiency of the buildings themselves, including the appliances used within them. The evaluation of some strategies developed by governments in this area has made it increasingly clear that the scope for fuel switching and the rate of investment in renewable energy sources between now and 2050 will not be sufficient to achieve sustainable carbon dioxide emission levels by 2050 (e.g. Carbon Trust, 2002). This has led to the recognition of the central role of energy efficiency measures in any strategy and the introduction of such ground-breaking measures as the Energy Performance in Buildings Directive from the European Union (OJ, 2003).

The delivery of more energy efficient buildings and appliances is a complex issue. A recent cost-abatement analysis conducted by the Building Research Establishment in the UK (Pout *et al.*, 2002) revealed that over 100 *bona fide* measures exist to improve the energy performance of buildings alone. The IPCC has made similar observations earlier (IPCC, 2001). Accordingly, solutions need to be tailored to the circumstances

Table 7.1. Comparing the respective benefits and limitations of different insulation types.

Mineral fibre		Cellular polymers	
Benefits	*Limitations*	*Benefits*	*Limitations*
Initial cost	Air-based thermal properties	Blowing-agent-based thermal properties	Fire performance (organic)
Availability	Moisture resistance*	Moisture resistance	Limited maximum temperature
High maximum temperature	Low structural integrity	Structural integrity	First cost (in some cases)
Fire performance		Lightweight	

* Potentially affecting long-term thermal performance

Table 7.2. Assessment of application areas by thermal insulation type.

Foam Type		Application Area							
		Refrigeration & Transport			**Buildings & Building Services**				
		Domestic Appliances	Other Appliances	Reefers & Transport	Wall Insulation	Roof Insulation	Floor Insulation	Pipe Insulation	Cold Stores
Polyurethane	Injected/ P-i-P	□□□	□□	□□□	□□□	□□□	□□□	□□□	□□□
	Boardstock				□□□	□□□	□		
	Cont. Panel			□□	□□□	□□□			□□□
	Disc. Panel			□□	□□□	□□□			□□□
	Cont. Block			□□					
	Disc. Block			□□				□□	□□
	Spray		□□		□□	□□□		□	□
	One-Component				□□	□			□
Extruded Polystyrene	Board			□□	□□□	□□□	□□□	□	□□
Phenolic	Boardstock				□□□	□□□			
	Disc. Panel				□□	□			□□□
	Disc Block							□□□	□□
Polyethylene	Board						□		
	Pipe							□□□	
Mineral Fibre		□	□	□	□□□	□□□		□□□	□

□□□ = Major use of insulation type   □□= Frequent use of insulation type   □= Minor use of insulation type

Table 7.3. Assessment of the use of foams by type in non-insulation applications.

Foam Type		Application Area					
		Transport		**Comfort**		**Packaging**	**Buoyancy**
		Seating	Safety	Bedding	Furniture	Food & Other	Marine & Leisure
Polyurethane	Slabstock	□□		□□□	□□□	□□	
	Moulded	□□□			□□	□□	
	Integral Skin		□□□		□	□	
	Injected/ P-i-P						□□□
	Cont. Block						□
	Spray						□
Extruded Polystyrene	Sheet					□□□	
	Board						□□□
Polyethylene	Board					□□	□□

□□□ = Major use of insulation type   □□= Frequent use of insulation type   □= Minor use of insulation type

and the requirements of individual buildings. This can include aspects involving building services (e.g. pipe insulation) as well as measures associated with the fabric of the building itself. The approach adopted is necessarily different for new buildings than for existing buildings and there always has to be a balance between 'passive', fabric-based measures such as double- or triple-glazed windows and 'active' measures such as improved controls for the internal climate.

There have been several studies of the potential role of thermal insulation in the achievement of these objectives. One recent study (Petersdorff *et al.*, 2002) has demonstrated that the technical saving potential through the more widespread use of thermal insulation in Europe alone is more than 350 Mtonnes CO_2-eq/yr. This contrasts with earlier estimates (Ashford, 1998) which took more account of other competing measures, but concluded that the annual savings potential for the same region was 150 Mtonnes CO_2-eq/yr. In any event, the role of thermal insulation in the future reduction of carbon dioxide emissions is undeniable.

One of the major challenges for governments is the identification of appropriate policies and measures for implementing and encouraging improved energy efficiency in buildings. The construction of new buildings is typically well-regulated and provides a natural opportunity for increased energy efficiency standards. However, targeting new building standards alone will not deliver the required emissions savings, since most new building activity is undertaken in response to overall economic growth and social changes such as ageing populations and population migration. The daunting prospect for many governments is that the demolition rates for existing building stock are less than 0.5% per year and in some cases less than 0.2% (UNECE (2002) Database). Accordingly, unless there is a major demolition and rebuilding programme, the focus of attention will need to shift to measures required to renovate existing buildings. This is not easy because governments cannot generally impose requirements on existing owners. The problem is compounded because many owners are not the occupiers (or the payers of the energy bills), and costs and benefits are therefore not linked. The assessment of cost-effectiveness criteria therefore becomes critical, with discussion focusing on matters such as investment costs, payback periods, discount rates and the like. In the end, the questions to be answered are 'who pays?', 'when?' and 'how much?'. Government intervention, particularly in the domestic household sector, is expected to be a major part of many national strategies and the costs need to be considered carefully in order to get best value for investments (Ashford and Vetter, 2004).

In summary, it is clear that building renovation requires specific innovative products, supported by equally innovative policies and measures to improve the overall energy efficiency of existing building stock. The role of thermal insulation has already been highlighted by several studies, but it is anticipated that insulating foams will have a particularly important part to play because of their space-saving qualities and ease of pre-fabrication and/or direct application (e.g. spray foams). One of the purposes of this chapter is to identify the strengths and weaknesses of HFC-based foams and their alternatives in order to assess their likely uptake and possibilities for sound management. This can only be done in the wider context of this over-arching role for energy efficient products and buildings.

Apart from specific insulation foams, one-component foams (OCF) are very widely used in the building industry as gap fillers around doors and windows, as well as in plumbing applications. They provide excellent seals against draughts, thereby reducing the heat loss or gain by the building through improved air-tightness. In addition, they have major economic benefits because of their speed of application compared to any alternative method. The blowing agent acts both as a frothing agent and as a propellant. Initially, CFC-12 was used, but this has been replaced by a variety of alternatives including HCFC-22, hydrocarbons such as butane, propane and dimethyl ether, or HFC-134a and HFC-152a. This application is emissive and a recent European Union Proposal for a Regulation (COM (2003) 492) will severely restrict the use of HFCs in this application (UNEP-FTOC, 2003).

Finally, insulating foam is increasingly being used in pre-insulated pipes used to transport hot water underground in district heating systems. The foam provides the combination of thermal performance, moisture resistance and heat resistance required for this demanding environment.

7.1.2.2   Domestic refrigerators

The appliance market is highly competitive in all regions of the world and subject to stringent energy regulations in some places. The cost and performance characteristics of the insulating material are therefore both important parameters. Furthermore, the best possible insulation value is required in appliances to reduce wall thickness and maximize internal space, while preventing the settlement which could occur with other non-foam insulation types. The use of polyurethane foam to insulate domestic appliances started in Europe in the late 1950s as an alternative to mineral fibre, thereby revolutionizing production and reducing energy consumption. The technology is now universally employed (although the transition was only completed in the 1990s in developing countries) and has been a key factor in achieving dramatic improvements in appliance efficiency.

In response to shortages of electricity and to limit power-plant emissions, several countries have implemented regulations that require the labelling of products to indicate their relative energy efficiency, and some have established limits on the energy consumption allowed for refrigerators and freezers. Mandatory standards were implemented nationwide in the US in 1990. Limits were tightened in 1993, and again in 2001. In Europe, mandatory standards were implemented in 1999 and they are complemented by a progressive labelling programme. Several other countries have also implemented mandatory standards and labels. These standards, along with other programmes that provide incentives for manufacturers to design and market high-efficiency products, have resulted in thicker walls, requiring more foam, and may influence the choice of foam-blowing agent. The Appliance Research Consortium sponsored studies to determine the relative performance of leading candidates for the replacement of CFC-11 and HCFC-141b as blowing agents for refrigerator and freezer insulating foam. The results of these studies indicated that the highest insulating value was achieved using HFC-245fa. (Wilkes *et al.*, 2003). In Japan, particularly stringent energy regulations have forced the limited introduction of vacuum panels, although cost penalties are preventing more widespread use.

Apart from its energy contribution, the foam also provides strength as the core of sandwich elements, therefore reducing the consumption of other materials previously needed for structural purposes. An example is the use of plastic internal liners instead of enamelled steel, with the resulting weight reduction. Since the foam is important for the performance and durability of the product, it must maintain both its insulation effectiveness and structural properties throughout the design life of the appliance.

7.1.2.3 Others

There are several additional applications in which rigid insulation foams act both as an insulating medium and as a strength-giving core material. These foams are generally polyurethane-based, although extruded-polystyrene foams are also used. Such foams are universally used in refrigerated containers (reefers) and in truck bodies. The best possible insulation value is needed in these applications to reduce wall thickness and maximize internal space, while preventing the settlement that could occur with other non-foam insulation types.

Rigid polyurethane foam is used for the same efficiency reasons in commercial refrigerator and freezer units, drink vending units and in domestic/commercial water heaters. The latter in particular are again subject to energy consumption limits.

CFC-11 has been used historically in all of these polyurethane systems, while CFC-12 has been used in extruded-polystyrene applications. Their main replacements globally have been HCFC-141b and HCFC-142b respectively. Pentane and HFCs are now being increasingly used as HCFC-141b is phased out in developed countries while, in the case of extruded polystyrene, HFC-134a based systems have been used to replace HCFC-142b owing to the high thermal insulation requirements.

7.1.3 Safety/resilience

Integral skin polyurethane foams are widely used in the transportation/automotive industries for in-cabin applications such as steering wheels, arm rests and gear shift knobs. They combine aesthetics and durability with the degree of softness required to minimize impact damage in the event of an accident. The components are produced by a reaction injection moulding technique. A low percentage of blowing agent is added to give a solid skin and a lower density core. The mouldings are produced in both developed and developing countries. Initially, the blowing agent in use was CFC-11 but options include HCFCs,

HFCs, pentane and CO_2 (water). The use of HCFCs has been phased out in developed countries (UNEP-FTOC, 2003).

Closed cell polyurethane rigid and semi-rigid foams are used for buoyancy aids and life jackets. The buoyancy aids for boats are usually foamed in place. The blowing agent was initially CFC-11. This has been replaced with a variety of alternatives including HCFC-141b, and the future use of HFCs is envisaged (UNEP-FTOC, 2003).

7.1.4 Comfort

The largest polyurethane foam sector by overall volume is flexible foam for furniture, mattresses and transportation components such as seats. This foam is used globally and a number of different blowing-agent technologies have been deployed to replace CFC-11, which was used initially as the auxiliary blowing agent. Neither HCFCs nor HFCs are used in this sector.

Integral skin polyurethane foams are also used for furniture components such as armrests, and for footwear, especially sports footwear. One of their key features is very high durability. The components are produced by a reaction injection moulding technique. A low percentage of blowing agent is added to give a solid skin and a lower density core. The mouldings are produced in both developed and developing countries. Initially, the blowing agent in use was CFC-11 but options include HCFCs, HFCs, pentane and CO_2 (water). The use of HCFCs has been phased out in developed countries (UNEP-FTOC, 2003).

7.1.5 Market growth factors

There are three distinct factors involved in projecting market changes in the insulating foam sector. These can be listed as follows:

(1) Overall economic activity based on GDP (as a default) or on the number of units being produced in a given year. Considering the major uses of foams, this would be the number of buildings built/refurbished or the number of appliances (e.g. domestic refrigerators or freezers) being manufactured.

(2) The amount of insulation used per unit. This has tended to increase as energy standards for appliances have been raised or as building regulations have called for more insulation in new buildings to meet energy efficiency requirements. The expression used for this measure is the 'foam-volume ratio'. Increasing energy costs can also encourage greater use of insulation, but the driver is relatively weak in most regions because energy costs have dropped in real terms over recent years.

(3) The third driver for market growth in the foams sector is the relative share of the insulation market accounted for by foams. Although the trend may be in either direction depending on product priorities (e.g. fire performance, thermal efficiency, moisture resistance), the general trend

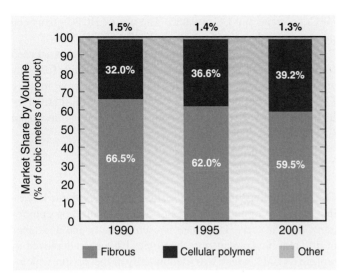

Figure 7.1. Variation of insulation by product type (1990–2001) in Europe

in Europe over the last ten years has been moving gradually towards foams (see Figure 7.1). Of course, this is unlikely to be consistent across all foam sub-sectors or regions.

These three factors have, in principle, constituted the overall market-growth drivers for the assessment of blowing-agent consumption from 1960 onwards. However, in practice, it has been more relevant to combine factors (2) and (3) into an overall 'foam-volume ratio' growth factor, since it is not normally possible to distinguish, at the regional level, overall insulation market growth 'per unit' from changes in market share of the types of insulation used.

For other types of non-insulating foams, the growth parameters are different and can be linked to indices such as global automobile production or fast-food sales. However, since these applications are unlikely ever to rely on HFCs, these parameters have not been considered quantitatively in this review.

7.2 Phase-out of ozone-depleting substances

The major classes of alternative blowing agents in use by sector are shown in Tables 7.4 and 7.5. In addition, there are a number of emerging blowing agents such as methyl formate and formic acid which are finding some acceptance. However, their use is not yet sufficiently widespread to justify inclusion in these tables.

7.2.1 Transitional use of HCFCs worldwide

The use of HCFCs in foam blowing in both developed and developing countries is generally limited to insulation and integral skin applications. Most foams for comfort and food applications use HCs (extruded-polystyrene sheet and polyethylene foams), liquid CO_2 (as a co-blowing agent) or methylene chloride (flexible, low-density PU foams).

7.2.1.1 Developing Countries

HCFC-141b is the most widely used CFC substitution technology for polyurethane foams in developing countries. The main reasons are the modest transition costs, easy implementation, similarity of end product performance (including energy) and wide availability. HCFC-142b and HCFC-22 are used for extruded polystyrene, although the uptake of this technology is limited to areas where the markets are sufficiently large to support the capital investment.

With limited further funding available to those who have already made the transition, HCFCs are likely to remain in widespread use in developing countries for a long time to come, possibly until 2040. This is particularly true of those who faced the challenge of CFC transition recently in stagnant economic conditions and favoured HCFCs as the most cost-effective option. Similar transition challenges face smaller enterprises in developed countries but, in this instance, the transition will be forced by regional regulation no later than 2015.

Other reasons for the continued use of HCFCs vary according to the application. The use of HCFC-141b as a blowing agent for the insulation of domestic refrigerators is associated with a potential lowering of energy consumption, which can assist in limiting energy demand in domestic markets where there are energy supply constraints. However, there may be constraints on trade for HCFC-containing articles in some regions (e.g. the European Union). Accordingly, HCFC-based technologies may simultaneously constitute an advantage and a disadvantage when multiple geographic markets are served. In summary, an appliance manufacturer's decision to use HCFCs, HFCs or hydrocarbons depends upon power supply and use, market forces, and existing regulations relating to energy and ozone-depleting substances in the targeted countries.

Similar considerations apply to sectors such as commercial refrigeration and other insulation applications. It is important to bear in mind that most industrial CFC conversions financed by the Multilateral Fund could use equipment that supports non-HCFC technologies such as CO_2 (water) and hydrocarbons. However, the degree of penetration of hydrocarbon technologies has been limited by the availability of hydrocarbons of an appropriate quality and by investment thresholds that set a cap on the funding available to save a kilogram of CFC.

HFCs, HCFCs, carbon dioxide (water) and HCs are currently used for automotive integral skin applications in developing countries in response to customer requirements, corporate policies or both.

In the case of extruded polystyrene, developing countries installing new capacity are likely to use HCFCs (HCFC-142b and/or HCFC-22) because HCFC equipment remains readily available and little equipment is available as yet for the installation of turnkey operations capable of using blowing agents with zero ozone-depleting potential.

In summary, as long as HCFCs remain available, HCs and HFCs will be used by developing countries only as widely as the additional costs associated with these technologies can be afforded or where corporate policy dictates. Some further ratio-

Table 7.4. Blowing-agent selection by product type and application area – thermal insulation.

Foam Type		Application Area							
		Refrigeration & Transport			**Buildings & Building Services**				
		Domestic Appliances	Other Appliances	Reefers & Transport	Wall Insulation	Roof Insulation	Floor Insulation	Pipe Insulation	Cold Stores
Polyurethane	Injected/ P-i-P	[HFC], HFC, HC, CO_2 (water)			HC, HFC			HC	
	Boardstock				Hydrocarbon, HFC				
	Cont. Panel			HC	HC, HFC				
	Disc. Panel			[HCFC] HC,HFC	[HCFC], HC,HFC				[HCFC] HC,HFC
	Cont. Block			HC, HFC				HC, HFC	
	Disc. Block			[HCFC] HC,HFC				[HCFC], HC,HFC	
	Spray	[HCFC] HC,HFC			[HCFC], HC,HFC				
	One-Component				HC, HFC				
Extruded Polystyrene	Board			HC, HFC	HCFC, CO_2, HFC, HC				HCFC, CO_2, HFC
Phenolic	Boardstock								
	Disc. Panel				HFC				HFC
	Disc Block							HC, HFC	
Polyethylene	Board						HC		
	Pipe							HC	

[ ] denotes mainly developing countries

nalization of HCFC usage patterns may be required post-2015 to comply with the anticipated freeze in consumption under the Montreal Protocol. The maturity of alternative technologies and conversion costs are likely to be decisive factors in prioritizing further substitution.

7.2.1.2 Developed Countries

In developed countries, ongoing HCFC use in polyurethane applications is limited to the smaller contractors (e.g. those applying spray foam). HCFCs also continue to be used for extruded polystyrene, particularly in North America. These uses are still in line with the Montreal Protocol. However, in some regions (e.g. the European Union), HCFCs have been phased out early and replaced by hydrocarbons, HFCs (HFC-134a, HFC-152a, HFC-245fa and HFC-365mfc/HFC-227ea blends) or CO_2-based blowing-agent systems, where these are appropriate. The approach adopted by different countries varies significantly, with some governments pursuing bans on use in specific foam sub-sectors, and others pursuing either a substance-based approach or relying on the Montreal Protocol timetable to ratchet down supply. There are often combinations of any two of these approaches and, sometimes, even all three. Although there have been regional difficulties, reductions in HCFC use are occurring in most developed countries. There is a slight complication, however, with the potential import of polyurethane systems containing HCFCs when developing countries are the source, since these are not discouraged by the Montreal Protocol itself.

Table 7.5. Blowing-agent selection by product type and application area – non-insulation.

Foam Type		Application Area					
		Transport		Comfort		Packaging	Buoyancy
		Seating	Safety	Bedding	Furniture	Food & Other	Marine & Leisure
Polyurethane	Slabstock	CO_2 (water LCD*) CH_2CL_2		CO_2 (water, LCD) CH_2CL_2		CO_2 (water, LCD) CH_2CL_2	
	Moulded	CO_2 (water LCD, GCD)			CO_2 (water LCD, GCD)	CO_2 (water LCD, GCD)	
	Integral Skin		HCFCs, HCs, HFCs, CO_2 water			HCFCs, HCs, HFCs, CO_2 water	
	Injected/ P-i-P						CO_2 (water)
Extruded Polystyrene	Sheet					HCs, CO_2, (water)	
	Board						HCFCs CO_2, HFCs
Polyethylene	Board					HCs	HCs

* liquid carbon dioxide technology

7.2.2 CO_2 (including CO_2(water))

CO_2 and CO_2 (water) are widely used as blowing agents in many non-insulating polyurethane foam sectors. These include flexible slabstock, moulded foam for transportation, footwear and integral skin foams for transport and furniture applications. In many cases, this blowing-agent technology has been the direct replacement for CFC-11.

This technology is not widely used for insulating polyurethane foams since it gives a lower thermal performance. However, there is emerging work in Japan (Ohnuma and Mori, 2003) on the potential use of super-critical CO_2 for some polyurethane spray foam formulations. The commercialization of this technology is being monitored closely, but most recent reports suggest that some subsidy for equipment investment may be required to stimulate market acceptance.

As a sole blowing agent or in conjunction with co-blowing agents (ethanol, hydrocarbons, acetone, isopropanol, water), CO_2 is principally used in XPS (extruded polystyrene) in regions which have already chosen to phase out HCFCs, and where the market considers the loss in thermal efficiency to be offset by the other inherent properties of XPS boards (moisture resistance, mechanical strength, compressive strength, dimensional stability and resistance to freeze-thaw deterioration). Higher densities or lower conversion may partially or wholly offset the low cost of CO_2 itself in some XPS applications. Furthermore, the use of CO_2-based blowing-agent systems is technically challenging, and not suitable for many existing manufacturing lines, particularly where such lines require substantial investment. These considerations put the technology out of reach for many small and medium enterprises.

Nevertheless, the use of CO_2-based technology for XPS is likely to continue to grow as long as the cost of alternative HFC solutions remains high.

7.2.3   Hydrocarbons

In rigid polyurethane insulating foams, the use of cyclopentane and various isomers of pentane to replace fluorocarbons has very rapidly gained wide acceptance in all regions. For building applications, normal pentane is the most widely used isomer. Normal pentane and cyclopentane/isopentane blends are typically used in rigid polyisocyanurate insulating foams in North America. In domestic refrigerators and freezers, cyclopentane was introduced in Europe in 1993 and it has gained acceptance in all regions except the USA, where existing VOC-emission regulations and stringent fire safety codes make the conversion of existing facilities uneconomical. The technology has been refined and optimized for cost-effectiveness and a blend of cyclopentane/isopentane is gaining prominence. Despite the higher thermal conductivity compared to foams based on HCFC-141b or HFC-245fa, appliances based on the hydrocarbon option attain the highest A+ energy efficiency category under the EU's stringent requirements. The technology provides good processing and other foam characteristics. The safety precautions for processing this flammable blowing agent are also well established. However, the cost of the necessary safety measures and the management infrastructure required can be uneconomical for small enterprises.

For rigid XPS insulation foams, hydrocarbons can be used as co-blowing agents with either CO_2- or HFC-based systems. Typically based on isomers of butanes or pentanes, they provide processability and in some cases contribute to the insulation value of the foams. The use of these flammable blowing agents is, however, limited due to their adverse influence on fire performance. Nonetheless, there is considerable uptake in Japan, where specific fire test configurations support the blowing agent's use.

7.2.4 HFCs

HFCs have been under consideration as replacements for HCFCs since the early 1990s. However, the absence of materials with an appropriate boiling point initially limited the potential for their use to gaseous processes such as those processes used to manufacture XPS (HFC-134a, HFC-152a). More recently, the development and commercialization of HFC-245fa and HFC-365mfc has enabled a wider range of processes to be encompassed by HFC technologies. However, the price of HFCs (particularly the newer materials) has ensured that HFCs are only used where they are perceived to be absolutely necessary – most notably where product liability, process safety and/or thermal performance benefits are evident.

As a consequence, it is now estimated that global HFC consumption in the foam sector is unlikely to exceed 75,000 tonnes annually in the period to 2015. This represents around 20% of the consumption that would have been associated with CFCs in the absence of the Montreal Protocol and around 50% of the uptake anticipated when the situation was first reviewed at the Petten Conference in 1999.

However, cost is a complex component, since it can be applied at both bulk substance level and at system formulation level. Box 7.1 illustrates the point.

7.3 Foam sub-sector analysis

7.3.1 Selecting blowing agents in applications where HFCs are considered

7.3.1.1 Thermal insulation foam
Building (polyurethane, phenolic, extruded polystyrene)
The construction industry is very cost-competitive and the use of HFCs in polyurethane rigid foams will naturally be restricted to those applications and circumstances where other blowing agents are unsuitable because of safety or performance factors. It is likely that the largest sector using HFCs will be the spray foam industry, because of safety constraints on the use of pentane and the product performance limitations (thermal conductivity and density) when using CO_2 (water). Another major sector, and one where small and medium enterprises are heavily involved, is the production of discontinuous sandwich panels. Here again, HFCs are expected to be the major blowing agent because of the high costs of flame-proofing such equipment for the use of HCs. The adoption of HFCs will only occur in the larger boardstock or continuously-produced sandwich panel sectors, where the end product has to meet very stringent flammability specifications. Although foams based on HFCs typi-

cally display superior insulation properties per unit thickness, this is only likely to be a decisive factor where there are significant space limitations[2]. One such sector may be the strategically important renovation market.

Box 7.2 illustrates the significance of different approaches to comparing thermal insulation and, in particular, the effect on lifetime CO_2 emissions of selecting different insulation materials with a constant thickness.

For phenolic systems used in the continuous boardstock sector and the discontinuous block and panel sectors, the considerations are very similar. However, because of the intrinsic fire properties of the foam matrix, the use of hydrocarbons has been found to be more feasible (c.f. progress in Japan with isobutane) than had been originally anticipated for foams used typically in fire-sensitive applications. A further factor influencing the non-selection of HFCs as blowing agents is the fact that phenolic chemistry does not offer the co-blowing options available to polyurethane systems. Accordingly, the blowing-agent component of formulations can be in excess of 10% in some instances, with significant cost implications. Nonetheless, the use of HFCs (particularly HFC-365mfc/HFC-227ea blends) is expected continue in a significant proportion of the phenolic foam market for the foreseeable future, particularly if fire performance standards continue to become more stringent, as is currently the case in Europe.

As with rigid polyurethane foams, the use of HFCs in rigid extruded polystyrene foams tends to be restricted to circumstances where other blowing agents are unsuitable (because of safety, processability and/or performance), as their price makes them too costly for many market segments. Systems based principally on HFC-134a can match HCFC-142b thermal conductivity, and are used in applications where insulation value and/or space are at a premium. HFC-134a is also long lived in the foam and provides excellent insulation performance over time (Vo and Paquet, 2004), making it suitable for applications with long lifetimes. HFC-152a, on the other hand, leaves the foam fairly quickly, but has improved processability. This makes it easier to use for the creation of large cross-sections, where the use of CO_2- or HFC-134a-based systems are technically challenging and beyond the capabilities of many small and medium enterprises today.

Appliances
Because of the challenging insulation and structural characteristics required for refrigerator and freezer insulation, no proven not-in-kind technologies are available at present that can meet all the requirements. Vacuum insulation panels have been used (as a supplement to the insulating foam) in limited quantities and are increasingly being used in critical locations in cabinets in Japan. However, high costs continue to limit their use to applications where thermal performance is of paramount impor-

---

[2] To facilitate the use of HFCs, BING has carried out studies (FIW, 2002) to define the ageing characteristics of the foams as required by CEN standards.

Box 7.1. Relationship between blowing-agent cost and foam cost

The thermal insulation market is a complex competitive environment but, in common with most suppliers to the construction industry, cost pressures in the insulation sector are ever-present. These pressures apply to all insulation types and the major driver is often the cost per unit of thermal resistance provided.

The blowing agent for thermal insulating foams is usually a significant contributor to thermal performance (Vo and Paquet, 2004). A more expensive blowing agent can therefore be justified if the benefit in thermal resistance can offset the additional cost. This has often been the case in traditional uses of CFCs. While more expensive than hydrocarbons and other alternatives, they provided cost-effective insulation because thinner boards could be used. Although the thermal performance of HCFC blown foams was initially slightly worse than their CFC blown equivalents (UNEP-FTOC, 1995) system improvements soon restored parity after the transition from CFCs (UNEP-FTOC, 1999). However, this was in a situation where HCFCs were trading at prices similar to traditional CFC levels.

For blowing agents with significantly higher prices, such as HFCs, there is no obvious way to provide a competitive insulation product in highly cost-sensitive markets other than reducing the amount of expensive blowing agent used (e.g. by using a blend with hydrocarbon) or reducing overall foam density without affecting the product properties (e.g. compressive strength). In practice, there is no easy route to lower-density products with the same performance levels. However, co-blowing with CO_2 (water) has been an option for several HFC-based technologies, albeit with some upper limits imposed by the performance requirements of the product.

In a traditional foam formulation (e.g. with HCFC-141b), blowing-agent levels of between 6% and 12% by weight are the norm. In the case of polyurethane foam blown with 8% blowing agent, the following scenario could be envisaged:

Polyurethane chemical costs	€1500 /tonne
Blowing-agent (1) cost	€1000 /tonne
Blowing-agent (2) cost	€4000/tonne

With an 8% loading and 32 kg/m³ density, the change in blowing agent from (1) to (2) would push up the cost of foam by 15%. However, if it were possible to reduce the expensive blowing-agent loading to 4% by co-blowing with CO_2 (water), the impact on cost would drop below 10%. These apparently small adjustments have been demonstrated to be critical in the competitive market place and the prudent use of expensive blowing agents has resulted in a reduction in the predicted uptake of HFCs worldwide.

Of course, there are insulation foam applications where the market is less sensitive to blowing-agent cost. This is particularly the case when a specific blowing-agent choice makes it possible to adopt a more economic building solution. This can be the case in some renovation markets as well as for exposed foam products such as pipe insulation.

---

tance. Polyurethane foam therefore continues to be used for the bulk of this application.

Although hydrocarbons are being successfully used as the blowing agent for refrigerator and freezer production in most regions of the world, there are some situations, most notably in North America, where manufacturers have chosen HFCs as the preferred replacement for ozone-depleting substances in this application. Their reasoning is that HFCs can provide the superior insulation performance needed to meet existing, stringent, energy standards while meeting customer demand for storage capacity, product dimensions and cost. Using HFCs as blowing agents also precludes the need for the prohibitively large investments needed to meet stringent safety codes and VOC-

emission regulations in existing facilities. These considerations can outweigh the advantage of lower material costs and lower direct GWP associated with the hydrocarbon alternatives. The use of HFCs as blowing agents for refrigerator foam was minor prior to 2003. However, it has since increased significantly in North America as HCFC-141b has been replaced, and may also increase in some other regions as HCFCs are ultimately phased out (UNEP-FTOC, 2003). The primary HFC used in North America is HFC-245fa.

With respect to the global warming implications of this choice, analyses of total equivalent warming impact (TEWI) and life cycle climate performance (LCCP) have indicated that, because of the superior insulating and ageing characteristics

---

Box 7.2. Choosing insulation for new building projects

The choice of insulation type can be impacted by many factors, including long-term thermal performance, efficient utilisation of space, cost and ease of installation. For most new buildings, a minimum energy performance is required for the building fabric (usually stated in terms of U-value or R-value). All insulation types can normally meet this minimum standard, but different thicknesses will often be required. Where space is constrained (constant thickness), this can pose a problem and energy penalties can be paid as shown below.

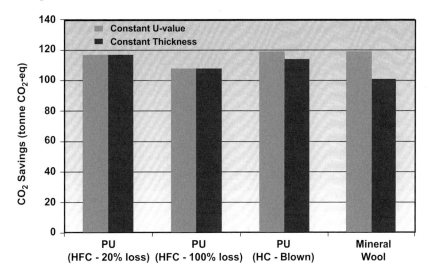

Figure Box 7.2. Lifetime CO_2 savings for a typical building fabric insulation measure (domestic - Netherlands)

However, where constant U-value can be maintained, blowing-agent choice is often the decisive factor. The graph illustrates the impact on total CO_2-equivalent savings of different loss assumptions for an HFC blowing agent. The impact of this direct emission is usually greater than any embodied energy consideration if any significant blowing-agent losses occur during the production, use or decommissioning of the foam.

Since the given parameters (e.g. thickness constraint) vary according to the application, it is vitally important that such an analysis is carried out on an application-by-application basis rather than at product level in order to make the most appropriate decision. Where the level of insulation permitted is generally low, the differences in overall lifetime CO2 savings can be substantial between different insulation materials.

---

when using HFC-245fa, the use of an HFC blowing agent can be at least neutral in most markets, or even beneficial when compared to the use of a hydrocarbon blowing agent (Johnson, 2004).

Some other appliances also rely on the insulation and structural benefits of insulating foam. Water heaters and commercial refrigerators and freezers (including display units), along with insulated picnic boxes (or coolers) and various flasks and thermoware, are substantial markets in this sector. The requirements and options are generally similar to those for domestic refrigerators and freezers. However, in some cases, the space requirements are not as stringent. The average operating temperature for water heaters is higher than for refrigerators and freezers and, at these temperatures, the energy disadvantage of

hydrocarbon blowing agents is not significant. Most manufacturers are therefore using hydrocarbons or CO_2 (from water) in such applications as substitutes with zero ozone-depleting potential.

7.3.1.2 Product and process safety
<u>Process Safety Issues: Toxicity and Flammability</u>
The production and use of foams involve important safety considerations. These often play a role at the manufacturing stage and include worker exposure and flammability issues.

In flexible polyurethane foams, a technically suitable alternative to CFC-11 as an auxiliary blowing agent is methylene chloride (dichloromethane). However, it is classified as Carcinogen category 3 in the EU and OSHA has established exposure standards in the USA. Nevertheless, HFCs are not

Box 7.3. LCCP analysis results for a refrigerator

A summary of the results of an LCCP analysis conducted on a European refrigerator is shown here. The two products compared were identical in all respects, except for the blowing agent (BA), which was HFC-245fa in one product and a cyclo/iso-pentane blend in the other product. This analysis assumed that 90% of the blowing agent would be recovered or destroyed at end of life by processing refrigerators in appropriate facilities as required by EU directives. In the table below, a second column was added, extending the analysis to include a case where only 50% of the remaining blowing agent was recovered or destroyed at end of life. Note the dominance of the CO_2 emissions due to energy consumption of the product in both cases. Emissions are in kg CO_2 equivalent.

	HFC-245fa (90% recycled)	HFC-245fa (50% recycled)	Pentane Mixture
BA Production Energy	4.5	4.5	1.2
BA Production Emissions	2.8	2.8	0
BA Transport Energy	0.17	0.17	0.05
Refrigerator Production: BA Emissions	18.7	18.7	0.1
Refrigerator Production Energy	5.0	5.0	5.0
Refrigerator Transport Energy	8.0	8.0	8.0
Refrigerator Use Energy	2697	2697	3198
Refrigerator Life BA Emissions	68.8	68.8	0.1
Refrigerator Shredding Emissions	17.0	85	0.1
Long-Term Emissions	67.9	339.5	0.2
Total Impact	2882	3230	3212

The analysis was extended to cases in which it was assumed that the product was used in several different regions of the world. Note that the net global warming impact is lower for the HFC-245fa product in all regions considered, except Brazil (where most of the power is from hydroelectric sources). In this analysis, it was assumed that 90% of decommissioned refrigerators in Europe would be recycled, with recovery or destruction of the blowing agent. The levels of recycling or destruction assumed for other countries were 50% for Australia, New Zealand, and the US; 95% for Japan; and 20% for China, India, and Brazil.

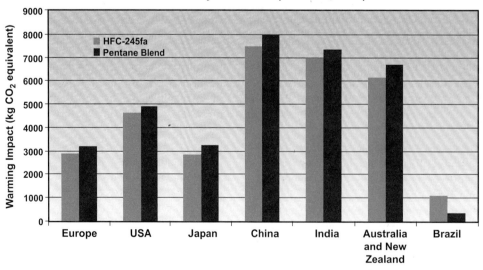

LCCP Analysis Results (Johnson, 2004)

considered to be alternatives to methylene chloride in this application (UNEP-FTOC, 2003).

The use of hydrocarbons in foam manufacture can be a major issue. The use of pentane for rigid polyurethane and polyisocyanurate foams has required significant re-engineering of production sites, as well as worker training to prevent explosive mixtures occurring. Although these investments can be amortized quickly over a large production volume in the larger enterprises, the conversion cost can be prohibitive for small and medium enterprises in both developed and developing countries. The safety record for the use of pentane has been good for over ten years and there have been very few incidents. By contrast, the use of highly flammable blowing agents in the extruded-polystyrene sheet sector to replace CFC-12 has resulted in several incidents, especially in developing countries. In the extruded-polystyrene-foam board sector, handling improve-

ments in, for example, Japan and Europe have resulted in the use of such flammable mixtures without incident (UNEP-FTOC, 2003). Significant concern exists in Europe and other areas about the safety of hydrocarbons in applications involving the use of polyurethane spray foam (Box 7.4).

Small and medium enterprises (SMEs) in both developed and developing countries can have particular difficulty with handling hydrocarbons. Whilst the unit costs and operating costs associated with hydrocarbons are low, the capital costs involved in converting manufacturing plants to comply with national regulations for the handling of flammable blowing agents may be prohibitive, particularly for SMEs in developed countries where no financial support is available. Even in developing countries, the Multilateral Fund limits the grant available for conversion by the adoption of a funding threshold, often leav-

---

Box 7.4. Process safety and renovation: The example of polyurethane spray foam in Spain

Although work continues on the safe use of hydrocarbon systems in polyurethane (PU) spray foams, the current commercial reality in Spain is that the only viable alternatives to HCFC-based systems for closed cell insulating foams are HFC-based. This has resulted in the need for well-coordinated transition programmes, not all of which have proceeded smoothly. Spray foam systems provide unique opportunities for the renovation of buildings, particularly roofs in both the residential and commercial/industrial sectors. The building style in Spain and other Southern European countries lends itself to the use of PU spray foam for upgrading the insulation of flat roofs. The key advantage is that there is no need to 'tear off' the existing roof material prior to the application of spray foam. This has a critical impact in the installed cost of a re-insulated roof, as shown in the table below:

Insulation option	Cost/m²
PU spray foam	€ 21.24
Mineral wool	€ 49.20
Extruded polystyrene (CO_2 blown)	€ 48.24
Expanded polystyrene	€ 47.16

In Spain, very few houses built prior to 1979 have any significant insulation in them. The estimated number of houses in this category is around 7.3 million. The policy objective of the Spanish government is understood as the renovation of at least 5% of this total between 2005 and 2012 (Energy Efficiency & Economy Strategy 2004–2012 – Buildings Sector E4 Working Document, July 2003). If the rate of renovation is constant, this will equate to just over 45,000 houses per year.

The budget set aside for the improvement of the building fabric (both residential and commercial/industrial) is € 2,780 million, although this will include the installation of new glazing and other fabric measures. This represents an average investment allocation of just over € 30 per m2. If the roofs of 45,000 houses per year were re-insulated with PU spray foam, the overall cost of the work, assuming an average roof size of 100m2, would be € 776 million (i.e. 28% of the total budget). This seems a reasonable proportion of funds to be allocated for this purpose bearing in mind the resulting CO_2 saving. However, if other insulation technologies were used as an alternative to PU spray foam, either the percentage of the total budget allocated to roof renovation would need to be increased to more than 50% of the total budget, thereby forfeiting other options, or the roofing area insulated per year would be less. The figure in this box illustrates the impact, given a constant budget allocation (28%), on CO_2 savings in lifecycle terms of the application of the different technologies. The lifetime of any roof renovation measure is assumed to be 15 years, given the durability of the waterproof membranes used, and it is the same for all options.

Box 7.4. Continued

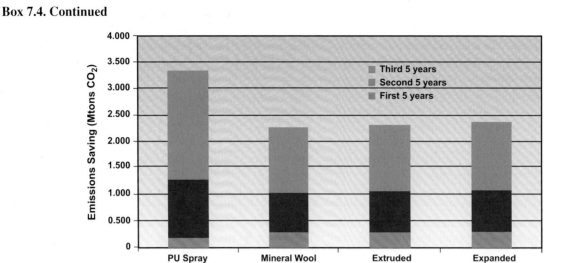

Figure Box 7.4. Cumulative renovation programme savings – five-year intervals [Insert from o/leaf]

The use of PU Spray foam involves the direct emission of HFCs (15%) during application, with the result that the CO_2 savings during the eight-year implementation period of the programme are offset, making them lower than for the other options. However, in the subsequent seven-year period, the increased market penetration of PU spray foam in the constant budget scenario makes the delivery of overall savings much more efficient. In the Spanish case, the prediction is that more than an additional 1 Mtonnes of CO_2 would be saved during the lifetime of the measures. These are important additional savings for a government faced with significant greenhouse-gas emission challenges and they explain why the Spanish are keen to see the retention of the HFC-based PU spray technology option.

(Source: Ashford and Vetter, 2004)

ing the enterprise significantly short of funding for a hydrocarbon route. Within the MLF system, hydrocarbons only become viable when the CFC-11 consumption reaches about 50 tonnes per annum. Smaller enterprises may therefore be obliged to use HFCs (or more likely HCFCs) despite their negative impact on operating costs and, potentially, the environment (UNEP-FTOC, 2003). In addition, the rigorous operating procedures required to store, handle and process flammable blowing agents safely may prove too much of a challenge for many SMEs, particularly in developing countries.

Product safety issues: End-product Fire Performance and Integral Skin Quality
The use of a flammable blowing agent can also have an effect on the fire properties of the end product. This is particularly important in the building/construction sector. Some of the most stringent fire product safety standards are not yet being met by hydrocarbon technology and the use of HFCs is envisaged for these applications (UNEP-FTOC, 2003).

There are other safety reasons why blowing-agent choice can be important. For example, in integral skin foams for the automotive sector, the combination of skin quality and impact absorbency is critical for steering wheel and facia applications. The use of in-mould coatings has been one approach used to combat the poorer quality skins created by some alternatives. However, there are drawbacks to this approach in terms of costs and processing. In some applications, HFCs are being used to produce skin quality more comparable with that once attained with CFCs and HCFCs.

7.3.2 *Introduction to tables*

Table 7.6 and 7.7 are intended to provide a reliable overview of the issues affecting the selection of insulation materials and, in particular, blowing agents. The tables are believed to be sufficiently comprehensive to provide sound guidance. However, the format does not lend itself to detailed reasoning in all cases. Accordingly, the reader is encouraged to refer to relevant text elsewhere in the chapter or to seek advice from sector experts if the information available requires further clarification or qualification.

Table 7.6. Sub-sector analysis – selection criteria.

Foam type	Not-in-kind options	Non-HFC zero-ODP options	HFC use criteria
Domestic appliances (domestic refrigerator and freezer).	Mineral fibre was widely used in this application until the mid-1970s but was replaced by polyurethane technology because of improved thermal performance and strength. Vacuum panels are emerging but are not used in isolation.	Hydrocarbons (HCs) (especially cyclopentane/iso-pentane blends)	HFCs are used where space savings and energy savings are paramount or where there are VOC and /or capital investment issues for hydrocarbons (e.g. USA). Benefits provided by (some) HFCs include: • higher insulating value (lower thermal conductivity) at operating temperature than other options, resulting in energy savings; • good thermal conductivity ageing characteristics;
Other appliances (water heaters, ice makers, display cases, vending machines…)	Again, mineral fibre was used historically, but was replaced by polyurethane technology in the mid-1970s. Mineral fibre is still used in some water heaters where moisture can be driven out routinely.	HCs (pentanes and cyclopentane) are more difficult to use because plants are sometimes set up for more customized manufacture. CO_2 (water) can also be used where foam plays a limited role. In limited applications methyl formate and formic acid in blends with HFC-134a.	• fire safety in manufacturing, often non-flammable atmospheres; • no local pollution effect (classified as non-VOC in some regions); • relatively low investment costs because of the benefits mentioned above. HFCs used are mostly HFC-245fa and some HFC-134a
Boardstock: polyurethane, polyisocyanurate, and phenolic	Mineral (glass) fibres, extruded- and expanded-polystyrene boards. Extruded-polystyrene and mineral (glass) fibres are widely used alternatives but are not suited to all building methods.	HCs (n-pentane and cyclopentane) are the predominant choice for cost reasons.	HFCs are used only when warranted by end-product fire performance or specific thermal performance. Benefits provided by some HFCs include: • higher insulating value (lower thermal conductivity) than other options; • fire safety in manufacturing and fire resistance of insulating boardstock. HFCs used include HFC-245fa and HFC-365mfc (and its blend with HFC-227ea).
Continuous and discontinuous panels	Mineral fibres compete as either built-up systems or as steel-faced panels. High densities (>150kg/m^3) are required to achieve strength as panels.	HCs (n-pentanes and cyclopentane) are the predominant choice for cost reasons. In limited applications, CO_2 (water), methyl formate and formic acid in blends with HFC-134a.	HFCs are used where end-product fire performance is an issue with insurers (e.g. LPC full-scale flammability test in UK) or where investment costs for HCs are prohibitive for small and medium enterprises (disc. panels). Benefits provided by some HFCs also include higher insulating value than other options at operating temperatures for applications such as walk-in coolers/cold storage. HFCs used include HFC-134a, HFC-245fa and HFC-365mfc (and its blend with HFC-227ea).

Table 7.6. continued (2)

Foam type	Not-in-kind options	Non-HFC zero-ODP options	HFC use criteria
Spray	No direct alternative	Attempts to commercialize HC technology in the USA but safety concerns remain among SMEs. CO_2 (water) can be used with equipment modification, but leads to the need for higher thickness (cost) to offset insulation and higher density (cost) to improve dimensional stability.	HFCs are currently the predominant option in this sector, providing safety for operators, insulation value and product performance (dimensional stability). HFCs used include HFC-365mfc-based blends, HFC-245fa and HFC-134a.
Blocks	No direct alternative. Mineral fibre is widely used in pipe insulation, although it requires greater space because of its higher thermal conductivity.	HCs (pentanes and cyclopentane) are used successfully for cost reasons. Plant safety can be an issue when making blocks discontinuously.	HFCs may be used where end-product fire performance is an issue. Blends of HFCs and blends of HFCs with pentanes can be a relevant compromise. HFCs used include HFC-245fa and HFC-365mfc-based blends.
Pipe (pipe-in-pipe)	Traditional insulation materials unsuitable for the environment – particularly in district heating (underground) applications.	Cyclopentane	Only rationale for HFC use is safety and/or investment constraints for small and medium enterprises. HFCs used include HFC-365mfc-based blends, HFC-245fa and HFC-134a.
Extruded-polystyrene board	Glass and mineral fibres, wood, cellular glass and other polymeric (urethane and phenolic) foams.	CO_2 or CO_2 with co-blowing agents such as ethanol, HCs, acetone and isopropanol. Most technologies are currently limited in product thickness.	HFCs are used because of more versatile processing. Fire and thermal performance can also have a bearing. HFCs used include HFC-134a and HFC-152a.
XPS panel	Mineral fibres compete as either built-up systems or as steel-faced panels. High densities (>150kg/m^3) are required to achieve strength as panels.	CO_2 or CO_2 with co-blowing agents such as ethanol, HCs, acetone and isopropanol. Most technologies are currently limited in product thickness.	HFCs are used because of more versatile processing. Fire and thermal performance can also have a bearing. HFCs used include HFC-134a and HFC-152a.
XPS pipe	Glass and mineral fibres. Silica, rock wool, slag wool and alumina silica fibres. Polymeric foams. Granular insulation.	None (CO_2-based systems have not been developed).	HFCs offer necessary process versatility together with dimensional stability and thermal benefits. HFCs used include HFC-134a and HFC-152a.

Table 7.6. continued (3)

Foam type	Not-in-kind options	Non-HFC zero-ODP options	HFC use criteria
Transportation (PU and XPS)	No viable alternatives owing to stringent requirements for insulation, structural strength and durability (resistance to vibration)	Hydrocarbons (pentanes and cyclopentane) for polyurethane. No non-HFC alternative for extruded polystyrene.	The thermal benefits can play a role where space is limited. Fire safety in manufacturing can also play a role in some operations. For extruded polystyrene, production versatility is the key. HFCs used include HFC-134a and HFC-152a (for extruded polystyrene). HFCs used include HFC-365mfc-based blend, HFC-245fa and HFC-134a for polyurethane.
Non-insulating foam one-component	Caulk	Hydrocarbons	No ongoing role for HFCs where HC technologies can be managed safely.
Extruded-polystyrene sheet	No real alternative	Hydrocarbons (butane and propane) in conjunction with CO_2.	Very limited use for safety applications. HFCs used include HFC-152a.
Automotive integral skin	PVC or leather skin over other substrates. Difficult to achieve same impact performance	CO_2 (water) with and without in-mould coating. Hydrocarbons for high volume and heavy duty productions	Only where safety and investment issues exist for small and medium enterprises. HFCs used include HFC-245fa, HFC-365mfc-based blend and HFC-134a.
Other integral skin	Flexibility and durability difficult to match with other substrates (e.g., EVA shoe soles have low resilience and durability)	CO_2 (water) is a viable option with in-mould coating.	Only where safety and investment issues exist for small and medium enterprises. HFCs used include HFC-245fa, HFC-365mfc-based blend and HFC-134a.
Buoyancy	PVC foams and inflatable buoyancy aids for life jacket	Hydrocarbons are technically feasible but not economically viable due to low production volume.	Only where safety and investment issues exist for small and medium enterprises.

Table 7.7. Sub-sector analysis – emission characteristics.

Foam type	Production/1st Year (%)	Use (%/yr)	End of life	Emissions functions
Domestic appliances (domestic refrigerator and freezer).	4 %	< 0.5% per year	Most HFCs will be present. The emission depends on the means of end-of-life management. HFCs may be incinerated (minimum emission) or shredded and sent to a landfill (maximum emission).	
Other appliances	5 %	< 0.5%	Most HFCs will be present. The emission depends on the means of end-of-life management. HFCs may be incinerated (minimum emission); or shredded and sent to a landfill (maximum emission).	
Polyurethane/polyisocyanurate/phenolic boardstock	6 %	0.5%	Significant portion of HFCs will be present at end of life. Separation for HFC recovery could be difficult and destruction will only be achieved if all combustible materials are treated together in a suitable incinerator.	
Panels	5–6 %	< 0.5 %	Most HFC content will be present. Can be reused, reprocessed through a recovery plant or incinerated.	
Spray	15 %	0.75%	Significant portion of HFCs will be present. Separation for HFC recovery will be difficult and destruction will only be achieved if all combustible materials are treated together in a suitable incinerator.	
Blocks	45% for pipe and 15% for slab.	0.75 % (with impermeable facing materials).	Significant portion of HFCs will be present. Separation for HFC recovery could be difficult in some slab applications (but not for pipe sections). Destruction may only be achieved if all combustible materials are treated together in a suitable incinerator.	
Pipe (pipe-in-pipe)	6 %	< 0.25 %	Most of HFCs will be present at end of life. It is feasible to recover the pipe section and treat the foam and other plastic components in a suitable incinerator.	
Extruded-polystyrene board	15–25% for HFC-134a based XPS. 15–25% for HFC-152a based XPS.	0.75 % +/- 0.25 for HFC-134a based XPS. 15% for 152a based XPS	Significant portion of HFC-134a will be present at end of life. The insulation boards can be re-used if not adhered to substrates. The boards with HFCs can be destroyed in a suitable incinerator.	
Extruded-polystyrene panel	25%	0.75% for permeable facer. Negligible for impermeable facer.	Most of HFC-134a will be present at end of life. The insulation panels can be re-used. The panels with HFCs can be destroyed in a suitable incinerator.	
Extruded-polystyrene pipe	25 %	0.66%	Significant portion of HFC-134a will be present at end of life. The insulation shells can be re-used. The shells with HFCs can be destroyed in a suitable incinerator.	
Transportation (PU and XPS)	5–6 % for PU, 25% for extruded polystyrene	0.5 % or negligible due to impermeable facers.	Most of HFCs will be present at the end of life. It is feasible to separate foam with its HFCs and destroy in a suitable incinerator.	
Non-insulating foam one-component	100%	N/A	N/A	
Extruded-polystyrene sheet	100%	N/A	N/A	
Automotive integral skin	100%	N/A	N/A	
Other integral skin	100%	N/A	N/A	
Buoyancy	10%	2%	Significant portion of HFCs will be present in the foam at end of life. Separation and recovery of foam and HFCs will be difficult because foam is part of composite.	

7.3.3 Summary of key themes

(1) The reasons given for the use of HFCs in Section 7.3.1 are replicated by sub-sector in the table.

They can be summarized as follows:
- better thermal performance, leading to enhanced capabilities in space-constrained environments;
- compliance with product fire classifications in applications and regions sensitive to 'reaction-to-fire' issues;
- minimization of risks in processing, particularly when risks cannot be specifically addressed by engineered solutions or when costs are prohibitive (e.g. SMEs);
- cost-effective application options, particularly important in the building renovation sector where costs may determine whether insulation is applied at all.

(2) Drawbacks to using HFCs remain:
- concerns about the ability to demonstrate objectively the environmental neutrality or benefit of using an HFC option, particularly for building applications;
- stakeholder concern about the growth of HFC use;
- cost implications of using HFCs in highly competitive markets where there is little, if any, return on the investment.

(3) Once an HFC technology has been selected, there are several strategies for minimizing emissions:
- optimization of the formulation to minimize HFC use (e.g. by using co-blowing agents);
- design of the product for low blowing-agent losses during the long use phase;
- design of the product for ease of processing at end of life (e.g. through greater use of pre-fabricated elements).

7.4 Blowing agents: use patterns and emissions

7.4.1 Quantifying consumption and establishing business as usual (BAU)

The anticipated growth drivers for foam use were discussed in Section 7.1.5. The three main growth factors driving foam use can be summarized as:
1. overall economic activity;
2. insulation usage per unit of activity (i.e. energy standards or costs);
3. changes in the respective market shares of foams and fibrous insulation.

Blowing-agent selection can have an influence on at least the two latter factors listed, as described explicitly in the sub-sector analysis. The summary of 'key themes' specifically explain the reasons why HFCs might be used. These include thermal performance, product/process safety and cost. It is important to note that cost appears as both a positive and a negative driver in

terms of HFC selection, reflecting the fact that, *in defined applications*, HFC use can have specific cost and environmental benefits, even though the basic cost of the HFC and its global warming potential would intuitively indicate otherwise. An additional factor with a potentially significant effect on use patterns is local regulation. However, one of the challenging aspects of determining the impact of such regulation is that most potential legislation is still in the future and the full implications on HCFC substitution are still to emerge. Accordingly, in the absence of detailed information, future regulatory impacts cannot therefore be considered to be part of the BAU scenario.

One-component foams have been a particularly high-profile use of HFCs (indeed, the primary use) in foams over recent years. Since OCFs are totally emissive in their application, the impact on short-term HFC emission assessments has been profound. However, alternative technologies do exist and have only been limited in their market penetration by initial safety concerns. It seems fairly certain that HFC use in this sector will be phased out within the next ten years, and the current draft EU regulation on fluorinated gases reflects this reality. Accordingly, this eventuality is included in the BAU case. Nonetheless, it is worth noting in passing that the role of one-component foams in improving the air tightness of buildings is possibly one of the largest single potential contributors to overall energy efficiency improvement.

For insulating foams themselves, the situation with respect to HFC usage for the European Polyurethane Foam industry is covered by Jeffs *et al.* (2004). Similar tables have been generated for the ten other regions of the world on the basis of the assessment carried out by the Foams Technical Options Committee (UNEP-FTOC, 2003). These have been further developed to account for all fourteen identified blowing-agent types in the insulated foam sector (Ashford *et al.*, 2004ab). These are shown in Table 7.8.

The resulting assessment of the global consumption of blowing agent is shown in the Figure 7.2, which illustrates the overall growth in consumption for the period to 2015 and how the newer blowing-agent options (HCs and HFCs) are replacing the ozone-depleting substances used previously (CFCs and HCFCs).

To generate substance-specific assessments, assumptions about the split between HFC-245fa usage and HFC-365mfc usage in competing markets are required. They have been made possible, at least in part, by the patent situation in North

Table 7.8. Identified blowing agents in the insulated foam sector.

CFCs	HCFCs	HFCs	Hydrocarbons
CFC-11	HCFC-141b	HFC-245fa	n-pentane
CFC-12	HCFC-142b	HFC-365mfc	isopentane
	HCFC-22	HFC-134a	cyclopentane
		HFC-152a	isobutane/LPG
		HFC-227ea	

Figure 7.2. Total blowing-agent consumption by type (1990–2015) - (aggregated)

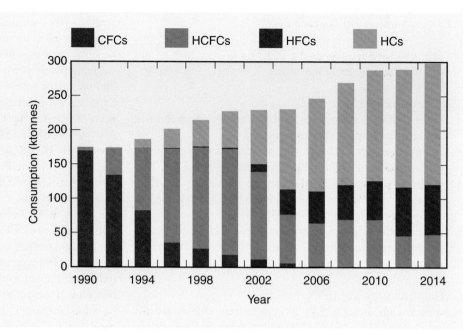

America, where the use of HFC-365mfc has been *de facto* prevented by the decision of the sole licence holder to focus entirely on HFC-245fa. This has further implications for the use of HFC-227ea in foam applications since it is used exclusively in blends with HFC-365mfc within the foam sector.

An additional element of uncertainty is the split between n-pentane, iso-pentane and cyclo-pentane usage in the hydrocarbon-based sector of the market. Where specific regional information is available, this has been included in the models. However, for the most part, assumptions have been made based on practices in other regions or similar usage sectors. In any event, the total hydrocarbon emissions and banks are unaffected by this approach and the uncertainty has little impact on greenhouse-gas emission assessments.

7.4.2   Discussion of business-as-usual scenarios

Overall HFC usage
At the Petten Conference (Ashford, 1999), it was estimated that the global uptake of HFCs in foams would be around 75,000 tonnes by 2004, with further growth to 115,000 tonnes in 2010. This was also reflected in emissions estimates for Europe at that time of 13 Mtonnes CO_2-eq (March, 1998). Since then, however, it has become increasingly apparent that the cost of the relevant HFCs has driven a high level of innovation for alternatives and, in particular, blends which make best use of HFC properties without being totally reliant upon them (e.g. co-blowing with CO_2 (water) for spray foam applications in the United States). The resulting estimate for 2010 is now approximately 57,000 tonnes, rising to just under **73,000 tonnes by 2015**, and the re-appraisal of European emissions to below 10 Mtonnes CO_2-eq (Preamble to draft EU F-Gas Regulation,

European Commission (2003)) is a further reflection of this re-alignment.

Ongoing HCFC usage in developed and developing countries
As a result of Montreal Protocol controls, the use of CFCs has dropped from more than 75,000 ODP tonnes in 1995 to around 10,000 ODP tonnes in 2002. The use of CFC blowing agents in developing countries is declining, but is likely to continue until 2005–2008. According to the TEAP HCFC Task Force Report (UNEP-TEAP, 2003), HCFCs are likely to play a continued role as 'transitional substances' in insulating and in integral skin foams. HCFC-141b is the most widely used blowing agent in the foam sector, with global demand at approximately 110,000 tonnes in 2001. Because it is banned for use in foams or scheduled for imminent phase-out in most developed countries, HCFC-141b consumption will drop significantly to approximately 28,000 tonnes in 2003–2004. However, HCFC-141b will still be considered a viable replacement for CFCs in developing countries and demand is expected to increase steadily to approximately 40,000 tonnes by 2015.

In the foam sector, demand for HCFC-142b and HCFC-22 is significantly lower than demand for HCFC-141b. Global demand for HCFC-142b was approximately 24,000 tonnes in 2001. That same year, global demand for HCFC-22 was approximately 9,000 tonnes. The demand for HCFC-142b is primarily based on the production of extruded-polystyrene foam in North America and Japan. HCFC-22 is also used, to a lesser degree, in polystyrene- and polyurethane-foam applications in North America. With the exception of HCFC-22 in Latin America, HCFC-142b and HCFC-22 consumption in developing countries is very low, although some use of HCFC-22 is occurring as a blend with HCFC-141b in integral skin foams. Once these chemicals are phased out in the foam sector in North America

(2010), their combined consumption in foams is expected to be less than 1,000 tonnes and to remain close to or below that level until 2015.

One of the key factors in defining the business-as-usual scenario related to the percentage of the phase-out of extruded polystyrene in the US foam sector in 2010 that would be dependent on the uptake of HFCs. This remains an extraordinarily difficult parameter to predict and a reasonably conservative estimate of 65% has therefore been adopted.

Finally, the business-as-usual scenario needs to address HFC demand assumptions beyond 2015 in order to establish a complete picture of the likely emissions scenario over the full lifecycle (up to 50 years) of foam products manufactured and installed in the period through to 2015. Beyond 2015, it is known that HCFC consumption in developing countries will be frozen under the Montreal Protocol and will be phased out by 2040. Accordingly, it is assumed that the decline is linear between 2030 and 2040. For HFC use beyond 2015, there is less certainty, but the degree of innovation that has marked the last 20 years would suggest that the foam industry is unlikely to be reliant on HFCs beyond 2030. Accordingly, the business-as-usual analysis proposes a freeze in consumption at 2015 levels and a linear decline in use between 2020 and 2030.

7.4.3 Business-as-usual emission projections and existing uncertainties

The development of interest in blowing-agent emissions from the foam sector was initially driven from within the Montreal Protocol process. Although this protocol operates in legal terms on a 'production/consumption' basis, the parties have never lost sight of the fact that three of the central clauses of the preamble to the protocol refer explicitly to controlling emissions (UNEP, 2003). Given this background, the parties have tended to encourage the technical and economic review of methods for reducing emissions from end uses involving ozone-depleting substances. In two key areas – refrigeration and foams – these opportunities are particularly significant because of the encapsulation and delayed release of the respective refrigerants and blowing agents. In promoting the consideration of these options, the protocol has been very careful not to mandate such emission reduction strategies, recognizing that these decisions are for individual parties or groups of parties.

The UNEP Foams Technical Options Committee (UNEP-FTOC, 1999) and, more latterly, the Task Force on Collection, Recovery and Long-term Storage (UNEP-TEAP, 2002ab) have addressed the issue of release profiles from foams. These have formed the basis for more recent modelling work (Ashford *et al.*, 2004ab). However, efforts continue to be made to update specific emission functions where new evidence supports such revision (see Table 7.7 for the latest emission functions). The recent work of the extruded-polystyrene industry has been particularly noteworthy in this respect (Vo and Paquet, 2004; Lee and Mutton, 2004). This assessment has re-evaluated emission losses during the use phase of extruded polystyrene at 0.75%

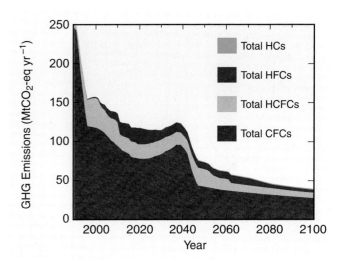

Figure 7.3. Business-as-usual emissions by blowing-agent type for the period 1990–2100

per annum against a previous estimate of 2.5% per annum.

Figure 7.3 emerges from the adoption of these emission functions for the business-as-usual scenario and the most recent scientific values for GWPs presented in Chapter 2.

Some significant trends emerge. The increased CFC emissions peak in the period between 2030 and 2050 reflects the impact of decommissioning CFC-containing products from buildings. The business-as-usual scenario does not envisage any recovery of the blowing agent from these sources. However, the continuation of CFC emissions beyond 2050 illustrates that considerable levels of CFCs are expected to be re-banked in landfill sites (other end-of-life options include re-use, shredding without blowing agent recovery and shredding with blowing agent recovery). The assumption in the case of landfill is that the initial losses from foams are not excessive. Work carried out by the Danish Technical University on losses from refrigerator foam during shredding and landfill suggests overall losses in the range of 20% (Scheutz and Kjeldsen, 2001). At first sight, this seems low. However, if one takes into consideration the fine-celled structure of these foams and the fact that some blowing agent is often dissolved in the matrix, it becomes more reasonable to assume losses of this order for most landfill processes. In cases where car shredders are used, there is a third potential factor, since the foam particles generated through such shredders tend to be 'fist-sized' and do not represent a substantial increase in surface area to volume. This contrasts with European and Japanese shredders, which are designed to maximize surface area and encourage the release of blowing agent under controlled conditions. The 'fist-sized' pieces are more likely to mirror the type of debris obtained from building demolition processes and it is for this reason that the parallel is drawn for modeling purposes. The assumption, however, involves a high level of sensitivity. Figure 7.4 illustrates this sensitivity by considering what the emissions forecast would look like if, in a worst-case scenario, 100% of blowing agent was emitted at the

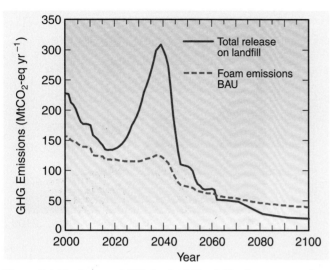

Figure 7.4. The impact of differing initial land-fill release assumptions on overall emission patterns

time of landfill.

This scenario, of course, means that the impact of blowing-agent emissions would be more concentrated and more severe during the period from 2030 to 2050. The uncertainty also indicates the need for further work in this area. What the adoption of a relatively conservative loss rate for landfill does ensure, however, is that any predicted savings from end-of-life measures are also conservative. The reality is probably somewhere in between.

Assuming the CFCs get there, consideration must be given to the potential for anaerobic degradation in the accumulated

CFC banks in landfills. The work of Scheutz, Fredenslund and Kjeldsen (2003) revealed the potential for both CFC-11 and CFC-12 breakdown in managed landfill environments. The believed need for 'management' places this potential emission reduction factor somewhere between a business-as-usual uncertainty and a potential emission reduction measure. However, since this approach generates breakdown products, including HFCs (notably HFC-41), further work still remains to be done in characterizing the potential consequential effects of the by-products before the approach can be classified as a genuine emission reduction option. Nevertheless, the overall global warming impact of the emissions would be reduced if such a process were to occur in practice.

It remains difficult to evaluate the impact of such a measure in quantitative terms. It would reduce the impact of blowing agents landfilled in the past or in the future. Since reliance on landfill is falling in many parts of the world, this reduction could be limited. While the current emission models used for this report allow, for the first time, the quantification of the banks that have already been landfilled (but not included in the bank estimates quoted in this report), the degree of mitigation that can be achieved will be dependent on the prevalence of appropriately designed and managed landfill sites. An assessment of this potential goes beyond the scope of this report.

In summary, the emissions projections are highly sensitive to emission function estimates because of the slow release rates involved and the size of the remaining banks. This is an important caveat when viewing detailed emission assessments. Table 7.9 illustrates the estimated uncertainties for key blowing-agent types. It should be noted that the uncertainties are more pronounced for annual estimates than they are for cumulative emis-

Table 7.9. Sources of uncertainty in emission estimates for commonly used blowing agents.

Blowing agent	Annual emission uncertainty (2002) (tonnes / % of bank)		Key sources in 2002	Uncertainty level
	Upper boundary	Lower boundary		
CFC-11	+10,000/0.6%	-10,000/-0.6%	• Domestic appliance (end of life) • Polyurethane boardstock (in-use) • Polyurethane spray (in-use)	High (for specific year) Low Low
HCFC-141b	+10,000/1.2%	-5,000/-0.6%	• Polyurethane boardstock (production/use) • Polyurethane spray (prod/use) • Domestic appliance (production/use)	Medium (transition year) High (transition plus production losses) Low
HCFC-142b	+5,000/2.2%	-0/0.0%	• Extruded polystyrene (production/use) • Polyethylene rigid (production/use)	Medium (production losses) High (activity)
HFCs	+2,500/21%	-2,500/-21%	• Extruded polystyrene (production/use)	High (HFC-152a emission rate)

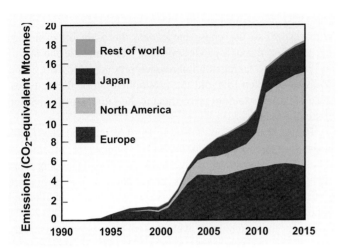

Figure 7.5. GWP-weighted global HFC emissions by region (1990–2015)

sions over longer periods. This reflects particular difficulties in predicting end-of-life activities (e.g. refrigerator disposal) on an annualized basis. It remains much easier to predict cumulative disposal over a ten-year period when the year-to-year variations offset one another.

The significant uncertainties in HFC emissions in 2002 as a proportion of bank size arise from the fact that this was a transition period (particularly in Europe) and the fact that HFC-152a emission rates were less certain. HFC uncertainties will continue to stabilize as the breadth of use develops and blowing-agent banks become established.

Focusing further on HFC emissions, it is clear that the vast majority will occur in the developed countries in the period to 2015, reflecting the fact that HCFCs will continue to be used during the period to 2040 in most developing countries. Figure 7.5 illustrates this fact.

This is important for the potential control of HFC emissions, since both Europe and Japan are already well advanced in end-of-life management options. The next section deals with these options as a sub-set of a review of emission reduction options in the foam sector.

7.5   Emission reduction options and their effects

As has already been noted elsewhere in the text, there are four areas of the product life cycle where emissions can be reduced. These are:

- selection of alternative technologies using non-GWP or lower GWP blowing-agents;
- production and first year (including installation losses);
- use phase;
- decommissioning and end of life.

The current 'best practice' and further options for improvement are discussed in the following text. 'Best practice' is not always easy to extrapolate from region to region because of significant differences in product types and processing methods.

7.5.1   *Production, first-year and use-phase options*

Current status and best practice
HFC selection, formulation and potential substitution
The selection of specific HFCs can have some bearing on the ultimate global warming impact of blowing-agent emissions. However, in practice, the two competing HFC technologies in most polyurethane applications are HFC-245fa (GWP 1020)[3] and HFC-365mfc/HFC-227ea blends (with GWPs of 950–1100 depending on blend composition). Accordingly, the distinction is minimal unless the formulations have significantly different loadings of HFC. This can be the case for HFC-245fa, which does have a higher blowing efficiency than HFC-365mfc/HFC-227ea. However, the difference in boiling points and flammability may offset this. In practice, therefore, there is little to be gained from switching HFCs to minimize impacts in the polyurethane sector.

For gaseous HFCs such as HFC-134a (GWP 1410) and HFC-152a (GWP 122), there would appear to be much more to gain by switching from HFC-134a to HFC-152a. However, this is far from being the case for two reasons. Both are related to the fact that the rate of diffusion of HFC-152a from foam cells is several times faster than that of HFC-134a. The implication is that the emission of HFC-152a is faster and less controllable than HFC-134a. This has the added implication that, where the foam is being used as thermal insulation, there is less blowing agent available to maintain that performance. Since long-term thermal performance is a critical characteristic of all insulation, the early loss of HFC-152a is a major drawback. As a result, HFC-134a is preferred in most insulation foam formulations. In practice (particularly in the case of extruded polystyrene), blends of HFC-134a and HFC-152a are used to obtain the best balance of properties.

With the possibility of HFC selection having only a superficial effect, and the formulation loadings being minimized in all cases because of cost, attention should rightly turn to potential substitution with non-GWP blowing-agent replacements. This subject was already covered in Sections 7.3.1 and 7.3.2 and it can be seen that, in most if not all cases, HFC selection is currently based on a well-grounded rationale. However, past experience shows that circumstances can change substantially with time, and technology breakthroughs in the areas of polyurethane spray foam, extruded-polystyrene foams or domestic appliances could have marked impacts on HFC consumption.

To reflect the fact that further technological development will be required and that experience with HFCs has only just started to be acquired in most instances, the emission reduction scenario considered in this analysis assumes that substitution can only realistically begin to take place after 2010. It is assumed

---

[3] The GWPs referred to in this section are the latest figures provided in Table 2.6 (from WMO (2003) and elsewhere). However, in the emission calculations carried out in this chapter, the appropriate SAR and TAR values have been used.

that up to 50% of the HFC use at that time could be replaced by alternative technologies but that this would occur progressively between 2010 and 2015, leaving a rump of foam types reliant on HFC-based technologies in 2015 which would, at best, freeze consumption until 2020, when further technological developments might be expected. The impact of such a scenario has been assessed under the heading of the '50% HFC reduction' option in the analysis that follows.

Production and first-year losses

Process management measures include the improved encapsulation of processing areas, particularly where unreacted chemical/blowing-agent mixtures are exposed at elevated ambient temperatures. The ability to capture the volatile blowing agent from the atmosphere will often depend on the amount of ventilation being used in the area (for safety and occupational health reasons, this is often significant) and the size of the processing area. The practicality and cost of such measures will vary significantly from process to process. For instance, in spray foams, there are few practical ways of recapturing blowing agent, although careful design of spray heads can limit over-spray and resulting blowing-agent loss.

In the extruded-polystyrene sector, process losses have also been under scrutiny within the industry and there is some evidence to suggest that losses in North America could already be lower than in Europe because of the product mix and process design. These differences have been documented earlier in a US EPA Notice of Data Availability (US EPA, 2001). The implication is that production losses could be as low as 10% in some scenarios. With current ranges being 15–25% for Europe and 10–25% for North America, the figure of 17.5% for North America and 20% for the rest of the world could justifiably be adopted in future as an alternative to the current default of 25% (Lee and Mutton, 2004). Nevertheless, a technical breakthrough is needed to handle the high-volume air flows and the very dilute blowing-agent concentrations generated from trim reprocessing (Cheminfo Services Inc., 2001) if major savings are going to be made in process emissions.

Another example of process management is the minimization of process waste. In the case of both phenolic and polyurethane block foams, there is an opportunity to recapture blowing agent from off-cuts. These can be significant in the case of fabricated pipe sections where block utilization rarely exceeds 55%. Similar waste minimization practices can also be encouraged during the installation phase of insulation products in buildings.

Taking these as examples of potential opportunities, this analysis has assumed that block-foam measures could be introduced as soon as 2005 and that efforts to improve processing and first-year losses could be implemented across other foam sectors from 2008. Given the technical and economic challenges, broader measures of this kind will not be expected to achieve a saving better than 20% on average.

Use-phase assessments

With respect to use-phase emissions, the main opportunities are provided through product design. They include the use of less permeable facings on boardstock products. However, care needs to be taken to avoid the misapplication of facing types, since breathable facings are critical to the satisfactory operation of some products.

The implementation of Article 16 of Regulation 2037/2000 in the European Union has increased the flow of information about the effectiveness of recovery of blowing agents from refrigerators (OJ, 2000). This information had already been established in several Member States and built into relevant standards (RAL GZ728, 2001; Draft DIN-8975-12, 2002). However, these standards were based on *minimum* recovery rates of approximately 70% of initial charge, implying considerably greater loss in the use phase than the assumptions used in the emissions models developed by the TEAP and others. This potentially higher loss rate has been supported by subsequent work by the Centre for Energy Studies in Paris (Zoughaib *et al.*, 2003). There is a case, therefore, for increasing the annual loss assumptions in the use phase from the 0.25% used in the TFCRS study (UNEP-TEAP, 2002ab) to 0.5% or 0.75%. The difference with use-phase losses for commercial refrigeration or other insulated panels could be rationally ascribed to the presence of a thermoplastic liner. However, it has been decided to wait for further investigation of aged refrigerators before adopting this change.

Conversely, as discussed earlier, measurements of effective blowing-agent-diffusion coefficients for laboratory and field-aged XPS samples by Vo and Paquet, and work by Lee and Mutton on the average thickness and installed use temperature of XPS boards, have given a better understanding of loss mechanisms and support a reduction of annual use-phase losses to 0.75% from the original TFCRS study value of 2.5% (UNEP-TEAP, 2002ab).

Since considerable uncertainties about emission rates in use exist and bearig in mind that general thicknesses of insulation are likely to increase (lower surface to volume ratio), leading to reductions in emission rates, it is naturally difficult, and perhaps inappropriate, to postulate a fully defined emission reduction scenario in this phase of the lifecycle. The reality is that the losses in use (i.e. after the first year) are low as a proportion of the total blowing-agent loading, and that changes in technology are unlikely to have any major impact.

7.5.2 Decommissioning and end-of-life options

Some foam sub-sectors (e.g. steel-faced panels, domestic and commercial appliances and some types of boardstock) have significant potential for the recovery of the blowing agent at end of life. To quantify this potential, it has been necessary to evaluate the condition of foam at the time of decommissioning to ensure that enough blowing agent is still available for recovery. Considerable work has been gone into the verification of these levels since the development and publication of the

TFCRS Report in 2001/2002 (UNEP-TEAP, 2002ab). This has included work by the Insulation Technical Advisory Committee of AHAM[4], the Danish Technical University (Scheutz and Kjeldsen, 2001), the Japan Technical Centre for Construction Materials (JTCCM) (Mizuno, 2003) and others. The work of the JTCCM in particular has involved the sampling of foams of varying ages from over 500 buildings and the conclusions have broadly endorsed the emission functions included in the TFCRS Report (UNEP-TEAP, 2002ab). Use-phase emission rates for extruded polystyrene are the exception here and these have been re-evaluated in the light of the recent work by Vo and Paquet (Vo and Paquet, 2004).

Much of the aforementioned work, as well as Fabian *et al.* (2004), has highlighted the need for a robust method of assessing blowing-agent content in foams and this has led to the development by JTCCM of a draft Japanese standard on the subject and a draft ASTM standard.

A second element of importance has been the establishment of possible end-of-life treatment methods for foams. The latest models used for this analysis (Ashford *et al.*, 2004ab) have incorporated four end-of-life scenarios:

1. re-use;
2. landfill;
3. shredding without recovery of blowing agent;
4. shredding with recovery of blowing agent.

It has been assumed that direct incineration of the foam results in the same loss profile as 'shredding with recovery of blowing agent' because of the similar front-end handling requirements. Both techniques have been used in Europe for refrigerators, although problems with incineration residue have curtailed the widespread commercialization of the incineration approach in this instance.

Technical options
The development of end-of-life management techniques for foams continues and further refinements to the mechanical recapture/recycle plants are emerging. One of the issues requiring improved management is the handling of hydrocarbon blown foams in refrigerators. At least two incidents have occurred in Europe in recent months and it has emerged that there is a need for increased awareness and risk management procedures, particularly in older plants.

As noted previously, the potential for the anaerobic degradation of ozone-depleting substances in landfill soils has also been investigated (Scheutz *et al.*, 2003) but at present it looks as though the 'fluorine-carbon' bond might be too strong to be broken down by this route, even though CFCs and HCFCs have been successfully degraded to HFCs. It may be that the microbial organisms responsible need further acclimatization and this will be evaluated further over time.

The direct incineration of foam continues to be a focus of attention. The MSWI co-combustion study (APME/ISOPA/EXIBA, 1996) paved the way for the incineration of building insulation. However, as noted above, problems have continued in Europe with the direct incineration of complete refrigerators. In other parts of the world, where foam and metals can be separated without controls, the incineration of the removed foam would be an option. In the UK, a study has been conducted on the partial dismantling of refrigerators prior to transporting (Butler, 2002) with the aim of minimizing cost and environmental impact. However, at present, refrigerators continue to be sent to mechanical recapture/recycle plants rather than municipal solid-waste incinerators.

There has been little progress in techniques for recovering blowing agent from insulation products used in buildings, although the ongoing trend towards the increased use of prefabricated building elements in Europe and elsewhere is assisting in making currently installed foams more accessible at end of life. The ongoing JTCCM project is spending its third year investigating the technical potential for recovering blowing agents from previously installed foams and the results should be reported internationally by mid-2005.

Economic considerations
The JTCCM study also involves the evaluation of the cost-effectiveness of recovering blowing agents from foams in buildings. This will be the subject of the fourth and final year of the project. Preliminary indications from previous work carried out in the field (Swedish EPA, 1996) show that recovery from traditional buildings may be uneconomical. However, the fact that steel-faced panels could be processed through recapture/recycle mechanical refrigerator plants may have an impact on future achievements.

All of these approaches have been stimulated under the Montreal Protocol framework, which does not mandate recovery. Under the Multilateral Fund, the finance made available for phase-out is typically capped at $15/kg (Jeffs *et al.*, 2004). At previous levels of activity, the mechanical recapture/recycling processes were handling refrigerators at a net cost of $15–20/unit (UNEP-TEAP, 2002ab), although more recent information from the market suggests that this may have even fallen as low as $10/unit. With typical recovery levels of 250–325 g per unit, the cost of recapture and destruction is $30–60 per kg of blowing agent.

The mathematics for HFC recovery will be based more specifically on climate change economics and the developing concept of the social cost of carbon (Clarkson and Deyes, 2002). These aspects are covered in more detail in Section 3.3.4.1. The cost of abatement calculations for HFC recovery are still in their infancy, but there is an expectation that if sufficient foam can reach such disposal facilities intact, the economics under a climate change scenario will be more favourable, even for foam from buildings. This would be particularly true if investments were to take into account the real value to the climate of the incremental destruction of ozone-depleting substances,

---

[4] AHAM – Alliance of Home Appliance Manufacturers – a US-based industry association

the GWPs of which are even higher than those of the HFCs replacing them. These issues will be addressed further in the forthcoming UNEP Task Force on end-of-life management for foam.

Expanded 'best practice'

The domestic refrigeration sector remains the major focus for end-of-life management in the short term. If the European approach to such management were to be extrapolated for use elsewhere in the world, the implications for HCFC emissions would be significant. This would be potentially achievable in developed countries from 2007, but would take longer to establish in developing countries. Under the end-of-life scenario developed under this analysis, it is assumed that all decommissioned refrigerators could be managed worldwide from 2010.

The situation is less clear with respect to the recovery of blowing agent from building insulation. However, it is apparent that some building elements (e.g. steel-faced panels) could be managed technically and economically. Work is continuing in the UK to confirm the economics of panel recovery and a report is expected in 2005. The specific advantage associated with panels is that they could be managed using the plants already established for refrigerators. This means that full recovery could commence as early as 2007.

For other types of building insulation, the situation is less clear. Even with favourable technology and economics, the development of an appropriate infrastructure would take time. The JTCCM project is expected to provide further information on this issue when it reports finally in 2005. For the sake of this assessment, however, it is assumed that no more than 20% of the remaining blowing agent will be technically and economically recoverable and, even then, this will not be possible before 2010.

7.5.3   *Further analysis of cost elements*

There have been few systematic reviews of abatement costs for the foam sector. The March Consulting Group (1998) reviewed potential emissions and abatement options for HFCs in the European Union. However, this review only included the costs of the options identified at that time for the reduction of potential extruded-polystyrene emissions. Ecofys picked up the agenda in the EU Sectoral Objectives Report (Harnisch and Hendriks, 2000) and conducted a more systematic review of abatement costs for both the extruded-polystyrene and polyurethane sectors. However, there were several drawbacks to this study. First of all, it was written before the transition to HFC blowing agents had taken place in Europe and before many of the technology choices had been finalized. As a consequence, it considerably over-estimated the uptake and consequent emissions of HFCs in the region. Secondly, it was only tasked to look at emission-abatement options that would be effective by 2010. This immediately ruled out the costing of end-of-life management options and forced the focus onto HFC-substitution technologies.

For the major polyurethane foam technologies, abatement costs in the range of $25–85 per tonne of CO_2-eq (1999 cost base) were estimated where replacement technologies were believed to exist (based approximately on a 50% emission reduction potential on average). If anything, it is anticipated that these costs are now on the low side, since many polyurethane-foam sectors have minimized the amount of HFCs used per unit of foam by the adoption of blends and other similar techniques, making abatement investments less effective.

Both March (1998) and Ecofys (Harnisch and Hendriks (2000)) gave particularly low estimates of abatement costs for measures for extruded polystyrene; both were in the range of $6–12 per tonne of CO_2-eq on the basis of transitions to CO_2 blowing options with investment amortization over a fifteen-year period. However, it should be noted that a transition to CO_2-based technologies is not currently possible for smaller producers and other regional markets (e.g. North America). The more recent IMAC Report (US EPA, 2004) has reassessed this option on a global basis and states an even lower estimate for the cost of transition. However, industry sources have been quick to point out that all of the emissions models used for these costing exercises have:
*   overstated in-life emissions (Vo and Paquet, 2004);
*   under-estimated the magnitude and costs of density increases;
*   under-estimated the significance of thickness constraints;
*   used investment cost information that is too low.

Estimates of polyurethane-foam technology abatement costs from the IMAC have also been lower than those from earlier European studies, but this may be due to the adoption of a 25-year amortization period for investments, which most commentators consider to be excessive for this industry.

The IMAC Report is the only study so far to provide a systematic appraisal of the costs of recovering blowing agents from appliances at end of life. The estimates, described by the authors as 'illustrative, rather than definitive', suggest costs in the order of $18–20 per tonne of CO_2-eq for automated processes and $48 per tonne of CO_2-eq for manually dismantled units with foam incineration. Anecdotal information from Europe suggests that ozone-depleting substances are currently being recovered at a price of around $40/kg. If transferred to HFC blowing agents, this would equate to an emission abatement cost of somewhere below $40 per tonne of CO_2-eq, depending on the profit margins applied. This would be considerably lower again for CFCs where GWPs are higher.

There is no data at present about blowing agents recovered from building elements and this is a subject for review in the UNEP TEAP Task Force Report scheduled for publication in mid-2005.

7.5.4   *Revised emission projections*

Figure 7.6 illustrates the effects of the three scenarios developed in this review. These are:

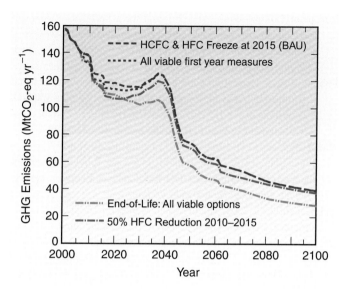

Figure 7.6. Summary of emission reduction scenarios for foams

deliver more overall savings in the period to 2100 if all blowing agent types are considered. The value is particularly significant for CFCs, where GWPs are significant and where there is an incremental effect of ozone depletion.

Table 7.10 sets out the cumulative savings as at 2015, 2050 and 2100 by blowing-agent type, with the aim of providing a clearer picture of the analysis conducted. However, it must be recalled that these savings (particularly in the end-of-life scenarios) could be conservative depending on the assumptions adopted for the business-as-usual losses for landfill.

Of course, in reality, any set of measures is likely to encompass elements from all three scenarios rather than to follow one exclusively at the expense of the others. The purpose of this assessment is only to indicate the relative significance of various options in order to provide a basis for prioritization.

1. a 50% reduction in HFC consumption between 2010 and 2015;
2. the implementation of production and first-year emission reductions in the block-foam sector from 2005 and in other sectors from 2008;
3. the geographic extension of end-of-life management for domestic refrigerators, together with a further extension to cover steel-faced panels and, where practicable, other building foams.

It can be seen that focusing on reducing HFC consumption provides the most significant saving in the period to 2015 and, if any such reduction is extrapolated to include use patterns beyond 2015, it offers the greatest 'HFC-specific' benefit to 2100 as well (see Table 7.10). In contrast, end-of-life measures deliver lower savings in the period to 2015, but have the potential to

Table 7.10. Summary of impacts of individual packages of measures by blowing agent type: cumulative emission reductions resulting under each scenario assessed.

Measure	*Year*	Cumulative Emission Reductions			
		CFCs (tonnes)	HCFCs (tonnes)	HFCs (tonnes)	CO_2-equivalents (MtCO$_2$-eq)
HFC consumption reduction (2010–2015)	2015	0	0	31,775	36
	2050	0	0	225,950	259
	2100	0	0	352,350	411
Production/installation improvements	2015	78	14,450	16,700	36
	2050	58	31,700	32,700	68
	2100	47	24,350	26,500	55
End-of-life management options	2015	8545	16,375	105	52
	2050	64,150	144,650	88,540	540
	2100	137,700	358,300	194,800	1200

References

APME/ISOPA/EXIBA, 1996: Co-combustion of Building Insulation Foams with Municipal Solid Waste – Final Report. Association of Plastics Manufacturers in Europe (APME) and European Isocyanate Producers Association (ISOPA), European Extruded Polystyrene Insulation Board Association (EXIBA), Brussels, Belgium, 1996.

Ashford, P., 1998: Assessment of Potential for the Saving of Carbon Dioxide Emissions in European Building Stock. Caleb Management Services, Bristol, United Kingdom.

Ashford, P., 1999: Consideration for the Responsible Use of HFCs in Foams. Proceedings of the Joint IPCC/TEAP Expert Meeting on Options for the Limitation of Emissions of HFCs and PFCs, L. Kuijpers, R. Ybema (eds.), 26-28 May 1999, Energy Research Foundation (ECN), Petten, The Netherlands (available online at www.ipcc-wg3.org/docs/IPCC-TEAP99).

Ashford, P., 2000: Final Report prepared for AFEAS on the Development of a Global Emission Function for Blowing Agents used in Closed Cell Foam. Caleb Management Services, Bristol, United Kingdom.

Ashford, P., D. Clodic, A. McCulloch and L. Kuijpers, 2004a: Emission profiles from the foam and refrigeration sectors, Comparison with atmospheric concentrations. Part 1: Methodology and data. *Int. J. Refrigeration*, **27**(7), 687-700.

Ashford, P., D. Clodic, A. McCulloch and L. Kuijpers, 2004b: Emission profiles from the foam and refrigeration sectors, Comparison with atmospheric concentrations. Part 2: Results and discussion, *Int. J. Refrigeration*, **27**(7), 701-716.

Ashford, P. and A. Vetter, 2004: Assessing the Role of HFC Blown Foam in European Building Renovation Strategies. Caleb Management Services, Bristol, UK. Proceedings of the 15th Annual Earth Technologies Forum, April 13-15, 2004, Washington, D.C., USA.

Butler, D.J.G, 2002: Determining Losses of Blowing Agent During Cutting of Refrigerators. Building Research Establishment (BRE), Garston, Watford, United Kingdom.

Carbon Trust, 2002: Submission to the UK Energy Policy Consultation (September 2002). Carbon Trust, London, United Kingdom, 42 pp.

Cheminfo Services Inc., 2001: Study on Alternatives to HCFCs in Foam Manufacturing, Draft Final Report (Januari 2001) to Environment Canada. Cheminfo Services Inc., Markham, Ontario, Canada.

Clarkson, R. and K. Deyes, 2002: Estimating the Social Cost of Carbon Emissions. Government Economic Service Working Paper No. 140, UK HM Treasury, London, United Kingdom, 59 pp.

DIN, 2002: Draft DIN-8975-12: Kälteanlagen - Sicherheitstechnische Grundsätze für Gestaltung, Ausrüstung und Aufstellung - Teil 12: Rückgewinnungssysteme. Deutsches Institut für Normung e.V., Berlin/Köln, Germany.

European Commission, 1996: European Energy to 2020 – A Scenario Approach. European Commission, DG XVII (ISBN 92-827-5226-7), Brussels, Belgium.

European Commission, 2003: Proposal for a European Parliament and Council Regulation on Certain Fluorinated Greenhouse Gases. Commission of the European Communities, COM(2003)492final, 2003/0189 (COD), August 11, 2003, Brussels, Belgium.

Fabian, B.A., S. Herrenbruck and A. Hoffee, 2004: The Environmental and Societal Value of Extruded Polystyrene Foam Insulation. Proceedings of the 15th Annual Earth Technologies Forum, April 13-15, 2004, Washington, D.C., USA.

Harnisch, J. and C. Hendriks, 2000: Economic Evaluation of Emission Reductions of HFCs, PFCs and SF_6 in Europe. Report prepared for the European Commission DG Environment, Ecofys, Cologne/Utrecht, Germany/Netherlands, 70 pp.

IAL Consultants, 2002: The European Market for Thermal Insulation Products. IAL Consultants, London, United Kingdom.

IPCC (Intergovernmental Panel on Climate Change), 2001: *Climate Change 2001 – Mitigation. Contribution of Working Group III to the Third Assessment Report of the Intergovernmental Panel on Climate Change* [Metz, B., O. Davidson, R. Swart and J. Pan (eds.)] Cambridge University Press, Cambridge, United Kingdom, and New York, NY, USA, 752 pp.

Jeffs, M., P. Ashford, R. Albach and S. Kotaji, 2004: Emerging Usage Patterns for HFCs in the Foams Sector in the Light of Responsible Use Principles. Proceedings of the 15th Annual Earth Technologies Forum, April 13-15, 2004, Washington, D.C., USA.

Johnson, R., 2004: The effect of blowing agent choice on energy use and global warming impact of a refrigerator. *International Journal of Refrigeration,* **27**(7), 794–799.

Kjeldsen, P., 2002: Determining the Fraction of Blowing Agent released from Polyurethane Foam used in Refrigerators after Decommissioning the Product. Proceedings of the Earth Technologies Forum, 25-27 March 2002, Washington, DC, USA.

Lee, S. and Mutton, J., 2004: Global Extruded Polystyrene Foams Emissions Scenarios. Proceedings of the 15th Annual Earth Technologies Forum, April 13-15, 2004, Washington, D.C., USA.

March (March Consulting Group), 1998: Opportunities to Minimise Emissions of Hydrofluorocarbons (HFCs) from the European Union. March Consulting Group, Manchester, UK, 121 pp.

McBride, M.F., 2004: Energy and Environmental Benefits of HCFCs and HFCs in Extruded Polystyrene Foam Insulation. Proceedings of the 15th Annual Earth Technologies Forum, April 13-15, 2004, Washington, D.C., USA.

Mizuno, K., 2003: Estimation of Total Amount of CFCs Banked in Building Insulation Foams in Japan. Proceedings of the 14th Annual Earth Technologies Forum, April 22-24, 2003, Washington, D.C., USA.

Ohnuma, Y. and J. Mori, 2003: Supercritical or Subcritical CO_2 Assisted Water-Blown Spray Foams. Proceedings of the Polyurethanes Expo 2003, September 30 to October 3, 2003, Orlando, Florida, USA.

OJ, 2000: Regulation (EC) No 2037/2000 of the European Parliament and of the Council of 29 June 2000 on Substances that Deplete the Ozone Layer. *Official Journal of the European Communities*, L244, pp. 0001-0024, 29 September 2000.

OJ, 2003: Directive 2002/91/EC of the European Parliament and of the Council of 16th December 2002 on the Energy Performance of Buildings. *Official Journal of the European Communities,* L1, pp. 0065-0071, 4 January 2003.

Petersdorff, C., T. Boermans, J. Harnisch, S. Joosen and Frank Wouters, 2002: The Contribution of Mineral Wool and other Thermal Insulation Materials to Energy Saving and Climate Protection in Europe. Report prepared by Ecofys for European Insulation Manufacturers Association (Eurima), Köln, Germany, 36 pp.

Pout, C.H., F. MacKenzie and R. Bettle, 2002: Carbon Dioxide Emissions from Non-Domestic Buildings – 2000 and beyond. Building Research Establishment (BRE), Garston, Watford, United Kingdom.

Price, L., L. Michaelis, E. Worrell and M. Khrushch, 1998: Sectoral trends and driving forces of global energy use and greenhouse gas emissions. *Mitigation and Adaptation Strategies for Global Change,* **3**, 263–319.

RAL, 2001: Quality Assurance RAL GZ 728: Demanufacture of Refrigeration Equipment Containing CFCs (2001 Update). RAL Deutsches Institut für Gütesicherung und Kennzeichnung e.V., Sankt Augustin, Germany.

Scheutz, C. and P. Kjeldsen, 2001: Determination of the Fraction of Blowing Agent Released from Refrigerator/Freezer Foam after Decommissioning the Product. Report prepared by Environment & Resources DTU – Technical University of Denmark for the Association of Home Appliance Manufacturers (AHAM), Lyngby, Denmark, 80 pp.

Scheutz, C., A.M. Fredenslund and P. Kjeldsen, 2003: Attenuation of Alternative Blowing Agents in Landfills. Report prepared by Environment & Resources DTU – Technical University of Denmark for the Association of Home Appliance Manufacturers (AHAM), Lyngby, Denmark, 66 pp.

Solvay, 2000: HFC-365mfc as Blowing and Insulation Agent in Polyurethane Rigid Foams for Thermal Insulation. (H. Krähling and S. Krömer), Solvay Management Support, Hannover, Germany.

Spanish Government, 2003: Estrategia de Ahorro y Eficiencia Energética 2004–2012 – Sector Edificación : Documento Trabajo 'E4' (Energy Efficiency and Economy Strategy 2004–2012 –Buildings Sector : Working Document 'E4'), 4th July 2003, Madrid, Spain.

Swedish EPA, 1996: Stena freonåtervinning Projekt SNVIII. Naturvardsverket (Swedish Environmental Protection Agency), Stockholm, Sweden (in Swedish).

UNEP (United Nations Environment programme), 2003: Handbook for the International Treaties for the Protection of the Ozone Layer – The Vienna Convention (1985), The Montreal Protocol (1987). Sixth edition (2003). [M. Gonzalez, G.M. Bankobeza (eds.)]. UNEP Ozone Secretariat, Nairobi, Kenya.

UNEP-FTOC (Foam Technical Options Committee), 1995: 1994 Report of the Rigid and Flexible Foams Technical Options Committee, UNEP Ozone Secretariat, Nairobi, Kenya.

UNEP-FTOC (Foam Technical Options Committee), 1999: 1998 Report of the Rigid and Flexible Foams Technical Options Committee, UNEP Ozone Secretariat, Nairobi, Kenya.

UNEP-FTOC (Foam Technical Options Committee), 2003: 2002 Report of the Rigid and Flexible Foams Technical Options Committee, UNEP, UNEP Ozone Secretariat, Nairobi, Kenya, 138 pp.

UNEP-TEAP (Technology and Economic Assessment Panel), 2002a: April 2002 Report of the Technology and Economic Assessment Panel – Volume 3A: Report of the Task Force on Collection, Recovery and Storage. [W. Brunner, J. Pons and S.O. Andersen (eds.)]. UNEP Ozone Secretariat, Nairobi, Kenya, 112 pp.

UNEP-TEAP (Technology and Economic Assessment Panel), 2002b: April 2002 Report of the Technology and Economic Assessment Panel – Volume 3B Report of the Task Force on Collection, Recovery and Storage. [S. Devotta, A. Finkelstein and L. Kuijpers (eds.)]. UNEP Ozone Secretariat, Nairobi, Kenya, 150 pp.

UNEP-TEAP (Technology and Economic Assessment Panel), 2003: May 2003 Report of the Technology and Economic Assessment Panel, HCFC Task Force Report. [L. Kuijpers (ed.)]. UNEP Ozone Secretariat, Nairobi, Kenya, 96 pp.

UNECE (United Nations Economic Commission for Europe), 2002: Environment and Human Settlement Statistics Database, Geneva, Switzerland.

US EPA (US Environmental Protection Agency), 2001: Protection of Stratospheric Ozone: Notice of Data Availability – New information concerning SNAP Program proposal on HCFC use in Foams (40-CFR-Part 82), Federal Register Vol. 66, No. 100, Wednesday May 23rd, pp. 28408-28410.

US EPA (United States Environmental Protection Agency), 2004:, 2004: Analysis of Costs to Abate International Ozone Depleting Substance Substitute Emissions. [D.S. Godwin (ed.)]. US EPA Report 430-R-04-006, US Environmental Protection Agency, Washington, DC, USA, 309 pp (available online at http://www.epa.gov/ozone/snap/).

Vo, C.V. and A.N. Paquet, 2004: An evaluation of the thermal conductivity for extruded polystyrene foam blown with HFC-134a or HCFC-142b. *Journal of Cellular Plastics,* **40**(3), 205-228.

Wilkes, K.E., D.W. Yarbrough and G.E. Nelson, 2003: Aging of Polyurethane Foam Insulation in Simulated Refrigerator Panels – Four-Year Results with Third-Generation Blowing Agents. Proceedings of the 14th Annual Earth Technologies Forum, April 22-24, 2003, Washington, D.C., USA.

Zoughaib, A., L. Palandre and D. Clodic, 2003: Measurement of remaining quantities of CFC-11 in Polyurethane Foams of Refrigerators and Freezers at End-of-Life. Centre for Energy Studies – Ecole des Mines de Paris, Paris, France. Proceedings of the 14th Annual Earth Technologies Forum, April 22-24, 2003, Washington, D.C., USA.

8

Medical Aerosols

Coordinating Lead Authors
Paul J. Atkins (United States), Ashley Woodcock (United Kingdom)

Lead Authors
Olga Blinova (Russia), Javaid Khan (Pakistan), Roland Stechert (Germany),
Paul Wright (United Kingdom), You Yizhong (China)

Contributing Authors
Ayman El-Talouny (Egypt)

Review Editors
David Fakes (United Kingdom), Megumi Seki (Japan)

Contents

EXECUTIVE SUMMARY

Asthma and COPD (Chronic Obstructive Pulmonary Disease) are major illnesses worldwide affecting over 300 million people. Inhaled therapy is, and is likely to remain, the gold standard for treatment. There are two main methods of delivering respiratory drugs for most patients: these are the metered dose inhaler (MDI) and the dry powder inhaler (DPI). The choice of the most suitable inhaler is a complex medical decision taken in consultation between the doctor and patient.

When the MDI was introduced in the mid-1950s, CFCs were used as propellants and these have been replaced in recent years with HFCs, although the complete phase-out of CFCs in MDIs is not expected to be before 2010. CFCs and HFCs are used in MDIs because of their chemical and physical properties and no other propellants have been developed to date that might constitute alternatives. Multidose DPIs, which do not require a propellant, have also become more widely available in the past decade as a not-in-kind replacement and this has mitigated the growth of the use of MDIs.

MDIs remain the dominant form of treatment for asthma and COPD worldwide in terms of units prescribed. In developed countries, the proportion of MDI to DPI varies substantially from country to country: from 9:1 MDI:DPI in the USA to 7:3 in UK and 2:8 in Sweden. The variation is accounted for by a number of factors, including availability (e.g. multidose DPIs only recently became available in the USA; by contrast, there is a local manufacturer and a long tradition of DPI use in Sweden), patient and physician choice, and relative cost. CFC MDIs are being phased out, and being replaced by HFC MDIs and DPIs. The use of DPIs in developing countries is negligible. Both MDIs and DPIs play an important role in the treatment of asthma/COPD and no single delivery system is universally acceptable for all patients. It is critical to maintain the range of therapeutic options.

The CFC transition under the Montreal Protocol is still in progress, with rates varying between countries. CFC inhalers are being replaced by both HFC MDIs and DPIs. The main impact on GWP is made by the transition from CFC to HFC MDIs. Recently, a number of new drugs have been launched in DPIs. DPIs are generally more expensive than MDIs (especially for salbutamol) and not all drugs are available in DPIs. As approximately 50% of CFC MDIs have historically contained salbutamol, it is likely that this rescue medication will continue to be given in MDIs and will necessitate the continued use of HFCs. As the development and regulatory time scales for new inhaled delivery systems are lengthy (10 years or longer) no major technical breakthroughs are expected to become available for patients for 10 to 15 years (Table 8a).

Predicting HFC usage in MDIs for developing countries is quite difficult and depends on a number of factors, some of which are under the control of national governments:

- the timing of CFC phase-out by local MDI manufacturers (including HFC intellectual property and manufacturing issues);
- nationally-produced DPIs will emerge, but it is likely these will still cost more per dose than HFC MDIs;
- disease trends and the continued uptake of multidose DPIs;
- general availability of affordable medicines.

Switching patients from reliable and effective medications has significant implications for patient health and safety. The provision of a range of safe alternatives is critical before enforcing change on environmental grounds. Any environmental policy measures for the future that could impact patient use of HFC MDIs would require careful consideration and consultation with physicians, patients, national health authorities and other health-care experts.

The overall use of HFCs for MDIs for asthma/COPD (rounded off to the nearest 100 tonnes) is predicted to be up to 15,000 metric tonnes (13,500 tonnes HFC-134a, 1500 tonnes HFC-227ea) by 2015. On the basis of these forecasts, the maximum environmental benefit of the hypothetical extreme case of switching all HFC MDIs to DPIs would be in the order of 23 million tonnes of CO_2 per year.

The estimated incremental cost of switching HFC MDIs to DPIs would be mainly related to inexpensive salbutamol HFC MDIs, and is of the order of an incremental and annually recurrent US$ 1.7–3.4 billion. This is equivalent to 150–300 US$/$tCO_2$-eq.

Small amounts of HFC are also used in non-inhalational topical aerosols, but most applications have non-HFC alternatives (mechanical pumps, hydrocarbon propellants, etc.).

Table 8a. Summary of estimates of global use of CFCs and HFCs in MDIs (IPAC, 2004; derived from UNEP-TEAP (2003), page 120).

Time scale	Annual CFC use in MDIs		Annual HFC use in MDIs	
	metric tonnes	$MtCO_2$-eq yr^{-1}	metric tonnes	$MtCO_2$-eq yr^{-1}
1987–2000	15,000	128	<500	<1
2001–2004	<8,000	<69	3,000–4,000	<5–6
2005–2015	<2,000	<17	13,000–15,000	23–26

8.1   Introduction

Medical aerosols include metered dose inhalers (MDIs) for asthma and COPD, and non-MDI medical aerosols.

8.1.1   *Metered dose inhalers*

Asthma and chronic obstructive pulmonary disease (COPD)
Asthma and chronic obstructive pulmonary disease (COPD) are the most common chronic diseases of the air passages (airways or bronchi) of the lung and are thought to affect over 300 million people worldwide. These illnesses account for high health-care expenditure, cause significant absence from work and school, and premature death. Modern treatment for these conditions involves the inhalation of aerosol medication with a specific particle size (1–5 micron), which is deposited into the airways of the lung (bronchi). This allows maximal local effect in the airways where it is needed and minimizes the side-effects of the drug elsewhere in the body.

Asthma
Asthma is a chronic condition with two main components: airway inflammation and narrowing. Most patients have symptoms every day, with more severe attacks intermittently, during which coughing and wheezing develop and the airways narrow, making it very difficult to breathe. Attacks of asthma may occur spontaneously or be triggered by many environmental factors or viral infections. Attacks of asthma may require urgent additional medication, they sometimes require hospitalization and they are occasionally fatal.

Asthma most often starts in childhood, and persists into adult life, causing frequent attacks, chronic ill health and incapacity. A recent international study of asthma in childhood has shown a prevalence of asthma that varies from approximately 1 percent in some countries such as Indonesia to over 30 percent in the United Kingdom, New Zealand, and Australia (ISAAC-SC, 1998). It is more common in affluent countries, but has been increasing rapidly in developing countries over the last two decades. This increasing prevalence is likely to be due to multiple factors including 'westernization', changes in house design, greater exposure to house dust mite, maternal smoking, diet, air pollution and/or tobacco smoking.

Chronic obstructive pulmonary disease (COPD)
COPD is a condition involving the narrowing and inflammation of the airways in conjunction with damage to the lung tissue (emphysema). COPD is caused primarily by cigarette smoking, with inhalation of occupational dusts or environmental air pollution as potential co-factors. COPD is persistent and progressive if the patient continues to smoke, and further deterioration can still occur even after smoking cessation. COPD ultimately leads to permanent disability and death. Acute exacerbations of COPD frequently require hospitalization.

The prevalence of COPD in many developed countries is around 4–17 percent in the adult population aged over 40 years (summarized in Celli *et al.*, 1999). Data are less certain in developing countries but figures as high as 26 percent have been quoted. Rates in men are generally higher than women and reflect smoking prevalence. Smoking is beginning to decline in some developed countries, but trends in developing countries indicate that both smoking and the prevalence of COPD are of increasing concern.

In the 1996 Global Burden of Disease Study sponsored by the World Health Organization, COPD was ranked 12 in terms of disability, but is projected to rank 5 in 2020, behind ischaemic heart disease, major depression, traffic accidents and cerebrovascular disease. In 1998, COPD was the fourth most common cause of death in the United States after heart disease, cancer and stroke. In most developed countries, the male death rate from COPD has been declining. By contrast, the female mortality rate is increasing and it is expected that mortality rates amongst females will overtake those men by about 2005.

Treatment
Modern treatment of asthma and COPD consists of inhaled therapy (Dolovich, 2000). This affords highly effective treatment with few side-effects.

There are two main categories of inhaled treatment for asthma and COPD: bronchodilators (also called acute relievers), and anti-inflammatory medication (also called controllers or preventers).

Bronchodilators (reliever medication)
Virtually all patients with asthma and COPD require short-acting bronchodilators. Bronchodilators reduce muscle tightening, which contributes to the narrowing in the airways. They are the key treatment for acute attacks and are lifesaving in severe attacks. In intervals between attacks, they may be needed frequently through the day, particularly in children for whom exercise-induced asthma is common.

Inhaled bronchodilators fall into three classes:
*   *Beta-agonists* – These are the main reliever treatment for asthma and COPD. *Short acting beta-agonists* include salbutamol (known as albuterol in the United States), terbutaline, and fenoterol. They act within a few minutes, and have an effect lasting approximately 4 hours.
*   *Long-acting beta-agonists* - salmeterol and formoterol have an effect that may last for up to 12 hours.
*   *Anti-cholinergics* including ipratropium bromide and tiotropium bromide. These are commonly used as first-line bronchodilator therapy in COPD.

Anti-inflammatory medication (controllers or preventers)
Inflammation of the airways is a fundamental part of asthma, and suppression of this inflammation is recommended in all but those with mild infrequent symptoms. Anti-inflammatory treatment stabilizes lung function and prevents acute exacerbation of asthma if used regularly, hence the term 'preventer'.

Preventers are commonly used in COPD, but are less effective in this condition. Inhaled preventers are usually one of two classes of drug:

- *inhaled steroids* (e.g. beclomethasone, budesonide, flunisolide, fluticasone or triamcinolone): these are the mainstay of preventer medication for asthma and COPD;
- *cromoglycate-like drugs* (e.g. sodium cromoglycate or nedocromil sodium): these are less effective than inhaled glucocorticosteroids.

Oral treatments

Oral treatments have a limited, but sometimes important, role in asthma/COPD. Theophyllines are an old inexpensive therapy, which is both relatively ineffective and has major side-effects. Newer oral leukotriene antagonists are safe but have generally been found to be less effective clinically and, in any event, they are not a substitute for inhaled steroids in asthma, and are ineffective in COPD.

8.1.2   Non-MDI medical aerosols

Before the Montreal Protocol phase-out of CFCs in non-Article 5(1) countries in 1996, CFCs were commonly used in aerosols outside the asthma/COPD indication, namely for:

- topical (skin) therapy for local anaesthesia and cooling for sports injuries;
- sub-lingual sprays for angina pectoris;
- nasal sprays for allergic rhinitis and sinusitis;
- vaginal foams for contraception;
- rectal foams for colitis.

All these uses have been replaced mainly by mechanical pump sprays, hydrocarbon, di-methyl ether and compressed gas, and a few by HFC pressurized aerosols. There are also some early technical developments of novel treatments formulated in DPIs or aqueous sprays by biotechnology companies (e.g. buccal insulin). At this point it can be concluded that it is unlikely that significant volumes of HFCs will be used in applications other than MDIs for asthma/COPD.

8.2   Technical performance characteristics

Inhalation aerosols have been the subject of significant investment in research and development for most of the past twenty years. The science and technology have taken leaps forward in response to both therapeutic (Hickey and Dunbar, 1997) and environmental needs (Molina and Rowland, 1974).

The requirement to phase out ozone-depleting chlorofluorocarbon propellants (in MDIs) has increased the interest in alternative systems, most notably dry powder inhalers (DPIs). Significant progress has been made in all aspects of their principles of operation, including formulation, metering and dispersion in the past decade.

Pressurized metered dose inhalers

A metered dose inhaler (MDI) is a complex system designed to provide a fine mist of medicament for inhalation directly to the airways as treatment for respiratory diseases such as asthma and COPD. The propellant-driven metered dose inhaler (MDI) has been the dominant inhaler for almost half a century. The combination of the drug dispersed in a high vapour pressure propellant, metered accurately in tens to hundreds of microgrammes and administered directly to the lungs was a powerful new tool for the treatment of pulmonary diseases (Hickey and Evans, 1996). The historical confluence of this new technology with potent new therapeutic agents revolutionized the treatment of asthma and COPD.

The main components of MDIs are:

- the active ingredient;
- excipient(s);
- the propellant (a liquefied gas);
- a metering valve;
- a canister;
- an actuator/mouthpiece.

Traditionally, MDIs have contained CFCs, primarily CFC-12 and CFC-11, and sometimes CFC- 114. However, because of the detrimental effect of CFCs on the Earth's ozone layer (Molina and Rowland, 1974), there was an extensive search in the 1990s for propellants that could be used as alternatives and had much less of an effect on the ozone layer and potentially on the environment in general. The hydrofluorocarbons 134a and 227ea are both now being used in the pharmaceutical industry as propellants for MDIs (Pischtiak *et al.*, 2001). These materials are pharmacologically inert and have similar properties to the CFC propellants they replaced (see Table 8.1). However, they

Table 8.1. Hydrofluorocarbon (HFC) propellants used in MDIs.

Substance	CFC-11	CFC-12	HFC-134a	HFC-227ea
Density (kg litre^{-1})	1.49	1.33	1.21	1.41
Vapour Pressure (at 20°C)				
in kPa	12.4	466	472	386
in psig	1.8	67.6	68.4	56.0
Boiling Point (°C)	23.7	-29.8	-26.5	-17.3

are so different from CFCs that they require a significant investment in formulation strategies for the range of drugs traditionally delivered by MDIs.

There are two types of MDI formulations: suspension formulations, in which micro-particulate drug (typically micronized material) is dispersed in a combination of propellants; and solution formulations, in which the drug freely dissolves in either the propellant or a combination of propellant and an acceptable co-solvent, typically ethanol (June *et al.*, 1994; Smith, 1995). Both types of formulation have inherent advantages and disadvantages. Traditionally, suspension formulations have been the more common dosage form, but with the advent of the hydrofluoroalkane propellants (HFC-134a; HFC-227ea), which have poor solvency characteristics, the use of co-solvents has become more common and solution formulations being used more (Leach *et al.*, 1998; Brown, 2002). Other potential propellants (e.g. hydrocarbons, pressurized gases) do not possess the safety profile of CFCs, or are impossible to reformulate.

Globally, there are a number of companies involved in developing HFC MDIs. They include: 3M Pharmaceuticals (USA); Aventis (France/Germany); Boehringer Ingelheim (Germany); Chiesi (Italy); Cipla (India); GlaxoSmithKline (UK); and Ivax Healthcare (USA/UK) (see Table 8.2). The development of HFC MDIs has provided many challenges for the pharmaceutical industry, requiring new formulations, novel surfactants and cosolvents, new valves and canisters, and new manufacturing plants. HFC MDIs are manufactured according to 'good manufacturing practices' regulated and inspected by government authorities, and efforts are made to ensure that fugitive HFC emissions and waste are strictly controlled (Medicines Control Agency, 1997).

An HFC MDI, for the widely prescribed short-acting beta-agonist salbutamol, was introduced in the United Kingdom for the first time in 1994. Today, there are over 60 countries where at least one HFC MDI containing salbutamol has been approved and marketed. In addition to the introduction of beta-agonist HFC MDIs, there are a growing number of controller medications (e.g. inhaled corticosteroids) available as HFC MDIs. It is estimated that there were approximately 100 million HFC MDIs produced globally in 2002 (UNEP-TEAP, 2002), representing approximately 25% of worldwide MDI production.

Nebulizers
Nebulizers for the delivery of solutions of drugs were in existence prior to the development of the MDI. These systems found their application in acute ambulatory care and domiciliary situations since the droplet size of the aerosol was slightly smaller than that of the MDI, could penetrate more deeply into the lungs of the patients and did not require a degree of co-ordination to deliver the drug effectively (Dalby *et al.*, 1996). Nebulizers continue to play a valuable role in severely compromised patients and for applications other than asthma, notably cystic fibrosis (Garcia-Contreras and Hickey, 2002).

Nebulizers can be used to generate aerosols for inhalation from liquid solutions or suspensions. Patient coordination of aerosol delivery with inhalation is not as critical as for MDIs or DPIs for achieving a therapeutic effect. In addition, aqueous solutions are often easily formulated for use in nebulizers (Niven, 1996). However, most nebulizers are bulky, inconvenient and too expensive for routine use.

Nebulizer systems typically fall into the categories of air jet or ultrasonic depending on the physical principle used for aerosol droplet generation. Jet nebulizers draw solution through a capillary tube using the Bernoulli effect and disperse droplets in air at high velocity. Ultrasonic nebulizers use high-energy ultrasonic vibration to create droplets suitable for inhalation. Until very recently, nebulizer use was limited to the hospital or home due to the energy requirements and poor portability of conventional systems. In general, the performance of nebulizers tends to vary significantly depending on the type and formulation.

A number of new 'portable' nebulizer technologies are being developed, although these may take some years to become commercially available (DeYoung *et al.*, 1998).

Dry powder delivery systems
DPIs contain dry powder formulations of inhalable drug, but do not use propellants. DPIs were used on a limited scale in the 1960s and 1970s, when the Spinhaler and Rotahaler were used

Table 8.2. Currently available formulations (by company) for the most widely prescribed inhaled drugs, salbutamol, beclomethasone, budesonide and cromoglycate.

Drug Compound	Formulation	Producer
Salbutamol	Ethanol/Surfactant/HFC-134a	3M Pharmaceuticals Ivax Healthcare
	HFC-134a alone	GlaxoSmithKline Cipla
Beclomethasone	Ethanol/HFC-134a	3M Pharmaceuticals Ivax Healthcare
	Ethanol/HFC-134a/Glycerol	Chiesi
Budesonide	Ethanol/HFC-134a/Glycerol	Chiesi
	HFC-134a alone	Cipla
Cromoglycate	HFC-227 only	Sanofi-Aventis

for the delivery of disodium cromoglycate and albuterol/salbutamol respectively (Dunbar *et al.*, 1998). These were relatively inconvenient single-dose capsule-based devices that delivered the drug directly into the inspiratory airflow of the patient in a manner considered passive since additional energy was not used to support dispersion. DPIs have continued to evolve. Two new devices for once- or twice-daily medications remain as unit dose devices, whereas several others are multidose blister or reservoir devices (Maggi *et al.*, 1999).

The powder formulation is a critical component of a DPI. Several key attributes are necessary for successful respiratory drug delivery. First, the active drug must be produced in the appropriate particle size range of 1–5 microns (Schuster *et al.*, 1998), whether these fine particles are delivered as pure drug or as a formulation with acceptable excipients. Second, the drug particles must be chemically and physically stable following their manufacture, storage, and subsequent processing when preparing the final drug product.

Powders of respirable size tend to have poor flow characteristics due to adhesive interparticle forces. To facilitate powder filling and dispersion, the drug particles can be formulated with additional excipients, usually lactose. Regardless of the composition, regulatory requirements mean that the final powder formulation must provide a stable and reproducible aerosol. This is mainly assessed by the total emitted dose and fine particle fraction (Hickey, 1992; US FDA, 1998). DPIs may be less suitable for some patients with low inspiratory flow rates, such as the elderly and young children.

Most DPI formulations are sensitive to moisture during processing and moisture ingress into packaging during storage. The presence of moisture in the ultimate product has been shown to reduce the de-aggregation of the powder, thereby decreasing the fine particle fraction of the aerosol (Boekestein *et al.*, 2002; Staniforth, 2002). One approach proposes the use of magnesium stearate as a lubricant and anti-caking agent to maintain an inhalation powder's fine particle fraction under extreme temperature and humidity conditions (Staniforth, 2002).

Sensitivity of asthmatics to excipients must be considered when developing new products. For example, a metaproterenol MDI product reformulated with a new surfactant was withdrawn shortly after launch due to escalating reports of coughing, gagging, and asthma exacerbation (Poochikian and Bertha, 2000). A powder's sensitivity to moisture can also be overcome by designing and manufacturing packages (multi- or single-dose) with a proper moisture barrier. Exploiting advances in packaging technology can improve inhalation powder stability while avoiding the risk of new excipients.

8.3   Health and safety considerations

Inhaled therapy is the mainstay of treatment for asthma and COPD. MDIs are currently the most widely used inhalation device and millions of patients around the world rely on these products to manage their chronic, lifetime illnesses effectively (Wright, 2002). In order to accomplish the phase-out of CFCs

under the Montreal Protocol, the MDI industry undertook an exhaustive search for an appropriate alternative aerosol propellant. An inhalation propellant must be safe for human use and meet several additional strict criteria relating to safety and efficacy: (i) liquefied gas, (ii) low toxicity, (iii) non-flammable, (iv) chemically stable, (v) acceptable to patients (in terms of taste and smell), (vi) appropriate solvency characteristics, and (vii) appropriate density (Tansey, 1997; Smith, 1995). It was extremely difficult to identify compounds fulfilling all of these criteria and in the end only two hydrofluorocarbons – HFC-134a and HFC-227ea – emerged as viable alternatives to CFCs. Two international consortia (IPACT-I and IPACT-II) were then established to conduct thorough toxicological testing and ensure that these propellants were safe for inhalation by humans (Tansey, 1997; Emmen, 2000).

Once suitable non-CFC propellants had been identified, the MDI industry undertook to reformulate the CFC MDIs so that they could use HFCs. The components and formulations had to be substantially modified to use the new HFC propellants. As drug products, MDIs are subject to extensive regulation by national health authorities to ensure product safety, product efficacy and manufacturing quality. The process for developing CFC-free MDIs was therefore essentially the same as the development of a wholly new drug product, involving full clinical trials for each reformulated MDI. Research and development for a new product is a lengthy, challenging, and resource-intensive process; typically, it takes about ten years to reach the prescribing doctor.

After identifying alternate medical propellants and developing safe, effective CFC-free MDIs, the final step in the phase-out of CFC MDIs is to switch millions of patients to reformulated MDIs and other CFC-free products. Switching patients from reliable and effective medications for an environmental, rather than therapeutic, reason is a large and unprecedented exercise with significant implications for patient health and safety (Wright, 2002; Yellen, 2003; Price, 2004). Patients depend on MDIs for the treatment of serious illnesses that frequently impact activities of daily life, and that have potentially life-threatening consequences. For these reasons, asthma and COPD patients may be particularly sensitive to a change in a trusted treatment regimen. Changes in medicine may also impact patient compliance with necessary therapy. Comprehensive educational programmes for patients and physicians are therefore important to ensure a smooth transition from CFCs to HFCs.

HFC MDIs play a central role in the timely and effective phase-out of CFC MDIs. These products came into existence under unique circumstances and solely because of an international environmental treaty. Any additional environmental policy measures taken in the future with a potential effect on patient use of HFC MDIs will necessarily raise significant health and safety issues. They will require careful consideration and consultation with physicians, patients, national health authorities, health-care experts and the pharmaceutical industry. In particular, it is important to bear in mind the realities of medical decision-making and the central role of physicians and patients when considering any measures that could impact MDIs. Therapeutic decisions, including the choice of delivery system, are made by the prescribing physician and are based primarily on the individual circumstances of each patient. The optimal therapeutic approach varies depending on a variety of factors, including a patient's symptoms, physiology and compliance patterns. Each of the existing inhalation delivery systems has an important role in the treatment of respiratory illnesses and no single delivery system is universally acceptable for all patients. DPIs do provide an alternative for patients with asthma/COPD, but are not available for some drugs. They do not use propellants, but are complex and usually disposable devices, which are generally more expensive than HFC MDIs. It is therefore critical to preserve a range of therapeutic options (ICF, 2003; UBA, 2004; US-EPA, 2004).

8.4   Cost issues

Cost of development

The development costs (technical, pharmaceutical and clinical) up to 1999 for the CFC–HFC transition were estimated to exceed US$1 billion (Everard, 2001) and significant additional investment is continuing as reformulation and clinical testing has expanded. Similar costs would be expected for the development of new DPI products.

Device/treatment costs

The cost of the HFC propellant is a negligible proportion (~ 5–10%) of the total cost of an MDI. There is general consensus in the literature that MDIs are less expensive than DPIs for asthma and COPD (Consumer Association, 2001; Dalby and Suman, 2003; Yellan, 2003). However, quantitative data are more limited. The UK is a good model for MDI/DPI usage since it is a market with high MDI use and a history of DPI usage. Salbutamol is also an important cost model since it is the most widely prescribed drug and it is off patent so there are a number of generic products competing on price (British National Formulary, 2003).

The UK Drug and Therapeutics Bulletin (Consumer Association, 2000) has compared salbutamol on the basis of cost per single dose; the result was US$0.030/0.035 for MDIs and US$0.078/0.143 for DPIs. For another widely prescribed generic drug, beclomethasone dipropionate, treatment costs for a year were approximately double for DPIs compared to MDIs. In a wider study, which included a comparison of costs across seven European countries, the average percentage increase was 160% for DPIs (Enviros March, 2000). No salbutamol DPI products are available at present in the USA. However, generic salbutamol CFC MDIs are less expensive than both branded CFC and HFC MDIs. Should a salbutamol DPI come to the market, it is anticipated that it would be substantially more expensive than the generic salbutamol CFC MDIs that are currently available.

In the treatment of asthma and COPD, it is less expensive, and at least as effective, to use an MDI with a spacer rather than a nebulizer (Barry and O'Callaghan 2003, Stark, 1999, Carnago and Kennedy, 2000). In developing countries, the cost of therapy is a prime consideration (Gupta, 1998), and DPIs are therefore used less (Fink, 2000).

Society costs

Asthma and COPD involve huge costs for society, both at a personal level for individual patients and also for society as a whole in the form of direct costs (hospitalization, for example), and indirect costs (absence from school or work). For example, in the USA in 1997, there were nearly 2 million emergency hospital visits (Carnago and Kennedy, 2000) for asthma and, according to the National Heart, Lung and Blood Institute, asthma cost an estimated US$11 billion in 1998. It should also be borne in mind that 26,000 people died of COPD in England and Wales in 1999.

Patient choice is important for compliance (Cochrane, 1999; Milgrom, 1996) and patients must therefore be satisfied with their MDI or DPI. Any restriction of patient choice may result in reduced compliance with medication, with an increase in the already substantial cost for society. Switching patients from reliable and effective medications has significant implications for patient health and safety. The provision of a range of safe alternatives is critical before enforcing change on environmental grounds. Any environmental policy measures for the future that could impact patient use of HFC MDIs would require careful consideration and consultation with physicians, patients, national health authorities and other health-care experts.

The future

The price of HFCs in the future is expected to vary with normal commercial factors and not to rise out of line with economies. This is supported by the present example of CFCs, where the price is only beginning to rise at the end of the transition period. In addition, pharmaceutical-grade HFC is already 2 to 3 times more expensive than normal technical-grade HFC due to purification costs, and is therefore less dependent on the supply cost of technical-grade HFC.

More sophisticated nebulizers (Smart, 2002) or effort-assisted DPIs will probably be even more expensive than the DPIs in use at present and there will therefore be an increased cost penalty compared to MDIs (Dalby, 2003). However, if they provide more effective delivery of the drug, treatment costs may be reduced. By contrast, some less complex DPIs currently under development (e.g. FlowCaps ™, Hovione; DirectHaler™ Pulmonary, DirectHaler) may prove less expensive, and over a longer time frame they become more comparable in cost to MDIs.

Reimbursment

It should be noted that, although some patients pay directly or indirectly via insurance for their medicines, many are reimbursed in some way by governments.

8.5 Regional considerations

MDIs are the dominant form of treatment for asthma and COPD worldwide. In *developed countries*, the proportion of MDI to DPI varies substantially from country to country: from 9:1 MDI:DPI in the USA to 7:3 in UK and 2:8 in Sweden This is linked to a number of factors, including availability (e.g. multidose DPIs only recently became available in the USA), patient and physician choice, and relative cost. CFC MDIs are being phased out and replaced by HFC MDIs, and increasingly by DPIs. The transition has proceeded slowly as alternatives become available and countries accept the overall increased cost, balancing the environmental benefits against the availability of affordable medications to patients. A transition to another new form of treatment could be both costly and time-consuming, without clinical benefits for patients. Whilst the proportion of HFC MDIs to DPIs is likely to decrease slowly because of availability, cost and patient choice, both forms of inhaled therapy will continue to be available.

In *developing countries*, inhaled therapy is mainly with salbutamol/albuterol, almost exclusively with pressurized MDIs from either multinationals or local manufacturers. CFCs are due to be phased out completely by 2010, and this is a major challenge for developing countries. The rates of both asthma and COPD are rising in these countries, and statistical data probably underestimate disease prevalence. Guidelines recommending the replacement of old and ineffective oral treatments with inhaled therapy are gaining increasing acceptance. Improved economic circumstances are likely to result in a substantial increase in the use of inhaled therapy. One company in India is locally producing and marketing a range of HFC MDIs. Affordable single dose DPIs are technically feasible and could be locally manufactured in developing countries. There would be significant pharmaceutical difficulties in hot and humid climates (Maggi *et. al.*, 1999), and DPIs would still be more expensive than MDIs on a cost per dose basis. If these became available and achieved a significant market share, they would reduce the future volumes of HFC needed for MDIs. Multidose DPIs from multinational companies are either unavailable or too expensive for many patients. MDIs are likely to remain the most affordable and acceptable form of inhaled therapy in the long term in developing countries. For all the above reasons, predictions of HFC needs for inhaled therapy in developing countries are uncertain.

8.6 Future developments and projections

It is always challenging to project market dynamics in areas like pharmaceutical use, where commercial and regulatory pressures have such a major impact. It is anticipated that global market growth in inhaled medicine in the coming decade will be approximately 7% (Enviros March, 2000) whereas growth in MDIs is lower at between 1.5 and 3%. Also, provided that the MDI transition in developing and industrialized countries proceeds according to schedule, it may be assumed that all MDIs produced globally will contain HFCs instead of CFCs by 2010.

Estimates have been made by IPAC (shown in Table 8.3) that provide a range for HFC consumption in both 2010 and 2015, based on various scenarios of market growth and dynamics within the market. Actual growth rates will vary depending on the rate of introduction of new therapies for asthma/COPD.

The above projections are based on certain assumptions. These include:
- Market growth will be 7% per annum and MDI growth within this will range from between 1.5 and 3%. The total number of MDI units will be 680 million in 2015.
- Salbutamol MDIs will still represent approximately 50% of the overall MDI market.
- Two HFCs will continue to be used in MDIs for asthma/COPD. These will be HFC-134a and HFC-227ea, in an approximate ratio of 90:10.

The above information makes it clear that the maximum projected usage of HFCs for MDIs in 2015 will be approximately 15,000 metric tones. This information allows us to calculate, on the basis of hypothetical scenarios, the potential mitigation cost for reducing the projected HFC use in MDIS.

There are essentially two approaches to projecting these costs. In the first case one can take the full amount of projected HFC use, assume a certain percentage will be attributable to salbutamol HFC MDIs and calculate the likely incremental annual costs to switch them to DPIs on the basis of published price differences in certain countries. In this example, we have used an average increase of US$5.0 per inhaler.

If we assume that approximately half of the HFC usage will be for salbutamol HFC MDIs (i.e. around 7,500 metric tonnes) and this equates to 340 million salbutamol inhalers, the economic burden to patients or healthcare payers will be around US$1.7 billion per annum for a reduction of 11.45 million tonnes of CO_2 This is equivalent to a mitigation cost of 148.5 US$/t$CO_2$-eq.

Alternatively, using the same set of assumptions, one can evaluate the change for a single canister of salbutamol and project the cost using a similar approach. In this case, for each salbutamol MDI, the amount of propellant in an individual canister is approximately 13.2 grammes of HFC-134a. As 134a has a GWP of 1300, the CO_2 equivalent of an individual HFC-134a MDI is 0.01716 tonnes (13.2 x 1300 / 1,000,000). With an incremental cost for each salbutamol inhaler converted from HFC MDI to DPI of US$ 5, this is equivalent to a mitigation cost of US$ 5/0.01716 tonnes or approximately 292 US$/t$CO_2$-eq.

In both cases, these estimates do not include any costs associated with the development of new multidose salbutamol dry powder inhalers, but these would be minimal with respect to the projected additional cost burden to health-care systems.

The calculations provide a range of potential cost impact on health-care systems for further HFC MDI mitigation. These estimates are of the same order as one available for Europe, based

Table 8.3. IPAC HFC volume projections (IPAC, 2004).

2010 PROJECTIONS (1999 IPAC ORIGINAL ESTIMATES)	2015 PROJECTIONS (1999 IPAC ORIGINAL ESTIMATES)
7,500 to 9,000 metric tonnes of HFCs *10.8 to 12.9 million metric tonnes* *of CO_2 equivalent* • This projection was developed by IPAC in 1999 and is primarily based upon a survey of IPAC member companies (including 3M) for 1998 data. The data were most reliable for IPAC companies' major markets (e.g. Europe and North America). The projections include an estimate for non-IPAC companies and the developing world. The range reflects assumed MDI market growth rates of 1.5% to 3.0%.	**9,000 to 10,500** metric tonnes of HFCs *12.9 to 15.0 million metric tonnes* *of CO_2 equivalent* • This range extrapolates from IPAC's 1999 estimated projections and, beginning with the 'high end' figure of 9,000, assumes a range of flat growth to 3% growth from a baseline of 9,000 tonnes of HFCs.

2010 PROJECTIONS (TEAP/IPAC 2001 DATA)	2015 PROJECTIONS (TEAP/IPAC 2001 DATA)
11,275 to 12,865 metric tonnes of HFCs *18.26 to 20.84 million metric tonnes* *of CO_2 equivalent* • In response to the request from the authors of the MDI chapter of the IPCC Special Report, IPAC recently reviewed its emissions projections and analyzed more recent data and information available since 1999. The range above is primarily based upon TEAP figures for the volume of CFCs used in the production of MDIs for 2001 and IPAC data on HFCs collected for 2001. • This range is somewhat higher than IPAC's original estimate given above. This is probably due to the fact that IPAC's 1999 estimate was based on hard data only from IPAC member companies. The 2001 TEAP figures are presumably more accurate for volumes used in the developing world and non-IPAC MDI companies. • IPAC's 2010 projections are also consistent with the 2010 emissions projection estimated by Enviros March (2000) in its December 2000 report, submitted to the European Climate Change Programme: 10,230 metric tonnes.	**12,865 to 14,915** metric tonnes of HFCs *20.84 to 24.16 million metric tonnes* *of CO_2 equivalent* • This range extrapolates from the 2010 estimated projection (TEAP/IPAC 2001) and, beginning with the 'high end' figure of 12,865, assumes a range of flat growth to 3% growth from a baseline of 12,865 tonnes of HFCs.

on data from the 1990s, of US$461 per tonne CO_2 (Enviros March, 2000). This figure has been quoted in environmental reports from a number of governments.

Conclusions
• Most reduction of GWP from MDIs will be achieved through the completion of the transition from CFC to HFC MDIs.

• No major breakthroughs for inhaled drug delivery are anticipated in the next 10–15 years given the current status of technologies and the development time scales involved.

• The health and safety of the patient is of paramount importance in treatment decisions and policymaking that might impact those decisions.

• If one assumes a hypothetical switch for the most widely used inhaled medicine (salbutamol) from HFC MDIs to DPI, the projected recurring annual costs would be on the order of US$ 1.7 billion with an effective mitigation cost of between 150–300 US$/t$CO_2$-eq.

• The environmental benefits of converting HFC MDIs to DPIs are small.

References

AAP (American Academy of Pediatrics), 1997: Policy statement: "Inactive" ingredients in pharmaceutical products. *Pediatrics*, **99**, 268–278.

Barry, P.W. and C. O'Callaghan, 2003: The influence of inhaler selection on efficacy of asthma therapies. *Advanced Drug Delivery Reviews*, **55**, 879–923.

BNF (British National Formulary), 2003 (September), British Medical Association and Royal Pharmaceutical Society of Great Britain, London, UK, pp. 134–147.

Boekestein, V.J., A.J. Hickey and T.M. Crowder, 2002: Uniform and reproducible delivery of albuterol from a variety of lactose powder blends using the Oriel active dispersion platform. In *Drug Delivery to the Lungs XIII*, T.A. Society, London, UK, pp. 107–110.

Brown, B.A.-S., 2002: 5 Myths about MDIs. *Drug Delivery Technology*, **2**, 52–59.

Carnago, C.A. and P.A. Kennedy, 2000: Assessing costs of aerosol therapy. *Respiratory Care*, **45**(6), 756–763.

Celli, B., J. Benditt and R.K. Albert, 1999: Chronic obstructive pulmonary disease. In *Comprehensive Respiratory Medicine*, R.K. Albert, S.G. Spiro and J.R. Jett (eds.), Mosby International Ltd., London, UK, pp. 371-395.

Cochrane, G.M., R. Home and P. Chanez, 1999: Compliance in asthma. *Respiratory Medicine*, **9**, 763–769.

Consumer Association UK, 2000: Inhaler devices for asthma. *Drug and Therapeutics Bulletin*, **38**(2), 9–14.

Consumer Association UK, 2001: Managing stable COPD. *Drug and Therapeutics Bulletin*, **39**(11), 81–85.

Dalby, R.N., A.J. Hickey and S.L. Tiano, 1996: Medical devices for the delivery of therapeutic aerosols to the lungs. In *Inhalation Aerosols*, A.J. Hickey (ed.), Marcel Dekker Inc., New York, NY, USA, pp. 441–473.

Dalby, R., 2003: Pulmonary devices – what's used. *European Pharmaceutical Review*, **8**(2), 70–73.

Dalby, R. and J. Sumas, 2003: Inhalation therapy: technological milestones in asthma treatment, *Advanced Drug Delivery Reviews*, **55**, 779–791.

DeYoung, L.R., F. Chambers, S. Narayan and C. Wu, 1998: The aerodose multidose inhaler device design and delivery characteristics. In *Respiratory Drug Delivery VI*, R.N. Dalby, P.R. Byron, and S.J. Farr (eds.), Interpharm Press Inc., Buffalo Grove, IL, USA, pp. 91–95.

Dolovich, M.A. and N.R. MacIntyre, 2000: Consensus statement: aerosols and delivery devices. *Respiratory Care*, **45**(6), 1–22.

Dunbar, C.A., A.J. Hickey and P. Holzner, 1998: Dispersion and characterization of pharmaceutical dry powder aerosols. *KONA*, **16**, 7–45.

Emmen,**H.H**, 2000: Human safety and pharmacokinetics of the CFC alternative propellants HFC 134a and HFC 227. *Regulatory Toxicology and Pharmacology*, **32**, 22–35.

Enviros March, 2000: Study on the Use of HFCs for Metered Dose Inhalers in the European Union. Report by Enviros March for the International Pharmaceutical Aerosol Consortium (IPAC), submitted to European Climate Change Programme, Washington, DC, USA, 45 pp.

Everard, M.L., 2001: Inhaler development: Where shall we be in 10 years'time? *Asthma Journal*, **6**(2), 84–88.

Fink, J.B., 2000: Metered-dose inhalers, dry powder inhalers and transitions. *Respiratory Care*, **45**(6), 623–635.

Garcia-Contreras, L. and A.J. Hickey, 2002: Pharmaceutical and biotechnological aerosols for cystic fibrosis therapy. *Adv. Drug Deliver. Rev.*, **54**, 1491–1504.

Gupta, S.K., K.S. Mazumdar, S. Gupta, A.S. Mazumdar and S. Gupta, 1998: Patient education programme in bronchial asthma in India: Why, how, what and where to communicate. *Indian Journal of Chest Dis. Allied Science*, **40**(2), 117–124.

Hickey, A.J., 1992: Methods of aerosol particle size characterization. In *Pharmaceutical Inhalation Aerosol Technology*, A.J. Hickey (ed.), Marcel Dekker, New York, USA, pp. 219–250.

Hickey, A.J. and R.M. Evans, 1996: Aerosol generation from propellent-driven Metered Dose Inhalers. In *Inhalation Aerosols: Physical and Biological Basis for Therapy*, A.J. Hickey (ed.), Marcel Dekker Inc., New York, USA, pp. 417–439.

Hickey, A.J. and C.A. Dunbar, 1997: A new millennium for inhaler technology. *Pharm. Tech.*, **21**, 116–125.

ICF, 2003: The Use of Alternatives to Synthetic Greenhouse Gases in Industries Regulated by the Montreal Protocol. Report prepared by ICF Consulting for the Australian Greenhouse Office, Canberra, Australia, 46 pp.

IPAC, 2004. International Pharmaceutical Aerosol Consortium. HFC usage projections in MDIs, (www.ipacmdi.com) and in Ensuring Patient Care, IPAC, Washington. Supplement Appendix B, 2004.

ISAAC-SC (International Study of Asthma and Allergies in Childhood Steering Committee), 1998: Worldwide variation in prevalence of symptoms of asthma, allergic rhino-conjunctivitis, and atopic eczema. *Lancet*, **351**, 1225–1232.

June, D.S., R.K. Schultz and N.C. Miller, 1994: A conceptual model for the development of pressurized metered-dose hydrofluro-alkane based inhalation aerosols. *Pharm. Tech.*, **17**, 40–52.

Keller, M. and R. Muller-Walz, 2000: Dry Powder for Inhalation. PCT Patent WO 00/28979, 25 May 2000.

Leach, C.L., P.J. Davidson and R.J. Boudreau, 1998: Improved airway targeting with the CFC-free HFC-beclomethasone metered-dose inhaler compared with CFC-beclomethasone. *Eur. Respir. J.*, **12**, 1346–1353.

Maggi, L., R. Bruni and U. Conte, 1999: Influence of moisture on the performance of a new dry powder inhaler. *Int. J. Pharm.*, **177**, 83–91.

Medicines Control Agency, 1997: Rules and Guidance for Pharmaceutical Manufacturers and Distributors 1997, The Stationery Office, London, UK.

Milgrom, H., B. Bender, L. Ackerson, P. Bowry, B. Smith and C. Rand, 1996: Non-compliance and treatment failure in children with asthma. *Journal of Allergy and Clinical Immunology*, **98**(6), 1051–1057.

Molina, M.J. and F. S. Rowland, 1974: Stratospheric sink for chlorofluoromethanes: chlorine atom catalyzed destruction of ozone. *Nature*, **249**, pp. 1810.

Niven, R.W., 1996: Atomization and nebulizers. In *Inhalation Aerosols: Physical and Biological Basis for Therapy*, A.J. Hickey (ed.), Marcel Dekker Inc., New York, USA, pp. 273–312.

Pischtiak, A.H., M. Pittroff and T. Schwartze, 2001: Characteristics, supply and use of the hydrofluorocarbons HFC 227ea and HFC 134a. *Aerosol Europe*, **9**(10), 44–47.

Poochikian, G. and C.M. Bertha, 2000: Inhalation drug product excipient controls: Significance and pitfalls. In *Respiratory Drug Delivery VII*, R.N. Dalby, P.R. Byron, S.J. Farr and J. Peart (eds.), Serentec Press, Inc., Raleigh, NC, USA, pp. 109–115.

Price, D., E. Valovirta and J. Fischer, 2004: The importance of preserving choice in inhalation therapy: the CFC transition and beyond. *Journal of Drug Assessment*, **7**, 45–61

Schuster, J., S. Farr, D. Cipolla, T. Wilbanks, J. Rosell, P. Lloyd and I. Gonda, 1998: Design and performance validation of a highly efficient and reproducible compact aerosol delivery system: AERx(TM). In *Respiratory Drug Delivery VI*, R.N. Dalby, P.R. Byron, and S.J. Farr (eds.), Interpharm Press Inc., Raleigh, NC, USA, pp. 83–90.

Smart, J., E. Berg, O. Nerbrink, R. Zuban, D. Blakey and M. New, 2002: Touchspray technology: Comparison of the droplet size measured with cascade impaction and laser diffraction. In *Respiratory Drug Delivery VIII (Volume 2)*, R.N. Dalby, P.R. Byron, J. Peart and S.J. Farr (eds.), Davis Horwood International Publishers Ltd., Surrey, UK, pp. 525–527.

Smith, I.J., 1995: The challenges of reformulation. *J. Aerosol Med.*, **8** (Suppl 1), S19–27.

Staniforth, J.N., 2002: Powders comprising anti-adherant materials for use in Dry Powder Inhalers. US Patent 6,475,523.

Tansey, I., 1997: The technical transition to CFC-free inhalers. *Br. J. Clin. Pract.*, **89** (suppl.), 22–27.

UBA, 2004: Fluorinated Greenhouse Gases in Products and Processes – Technical Climate Protection Measures. [K. Schwaab and W. Plehn (eds.)] Umweltbundesamt (German Federal Environmental Agency), Berlin, Germany, 240 pp.

UNEP-TEAP, 2003: 2002 Report of the Technology and Economic Assessment Panel (Pursuant to Article 6 of the Montreal Protocol). UNEP Ozone Secretariat, Nairobi, Kenya, 117 pp.

US EPA, 2004: Analysis of Costs to Abate International Ozone-Depleting Substance Substitute Emissions. US Environmental Protection Agency report 430-R-04-006, D.S. Godwin (ed.), Washington, D.C. 20460, USA, 309 pp.

US FDA, 1998: Draft Guidance for Industry: Metered Dose Inhaler (MDI) and Dry Powder Inhaler (DPI) Drug Products – Chemistry, Manufacturing and Controls Documentation, US Food and Drug Administration, Washington, DC, USA, 66 pp.

Voss, A. and W.H. Finlay, 2002: Deagglomeration of dry powder pharmaceutical aerosols. *Int. J. Pharm.*, **248**, 39–50.

Yellen, D. 2003: The role of HFC MDIs in the CFC transition: A case study in the European Union. Proceedings of the 14th Annual Earth Technologies Forum, April 22-24, 2003, Washington, D.C., USA.

9

Fire Protection

Coordinating Lead Author
Dan Verdonik (USA)

Lead Authors
H.S. Kaprwan (India), E. Thomas Morehouse (USA), John Owens (USA), Malcolm Stamp (United Kingdom), Robert Wickham (USA)

Review Editors
Stephen O. Andersen (USA), Megumi Seki (Japan)

Contents

EXECUTIVE SUMMARY

Halons are gaseous fire- and explosion-suppression agents that leave no damaging residue and are safe for human exposure. They are inexpensive to make, forgiving to use and applicable across a wide range of conditions. Despite these challenging benchmarks halons are no longer needed for over 95% of the applications that used halons before the Montreal Protocol. This transformation came about through a combination of national regulations, research, commercial product development and the approval of alternatives under fire protection regulations.

The ozone layer

The environmental goal of ozone protection is being achieved with a growing base of newly installed systems which now use halon alternatives. Seventy-five percent of the applications that originally used halons have shifted to agents with no climate impact. Four percent of the original halon applications continue to employ halons. The remaining 21% have shifted to HFCs and a small number of applications have shifted to PFCs and HCFCs. While this is a significant achievement, it is tempered by the reality that there remains a large base of installed halon systems and the eventual choice of alternatives for these systems remains unknown at this time.

Past practice (1987–2002)

Before the Montreal Protocol, HFCs, PFCs and HCFCs were not used in fire protection. Their current, and growing, usage is a direct result of their adoption as halon alternatives, despite being inferior to halons both in terms of cost and performance. Because fire protection is regulated in most countries, only those agents meeting minimum acceptable fire extinguishment and life-safety performance levels have been certified. About half of former halon users are now choosing to protect new installations with zero-ODP gaseous (in-kind) clean agents. Some have no significant effect on the climate system, while others have substantial GWPs. The other half are choosing non-gaseous (not-in-kind) agents, such as water, water mist, dry chemical, foam and aerosols, all of which produce no direct greenhouse-gas emissions. There are a small number of users that have adopted HCFC-123, HCFC-22 and HCFC-124 agent blends.

Emissions of gaseous halon alternatives are very small compared with those estimated for halon. This is the result of cooperation between industry and governments to implement standards for system maintenance, training and certification programmes, and equipment that minimizes or eliminates escape during transfer and detects leaks from storage. Emissions are estimated to range from 1 to 3% of the fixed-system bank and 2 to 6% of the portable extinguisher bank per year.

With the exception of civil aviation applications, where the added weight of halon alternatives produces indirect emissions through additional fuel use over the aircraft's life, the indirect greenhouse-gas emissions are negligible compared with the direct effects. Total equivalent warming impact (TEWI) and life cycle climate performance (LCCP) are therefore not particularly relevant to fire protection.

Recent modelling suggests the global fixed-system bank for HFCs, PFCs and HCFCs at the end of 2002 was 22,000 tonnes, with 1.1 MtCO$_2$-eq of emissions at a 2% emission rate. For portable extinguishers, the global bank of HFCs, PFCs and HCFCs was approximately 1,500 tonnes at the end of 2002, with 0.12 MtCO$_2$-eq of emissions at a 4% emission rate.

The most recent estimate from the Halons Technical Options Committee (HTOC) of the size of the existing halon bank and annual emission rates (2002) is 42,000 tonnes with 2,100 tonnes of emissions for halon 1301, and 125,000 tonnes with 17,000 tonnes of emissions for halon 1211 (HTOC, 2003). Atmospheric measurements in 2002 for halon 1301 suggest emissions of 1000–2000 tonnes and emissions of 7,000–8,000 tonnes for halon 1211. Halon 2402 was used mainly in the former Soviet Union and no information on banks or emissions was found in the literature.

Present practice (2003–2004)

In fixed systems where a clean agent is necessary, the alternatives currently available are carbon dioxide (lethal at concentrations that extinguish fires) and inert gases, HFCs, PFCs, HCFCs and more recently, a fluoroketone[1] (FK). Carbon-dioxide and inert-gas systems account for approximately half of the new clean-agent systems, with HFCs, PFCs, HCFCs and FK accounting for the other half. HFCs are playing an important role in the transition away from halon where the unique properties of this type of agent are required to achieve safe, fast fire extinguishing without causing residual damage. PFCs played an early role but current use is limited to the replenishment of previously installed systems. The current fixed system bank of HFCs, PFCs, HCFCs and FK was estimated to be 27,000 tonnes at the end of 2004 with 1.4 MtCO$_2$-eq of emissions at a 2% emission rate.

Compared to halon 1211, portable extinguishers using HFCs, PFCs and HCFCs have achieved very limited market acceptance, primarily because of their high cost. PFCs are limited to use as a propellant in one manufacturer's portable extinguisher agent blend. The portable extinguisher bank based on information from a producer is approximately 1,850 tonnes, with 0.16 MtCO$_2$-eq of emissions at a 4% emission rate at the end of 2004.

In the case of fixed clean-agent systems, four main factors contribute to the choice of replacement agents: performance, life safety, cost and environmental concerns. In addition, two other factors may be important in some instances: demonstrated special capabilities and multiple supply sources. For portable extinguishers, cost is the main factor. Table 9.1 compares the performance of the available alternatives for gaseous fire extinguishing systems and Table 9.2 does the same for portable fire extinguishers.

---

[1] Fluoroketone (FK) - An organic compound in which two fully fluorinated alkyl groups are attached to a carbonyl group (C=O).

The future (2005–2015)

During this period, the use of HFCs and FK for fire protection is expected to increase due to general economic expansion and improvements in technologies that allow these materials to displace current halon applications. The halon alternatives most likely to be employed through 2015 are identified in Tables 9-1 and 9-2.

HTOC (2003) estimates that significant quantities of halon will remain in the global bank in 2015. The figures for halon 1301 are as follows: in 2005, there will be a bank of 39,000 tonnes with 1,900 tonnes of emissions; in 2010, there will be a bank of 31,000 tonnes with 1,500 tonnes of emissions and, in 2015, there will be a bank of 24,000 tonnes with 1,300 tonnes of emissions. The figures for halon 1211 are: a bank of 83,000 tonnes with 16,000 tonnes of emissions in 2005, a bank of 33,000 tonnes with 6,000 tonnes of emissions in 2010 and a bank of 19,000 tonnes with 1,600 tonnes of emissions in 2015.

Modelling suggests an HFC/HCFC/PFC/FK fixed-system bank of 67,000 tonnes with annual emissions of 4 $MtCO_2$-eq in 2015 at an emission rate of 2%. In the absence of a change in emission rate, total annual emissions will change in proportion to changes in the installed base. For the portable extinguisher bank, projections of information provided by a producer are 4,000 tonnes in 2015 with 0.34 $MtCO_2$-eq of emissions at a 4% emission rate.

Clean-agent demand will be influenced by economic growth and decisions by regulators and halon owners regarding the disposition of agent from decommissioned systems. If decommissioned halon is destroyed, the demand for new clean agents will increase, probably in the same proportion as for new systems. In addition, clean-agent demand will be influenced by existing and future regulation. As research into new fire protection technologies continues, additional options will likely emerge. However, due to the lengthy process of testing, approval and market acceptance of new fire protection equipment types and agents, no additional options are likely to be available in time to have appreciable impact by 2015.

Since the Montreal Protocol, the fire protection community has become much more active in managing emissions. Industry practices of capturing, recycling and reusing halons have carried over to all high-value gaseous agents, including HFCs, HCFCs and PFCs. There is no reason to believe they will not also apply to any additional agents entering the market. Because of their more complex chemistries, halon replacements are unlikely to ever be as inexpensive as halons. As a result, there is both a market incentive as well as an industry culture that encourage the capture, recycling and reuse of fire-fighting agents. Adhering to these practices, including certification programmes, has been shown to minimize emissions and to limit the use of these agents to applications where their cleanliness is needed. In countries where high levels of regulation exist, the emission rates have been estimated at 1% or less; the figure is 2% where levels of regulation are average and approaches 3% where there is less regulation. Portable extinguishers have an emission rate of approximately 4%.

Table 9.1. Comparison table – clean-agent systems suitable for occupied spaces.

Fixed systems	Halon 1301 (reference)	HFC-23	HFC-227ea	HFC-125[1]	FK- 5-1-12	Inert Gas
Substance characteristics						
Radiative efficiency (W m^{-2} ppb^{-1})	0.32	0.19	0.26	0.23	0.3	N/A
Atmospheric lifetime (yr)	65	270	34.2	29	0.038	N/A
Direct GWP (100-yr time horizon)						
- This report	7030	14,310	3140	3450	not	N/A
- IPCC (1996)	5400	11,700	2900	2800	available[2]	
Ozone depletion potential	12	~0	-	~0	-	N/A
Technical data						
Demonstrated special capabilities	yes	yes[3]	yes[4]	yes[4]	note[6]	no
Weight (kg m^{-3}) [a]	0.8	2.3	1.1	1.1	1.2	4.3
Area (10^4 m^2/m^3) [b]	5.8	12.0	6.8	7.4	7.3	28.2
Volume (10^4 m^3/m^3) [c]	8.6	18.0	13.1	14.4	13.8	56.6
Emission rate [d]	2 ± 1%	2 ± 1%	2 ± 1%	2 ± 1%	2 ± 1%	2 ± 1%
Costs						
Investment cost (relative to halon 1301)	100%	535%	377%	355%	484%	458%
Additional service costs (US$ kg^{-1}) [e]	0.15	0.43	0.60	0.53	0.72	0.31
Additional recovery costs at end-of-life (US$ kg-1) [f] ( ) indicates income	(3.85)	(10.75)	(15.07)	(13.20)	(18.00)	0.00
HFC abatement costs (US$ per tCO$_2$-eq) [g]	-	-	-	-	21–22	14–27
Commercial considerations						
Multiple agent manufacturers	-	yes	yes	yes	no[7]	yes

Notes:

[a] Average weight of the agent storage containers and contents in kilogrammes per cubic metre of space protected.

[b] Average area of a square or rectangle circumscribing the agent cylinder bank expressed in square metres x 10^4 per cubic metre of volume protected.

[c] Average volume is the area multiplied by the height of the cylinders measured to the top of the valves expressed in cubic metres x 104 per cubic metre of volume protected.

[d] Total average in-service-life annual emissions rate including system discharges for fire and inadvertent discharges.

[e] Additional annual service costs are based on the replacement of 2% of the agent charge emitted per year.

[f] For the halocarbon agents, the end-of-life agent value is positive and represents a cost recovery equivalent to 50% of the initial cost of the agent as the agent is recovered, recycled and resold for use in either new systems or for the replenishment of existing systems.

[g] HFC abatement costs for FK-5-1-12 and inert gas are based on HFC-227ea, the predominant HFC, as the reference. The lower value reflects the cost in US$ per tonne of CO$_2$-equivalent at a discount rate of 4% and tax rate of 0%. The range includes both the lowest and highest of costs for the USA, non-USA Annex 1 and non-Annex 1 countries.

Explanation of special capabilities:

1. In some jurisdictions HFC-125 is not allowed for use in occupied spaces while in other jurisdictions that use is permitted under certain conditions.

2. Due to the short atmospheric lifetime, no GWP can be given. It is expected to be negligible for all practical purposes (Taniguchi *et al.*, 2003). See Section 2.5.3.3 'Very short-lived hydrocarbons' for additional information.

3. HFC-23 is effective at low temperatures (cold climates) and in large volumes due to its high vapour pressure.

4. HFC-227ea is effective in shipboard and vehicle applications due to extensive large-scale testing that has established the use parameters and demonstrated its specialized capabilities in these applications.

5. HFC-125 is effective in vehicle and aircraft engine applications as a result of extensive large-scale testing that has established the use parameters and demonstrated its specialized capabilities in these applications.

6. FK-5-1-12 is in the early stages of its product life cycle and has yet to be tested for special applications beyond those achieved through the conventional approval testing of the requirements in ISO and NFPA type standards.

7. While the agent FK-5-1-12 is a proprietary product of a single agent-manufacturer, the agent is available from multiple systems manufacturers.

Table 9.2. Comparison table – extinguishing agents for portable fire extinguishers.

Portable systems	Halon 1211 (reference)	HCFC Blend B	HFC-236fa	Carbon Dioxide	Dry Chemical	Water
Substance characteristics						
Radiative efficiency (W m^{-2} ppb^{-1})	0.3	Note [a]	0.28	See Ch. 2	-	-
Atmospheric lifetime (yr)	16	Note [a]	240	See Ch. 2	-	-
Direct GWP (100-yr time horizon)						
- This report	1860	<650 [a]	9500	1	-	-
- IPCC (1996)	not given	<730 [a]	6300	1	-	-
Ozone depletion potential	5.3	<0.02 a	-	-	-	-
Technical data						
Agent residue after discharge	no	no	no	no	yes	yes
Suitable for Class A fires	yes	yes	yes	no	yes	yes
Suitable for Class B fires	yes	yes	yes	yes	yes	no
Suitable for energized electrical	yes	yes	yes	yes	yes	no
Extinguisher fire rating [b]	2-A:40-B:C	2-A:10-B:C	2-A:10-B:C	10-B:C	3-A:40-B:C	2-A
Agent charge (kg)	6.4	7.0	6.0	4.5	2.3	9.5
Extinguisher charged weight (kg)	9.9	12.5	11.6	15.4	4.15	13.1
Extinguisher height (mm)	489	546	572	591	432	629
Extinguisher width (mm)	229	241	241	276	216	229
Emission rate [c]	4 ± 2%	4 ± 2%	4 ± 2%	4 ± 2%	4 ± 2%	4 ± 2%
Costs						
Investment costs (relative to halon 1211)	100%	186%	221%	78%	14%	28%
Additional service costs (US$ kg^{-1})	- [d]	- [d]	- [d]	- [d]	- [d]	- [d]
Additional recovery costs at end-of-life (US$ kg^{-1})	- [d]	- [d]	- [d]	0.00	0.00	0.00

Notes:

[a] HCFC Blend B is a mixture of HCFC-123, CF$_4$ and argon. While the ratio of the components is considered proprietary by the manufacturer, two sources report that HCFC-123 represents over 90% of the blend on a weight basis, with CF$_4$ and argon accounting for the remainder. The atmospheric lifetime of HCFC-123 is 1.3 years; this figure is 50,000 years for CF$_4$.

[b] Fire extinguisher rating in accordance with the requirements of Underwriters Laboratories, Inc. The higher the number, the more effective the extinguisher.

[c] This value is the total average in-service-life annual emissions rate, including both intentional discharges for fire and inadvertent discharges.

[d] This information is neither in the literature nor available from other sources, as it is considered confidential.

9.1 Introduction

Halons are halogenated hydrocarbons that contain bromine and exhibit exceptional fire-fighting effectiveness. They are known as "clean agents" because they are volatile liquids or gases, electrically non-conductive and leave no residue. As gases, when dispensed in air, they can extinguish hidden fires and those with complex geometries (three-dimensional fires). At least one halon (1301) is safe for human exposure at concentrations that will extinguish fires. In short, they are easy to use, versatile and inexpensive.

After their introduction in the early 1960s, the use of halons grew steadily worldwide until the Montreal Protocol required an end to their production by 1/1/94 in developed countries. Global production of halon 1211 and 1301 peaked in 1988 at 43,000 and 13,000 tonnes respectively (HTOC, 1991). In developing countries halon 1211 production began in the 1980s, and showed a growth curve similar to developed countries until Multilateral Fund projects began to reverse that trend in the 1990s (HTOC, 2003). A third halon, 2402, was mainly used in the former Soviet Union. No information on emissions or bank size was found in the literature.

The most recent estimate of the Halons Technical Options Committee (HTOC) of the existing halon bank (their 2002 assessment report (HTOC, 2003)) is as follows. For halon 1301: a bank of 39,000 tonnes with 1,900 tonnes of emissions in 2005, a bank of 31,000 tonnes with 1,500 tonnes of emissions in 2010 and a bank of 24,000 tonnes with 1,300 tonnes of emissions in 2015. For halon 1211: a bank of 83,000 tonnes with 16,000 tonnes of emissions in 2005, a bank of 33,000 tonnes with 6,000 tonnes of emissions in 2010 and a bank of 19,000 tonnes with 1,600 tonnes of emissions in 2015.

Atmospheric measurements in 2002 place emissions of halon 1301 between 1,000 and 2,000 tonnes. This is close to the HTOC estimates. Atmospheric measurements for halon 1211, however, are between 7,000 and 8,000 tonnes in 2002, considerably lower than estimated by the HTOC. It appears that recent actions by several parties may result in significant future adjustments to the estimated quantities and to the geographical distribution of halons. However, the HTOC estimate of future emissions is currently the best available in the literature.

Halons' unique combination of properties led to their selection for many fire protection situations: computer, communications and electronic equipment facilities; museums; engine spaces on ships and aircraft; ground protection of aircraft; flammable liquid storage and processing facilities, general office fire protection and industrial applications (HTOC, 1989). Wickham (2002), however, finds that the halocarbon alternatives are not in such general use now. In part, their high cost appears to be limiting their use to applications where a clean agent is a necessity. The development and use of halon alternatives are driven entirely by the Montreal Protocol. None of the available alternatives offer performance or cost advantages over halons.

9.1.1 Overview of the halon market before the Montreal Protocol

The market for halons and their alternatives is divided into two distinctly separate sub-sectors: portable extinguishers and fixed systems. Halon 1301 (ODP = 10, Montreal Protocol value) dominated the market for fixed gaseous systems while halon 1211 (ODP = 3) dominated the market for gaseous portable extinguishers. Halon 2402 (ODP = 6) is more toxic than the other halons and was used only in a small number of countries.

9.1.2 Progress since the Montreal Protocol

Research on alternatives to halons has been underway since at least 1988 (HTOC, 1989). In 1994, the HTOC (1994) reported that newly developed replacement products were commercially available for most new applications, but stated that retrofitting of existing systems to use new alternatives needed further evaluation.

As of 1999, it was estimated that only 4% of the former halon market still required halon in new systems. Applications with halon had successfully shifted to new systems with many different alternative agents and approaches, as shown below (IPCC/TEAP, 1999):
- not-in-kind (non-gaseous) agents   50%
- clean agents   50%
  - *carbon dioxide and inert gases*   *25%*
  - *halons*   *4%*
  - *PFCs*   *<1%*
  - *HFCs*   *20%*

In 2002, the HTOC (2003) concluded that halon was no longer necessary in virtually any new installations, with the possible exceptions of engine nacelles, passenger spaces and cargo compartments of commercial aircraft, and crew compartments of military combat vehicles. A new fluorinated agent, fluoroketone FK-5-1-12, was also introduced in 2002.

For all practical purposes, PFC use has ceased and been replaced by HFCs. PFC use is currently limited to the replenishment of existing systems and as a minor component in a streaming agent identified as HCFC Blend B.

9.1.3 Emission characteristics of fire protection applications

Any fire protection system or extinguisher must be available for discharge at the moment a fire event occurs. Fixed systems and extinguishers should therefore be designed, produced and maintained in a manner that eliminates loss of agent through leakage and, in the case of fixed systems, inadvertent or unwanted discharges. As a result, fire protection codes and standards establish minimum levels of design, require periodic maintenance, and require regulatory authorities to adhere to these minimum levels (NFPA, 2000, 2004).

Since the Montreal Protocol, the fire protection community has become much more active in managing emissions. For example, the highly emissive equipment testing and personnel training practices were eliminated (HTOC, 1991, 1994) and as a result emissions have decreased significantly. Until 2003, the only estimates of emission rates for halons and their alternatives were based on expert opinion. The 1996 IPCC Revised Guidelines for National Greenhouse Gas Inventories recommended using the estimates for halons developed by McCulloch (1992) to estimate HFC and PFC emissions (IPCC, 1997). In Volume 3, however, it is also noted that the emissions of HFCs and PFCs could be reduced by more than 50% in recognition of the changes being implemented by the fire protection community. The HTOC (1997) also estimated that these new procedures reduced the non-fire emissions of halons by up to 90%. Ball (1999) reported that many of the procedures originally developed for halons are now being applied to alternatives, reducing emissions far below pre-Montreal Protocol levels. Ball opined that emissions for HFCs and PFCs could be as low as 1% of in-use quantities. On the basis of consultation with industry, a study prepared for the United Kingdom Department of Environment, Food and Rural Affairs (UK DEFRA) estimated current annual HFC emissions at approximately 5% of the installed base per annum (AEAT, 2003).

Recent data from the Halon Recycling and Banking Support Committee in Japan indicate a 0.12% emission rate for halon 1301 systems, with the exception of ships, aircraft and military systems (Verdonik and Robin, 2004). Verdonik and Robin (2004) evaluated the publicly available data relating to the production and emissions of HFCs and PFCs from fire suppression applications and developed three independent approaches to determine emission rates. The study derived an average emission rate of 2% of the installed base, with an uncertainty range of 1 to 3% (i.e. 2% ±1%).

9.1.4 Estimates of direct greenhouse-gas emissions

The actual global production or consumption of halon alternatives in fire protection applications is not known. This is because each is produced by a limited number of companies and each company regards their sales-related data as proprietary. Actual emissions are therefore reported as an aggregated value using the GWPs in the IPCC Second Assessment Report (SAR) (IPCC, 1996). All bank and emission estimates presented in this report are based on the SAR 100-yr GWP values. It is anticipated that the new 100-yr GWPs for fire protection agents provided in this report would change these estimates by less than 10%, which is well within the range of results presented and significantly less than the uncertainty of these estimates. As such, no adjustments to the SAR 100-yr GWP values have been included in the estimates in this chapter.

The UK fire industry collected and reported aggregate emissions of HFCs and PFCs from fire protection applications for the years 1997–1999 as part of their voluntary agreement with the government (FIC, 1997–1999). The results were: 1997 –

0.010 $MtCO_2$-eq; 1998 – 0.012 $MtCO_2$-eq; and 1999 – 0.014 $MtCO_2$-eq. The US fire protection industry collected and reported aggregate emissions of HFCs and PFCs for 2002 under their voluntary agreement with the government (Cortina, 2004). The result was 0.56 $MtCO_2$-eq. These methodologies estimate emissions on the basis of the amount of agent sold to recharge systems. The data may therefore slightly understate emissions where a system that has been discharged has, for any reason, not been recharged.

Verdonik (2004) used these UK and US emissions data and other publicly available data to update the Fire Protection Emissions Model developed under the Greenhouse Gases Emission Estimating Consortium. Estimates of emissions of HFCs and PFCs from fire protection applications and the fixed-system bank of HFCs and PFCs were developed using the 1 to 3% emission rate range developed by Verdonik and Robin (2004). The results are provided in Table 9.3 where the composite gas is defined as consisting of 97.5% HFC-227ea and 2.5% HFC-23. As fixed PFC systems are no longer installed, only a small difference results when PFC is included in the calculation. Information provided by a producer of HCFCs for fixed systems places the bank in 2002 at 3,400 and in 2004–2015 at 3,600 tonnes with annual emissions of 0.09 $MtCO_2$-eq. The fixed-system bank of HCFCs is assumed to be 85% HCFC-22, 10% HCFC-124 and 5% HCFC-123.

Additional modelling using the methodology of Verdonik (2004) suggests the global fixed system bank for HFCs, PFCs and HCFCs at the end of 2002 was 22,000 tonnes with 1.1 $MtCO_2$-eq of emissions at a 2% emission rate assuming (1) that the fixed system bank of HFCs/PFCs is approximated in 2002 by 95.1% HFC-227ea and 2.45% each of HFC-23 and PFC-3-1-10, and (2) that the fixed-system bank of HCFCs is comprised of 85% HCFC-22, 10% HCFC-124 and 5% HCFC-123 (henceforth all referred to as the fixed-system composite bank). PFCs played an early role but current use is limited to the replenishment of previously installed systems. The annual addition to the fixed-system composite bank consists of 97.5% HFC-227ea and 2.5% HFC-23. The current fixed-system composite bank is estimated at 27,000 tonnes at the end of 2004 with 1.4 $MtCO_2$-eq of emissions at a 2% emission rate.

Looking to the future, modelling suggests a fixed-system bank (including 3,600 tonnes of HCFCs and some use of FK) of 67,000 tonnes with annual emissions of 4 $MtCO_2$-eq in 2015 at an emission rate of 2%. In the absence of a change in emission rate, total annual emissions will change in proportion to changes in the installed base.

In countries where high levels of regulation exist, the emission rates have been approximated as 1% or less. The estimated emission rate is 2% where there are average levels of regulation, approaching 3% at lower levels of regulation. On the basis of the 2000 estimate from HTOC (2003a), portable extinguishers are thought to have an emission rate of approximately twice that of fixed systems. Applying that factor provides an uncertainty range of 2 to 6% (i.e. 4% ±2%). Looking forward, one might expect that the continuation of efforts to minimize emissions

Table 9.3. Estimation of global emissions and bank of HFC/PFCs in fixed-system fire protection.

		1995	2000	2005	2010	2015
1% Emission Rate						
Fixed-system emissions	$MtCO_2$-eq	0.06	0.60	1.20	1.99	2.87
	MtC-eq	0.02	0.16	0.33	0.54	0.78
Fixed-system bank	tonnes composite gas	1,805	19,325	38,599	63,685	92,052
2% Emission Rate						
Fixed-system emissions	$MtCO_2$-eq	0.11	0.85	1.64	2.74	3.95
	MtC-eq	0.03	0.23	0.45	0.75	1.08
Fixed-system bank	tonnes composite gas	1,684	13,576	26,360	43,968	63,315
3% Emission Rate						
Fixed-system emissions	$MtCO_2$-eq	0.11	1.10	2.13	3.46	4.91
	MtC-eq	0.03	0.30	0.58	0.94	1.34
Fixed-system bank	tonnes composite gas	1,127	11,726	22,711	37,003	52,415

further would drive the emission rates toward the 1% level for fixed systems and 2% for portable extinguishers; conversely, a relaxation of this goal would likely increase those emissions to the 3% level for fixed systems and 6% for portable extinguishers.

Products have been available for portable extinguisher applications since the early 1990s. No information was found in the literature about the quantities of HCFCs, PFCs or HFCs used for portable extinguisher applications. Information provided by a producer, combined with modelling, results in an estimate for the portable extinguisher composite bank of HCFCs, HFCs, and PFCs of approximately 1,500 tonnes at the end of 2002 with 0. 12 $MtCO_2$-eq of emissions and approximately 1,900 tonnes at the end of 2004 with emissions of 0.16 $MtCO_2$-eq at a 4% emission rate. The portable extinguisher composite bank is assumed to be comprised of 68% HCFC-123, 30% HFC-236fa and 2% PFC-14. Assuming an annual 3% growth rate, the estimated size of the portable extinguisher composite bank in 2015 will be 4,000 tonnes with 0.34 $MtCO_2$-eq of emissions at a 4% emission rate.

9.1.5 Regulatory and approval processes – hurdles to introducing new technologies

Most countries regulate the requirements for fire protection and the type provided. These controls can take the form of requiring fire protection in specific situations, such as sprinklers in hotels and offices. They can also take the form of required approval for the design and installation of specific fire protection systems. For example, organizations such as the National Fire Protection Association (NFPA), the Comité Européen de Normalisation (CEN) and the International Organization for Standardization (ISO) publish standards for specific types of fire protection systems, such as fixed systems using gaseous agents (NFPA, 2000; ISO, 2000). Some countries simply adopt NFPA or ISO standards, while others have their own standard-making bodies.

For example, the Brazilian Association for Technical Standards (ABNT) includes a committee for Fire Protection (CB24) (UNEP-TEAP, 1999) and there are some 12 Indian standards for halon alternatives. Furthermore, testing organizations such as Underwriters Laboratories (UL) or Factory Mutual Research Corporation (FM) test specific manufacturers' products to validate their performance against the standards written by the standard-setting organizations.

Such regulatory and approval processes can limit the introduction of new agents and techniques to those that have demonstrated acceptable levels of performance in two, and sometimes three, key areas: fire extinguishing effectiveness, safety for personnel/life safety and possible environmental considerations. The United Nations Environment Programme (UNEP) has prepared two documents that compile examples of such environmental regulations (UNEP 2000, 2001). The reader is encouraged to review these documents for further information. Only alternative fire protection agents and techniques that have satisfied or are undergoing nationally or internationally recognized regulatory approval processes are discussed in this section.

9.2 Reducing emissions through the choice of agents in fixed systems

9.2.1 Agents and systems with the potential to replace halons

With the halt in production of halon total flooding agents, fire protection professionals had two choices: (1) to begin using not-in-kind alternatives, accepting their deficiencies or (2) to wait for the development of new alternatives and techniques as they come available. Different market sectors were able to accept different choices based on the particular fire threat and risk acceptance.

9.2.1.1 Agents existing at the time of the Montreal Protocol

In its 1989 report, HTOC discussed a technique called the selection matrix. Basically, it places a value on performance parameters for several fire extinguishing system types, helping end users and authorities select the most appropriate alternative for a given application. The parameters in the matrix are: (1) low space and weight (of the agent storage containers), (2) damage limiting (speed of extinguishment and no agent residue), (3) ability to permeate (works around obstructions), (4) occupant risk (safe at concentrations used), (5) flammable liquid extinguishing capability, (6) system efficacy, (7) energized electrical equipment capability (electrically non-conductive) and (8) installed cost.

None of the new halon alternatives (HCFCs, HFCs, PFCs, FK, inert gases, water mist, etc.) had been developed at the time of the 1989 report. The types of systems directly considered in the matrix at that time were: automatic water sprinklers, fast response water sprinklers, pre-action water sprinklers, total flooding dry-chemical systems, total flooding carbon-dioxide systems, deluge water-spray systems, low-expansion foam and high-expansion foam.

The HTOC matrix effectively illustrates the individual strengths and weaknesses of the various systems, and reveals that all of the alternatives available at that time had one or more shortcomings that would prevent them from fully replacing halon in every application.

9.2.1.2 Agents developed since the Montreal Protocol

The conclusion drawn from the HTOC 1989 matrix – that no single suitable alternative for halon 1301 existed at that time – served as the incentive for the development of fire extinguishing systems using agents in four broad categories:

- halocarbon gaseous chemical agents including HCFCs, PFCs, HFCs and FK;
- inert gases, such as nitrogen or argon, or blends of those inert gases;
- water mist;
- fine powder aerosol.

Alternatives to halon fall into two broad categories: (1) clean agents similar to halon 1301 and (2) additional not-in-kind agents and systems similar to the other pre-Montreal Protocol alternatives.

9.2.2 Not-in-kind technologies

Not-in-kind technologies include the agents and systems considered in the 1989 HTOC matrix and the more recently developed water-mist and fine-powder-aerosol systems. When compared to halon, all of the not-in-kind alternatives continue to have the same fire protection shortcomings described in 1989. However, each has an appropriate place within fire protection. Not-in-kind alternatives offer the advantage of generating no direct greenhouse-gas emissions and are currently being used in about half of the applications that used halon before the Montreal Protocol (IPCC/TEAP, 1999).

9.2.2.1 Water-based systems

Water-based systems include automatic, fast response and pre-action water sprinklers, deluge water spray and water mist. In some applications, additives protect against freezing. Water-based systems have a limited ability to permeate obstructions, poor flammable liquid extinguishing capability and use limitations around energized electrical equipment (HTOC, 1989). Water itself may damage the items to be protected, i.e. it is not a clean agent. By contrast with gaseous agents, there is also an agent residue after the systems discharge. Water mist technology has achieved limited market acceptance, primarily in turbine and diesel-powered machinery and, somewhat less frequently, machinery spaces aboard ships. Water mist faces two obstacles (Wickham, 2002). First, the systems are not good at extinguishing small fires in large volumes, even to the point of failing to extinguish them. Second, applications are limited (in size and characteristics) to those where fire test protocols have been developed in which system performance has been determined empirically. This drives up costs higher than for other technologies.

9.2.2.2 Total flooding dry-chemical and aerosol systems

These systems use compounds such as sodium bicarbonate, ammonium phosphate and potassium bromide. They have a limited ability to permeate obstructions and leave a residue that normally precludes use for the protection of electronic equipment spaces. With aerosol systems, the technology is so new that suitable standards for approval testing and application requirements are still under development by the major standards organizations, including the NFPA, CEN and ISO.

9.2.2.3 Foam systems

Systems using foam employ water-based formulations containing surfactants to produce semi-stable foams. These systems leave agent residue after discharge and are larger and heavier than other options. They have limited ability to permeate obstructions and are not used around energized electrical equipment. Foams also have other environmental impacts on water quality and aquatic life (Ruppert and Verdonik, 2001). One manufacturer's version of Aqueous Film Forming Foam (AFFF), the best performing foam, is no longer produced in the US due to persistence, bioaccumulation and toxicity concerns (Dominiak, 2000).

Continuing research and development is underway with water mist and aerosol systems, as are efforts to develop and improve test methods and application standards (IMO, 1996, 1999, 2001).

9.2.3 Clean agents

Clean agents include inert gases (nitrogen, argon or blends of these two sometimes incorporating carbon dioxide as a third component), HCFC blends, HFCs, PFCs, FK and in some definitions carbon dioxide also (NFPA, 2000). With the exception of carbon dioxide, all clean-agent alternatives have been de-

Table 9.4. Clean-agent alternatives to halon 1301 (ISO, 2000; NFPA, 2004).

Generic name	Group	Comment
HFC-23	HFC	
HFC-125	HFC	
HFC-227ea	HFC	
HFC-236fa	HFC	Not commercialized in fixed systems
HCFC Blend A	HCFC Blend	Unsuitable for occupied spaces
HCFC-124	*HCFC*	*Being withdrawn from ISO standard. Unsuitable for occupied spaces.*
FC-2-1-8	*PFC*	*Being withdrawn from ISO and withdrawn from NFPA standard(s).*
FC-3-1-10	*PFC*	*Being withdrawn from ISO standard.*
FIC-13I1	FIC	Unsuitable for occupied spaces
FK-5-1-12	FK	New agent in the NFPA and expected in ISO referenced standards
Inert Gases	IG	Argon, nitrogen or blends of the two, sometimes with carbon dioxide

veloped since the Montreal Protocol and are used when not-in-kind technologies do not offer the required level of performance. While carbon dioxide systems have been used for many years and the technology is well developed, its use as a total flooding agent in occupied spaces involves significant life safety considerations. This is because the concentration needed to extinguish a fire is lethal. Numerous incidents of fatalities have been reported (Wickham, 2003). Carbon dioxide systems are therefore not included in the detailed sections below. They may be appropriate for use in some applications but only where personnel cannot be exposed.

PFC systems were initially used to help in the transition away from halons, but because of their impact to the climate system and lack of performance advantage over other alternatives new systems are no longer installed. The original agent manufacturer is supporting existing systems where necessary.

9.2.3.1   Progress with clean-agent systems

Market acceptance of the halocarbon and inert gas systems was generated relatively quickly. This was made possible by the early development of standards and the willingness of end users to accept these agents (NFPA, 2000; ISO, 2000; and IMO, 1998a,b). However, the path to market for new fire extinguishing agents and systems is laborious (Wickham, 2002). In most instances, the suitability of an agent is determined through data submittals, reviews and often testing involving the following: national health authorities to approve safe usage levels, national environmental authorities to assure compliance with laws and treaties, national and international standard-making organizations to write rules for safe use, national testing laboratories to test the chemicals, national testing laboratories to test and approve agent systems, national and international certification bodies for approvals.

This process is lengthy, expensive and often has to be repeated country-by-country to meet different national standards. While it may be onerous, it is important from the points of view of both fire protection and the environment. Countries and regions with high levels of regulatory supervision tend to avoid unapproved products, while those with less regulation have experienced difficulties with agents of questionable safety and effectiveness. Table 9.4 provides a listing of the clean extinguishing agents identified as suitable for use, within limitations, in ISO Standard 14520-1 and NFPA Standard 2001. This list includes all agents that are known to have been subjected to the appropriate approval processes at the time the list was compiled. Using an agent for an application for which it has not been approved can result in significant loss of life and property. From an environmental point of view, any use of unapproved agents that are greenhouse gases may result in emissions that otherwise could have been avoided

9.2.3.2   Refining the list of available gaseous alternatives to halon 1301

While Table 9.4 shows numerous halocarbon alternatives, the TEAP concluded that only three were commercially viable in 1999: HFC-23, HFC-227ea and HFC-236fa (UNEP-TEAP, 1999). A key characteristic of halon is its safety for use in occupied areas, but TEAP concluded that HFC-125, HCFC Blend A, HCFC-124 and FIC-13I1 were unsuitable due to unacceptable toxicity. However, new exposure guidelines are being adopted in some jurisdictions that will allow the use of HFC-125 in occupied spaces under certain conditions (NFPA, 2003).

Since 1999, FK-5-1-12 has become available. It has been listed as an acceptable alternative to halon under the US Environmental Protection Agency's Significant New Alternatives Policy (SNAP) programme (US EPA, 1994, 2002). It has also been included in the latest revision of the NFPA clean-agent standard (NFPA, 2004) and is expected to be in the next revision of the ISO standard (ISO, 2003). Systems have been approved by several testing and approval organizations and are now available. Additional technical information on FK-5-1-12 and the other clean-agent alternatives to halon 1301 is available in the referenced ISO and NFPA documents.

Table 2.6 (see chapter 2) lists the environmental properties of the halocarbon agents deemed acceptable for safe human exposure for use in normally occupied spaces: HFC-23, HFC-125,

Table 9.5. Remaining gaseous alternatives to halon 1301.

Generic name	Group	Comment
HFC-23	HFC	
HFC-125	HFC	
HFC-227ea	HFC	
FK-5-1-12	FK	New agent that went into commercial use in 2002.
Inert Gases	IG	Argon, nitrogen or blends of the two, sometimes with carbon dioxide

HFC-227ea, HFC-236fa, FC-2-1-8, FC-3-1-10 and FK-5-1-12. It is important to note that:

- HFC-23's high vapour pressure and low boiling point make it a unique replacement for halon 1301 in large-volume, low-temperature applications such as those found on the oil and gas industry on the North Slope of Alaska (Catchpole, 1999);
- HFC-236fa has never been commercialized as a fire extinguishing agent in fixed systems;
- FC-2-1-8 and FC-3-1-10 are being withdrawn from the ISO standard and all PFCs are prohibited by the International Maritime Organization (IMO) for shipboard fire extinguishing systems (SOLAS, 2000).

9.2.3.3   Available clean agents

After elimination of the agents either found unsuitable for occupied spaces, withdrawn from standards, prohibited by IMO or not commercialized in fire extinguishing systems, the list of potential gaseous total flooding agents is reduced considerably, as shown in Table 9.5.

Comparing the list from Table 9.5 against the performance parameters in the 1989 HTOC report shows that the agents listed in Table 9.5 are quite similar in terms of several parameters. For example, they are all gaseous agents that readily permeate obstructions, they are effective at concentrations safe for human exposure, they are effective for flammable liquids, they are non-conductive and can be used around energized electrical equipment. The real differentiation in system performance is (1) space and weight, (2) cost, (3) greenhouse-gas effect and (4) speed of extinguishing. In addition, two other factors may be important in some instances: (5) demonstrated special capabilities and (6) multiple supply sources.

- Space, weight and cost: A comparison was made of the agent storage container weights, floor area occupied (footprint), volume occupied (cube) and costs of several types of systems to protect volumes of 500, 1,000, 3,000 and 5,000 m^3 under the rules of the IMO for the protection of shipboard machinery spaces (Wickham, 2003). The results are shown in Table 9.6, using a halon 1301 system as the basis for comparison. The following definitions apply to Table 9.6:
  a. Weight includes the storage containers and contents but not the weight of piping, hangers, etc.
  b. Footprint is that area occupied by the agent containers defined by a square or rectangle circumscribing the agent cylinder bank.
  c. Cube is that volume occupied by the agent containers defined as the area multiplied by the height of the agent cylinders measured to the top of the valves.
  d. System cost is the average selling price of the system components charged by a manufacturer to a distributor or installer, and includes agent, agent storage containers, actuators, brackets, discharge and actuation hoses, check valves, stop valves and controls, time delay, manually-operated stations, predischarge alarms, pilot cylinders and controls. Cost does not include agent distribution piping and fittings, pipe supports and hangers, actuation tubing and fittings, electrical cables and junction boxes or labor to install, packing or freight.

- Greenhouse-gas effect: The GWPs of the agents (as listed in Table 2.6) provide a relative comparison of the direct greenhouse-gas emissions of fire protection systems and do not take into account any effects from indirect emissions. For most applications, the indirect effects are negligible compared with the direct effects. By contrast with

Table 9.6. Comparisons of average values in the 500 to 5,000 m^3 range (per cubic metre of protected volume at the concentration indicated).

		Halon 1301	HFC-23	HFC-227ea	HFC-125	FK 5-1-12	Inert gas
Concentration	Vol. %	6.0	19.5	8.7	12.1	5.5	40.0
Weight	kg/m^3	0.8	2.3	1.1	1.1	1.2	4.3
Footprint	10^4 m^2/m^3	5.8	12.0	6.8	7.4	7.3	28.2
Cube	10^4 m^3/m^3	8.6	18.0	13.1	14.4	13.8	56.6
System cost	US\$/m^3	7.43	39.77	28.05	26.37	35.98	34.07
Agent cost	US\$/m^3	3.34	18.33	19.08	16.81	26.59	9.62

other sectors, the amount of energy required to operate fire protection systems is trivial and largely unaffected by the agent used. This includes the impact of heating and cooling the systems installed in buildings, as this is quite small compared to the direct effect of an annual emission rate of even 1%. Techniques such as TEWI and LCCP identify only negligible levels of change compared to evaluations based only on the GWPs of the agents. The notable exception is aviation. The added weight of any of the halon alternatives has an indirect emissions effect because of the additional fuel use during the life of an aircraft.

- Speed of suppression: The discharge time for inert gas systems is in the order of 60 seconds or more. This is significantly longer than the discharge time of 10 seconds for halocarbon systems. Inert gas systems are therefore not recommended for areas where a rapidly developing fire can be expected (Kucnerowicz-Polak, 2002).
- Demonstrated special capabilities: There can be other performance requirements that apply to specific applications for which alternatives may not have been tested. For example, Catchpole (1999) described the unique capabilities of HFC-23 for deployment in high bay areas and in low ambient temperature conditions. Wickham (2002) described multiple special applications for HFC-227ea in military shipboard and vehicle applications and the choice of HFC-125 for both military vehicle and high-performance aircraft engine protection. The comprehensive testing and evaluation of HFC-23, HFC-227ea and HFC-125 for these applications and others are well documented in the publications of the US National Institute of Standards and Technology (NIST, 2003). FK-5-1-12 became commercially available in 2002 but has not yet been tested for some special applications. It has satisfied regulatory review under US EPA SNAP, NFPA and VdS, (VdS Schadenverhütung GmbH) and is pending at ISO. Market acceptance will depend on the comparative cost of the competing options. No estimates are available about how further restrictions on ozone-depleting substances and greenhouse-gases would affect agent choice in the future.
- Multiple supply sources: HFC-23, HFC-125 and HFC-227ea are manufactured by several companies in the US, Europe and Asia, and inert gas agents and systems are available worldwide from several competitive sources. FK-5-1-12 is available from one manufacturer in the US. Fixed systems utilizing these agents are available from several manufacturers.

9.2.3.1   Abatement costs

The abatement costs for assessing fire protection alternatives vary greatly depending upon specific assumptions that may or may not be applicable for any given situation. Harnish *et al.* (2001) assessed abatement costs for leakage reduction and recovery assuming a cost structure comparable to commercial refrigeration. The abatement cost was US$ 158,000 per ton of abated substance based on a projected 30% reduction of these emissions in 2010 and a 60% reduction in 2020. Harnish and Schwartz (2003) calculated an abatement cost to retrofit PFC 3-1-8 fixed systems with HFC-227ea in the EU assuming a 7% emission rate. The results were approximately 26 US$/tCO$_2$-eq abated (27.53 /tCO$_2$-eq). With a 2% emission rate, their result changes to approximately 130 US$/t CO$_2$-eq abated. Godwin (2004) proposed three abatement options for future systems: (1) replacing up to 30% of HFC-227ea with inert gases by 2020; (2) replacing all HFC-227ea in Class B (flammable liquids) applications with water mist by 2020; and (3) replacing up to 50% of HFC-227ea with FK 5-1-12 by 2020. The specific abatement costs range from approximately 14 to 57 US$/tCO$_2$-eq (4% discount rate) per year over the twenty-year lifetime. A fifteen-year lifetime leads to a range of 17–70 US$/tCO$_2$-eq per year.

While Godwin (2004) provides these possible scenarios, they must be viewed as hypothetical. HFCs and inert gases have evolved as the most commonly used agents, having achieved a degree of equilibrium in terms of market applications and share. FK-5-1-12 has been commercialized and is now available but there is no basis for predicting the rate of its market acceptance or its effect on the already established equilibrium. Under these circumstances, it is impossible to quantify the reduction in the use or emissions of HFCs in fire protection through 2015.

9.3   Reducing emissions through the choice of portable extinguisher agents

9.3.1   *Regulatory requirements for portable extinguishers*

Users almost always purchase hand portable fire extinguishers to comply with fire codes. In general, depending on the type of occupancy, fire codes and regulations describe the hand portable fire extinguisher requirements in terms of (1) either the charged weight or fire test rating of the extinguisher and (2) the number of extinguishers required based on the floor area of the facility and the maximum travel distance to an extinguisher. Furthermore, most codes require extinguishers capable of extinguishing fires in Class A and B fuels, often with the additional requirement that the extinguisher can be used safely around energized electrical equipment (NFPA, 1998 and UL, 2000). Extinguishers suitable for Class A fuels are tested by approval laboratories on wood cribs, wood panels and excelsior material. Extinguishers suitable for Class B fuels are tested by approval laboratories on flammable liquids in metal pans. The code requirements can be met either by employing an extinguisher capable of both Class A and Class B fires or separate extinguishers, one for each type of fire (Wickham, 2002).

9.3.2   *Agents existing at the time of the Montreal Protocol*

The HTOC (1989) also developed an agent selection matrix for portable fire extinguishers as follows: (1) effectiveness on ordinary combustibles, (2) effectiveness on flammable/combustible liquid fires, (3) electrical conductivity, (4) ability to permeate, (5) range, (6) effectiveness to weight ratio, (7) secondary damage (by extinguishing agent residue) and (8) cost.

Since none of the new halon alternatives were developed at the time of the 1989 report, the types of portable extinguishers considered in the matrix are regarded as not-in-kind technologies: carbon dioxide, multipurpose dry chemical, aqueous film forming foam and water.

The matrix illustrates the strengths and weaknesses of the various types of portables, and reveals that each one has shortcomings that would limit widespread use as halon alternatives.

- Water can be used for Class A fuels (ordinary combustibles) but has limited ability to permeate obstructions, poor flammable liquid extinguishing capability (Class B fuels) and limitations of use around energized electrical equipment (Class C fires). Water leaves a residue but produces no direct greenhouse-gas emissions (HTOC, 1999b).
- Carbon dioxide is effective on Class B and Class C fuels, but has poor effectiveness on ordinary combustibles (Class A fires), (HTOC, 1999a), poor range and high weight to effectiveness ratio. Its direct greenhouse-gas emissions are negligible (HTOC, 1999b).
- Multipurpose dry-chemical extinguishers, such as ammonium phosphate-based powder, are rated for use on Class A, B and C fires. They have a limited ability to permeate obstructions and can produce significant secondary damage from agent residue. They produce no direct greenhouse-gas emissions (HTOC, 1999b).
- Aqueous –film forming foam (AFFF) extinguishers use water-based formulations that contain surfactants to create a semi-stable foam. They are effective on Class A and B fuels, but have a limited ability to permeate and there are limitations on use around energized electrical equipment. In addition, AFFF extinguishers leave a residue after discharge. They produce no direct greenhouse-gas emissions (HTOC, 1999b).

9.3.2.1   Agents developed since the Montreal Protocol

Table 9.7 is an illustration of the clean agents developed for use in portable extinguishers since the Montreal Protocol. According to Wickham (2002), only HCFC Blend B and HFC-236fa have achieved any significant level of commercialization and the

manufacture of PFC-5-1-14 has been discontinued. FK 5-1-12 is a newly developed agent capable of extinguishing Class A, B and C fires. Extinguishers using FK-5-1-12 have negligible direct greenhouse-gas emissions (Taniguchi *et al.*, 2003). Other blends of HCFCs are in use in portable extinguishers but the literature contains no information about commercial acceptance. All of the agents in Table 9.7 have had use limitations applied to them by the US EPA, which has found the agents "acceptable in non-residential uses only". In effect, this precludes their use in residential applications (US EPA, 2003). The EPA uses the expression "residential" to differentiate from "commercial" applications.

9.3.2.2   Options for prospective owners of portable extinguishers

An end user is faced with making trade-offs between effectiveness, cleanliness and cost because, given equal size, (1) halocarbons are most expensive and least effective and (2) multipurpose dry chemicals cost least but are most effective.

End users who would have purchased halon 1211 portable extinguishers to protect their facilities or equipment fifteen years ago have three options today. (1) They can use an "in-kind" halon 1211 alternative such as one of the new halocarbon agents. (2) They can use two extinguishers (to do the job of one halon 1211 unit): water for Class A fires and carbon dioxide for Class B fires and those around electrically energized equipment. (3) They can use a multipurpose dry-chemical extinguisher in situations where the agent residue can be tolerated.

9.3.2.3   Progress with halocarbon portable extinguishers

Wickham (2002) compared the agent charge, fire rating and average selling price of several types of extinguishers. This is shown in Table 9.8. The available data for extinguishers with FK-5-1-12 are limited for this comparison, as few portable extinguishers have been developed using this agent to date.

As an example, a commonly specified extinguisher is 2-A:10-B:C rating, where "A" indicates it is suitable for use on Class A fires and the preceding number indicates its degree of effectiveness on those types of fires as determined in testing by approval laboratories; "B" indicates it is suitable for Class B fires and the preceding number indicates its degree of effectiveness on those

Table 9.7. Gaseous alternatives to halon 1211 in portable extinguishers (HTOC, 1999).

Generic name	Group	Comment
HCFC Blend B	HCFC+ PFC + inert gas	Blend of $CHCl_2CF_3$, CF_4 and argon
HCFC Blend E	HCFC+ HFC + hydrocarbon	Blend of $CHCl_2CF_3$, CF_3CHF_2 and $C_{10}H_{16}$
HCFC-124	HCFC	$CHClFCF_3$
HFC-236fa	HFC	$CF_3CH_2CF_3$
HFC-227ea	HFC	CF_3CHFCF_3
FC-5-1-14	PFC	C_6F_{14}
FK-5-1-12	FK	$CF_3CF_2C(O)CF(CF_3)_2$

Table 9.8. Cost comparisons for portable extinguishers.

Type	Agent charge		Fire rating	Average selling price (US$)
Halon 1211*	14.0 pounds	6.35 kg	2-A:40-B:C	223
Multipurpose dry chemical	5.0 pounds	2.27 kg	3-A:40-B:C	30
HFC-236fa	13.3 pounds	6.03 kg	2-A:10-B:C	493
HCFC Blend B	15.5 pounds	7.03 kg	2-A:10-B:C	415
Carbon dioxide	10.0 pounds	4.54 kg	10-B:C	175
Water	2.5 gallons	9.5 litre	2-A	63

*The halon 1211 price information is from 1993 before the halt of production of new halon 1211.

types of fires; and "C" indicates it is safe for use around energized electrical equipment. Table 9.8 shows end users have a choice of using the five-pound multipurpose dry-chemical unit for an average end user cost of US$ 30 each, the 13.3-pound HFC-236fa unit for US$ 493, the 15.5-pound HCFC Blend B unit for US$ 415, or both a ten-pound carbon dioxide unit and a 2.5 US gallon water extinguisher with a combined average end-user cost of US$ 238. This is compared to a cost of US$ 223 for equivalent halon 1211 units prior to 1994 when they were still being manufactured.

History has shown that a large portion of the market place was willing to pay over seven times more to get a clean-agent halon 1211 unit rather a not very clean dry-chemical extinguisher (US$ 223/30=7.43). The current cost multiple of 13 to 16 for the HCFC Blend B and HFC agents appears to be limiting market acceptance to those applications where users consider cleanliness a necessity.

9.4 Additional abatement options for emissions

9.4.1 Responsible agent management

The significantly reduced emission rates were achieved through a variety of actions within the fire protection community. For example, both the United Kingdom and the United States have developed voluntary agreements between the government and the fire protection industry that include specific actions by the fire protection community and also include the specific codes and standards that must be followed (DETR and FIC, 1997; FEMA *et al.*, 2002). While the specific form of implementation will differ according to the country or region depending on culture and legal traditions, the technical practices needed to minimize emissions will be the same. An explanation of these kinds of practices may be found in the agreements themselves and in the HTOC Technical Note #2 (1997), which also include important provisions for stored and stockpiled halons.

Another method chosen by some authorities is to use regulation to reduce emissions. For example, the International Convention for the Prevention of Pollution from Ships (Marpol 73/78) prohibits deliberate emissions of ozone-depleting substances and the European Union requires that all halon systems

and extinguishers, except those on a critical list, be decommissioned.

9.4.2 Importance of applying and enforcing codes and standards to minimize emissions

The UK and US agreements, and HTOC Technical Note #2, all make reference to the specific codes and standards that need to be followed in carrying out these provisions. Many, but not all, of these codes and standards have been discussed earlier in this chapter and have been an integral part of successful national programmes for reducing non-fire emissions. The specific codes or standards are different in the UK and US agreements, but their use was key to the success of both countries' programmes. The reader is encouraged to review the cited references to gain a better understanding of the types of codes and standards available, how they were applied and how their use produced emissions reductions.

9.4.3 End-of-life considerations for clean-agent systems and extinguishers

There are three distinct end-of-life considerations for clean-agent fixed systems and extinguishers: (1) end of useful life of the fire protection application, (2) end of useful life of the fire protection equipment and (3) end of useful life of the fire protection agent.

Fixed systems protect specific volumes (rooms), often telecommunications suites. Over the typical system life of 15 to 20 years, the protected equipment may change many times, with the fire protection system remaining in place. In specialized applications, such as aircraft and military systems, systems can remain in use for 25 to 35 years or longer (HTOC, 1994). Similarly, portable extinguishers meet a particular fire code or level of safety and are not replaced unless the use of the space changes in a way that requires a different fire protection capability. Portable extinguishers are generally required to undergo periodic pressure testing that involves removing the agent from the cylinder (NFPA 10), which is then recycled and redeployed. Estimates of the useful lifetime of portable extinguishers are between five to 25 years, depending upon regional factors

(HTOC, 1994).

The current availability of halons is the direct result of a strategy to recycle, effectively manage, and allow trade of the existing supplies of halons. The objectives were twofold: to enable a production phase-out in advance of alternatives to important national security and public safety applications, and to preclude the need for any future new production and consumption. The current global trade in halons produced before the phase-out is neither accidental nor a situation arising from oversupply. No exemptions to the halon production and consumption phase-out have been requested, or are expected to be granted by the parties (HTOC, 1994). Halon's positive market value provides a financial incentive to minimize emissions. Alternatively, because halon released to the atmosphere is untraceable, policies that diminish halon's value or make ownership a liability could result in increased emissions (UNEP-TEAP, 1998).

In countries or regions that have also placed export restrictions on the stores of halon, the only solution proposed has been destruction, often through a plasma arc or incineration process. To date, there has been only one known potentially economically feasible process described in the literature to convert either halons or the halocarbon alternatives to other useful products (Uddin *et al.*, 2004).

To date, there are no fire-protection-specific proposals in the literature for managing the end of life of halocarbon agents other than halons (HTOC, 2002, page 6). However, the industry practices of capturing, recycling and reusing halons have carried over to all high value gaseous agents, including HFCs, HCFCs and PFCs. There is no reason to believe they will not also be applied to FK agents as they enter the market. In addition, because of their more complex chemistries, it appears unlikely that halon replacements will ever be as inexpensive as halons. As a result, there is both a market incentive as well as an industry culture favouring capture, recycle and reuse. Adherence to these practices, including certification programmes, has been shown to minimize emissions and to limit the use of these agents to applications where their cleanliness is needed.

References

AEAT (AEA Technology), 2003: Emissions and Projections of HFCs, PFCs and SF$_6$ for the UK and Constituent Countries. Report prepared for the Global Atmosphere Division of UK Department of Environment, Food and Rural Affairs, AEA Technology Environment, Oxfordshire, UK, 159 pp.

Ball, D., 1999: Keeping the Options Open. Proceedings of the Halon Options Technical Working Conference, Albuquerque, NM, USA, 1999, pp. 19-24.

Catchpole, D.V., 1999: Fire Suppression In Cold Climates – A Technical Review. Report prepared for the U.S. Environmental Protection Agency, Washington, DC, USA, 29 pp.

DETR (UK Department of the Environment, Transport and the Regions), 1997: Voluntary Agreement Between UK Government and the Fire Industry Concerning the Use of HFC and PFC Fire Fighting Agents, UK DETR and the Fire Industries Council (FIC).

Dominiak, M., 2000: Phasing Out a Problem: Perfluorooctyl Sulfonate (PFOS). Minutes of the DOD AFFF Environmental Meeting, US Naval Research Laboratory Report No. 6180/0394A:FWW, Washington DC, USA.

FEMA (Fire Equipment Manufacturers Association) *et al.*, 2002: Voluntary Code of Practice for the Reduction of Emissions of HFC & PFC Fire Protections Agents. Fire Equipment Manufacturers Association (FEMA), Fire Suppression Systems Association (FSSA), Halon Alternatives Research Corporation (HARC), National Association of Fire Equipment Distributors (NAFED) and US Environmental Protection Agency (US EPA), March 2002 (available online at http://ww c.org/vcopdocument.pdf).

Fire Industry Confederation (FIC), 1997-1999: Reports under the Voluntary Agreement between UK Government and the Fire Industry Concerning the Use of HFC and PFC Fire Fighting Agents, Kingston upon Thames, Surrey, UK.

Godwin, D.S., 2004: Analysis of Costs to Abate International Ozone Depleting Substance Substitute Emissions. US EPA Report 430-R-04-006, US Environmental Protection Agency, Washington, DC, USA, 309 pp (available online at http://www.epa.gov/ozone/snap/).

Harnisch, J., O. Stobbe and D. de Jager, 2001: Abatement of Emissions of Other Greenhouse Gases: Engineered Chemicals. Report prepared for the IEA Greenhouse Gas R&D Programme, Ecofys, Utrecht/Cologne, The Netherlands/Germany, 85 pp.

Harnisch, J. and W. Schwarz, 2003: Costs and the Impact on Emissions of Potential Regulatory Framework for Reducing Emissions of Hydrofluorocarbons, Perfluorocarbons and Sulphur Hexafluoride. Report prepared for the European Commission, Ecofys/Öko-Recherche, Nürnberg, Germany, 57 pp (available online at http://europa.eu.int/comm/environment/climat/pdf/ecofys_oekorecherchestudy.pdf).

Cortina, T., 2004: HFC Emissions Estimating Program (HEEP). Proceedings of the 15th Annual Earth Technologies Forum, April 13-15, 2004, Washington, DC, USA, 3 pp.

HTOC (Halon Technical Options Committee), 1989: Assessment Report of the Halons Technical Options Committee. UNEP Technology and Economic Assessment Panel / UNEP Ozone Secretariat (Secretariat to the Vienna Convention for the Protection of the Ozone Layer and the Montreal Protocol on Substances that Deplete the Ozone Layer), Nairobi, Kenya, 101 pp.

HTOC (Halon Technical Options Committee), 1991: 1991 Assessment Report of the Halons Technical Options Committee. UNEP Technology and Economic Assessment Panel / UNEP Ozone Secretariat, Nairobi, Kenya, 166 pp.

HTOC (Halon Technical Options Committee), 1994: Assessment Report of the Halons Technical Options Committee. UNEP Technology and Economic Assessment Panel / UNEP Ozone Secretariat, Nairobi, Kenya, 186 pp.

HTOC (Halon Technical Options Committee), 1997: Technical Note #2 – Halon Emission Reduction Strategies. UNEP Technology and Economic Assessment Panel / UNEP Ozone Secretariat, Nairobi, Kenya, 18 pp.

HTOC (Halon Technical Options Committee), 1999a: 1998 Assessment Report of the Halons Technical Options Committee. UNEP Technology and Economic Assessment Panel / UNEP Ozone Secretariat, Nairobi, Kenya, 222 pp.

HTOC (Halon Technical Options Committee), 1999b: Technical Note #1 – New Technology Halon Alternatives – Revision 2. UNEP Technology and Economic Assessment Panel / UNEP Ozone Secretariat, Nairobi, Kenya, 30 pp.

HTOC (Halon Technical Options Committee), 2003: 2002 Assessment Report of the Halons Technical Options Committee. UNEP Technology and Economic Assessment Panel / UNEP Ozone Secretariat, Nairobi, Kenya, 69 pp.

IMO (International Maritime Organization), 1996: Amendments to the Test Method for Equivalent Water-Based Fire-Extinguishing Systems for Machinery Spaces of Category A and Cargo Pump-Rooms, Contained in MSC/Circular 668 Annex, Appendix B. MSC/Circular 728, Maritime Safety Committee, International Maritime Organization (IMO), London, UK.

IMO (International Maritime Organization), 1998a: Revised Guidelines for the Approval of Equivalent Fixed Gas Fire-Extinguishing Systems, As Referred to in SOLAS 74, for Machinery Spaces and Cargo Pump-Rooms (Annex). MSC/Circular 848, Maritime Safety Committee, International Maritime Organization (IMO), London, UK.

IMO (International Maritime Organization), 1998b: International Code for Fire Safety Systems, Chapter 5: Fixed Gas Fire Extinguishing Systems, Paragraph 2.5: Equivalent Fixed Gas Fire-Extinguishing Systems For Machinery Spaces And Cargo Pump Rooms, International Maritime Organization (IMO), London, UK.

IMO (International Maritime Organization), 1999: Guidelines for the Approval of Fixed Water-Based Local Application Fire-Fighting Systems for Use in Category A Machinery Spaces. MSC/Circular 913, Maritime Safety Committee, International Maritime Organization (IMO), London, UK

IMO (International Maritime Organization), 2001: Guidelines for the Approval of Fixed Aerosol Fire-Extinguishing Systems Equivalent to Fixed Gas Fire-Extinguishing Systems, As Referred to in SOLAS 74, for Machinery Spaces. MSC/Circular 1007, Maritime Safety Committee, International Maritime Organization (IMO), London, UK

IPCC (Intergovernmental Panel on Climate Change), 1996: *Climate Change 1995: Impacts, Adaptation and Mitigation of Climate Change: Scientific Technical Analyses. Contribution of Working Group II to the Second Assessment Report of the Intergovernmental Panel on Climate Change* [Watson, R.T., M.C. Zinyowera and R.H. Moss (eds.)]. Cambridge University Press, Cambridge, United Kingdom, and New York, NY, USA, 878 pp.

IPCC (Intergovernmental Panel on Climate Change), 1997: *Revised 1996 Guidelines for National Greenhouse Gas Inventories – Reference Manual* [Houghton, J.T., L.G. Meira Filho, B. Kim, K. Treanton, I. Mamaty, Y. Bonduki, D.J. Griggs and B.A. Callender (eds.)]. Published by UK Meteorological Office for the IPCC/ OECD/IEA, Bracknell, United Kingdom.

IPCC (Intergovernmental Panel on Climate Change), 2001: *Climate Change 2001 – Mitigation. Contribution of Working Group III to the Third Assessment Report of the Intergovernmental Panel on Climate Change* [Metz, B., O. Davidson, R. Swart and J. Pan (eds.)] Cambridge University Press, Cambridge, United Kingdom, and New York, NY, USA, 752 pp.

IPCC/TEAP (Technology and Economic Assessment Panel), 1999: Meeting Report of the Joint IPCC/TEAP Expert Meeting on Options for the Limitation of Emissions of HFCs and PFCs, [L. Kuijpers and R. Ybema (eds.)], 26-28 May 1999, Energy Research Foundation (ECN), Petten, The Netherlands, 51 pp (available online at www.ipcc-wg3.org/docs/IPCC-TEAP99).

ISO (International Organization for Standardization), 2000: ISO 14520-1:2000 to -15:2000, Gaseous Fire-Extinguishing Systems – Physical Properties and System Design. (Part 1: General requirements, Part 2: CF_3I extinguishant, Part 3: FC-2-1-8 extinguishant, Part 4: FC-3-1-10 extinguishant, Part 5: FK-5-1-12 extinguishant, Part 6: HCFC Blend A extinguishant, Part 7: HCFC 124 extinguishant, Part 8: HCFC 125 extinguishant, Part 9: HFC 227ea extinguishant, Part 10: HFC 23 extinguishant, Part 11: HFC 236fa extinguishant, Part 12: IG-01 extinguishant, Part 13: IG-100 extinguishant, Part 14: IG-55 extinguishant and Part 15: IG-541 extinguishant). International Organization for Standardization (ISO), Geneva, Switzerland.

ISO (International Organization for Standardization), 2002: Report of 7th Meeting of ISO/TC 021 / SC 08 Held In New Orleans, LA, USA, 10-11 September 2002 – Resolution 60, ISO/TC 21/SC 8/N 184, International Organization for Standardization (ISO), Geneva, Switzerland (available online at http://isotc.iso.org/iso-tcportal/index.html).

ISO (International Standards Organization), 2003: Gaseous Fire Extinguishing Systems – Physical Properties – Part 5: FK-5-1-12mmy2 Extinguishant, ISO/Working Draft 14520-5, ISO/TC 021 / SC 08 N 204, 11 July 2003. International Organization for Standardization (ISO), Geneva, Switzerland (available online at http://isotc.iso.org/isotcportal/index.html).

Kucnerowicz-Polak, B., 2002: Halon Sector Update. Presentation at the 19th Meeting of the Ozone Operations Resource Group (OORG), The World Bank, Washington, DC, USA (available olnline at http://www.worldbank.org/montrealprotocol).

McCulloch, A., 1992: Global production and emissions of bromochlorodifluoromethane and bromotrifluoromethane (halons 1211 and 1301). *Atmospheric Environment*, **26**, 1325-1329.

NFPA (National Fire Protection Association), 1998: NFPA 10 – Standard for Portable Fire Extinguishers. National Fire Protection Association (NFPA), Quincy, MA, USA.

NFPA (National Fire Protection Association), 2000: NFPA 2001 – Standard on Clean Agent Fire Extinguishing Systems. National Fire Protection Association (NFPA), Quincy, MA, USA.

NFPA (National Fire Protection Association), 2004: NFPA 2001 – Standard on Clean Agent Fire Extinguishing Systems – 2004 Edition. National Fire Protection Association (NFPA), Quincy, MA, USA.

NIST (National Institute of Standards and Technology), 2003: Papers from the 1991-2003 Halon Options Technical Working Conference (HOTWC), NIST Special Publication 984-1, National Institute of Standards and Technology, US Department of Commerce, Washington, DC, USA.

Rivers, P.E., 2000: A Novel Class of Low GWP Compounds as In-Kind Clean Agent Alternatives to HFCs and PFCs. Proceedings of the Halon Options Technical Working Conference, Albuquerque, NM, National Institute of Standards and Technology, Washington, DC, USA, pp. 186 - 192.

Ruppert, W.R., and D.P. Verdonik, 2001: Environmental Impacts of Aqueous Film Forming Foam. Proceedings of the AFFF and the Environment Technical Working Conference, Panama City, FL, USA, October 16-19, pp 24-34.

SOLAS (International Convention for the Safety of Life at Sea), 2000: International Convention for the Safety of Life at Sea, 1974; The 1988 Protocol; 2000 Amendments (Effective January and July 2002), Chapter II-2, Regulation 10, Paragraph 4.1.3, International Maritime Organization (IMO), London, UK.

Taniguchi, N., T.J. Wallington, M.D. Hurley, A.G. Guschin, L.T. Molina and M.J. Molina, 2003: Atmospheric chemistry of $C_2F_5C(O)CF(CF_3)_2$: Photolysis and reaction with Cl Atoms, OH radicals, and ozone. *J. Physical Chemistry A*, **107**(15), 2674-2679.

Uddin, M.A., E.M. Kennedy, H. Yu and B.Z. Dlugogorski, 2004: Process for Conversion of Surplus Halons, CFCs and Contaminated HFCs into Fluoroelastomer Precursors. In *Papers from 1991-2004 Halon Options Technical Working Conferences (HOTWC)*, [R.G. Gann, S.R. Burgess, K. Whisner and P.A. Reneke (eds.)], CD-ROM, NIST SP 984-2, National Institute of Standards and Technology, Gaithersburg, MD, USA.

UNEP (United Nations Environment Programme), 2000: Regulations to Control Ozone-Depleting Substances: A Guidebook. United Nations Environment Programme, Earthprint Ltd., Stevenage, Hertfordshire, UK, 385 pp.

UNEP (United Nations Environment Programme), 2001: Standards and Codes of Practice to Eliminate Dependency on Halons: Handbook of Good Practices in the Halon Sector. United Nations Environment Programme (OzonAction Programme), Multilateral Fund for the Implementation of the Montreal Protocol and Fire Protection Research Foundation, Paris, France, 74 pp.

UNEP-TEAP (Technology and Economic Assessment Panel), 1998: 1998 Assessment Report of the Technology and Economic Assessment Panel. UNEP Ozone Secretariat, Nairobi, Kenya, 300 pp.

UNEP-TEAP (Technology and Economic Assessment Panel), 1999: April 1999 Report of the Technology and Economic Assessment Panel. [L. Kuijpers (ed.)]. UNEP Ozone Secretariat, Nairobi, Kenya, 245 pp.

US EPA (US Environmental Protection Agency), 1994: Title 40, Code of Federal Regulations, Part 111.59, Sub-Chapter J, Significant New Alternatives Policy (SNAP) Program, Federal Register, Volume 59, pp. 13044.

US EPA (US Environmental Protection Agency), 2002: Title 40 Code of Federal Regulations, Part 82, Protection of Stratospheric Ozone: Notice 17 for Significant New Alternatives Policy (SNAP) Program, Federal Register, Volume 67, pp. 77928.

US EPA (US Environmental Protection Agency), 2003: Acceptable Substitutes for Halon 1211 Streaming Agents Subject to Narrowed Use Limits under the Significant New Alternatives Policy (SNAP) Program as of August 21, 2003, US EPA, Washington, DC, USA (available online at http://www.epa.gov/ozone/snap/fire/halo.pdf).

UL (Underwriters Laboratories), 2000: UL Standard for the Rating and Fire Testing of Fire Extinguishers – UL 711 (Fifth Edition). Underwriters Laboratories Inc., Northbrook, IL, USA.

Verdonik, D.P., 2004: Modelling Emissions of HFCs and PFCs in the Fire Protection Sector, Proceedings of the 15th Annual Earth Technologies Forum, April 13-15, 2004, Washington, DC, USA, 13 pp.

Verdonik, D.P. and M.L. Robin, 2004: Analysis of Emission Data, Estimates, and Modeling of Fire Protection Agents. Proceedings of the 15th Annual Earth Technologies Forum, April 13-15, 2004, Washington, DC, USA, 11 pp.

Wickham, R.T., 2002: Status of Industry Efforts to Replace Halon Fire Extinguishing Agents. Report prepared for the U.S. Environmental Protection Agency, Wickham Associates, Stratham, NH, USA, 46 pp.

Wickham, R.T., 2003: Review of the Use of Carbon Dioxide Total Flooding Fire Extinguishing Systems. Report for the U.S. Environmental Protection Agency, Wickham Associates, Stratham, NH, USA, 89 pp.

10

Non-Medical Aerosols, Solvents, and HFC 23

Coordinating Lead Authors
Sally Rand (USA), Masaaki Yamabe (Japan)

Lead Authors
Nick Campbell (United Kingdom), Jianxin Hu (China), Philip Lapin (USA), Archie McCulloch (United Kingdom), Abid Merchant (USA), Koichi Mizuno (Japan), John Owens (USA), Patrice Rollet (France)

Contributing Authors
Wolfgang Bloch (Germany)

Review Editors
Yuichi Fujimoto (Japan), José Pons (Venezuela)

Contents

EXECUTIVE SUMMARY

Non-medical aerosols

Prior to the Montreal Protocol, CFCs, particularly CFC-11 and CFC-12, were used extensively in non-medical aerosols as propellants or solvents. In developed countries today, more than 98% of non-medical aerosols now use non-ozone-depleting, low-GWP propellants (hydrocarbons, dimethylether, CO_2 or nitrogen). Some of the non-medical applications that previously used CFCs have also converted to not-in-kind technologies (spray pumps, rollers, etc.). These substitutions led to a total reduction of greenhouse-gas emissions of aerosol origin by over 99% between 1977 and 2001 (ADL, 2002; FEA, 2003). Although there are no technical barriers to the phase-out of CFCs, estimated consumption of CFCs in developing and CEIT countries was 4,300 tonnes in 2001 (UNEP, 2003a).

The remaining aerosol products use either HCFCs or HFCs because these propellants provide a safety, functional or health benefit for the users. In 2003, HFC use in aerosols represented total emissions of approximately 22.5 $MtCO_2$-eq. Both the US and European aerosol trade associations and the Japanese Ministry of Economy, Trade and Industry project low growth over the next decade.

Technical aerosols are a group of pressurized products used to clean, maintain, fix, test, manufacture, disinfect, or apply lubricant and mould release agents to various types of equipment. Technical aerosols are used in a number of industrial and commercial processes. In technical aerosols, HFCs are used most in dusters, where the substitution of HFC-134a by HFC-152a is a leading factor in reducing greenhouse-gas emissions. The substitution of HCFC-141b solvent by HFEs and HFCs with lower GWPs makes additional emission reductions possible in other technical aerosols.

Safety aerosols and insecticides for planes and restricted areas continue to rely on HFC-134a because of its non-flammability. Novelty aerosols generally use hydrocarbons, but some applications have been reformulated to HFC-134a in response to safety considerations. In consumer aerosols, the use of HFC-152a is likely to increase in the US due to regulations for the control of ground-level ozone formation caused by the emission of hydrocarbons from these products. However, on a global basis, such HFC-forcing regulation is not common.

Solvents

The primary ozone-depleting substances (ODSs) traditionally used in solvent applications are CFC-113, 1,1,1-trichloroethane and, to a lesser extent, CFC-11 and carbon tetrachloride. The ozone-depleting solvents have a variety of end uses. However, this report focuses on the largest application – cleaning – which includes precision, electronics and metal cleaning. It is estimated that 90% of ozone-depleting-solvent use had been reduced through conservation and substitution with not-in-kind technologies by 1999 (no-clean flux, aqueous or semi-aqueous cleaning, and hydrocarbon solvents). The remaining 10% of solvent uses are shared by several organic solvent alternatives.

The primary in-kind substitutes for 1,1,1-trichloroethane and carbon tetrachloride are non-ozone-depleting chlorinated solvents such as trichloroethylene, perchloroethylene and methylene chloride. More recently, n-propyl bromide has begun to be used in cleaning and adhesive applications, although there are concerns and uncertainties about its safety and ozone-depleting potential.

The in-kind substitutes for CFC-113 and CFC-11 are fluorinated alternatives such as hydrochlorofluorocarbons (HCFC-141b and HCFC-225ca/cb), perfluorocarbons (primarily C_6F_{14}), hydrofluorocarbons (HFC-43-10mee, HFC-365mfc and HFC-245fa, HFC-c447ef) and hydrofluoroethers (HFE-449s1, HFE-569sf2). These in-kind solvent substitutes cost more than the alternatives, and so they are primarily used in applications where safety, performance (such as compatibility, selective solvency and stability) or other environmental considerations are crucial. In most of the cleaning applications, the fluorinated solvents are used as blends with alcohols and/or chlorocarbons, thereby enhancing cleaning performance and reducing the overall cost of the solvents.

While HCFC-141b use in developed countries is rapidly decreasing, its use in developing countries and countries with economies in transition continues to increase due to low cost and widespread availability. HCFC-225ca/cb is currently used mainly in Japan and the US, but its use is expanding in developing countries.

PFC solvent use is constrained to a few high-performance niche applications due to its very limited solvency and high cost. Consumption is known to have decreased since the mid-1990s as a result of replacement with lower-GWP solvents.

Although HFCs are available in all regions, they have been used as solvents primarily in developed countries because of their high cost and the concentration of use in high-tech industries. With increasing concern about their GWP, uses tend to be focused in critical applications with no other substitutes. It is thought that use has now peaked at current levels and may even decline in the future.

Emission reduction options for solvent applications fall into two categories:

(1) Improved containment in existing uses: New and retrofitted equipment can significantly reduce emissions of all solvents. Optimized equipment can reduce solvent consumption by as much as 80% in some applications. Due to their high cost and ease of recycling, the fluorinated solvents can be recovered and reused by the end-users or their suppliers.

(2) Alternative fluids and technologies: A variety of organic solvents can replace HFCs/PFCs and ODSs in many applications. Caution is warranted prior to the adoption of any alternatives for which the toxicity profile is not complete. These alternative fluids include compounds with lower GWPs such as traditional chlorinated solvents, hydrofluoroethers (HFEs), and n-propyl bromide (nPB)

if its use continues to be allowed under the Montreal Protocol. The numerous not-in-kind technologies, including hydrocarbon and oxygenated solvents, are also viable alternatives in some applications. In a limited number of applications, no substitutes are available due to the unique performance characteristics of the HFC or PFC for those applications.

HFC-23 emissions from HCFC-22 production

Trifluoromethane (HFC-23 or CHF_3) is generated as a byproduct during the manufacture of chlorodifluoromethane (HCFC-22 or $CHClF_2$). HCFC-22 is used as a refrigerant in several different applications, as a blend component in foam blowing and as a chemical feedstock for manufacturing fluoropolymers such as polytetrafluoroethylene (PTFE). While the Montreal Protocol will eventually phase out the direct use of HCFC-22, its use as a feedstock will be permitted indefinitely because it does not involve the release of HCFC-22 to atmosphere.

HCFC-22 is produced in several developed and developing countries. The US, EU, Japan, China, India, and Korea are major producers of HCFC-22. HFC-23 is formed at the reactor stage of the manufacture of HCFC-22 as a result of over-fluorination. Its formation is dependent upon the conditions used in the manufacturing process and varies from 1.4 to 4.0 % of the production of HCFC-22 so that, while it is possible to reduce its formation by optimizing process conditions, it is not possible to eliminate the production of HFC-23. In a number of plants, HFC-23 is destroyed by thermal oxidation. The cost of reducing emissions varies significantly depending on the option used and will vary from plant to plant depending on the particular baseline situation.

Global emissions of HFC-23 increased by an estimated 12% between 1990 and 1995 as a result of a similar increase in global production of HCFC-22. However, due to the widespread implementation of process optimization and thermal destruction, this trend in HFC-23 emissions from developed countries has not continued and, since 1995, emissions have decreased as a proportion of production levels in those countries. On the other hand, HCFC-22 production has increased very rapidly in China and India in the last few years and global emissions of HFC-23 are still increasing.

Options for further emission reduction include:

(1) Process optimization at additional facilities: In fully optimized processes, average HFC-23 generation can be reduced to as little as 1.4% of production. However, actual achievements vary for each facility and it is not possible to completely eliminate HFC-23 emissions by this means.
(2) Capture and destruction: Thermal oxidation is an effective option for the abatement of HFC-23 byproduct emissions. Destruction efficiency can be >99.0% with measurements at one facility indicating the destruction of 99.9% of the HFC-23 but the impact of the "down-time" of thermal oxidation units on emissions needs to be taken into account. Several facilities in the EU, Japan and the US have already implemented this option.

Destruction of byproduct emissions of HFC-23 from HCFC-22 production has a reduction potential of up to 300 $MtCO_2$-eq per year by 2015 and specific costs below 0.2 US$/$tCO_2$-eq according to two European studies in 2000. Reduction of HCFC-22 production due to market forces or national policies, or improvements in facility design and construction also could reduce HFC-23 emissions.

10.1 Introduction

Chapter 10 assesses the remaining or "miscellaneous" ODS-replacement applications where greenhouse-gas emissions have been identified. These applications are in the non-medical aerosols and solvents sectors. The majority of substitutes in these sectors are ozone- and climate-safe alternatives, except for some limited use of HFCs, HFEs and PFCs for technical or safety purposes. A few additional niche applications of ODSs with possible greenhouse-gas replacements are also identified but there is insufficient literature available to evaluate the global status of replacements, or of current or future emissions. Emissions from fluorocarbon production are considered, and a detailed evaluation is made of HFC-23 byproduct emissions from HCFC-22 production. The use of PFCs and SF6 as ODS substitutes is also considered but the vast majority of emissions are found in sectors or applications that have never used ODSs.

10.2 Non-medical aerosols (including technical, safety, consumer and novelty aerosol products)

Aerosol products use liquefied or compressed gas to propel active ingredients in liquid, paste or powder form in precise spray patterns with controlled droplet sizes and amounts. Some aerosol products use the propellant as the only active ingredient, an example being dusters. Typical products use an aerosol propellant that is a gas at atmospheric pressure but is a pressurized liquid in the can. The phase change from liquid to gas allows a high level of active ingredient per unit of propellant to be expelled from the aerosol can.

The first aerosol product was developed during World War II as a result of efforts to control tropical disease. Controlled-release aerosol products distributed DDT insecticide to small enclosed areas (e.g. rooms, tents, etc.) to control mosquitoes.

The aerosol sector grew extensively worldwide after 1945. In the 1960s and 1970s, the vast majority of these spray products used CFC-12 as a propellant or CFC-11 as a solvent. The publication of the Rowland-Molina theory in 1974, the banning of CFCs in most aerosols in the United States, Denmark, Sweden and Norway in the late 1970s and early 1980s, and the ratification and implementation of the Montreal Protocol have led to aerosol reformulations without CFCs. More than 10 billion aerosol products were used worldwide in 2001 (UNEP, 2003a).

In developed countries, more than 98% of the conversions of non-medical aerosols encouraged by the Montreal Protocol have been to ozone- and climate-safe propellants: hydrocarbons (HC), dimethylether (DME), CO_2, or nitrogen (N_2). (In the aerosol industry, the terms "hydrocarbons" or "HAPs" (hydrocarbon aerosol propellants) refer more specifically to the natural gas liquids: n-butane, propane, and iso-butane.) In addition, where this is feasible and where they are effective in use, many aerosol products have been replaced by not-in-kind substitutes. Over time, these have included finger or trigger spray pumps,

sticks (deodorants, insect repellents), rollers, brushes, and cloth. This substitution process led to a total reduction of greenhouse-gas emissions from aerosols by over 99% in the US between 1977 and 2001 (ADL, 2002), and the same reduction in Europe between 1988 and 2002 (FEA, 2003).

Those remaining products made with either HCFCs (in developing countries where the use of HCFC-22 and HCFC-142b is allowed until 2040) or HFCs (HFC-152a, -134a) are niche products where safety or other environmental considerations such as tropospheric ozone are a crucial issue. HCFC-22 and HCFC-141b, which were once considered to be replacements for CFCs, are in minimal use today in aerosol production (UNEP, 2003a). However, developing countries in general do not have prohibitions on the use of HCFCs in aerosols. HCFC-22 and HCFC-141b therefore have some limited applications in non-medical aerosol products requiring low flammable propellants or solvents. HCFC-22 is much cheaper than 134a. HCFC-141b, however, is a strong solvent so it is usually used in blends. Table 10.1 provides an overview of the non-medical aerosol propellant alternatives.

Although there are no technical barriers to the transition from CFCs to alternatives for non-medical aerosol products, the estimated consumption of CFCs in developing countries and countries with economies in transition was still 4,300 tonnes in 2001 (UNEP, 2003a).

Most of the remaining aerosol products use either HFCFs or HFCs because these propellants provide an important safety, functional, or health benefit for the users. The Code of Practice on HFC Use in Aerosols of the FEA (European Aerosol Federation) and the responsible use practices of the trade group for the US aerosol industry (the Consumer Specialty Products Association) insist that HFCs should only be used in aerosol applications where there are no other safe, practical, economic or environmentally acceptable alternatives. The use of HFCs in aerosols is also limited further by cost. HFCs are between five and eight times more expensive than HCs. Table 10.2 contains estimates of total current and projected use of HFCs by region.

In 2003, HFC use in non-medical aerosols represented total emissions in the range of around 22 $MtCO_2$-eq: approximately 9 $MtCO_2$-eq for the US, 6 $MtCO_2$-eq for the EU, and 2.5 $MtCO_2$-eq for Japan (ADL, 2002; FEA, 2003; METI, 2004). It is estimated that emissions in 2010 will be around 23 $MtCO_2$-eq (ADL, 2002; FEA, 2003, METI, 2004). Both American and European aerosol trade associations and the Japanese Ministry of Economy, Trade and Industry expect low levels of growth in overall GWP-weighted emissions as a result of the introduction of HFC-152a in a number of applications currently using HFC-134a. Banking is not an issue for the aerosol sector. Products are generally used within 2 years or less after filling.

Key products using HFCs and HCFCs are listed in the next section, with a discussion of each product's efficacy, safety/health impact, volumes, growth pattern and substitution or reformulation potentials.

Table 10.1. Overview of non-medical aerosol propellant alternatives.

	HCFC-22	HFC-134a	HFC-152a	Dimethylether	Isobutane[1]
Substance characteristics					
Radiative efficiency (W m^{-2} ppb^{-1})	0.20	0.16	0.09	0.02	0.0047
Atmospheric lifetime (yr)	12	14	1.4	0.015	0.019
GWP (100-yr time horizon)					
- This report	1780	1410	122	1	n/a
- IPCC (1996)	1500	1300	140	1	
ODP	0.05	~0	-	-	-
Ground-level ozone impact					
- MIR[2] (g-O$_1$/g-substance)	<0.1	<0.1	<0.1	0.93	1.34
- POCP[3] (relative units)	0.1	0.1	1	17	31
Flammability (based on flashpoint)	None	None	Flammable	Flammable	Flammable
Technical data					
Stage of development	Commercial	Commercial	Commercial	Commercial	Commercial
Type of application:					
- Technical aerosols	X	X	X	X	X
- Safety aerosols	X	X			
- Consumer products	Phased out in industrialized countries		X	X	X
Emissions	Use totally emissive in all cases				
Costs					
Additional investment costs			Special safety required at filling plant	Special safety required at filling plant	Special safety required at filling plant

[1] Listed values refer to isobutane only. Additional hydrocarbon propellants are used in non-medical aerosol applications as indicated in section 10.2.
[2] Maximum Incremental Reactivity (See Chapter 2, Table 2.11 for more details)
[3] Photochemical Ozone Creation Potential (See Chapter 2, Table 2.11 for more details)

Table 10.2. HFC use in aerosols (MtCO$_2$-eq).

Region	HFC use in aerosols (MtCO$_2$-eq)		
	2000	**2003**	**2010**
EU	6.0	6.25	5.8
Japan	2.8	2.5	2.5
US	9.9	9.0	8.6
Rest of World	3.2	4.8	6.5
Total	**21.9**	**22.5**	**23.4**

10.2.1 Technical aerosols

Technical aerosols are a group of pressurized products used to clean, maintain, fix, test, manufacture, disinfect, or apply lubricant and mould release agents to various equipment. They are also used in a number of industrial processes.

10.2.1.1 Dusters

Compressed gas "dusters" generate small jets of "clean" gas as a non-contact method for removing dust and other particulate matter. Dusters are packaged in the same manner as aerosols, but the propellant is the only active ingredient in the can. As a result, there is usually considerably more gas in these items than in other aerosol products. Dusters are formulated in such a way that they leave no residue. The fact that there can be no flame suppression ingredients, e.g. water, increases the risk of ignition.

Dusters are used to clear dust and lint and other light particulate material from delicate surfaces (photographic negatives and lenses, substrates in semiconductor "chip" manufacture, computer/tv screens, high quality optics, specimens for electron microscopy) and difficult-to-reach places (the inside of Central Processing Unit "towers", computer keyboards, the inside of cameras, photocopiers, ATMs). These products are used in laboratories, darkrooms, offices (both in homes and on the commercial scale) and workbenches. Dusters are very widely used by industrial entities for a wide range of applications. Dusters are also used by consumers in a similar broad set of applications in offices, homes and workshops.

Dusters have been reformulated in the US, Europe, and Japan, switching from CFC-12 to HCFC-22 (1989/1990) and later to HFC-134a or HFC-152 (1993/1994). In some countries, where its use as an aerosol propellant is not prohibited, a small amount of HCFC-22 is still used in dusters. It is estimated that dusters account for most HFC use in technical aerosols worldwide (excluding metered dose inhalers). The replacement of HFC-134a by HFC-152a, which has started in the US and is expected to develop in the EU and Japan, is a key factor in the stabilization of the reduction of the CO_2-eq emission level in 2010. In Japan, emissions of HFC-134a from dusters have decreased from 2,137 metric tonnes in 2000 to 1,851 metric tonnes in 2003 as a result of the introduction of HFC-152a. HFC-152a use increased from 18 metric tonnes in 2000 to 371 metric tonnes in 2003 (METI, 2004).

HFC-152a is far less ignitable than hydrocarbons (HCs) and can be used in less fire-sensitive applications. Generally, this use is acceptable in the office environment where usage would be characterized by short blasts emitting small amounts of gas, and where contact with ignition sources is very limited. When used in machinery or equipment where an ignition source is present, the power should be turned off. In the US, Consumer Products Safety Commission (CPSC) standards do not require HFC-152a in a duster to be labelled as flammable.

Reduction options for dusters
The expected decrease in emissions from HFCs in aerosols comes as a result of the replacement of HFC-134a by HFC-152a in a number of technical aerosol applications. In addition, there are some cases of substitution in Europe by HCs in these products. This is a recent development and it remains to be seen how the safety issues with these gases will develop.

Some reduction of greenhouse gas emissions can still be realized through the partial replacement of HFC-134a by HFC-152a or dimethyl ether. Mixtures of these propellants can have acceptable flammability characteristics for use in applications where flammability is of less concern.

Some potential exists for the use of carbon dioxide (CO_2) but this is very expensive. Compressed CO_2 has the disadvantage of providing fewer blasts per can and CO_2 products require expensive high-pressure canisters. The CO_2 systems in the market today use very small aerosol cans (less than 20 gr). According to a recent analysis entitled "Dustometer Study", it takes approximately twenty of the small CO_2 aerosol cans to replace the cleaning power of one HFC-152a-based aerosol can (12 oz/340 gr) (DuPont, 2004). The assumption is that the emissions from the energy use needed to make the steel CO_2 canisters would at least equal the inherent emissions from an HFC duster.

There is one not-in-kind solution for high-pressure industrial applications: electrical air compressors. They are used in large high-volume production environments. Their cost and total efficacy make these systems useful in only a small number of situations. Rubber bulb brushes, if kept clean, are an alternative to dusters for low-pressure applications such as dusting lenses and dislodging dust from camera interiors.

10.2.1.2 Other technical aerosols
This category of products accounts for considerably less HFC use and is unlikely to grow in the future.

Freeze sprays are used to test the electrical conductivity of components on circuit boards. They are used for thermally locating intermittently operating electrical components. They are used in static-sensitive areas and therefore need to be non-flammable. Freeze sprays are like dusters in that they are cans filled with propellant only. These cannot be anything other than non-flammable propellants since they are mainly used to test electrically charged circuits that could ignite flammable gases.

Flux remover is a product used to clean excess flux and solder residue from circuit boards and other electronic components, especially in the repair of components and printed circuit boards. Since these are usually used near soldering irons or other sources of ignition, the formulations are made to be non-flammable in order to safeguard worker safety.

Mould release agents are used to release products moulded in plastic or synthetic fibre from their moulds without damage to the end product. These aerosols are in the process of being re-formulated due to the phase-out of HCFC-141b in all aerosol solvent uses in developed countries. The substitution of HFCs by HCs is technically feasible for some applications when appropriate consideration is given to flammability risks.

Electronic contact cleaners are used to dissolve and remove oil, grease, flux, condensation and other contaminants quickly from delicate electronic circuitry and expensive electronic instrumentation. Contact cleaners combine quick drying and high dielectric solvency for residue-free cleaning. They are used to clean switchboards or circuitry, magnetic tape, printed circuit boards, electronic relays, and security equipment. Contact cleaning needs a completely non-flammable formulation for use while equipment is in operation. It is therefore necessary to use pressurized cleaners containing non-flammable solvents as well as non-flammable propellants (in most cases HFC-134a).

Reduction options for other technical aerosols
For cleaners (contact cleaners, flux removers, etc.) and mould release agents, the substitution of the HCFC-141b solvent by hydrofluoroethers (HFEs) or HFCs with lower GWPs would reduce emissions without giving rise to substantial technical issues. The current cost of HFEs is higher than that of HFC-134a or HFC-152a. Mixes of HFC-152a, or dimethyl ether in smaller percentages, with HFC-134a are also a practical solution. Compressed gas is also an emission reduction option but with an important technical limitation. Compressed gas does not ensure constant pressure level within the can and its use is therefore impractical once two-thirds of the product has been expelled. Some not-in-kind solutions (manual or mechanical sprays) have been tested without success due to their reduced efficiency; new solutions like bag-on-bag containers appearing in Japan could be adapted for uses not requiring high-pressure sprays.

10.2.2   Safety aerosols

Tyre inflators

Tyre inflators are made with latex sealants, a solvent acting as a diluent, and propellant. They are a safe, effective, and a far less strenuous alternative to changing a flat tyre for people who cannot change a tyre or do not wish to risk doing so for safety reasons (e.g. busy traffic or personal safety late at night and on dark roads). Many of the tyre inflators on the market initially used dimethyl ether or HCs as the propellant. However, after a series of accidents and fatalities while repairing tyres containing HC in an enclosed garage, many manufacturers of these products choose to reformulate and switch to the more expensive HFC-134a and away from flammable propellants.

It is estimated that more than 80% of inflators have now been converted to HFC-134a because of its non-flammability, and sometimes because of local regulations.

Safety signal horns

These products are made and used for safety purposes such as signalling on boats, and signalling in industrial plants and offices for emergency evacuations. Performance requirements include portability, and easy and non-electrical operation. In the United States, horns designed to meet US Coast Guard (USCG) specifications must be audible one mile (1.6 kilometres) from the point of activation.

Insecticides for planes or restricted areas

The non-flammability requirement for products used in certain areas (planes, source of energized equipment like transformers etc.) leads to the use of HFC-134a, particularly for insecticides.

Reduction options for safety aerosols

Technically, HCs could be used to operate safety horns. However, because hydrocarbons are highly flammable, they constitute an additional risk element in emergency situations or where flame or ignition sources are frequently present. Whistles and mouth horns are substitutes but are not as loud as safety horns and therefore do not perform as well, especially in an emergency on the water or in burning buildings. Whistles and mouth horns can also be difficult to use in smoke-filled environments.

10.2.3   Consumer aerosols (cosmetic and convenience aerosols)

In the US, in response to mandates from the Federal Clean Air Amendments of 1990, state regulations in 16 states have been adopted to limit the content of volatile organic compounds (VOCs) in consumer products such as hair spray and deodorants. A comparatively small number of aerosol products (less than 2%) have needed to utilize HFC-152a in order to comply with these standards. Unlike hydrocarbons and dimethyl ether, HFC-134a and HFC-152a are exempt from VOC controls because they do not participate in the reaction to form ground-level ozone.

While there will be an increase in the use of HFC-152a in US consumer aerosol products, overall the replacement of HFC-134a by HFC-152a in dusters will result in the net decrease in emissions shown in Table 10.2.

Currently, VOC controls in Europe do not exempt HFCs because of the broad definition of VOCs (boiling point < 250°C under standard pressure/temperature conditions). No transition from HCs to HFCs for consumer products is therefore expected in the EU. No other VOC regulations identified elsewhere in the world restrict the use of HCs in aerosols.

Reduction options for formulated consumer products

HFCs are only used in consumer aerosol products in the US where no other options exist to meet product safety or VOC regulatory needs. Reformulation options are therefore very limited. In addition, since the overall world market for aerosols involves only very limited HFC use (less than 2% of all aerosol products), such use represents only a very small fraction of total HFC use, limiting the potential for substitution.

10.2.4   Novelty products

Products in this category include artificial snow, silly string (spray of solid polymer material), and noise-makers. No data about global or regional HFC use are available for these applications. However, in the EU, this category could represent up to 25% of total HFC emissions for non-medical aerosols (Harnisch and Schwartz, 2003). Noise-makers are a variation of the safety horns discussed in section 10.1.2, but noise-makers are different in that they emit a less loud, deeper pitched sound. Additionally, the large majority (>80%) of horns marketed as noise-makers are made with HCs and not HFCs. Noise-makers with HFCs account for a very small percentage of the overall market. Artificial snow and string novelties were once marketed using flammable propellants but, after highly publicized safety incidents, were reformulated to HFC-134a in many developed countries. Conversions to lower-GWP propellant mixes or HC propellants should be considered where safety permits.

10.2.5   Constraints on emission reduction potential

In the US, in order to comply with limits on tropospheric ozone levels pursuant to the Federal Clean Air Act, some state-level VOC emission restrictions are leading to the reformulation of hydrocarbon-propellant aerosol products, e.g. consumer aerosols, to HFC-152a. In the EU, the recently proposed regulation on fluorinated gases has not restricted HFC use in technical or safety aerosols but has proposed a ban on use in novelty products (according to the definition of novelty in Directive 94/48/EC; for example artificial snow, silly string, decorative foams etc.).

10.2.6 *Emission reduction potential and abatement cost*

Three separate studies estimate that the potential reduction in HFC non-medical aerosol emissions ranges from 10% to 88% (see Table 10.3 for more details) depending on the country and substitution costs (Harnisch and Gluckman 2001; METI, 2004; US EPA, 2004b).

The abatement costs associated with using the HFC aerosol substitutes vary greatly from negative costs to over 150 US$$_{2002}$/tCO$_2$-eq depending upon the applications and the required level of final safety (flammability).

A European study covering only 15 EU member states was conducted by Harnisch and Gluckman (2001) and projects an emissions reduction in 2010 of 39% for a cost of less than below 23 US$$_{2002}$/tCO$_2$-eq /tCO$_2$-eq and an additional reduction of 39% for a cost of between 23 and 57 US$$_{2002}$/tCO$_2$-eq. Another European study by Harnisch and Schwartz (2003) is not referred to as it focuses only on novelty aerosols, which represent 25% of total HFC aerosol emissions in the EU.

The Japanese study by METI (2004) projects a 10% emission reduction potential for non-medical aerosols used in Japan, mainly as a result of the substitution of HFC-134a by HFC-152a. This study does not give any cost estimates.

The US EPA's emission-abatement report (2004) has estimated emission reduction potentials in 2020 in the US, non-US Annex I and non-Annex I countries. This study considered three abatement options:

- replacement of HFC-134 as a propellant by HFC-152a;
- substitution of HFCs by hydrocarbon propellants (with safety concerns being properly handled); and
- substitution by not-in-kind packaging such as triggers, pumps, or sticks.

These substitutions represent a 57% reduction potential of total emissions in 2020, with reductions of 37%, 10% and 10% for each category respectively.

10.3 Solvents

On an ozone-depletion-weighted basis, solvents constituted approximately 15% of the market for chemicals targeted for phase-out under the Montreal Protocol. In the case of the four ozone-depleting substances used as solvents – CFC-113, CFC-11, carbon tetrachloride and 1,1,1-trichloroethane (also known as methyl chloroform) – most use by far involved CFC-113 and 1,1,1-trichloroethane in industrial applications and in industrial cleaning equipment. Carbon tetrachloride was predominately for dry-cleaning textiles, or in hand or maintenance cleaning. The three main end uses of CFC-113 and 1,1,1-trichloroethane in the past were metal cleaning, electronics cleaning, and precision cleaning. Metal cleaning usually involves removing cutting oils and residual metal filings from metal surfaces, and the maintenance and repair of equipment and machinery. This sector relied principally on 1,1,1-trichloroethane. Precision and electronics cleaning mostly used CFC-113. Precision cleaning mainly involves products that require a high level of cleanliness and generally have complex shapes, small clearances, and other cleaning challenges (UNEP, 1998). Electronics cleaning, or defluxing, consists mainly of the removal of the flux residue left after a soldering operation on printed circuit boards and other contamination-sensitive electronics applications. Tables 10.4 and 10.5 provide an overview of CFC-113 and 1,1,1-trichloroethane solvent cleaning applications and market estimates.

The Solvent Technical Options Committee (STOC) of TEAP estimates that 90% of ozone-depleting solvent use (based on the peak consumption of 1994-95) has been reduced by switching to not-in-kind technologies (UNEP, 1998). The remaining 10% of the market for ozone-depleting substances is shared by several organic solvent alternatives. In developed countries, almost all ozone-depleting solvents (CFC-113, CFC-11, carbon tetrachloride and 1,1,1-trichloroethane) have been replaced by either not-in-kind technology (no-clean flux, aqueous, semi-aqueous or hydrocarbon solvents) or other organic solvents. HCFC use is declining and is expected to be

Table 10.3. Summary of emission reduction potential and abatement cost for non-medical aerosols applications.

Reference	Harnisch and Gluckman (2001)	US EPA (2004)		METI (2004)
Region **Year**	**European Union** **2010**	**USA** **2020**	**non-USA** **2020**	**Japan** **2010**
Total emission projection (MtCO$_2$-eq)	5.1 (2010)	4.0 (2020)	6.7 (2020)	2.5 (2010)
Emission abatement (MtCO$_2$-eq)	a) 2.0 b) 2.0 c) 0.5	2.3	3.8	0.3
Abatement cost (US$$_{2002}$/tCO$_2$-eq)	a) <23 b) 23 to 57 c) 57 to170	-25 to 2.5	-25 to 2.5	n/a

Table 10.4. CFC-113 uses.

Applications	Electronics cleaning	Metal cleaning	Drying	Particle removal	Carrier fluid	Dielectric fluid	Medical	Misc.
Percentage	40	30	8	4	4	2	3	9

Table 10.5. 1,1,1-trichloroethane uses.

Applications	Electronics cold cleaning	Degreasing and	Adhesives	Aerosols	Coating	Chemical process intermediate
Distribution	4%	53%	13%	13%	10%	7%

eliminated by 2015. In developing countries and countries with economies in transition, the use of ozone-depleting solvents, especially HCFC-141b, continues. Although many of the larger international firms with operating facilities located in those countries have switched to non-ozone-depleting technology, few small or medium enterprises have completed the transition.

The primary in-kind substitutes for 1,1,1-tricholorethane have been the chlorocarbon alternatives such as trichloroethylene, perchloroethylene and methylene chloride. These substitutes have very low (0.005-0.007) ozone-depletion potential, and similar cleaning properties (UNEP, 1998). These chemicals were widely used for cleaning in the past until toxicity concerns resulted in switches to ozone-depleting substances. In addition, a brominated solvent – n-propyl bromide – was recently introduced. It has similar properties and has acquired a definite market share of organic solvents for defluxing, general cleaning and adhesives applications.

The in-kind substitutes for CFC-113 and CFC-11 are fluorinated alternatives such as hydrochlorofluorocarbons (HCFC-141b and HCFC-225ca/cb), perfluorocarbons (primarily C6F14), hydrofluorocarbons (HFC-43-10mee, HFC-365mfc and HFC-245fa, HFC-c447ef) and hydrofluoroethers (HFE-449s1, HFE-569sf2). HCFC-141b and HCFC-225ca/cb have properties similar to CFC-113 and CFC-11 and have therefore been preferred as alternatives where in-kind alternatives are required. In particular, the low cost and good solvency of HCFC-141b initially allowed it to capture a significant market share of CFC solvents. However, HCFCs are transitional substances due to their ozone-depleting potential. Their use is declining in developed countries but still increasing in developing countries (UNEP-TEAP, 2003).

Most of the electronic cleaning market converted to not-in-kind alternatives such as no-clean processes, aqueous and semi-aqueous cleaning. Oxygenated solvents (e.g. alcohols, glycols, ethyl lactate) have also replaced CFC-113 in some portions of the electronics cleaning market. A very small segment of the market went to the HCFC alternatives and their azeotropic blends. Some of the metal-cleaning, particulate removal, medical and miscellaneous applications also switched to

aqueous, semi-aqueous, HCFC and chlorocarbon alternatives. Recently, some of these applications have moved to a newer brominated solvent – normal propol bromide (nPB) – an ozone-depleting substance not yet controlled by the Montreal protocol. However, uncertainty about the toxicity of nPB has limited its development in certain countries. Most of the displacement drying applications switched to either hot-air drying or alcohol drying systems. Some of the drying applications also switched to HCFCs. The cleaning applications (electronics, metal and particulate removal) and carrier solvents are the focus of this section.

In the late 1980s and early 1990s, PFCs (primarily C_6F_{14}) entered into some of the CFC-113 markets, with limited use in precision and electronics cleaning applications. Use was mostly limited to high-performance, precision-engineered applications where no other alternatives were technically feasible due to performance or safety requirements. Other alternatives such as HFCs and HFEs have been developed with better cleaning performance, significantly lower global warming potential and comparable cost. PFCs are no longer considered technically necessary for most applications.

In the long-term, HFCs and PFCs are expected to acquire a very small share of the ozone-depleting-solvent market. The IPCC's TAR has produced projections indicating that less than 3% of the possible CFC solvent demand will be replaced by HFCs and PFCs (IPCC, 2001b). Some of the reasons for these low uses are the higher cost of HFCs and PFCs, the ready availability of other low cost and acceptable alternatives, and the limited solvency of the common soils in these solvents. Because of this limited use of HFC and other fluorinated alternatives, the EU Commission did not include any restriction on HFC solvent use in its recent proposed regulation on certain fluorinated gases (CEC, 2003).

10.3.1   *Current uses – HFCs, PFCs, HCFCs*

The solvent applications that currently employ HCFCs, HFCs, and PFCs are shown in Table 10.6, together with an overview of the environmental and technical data for each substance.

Table 10.6. Overview of HFCs, PFCs and HCFCs in solvent applications.

	HCFC-141b	HCFC-225ca/cb	HFC-43-10mee	HFC-365mfc (C_6F_{14})	PFC–51-14
Substance characteristics					
Radiative efficiency (W m^{-2} ppb^{-1})	0.14	0.2/0.32	0.4	0.21	0.49
Atmospheric lifetime (yr)	9.3	1.9/5.8	15.9	8.6	3,200
GWP (100-yr time horizon)					
- This report	713	120/586	1,610	782	9,140
- IPCC (1996, 2001a)	600	180/620	1,300	890	7,400
ODP	0.12	0.02/0.03	-	-	-
Ground-level ozone impact					
- MIR$^{(1)}$ (g-O$_3$/g-substance)	<0.1	<0.1	n/a	n/a	n/a
- POCP$^{(2)}$ (relative units)	0.1	0.2/0.1	n/a	n/a	n/a
Ground-level ozone impact	None	None	None	None	None
Flammability (based on flashpoint)	None	None	None	Flammable	None
Technical data					
Stage of development	Commercial	Commercial	Commercial	Commercial	Commercial
Type of application:					
- Electronics cleaning	X	X	X		
- Precision cleaning	X	X	X	X	X
- Metal cleaning	X	X	X		
- Drying	X	X	X	X	
- Carrier solvent	X	X	X		X

[1] Maximum Incremental Reactivity (See Chapter 2, Table 2.11 for more details)
[2] Photochemical Ozone Creation Potential (See Chapter 2, Table 2.11 for more details)

10.3.1.1 HFCs

Two HFC solvent alternatives are available commercially: HFC-43-10mee (C5H2F10) and HFC-c447ef (heptafluorocyclopentane; C5H3F7). Two other HFCs are coming onto the solvent markets.

HFC-43-10mee is a non-flammable, non-VOC (US definition) solvent with low toxicity (US EPA, 1998). The atmospheric life of HFC-43-10mee is 15.9 years; it has a GWP of 1610 (Chapter 2, Table 2.6). HFC-43-10mee readily forms azeotropes with many desirable alcohols, chlorocarbons and hydrocarbons to give blends enhanced cleaning properties (Merchant, 2001). It has better solvent properties than PFCs. The neat material is used as a carrier fluid, and in particulate removal and other miscellaneous applications. The blends are used in applications such as precision cleaning, defluxing flip chips and printed wiring board (PWB), and oxygen system cleaning.

HFC-c447ef (C$_5$H$_3$F$_7$) is a non-flammable, non-VOC solvent (US definition) with a boiling point of 82°C (Zeon Corporation, 2004). Its atmospheric lifetime is reported to be 3.4 years with a GWP of 250 (UNEP-TEAP, 2003)[1]. This is lower than most HFCs and some HFEs. However, because of its high freezing point (21.5°C), its neat use is limited. Some azeotropic mixtures with alcohol and ketone, or mixtures with terpenes, are provided for the degreasing metal parts and flux cleaning of PWBs. The levels of use of this solvent are low and most current users are in Japan, but marketing has started in the US.

Two other HFC candidates, although primarily developed for foam-blowing applications, have been promoted in some solvent applications. They are HFC-365mfc (pentafluorobutane) (Solvay, 2005) and HFC-245fa (pentafluoropropane) (Honeywell, 2004).

HFC-365mfc is a non-VOC solvent (US definition) with an atmospheric life of 8.6 years, a GWP of 782 (Table 2.6) and very low toxicity (exposure guide of 500 ppm). The manufacturer's data sheet (Solvay, 2005) indicates that it is flammable (flashpoint of –27°C) and it is therefore used primarily in blends with other non-flammable HFCs or HFEs to suppress its flammability. The current cost of HFC-365mfc is lower than that of other HFCs and HFEs and so it is being increasingly used in low-cost non-flammable blends (with HFCs and HFEs) for some applications.

HFC-245fa has a relatively low boiling point (15°C) and this limits its use in conventional degreasers. At present, its use has been limited to aerosol blends; its use in cold cleaning is also envisaged, though this is also limited due to the rapid evaporation rate.

Although HFCs are available in all regions, they have been primarily used in developed countries due to their relatively

---

[1] Note that HFC-c447ef is not included in the assessment of literature on chemical and radiative effects in this report (in chapter 2).

high cost and the concentration of use in high-tech industries. In addition, given the increasing concern about their GWPs, they now tend to be used in critical applications with no other substitute. Growth is therefore expected to be minimal (Merchant, 2001).

10.3.1.2　PFCs

PFC solvents such as C_6F_{14} (PFC-5-1-14) have physical properties similar to CFC-113. The atmospheric lifetimes and GWPs of PFCs are very high (C_6F_{14} has an atmospheric lifetime of 3200 years and a GWP of 9140 (Table 2.6)). However, the very limited solubility of organic soils in these materials has restricted their use in cleaning to particulate removal and the cleaning of fluorinated oils and greases (Agopovich, 2001). Due to the high cost, the availability of substitutes, and regulations restricting or banning use, PFC use has been limited primarily to niche applications in developed countries (Japan, US and Western Europe) with virtually no use in developing countries or countries with economies in transition (IPCC/TEAP, 1999). For example, in the US, PFC use is restricted to only those applications where no other alternative meets performance or safety requirements. In Germany, the use of CFCs, HCFCs, HFCs and PFCs is prohibited in solvents, except the pertinent authority can allow for high-quality applications provided negative environmental impacts and impacts on the climate are not expected and only if state-of-the-art technology does not allow the use of any fluorine-free solvent (Germany, 1990).

The published data about the consumption of C_6F_{14} are limited. Volumes are known to have decreased since the mid-1990s due to replacement with lower-GWP compounds (IPCC/TEAP, 1999).

10.3.1.3　HCFCs

The only HCFC solvents used are HCFC-141b and HCFC-225ca/cb, with ODPs of 0.12 and 0.02/0.03 (Chapter 1, Table 1.2) and GWPs of 713 and 120/586 respectively (Chapter 2, Table 2.6). HCFC-141b is mainly used in foam blowing, with use as a solvent representing less than 10% of global use in 2002 (UNEP-TEAP, 2003).

The use of HCFC-141b as a solvent is widely banned in developed countries, with a few derogations for technical reasons. Use from existing stockpiles is, however, allowed in the US.

In developing countries, the use of HCFC-141b is still increasing, especially in China, India, and Brazil. As economic growth rates are high, use could have exceeded 5000 tonnes in 2002, even with process containment and recycling (AFEAS, 2001). HCFC-141b is often the most cost-effective replacement for 1,1,1-trichloroethane or CFC-113. The decrease in developed countries, especially as use for foam blowing is phased out, will lead to production restructuring which could create shortages in 2010–2015 (UNEP-TEAP, 2003).

HCFC-225ca/cb use has always focused on niche applications in precision cleaning and use as a carrier solvent because of its ozone-depleting potential and phase-out schedule; it is being gradually replaced by HFC, HFE and not-in-kind alternatives. Most sales are in Japan and the US (around 4,000 tonnes in 2002); sales are increasing in developing countries (2,000 tonnes in 2002) (UNEP-TEAP, 2003).

The HCFC total consumption phase-out for developing countries is scheduled for 1 January 2040, with a freeze in consumption in 2015. But some countries are accelerating this schedule, especially for solvents (e.g., Malaysia, Thailand). Most of the uses of HCFC-141b and HCFC-225ca/cb can be technically replaced by partly introducing zero-ODP HFC or HFE, with a lengthy qualification transition time for complex and crucial applications and sometimes at a high cost.

10.3.2　*Projected consumption / emissions*

Most solvent uses are emissive in nature, with a short inventory period of a few months to two years (IPCC, 2000). Although used solvents can and are distilled and recycled on site, all quantities sold are eventually emitted. The IPCC Good Practice Guidance recommends a default emission factor of 50% of the initial solvent charge per year (IPCC, 2000). A report by the US Environmental Protection Agency (US EPA) assumes that 90% of the solvent consumed annually is emitted to the atmosphere (US EPA, 2004a). Consequently, the distinction between consumption and emission is typically not significant for these applications.

The US EPA has projected global emissions of HFCs and PFCs from solvent uses as shown in Table 10.7 (US EPA, 2004a). PFC emissions are assumed to decline linearly until they are essentially no longer used in solvent applications in 2025.

Table 10.7. Projected HFC and PFC emissions from solvent uses (MtCO$_2$-eq) (US EPA, 2004).

Region	HFC and PFC emissions from solvent use (MtCO$_2$-eq)			
	2005	**2010**	**2015**	**2020**
United States	1.65	1.80	1.91	2.09
Developed countries (non-US)	2.05	2.09	2.09	2.13
Developing countries	0.22	0.33	0.37	0.44
Total	**3.92**	**4.18**	**4.40**	**4.62**

Note: Totals may not tally due to independent rounding-off.

10.3.3   *Emissions reduction options*

Most of the new solvent handling equipment is designed to ensure very low emissions, partly because of the higher cost of the HFC, PFC and HCFC solvents and partly due to increased environmental awareness and regulations in some countries for all solvents requiring zero or low emissions. The new tight machines generally have higher freeboards, dual temperature cooling coils, automated work transport facilities, welded pipe joints, hoods or sliding doors, and superheated vapour drying systems. Likewise, it is possible to reduce emissions to a minimal level by implementing good handling practices, such as reducing drag-out losses of solvent from the system by keeping the workload in the vapour zone long enough to drain and drop any entrapped or remaining solvent (UNEP, 1998). It is also possible to minimize evaporative losses by improving the design of solvent bath enclosures and of vapour-recovery condensing systems (March, 1998).

10.3.3.1   *HFCs, PFCs, HCFCs emission reduction*
In tests with both new degreasers and retrofitted degreasers with specified enhancement features, measurements of HFC, PFC and HCFC solvent emissions showed that there was a decrease of as much as 80% (Ramsey, 1996). Another study indicated that retrofitting a vapour degreaser with an open-top area of 3.6 meters square (13 square feet) can, in combination with proper maintenance, reduce emissions from a solvent process by 46 to 70 percent, depending on the specific retrofit methods chosen (Durkee, 1997). With higher-cost fluorinated solvents, the pay-off from upgrading an old degreaser to a new degreaser standard is estimated to be > 62% (rate of return with break-even in less than 2 years) (Ramsey, 1996).

10.3.3.2   *Alternative fluids (chlorinated solvents, HFE-449s1, HFE-569sf2 HFE-347pcf, n-propyl bromide)*
In some cleaning applications, lower-GWP solvents can be used to replace higher-GWP solvents. These options include the use of lower-GWP HFCs to replace the use of PFCs and higher-GWP HFCs. Similarly, with very few exceptions, the non-ozone-depleting solvents can be used as alternatives to HCFCs.

Cleaning processes using halogenated alternative fluids, with the exception of n-propyl bromide (due to the higher vaporization temperature compared to chlorocarbons), are typically similar to the processes they replace with respect to energy use (ADL, 2002). Consequently, these aspects are not assessed for each option.

Fluids that are alternatives to both HFCs/PFCs and ODSs include:

Chlorinated solvents
Chlorinated solvents such as dichloromethane (methylene chloride), trichloroethene (trichloroethylene) and tetrachloroethene (perchloroethylene) have been used for decades in cleaning applications (Risotto, 2001). The implementation of the Montreal Protocol motivated a return to the use of these traditional solvents. Chlorinated solvents are used as replacements for CFCs, 1,1,1-trichloroethane and HCFCs in a variety of cleaning applications due to their high solvency. While the higher boiling points of trichloroethene (87°C) and tetrachloroethene (121°C) limit their use in some applications, they also provide an advantage in cleaning some soils such as resins and waxes (Risotto, 2001). However, in some instances, the solvent strength of the chlorinated solvents may cause incompatibility with the components to be cleaned.

Direct greenhouse-gas emissions from the use of chlorinated solvents are much lower than those from the products they replace due to their very low GWPs.

Chlorinated solvents are non-flammable but have toxicological profiles that require low exposure guidelines. The 8-hour exposure guidelines for dichloromethane, trichloroethene and tetrachloroethene are 50 ppm, 50 ppm and 25 ppm respectively (Risotto, 2001). Worker exposure to these solvents needs to be minimized in order to use them safely. Many regions have placed restrictions on the handling, use and disposal of chlorinated solvents (Risotto, 2001). The EU has reclassified trichloroethylene as carcinogenic (OJ, 2003).

Chlorinated solvents are among the lowest-cost solvents and are widely available in virtually all regions (UNEP, 2003b).

Hydrofluoroethers
HFE-449s1, HFE-569sf2, HFE-449s1 and HFE-569sf2 are segregated hydrofluoroethers with the ether oxygen separating a fully fluorinated and a fully hydrogenated alkyl group. Both of these compounds are used as replacements for CFCs and HCFCs. The pure HFEs are limited in utility in cleaning applications due to their mild solvent strength. However, HFEs are also used in azeotropic blends with other solvents (such as alcohols and trans-1,2-dichloroethylene) and in co-solvent cleaning processes, giving them broader cleaning efficacy (Owens, 2001).

As a result of the low GWPs of these HFEs, the direct greenhouse-gas emissions from them are significantly reduced compared to the products they replace (GWP of 56 and 397 for HFE-569sf2 and HFE-449s1 respectively; see Table 2.6 in Chapter 2).

HFE-449s1 and HFE-569sf2 are non-flammable and have relatively high exposure guidelines (750 ppm and 200 ppm respectively), allowing them to be used safely in many cleaning applications (Owens, 2001).

These HFE compounds are available in most regions. The relatively high cost of these materials limits their use compared to lower-cost solvents such as the chlorinated solvents and hydrocarbons (UNEP, 2003b).

HFE-347pcf
This compound is a non-segregated hydrofluoroether with oxygen separating two partially fluorinated alkyl groups. The material is a new compound and has only recently become commercially available. Very little information is available regarding the performance of this material in cleaning applications.

Normal-propyl bromide
Normal-propyl bromide (nPB) is a non-flammable, brominated alkane. Its high solvent strength makes it effective in a variety of cleaning applications, including both vapour degreasing and cold cleaning (Shubkin, 2001). In some instances, n-propyl bromide may be incompatible with the components to be cleaned (Shubkin, 2001).

The direct greenhouse-gas emissions from n-propyl bromide are very low due to its very low GWP. The ODP of n-propyl bromide varies depending upon the latitude at which it is emitted. The current calculations indicate a range of ODPs from 0.013 to 0.105 (WMO, 2002).

Normal-propyl bromide has low acute toxicity but its complete toxicological profile necessitates a low exposure guideline. Current exposure guidelines set by n-propyl bromide manufacturers range from 5 to 100 ppm (US EPA, 2003b). However, recent toxicological studies continue to raise concerns regarding its chronic toxicity (both reproductive and central nervous system effects), causing some organizations to suggest that even lower exposure guidelines are required to protect workers (CDHS, 2003). The EU has decided to reclassify its toxicity as altering fertility and dangerous for the foetus (OJ, 2004).

Normal-propyl bromide is available in most regions, including Asia, with local production in China (UNEP, 2003b). World production capacity for the manufacture of nPB is increasing, although consumption is currently stable. This is resulting in a lowering of bulk prices to a level comparable to the upper range of chlorinated solvents (UNEP, 2003b).

10.3.3.3   Not-in-kind options
These are numerous, and include no-clean, aqueous, hydrocarbon and oxygenated solvents, and they correspond to the fragmented market with a large number of small users in very diverse industries. Each presents "trade-offs" between effectiveness in use, cost, safety, and environmental properties.

No-clean applications
The use of no-clean flux is the preferred option in electronics cleaning and can be used for practically all types of printed circuit boards, even for most applications in the aerospace, aeronautics and military sectors (given the crucial reliability and safety issues). It is even used in the vast majority of high-tech printed circuit boards for automotive controls (>70%). But this approach requires very tight control of incoming materials, adapted chemicals for soldering and skilled operators to master the process. Overall, industry experts report 85% of printed circuit boards are assembled with a no-clean process, in developed as well as in developing countries. Total process costs are reduced and environmental impact is very low with no direct greenhouse-gas emissions.

Hydrocarbon solvent cleaning
This process has proven a good solution with paraffin hydrocarbon formulations; cleaning is efficient but the non-volatile or less-volatile residue can be incompatible with some downstream manufacturing or finishes. Some cases of materials incompatibility are also known. Environmental impact is low (low GWP, no ODP) but these solvents are generally classified as VOCs. Their toxicity is also low. Due to their flammability (flashpoint > 55°C), they have to be used in open tank equipment at a temperature at least 15°C below their flashpoint. After cleaning, rinsing has to be done using fresh solvent. To reduce solvent consumption, on-site recycling with vacuum distillation is recommended. In some cases, safety considerations and emissions reduction require a closed machine with a vacuum system. This is not always compatible with high-throughput processes. Use in precision mechanics (injectors, bearings etc.) has developed in the last 3 years, but it involves a drastic change of process, and a medium to high investment, as well as good operator training.

Oxygenated solvent cleaning
Oxygenated organic solvents are compounds based on hydrocarbons containing appendant oxygen (alcohols and ketones), integral oxygens (ethers), or both (esters). These substances have been used for many years in diverse cleaning applications. Their cost and environmental impact are low (low GWP, no ODP), except they are classified as VOCs and may contribute to ground-level ozone pollution. Some of the oxygenated compounds are flammable as neat product or the mixture may become flammable in use. Precautions for handling flammables must therefore be considered, e.g. explosion-proof equipment.

Aqueous cleaning
These processes can be good substitutes for metal degreasing or even precision cleaning when corrosion of the materials is not an issue. For some tasks, where finish is a crucial issue, the number of required washing and rinsing baths, and the quality of the water, might be a deterrent. The availability of good-quality water and water disposal issues need to be addressed from the outset of process design. Aqueous cleaning processes have low environmental impact (not VOCs, low GWP, and no ODP) and low toxicity. The total TEWI can be equivalent to low-GWP solvent processes, as energy for drying has to be included in the total evaluation (ADL, 2002). It should be noted that a change from solvent to water cleaning does require good training of the workforce. The process is generally not forgiving; any error in operation or maintenance could result in corrosion and other quality problems. Investment costs can be high but operating costs are generally lower than those with solvent alternatives.

Table 10.8 provides an overview of the alternative fluids and not-in-kind technologies, including environmental and technical data.

Like other solvents, the hydrocarbon and oxygenated solvents lead to emissions that vary with the equipment used as well as with storing, handling and recovery procedures. The emission rate (per kg of parts cleaned for example) can decrease from 10 to 1 if containment in the equipment (upgrading or new

Table 10.8. Overview of alternative fluids and not-in-kind technologies in solvent applications.

	CH_2Cl_2[(1)]	HFE-449s1[(2)]	n-propyl bromide	No Clean	Hydro-carbon / oxygenated	Aqueous / semi-aqueous
Substance characteristics						
Radiative efficiency (W m^{-2} ppb^{-1})	0.03	0.31	0.3	n/a		n/a
Atmospheric lifetime (yr)	0.38	5	0.04	n/a		n/a
GWP (100-yr time horizon)						
- This report	10	397	n/a	n/a		n/a
- IPCC (1996)	9	not given				
ODP	-	-	-	-	-	-
Ground-level ozone impact						
- MIR[(3)] (g-O$_1$/g-substance)	0.07	n/a	n/a			
- POCP[(4)] (relative units)	7	n/a	n/a			
Ground-level ozone impact	Low to moderate	None	Low to moderate	None	Low to moderate	None
Flammability (based on flashpoint)	None	None	None	n/a	Flammable	n/a
Technical data						
Stage of development	Commercial	Commercial	Commercial	Commercial	Commercial	Commercial
Type of application:						
- Electronics cleaning		X	X	X	X	X
- Precision cleaning		X	X		X	X
- Metal cleaning	X	X	X		X	X
- Drying					X	
- Carrier solvent	X	X	X		X	

[(1)] Listed values refer to CH$_2$Cl$_2$ only. Additional chlorinated solvents are used in these applications as indicated in section 10.3.3.2.
[(2)] Listed values refer to HFE-449s1 only. Additional HFE solvents are used in these applications as indicated in section 10.3.3.2.
[(3)] Maximum Incremental Reactivity (See Chapter 2, Table 2.11 for more details.)
[(4)] Photochemical Ozone Creation Potential (See Chapter 2, Table 2.11 for more details.)

installation) is optimized and if care is exercised throughout the implementation of the process.

The flammability of oxygenated and, to a lesser extent, hydrocarbon solvents generally requires the use of higher-containment, flammable-rated equipment, which leads to reduced emissions. In the case of hydrocarbons, the high boiling point induces lower volatility at the normal use temperature and, accordingly, low emissions. In many countries, environmental and VOC regulations have further pushed the emissions reduction from all solvent use and in particular from cleaning with organic or hydrocarbons solvents.[2]

No-clean and aqueous cleaning processes do not emit any volatile compounds; however, for aqueous cleaning, total energy consumption (heating and drying) is a source of CO_2 emissions (ADL, 2002).

10.3.4 Emission reduction potential and abatement cost

Reduction potential
Four separate studies (two European, one Japanese and one from the US EPA) state a range for the potential reduction in HFC/PFC solvent emissions of 0% to 67%. See Table 10.9 for more details. The two European studies, which covered EU countries, were conducted by Harnisch and Gluckman (2001) and Harnisch and Schwartz (2003). The first study stated an estimate for solvent emission reductions of 0.3 to 0.5 MtCO$_2$-eq (15% to 25%) depending upon the level of investment (abatement cost). The estimate in the second study was a solvent emission reduction potential of 60% (0.3 MtCO$_2$-eq) by 2010 using substitution with lower-GWP solvents. For Japan, a recent study concludes there is no emission reduction potential for solvent uses below projected business as usual quantities (METI, 2004).

The US EPA's recently published emission-abatement report (2004) has estimated emission reduction potentials in 2020 in the US, non-US Annex I and non-Annex I countries. This study looked at several abatement options and concluded that the global reduction potential was 58% (2.7 MtCO$_2$-eq in 2020). Based on these abatement options, the reduction potentials are:

Retrofit	0% to 10.5%
Conversion to low-GWP solvents such as HFE	21% to 50%
Conversion to not-in-kind semi-aqueous	5%-10%
Conversion to not-in-kind aqueous	10%-20%

---

[2] For example EU Directive 1999/13/EC on industrial solvent cleaning (OJ, 1999).

Table 10.9. Summary of emission reduction potential and abatement cost for solvent applications.

Reference Region Year	Harnisch and Gluckman (2001) European Union 2010	Harnisch and Schwartz (2003) European Union 2010	US EPA (2004)		METI (2004) Japan 2010
			USA 2020	non-USA 2020	
Total emission projection ($MtCO_2$-eq)	2.0	0.5	2.1	2.6	3.6
Emission abatement ($MtCO_2$-eq)	a) 0.3 b) 0.5 c) 0.5	0.3	1.4	1.3	0
Abatement cost ($US\$_{2002}/tCO_2$-eq)	a) <23 b) 23 to 57 c) 57 to170	0	-37 to 2	-37 to 2	n/a

Abatement costs

The abatement costs associated with using the solvent alternatives vary greatly: from increased costs for adopting a new technology to cost savings due to reduced solvent consumption. Some options are cost-neutral.

Harnisch and Gluckman (2001) assumed that abatement costs could range from <20 €$_{1999}$ (23 US\$$_{2002}$) to a range of 50 to 150 €$_{1999}$ (57 to 170 US\$$_{2002}$), depending on the level of abatement achieved.

Harnisch and Schwartz (2003) estimate abatement costs for HFC/PFC solvent emission reduction in the EU to be cost-neutral based upon substitution with lower GWP solvents (0 €$_{2000}/tCO_2$-eq [0 US\$$_{2002}/tCO_2$-eq] using a ten-year depreciation period and 10% discount rate).

The US EPA (2004) has estimated the abatement costs for several alternatives using a 4% discount rate:
Equipment Retrofit
    –36 US\$$_{2000}/tCO_2$-eq   [–37 US\$$_{2002}/tCO_2$-eq]
Conversion to lower GWP solvents such as HFE
    0 US\$$_{2000}/tCO_2$-eq   [0 US\$$_{2002}/tCO_2$-eq]
Conversion to not-in-kind semi-aqueous
    0.2 US\$$_{2000}/tCO_2$-eq   [0.2 US\$$_{2002}/tCO_2$-eq]
Conversion to not-in-kind aqueous
    2 US\$$_{2000}/tCO_2$-eq   [2 US\$$_{2002}/tCO_2$-eq]

10.4   HFC-23 from HCFC-22 production

In addition to emissions of a chemical during its use and distribution, there may also be emissions of the material that is being produced, the so-called "fugitive emission" or emissions of other byproduct chemicals generated during the production of a related chemical (e.g. HFC-23 from HCFC-22). Emissions from fluorocarbon production processes have been estimated at approximately 0.5% of the total production of each compound (AFEAS, 2004). Emissions may vary because of the process used or destruction technology employed. However, because manufacturers have an incentive to minimize lost product, and emission rates are already reported to be very low, fugitive emissions are assumed to be insignificant compared to emissions from consumption. In contrast, byproduct emissions of HFC-23 from HCFC-22 are significant and well documented.

Trifluoromethane (HFC-23 or CHF_3) is generated as a byproduct during the manufacture of chlorodifluoromethane (HCFC-22 or $CHClF_2$). HCFC-22 is used as a refrigerant in several different applications, as a blend component in foam blowing and as a chemical feedstock for manufacturing synthetic polymers. In developed countries, the phase-out of HCFC-22 consumption scheduled under the Montreal Protocol will occur by 2020 because of its stratospheric ozone-depleting properties. Feedstock production, however, will be allowed to continue indefinitely because it does not involve the release of HCFC-22 to atmosphere in the same way as the dispersive uses. Consumption in developing countries will be phased out over a longer time period, using consumption (and production) in the year 2015 as a baseline.

HCFC-22 is produced in several developed and developing countries. The USA is a major world producer of HCFC-22, with three plants and approximately one-third of all developed-country production. There are ten manufacturing plants for HCFC-22 in the EU, with total capacity estimated at 184,000 metric tonnes; the reported levels of production are very close to this capacity (Irving and Branscombe, 2000; EFCTC, 2001; UNEP, 2002). Japan is another large producer of HCFC-22, with a reported production of 77,310 metric tonnes in 2003, and production in other Asian countries (notably China, India and Korea) has increased dramatically in recent years. There are now 19 HCFC-22 producers in China with a total capacity of 200,000 metric tones. They produced 177,000 metric tonnes in 2003 (CAOFSMI, 2003). Table 10.10 shows the historic and projected time series for global HCFC-22 production.

Trifluoromethane (HFC-23) is formed at the reactor stage of the manufacture of chloro-difluoromethane (HCFC-22) as a result of over-fluorination. Its formation depends upon the conditions used in the manufacturing process and varies from 1.4% to 4.0% of the production of HCFC-22 so that, while it is possible to reduce its formation by optimizing process conditions, it is not possible to eliminate the production of HFC-23.

Table 10.10. Business-as-usual and reduced emissions forecasts for HFC-23 (given current best practice).

Year	HCFC-22 Production BAU	HFC-23 emissions BAU		HFC-23 emissions Current best practice		Emission reduction potential
	ktonnes	ktonnes	MtCO$_2$-eq	ktonnes	MtCO$_2$-eq	MtCO$_2$-eq
1990	**341**	**6.4**	**92**	**6.4**	**92**	
1991	373	7.0	100	7.0	100	
1992	385	7.3	104	7.3	104	
1993	382	7.2	103	7.2	103	
1994	387	7.2	103	7.2	103	
1995	**385**	**7.3**	**104**	**7.3**	**104**	
1996	430	8.6	123	8.6	123	
1997	408	8.4	120	8.4	120	
1998	448	9.7	139	9.7	139	
1999	458	10.4	149	10.4	149	
2000	**491**	**11.5**	**165**	**11.5**	**165**	
2001	521	12.6	180	12.6	180	
2002	551	13.6	195	13.6	195	
2003	581	14.7	210	14.7	210	
2004	520	14.1	202	14.1	202	
2005	**550**	**15.2**	**218**	**13.8**	**197**	**20**
2006	580	16.2	232	13.3	190	41
2007	610	17.3	248	12.6	180	67
2008	640	18.4	263	11.8	169	94
2009	670	19.4	278	10.7	153	124
2010	**622**	**19.0**	**272**	**8.8**	**126**	**146**
2011	652	20.1	288	7.4	106	182
2012	682	21.2	303	5.9	84	219
2013	712	22.3	319	4.2	60	259
2014	742	23.3	333	2.3	33	301
2015	**707**	**23.2**	**332**	**2.3**	**33**	**299**

In a number of plants, HFC-23 is destroyed by thermal oxidation, and emission tests have shown this to be highly effective (Irving and Branscombe, 2000). The cost of reducing emissions varies significantly depending on the option used and will vary from plant to plant depending on the particular situation.

Business-as-usual emissions prior to 2015

Global emissions of HFC-23 increased by an estimated 12% between 1990 and 1995, due to a similar increase in global production of HCFC-22 (Oram *et al.*, 1998). However, the widespread implementation of process optimization and thermal destruction has brought an end to this trend in HFC-23 emissions and, since 1995, the emission trend has fallen below the increase in production (US EPA, 2003a).

While such abatement measures serve to reduce or eliminate emissions, the quantity of HFC-23 actually produced is directly related to the production of HCFC-22 and so emission projections require a scenario for future HCFC-22 production volumes. These will depend on the consumption of HCFC-22 in developed countries, which is declining, and the consumption in developing countries and global demand for fluoropolymers feedstock (UNEP-TEAP, 2003), both of which are increasing.

The introduction in developed countries of additional controls on HCFC production after 2005 (over and above the required freeze) may have a neutral effect on production globally because it might stimulate investment plans for further HCFC-22 capacity in developing countries (UNEP-TEAP, 2003).

Consumption in developed countries is expected to decrease as a consequence of increasingly stringent national and regional regulations. However, a scenario that followed the requirements of the Montreal Protocol exactly in these countries was adopted as the best estimate (business-as-usual case) in the most recent Scientific Assessment of Ozone Depletion (Fraser, Montzka *et al.*, 2003). This envisages that, by 2015, consumption of HCFC-22 in these countries will have fallen by a factor of 10 from the level in 2000-2003.

In the same region, over the 1990s, growth in demand for fluoropolymer feedstock increased linearly at a rate of 5800 tonnes yr^{-1}, with a coefficient of variance (R^2) of 0.84 (Cefic, 2003; SRI, 1998). In the absence of other influences, this rate of increase could continue, leading to a doubling of feedstock demand for HCFC-22 in developed countries by 2015.

In the remaining countries of the world, production of HCFC-22 for both dispersive and feedstock uses has grown rapidly in recent years. Over the period 1997 to 2001, production for dispersive uses grew linearly at 20,000 tonnes yr^{-1}, with a coefficient of variance (R^2) of 0.95, and feedstock use increased by 4100 tonnes yr^{-1} (R^2 equal to 0.99) (Bingfeng *et al.*, 2000; CCR, 2002; UNEP, 2002; UNEP-TEAP, 2003). Projected at these rates until 2015, the total global requirement for HCFC-22 could become about 710,000 tonnes yr^{-1}, about 40% of which would be for feedstock, compared with a total of 490,000 tonnes yr^{-1} in the year 2000.

Up to 1995, records of atmospheric concentrations of HFC-23 (Oram *et al.*, 1998) and global HCFC-22 production (McCulloch *et al.*, 2003) indicate an emission intensity for HFC-23 of 20 kg per tonne of HCFC-22 production, or 2%. This suggests that there was substantial abatement of HFC-23 emissions at that time, either by process optimization or by destruction; unabated emissions would be expected to be of the order of 4% (Irving and Branscome, 2000). In the "business-as-usual" case for the period prior to 2015, it has been assumed that emissions from existing capacity will continue at 2% of HCFC-22 production and that new capacity will emit HFC-23 at a rate of 4%. Consequently, emissions of HFC-23 could increase by 60% between now and 2015, from about 15,000 tonnes yr^{-1} in 2003 to 23,000 tonnes yr^{-1}.

10.4.1   Relevant practices during life cycle

In the most commonly used process, chlorodifluoromethane (HCFC-22) is produced by the reaction of chloroform ($CHCl_3$) and anhydrous hydrogen fluoride (HF) in the presence of an antimony pentachloride ($SbCl_5$) catalyst. The reaction takes place in a continuous-flow reactor full of boiling liquid, products being removed from the reaction system as vapours. Two molecules of HF react with one molecule of chloroform to yield chlorodifluoromethane, and reaction conditions are maintained to optimize HCFC-22 production but overfluorination (reaction of HCFC-22 with HF) can occur, yielding HFC-23 (CHF_3). Consequently, the vapour stream leaving the reactor contains, in addition to HCFC-22 ($CHClF_2$), HCFC-21 ($CHCl_2F$), $CHCl_3$, HF and some entrained catalyst (which are recycled to the reactor) and also HFC-23 (CHF_3) and HCl. Separation of the hydrocarbon compounds is facilitated by the differences in volatility, which makes it possible to condense the chloroform and HCFC-21 for recycling and conversion to HCFC-22. It is technically feasible but not economical to convert HFC-23, once formed, back into HCFC-22 (Merchant, 2001). Most of the HFC-23 produced is released from the reaction system at the control valve used to maintain the system pressure (the "condenser vent") and, unless separated for collection, is then emitted to the atmosphere.

This production process is relatively old (over 30 years) and has been extensively researched. However, the optimum operating conditions dictated by business economics are not necessarily the conditions required to minimize HFC-23 production. The upper limit for HFC-23 emissions is of the order of 3 to 4 percent of HCFC-22 production but the actual quantity of HFC-23 produced depends in part on how the process is operated and the degree of process optimization. There are a number of factors that affect halogen exchange of chlorine to fluorine and thus affect the generation of HFC-23 in the reactor, and a significant reduction in HFC-23 formation can be achieved by adjusting process operating conditions, including modifying process equipment.

One variation on the liquid phase process is the so-called "swing" plant, designed to be capable of manufacturing either CFC-12 or -11 from carbon tetrachloride feedstock or HCFC-22 from chloroform. Reactor designs for the two products are basically very similar, with differences only in operating pressure, operating temperature and heat load. However, the chemical engineering design of the whole plant is, of necessity, a compromise between the ideals for the two products and, consequently, optimization of the process to minimize HFC-23 production when producing HCFC-22 is intrinsically more difficult than in a plant designed for HCFC-22 specifically. It is to be expected that swing plants will operate at the higher end of the HFC-23 production range (say 3 to 4% of HCFC-22 by mass).

The major exit point for HFC-23 from the reaction system in the HCFC-22 production process is the condenser vent, whence it can be discharged to the atmosphere after separation from useful products (HCFC-22 and HCl). There are four additional sources of HFC-23 emissions, three of which are inconsequential:

- Fugitive emissions from leaking compressors, valves and flanges. Because the process is maintained under relatively high pressures, there is a potential for fugitive emissions from leaking equipment. However, there is a strong incentive to prevent leaks because of the noxious nature of the reactor contents and the potential for loss of valuable product (HCFC-22). Any leaks can usually be identified and repaired quickly and effectively. Fugitive emissions are a minor source; in two plants where they were measured, the HCFC-22 lost in this way amounted to less than 0.1% of plant production (Irving and Branscombe, 2000).

- Removal with the HCFC-22 product and subsequent emission. Industry standard specifications limit the concentration of HFC-23 allowed in the product to a small fraction of a percent. Consequently, removal with the product also accounts for only a very small portion of the HFC-23 that is generated.

- Vents from product storage. These are not significant sources, even when plants capture HFC-23 (by condensation at high pressures and low temperatures) and subsequently store it as pure material. The storage tanks are enclosed vessels, not normally venting to atmosphere, and only a very small fraction of the plants producing HCFC-22 recover HFC-23. This is clear from the small quantity of HFC-23 that is sold.

• The most important of the additional loss routes is in the aqueous phases from caustic and water scrubbers used in the process. These may yield useful product, such as hydrochloric acid (aqueous HCl), or be destined for waste treatment, but in any case volatile organics in such aqueous streams are readily emitted when the streams are subsequently managed in open systems, such as wastewater treatment processes. HFC-23 concentrations in the aqueous effluent streams are only a few ppm at the most but the concentration in aqueous HCl can be several hundred ppm and this can represent an important release vector if the acid is used untreated. However, compared to the condenser vent, the aqueous HCl source is minor. Furthermore, the HFC-23 can be removed from the aqueous HCl by air-stripping for subsequent treatment in the thermal oxidizer.

10.4.2   Emission reduction options

Process optimization
In fully optimized processes, the likely range of emissions is about 1.5 to 3 percent of production, with 2 percent being a reasonable average estimate. Actual emissions depend on the age and design of the facility as well as the process management techniques applied. However, it is not possible to eliminate HFC-23 production completely by this means.

Capture and destruction
Further reductions in emissions beyond what is technically achievable through process optimization require additional equipment and, given that there is one main outlet for HFC-23 from the process, the most favourable option for eliminating emissions is to collect and treat the vent gases. Thermal oxidation is an effective treatment but the system must be designed to cope with, and render harmless, the halogen acids (HF and HCl) produced as a result of combustion. Depending on cost and availability, treatment can be on- or off-site. Emission tests at one plant showed that thermal oxidation destroys over 99.996 percent of the HFC-23 (Irving and Branscombe, 2000), making this a highly effective treatment option. However, the impact of the down-time of thermal oxidation units on the emissions of HFC-23 needs to be considered (EFCTC, 2001).

In the EU, thermal destruction was available from mid-2000 onwards, either on- or off-site, for six of the ten HCFC-22 manufacturing facilities. These facilities make up approximately 80% of EU HCFC-22 production but manufacturing plants in Spain (2), UK (1) and Greece (1) remain without such facilities. The cost of a typical unit is 3 million (about US$4 million) to destroy 200 metric tonnes of HFC-23 per year, plus 200,000 yr^{-1} (US$250,000) in operating costs (EFCTC, 2001; Harnisch and Hendriks, 2000). Assuming a technological lifetime of 15 years, it is possible to calculate that specific abatement costs will be <0.2 US$/t$CO_2$-eq.

In 1995, European emissions amounted to 3,150 metric tonnes. The potential emission in the year 2000 could have been up to 7,340 tonnes per annum, depending upon the rate of for-

mation in the individual manufacturing plants, but these were reduced to actual emissions of less than 2,025 metric tonnes (EFCTC, 2001).

The US EPA estimates that US emissions of HFC-23 in 2002 amounted to 1,690 metric tonnes of gas. Annual emissions have fluctuated since 1990, before dropping by 45 percent/tonne of HCFC-22 produced in 2002 (US EPA, 2004b).

In Japan, the production of HCFC-22 was reported at 72,787 metric tonnes for 2002, with the byproduction of 1,124 tonnes of HFC-23, of which 253 tonnes were destroyed. In 2003, 77,310 tonnes of HCFC-22 were produced, with the byproduction of 1,277 tonnes of HFC-23, 367 tonnes of which were destroyed (METI, 2004).

10.4.3   Reduced emissions in the period prior to 2015 given current best practice

It is apparent from historical trends, as described in *Business-as-usual emissions prior to 2015* above, that the amount of HFC-23 co-produced with HCFC-22 in the developed world will be reduced by approximately half. There is at present no published information about current HFC-23 global emissions and, for the reduced emissions case, it is assumed that destruction technology will be progressively introduced for all new and non-abated existing capacity from 2005 onwards. Destruction technology is assumed to be 100% efficient and to operate for 90% of the on-line time of the HCFC-22 plant.

Reduced emissions were calculated for the same activity (in the form of assumed future HCFC-22 production) as in the business-as-usual case. The difference between the two HFC-23 forecasts is therefore solely due to the extent of the deployment of destruction technology. The forecasts represent potential extreme cases and future changes in activity will tend to increase the probability of one or the other. Table 10.10 lists values for the forecasts and the HCFC-22 production on which they are based.

Implications of HCFC-22 production phase-out
In view of the uncertainties of scenario generation, the business-as-usual scenario uses consumption forecasts based on adherence to the Montreal Protocol, with no further controls. This is a fixed scenario (for the developed world) and amounts to a 90% reduction in the production and consumption of HCFCs for non-feedstock uses there by 2015, although many countries already have, or are planning, more stringent regulations. In 2010 in the US, the production and imports of HCFC-22 will be banned under the US Clean Air Act (US EPA, 1993), except for use in equipment manufactured before 2010. Within the EU, it is estimated that levels of HCFC-22 production for non-feedstock uses will decrease by 30% by the year 2010 under EU Regulation 2037/2000 on Substances that Deplete the Ozone Layer (EFCTC, 2001). Such legislation is well documented and the current effect on production of HCFC-22 is recorded in the data reported by AFEAS (2003), which shows a fall of 54,000 tonnes yr^{-1} by the year 2001 from the peak production

of 271,000 tonnes yr^{-1} that occurred in 1996. Potential dispersive consumption in the developed world is therefore already substantially below that in the business-as-usual scenario and will fall faster, skewing the probability of HFC-23 emissions towards the reduced emissions case.

Under the Montreal Protocol, the baseline year for developing countries is 2015, so that the regulatory maximum has yet to be established and production must be extrapolated from prior trends. In China, the production of HCFC-22 for dispersive use grew over the four years from 1998 to 2001 at a linear rate of 18,100 tonnes yr^{-1} ($R^2 = 0.95$), or 26% in the year 2000 (CCR, 2002). When the other producers from developing countries and countries with economies in transition are included in the calculation, the growth rate reaches 20,000 tonnes yr^{-1} and this rate has been assumed for both business-as-usual and reduced emissions cases. Any changes made to the assumed production rates will have a direct effect on the forecast HFC-23 emissions.

Use of HCFC-22 as feedstock
Historically, the demand for HCFC-22 in fluoropolymers has been growing linearly in the developed world at a rate that approximated to 3% in the year 2000, or absolute growth rates of 2200 tonnes yr^{-1} in the US, 1600 tonnes yr^{-1} in Japan and 2000 tonnes yr^{-1} in the EU (Cefic, 2003; SRI, 1998). In 2001 in China, the demand for HCFC-22 in the production of fluoropolymers was 20,300 metric tonnes and the linear growth rate was 4100 tonnes yr^{-1}, with an R^2 value of 0.99 over four years (CCR,

2002). There is every reason to expect that this demand will continue to grow and there is no evidence for a future change in the growth rate. Hence, for both cases, demand for HCFC-22 in fluoropolymers was extrapolated using linear growth at these rates. Any changes made to the assumed production rates will have a direct effect on the forecast HFC-23 emissions.

Alternative processes for HCFC-22 production
Although it is technically feasible to manufacture HCFC-22 in a reaction involving chloroform and HF in the vapour phase over a solid catalyst, most, if not all, modern processes use the liquid-phase route described above (Hoechst, 1962). Furthermore, it is intrinsically more difficult to control a vapour-phase process for minimizing the formation of HFC-23 than to control a liquid-phase process. In the latter, conditions can be adjusted so that the boiling liquid in the reactor contains little HCFC-22, therefore minimizing the possibilities of fluorinating it to produce HFC-23. In the vapour phase, on the other hand, all HCFC-22 that is produced co-exists towards the exit of the catalyst with the excess of HF. In a vapour-phase process it is not impossible to adjust conditions to minimize HFC-23; it is only more difficult to do so than in the liquid phase.

Emission reduction potential
Table 10.10 shows the emissions under the mitigation scenario of this report, assuming implementation of current best practices. Destruction of byproduct emissions of HFC-23 from HCFC-22 production has a reduction potential of up to 300

Table 10.11. Niche applications of ozone-depleting substances with greenhouse-gas replacements.

Application	Comment
CFC-11 as a flushing agent for refrigerant and air-conditioning systems	Refrigeration and air-conditioning systems suffer occasional failures that result in contamination. The most common failure is a compressor burnout. Flushing systems with CFC-11 were effective, fast and safe. Alternatives include HCFC-141b, HCFs, terpenes, glycols and hydrocarbons. Flammability and system compatibility must be addressed.
CFC-11, CFC-113 for non-mechanical heat transfer	Cooling systems that rely on convection to remove heat from an area, rather than relying on mechanical refrigeration. A broad range of alternatives including not-in-kinds, HFCs and HFEs have been identified for retrofits and new systems. However, for applications where no other alternatives are technically feasible, PFCs are used. Most are used in closed systems with minimal emissions. Other applications have evaporative losses. For example, the electronics industry employs PFCs for cooling certain process equipment, during testing of packaged semiconductor devices and during vapour-phase reflow soldering of electronic components to circuit boards.
CFC-11 as an extraction solvent for perfumes, flavours, decaffeinated coffee	Used in a closed system. HFC-134a and not-in-kind replacements.
CFC-11, CFC-113 for cold cleaning and degreasing, hand cleaning	CFC used in buckets, as a wipe or spray to degrease parts, wash down equipment, clean oil spills, etc. No technical need for CFCs, HFCs or PFCs, but CFC use may continue if permitted by local regulation and if it is low cost and readily available.
CFC-114 for uranium isotope separation	This process requires operation of a heat transfer cycle to cool uranium isotope separation. Substitutes must meet an extremely rigorous set of criteria to be applicable to this end use. PFCs are used as a substitute.

$MtCO_2$-eq per year by 2015 and specific costs below 0.2 US$/ tCO_2-eq according to two European studies in 2000. Reduction of HCFC-22 production due to market forces or national policies, or improvements in facility design and construction also could reduce HFC-23 emissions.

10.5   Other miscellaneous uses

A few additional niche applications of ODSs with possible HFC, PFC or other greenhouse-gas replacements were identified. ODS use at the time of the Montreal Protocol was very low and there is insufficient literature available to evaluate the status of replacements, current or future emissions. It may also be the case that ODS use continues in some of these applications, especially where the source of the ODS is a common use such as a refrigerant and is redirected for convenience to an application such as cold cleaning or flushing. The consumption of ODSs or replacements in these applications is generally assumed to be insignificant but documentation does not exist. Although not comprehensive, Table 10.11 lists applications identified through expert judgment.

At the time of the Montreal Protocol, CFCs, PFCs and SF_6 were used simultaneously in a few applications, including wind tunnel tests, military dielectric applications, adiabatic applications, leak detection and tracer gases. Although there is little published information, consumption and emissions are not currently significant and not expected to grow significantly in the future.

Finally, other significant global sources of emissions of PFCs and SF_6 were considered and found to be from processes that never used ODSs or for new applications that would not have used ODSs. This includes PFCs from primary aluminium production, and semiconductor manufacture, and SF_6 used in high-voltage electrical equipment.

References

Agopovich, J, 2001: Review of Solvents for Precision Cleaning. Chapter 1.4 in *Handbook of Critical Cleaning*, B. Kanegsberg and E. Kanegsberg (eds.), CRC Press, Boca Raton, USA, p. 59-73.

ADL (A.D. Little, Inc.), 2002: Global Comparative Analysis of HFC and Alternative Technologies for Refrigeration, Air Conditioning, Foam, Solvent, Aerosol Propellant, and Fire Protection Applications. Final Report to the Alliance for Responsible Atmospheric Policy, March 21, 2002 (available online at www.arap.org/adlittle/toc.html), Acorn Park, Cambridge, MA, USA, 150 pp.

AFEAS (Alternative Fluorocarbons Environmental Acceptability Study), several years: Alternative Fluorocarbons Environmental Acceptability Study - Production, Sales and Atmospheric Releases of Fluorocarbons, AFEAS Arlington, VA, USA (available online at www.afeas.org).

Bingfeng Y., W. Yezheng and W. Zhigang, 2000: Phase-out and replacement of CFCs in China. *Bulletin of the International Institute of Refrigeration*, 1/2000, pp. 3-11.

CCR (China Chemical Reporter), 2002: Market Report: Fluorochemical develops rapidly in China. *China Chemical Reporter,* **13**, September 6, 2002.

CDHS (California Department of Health Services), 2003: Health Hazard Alert : 1-bromopropane. Hazard Evaluation System & Information Service, California, USA.

CEC (Commission of the European Communities), 2003: Proposal for a European Parliament and Council Regulation on Certain Fluorinated Greenhouse Gases. Commission of the European Communities, COM(2003)492final, 2003/0189 (COD), August 11, 2003, Brussels, Belgium.

Cefic (European Council of Chemical Industry Federations), 2003: Data submitted to Commission of the European Communities under Community Regulation EC 2037/2000, Cefic, Brussels, Belgium.

CAOFSMI (China Association of Organic Fluorine and Silicone Material Industry), 2003: Survey Report on HCFC-22 production in China, http://www.sif.org.cn (in Chinese).

DuPont (J. Creazzo) and Falcon, 2004: Dustometer Study. DuPont Fluorochemicals and Falcon Safety Products, Wilmington, DE, USA.

Durkee, J.B., 1997: Chlorinated Solvents NESHAP – Results to Date, Recommendations and Conclusions. Proceedings of the International Conference on Ozone Protection Technologies, November 12-13, 1997, Baltimore, MA, USA.

EFCTC (European Fluorocarbon Technical Committee), 2001: HFC-23 from HCFC-22 Manufacturing. In *Annex I to Final Report on the European Climate Change Programme, Working Group Industry: Work Item Fluorinated Gases*, February 2001, Brussels, Belgium.

FEA (Fédération Européenne des Aérosols or European Aerosol Federation), 2003: FEA Statiustics report – 2003. European Aerosol Federation, Brussels, Belgium.

Germany, 1990: Zweite Verordnung zur Durchführung des Bundes-Immissionsschutzgesetzes, Verordnung zur Emissionsbegrenzung von leichtflüchtiger Halogenkohlenwasserstoffen (Second Ordinance on Emission Abatement of Volatile Halogenated Organic Compounds). *BGBl. I S.* 2694 (December 10, 1990), adapted *BGBl. I S.* 2180 (August 21, 2001).

Harnisch, J. and R. Gluckman, 2001: Final Report on the European Climate Change Programme Working Group Industry, Work Item Fluorinated Gases. Report prepared for the European Commission (DG Environment and DG Enterprise), Ecofys and Enviros, Cologne, Germany, 58 pp.

Harnisch, J. and C. Hendriks, 2000: Economic Evaluation of Emission Reductions of HFCs, PFCs and SF_6 in Europe. Report prepared for the European Commission DG Environment, Ecofys, Cologne/Utrecht, Germany/Netherlands, 70 pp.

Harnisch, J. and W. Schwarz, 2003: Costs and the Impact on Emissions of Potential Regulatory Framework for Reducing Emissions of Hydrofluorocarbons, Perfluorocarbons and Sulphur Hexafluoride. Report prepared for the European Commission, Ecofys/Öko-Recherche, Nürnberg, Germany, 57 pp (available online at http://europa.eu.int/comm/environment/climat/pdf/ecofys_oekorecherchestudy.pdf).

Hoechst AG, 1962: Patent Japan-Kokai 62-186945, 1962.

Honeywell, 2004: Honeywell Corporation Data Sheet for HFC-245fa (1,1,1,3,3-pentafluoropropane, $CHF_2CH_2CF_3$, CAS 460-73-1. Honeywell Corporation (available online at www.honeywell.com).

IPCC/TEAP, 1999: Meeting Report of the Joint IPCC/TEAP Expert Meeting on Options for the Limitation of Emissions of HFCs and PFCs, [L. Kuijpers and R. Ybema (eds.)], 26-28 May 1999, Energy Research Foundation (ECN), Petten, The Netherlands, 51 pp (available online at www.ipcc-wg3.org/docs/IPCC-TEAP99), p. 22.

IPCC (Intergovernmental Panel on Climate Change), 1996: *Climate Change 1995: The Science of Climate Change. Contribution of Working Group I to the Second Assessment Report of the Intergovernmental Panel on Climate Change* [Houghton, J.T., L.G. Meira Filho, B.A. Callander, N. Harris, A. Kattenberg and K. Maskell (eds.)]. Cambridge University Press, Cambridge, United Kingdom, and New York, NY, USA, 572 pp.

IPCC (Intergovernmental Panel on Climate Change), 2000: *Good Practice Guidance and Uncertainty Management in National Greenhouse Gas Inventories* [Penman, J., M. Gytarsky, T. Hiraishi, T. Krug, D. Kruger, R. Pipatti, L. Buendia, K. Miwa, T. Ngara, K. Tanabe and F. Wagner (eds.)]. Published by the Institute for Global Environmental Strategies (IGES) for the IPCC, Hayama, Kanagawa, Japan.

IPCC (Intergovernmental Panel on Climate Change), 2001a: *Climate Change 2001: The Scientific Basis. Contribution of Working Group I to the Third Assessment Report of the Intergovernmental Panel on Climate Change* [Houghton, J. T., Y. Ding, D. J. Griggs, M. Noguer, P. J. van der Linden, X. Dai, K. Maskell, and C. A. Johnson (eds.)]. Cambridge University Press, Cambridge, United Kingdom, and New York, NY, USA, 944 pp.

IPCC (Intergovernmental Panel on Climate Change), 2001b: *Climate Change 2001 – Mitigation. Contribution of Working Group III to the Third Assessment Report of the Intergovernmental Panel on Climate Change* [Metz, B., O. Davidson, R. Swart and J. Pan (eds.)] Cambridge University Press, Cambridge, United Kingdom, and New York, NY, USA, 752 pp.

Irving, W.N. and M. Branscombe, 2002: HFC-23 Emissions from HCFC-22 Production. In *Background Papers – IPCC Expert Meetings on Good Practice Guidance and Uncertainty Management in National Greenhouse Gas Inventories*, IPCC/OECD/IEA Programme on National Greenhouse Gas Inventories, published by the Institute for Global Environmental Strategies (IGES), Hayama, Kanagawa, Japan, pp. 271-283 (available online at www.ipcc-nggip.iges.or.jp/public/gp/gpg-bgp.htm).

March (March Consulting Group), 1998: Opportunities to Minimise Emissions of Hydrofluorocarbons (HFCs) from the European Union. March Consulting Group, Manchester, UK, 121 pp.

McCulloch, A., P.M. Midgley and P. Ashford, 2003: Releases of refrigerant gases (CFC-12, HCFC-22 and HFC-134a) to the atmosphere. *Atmos. Environ.,* **37**(7), 889-902.

Merchant, A, 2001: Hydrocarbons. Chapter 1.6 in *Handbook of Critical Cleaning*, B. Kanegsberg and E. Kanegsberg (eds.), CRC Press, Boca Raton, USA, p. 89-110.

METI (Japanese Ministry of Economy, Trade and Industry), 2004: Report on Japanese Voluntary Action Program in Industry. Japanese Ministry of Economy, Trade and Industry, Tokyo, Japan, p. 10, 15, 19 (available online at http://www.meti.go.jp/committee/downloadfiles/g41108b31j.pdf) (in Japanese).

Montzka, S.A., P.J. Fraser, J.H. Butler, P.S. Connell, D.M. Cunnold, J.S. Daniel, R.G. Derwent, S. Lal, A. McCulloch, D.E. Oram, C.E. Reeves, E. Sanhueza, L.P. Steele, G.J.M. Velders, R.F. Weiss and R.J. Zander, 2003: Controlled Substances and Other Source Gases. Chapter 1 in *Scientific Assessment of Ozone Depletion: 2002*. Global Ozone Research and Monitoring Project – Report No. 47, World Meteorological Organization, Geneva.

OJ (Official Journal of the European Community), 1999: Council Directive 1999/13/EC of 11 March 1999 on the limitation of emissions of volatile organic compounds due to the use of organic solvents in certain activities and installations. *Official Journal of the European Communities*, L 085 pp. 0001-0022, 29 March 1999.

OJ (Official Journal of the European Community), 2004: Directive 2003/36/EC of the European Parliament and of the Council of 26 May 2003 amending, for the 25th time, Council Directive 76/769/EEC on the approximation of the laws, regulations and administrative provisions of the Member States relating to restrictions on the marketing and use of certain dangerous substances and preparations (substances classified as carcinogens, mutagens or substances toxic to reproduction – c/m/r). *Official Journal of the European Communities*, L 156, pp. 0026-0030, 25 June 2003.

OJ (Official Journal of the European Community), 2003: Commission Directive 2004/73/EC of 29 April 2004 adapting to the technical progress for the twenty-ninth time Council Directive 67/548/EEC on the approximation of the laws, regulations and administrative provisions relating to the classification, packaging and labelling of dangerous substances. *Official Journal of the European Communities*, L 152, pp. 0001-0311, 30 April 2004.

Oram, D.E., W.T. Sturges, S.A. Penkett, A. McCulloch and P.J. Fraser, 1998: Growth of fluoroform (CHF_3, HFC-23) in the background atmosphere. *Geophys. Res. Lett.*, **25**(1), 35-38.

Owens, J, 2001: Hydrofluoroethers. Chapter 1.5 in *Handbook of Critical Cleaning*, B. Kanegsberg and E. Kanegsberg (eds.), CRC Press, Boca Raton, USA, p. 75-88.

Ramsey, R.B. and A.N. Merchant, 1995: Considerations for the Selection of Equipment for Employment with HFC-43-10. Proceedings of the International CFC and Halon Alternatives Conference, Washington, DC, USA, October 1995.

Risotto, S., 2001: Vapor Degreasing with Traditional Chlorinated Solvents. Chapter 1.8 in *Handbook of Critical Cleaning*, B. Kanegsberg and E. Kanegsberg (eds.), CRC Press, Boca Raton, USA, p. 133-145.

Solvay, 2005, Solvay Fluor Data Sheet for HFC-365mfc (Solkane® 365mfc). Solvay Fluor (available online at http://www.solvay-fluor.com).

SRI (Stanford Research Institute, International), 1998: Fluorocarbons. Sections 543.7001 to 543.7005 in *Chemical Economics Handbook*, SRI International, Menlo Park, USA.

Shubkin, R., 2001: Normal-Propyl bromide. Chapter 1.7 in *Handbook of Critical Cleaning*, B. Kanegsberg and E. Kanegsberg (eds.), CRC Press, Boca Raton, USA, p. 111-131.

UNEP (United Nations Environment Programme), 1998: 1998 Report of the Solvents, Coatings and Adhesives Technical Options Committee (STOC) – 1998 Assessment (STOC 1998). [M. Malik (ed.)]. UNEP Ozone Secretariat (Secretariat to the Vienna Convention for the Protection of the Ozone Layer and the Montreal Protocol on Substances that Deplete the Ozone Layer), Nairobi, Kenya, 242 pp.

UNEP (United Nations Environment Programme), 2002: Production and Consumption of Ozone Depleting Substances under the Montreal Protocol, 1986-2000. UNEP Ozone Secretariat, Nairobi, Kenya.

UNEP (United Nations Environment Programme), 2003a: 2002 Report of the Aerosols, Sterilants, Miscellaneous Uses and Carbon Tetrachloride Technical Options Committee – 2002 Assessment (ATOC 2002). [H. Tope (ed.)]. UNEP Ozone Secretariat, Nairobi, Kenya, 100 pp.

UNEP (United Nations Environment Programme), 2003b: 2002 Report of the Solvents, Coatings and Adhesives Technical Options Committee – 2002 Assessment (STOC 2002). [B. Ellis (ed.)]. UNEP Ozone Secretariat, Nairobi, Kenya, 111 pp.

UNEP-TEAP (Technology and Economic Assessment Panel), 2003: May 2003 Report of the Technology and Economic Assessment Panel, HCFC Task Force Report. [L. Kuijpers (ed.)]. UNEP Ozone Secretariat, Nairobi, Kenya, 96 pp.

US EPA (US Environmental Protection Agency), 1993: ODS Phase-out Schedule Final Regulation. US Federal Register, Vol. 58, No. 236, p. 65018, December 10, 1993.

US EPA (US Environmental Protection Agency), 1998: Air Quality: Revision to Definition of Volatile Organic Compounds – Exclusion of Methyl Acetate, US Federal Register, Vol. 63, No. 68, p. 17331, April 9, 1998 .

US EPA, (US Environmental Protection Agency), 2003a: Inventory of U.S. Greenhouse Gases and Sinks: 1990-2001. Report EPA 430R03004, US Environmental Protection Agency, Washington, DC, USA.

US EPA (US Environmental Protection Agency), 2003b: Protection of Stratospheric Ozone: Listing of Substitutes for Ozone-Depleting Substances – n-Propyl Bromide. US Federal Register, Vol. 68, No. 106, pp. 33283-33316, June 3, 2003.

US EPA, (US Environmental Protection Agency) 2004a: Analysis of International Costs to Abate HFC and PFC Emissions from Solvents. US Environmental Protection Agency, Washington, DC, USA.

US EPA (United States Environmental Protection Agency), 2004b. 2004 Global Emissions Report for Non-CO_2 Gases. [E. Scheehle (ed.)], US Environmental Protection Agency, Washington, DC, USA.

WMO (World Meteorological Organization), 2002: Scientific Assessment of Ozone Depletion: 2002, Global Ozone Research and Monitoring Project – Report No. 47. World Meteorological Organization, Geneva.

Zeon Corporation, 2004: Data Sheet for Zeorora™ H (HFC-c447ef ($C_5H_3F_7$)). Zeon Corporation, Japan (online available at www.zeon.co.jp).

11

HFCs and PFCs: Current and Future Supply, Demand and Emissions, plus Emissions of CFCs, HCFCs and Halons

Coordinating Lead Authors
Nick Campbell (United Kingdom), Rajendra Shende (India)

Lead Authors
Michael Bennett (Australia), Olga Blinova (Russia), Richard Derwent (United Kingdom), Archie McCulloch (United Kingdom), Masaaki Yamabe (Japan), James Shevlin (Australia), Tim Vink (The Netherlands)

Contributing Authors
Paul Ashford (United Kingdom), Pauline Midgley (United Kingdom), Mack McFarland (USA)

Review Editors
Garry Hayman (United Kingdom), Lambert Kuijpers (The Netherlands)

Contents

EXECUTIVE SUMMARY

- Current production capacity for HFCs and PFCs exceeds current demand. There are a number of plants in developed countries and one plant in a developing country. The estimated production capacity for HFC-134a in 2002 was 185,000 tonnes. No published data are available to forecast future production capacity. However, as there are no technical or legal limits to production, it can be assumed that production will continue at a rate necessary to satisfy demand over time, although short-term fluctuations are possible. Future production has therefore been estimated by aggregating sectoral demand.

- Major drivers of demand for HFCs and PFCs include the Montreal Protocol and the expected worldwide increase in disposable income. HCFCs, HFCs and PFCs, together with not-in-kind alternatives, have contributed to the phase-out of CFCs and halons. The Multilateral Fund of the Montreal Protocol has been instrumental in introducing CFC replacements in the markets of the developing countries. Nearly US$ 130 million has been disbursed to the developing countries for the conversion to HFCs, with another US$ 250 million being disbursed for the conversion to HCFCs.

- Emissions of CFCs have fallen significantly over the period 1990–2000. HCFC emissions have grown significantly during this period. HFC emissions are also increasing. Demand for HCFCs, and hence emissions, are expected to continue to rise significantly between 2000 and 2015, especially in developing countries. The fall in CFC emissions will not necessarily be accompanied by a directly proportional increase in emissions of HCFCs and HFCs because of the switch to alternatives, improvements in design, containment, and improved recovery and disposal practices.

- Despite the fall in the production of CFCs, the existing bank of CFCs is over 1.1 million tonnes and is therefore a significant source of potential future emissions. Banks of HCFCs and HFCs are being established as use increases. The management of CFC and HCFC banks is not controlled by the Montreal Protocol or taken into account under the United Nations Framework Convention on Climate Change (UNFCCC).

- Byproduct emissions of HFC-23 are expected to rise globally by 60% during this period – especially in developing countries – but these emissions can be significantly reduced with best-practice capture and the destruction of vent gases.

- The estimates of aggregate emissions for the business-as-usual scenario differ from other emission estimates (the IPCC (2000) SRES scenarios, for example), largely as a result of the more detailed analyses of the current applications that have been carried out for the purposes of this report. The SRES scenarios relied mainly on the application of regional economic growth factors to the baseline consumption of individual compounds; furthermore, little distinction was made between annual consumption and annual emission.

- Total direct emissions amount to about 2.5 $GtCO_2$-eq yr^{-1} (2.0 $GtCO_2$-eq yr^{-1} using SAR/TAR values), which is similar to the estimate derived in Chapter 2 from atmospheric measurements. However, as shown in Table 11.6 and in Figure 2.4, there are significant differences between calculated and observed emissions of individual substances. For CFC-11, HCFC-141b and HCFC-142b in particular, the observations indicate significantly higher emissions but the resolution of these systematic errors, while important and urgent, is outside the scope of this report. The errors affect the results quantitatively, but qualitative conclusions are sound.

- There are opportunities to reduce direct emissions significantly through the global application of best practices and recovery methods, with a reduction potential of about 1.2 $GtCO_2$-eq yr^{-1} of direct emissions by 2015, as compared to the business-as-usual scenario. The potential emission reductions involve a broad range of costs: from net zero to US$300 per tonne of CO_2 equivalent. These estimates are based on a mitigation scenario which assumes the global application of best practices for the use, recovery and destruction of these substances. About 60% of this potential is HFC emission reduction; HCFCs and CFCs contribute about 30% and 10% respectively.

- Recovery, recycling and reclamation, and the destruction of refrigerant not suitable for reprocessing, may reduce emissions by as much as 20%. Licensing and compliance programmes may have a significant impact, with average leakage rates being achievable of less than 5% for stationary equipment and 12% for mobile systems. The costs for units used for recovery and recovery/recycling start at approximately US$500. Recovery, recycling and reuse costs may be offset by the savings in new refrigerant purchases. The destruction of contaminated and unwanted refrigerant can be achieved for less than US$3.

- In some sectors, action to reduce indirect emissions (for example, through the improved energy efficiency of appliances) will have a significantly higher impact than focusing exclusively on the chemicals used.

11.1   Introduction

11.1.1   Objectives

This chapter has three major objectives. The first is to produce global estimates for current and future demand by aggregating "bottom-up" sectoral information, and to compare this with information about supply from known chemical production facilities. The second is to compare the associated "bottom-up" sectoral estimates of emissions with the "top-down" estimates derived from observations of atmospheric concentrations of the selected chemical species presented in Chapter 2. The third objective is the identification of the global extent of the opportunity for additional emission reductions through the implementation of additional sectoral measures beyond business as usual. Data have been presented in both metric tonnes and also in CO_2-equivalent-weighted tonnes. These data have been calculated using the global warming potentials that appear in Section 2.5 of this report and also the global warming potentials in the IPCC 2nd and 3rd Assessment Reports (IPCC, 1996, 2001). These SAR and TAR values are those used by parties to the UNFCCC for reporting inventory data and also adopted (Decision 1CP/5) in the Kyoto Protocol[1]. Use of these values facilitates a comparison against the regulatory commitments adopted within these agreements.

11.1.2   Structure

Following a brief explanation of the methodology used, the chapter identifies current and planned production facilities for the supply of HFCs, and compares this capacity with estimates of aggregate current demand and emissions (based on the sectoral information provided in Chapters 4 to 10). A comparison is then made between these "bottom-up" emission estimates and the "top-down" observed atmospheric concentrations presented

---

[1] These GWPs for HFCs, PFCs and SF_6 are the values adopted for the Kyoto Protocol pursuant to paragraph 3 of Article 5 of Decision 1/CP.3 (FCCC/CP/1997/7/Add.1). "3. The global warming potentials used to calculate the carbon dioxide equivalence of anthropogenic emissions by sources and removals by sinks of greenhouse gases listed in Annex A shall be those accepted by the Intergovernmental Panel on Climate Change and agreed upon by the Conference of the Parties at its third session. Based on the work of, inter alia, the Intergovernmental Panel on Climate Change and advice provided by the Subsidiary Body on Scientific and Technological Advice, the Conference of the Parties serving as the meeting of the Parties to this Protocol shall regularly review and, as appropriate, revise the global warming potential of each such greenhouse gas, taking fully into account any relevant decisions by the Conference of the Parties. Any revision to a global warming potential shall apply only to commitments under Article 3 in respect of any commitment period adopted subsequent to that revision." and Paragraph 3 of Decision 2/CP.3 (FCCC/CP/1997/7/Add.1): "3. Reaffirms that global warming potentials used by Parties should be those provided by the Intergovernmental Panel on Climate Change in its Second Assessment Report ("1995 IPCC GWP values") based on the effects of the greenhouse gases over a 100-year time horizon, taking into account the inherent and complicated uncertainties involved in global warming potential estimates. In addition, for information purposes only Parties may also use another time horizon, as provided in the Second Assessment Report."

in Chapter 2. The chapter then discusses the drivers of future demand and supply, and provides estimates of demand and emissions in 2015 under both a business-as-usual and a mitigation scenario. It concludes with a comparison of these estimates with those derived under previous scenarios, and the identification of the aggregate extent of the additional emission reduction opportunities that may be available through the implementation of additional measures.

11.2   Methodology for estimations

In estimating the supply of HFCs and PFCs, the basic assumption is that supply will match demand. In addition, for current production, the aggregate supply of some individual substances can be verified using published capacity and production data. For projections, a check of this kind is not possible. Demand can exceed supply in some circumstances because chemical producers may be unwilling to invest in new plants due to uncertainties resulting from the threat of government regulation. This, in turn, may result in tight supply, which will increase prices and may increase the use of alternatives. If this happens, actual use and emissions of HFCs and PFCs will be lower than projected in this chapter.

Two scenarios have been developed for demand (see Section 11.5), the first one based on business-as-usual assumptions (i.e. that all existing measures, including the Montreal Protocol, will continue and that no new measures will be introduced), and another mitigation scenario based on an increased effort to achieve emission reductions through the global application of current best-practice emission reduction techniques (e.g. containment, recovery, recycling and destruction). The precise assumptions included in the mitigation scenario vary between chapters – and therefore these assumptions are discussed in detail in each of the chapters. Because the Kyoto Protocol does not impose specific emission limitations on a substance-by-substance basis, the entry into force of the Kyoto Protocol does not directly alter either of the scenarios. However, national measures related to fluorocarbons have been included where they have been introduced or announced.

Wherever possible, the "bottom-up" methodology has been used to develop demand and emission estimates; the "top-down" approach often has to rely on a limited number of data points since the substances concerned are often in their early stage of market introduction (see also section 11.4).

11.3   Current supply, demand and emissions situation for HFCs and PFCs, plus emissions for CFCs, HCFCs and halons

11.3.1   Current and planned global supply of HFCs and PFCs – production and number of facilities

From a number of data sources, including UNFCCC, AFEAS and the World Bank, it is possible to make estimates of the global capacities for the production of HFC-134a in different

Table 11.1. Estimated HFC production capacity for HFC-134a (in ktonnes) in 2003 and number of production facilities (in brackets).

	HFC-134a	HFC-125	HFC-143a	HFC-23	HFC-32	HFC-152a	HFC-227ea	HFC-245fa	HFC-365mfc
European Union	40 (4)	(2)	(2)		(1)				(1)
USA	100 (4)	(1)	(2)		(1)	(1)	(1)	(1)	-
Japan	40 (3)	(2)	(1)		(1)	-	-	-	-
Russian Federation		(1)	(1)	(1)	-	(1)	(2)	-	-
China	5 (1)		(1)		(1)	(1)			
Korea					(1)				

regions, and the companies and the number of production facilities that produce the other main HFCs (see Table 11.1 and Figure 11.1). A complicating factor is that, apart from HFC-134a, -125 and -143a, the other individual HFCs are produced by three producers or less. Consequently, the actual production volumes are not publicly available for those substances under anti-trust rules. Moreover, AFEAS data do not cover all producing companies, but only those established in Japan, the EU and the USA. Facilities in China, Korea and the Russian Federation, for example, are not covered under AFEAS.

At present, global capacity just exceeds aggregate demand, since fluorocarbon producers have made investments in anticipation of the likely future needs for, among other substances, HFCs and PFCs triggered by phase-out under the Montreal Protocol.

There is no available information about supply that enables a forecast of future production capacities for different HFCs or PFCs. However, as there are no technical or legal limits to the production of these chemicals, it has been assumed that production will continue to occur at the rate necessary to satisfy demand, and it has been assumed that construction lead time on the year-to-year availability of product and the perception of potential legislative controls on investors' confidence have no effects. Future production has been estimated by the simple aggregation of sectoral demand projections.

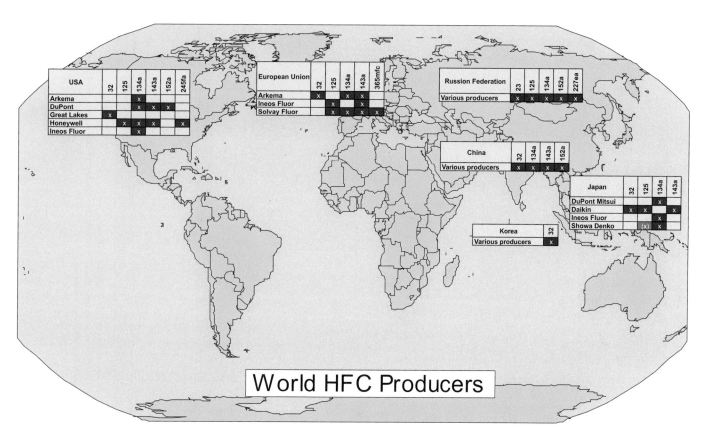

Figure 11.1. Global HFC production.

Table 11.2. Demand for CFCs, halons, HCFCs, HFCs and PFCs (metric tonnes) in 2002, in the business-as-usual projection for 2015, and in the mitigation scenario for 2015.

	2002 demand — Refrigeration, SAC, MAC[1] (kt yr⁻¹)	Foams (kt yr⁻¹)	Medical aerosols (kt yr⁻¹)	Fire protection & other[2] (kt yr⁻¹)	Total (kt yr⁻¹)	2015 BAU — Refrigeration, SAC, MAC[1] (kt yr⁻¹)	Foams (kt yr⁻¹)	Medical aerosols (kt yr⁻¹)	Fire protection & other[2] (kt yr⁻¹)	Total (kt yr⁻¹)	2015 MIT — Refrigeration, SAC, MAC[1] (kt yr⁻¹)	Foams (kt yr⁻¹)	Medical aerosols (kt yr⁻¹)	Fire protection & other[2] (kt yr⁻¹)	Total (kt yr⁻¹)
CFC-11	6	11	3		19	2		1		2	1				1
CFC-12	132	1	5		137	8		1		9	6				6
CFC-113					n.a.					n.a.					n.a.
CFC-114			0.4		0.4			0.1		0.1			0		
CFC-115	12				12	2				2	1				1
Halon-1211				3	3										
Halon-1301				1	1										
HCFC-22	346	7		0.3	353	482	0.4	*	0.1	482	324	0	*	0	324
HCFC-123	8			0.2	8	*	*	*		3	*	*	*		1
HCFC-124	3			0.04	3	*	*	*		1	*	*	*		1
HCFC-141b		97		5	102		49		10	59		49		10	59
HCFC-142b		25			25		1			1		1			1
HCFC-225				6	6				6	6				6	6
HFC-23				0.1	0.1	*	*	*	*	0	*	*	*	*	0.1
HFC-32	3				3	25				25	28				28
HFC-125	23				23	86				86	58				58
HFC-134a	133	6	3		142	408	27	10		446	318	7	12		337
HFC-143a	28				28	72				72	35				35
HFC-152a	1	3			4	0.4	3			4	0	2			2
HFC-227ea		0.1	1	3	4	*	*	*	*	7	*	*	*	*	7
HFC-245fa		2			2		32			32		15			15
HFC-236fa				0.1	0.1	*	*		*	0	*	*		*	0.1
HFC-365mfc		0			0		9			9		5			5
HFC-43-10mee				1	1				1	1				1	1
Other HFCs				p.m.	p.m.	*	*	*	*	p.m.	*	*	*	*	p.m.
PFC-14	0			0	0	*	*	*	*	0.005	*	*	*	*	0
PFC-3-1-10				0	0					0.01					0
PFC-5-1-14				0	0				0.01	0.01				0	0
Per group															
Halons				4	4				*	0				*	0
CFCs	149	11	8		169	12	*	2	*	14	8	0		*	8
HCFCs	357	128		12	496	485	50		16	551	325	50		16	391
HFCs	190	11	4	4	209	591	73	13	5	681	440	30	15	5	490
PFCs				0.1	0.1				0.01	0.02				0	0.02
Total	696	150	12	20	877	1,087	123	15	21	1,246	773	80	15	21	889

* Data not available due to commercial confidentiality of data.

[1] "Refrigeration" comprises domestic, commercial, industrial (including food processing and cold storage) and transportation refrigeration. "Stationary air-conditioning" (SAC) comprises residential and commercial air conditioning and heating; MAC stands for mobile air-conditioning.

[2] "Other" includes solvents, but excludes non-medical aerosols. Estimates expressed in MtCO₂-eq are available for non-medical aerosols only.

11.3.2 Current demand for HFCs and PFCs

Sectoral demand: past trends and current situation in developed and developing countries

When the Montreal Protocol was signed in 1987, alternatives to CFCs and halons were not commercially available for most applications. By 1989, the first reports of the UNEP Technology Assessment Panels of the Montreal Protocol (UNEP-TEAP, 1989, Chapter 3.10) concluded that, while non-fluorocarbon alternatives including not-in-kind alternatives would be important, HCFCs and HFCs would be necessary in order to make substantial and timely reductions in CFC and halon consumption. In 1990 the Montreal Protocol was amended to require a phase-out of CFCs, halons, carbon tetrachloride and 1,1,1-trichloroethane.

Technical advances subsequently accelerated the commercial availability and use of the new alternatives, including not-in-kind technologies. Consistent with these advances, the Montreal Protocol was further amended in 1992 to require a phase-out of HCFCs. By 1998, the TEAP reports were noting a rapid development of technologies that were no longer reliant on fluorocarbons (UNEP-TEAP, 1998).

Table 11.2 shows the aggregated demand for CFCs, HCFCs and halons, as well as HFCs and PFCs. The data are derived primarily from Ashford *et al.* (2004) and supplemented by data from Chapters 4 – 10 in this report, UNEP (2004b) and AFEAS (2004). These data serve as the baseline for the projections of 2015 demand and emissions in both the business-as-usual and mitigation scenarios.

In the absence of the Montreal Protocol, the use of CFCs would have increased significantly (Prather *et al.*, 1996). However, as a result of its adoption, the use and production of CFCs and halons have almost completely ceased in developed countries (UNEP, 2004b), and are rapidly falling in developing countries, as demanded by the schedule in the protocol. The use of HFCs has facilitated this rapid phase-out in developed countries, particularly (but not solely) in applications where

other alternatives were not available. In many cases, HFCs are still important for the safe and cost-effective phase-out of CFCs and halons in countries with economies in transition and in developing countries (UNEP-TEAP, 2004). Developed countries have replaced about 8% of projected CFC use with HFCs and 12% with HCFCs. They have eliminated the remaining 80% by controlling emissions, making specific use reductions, or by using alternative technologies and non-HCFC and HFC fluids, including ammonia, carbon dioxide, water and not-in-kind options. Approximately 20% of halon use has been replaced by HFCs, less than 1% by PFCs and 50% by non-gaseous agents (UNEP-TEAP, 1999b). Some of the options are shown in Table 11.3.

11.3.3 Current emissions of CFCs, halons, HCFCs, HFCs, and PFCs

Emissions occur during both the production and use of these substances. During production, the emissions are largely inadvertent, whereas emissions arising from use of the substances can be either intentional or inadvertent.

11.3.3.1 Production emissions

Emissions of HFCs and PFCs occur during the production of fluorocarbons, either as undesired byproducts or as losses of useful material, for example during the filling of containers or when production equipment is opened for maintenance. In general, the emissions during production are much smaller than the emissions during the use of the products. Nevertheless, techniques and procedures to minimize emissions have been developed, including the thermal oxidation of gas streams before they are released into the atmosphere. Calculating emissions from such sources requires data about not only the quantities of materials produced (the activity) and the rate of emission (which is influenced by the process design and operating culture) but also about the extent to which emissions are abated. This is particularly important in determining the uncertainty of

Table 11.3. Replacement of CFCs and halons in developed countries (derived from UNEP-TEAP, 2000).

Application	Replacement rate of CFCs and halons by HCFCs and HFCs	Dominant alternative fluid or technology
Refrigeration and air-conditioning	30%	Emission reductions (containment, recycle, recovery), hydrocarbons and ammonia, reduced charge size, design improvements (e.g. heat exchangers)
Closed-cell foam	<45-50%	hydrocarbons and carbon dioxide
Open-cell foam	15%	water and hydrocarbons
Aerosol propellant (not MDIs)	3%	hydrocarbons and alternative dispensing technologies
Fire extinguishants	20-25%	conservation, water mist, sprinklers, foam, carbon dioxide, inert gases
Other (primarily solvents)	3%	"no-clean technologies", water-based systems and chlorocarbons
All applications	HCFCs 12% HFCs 8%	

estimates of future emissions. The majority of these estimates are made using standardized tables for emissions that cover valves, joints, pumps and other processing parts.

Byproduct emissions of HFC-23

The most significant of the byproducts is HFC-23 (fluoroform). It is produced during the manufacture of HCFC-22. Global emissions of HFC-23 increased by an estimated 12% between 1990 and 1995, due to a similar increase in the global production of HCFC-22 (see Table 10.10 in Chapter 10, and the description of chemical routes of manufacture in Appendix 11A).

The quantity of HFC-23 that is produced (and that, potentially, may be emitted) is directly related to the production of HCFC-22 and so emission forecasts require a scenario for future HCFC-22 production volumes. These will depend on the consumption of HCFC-22 in developed countries, which is declining, and the consumption in developing countries and global demand for fluoropolymers feedstock, both of which are increasing (see Section 10.4). After thorough investigation, the US EPA concluded that process optimization reduces, but does not eliminate, HFC-23 emissions. To reduce the emissions below the 1% level, thermal oxidation is required (Irving and Branscombe, 2002).

On the basis of a business-as-usual scenario that assumes that consumption of ozone-depleting substances will match the maximum levels permitted under the Montreal Protocol, the consumption and production of non-feedstock HCFC-22 will fall by a factor of almost 10 by 2015 from the average level in 2000-2003 in developed countries (including the 15% production allowance to meet the basic domestic needs of developing countries). In developed countries, demand for fluoropolymer feedstock is projected to continue increasing linearly, leading to a doubling in feedstock demand for HCFC-22 in those countries by 2015 (see Section 10.4). In developing countries, production of HCFC-22 for both feedstock and non-feedstock uses has grown rapidly in recent years; over the period 1997 to 2001, production for potentially emissive (or non-feedstock) uses grew linearly at 20,000 tonnes yr^{-1} and feedstock use grew at 4,100 tonnes yr^{-1} (see Chapter 10) Assuming that these rates continue until 2015, the total global requirement for HCFC-22 will be about 707,000 tonnes yr^{-1}, about 40% of which would be for feedstock, compared to a total of 490,000 tonnes yr^{-1} in the year 2000 (Table 10.10).

In the business-as-usual case prior to 2015, it has been assumed that emissions from existing capacity will continue at 2% of HCFC-22 production (reflecting the average of plants in developed and developing countries, including current levels of abatement) and that new capacity (mainly in developing countries) will emit HFC-23 at a rate of 4% (see Chapter 10). Consequently, emissions of HFC-23 could grow by 60% between now and 2015, from about 15,000 tonnes yr^{-1} in 2003 to 23,000 tonnes yr^{-1} (Table 10.10).

In the variant for this scenario, the current best-practice technology, which includes the capture and thermal oxidation of the "vent gases", is introduced into all facilities progressively from 2005 onwards. Destruction technology is assumed to be 100% efficient and to operate for 90% of the on-line time of HCFC-22 plants. Reduced emissions were calculated for the same activity (in the form of assumed future HCFC-22 production) as the business-as-usual case. The difference between the two HFC-23 forecasts is therefore solely due to the extent of the deployment of destruction technology. The forecasts represent potential extreme cases, and future changes in activity will tend to increase the probability of one or the other.

The reduction of HCFC-22 production due to market forces or national policies, or improvements in facility design and construction to reduce HFC-23 output, could further influence these scenarios. An additional factor is the possible implementation of Clean Development Mechanism projects to install destruction facilities in non-annex I countries. This could result in a more rapid reduction of HFC-23 emissions than would otherwise have occurred.

11.3.3.2   Emissions arising from use

As noted under 11.3.2, the production and use of CFCs and halons have almost completely ceased in developed countries, and are rapidly falling in developing countries (Ashford *et al.*, 2004). HFCs and HCFCs have replaced about 8% and 12% of projected CFC use respectively in developed countries (UNEP-TEAP, 1999b). Similarly, about 20 % of halon use has been replaced by HFCs and less than 1% by PFCs (UNEP-TEAP, 1999b). The resulting changes in historic emissions of CFCs and HCFCs are shown in Table 11.4 (Ashford *et al.*, 2004).

Despite an equivalent fall in production of CFCs, the bank of material that remains in equipment and has not yet been emitted remains significant. For example, in the year 2002, the CFC bank amounted to over 2.4 million tonnes, consisting mainly of CFC-11 in foams (UNEP-TEAP, 2002ab). In addition to the emissions of CFCs and HCFCs from future consumption in developing countries, it is expected that a substantial proportion of these banks will eventually be released into the atmosphere under a business-as-usual scenario. There are opportunities

Table 11.4. Estimated historic emissions of ozone-depleting substances (tonnes yr^{-1}).

Year	CFC-11	CFC-12	CFC-113	HCFC-123	HCFC-124	HCFC-141b	HCFC-142b	HCFC-22
1990	258,000	367,000	215,000	0	0	0	2,100	217,000
1995	106,000	256,000	28,000	2,100	1,600	25,400	9,500	252,000
2000	75,000	134,000	2,700	4,200	2,900	25,600	11,100	286,000

to make significant reductions in these emissions through the global application of best-practice handling and eventual destruction of these gases (UNEP-TEAP, 2002ab).

Tables 11.5a and 11.5b show the development of banks of some fluorocarbons.

Refrigeration, air conditioning (particularly air conditioning in vehicles) and heat pumps are the largest source of emissions of HFCs. In the future, improved design, tighter components, and recovery and recycling during servicing and disposal could reduce HFC emissions at moderate to low costs (see Chapter 6). The impact of improved practices will vary between regions and countries.

Insulating foams are expected to become the second-largest source of HFC emissions, and HFC use is expected to grow rapidly as CFCs and HCFCs are replaced with HFC-134a, HFC-

227ea, HFC-245fa and HFC-365mfc (see Chapter 7). In replacing CFC-11 use in insulating foams in the refrigeration sector, around 66% of the ozone-depleting substances (ODSs) were replaced by hydrocarbons in investment projects approved by the Multilateral Fund of the Montreal Protocol. In non-refrigeration uses, 30% of the ODSs were replaced by hydrocarbons, 25% by HCFCs and 45% by zero-ODP/zero-GWP alternatives (UNEP-TEAP, 1999b, Chapter 24).

Other sources of HFC emissions are industrial-solvent applications, medical aerosol products, other aerosol products, fire protection and non-insulating foams. Increased containment, recovery, destruction and substitution can reduce emissions. In some applications in all these sectors, there are zero- or low-GWP options.

Limited data are available about PFC emissions and use.

Table 11.5a. Evolution of banks (in metric ktonnes).

2002	Banks (ktonnes of substance)							
	Refrigeration[1]	Stationary air-conditioning[2]	Mobile air-conditioning	Foams	Medical aerosols[4]	Fire protection	Other[3,4]	Total
Halons					-	167		167
CFCs	330	84	149	1,858				2,430
HCFCs	461	1,028	20	1,126	8	4	11	2,651
HFCs	180	81	249	12		19	p.m.	543
PFCs					3.5	0.5	0.1	1
Total	971	1,192	418	2,996		191	11	**5,791**

2015 Business-as-usual scenario	Banks (ktonnes of substance)							
	Refrigeration	Stationary air-conditioning[2]	Mobile air-conditioning	Foams	Medical aerosols[4]	Fire protection	Other	Total
Halons	-	-	-	-	-	43		43
CFCs	64	27	13	1,305	2			1,411
HCFCs	891	878	23	1,502		6	16	3,317
HFCs	720	951	635	566	13	64	p.m.	2,949
PFCs						1	0.01	1
Total	1,675	1,856	671	3,374	15	114	16	**7,722**

2015 Mitigation scenario	Banks (ktonnes of substance)							
	Refrigeration	Stationary air-conditioning[2]	Mobile air-conditioning	Foams	Medical aerosols	Fire protection	Other	Total
Halons	-	-	-	-	-	43		43
CFCs	62	27	13	1,305	0			1,407
HCFCs	825	644	23	1,502		6	16	3,017
HFCs	568	1,018	505	443	15	64	p.m.	2,612
PFCs						1	0.01	1
Total	1,455	1,689	541	3,250	15	114	16	**7,080**

[1] "Refrigeration" comprises domestic, commercial, industrial (including food processing and cold storage) and transportation refrigeration.
[2] "Stationary air-conditioning" comprises residential and commercial air conditioning and heating.
[3] "Other" includes solvents, but excludes non-medical aerosols. Estimates expressed in MtCO$_2$-eq are available for non-medical aerosols only.
[4] Emissive use applications are assumed to have banks that are equal to annual emissions.

Table 11.5b. Evolution of banks (in $MtCO_2$-eq).

2000[2]	Banks ($MtCO_2$-eq)							
	Refrigeration	Stationary air-conditioning[2]	Mobile air-conditioning	Foams	Medical aerosols[5]	Fire protection	Other[3,5]	Total
Halons						[531 (391)]		[531 (391)]
CFCs	3,423 (2,641)	631 (489)	1,600 (1,209)	10,026 (8,008)	69 (53)	0		15,749 (12,400)
HCFCs	810 (682)	1,755 (1,480)	36 (31)	1,229 (1,009)		5 (5)	6 (6)	3,841 (3,212)
HFCs	518 (446)	123 (111)	350 (323)	16 (14)	6 (6)	65 (59)	25 (25)	1,103 (984)
PFCs						4 (3)	1 (1)	5 (4)
Total[4]	**4,751 (3,769)**	**2,509 (2,079)**	**1,987 (1,563)**	**11,270 (9,031)**	**75 (59)**	**74 (67)**	**32 (31)**	**20,698 (16,600)**

2015 Business-as-usual scenario	Banks ($MtCO_2$-eq)							
	Refrigeration	Stationary air-conditioning[2]	Mobile air-conditioning	Foams	Medical aerosols[4]	Fire protection	Other	Total
Halons	-	-	-	-	-	[206 (156)]		[206 (156)]
CFCs	653 (510)	208 (161)	138 (104)	7,286 (5,798)	17 (13)	0		8,302 (6,587)
HCFCs	1,582 (1333)	1,536 (1295)	42 (35)	1,696 (1,391)		6 (5)	9 (9)	4,871 (4,068)
HFCs	1,922 (1661)	1,488 (1333)	896 (826)	644 (612)	23 (21)	226 (204)	27 (27)	5,227 (4,683)
PFCs						4 (4)	0.1 (0.1)	4 (4)
Total[4]	**4,157 (3,504)**	**3,232 (2,788)**	**1,076 (965)**	**9,626 (7,801)**	**40 (34)**	**236 (212)**	**37 (36)**	**18,404 (15,341)**

2015 Mitigation scenario	Banks ($MtCO_2$-eq)							
	Refrigeration	Stationary air-conditioning[2]	Mobile air-conditioning	Foams	Medical aerosols[4]	Fire protection	Other	Total
Halons	-	-	-	-	-	[206 (156)]		[206 (156)]
CFCs	627 (491)	208 (160)	138 (104)	7,286 (5,798)	0	0		8,258 (6,553)
HCFCs	1,466 (1,235)	1,134 (956)	41 (34)	1,696 (1,391)		6 (5)	9 (9)	4,352 (3,630)
HFCs	1,455 (1,255)	1,586 (1,422)	712 (656)	494 (471)	26 (24)	226 (204)	27 (27)	4,527 (4,059)
PFCs						4 (4)	0.1 (0.1)	4 (4)
Total[4]	**3,548 (2,980)**	**2,928 (2,539)**	**891 (795)**	**9,475 (7,659)**	**26 (24)**	**236 (212)**	**37 (37)**	**17,141 (14,247)**

Note: Direct GWPs for a 100-year time horizon were used from IPCC 2001 and WMO 2003 (as listed in Table 2-6). Data in brackets use the direct GWPs for a 100-year time horizon from IPCC 1996 (SAR) and IPCC 2001 (TAR).

[1] "Refrigeration" comprises domestic, commercial, industrial (including food processing and cold storage) and transportation refrigeration.

[2] "Stationary air-conditioning" comprises residential and commercial air conditioning and heating.

[3] "Other" includes non-medical aerosols and solvents.

[4] Halons cause much larger negative indirect than positive direct radiative forcing and, in the interests of clarity, their positive direct effects are not included in the totals.

[5] Emissive use applications are assumed to have banks that are equal to annual emissions.

These substances are predominantly used in fire protection and in solvent applications. In the case of fire protection, use is emissive but products remain in the equipment until use once it has been installed and the options for reducing these emissions are therefore limited (see Chapter 9). PFC solvents are no longer considered technically necessary for most applications and use has decreased since the mid-1990s due to replacement with lower-GWP solvents (see Chapter 10).

11.3.4 Comparison of emissions estimated via a bottom-up approach with those calculated from observed atmospheric concentrations of CFCs, HCFCs and HFCs

As noted in Chapter 2, one approach to evaluating the total global emissions of long-lived greenhouse gases uses observations of their atmospheric concentrations. By contrast, Table 11.6 presents the breakdown of direct emissions of each halocarbon for the year 2002, as derived via the bottom-up approach described in Chapters 4 to 10. Total direct emissions amount to

Table 11.6. Emissions for CFCs, halons, HCFCs, HFCs and PFCs in 2002, in the business-as-usual projection for 2015, and in the mitigation scenario for 2015.

	2002 emissions								2015 business-as-usual (BAU) emissions							2015 mitigation scenario (MIT) emissions						
	Refr. Etc.[3] (kt yr^{-1})	Foams (kt yr^{-1})	Med. Aerosols (kt yr^{-1})	Fire Prot. & Other[4] (kt yr^{-1})	Total (kt yr^{-1})	Global Estimate (Ch.2) (kt yr^{-1})	Total (direct GWP-weighted) A[1] (MtCO$_2$-eq yr^{-1})	B[2]	Refr. Etc.[3] (kt yr^{-1})	Foams (kt yr^{-1})	Med. Aerosols (kt yr^{-1})	Fire Prot. & Other[4] (kt yr^{-1})	Total (kt yr^{-1})	Total (direct GWP-weighted) A[1] (MtCO$_2$-eq yr^{-1})	B[2]	Refr. Etc.[3] (kt yr^{-1})	Foams (kt yr^{-1})	Med. Aerosols (kt yr^{-1})	Fire Prot. & Other[4] (kt yr^{-1})	Total (kt yr^{-1})	Total (direct GWP-weighted) A[1] (MtCO$_2$-eq yr^{-1})	B[2]
CFC-11	7	20	3		30	70-90	140	113	3	14	1		19	87	71	2	14			15	71	58
CFC-12	127	2	5		134	110-130	1,433	1,083	19	2	1		22	235	177	11	2			13	140	105
CFC-113						5-12																
CFC-114			0.4		0	-	4	4			0.1		0.1	1	1							
CFC-115	10				10	-	75	74	2				2	15	15	1				1	10	10
Halon-1211				17	17	7-8	32	23				2	2	3	2				2	2	3	2
Halon-1301				2	2	1-2	14	11				1	1	9	7				1	1	9	7
HCFC-22	229	2		0.1	231	240-260	411	346	448	1		0.1	449	800	674	257	1		0.1	258	459	387
HCFC-123	4			0.04	4	-	0	0	6			0.1	6	0.4	1	2			0.1	2	0.1	0.2
HCFC-124	3			0.01	3	-	2	1	1			0.01	1	1	0.5	0.4			0.01	0.4	0.3	0.2
HCFC-141b		14		5	19	55-58	14	11		17		10	27	19	16		14		10	24	17	14
HCFC-142b		8			8	25	19	15		3			3	6	5		3			3	6	5
HCFC-225				6	6	-	2	3				6	6	2	3				6	6	2	3
HFC-23				14	14	13	195	159				23	23	332	272				2	2	33	27
HFC-32	1				1	-	0	0	10				10	7	6	8				8	5	5
HFC-125	10				10	9-10	34	28	51				51	175	142	23				23	81	66
HFC-134a	74	2	3		79	96-98	111	102	249	8	10		267	376	347	105	3	12		120	169	156
HFC-143a	15				15	-	65	56	49				49	215	186	19				19	82	71
HFC-152a	1	3			4	21-22	1	1	0.4	3			4	0.5	1	0.2	2			2	0.3	0.3
HFC-227ea		0.01	1	0.3	1	-	3	3		0.2	3	1	4	12	11		0.2	3	1	4	13	12
HFC-245fa		0.2			0.2	-	0.2	0.1		5			5	5	5		3			3	3	3
HFC-236fa				0.02	0.02	-	0.2	0.1				0.05	0.05	0.5	0.3				0.05	0.05	0.5	0.3
HFC-365mfc		0.1			0.1	-	0.1	0.1		2			2	1	2		1			1	1	1
HFC-43-10mee				1	1	-	2	1				1	1	2	1				1	1	2	1
Other HFCs				p.m.	p.m.	-	23	23				p.m.	p.m.	26	26				p.m.	p.m.	26	26
PFC-14				0.001	0.001	-	0.005	0.005				0	0.002	0.01	0.01				0	0.002	0.01	0.01
PFC-3-1-10				0.01	0.01	-	0.1	0.1				0	0.01	0.1	0.1				0	0.01	0.1	0.1
PFC-5-1-14				0.10	0.10	-	1	1				0	0.01	0.1	0.1				0	0.01	0.1	0.1
Per group[5]																						
Halons[5]				19	19	8-10	[47]	[34]				3	3	[12]	[9]				3	3	[12]	[9]
CFCs	144	22	8		174	185-232	1,651	1,274	25	16	2		43	338	264	14	15			30	221	173
HCFCs	236	24		11	271	310-333	447	377	455	21		16	492	828	699	259	17		16	292	484	408
HFCs	100	5	4	15	124	133-137	434	374	359	18	13	25	415	1,153	999	155	9	15	4	184	416	369
PFCs				0.10	0.1	-	1	1					0.02	0.2	0.2					0.02	0.2	0.2
Total[5]	481	51	12	46	589	652-728	2,534	2,026	838	55	15	44	953	2,319	1,961	429	42	15	24	509	1,121	951

[1] A: Direct GWPs for a 100-year time horizon were used from IPCC 2001 and WMO 2003 (as listed in Table 2-6)

[2] B: Direct GWPs for this column were taken from the Intergovernmental Panel on Climate Change Second Assessment Report (SAR) (Climate Change 1995: The Science of Climate Change, J.T. Houghton, L.G. Meira Filho, B.A. Callandar, N. Harris, A. Kattenberg, and K. Maskell (eds.), Cambridge University Press, Cambridge, UK, 1996). Where no value was published in the SAR, the value was taken from Scientific Assessment of Ozone Depletion: 1998, World Meteorological Organisation Global Ozone research and Monitoring Project report No. 44, Geneva, 1999).

[3] "Refr. Etc." comprises refrigeration, stationary air-conditioning (SAC) and mobile air-conditioning (MAC) (see Table 11.2)

[4] "Other" includes solvents, but excludes non-medical aerosols. Estimates expressed in MtCO2-eq are available for non-medical aerosols only. These are included in the columns "Total (direct GWP-weighted)".

[5] Halons cause much larger negative indirect than positive direct radiative forcing and, in the interests of clarity, their positive direct effects are not included in the totals in the columns A and B.

Table 11.7. Estimated historic emissions of HFCs (tonnes yr^{-1}).

Year	HFC-125	HFC-134a	HFC-143a	HFC-152a	HFC-227ea	HFC-245fa	HFC-32	HFC-365mfc
1990	0	180	0	12	0	0	0	0
1995	200	17,500	1,110	7,340	100	0	0	0
2000	5,150	73,700	9,180	15,200	1,950	0	230	0

about 2.5 GtCO$_2$-eq yr^{-1} (2.0 GtCO$_2$-eq yr^{-1} using SAR/TAR values), which is similar to the estimate derived in Chapter 2 from atmospheric measurements. However, as shown in Table 11.6 and in Figure 2.4, there are significant differences for individual substances between calculated and observed emissions, notably for CFC-11, HCFC-141b and HCFC-142b. The possible reasons for such differences are examined in Appendix 11B, together with suggestions for further work in this area that could clarify the sources. The lack of information on use patterns for these substances makes it difficult to assess the contribution to observed emissions from current production and use.

The differences between emissions estimates constitute a long-standing problem that has yet to be resolved by the research community. Nevertheless, qualitative conclusions can be drawn for all gases and for some gases, notably the HFCs, atmospheric concentrations match emissions estimates and quantitative conclusions are justified. Current CFC emissions are dominated by delayed releases from banks (such as material contained in existing foams, air conditioning, refrigeration, and other applications). Calculating the magnitude of these emissions is therefore a complicated process requiring massive amounts of data and there is significant scope for systematic error. Observations of the changes in concentrations of CFCs can provide direct insights into their emissions. The observed rates of change differ among CFCs, depending mainly upon the gas lifetimes and emissions (see Chapter 2). The concentrations of some CFCs have peaked and a slow decline is now being observed, while others are expected to decline in the future. CFC-11 concentrations are decreasing about 60% slower than they would in the absence of emissions, while CFC-12 has only now stabilized. These changes show that CFC-12 emissions, largely from the banks, remain similar in magnitude to the quantities of these gases that are naturally destroyed in the atmosphere.

Table 11.7 shows the historical emissions arising from the use of HFCs, as derived from production and sales data, and sector-specific emission factors. For HFC-134a, these concur well with measured atmospheric concentrations where known. For example, Figure 11.2 compares concentrations of HFC-134a measured at Mace Head, Ireland with northern hemispherical concentrations calculated from the emission data.

The good historic match for HFC-134a is found for emissions calculated using both top-down methodology (as in Table 11.7) and bottom-up methods (that rely on sector-specific activities and emission functions).

The EDGAR database contains other historical data for 1990 and 1995 (http://arch.rivm.nl/env/int/coredata/edgar/). This database relies on calculations described in Olivier (2002) and Olivier and Berdowski (2001). Although the values quoted for emissions of HFC-134a and HFC-23 agree with the data used in this chapter, those for HFC-125, -143a, -152a and -227ea are markedly different. The lack of exact information about production volumes (see 11.3.1) introduces major uncertainties with regard to top-down estimates of the future emissions based on those data.

11.4 Discussion of demand drivers, including differentiation between developed and developing countries

11.4.1 Demand drivers for HFCs and PFCs

By far the most important factor in the demand for HFCs and to a lesser extent PFCs has been the adoption of the Montreal Protocol and the subsequent amendments to it. Moreover, a number of developed countries have accelerated the phase-out of ozone-depleting substances under the Montreal Protocol, thus providing extra impetus for demand. These substitution effects have been discussed extensively in the previous chapters.

Perhaps the single most important demand factor after the

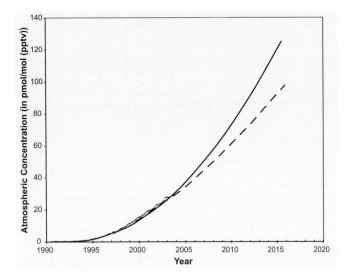

Figure 11.2. Comparison of measured HFC-134a concentrations at Mace Head, Ireland (crosses) with concentrations calculated from emissions (solid line: historic data and BAU scenario for the future; dashed line: reduced emissions). The future trends are simply illustrative. (The scenarios are described in Ashford *et al.* (2004)).

implementation of the Montreal Protocol is the expected increase in disposable income in the developing world and the acquisition by families of more than one appliance of the same kind, as noted in various chapters above. In the next decade, hundreds of millions of families can be expected to pass the wealth threshold enabling them to purchase products that may contain or be manufactured with HFCs or PFCs. As noted in SRES, world population growth and GDP growth are also major factors in future global demand for HFCs and PFCs.

Other factors that are expected to have a positive impact on demand for HFCs and PFCs are:

- energy efficiency requirements which may in some cases encourage the use of fluids with superior thermodynamic properties or insulation values;
- safety requirements (including standards) which exclude the use of flammable or toxic substances or fluids with very high operating pressures;
- other environmental legislation, e.g. VOC rules that set a ceiling on emissions, or that prohibit their use in particular applications;
- especially in the USA, product liability concerns encourage the use of HFCs rather than flammable or toxic substances;
- in the mobile air-conditioning sector, the primary replacement for CFC-12 is HFC-134a. There has been a steep rise in the use of these systems in Europe and developing countries in recent years and this is predicted to continue;
- financial mechanisms that enable the compliance of developing countries and of countries with economies in transition under the Multilateral Fund of the Montreal Protocol and the Global Environment Facility respectively have had a positive impact on demand for HFCs, but this impact may decline in the future. A number of the projects that were supported by these mechanisms converted CFCs to HCFCs and HFCs.

Other developments may reduce demand:

- in the context of their commitments under the Kyoto Protocol, some countries have adopted legislation restricting the use of fluorinated greenhouse gases in some applications;
- the establishment of a recovery, recycling, reclamation and destruction infrastructure for ozone-depleting substances will prove useful for HFCs and PFCs as well, thus reducing demand for virgin material;
- because of their complex chemistry, HFCs and PFCs tend to be an order of magnitude more expensive than ozone-depleting substances, thus further encouraging recovery/reuse and the switch to not-in-kind technology;
- intellectual property rights for particular uses may equally encourage the switch to not-in-kind technology;
- other countries tax these gases according to their global warming potential and/or require their recovery when equipment is serviced or disposed;
- awareness among the governments and public in general of the linkages between ozone layer protection and climate change and impacts of HFCs/PFCs on the climate system

---

Box 11.1 The Multilateral Fund of the Montreal Protocol

The Multilateral Fund of the Montreal Protocol funded projects totalling approximately US$ 130 million that have contributed to the replacement of nearly 13,000 tonnes per year of CFCs by HFC-134a in developing countries (10,000 tonnes of conversion under these projects has already been completed). This represents approximately 30% of the CFCs phased out in the refrigeration sector under the Multilateral Fund. Most (99%) of the replacement by HFCs is in the refrigeration sector, in other words chillers, commercial refrigeration, compressors, domestic refrigeration and mobile air-conditioning. The remaining introduction of HFCs was in the solvents, aerosols and foam sectors. The introduction of HFC-152a and R-404A as replacements for CFCs in developing countries has been minimal.

The Multilateral Fund has also supported projects in developing countries totalling nearly US$ 250 million for the replacement of approximately 32,000 tonnes per year of CFCs by HCFCs (20,000 tonnes of conversion under these projects has already been completed). Eighty-five percent of this replacement has involved HCFC-141b (majority of which in the foam sector) and the remainder HCFC-22 in the refrigeration sector.

The Multilateral Fund has provided incremental costs for the shut-down of CFC production facilities in developing countries. The flexibility provided has enabled China, for example, to use part of the funds to establish the small-scale production of HFC-134a.

There is no inventory system for GEF-funded projects in CEITs that enables a clear determination of the overall share of ODS replacement by HCFCs and HFCs.

(Calculations based on information in UNEP (2004a).)

as a result of the decisions in the Conference of the Parties (Climate Convention)/ Meeting of the Parties.

The impact of technology is more uncertain. During the phase-out of ozone-depleting substances, many industries had significant technical hurdles to overcome, as described in previous chapters. This has led, in many cases, to the introduction of new technology and has significantly reduced equipment leakage rates through improved containment. Furthermore, energy efficiency has improved in many cases. Consequently, energy use and associated indirect emissions have grown less than they would have otherwise.

11.4.2 Discussion of drivers of supply for HFCs and PFCs

Drivers that will influence the business-as-usual scenario for the supply of HFCs and PFCs include the impact of current and anticipated policies and measures (including impact on demand), mitigation options, the economics of manufacturing HFCs and PFCs (both direct and indirect), technological change, supply capacity, and the timing and nature of the transition from ozone-depleting substances.

As indicated above, the basic assumption is that supply will follow demand. Individual manufacturers will make their own assessment of the market potential (=demand) and take investment decisions accordingly. In doing so, several considerations will be taken into account:

- the manufacture of HFCs and PFCs is increasingly complex and often surrounded by measures for protecting intellectual property;
- using existing infrastructures (HCFC plants, for example) will reduce the required capital expenditure and construction time;
- economies of scale are more important for HFCs because of the large capital investment required for these plants due to their complexity (see appendix 1A);
- the reluctance of downstream users to be dependent on a single source of supply;
- access to, and the availability of, key raw materials are gaining importance;
- new recovery and reclamation requirements will result to some extent (albeit an uncertain extent) in a switch away from newly produced substances;
- uncertainties relating to future policies and measures for greenhouse gases will add a risk premium to the minimum acceptable rate of return;
- the disposal of carbon tetrachloride co-generated from chloromethane plants may drive some producers in developing countries to use this product to start the manufacture of HFC-245fa and HFC-365mfc;
- producers of HFCs/PFCs may not make investment decisions due to uncertainties about the future of HFCs/PFCs arising out of the assessments and also due to impending regulations at the national, regional and global levels.

As a result, investments are likely to be made in existing sites and the number of production sites can be expected to remain limited. Greenfield investment will most likely occur in regions with substantial market potential, particularly China. Equally, supply can be expected to follow demand more closely than has been the case for CFCs and HCFCs. The perceived regulatory risks, and the existence of valid process patents, may discourage the conversion of current production facilities for ozone-depleting substances in developing countries to HFC and PFC production (with the latter being less likely). Furthermore, due to the complexity of the processes, most production facilities for ozone-depleting substances are not easily convertible to HFC production.

With CFCs phased out in developed countries, the production of HFCs and PFCs for supply is now focused upon maintaining the existing market, satisfying market growth, exports to other markets and developing capacity to cater for the demand for replacing HCFCs. Rapidly expanding markets in developing countries, in particular for CFC replacements, are resulting in new capacity for fluorinated gases, currently through the expansion of HCFC-22 and -141b capacity. Much of this is being achieved through local production in developing countries and joint ventures.

11.4.3 Impact of recovery, recycling and reclamation of CFCs, HCFCs, halons and HFCs

11.4.3.1 Current situation

Emissions may be reduced by cutting back on demand and destroying refrigerant at end of life. Demand may be reduced by the reuse of refrigerant affected by recovery programmes, and the prevention of leakage and losses through improved work practices and higher standards. These measures may be voluntary or mandatory and experience indicates that a co-regulatory approach, and cooperation between industry and governments resulting in mutually agreed regulation, provide a genuine and sustainable reduction in emissions.

Recovery, recycling and reclamation take place on a number of levels in developed countries. In many developed countries and regions – Australia, the European Union, United States and Japan for example – there are regulatory requirements that mandate the removal of "used" fluorinated gases from equipment during dismantling and destruction, and/or regulations that prohibit the venting of fluorinated gases. These requirements have been extended, or are in the process of being extended, to include HFCs and PFCs (Snelson and Bouma, 2003). In certain regions, the regulatory requirements are supported by industry voluntary schemes. In developing countries, UNEP has launched initiatives to encourage the recovery and recycling of CFCs and HCFCs although, at present, these have not been extended to HFCs and PFCs (UNEP, 2003).

Product stewardship schemes have been established by industry and governments in numerous countries to take back, and either reclaim or destroy, recovered refrigerant. These schemes may provide incentives for technicians to recover and return refrigerant, although the potential to charge for the service exists.

Current recovery and reuse practices include: service companies and technicians recovering and reinstalling used fluorinated refrigerants; major users (supermarkets, buildings etc.) recovering and reinstalling fluorinated gases in their own plant; large distributors of fluorinated gases collecting and either recycling or reclaiming fluorinated gases for resale; and producers of the fluorinated gases recovering and reusing refrigerants, sometimes as feedstock. In the EU, where used refrigerants are classed as "hazardous waste", transportation is a costly bureaucratic process that acts as a disincentive, particularly when the used refrigerants must be moved across national borders. In the EU, it is mandatory to destroy recovered CFCs under Regulation 2037/2000.

Similar recovery, recycling and reclamation processes take place in the fire-fighting industry. For the foam industry, the recovery of the foam products during the dismantling of foam-containing equipment and buildings is possible and, in some developed countries, mandatory. Fluorinated gases that are recovered will primarily be sent for destruction, as the presence of contaminants makes recycling and reuse non-economic. In the case of solvent use, recycling is an integral part of the process in which they are used: the solvents are boiled constantly in the equipment in which they are used and the vapour is condensed and reused. Solvents are also recovered from equipment and recycled externally from that equipment; this work is mainly done by the user of the solvent. Once the solvent reaches a level of contamination where recycling is not possible, it is destroyed. During the manufacture of metered dose inhalers, fluorinated gases are captured during the filling process and recycled in the filling equipment, thereby reducing emissions. Recycling of fluorinated gases externally is not often seen as an option because of contamination with pharmaceutical residues.

As a result of the majority of recovery, recycling and reclamation being carried out either on the premises of the user of the fluorinated gases or by independent distributors of the gases, there is only limited information on which to base estimates of the quantities of gases recovered, recycled or reclaimed. The cost of the gases is a major driver for recovery, recycling and reclamation, and some countries – notably Australia, the Netherlands and Norway – have a system that encourages the return of gases through a payment (or equivalent compensation) based upon quantity and quality.

11.4.3.2   Future opportunities

Recovery and reuse or destruction are important components of emission reduction. Whilst it is difficult to assess the volume of refrigerant that may be recovered, there has been some quantification work. A survey undertaken by Refrigerant Reclaim Australia (RRA) (Bennett, 2001) suggests that the equivalent of 25% of refrigerant sales may be recovered in any given year. The amount of recovered refrigerant recycled, reclaimed or destroyed translates directly to an equivalent reduction in demand and emissions. The survey indicated that the equivalent of 20%

of sales may be recovered from service and maintenance applications, and that the equivalent of 5% of sales may be recovered from decommissioned plants.

The RRA study, which included two hundred direct interviews with service companies, sought to establish a picture of the applications for new refrigerant first in order to ascertain the potential for recovery. Figure 11.3 shows the uses for new refrigerant.

The quantification project used the methodology outlined in Figure 11.3. Essentially, this process tracks imported product through the sales and application chain to determine the amount of product that may be returned. A number of the data points in the graphic were determined through research and survey work. The balance of required information was collected through surveys of contractors.

Figure 11.4 displays the applications using purchased new refrigerants. These results refer only to refrigerant sold to contractors and not refrigerant purchased by original equipment manufacturers (OEMs).

The results from the survey indicate that only about 20% of new refrigerant applications, retrofitting and contaminations lead to the controlled recovery of existing refrigerant.

The other application from which installed refrigerant is recovered is the decommissioning of existing aged plants. Industry experience indicates that approximately 5% of sales of new refrigerant are recovered annually from decommissioned plants. This figure does not include domestic refrigeration and air conditioning, as coordinated recovery and recycling programmes for electrical goods have not commenced in Australia as yet.

The second section of the survey looked at the processes contractors and technicians applied to the refrigerant they recovered. The responses can be found in Figure 11.5.

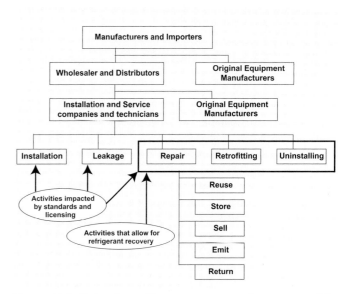

Figure 11.3. Estimating refrigerant recovery and reuse potentials: methodology of a survey in Australia (Bennett, 2001).

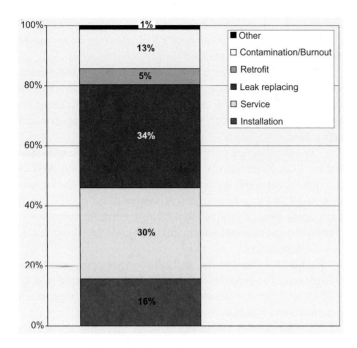

Figure 11.4. Applications to which purchased new refrigerants are applied: results of a survey by Refrigerant Reclaim Australia (Bennett, 2001).

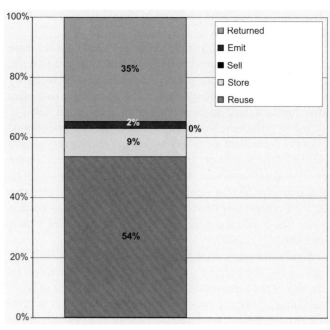

Figure 11.5. Destiny of recovered refrigerants: results of a survey by Refrigerant Reclaim Australia (Bennett, 2001).

The results are as follows:

- Reuse/Store: Approximately 63% of all refrigerant recovered was recycled and reused or stored by contractors for later use. This is an encouraging result, as recycling was not an activity undertaken by the industry in Australia only some years before. This activity impacts directly on emission levels by reducing the volume of refrigerant being imported.
- Emit/Sell: A very small percentage of contractors admitted to contravening the law by emitting refrigerant or selling impure recovered refrigerant.
- Returned: Contractors indicated that approximately 35% of all refrigerant recovered was returned to RRA for safe disposal.

The key results are firstly that a volume equivalent to 25% of sales of new refrigerant may be recovered in any year, and secondly that approximately 35% of the amount recovered – about 9% – may be expected to be returned for destruction.

Impact of financial incentives on recovered volumes

The impact of financial incentives on the volume of refrigerant recovered and returned may be significant. In early 2003, RRA doubled the financial incentive paid to service companies for returned refrigerant. This measure, in conjunction with promotional and information programmes highlighting the environmental and financial benefits of recovery, produced an increase of more than 50% of refrigerant returned to RRA (RRA Annual Report, 2004), see Figure 11.6.

Other financial incentives

Some countries tax imports and/or sales of HFC and PFCs, thereby increasing their price and encouraging reuse. Norway and Denmark use taxation based on the global warming potential of the refrigerant (Chapter 4), and then provide a rebate for

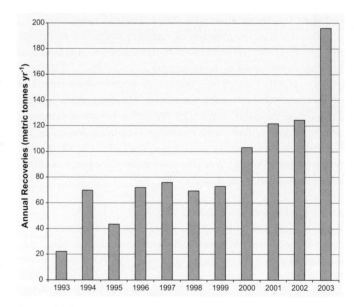

Figure 11.6. Impact of financial incentives on recovered volumes of refrigerants: doubling the financial inventive paid to service companies for returned refrigerant in 2003 in Australia resulted in an increase of more than 50% of refrigerant returned (RRA, 2004).

the return of recovered refrigerant. In the Netherlands, a subsidy of 50% on the cost of collection and destruction of halons and CFCs was provided to discourage venting.

Whilst there are no comparative studies of programmes available, logic and general behaviour would suggest that programmes that provide financial incentives, such as payments for return, will perform better compared to those that charge for the return of recovered refrigerant.

Funding of recovery programmes

There are numerous methods for funding refrigerant recovery programmes (Snelson and Bouma, 2003). In general, the ultimate user of the refrigerant, the equipment owner, will bear the cost in accordance with the polluter-pays principle. Where there is government funding (USA) and joint government and industry funding (Japan), the most common method of funding is the application of an industry-wide levy on sales of new refrigerant.

A good example is Australia, where a levy of Australian $1.00 (0.78 US$) per kilogramme was applied in 1993 (Bennett, 2001). The collected funds are held in a special purpose trust. Initially, the levy applied only to ozone-depleting refrigerants but it was extended in 2004 to "synthetic greenhouse gas" refrigerants to fund the expansion of the recovery programme to take back those products. Anti-trust issues were managed through application to, and approval from, the Australian Competition Commission. An industry-wide levy has numerous advantages, including cost sharing as a result of the contribution from all of industry, guaranteed revenue over the long term, and the flexibility to allow for increased recovery and subsequent costs.

Licensing, accreditation and standards

Improvements in technical standards and their conscientious application will result in significant reductions in direct and indirect emissions. The application of technical standards can be effectively achieved within a licensing or accreditation programme. The benefits of such programmes are: improved standards of workmanship; more efficient installations; lower leakage rates; improved and more regular maintenance practices; and increased rates of recovery, recycling, reuse and safe disposal. It should be noted that a common element of successful licensing schemes is a robust compliance programme.

The two best-known programmes of this type are the KMO programme in Denmark, and the STEK programme in the Netherlands. In the Netherlands, emissions have been significantly reduced through national mandatory regulations established in 1992 for CFCs, HCFCs, and HFCs, assisted by an industry-supported certification model. STEK is the abbreviation for the institution that certifies practices for installation companies in the refrigeration business. Elements of the regulation are detailed by Van Gerwen and Verwoerd (1998). They include, among other requirements, technical requirements to improve tightness, system commissioning to include pressure and leakage tests, refrigerant record keeping, periodic system inspections for leak tightness, maintenance and installation work by

certified companies and servicing personnel.

The success of the Dutch regulations and the STEK organization in reducing refrigerant emissions was demonstrated with results from a detailed study in 1999 of emission data from the refrigeration and air-conditioning sectors. For commercial refrigeration, annual refrigerant emissions (emissions during leakage plus disposal) were 3.2% of the total bank of refrigerant contained in this sector (Hoogen *et al.*, 2002). The overall loss rate from stationary equipment has been reduced to less than 5%, whilst the loss rate for automotive air conditioning has been reduced to less than 12%.

Influence of associated product-stewardship programmes

Product-stewardship programmes, sometimes known as producer responsibility, have been established in numerous countries for a range of products that contain refrigerants and foam agents. The best examples are the WEEE and EOL Vehicle Directives in Europe, and the Japanese Appliance and Recycling Act. The recovery and recycling of refrigerators, air-conditioning units and motor vehicles provides for the extraction of refrigerant from systems, and also for the recovery of the blowing agent from foam. In Japan, where 1,800 tonnes of fluorocarbons were recovered in 2003, electrical goods and automotive recycling programmes contributed 1,000 tonnes (RRC, 2003).

Foam

It is uncertain whether recovery, recycling or reclamation of the actual fluorinated gases from foam products will increase in the future and thereby have a substantial impact on emissions. For CFC-11, recovery from foams is discouraged by the chemical's low market value and by laws requiring the destruction of CFCs. For HFCs, recovery may be more attractive due to the higher market values of the chemicals. Even if the chemicals are not recovered, however, a major abatement option will be the recovery of used foam products and their subsequent destruction. Either destruction or chemical recovery could substantially reduce emissions of CFCs, HCFCs and HFCs from the banks in foams (see Table 7.10).

11.4.4 Destruction technologies and capacities

11.4.4.1 Current situation

The decomposition or destruction of ozone-depleting substances, HFCs and PFCs can be achieved using various technologies. The commercially available and successfully demonstrated technologies are listed by region in Table 11.8. A more detailed overview of destruction technologies and capabilities can be found in Appendix 11C.

The costs for recovery and recovery/recycling units start at approximately US$ 500 per unit. Recovery, recycling and reuse costs may be offset by the savings in new refrigerant purchases. Contaminated and unwanted refrigerant can be destroyed for less than US$ 3 per kilogram.

At the present time, the quantities of fluorinated gases destroyed by the techniques described above are extremely low;

Table 11.8. Commercially available destruction technologies by region.

Region	Country	Technology	Capacity
Australasia	Australia	Argon plasma	60 kg/hr fluorocarbons 120 kg/hr halons
Europe	France Germany Denmark Norway Switzerland Russian Federation	Reactor cracking Reactor cracking Incineration Cement kiln Rotary kiln Plasma chemical incineration	200 kg/hr fluorocarbons 200 kg/hr fluorocarbons
North America	Canada United States	Rotary kiln Reactor cracking Gas/fume oxidation Rotary kiln Cement kiln ICRF plasma	
Asia	Japan	Gas/fume oxidation ICRF plasma Rotary kiln Cement kiln Liquid injection	165 kg/hr fluorocarbons

potentially, they do not exceed a few thousand tonnes. In the European Union, the EU Regulation mandates the destruction of CFCs following their recovery; this also applies to foam insulation when recovered from dismantled equipment. Similarly, CFCs recovered from refrigeration and air-conditioning equipment must also be destroyed.

The different technologies involve varying costs, effluents and emissions, energy usage, and destruction efficiencies.

11.4.4.2 Future opportunities

Considerable scope exists to increase the quantities of CFCs, HCFCs , HFCs and PFCs recovered and then destroyed at the end of their useful life. This particularly applies to the quantities that are "banked" in foam insulation products. Recovery and destruction could have a substantial impact on emissions of the fluorinated gases. Further work is required to estimate whether the current number of destruction facilities globally could cope with the quantities of fluorinated gases recovered and destroyed. Another concern would apply to regions and countries where destruction facilities do not exist and where transport regulations for hazardous wastes prevent their transfer to such facilities in other countries.

11.5   Scenarios for 2015

11.5.1   Business-as-usual projections

The previous chapters have developed business-as-usual (BAU) projections for the use and emissions of CFCs, HCFCs, HFCs

and some PFCs (where these are used as replacements for ozone-depleting substances). These projections have assumed that all existing measures will remain in place, including the Montreal Protocol (phase-out) and relevant national regulations. It is also assumed that the usual practices and emission rates will remain unchanged up to 2015 and that recovery efficiency will not increase. In order to facilitate the calculations of emissions, estimates have been made of the size of the bank of ozone-depleting substance and fluorinated gases in equipment and applications in 2002. Table 11.9 summarizes the key assumptions of the business-as-usual projections.

The activities underlying emissions of fluorocarbons are expected to expand significantly between now and 2015. These activities (such as the requirements for refrigeration, air conditioning and insulation) will involve a number of technologies, including CFCs and HCFCs. In industrialized countries, the use and emissions of CFCs and HCFCs will decline as obsolete equipment is retired. In developing countries, ozone-depleting substances (particularly HCFCs) may be used for most of the first half of this century and significant growth is expected. These changes, and their impacts as developed by previous chapters, are reflected in the data in Table 11.2 (demand), 11.5 (banks), and 11.6 (emissions).

The fall in CFC emissions is not accompanied by a similar increase in emissions of HFCs because of continuing trends towards non-HFC technology and substitutes with lower GWPs. Additional factors not included in the BAU scenario are the capture and safe disposal of materials that would historically have been emitted. The business-as-usual case assumes the

Table 11.9. Key assumptions in the business-as-usual (BAU) and mitigation (MIT) scenarios.

Sector (Refrigeration SAC and MAC)	Annual market growth 2002-2015 (both in BAU and MIT)[1] (% yr⁻¹)				Best-practice assumptions								
	EU % yr⁻¹	USA % yr⁻¹	Japan % yr⁻¹	DCs[1] % yr⁻¹	Type of reduction	EU BAU	EU MIT	USA BAU	USA MIT	Japan BAU	Japan MIT	DCs[1] BAU	DCs[1] MIT
Domestic refrigeration	1	2.2	1.6	2–4.8	Substance	HFC-134a / HC-600a	HC-600a	HFC-134a	HFC-134a / HC-600a (50%)	HFC-134a	HC-600a	CFC-12 / HFC-134a	Plus HC-600a (50% in 2010)
					Recovery	0%	80%	0%	80%	0%	80%	0%	50%
Commercial refrigeration	1.8	2.7	1.8	2.6–5.2	Substance	R-404A R-410A	R-404A / R-404A (50%)	HCFC-22 / R-404A /	R-404A / R 410A (50%)	HCFC / R-404A	R-404A R-410A (50%)	CFC / HCFC	R-404A / R 410A (50%)
					Recovery	50%	90%	50%	90%	50%	90%	25%	30%
					Charge		-30%		-30%		-30%		-10%
Industrial refrigeration	1	1	1	3.6–4	Substance	HFC-NH₃ (35%)	HFC-NH₃ (70%)	HCFC / HFC-NH₃ (60%)	HCFC / HFC-NH₃ (80%)	HCFC / HFC-NH₃ (35%)	HCFC / HFC-NH₃ (70%)	CFC / HCFC-22	NH₃ (40–70%)
					Recovery	50%	90%	50%	90%	50%	90%	15–25%	50%
					Charge		-40%		-40%		-40%		-10%
Transport refrigeration	2	3	1	3.3–5.2	Substance	HFCs	HFCs	HCFCs / HFCs	HCFCs / HFCs	HCFCs / HFCs	HCFCs / HFCs	CFC / HCFC-22	Plus HFCs, up tp 30%
					Recovery	50%	80%	50%	70%	50%	70%	0%	20–30%
Stationary AC	3.8	3	1	5.4–6	Substance	HFCs	HFCs	HCFCs / HFCs	HCFCs / HFCs	HCFCs / HFCs	HCFCs / HFCs	CFC / HCFC-22	CFC / HCFC-22 (HFCs 30% in some DCs)
					Recovery	50%	80%	30%	80%	30%	80%	0%	50%
					Charge		-20%		-20%		-20%		
Mobile AC	4	4	1	6–8	Substance	HFC-134a / CO₂ (10%) as of 2008	HFC-134a / CO₂ (50%) as of 2008	HFC-134a	HFC-134a / CO₂ (30%) as of 2008	HFC-134a	HFC-134a / CO₂ (30%) as of 2008	CFC / HCFC-134a	CFC / HCFC-134a
					Recovery	50%	80%	0%	70%	0%	70%	0%	50%
					Charge	700 g	500 g	900 g	700 g	750 g	500 g	750–900 g	750–900 g

Table 11-9 (continued)

Sector	Annual market growth 2002-2015 (both in BAU and MIT)[1] (% yr[-1])	Best-practice assumptions	
Foams	About 2% yr[-1]	BAU	
		MIT	Assumptions on substance use (see Chapter 7) HFC consumption reduction: A linear decrease in use of HFCS between 2010 and 2015 leading to 50% reduction by 2015. Production / installation improvements: The adoption of production emission reduction strategies from 2005 for all block foams and from 2008 in other foam sub-sectors. End-of-life management options: The extension of existing end-of-life measures to all appliances and steel-faced panels by 2010 together with a 20% recovery rate from other building-based foams from 2010.
Medical aerosols	1.5–3% yr[-1]	BAU	Partial phase-out of CFCs
		MIT	Complete phase-out of CFCs
Fire protection	−4.5% yr[-1] (all substances) +0.4% yr[-1] (HCFCs/HFCs/PFCs)	BAU	Phase-out of halons
		MIT	Not quantifiable
HFC-23 byproduct	2.5% yr[-1]	BAU	HFC-23 emissions of existing production capacity: 2% of HCFC-22 production (in kt) HFC-23 emissions of new production capacity: 4% of HCFC-22 production (in kt)
		MIT	100% implementation of reduction options (90% emission reduction)
Non-medical aerosols	16% increase period in total CO-weighted emissions over 2002–2015	BAU	See Chapter 10
		MIT	Not quantifiable

[1] BAU: Business-As-Usual Scenario; MIT: Mitigation Scenario; DCs: developing countries

continuing application of all existing measures and the alternative scenario(s) embody improvements that could be implemented assuming global application of current best-practice emission reduction techniques.

Future emissions from use

Table 11.6 above shows, for the business-as-usual case, the expected future emissions arising from the use of HFCs in all sectors, predominantly refrigeration, foam blowing, solvents and aerosols. Sectoral emission functions that reflect the current levels of containment were applied to the activities projected for these uses so that growth in emissions reflects growth in use. These emissions will be accompanied by significant changes in the banks of fluorocarbons remaining in equipment and not yet released. In the case of HFC-134a, the quantity retained in equipment is expected to increase in this scenario to 2.2 million metric tonnes by 2015 from the 2002 level of 517,000 metric tonnes (Ashford *et al.*, 2004).

Growth was calculated as a projection of current demand patterns and has no connection to the global capability to produce HFCs. The most important of these materials continues to be HFC-134a. However, current production capacity for this material is about 180,000 tonnes yr^{-1}, so that the increase in demand (and therefore emissions) after 2008 cannot be sustained and would require the construction of new plants.

Future atmospheric concentrations of HFC-134a that would result from these emissions (solid line) are indicated in Figure 11.2. In the second scenario (dashed line), the current best practices for containment are implemented in all new equipment, recovery efficiency and the extent of recovery are improved, and the quantities in systems are reduced. This has the effect of reducing the atmospheric concentration (and also the climate impact) in 2015 by some 23% but still requires additional capacity. The projected emissions are given in Table 11.6 as well. In this case, no expansion of current capacity would be needed.

Byproduct emissions of HFC-23

Emissions of HFC-23 from the production of HCFC-22 have significant climate impact. As discussed above, these emissions may be reduced by changes to process procedures or eliminated by thermal oxidation without compromising the ability to produce HCFC-22, which will continue to be required as a fluoropolymer feedstock after its commercial uses have been phased out. The emissions of HFC-23 expected under the business-as-usual case to 2015 are shown in Table 10.10. It has been assumed that emissions from existing capacity (in both developed and developing countries) will continue at 2% of HCFC-22 production and that new capacity (mainly in developing countries) will emit HFC-23 at a rate of 4%. Consequently, emissions of HFC-23 could increase by 60% between now and 2015 from about 15,000 tonnes yr^{-1} in 2003 to 23,000 tonnes yr^{-1}.

The US EPA (2004) has developed a business-as-usual scenario for HFC-23, projecting emissions of 100 $MtCO_2$-eq in 2000, 115 $MtCO_2$-eq in 2010 and 130 $MtCO_2$-eq in 2020. The difference between the EPA projections and those in this report is primarily due to EPA's use of a lower emission factor for developing countries (3% instead of 4%).

11.5.2 Mitigation scenario for supply and demand through to 2015

As shown above, previous chapters have identified the potential impact on demand (and supply) under BAU assumptions (i.e. all existing measures continue to be applied). These chapters have also considered the potential impact of the global application of current best-practice emission reduction techniques. The key assumptions of the mitigation scenario are summarized again in Table 11.9, and Tables 11.2, 11.5, and 11.6 present the implications for demand, banks, and emissions.

Byproduct emissions of HFC-23

The alternative scenario for byproduct emissions of HFC-23 assumes that the current best-practice technology, comprising the capture and thermal oxidation of "vent gases", is introduced into all facilities progressively from 2005 onwards. Destruction technology is assumed to be 100% efficient and to operate for 90% of the on-line time of the HCFC-22 plant. Reduced emissions were calculated on the basis of the same level of activity (in the form of assumed future HCFC-22 production) as the business-as-usual case. The difference between the two HFC-23 forecasts is therefore solely due to the extent of deployment of destruction technology. The forecasts represent potential extreme cases and future changes in activity will tend to increase the probability of one forecast or the other.

11.5.3 Comparison with other scenarios

Future global HFC and PFC use and/or emissions as substitutes for ozone-depleting substances have been separately estimated by IPCC (2000), Midgley and McCulloch (1999), McCulloch (2000) and Madronich *et al.* (1999). Midgley and McCulloch (1999) projected carbon-equivalent emissions of HFCs and PFCs (excluding unintended chemical byproduct emissions) at 60 MtC-eq in 2000, 150 MtC-eq in 2010 and 280 MtC-eq in 2020 (220, 550 and 1030 $MtCO_2$-eq, respectively). Considering that emissions lag behind consumption by many years, the Midgley and McCulloch figures are much larger than the values in Madronich *et al.* This discrepancy is consistent with the Midgley and McCulloch scenario which was constructed to represent plausible upper limits to future emissions (McFarland, 1999).

The Intergovernmental Panel on Climate Change published in its Special Report on Emissions Scenarios (IPCC, 2000), scenarios for emissions of HFCs through to 2100 (Figure 2.9 and Table 11.10). Scenarios are alternative images of how the future might unfold and are used to analyze the effects of different driving forces on emissions outcomes and to allow contingency planning; they are not predictions or projections of future emissions. The emission values contained within the SRES report

Table 11.10. Global anthropogenic emissions (kt) projections for the years 2000, 2010 and 2020 for ODS, HFC and PFC emissions in the four marker scenarios (from SRES, 2000, Table 5-8, and http://sres.ciesin.org), and comparison with the 2002 and 2015-BAU results from this report.

Emissions (kt)		SRES Marker Scenario				SROC-BAU
		A1	A2	B1	B2	(This report)
2000 (SRES) /	ODSs			842			569
2002 (SROC)	HFC-23	12.6				
	HFC-134a	80.0				
	HFCs – Total	93	124			
	PFCs – Total	14	0.1			
2010	ODSs			786		
	HFC-23	14.8	14.8	14.8	14.8	
	HFC-32	3.7	3.5	3.2	3.3	
	HFC-43-10	7.2	6.8	6.2	6.3	
	HFC-125	12.1	11.4	10.9	11.1	
	HFC-134a	175.7	165.8	162.9	166.4	
	HFC-143a	9.3	8.8	8.0	8.1	
	HFC-227ea	13.1	11.8	13.2	13.6	
	HFC-245ca	62.1	58.6	59.7	61.4	
	HFCs – Total	**298**	**282**	**279**	**285**	
	PFCs – Total	**17**	**22**	**16**	**23**	
2015	ODSs			519			538
(interpolation	2000/2-2015 growth rate			(-3% yr⁻¹)			(1% yr⁻¹)
for SRES)	**HFCs – Total**	**408**	**342**	**340**	**354**	**415**
	2000/2-2015 growth rate	(10% yr⁻¹)	(9% yr⁻¹)	(9% yr⁻¹)	(9% yr⁻¹)	(10% yr⁻¹)
	PFCs – Total	**20**	**25**	**17**	**26**	**0.02**
	2000/2-2015 growth rate	(2% yr⁻¹)	(4% yr⁻¹)	(1% yr⁻¹)	(4% yr⁻¹)	(-12% yr⁻¹)
2020	ODSs	253				
	HFC-23	4.9	4.9	4.9	4.9	
	HFC-32	8.3	6.4	6.0	6.2	
	HFC-43-10	8.8	7.6	6.9	7.2	
	HFC-125	27.1	20.7	20.6	21.5	
	HFC-134a	325.5	252.2	248.8	261.9	
	HFC-143a	20.6	16.0	15.0	15.6	
	HFC-227ea	22.2	16.6	18.5	19.7	
	HFC-245ca	100.5	78.7	80.3	85.4	
	HFCs – Total	**518**	**403**	**401**	**422**	
	PFCs – Total	**23**	**28**	**17**	**30**	

are scenarios of potential emissions to the year 2100 assuming that no action is taken to limit emissions based on concerns over climate change. The scenarios are based upon historical market information and expected developments in use and emissions. Projected emissions in SRES are lower than the reported 2015 emissions in the BAU scenario of this report, although annual growth rates are more or less the same (Table 11.11). These differences are the consequence of differences in scenario methodologies and assumptions.

The Mitigation Scenario demonstrates that emission levels can be significantly reduced below those of business as usual leading to reductions of over 200,000 metric tonnes of HFCs (Table 11.11). This clearly demonstrates that potential benefits

could be gained by implementing measures such as improved containment, recovery and recycling and end of life treatment of HFCs. Such measures would reduce the environmental impact of HFCs and PFCs and allow the benefits of their use in many applications (such as energy efficiency, non-flammability etc.) to be maintained.

A paper by Harnisch and Höhne (2002) contains estimates of historic emissions based on atmospheric concentrations. These are consistent with the data in this chapter and in AFEAS (2004) and Madronich *et al.*, 1999. A similar paper on halogenated compounds and climate change (Harnisch *et al.*, 2002) contains graphical estimates of future emissions in which HFC-23 levels are expected to fall markedly (potentially a factor of 5

Table 11.11. Summary of direct emission projections for 2015 for CFCs, halons, HCFCs, HFCs and PFCs.

	2015 business-as-usual		2015 mitigation scenario		2015 reduction potential	
	ktonnes yr$^{-1\,a)}$	MtCO$_2$-eq $^{b)}$	ktonnes yr$^{-1\,a)}$	MtCO$_2$-eq $^{b)}$	ktonnes yr$^{-1\,a)}$	MtCO$_2$-eq $^{b)}$
Halons$^{c)}$	3	[12 (9)]	3	[12 (9)]	-	-
CFCs	43	338 (264)	30	221 (173)	13	117 (91)
HCFCs	492	828 (699)	292	484 (408)	200	344 (290)
HFCs	415	1,153 (999)	184	416 (369)	231	737 (630)
PFCs	0.02	0.2 (0.2)	0.02	0.2 (0.2)	-	-
Total$^{c)}$	**953**	**2,319 (1,961)**	**509**	**1,121 (951)**	**444**	**1,198 (1,011)**

Note: Direct GWPs for a 100-year time horizon were used from IPCC 2001 and WMO 2003 (as listed in Table 2-6). Bracketed data use direct GWPs for a 100-year time horizon from SAR/TAR.

$^{a)}$ Excluding non-medical aerosols.

$^{b)}$ Including non-medical aerosols and solvents.

$^{c)}$ Halons cause much larger negative indirect than positive direct radiative forcing and, in the interests of clarity, their positive direct effects are not included in the totals showing total CO$_2$-equivalents.

to 10 between 1996 and 2020). This is consistent with the HFC-23 mitigation scenario rather than the business-as-usual scenario. The same seems to apply to future emissions of HFCs:

- The total HFC emissions in the graph in Harnisch *et al.* (2002) are approximately 200, 450 and 700 MtCO$_2$-eq yr^{-1} for the years 1996, 2010 and 2020 (IPCC (1996) SAR values for GWPs).
- In this report (using 1996 SAR/ 2001 TAR GWP values for consistency), the values for the business-as-usual scenario are 375 MtCO$_2$-eq yr^{-1} in 2002 (159 MtCO$_2$-eq of which are from HFC-23), increasing to 1001 MtCO$_2$-eq in 2015 (272 MtCO$_2$-eq of which are from HFC-23).
- In the mitigation scenario, these emissions are reduced to 368 MtCO$_2$-eq yr^{-1} in 2015 (27 MtCO$_2$-eq of which are from HFC-23).

The Harnisch *et al.* (2002) numbers are, therefore, consistent with this report up to 2010; from then on they continue to increase, whereas the mitigation scenario envisages emissions under better control. Similarly, the values calculated for developed countries by the US EPA (2001) are also consistent with those considered here (as the sub-set of the same countries in the global databases).

11.5.4 Aggregate additional emission reduction opportunities

Tables 11.11 and 11.12 summarize the projections of the business-as-usual and mitigation scenario emissions for 2015. In the business-as-usual scenario, total direct emissions are projected to represent approximately 2.3 GtCO$_2$-eq yr^{-1} in 2015 (2.0 GtCO$_2$-eq yr^{-1} using SAR/TAR values), as compared to about 2.5 GtCO$_2$-eq yr^{-1} in 2002 (2.0 GtCO$_2$-eq yr^{-1} using SAR/TAR values).

Refrigeration applications, and stationary and mobile air-conditioning (MAC) contribute most to global direct green-house-gas emissions. A significant reduction in CFC emissions, notably from refrigeration banks, can be expected during the next decade (CFC emissions fall to approximately 0.3 GtCO$_2$-eq yr^{-1} in 2015). Most greenhouse-gas emissions from foams are expected to occur after 2015. HCFC emissions are projected to increase by a factor of two between 2002 and 2015, owing to a steep increase in the use in refrigeration and air-conditioning (AC) applications, in particular in commercial refrigeration and stationary air-conditioning (commercial refrigeration represents about 25% of the total of all refrigeration, AC and MAC systems). HFC emissions are projected to increase by about a factor of three, particularly in refrigeration, the stationary AC and MAC sectors, and due to byproduct emissions of HFC-23 during HCFC-22 production. It is expected that these projected increases will be accompanied by a build-up of banks and a corresponding added potential for future radiative forcing, depending on bank management and, notably, end-of-life measures. Action in these sub-sectors could therefore have a substantial influence on future emissions of HCFCs and HFCs.

There are opportunities to reduce direct emissions significantly through the global application of best practices and recovery methods, with a reduction potential of about 1.2 GtCO$_2$-eq yr^{-1} (1.0 GtCO$_2$-eq yr^{-1} using SAR/TAR values) of direct emissions by 2015, as compared to the BAU scenario. About 60% of this potential is HFC emission reduction; HCFCs and CFCs contribute about 30% and 10% respectively. A key factor determining whether this potential will be realized are the costs associated with the implementation of the measures to achieve the emission reduction. These vary considerably from a net benefit to 300 US$/tCO$_2$-eq (Chapters 4–10 of this report, US- EPA, 2004). Chapters 4–10 of this report have identified various reduction opportunities. The major opportunities for reducing greenhouse-gas emissions in the period prior to 2015 include:

Refrigeration: In refrigeration applications, direct green-house-gas emissions can be reduced by 10% to 30%. For the refrigeration sector as a whole, the mitigation scenario indicates an overall direct-emission reduction of about 490 MtCO$_2$-eq

yr^{-1} by 2015 (411 $MtCO_2$-eq yr^{-1} using SAR/TAR values), with most being predicted for commercial refrigeration. Specific costs are in the range of 10 to 300 $US\$/tCO_2$-eq[2]. Improved system energy efficiencies can also significantly reduce indirect greenhouse-gas emissions. In full supermarket systems, reductions of values for life cycle climate performance (LCCP) of up to 60% can be obtained by using direct expansion systems with alternative refrigerants, improved containment, distributed systems, indirect systems or cascade systems.

The emission reduction potential in domestic refrigeration is relatively small. Indirect emissions of systems using either HFC-134a or isobutane (HC-600a) dominate total emissions, practically regardless of the carbon intensity of electric power generation. The difference between the LCCP of HFC-134a and isobutane systems is small and end-of-life recovery can further reduce the magnitude of the difference.

Residential and commercial air conditioning and heating: Direct greenhouse-gas (GHG) emissions of residential and commercial air conditioning and heating equipment can be reduced by about 200 $MtCO_2$-eq yr^{-1} (169 $MtCO_2$-eq yr^{-1} using SAR/TAR values) relative to the BAU scenario (2015). Specific costs range from –3 to 170 $US\$/tCO_2$-eq[2]. Improved system energy efficiencies can significantly reduce indirect GHG emissions, leading in some cases to overall savings of 75 $US\$/tCO_2$-eq. Opportunities to reduce direct GHG (i.e. refrigerant) emissions can be found in (i) more efficient recovery of refrigerant at end-of-life (in the mitigation scenario, this is assumed to be up to 50% and 80% for developing and developed countries, respectively); (ii) charge reduction (up to 20%); (iii) better containment and (iv) the use of non-fluorocarbon refrigerants in suitable applications.

Mobile air-conditioning: In mobile air-conditioning, a reduction potential of 179 $MtCO_2$-eq yr^{-1} (162 $MtCO_2$-eq yr^{-1} using SAR/TAR values) could be achieved by 2015 at a cost of 20 to 250 $US\$/tCO_2$-eq[2]. Specific costs differ per region and per solution. Improved containment, and end-of-life recovery (both of CFC-12 and HFC-134a) and recycling (of HFC-134a), could reduce direct GHG emissions by up to 50%, and total (direct and indirect) GHG emissions of MAC units by 30 to 40%. New systems with either CO_2 or HFC-152a are likely to penetrate the market in the coming decade, leading to total GHG system emission reductions estimated at 50 to 70% in 2015. Hydrocarbons are in use as service refrigerants in several countries, despite manufacturer recommendations.

Foams: Because of the long life span of most foam applications, only a limited emission reduction of about 17 $MtCO_2$-eq yr^{-1} (about 15 $MtCO_2$-eq yr^{-1} using SAR/TAR values) by 2015 is projected, with costs ranging from 10 to 100 $US\$/tCO_2$-eq[2]. There are two key areas of potential emission reduction in the foams sector. The first is a potential reduction in halocarbon

use in newly manufactured foams. However, the enhanced use of blends and the further phase-out of fluorocarbon use both depend on further technology development and market acceptance. Action to reduce HFC use by 50% between 2010 and 2015 would result in an emission reduction of about 10 $MtCO_2$-eq yr^{-1} (about 9 $MtCO_2$-eq yr^{-1} using SAR/TAR values), with further reductions thereafter, at a cost of 15 to 100 $US\$/tCO_2$-eq[2].

The second opportunity can be found in the worldwide banks of halocarbons contained in insulating foams in existing buildings and appliances (about 9 and 1 $GtCO_2$-eq for CFC and HCFC respectively in 2002 (about 7 and 1 $GtCO_2$-eq using SAR/TAR values)). Although recovery effectiveness is yet to be proven, particularly in the buildings sector, commercial operations are already recovering halocarbons at 10 to 50 $US\$/tCO_2$-eq[2] for appliances. Emission reductions may be about 7 $MtCO_2$-eq yr^{-1} (about 6 $MtCO_2$-eq yr^{-1} using SAR/TAR values) in 2015. However, this potential could increase significantly in the period between 2030 and 2050, when large quantities of building insulation foams will be decommissioned.

Medical Aerosols: The reduction potential for medical aerosols is limited due to medical constraints, the low emission level and the higher costs of alternatives. The major contribution (14 $MtCO_2$-eq yr^{-1} by 2015 compared to a BAU emission of 40 $MtCO_2$-eq yr^{-1} (using SAR/TAR values, this is 10 as compared to 34 MtCO2-eq yr^{-1}) to a reduction of GHG emissions for metered dose inhalers (MDIs) would be achieved by the completion of the transition from CFC to HFC MDIs beyond what is already assumed to be BAU. The health and safety of the patient is of paramount importance in treatment decisions, and there are significant medical constraints on the use of HFC MDIs. If salbutamol MDIs (approximately 50% of total MDIs) were to be replaced by dry powder inhalers (this is not an assumption in the mitigation scenario) this would only result in an annual emission reduction of about 10 $MtCO_2$-eq yr^{-1} by 2015, at a project cost in the range of 150 to 300 $US\$/tCO_2$-eq.

Fire protection: In fire protection, the potential reduction by 2015 is small due to the relatively low emission level, the significant shifts to not-in-kind alternatives in the past and the lengthy procedures for introducing new equipment. Direct GHG emissions for the sector are estimated at about 5 $MtCO_2$-eq yr^{-1} (about 4 $MtCO_2$-eq yr^{-1} using SAR/TAR values) in 2015 (BAU). Agents with no climate impact have replaced 75% of original halon use . Four percent of the original halon applications continue to employ halons. The remaining 21% has been replaced by HFCs, with a small number of applications switching to HCFCs and to PFCs. PFCs are no longer needed for new fixed systems and are limited to use as propellants in one manufacturer's portable-extinguisher agent blend. Due to the lengthy process of testing, approval and market acceptance for new types of equipment and agents for fire protection, no additional options are likely to have an appreciable impact by 2015. With the introduction of a fluoroketone (FK) in 2002, additional reductions at increased cost will be possible in this sector through 2015. Currently, those reductions are estimated to

---

[2] The presented cost data relate to direct emission reductions only. Taking into account energy efficiency, improvements may even result in net negative specific costs (savings).

be small compared to other sectors.

For *non-medical aerosols and solvents*, the reduction potentials are likely to be rather small because most remaining uses are critical to performance or safety. The projected BAU emissions by 2015 for solvents and aerosols are about 14 and 23 MtCO$_2$-eq yr^{-1} respectively (the figures are about the same with SAR/TAR values). The replacement of HFC-134a by HFC-152a in technical aerosol dusters is a leading option for reducing GHG emissions. For contact cleaners and mould release agents for plastic casting, the substitution of HCFCs by HFEs and HFCs with lower GWPs offers an opportunity. Some countries have banned HFC use in novelty aerosol products, although HFC-134a continues to be used in many countries for safety reasons.

A variety of organic solvents can replace HFCs, PFCs and ozone-depleting substances in many applications. These alternative fluids include lower-GWP compounds such as traditional chlorinated solvents, HFEs, and n-propyl bromide. Many not-in-kind technologies, including hydrocarbon and oxygenated solvents, are also viable alternatives in some applications.

The destruction of byproduct emissions of HFC-23 from HCFC-22 production has a reduction potential of up to 300 MtCO$_2$-eq yr^{-1} by 2015 (245 MtCO$_2$-eq yr^{-1} using SAR/TAR values) and specific costs below 0.2 US$/tCO2-eq according to two European studies in 2000. Reduction of HCFC-22 production due to market forces or national policies, or improvements in facility design and construction also could reduce HFC-23 emissions.

References

AFEAS (Alternative Fluorocarbons Environmental Acceptability Study), 2004: Alternative Fluorocarbons Environmental Acceptability Study – Production, Sales and Atmospheric Releases of Fluorocarbons through 2002. RAND ES and P Center, Arlington, VA, USA (available online at www.afeas.org).

Ashford, P., D. Clodic, L. Palandre, A. McCulloch and L. Kuijpers, 2004: Determination of Comparative HCFC and HFC Emission Profiles for the Foam and Refrigeration Sectors until 2015. Part 1: Refrigerant Emission Profiles [L. Palandre and D. Clodic, Armines, Paris, France, 132 pp.], Part 2: Foam Sector [P. Ashford, Caleb Management Services, bristol, UK, 238 pp.], Part 3: Total Emissions and Global Atmospheric Concentrations [A. McCulloch, Marbury Technical Consulting, Comberbach, UK, 77 pp.]. Reports prepared for ADEME and US EPA.

Bennett, M., 2001: Refrigerant Reclaim Australia. Proceedings of the IIR conference 'Refrigerant Management and Destruction Technologies of CFCs', August 29-31, 2001, Dubrovnik, Croatia. International Institute of Refrigeration (IIR/IIF), Paris, France.

Gerwen, R.J.M. van, M. Verwoerd, 1998: Dutch Regulations for Reduction of Refrigerant Emissions: Experiences with a Unique Approach over the period 1993-1998. ASERCOM Symposium: Refrigeration/Air conditioning and Regulations for Environment Protection – A Ten Years Outlook for Europe, 7 October 1998, Nürnberg, Germany.

Harnisch, J. and N. Höhne, 2002: Comparison of emissions estimates derived from atmospheric measurements with national estimates of HFCs, PFCs and SF$_6$. *Environ. Sci. & Pollut. Res.*, **9**(5), 315-320.

Harnisch, J., D. de Jager, J. Gale and O. Stobbe, 2002: Halogenated compounds and climate change. *Environ. Sci. & Pollut. Res.*, **9**(6), 369-374.

Hoogen, B. van den, and H. van der Ree, 2003: The Dutch Approach to Reduce Emissions of Fluorinated Greenhouse Gases. Proceedings of the 21st International Congress of Refrigeration, Washington, DC, USA, 17-22 August 2003. International Institute of Refrigeration (IIR/IIF), Paris, France.

IPCC (Intergovernmental Panel on Climate Change), 1996: *Climate Change 1995: The Science of Climate Change. Contribution of Working Group I to the Second Assessment Report of the Intergovernmental Panel on Climate Change* [Houghton, J.T., L.G. Meira Filho, B.A. Callander, N. Harris, A. Kattenberg and K. Maskell (eds.)]. Cambridge University Press, Cambridge, United Kingdom, and New York, NY, USA, 572 pp.

IPCC (Intergovernmental Panel on Climate Change), 2001: *Climate Change 2001: The Scientific Basis. Contribution of Working Group I to the Third Assessment Report of the Intergovernmental Panel on Climate Change* [Houghton, J. T., Y. Ding, D. J. Griggs, M. Noguer, P. J. van der Linden, X. Dai, K. Maskell, and C. A. Johnson (eds.)]. Cambridge University Press, Cambridge, United Kingdom, and New York, NY, USA, 944 pp.

IPCC (Intergovernmental Panel on Climate Change), 2000: *Emissions Scenarios.* Special Report on Emissions Scenarios to the

Intergovernmental Panel on Climate Change [Nakicenovic, N. and R. Swart (eds.)]. Cambridge University Press, Cambridge, United Kingdom, and New York, NY, USA, 570 pp.

Irving, W.N. and M. Branscombe, 2002: HFC-23 Emissions from HCFC-22 Production. In *Background Papers – IPCC Expert Meetings on Good Practice Guidance and Uncertainty Management in National Greenhouse Gas Inventories*, IPCC/OECD/IEA Programme on National Greenhouse Gas Inventories, published by the Institute for Global Environmental Strategies (IGES), Hayama, Kanagawa, Japan, pp. 271-283 (available online at www.ipcc-nggip.iges.or.jp/public/gp/gpg-bgp.htm).

Madronich, S., G.J.M. Velders, J.S. Daniel, M. Lal, A. McCulloch and H. Slaper, 1999: Halocarbon Scenarios for the Future Ozone Layer and Related Consequences. Chapter 11 of *Scientific Assessment of Ozone Depletion: 1998*. D.L. Albritton, P.J. Aucamp, G. Megie and R.T. Watson (eds), World Meteorological Organization Global Ozone Research and Monitoring Project, Report No 44, WMO, Geneva, Switzerland.

McCulloch, A., 2000: Halocarbon Greenhouse Gas Emissions During The Next Century. In: *Non-CO$_2$ Greenhouse Gases: Scientific Understanding, Control and Implementation. Proceedings of the Second International Symposium, Noordwijkerhout, The Netherlands, September 8-10, 1999*, J. van Ham, A.P.M. Baede, L.A. Meyer and R. Ybema (eds.), Kluwer Academic Publishers, Dordrecht, The Netherlands, pp. 223-230.

McFarland, M., 1999: Applications and Emissions of Fluorocarbon Gases: Past, Present and Prospects for the Future. Proceedings of the Joint IPCC/TEAP Expert Meeting on Options for the Limitation of Emissions of HFCs and PFCs, L. Kuijpers, R. Ybema (eds.), 26-28 May 1999, Energy Research Foundation (ECN), Petten, The Netherlands (available online at www.ipcc-wg3.org/docs/IPCC-TEAP99).

Midgley, P and A. McCulloch, 1999: Properties and Applications of Industrial Halocarbons. Chapter 5 of *Reactive Halogen Compounds in the Atmosphere, Vol. 4E of The handbook of Environmental Chemistry*, O. Hutzinger (ed.), Springer-Verlag, Berlin, Germany.

Olivier, J.G.J., 2002: Greenhouse gas emissions. Part III in *CO$_2$ Emissions from Fuel Combustion 1971-2000* (2002 Edition). International Energy Agency (IEA), Paris, France, pp. III.1-III.31.

Olivier, J.G.J. and J.J.M. Berdowski, 2001: Global emissions sources and sinks. In *The Climate System*, J. Berdowski, R. Guicherit and B.J. Heij (eds.). A.A. Balkema Publishers/Swets & Zeitlinger Publishers, Lisse, The Netherlands, pp. 33-78.

Prather, M., P. Midgley, F.S. Rowland and R. Stolarski, 1996: The ozone layer: the road not taken. *Nature*, 381, 551-554.

RRA (Refrigerant Reclaim Australia), 2004: Annual Report 2004. Refrigerant Reclaim Australia, Canberra, Australia.

RRC (Refrigerants Recycling Promotion and Technology Center), 2003: Recovery, Reclamation, Reuse and Destruction Technology of Fluorocarbons used as Refrigerants, Toky, Japan.

Snelson, K. and J. Bouma (eds.), 2003: Refrigerant Recovery, Recycling, Reclamation and Disposal – An International Assessment, Parts 1 and 2. IEA Heat Pump Centre, Sittard, Netherlands 2003.

UNEP (United Nations Environment Programme), 2003: Extended Desk Study on RMP Evaluation (UNEP/OzL.Pro/ExCom/39/14, March 2003), Multilateral Fund for the Implementation of the Montreal Protocol, Montreal, Canada.

UNEP (United Nations Environment Programme), 2004a: Report on Implementation of Approved Projects with Specific Reporting Requirements (UNEP/OzL.Pro/ExCom/42/14 and Corr.1), Multilateral Fund for the Implementation of the Montreal Protocol, Montreal, Canada.

UNEP (United Nations Environment Programme), 2004b: Information Provided by Parties in Accordance with Article 7 of the Montreal Protocol on Substances that Deplete the Ozone Layer (UNEP/OzL.Pro.16/4, 18 October 2004). Sixteenth Meeting of the Parties to the Montreal Protocol on Substances that Deplete the Ozone Layer, Prague, 22–26 November 2004. UNEP Ozone Secretariat, Nairobi, Kenya, 110 pp.

UNEP-TEAP (Technology and Economic Assessment Panel), 1998: 1998 Assessment Report of the Technology and Economic Assessment Panel. UNEP Ozone Secretariat, Nairobi, Kenya, 300 pp.

UNEP-TEAP (Technology and Economic Assessment Panel), 1999b: The Implications to the Montreal Protocol of the Inclusion of HFCs and PFCs in the Kyoto Protocol. October 1999 Report of the TEAP HFC and PFC Task Force. [J. Phillips (ed.)]. UNEP Ozone Secretariat, Nairobi, Kenya, 204 pp.

UNEP-TEAP (Technology and Economic Assessment Panel), 2000: April 2000 Report of the Technology and Economic Assessment Panel. [W. Brunner, S. Machado Carvalho and L. Kuijpers (eds.)]. UNEP Ozone Secretariat, Nairobi, Kenya, 193 pp.

UNEP-TEAP (Technology and Economic Assessment Panel), 2002a: April 2002 Report of the Technology and Economic Assessment Panel – Volume 3A: Report of the Task Force on Collectrion, Recovery and Storage. [W. Brunner, J. Pons and S.O. Andersen (eds.)]. UNEP Ozone Secretariat, Nairobi, Kenya, 112 pp.

UNEP-TEAP (Technology and Economic Assessment Panel), 2002b: April 2002 Report of the Technology and Economic Assessment Panel – Volume 3B Report of the Task Force on Collectrion, Recovery and Storage. [S. Devotta, A. Finkelstein and L. Kuijpers (eds.)]. UNEP Ozone Secretariat, Nairobi, Kenya, 150 pp.

UNEP-TEAP (Technology and Economic Assessment Panel), 2004: May 2004 Report of the Technology and Economic Assessment Panel – Progress Report. [L. Kuijpers (ed.)]. UNEP Ozone Secretariat, Nairobi, Kenya, 115 pp.

US EPA (US Environmental Protection Agency), 2001: Non-CO$_2$ Greenhouse Gas Emissions from Developed Countries: 1990-2010. [E. Scheehle (ed.)]. Report No. EPA-430-R-01-007, US Environmental Protection Agency, Office of Air and Radiation, Washington, DC, USA, 132 pp.

US EPA (US Environmental Protection Agency), 2004: Analysis of Costs to Abate International Ozone Depleting Substance Substitute Emissions. [D.S. Godwin (ed.)]. US EPA Report 430-R-04-006, US Environmental Protection Agency, Washington, DC, USA, 309 pp (available online at http://www.epa.gov/ozone/snap/).

Appendix 11A Chemical routes of manufacture

Manufacturing HFCs and PFCs involves complex processes, and the handling of the highly toxic chemical, anhydrous hydrogen fluoride (HF). This has resulted in low global numbers of manufacturers and production units being situated where there is relatively straightforward access to raw materials and the basic infrastructure to handle toxic chemicals. In general, companies that have, or are currently, manufacturing CFCs and/or HCFCs are now producing HFCs. PFC manufacture requires more complex manufacturing processes and is generally carried out by specialist companies on a smaller scale than HFC production.

The main routes for the manufacture of HFCs, HCFCs and PFCs are shown in Figure 11A.

Production pathways of HFCs

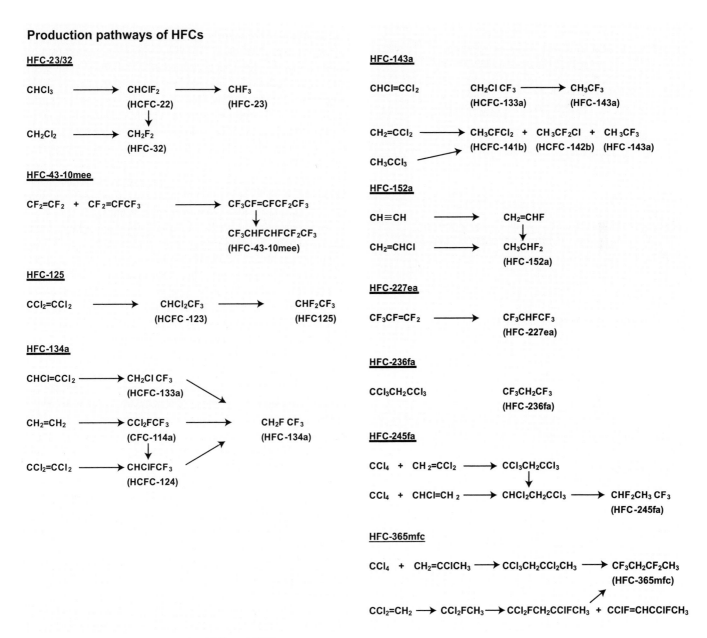

Figure 11A. Manufacturing routes for HFCs and PFCs.

Production pathways of HCFCs

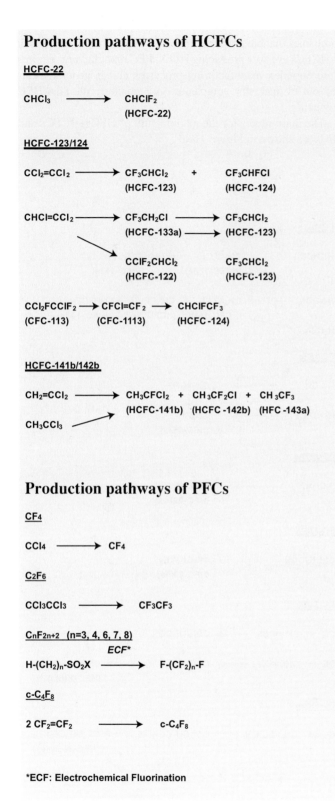

HCFC-22

$CHCl_3 \longrightarrow CHClF_2$
(HCFC-22)

HCFC-123/124

$CCl_2=CCl_2 \longrightarrow CF_3CHCl_2 \quad + \quad CF_3CHFCl$
(HCFC-123)   (HCFC-124)

$CHCl=CCl_2 \longrightarrow CF_3CH_2Cl \longrightarrow CF_3CHCl_2$
(HCFC-133a) $\longrightarrow$ (HCFC-123)

$CClF_2CHCl_2 \quad CF_3CHCl_2$
(HCFC-122)   (HCFC-123)

$CCl_2FCClF_2 \longrightarrow CFCl=CF_2 \longrightarrow CHClFCF_3$
(CFC-113)   (CFC-1113)   (HCFC-124)

HCFC-141b/142b

$CH_2=CCl_2 \longrightarrow CH_3CFCl_2 + CH_3CF_2Cl + CH_3CF_3$
(HCFC-141b) (HCFC-142b) (HFC-143a)

CH_3CCl_3

Production pathways of PFCs

CF₄

$CCl_4 \longrightarrow CF_4$

C₂F₆

$CCl_3CCl_3 \longrightarrow CF_3CF_3$

C_nF_{2n+2} (n=3, 4, 6, 7, 8)

ECF*
$H-(CH_2)_n-SO_2X \longrightarrow F-(CF_2)_n-F$

c-C₄F₈

$2\ CF_2=CF_2 \longrightarrow c\text{-}C_4F_8$

*ECF: Electrochemical Fluorination

Figure 11A. Manufacturing routes for HFCs and PFCs.

Appendix 11B Analysis of the potential reasons for discrepancies in calculated emissions and the means of reconciling data sets

The origins of discrepancy

The average residence time of a fluorocarbon in equipment or other products varies from less than a year to many decades, depending on the application. At the time of the publication of the ozone-depletion hypothesis by Molina and Rowland (1974), fluorocarbons were mainly used as propellants in aerosols, resulting in rapid emissions of these gases to the atmosphere within a year of sale. Use as propellants, and other applications with rapid emissions, have largely disappeared in developed countries (AFEAS, 2004; Montzka and Fraser, 2002) and are gradually being minimized in developing countries. Current fluorocarbon consumption is dominated by their continuing use in applications with relatively slow emissions to the atmosphere such as refrigerant and closed-cell foam applications. The slow emission of refrigerants and blowing agents establish banks, which can emit fluorocarbons for prolonged periods and long after consumption of the chemical has ceased.

When calculating banks of fluorocarbon and other chemicals in refrigeration and in foams, it is necessary to study sales of equipment and/or products, lifetime aspects, leakage during use, and end-of-life issues. This bottom-up approach is based on data from surveys, which must be well designed to ensure complete coverage. However, there are usually cross-checks to ensure completeness. Emission factors for refrigeration equipment and foams have been assessed over several years. These vary significantly according to the application and good market analysis is required to ensure the use of appropriately weighted emission factors. The differences between consumption and forecast emissions using these functions are assigned to banks, which accrue and diminish with time. Verification of the physical existence of these banks (e.g. Mizuno, 2003) has largely served to confirm the appropriateness of the emission functions adopted, although year-to-year emission projections can have significant error bars for banks that release slowly (e.g. foams), particularly where emission projections are largely based on end-of-life practices. These year-to-year errors tend to be offset with time, making the level of confidence greater over a multi-year period[3].

Observations of concentrations of long-lived species in the atmosphere can be used to estimate past emissions. If the lifetime of the chemical is long enough and the rate of mixing is sufficient to ensure a more or less homogeneous concentration in the whole lower troposphere, then past emissions can be derived by simple inverse modelling using

---

[3] For instance, it is difficult to assess with any accuracy the number of domestic refrigerators that will be decommissioned in a given year, but it is much easier to assess the number that will be decommissioned in a ten-year period based on the knowledge of the age profile of the global stock.

the change in observed concentration and the lifetime of the species (see Table TS-2 in the Technical Summary).

Measurements of long-lived species are performed frequently and with high accuracy in several measurement networks. The uncertainty in derived emissions depends on the uncertainty in the trend in observed concentration and on the uncertainty in the lifetime. Both are relatively small for most species. Using different measurement networks may also yield different emissions and therefore information about the uncertainty. On this basis, the uncertainty in derived emissions is about 10% for CFC-11 and between 1% and 6% for other long-lived halocarbons.

The determination of emissions via the bottom-up method has revealed substantial discrepancies with the emissions calculated from atmospheric measurements of the same chemicals. However, closer analysis of these discrepancies has found that even relatively small levels of consumption of fluorocarbons in rapidly emitting uses in developing countries can dominate emissions for specific chemicals (e.g. CFC-11) in specific years. This implies that there is a need to investigate production data, both from producers as reported to AFEAS and from countries as submitted to UNEP under the Montreal Protocol (Article 7 on data reporting), in order to determine where such emissions might appear. However, the key weakness in the UNEP data set is that it makes no provision for the recording of use patterns. Nonetheless, the quantification of the gap between UNEP production/consumption data and AFEAS data is essential.

Production and consumption issues

Since 1987, countries have reported production and consumption data as well as exports and imports for each of the relevant

CFC and HCFC chemicals to UNEP. UNEP publishes aggregated data for production and consumption on an annual basis (in ODP tonnes) for developed (i.e., non-Article 5(1)) and developing (i.e., Article 5(1)) countries. Annual production data for certain groups of countries can also be obtained, although data at the country level are confidential.

Since 1976, the chemical industry has voluntarily reported the production and sales of fluorocarbons through a survey compiled by an independent accountant. The purpose is to provide the scientific community with data estimating the atmospheric release of CFCs and the alternative fluorocarbons. Data are currently available up until 2002 for all relevant CFCs, HCFCs and HFCs. The sum of all CFC production reported as above to AFEAS in 2002 was only 3% of the total in the peak year, 1988 (AFEAS, 2004). The alternatives initially grew rapidly after their introduction to replace CFCs but currently have varied growth rates, with most levelling off as they become more mature products (Table 11B).

The companies surveyed by AFEAS include subsidiaries and joint ventures in all developed and many developing countries. Some production in developing countries, for example in China, India and Korea, is not included in the survey. The data collected by AFEAS for 2002 are thought (by comparison with the UNEP totals for production) to cover about 30-35% of global CFC production. Global coverage is much greater in the AFEAS survey for the HCFCs and HFCs. The AFEAS data currently cover 73% of all non-feedstock HCFC production, and are thought to represent at least 97% of global HFC production (AFEAS, 2004). Table 11B shows the decreasing trend in the coverage of the production reported to AFEAS in the total.

With respect to the accuracy of reporting, the numbers re-

Table 11B. AFEAS and UNEP production data (for the same group of countries/ manufacturers reporting to AFEAS and those not reporting, for CFC-11, HCFC-22 and -141b)

Year/Chemical	1989	1995	1997	1999	2000	2001	2002
CFC-11 – AFEAS	*302,489*	*32,683*	*18,577*	*12,871*	*9,900*	*8,311*	*6,795*
CFC-11 – UNEP*	314,749	36,098	27,798	20,616	15,064	11,831	10,336
CFC-11 – UNEP°	39,251	36,495	34,675	33,370	28,960	20,353	21,487
CFC-12 – AFEAS	*379,778*	*82,822*	*32,900*	*27,132*	*24,564*	*20,873*	*20,181*
CFC-12 – UNEP*	378,992	70,419	30,906	32,814	22,016	22,556	22,666
CFC-12 – UNEP°	98,350	86,186	59,069	54,707	61,389	43,049	36,505
HCFC-22 – AFEAS	*219,537*	*243,468*	*251,108*	*252,375*	*243,847*	*217,465*	*198,208*
HCFC-22 – UNEP*	204,460	297,981	214,717	235,416	220,260	196,467	181,483
HCFC-22 – UNEP°	36,602	30,075	41,044	86,016	118,533	140,589	144,757
HCFC-141b – AFEAS	-	*113,154*	*122,356*	*132,355*	*134,393*	*123,565*	*118,406*
HCFC-141b – UNEP*	287	111,319	109,120	132,667	128,385	117,531	123,931
HCFC-141b – UNEP°	445	0	5,112	11,605	11,975	13,360	24,697

Notes:

* non-Article 5(1) and Article 5(1) countries reporting to AFEAS

° countries not reporting to AFEAS

ported to AFEAS and the ones reported by the same countries to UNEP concur quite well in the early 1990s for CFC-11 and CFC-12. The differences increase after 1995, particularly for CFC-11, where there seems to be a systematic over-reporting to UNEP, or under-reporting to AFEAS, whilst the differences for CFC-12 remain more or less the same, also because quantities are larger. Reporting for HCFC-22 generally shows significant under-reporting by UNEP compared to AFEAS, with differences in the order of 10% or more. In the case of HCFC-22 it can be clearly seen that the quantity produced in 2002 by countries not reporting to AFEAS is more than 40% of the total. In the case of HCFC-141b, the agreement between AFEAS and UNEP reporting is remarkably good, with annual differences smaller than about 5%.

Determining use patterns and emissions

Annual production and sales data are reported to AFEAS for each of the 12 fluorocarbons surveyed. Sales are divided into use categories – refrigeration, foam blowing distinguishing between open-and closed-cell foams, aerosols, and all other uses – to the best knowledge of the fluorocarbon producers. Some degree of geographical breakdown is also provided but the AFEAS survey does not distinguish between developing and developed countries. Emissions of the individual fluorocarbons to the atmosphere are estimated by applying emission factors to the sales reported in each use category. These factors are derived from a consideration of the respective emission patterns and release delays for each application but are not always as sophisticated as those used in the bottom-up assessments. By way of an example, the allocation of all HCFC-141b foam sales to "closed-cell foams" may be inappropriate in view of uses in integral skin and other more rapid-release foam applications.

This said, estimates of atmospheric concentrations based on production statistics provided by manufacturers and national governments coupled with historical emission factors were, until recently, a good match with observations (McCulloch *et al.*, 2001, 2003), largely because of the dominance of rapid-release applications in developed countries. However, as the detailed assessment of delayed-release applications has become more important in developed countries and the gap between UNEP data and AFEAS data has increased, reflecting higher levels of activity in developing countries, the lack of use pattern data via the UNEP reporting structure is now a key impediment, particularly where usage might be occurring in rapid-release applications.

The improvement of knowledge about the use of the individual chemicals, both in the developed and in the developing countries, as well as uncertainties in emissions factors from banks following the bottom-up method, is unlikely to bear fruit within the time scale of this report. This has implications for the calculation of emission scenarios for the period 2002-2015, in particular the contributions from developing countries. It may therefore be the case that the emissions are significantly larger than calculated via the bottom-up method for refrigeration and foams only because of additional contributions from rapid-release applications.

In addition, the emission factors applied to HCFCs and HFCs were derived from those developed earlier for CFCs and HCFC-22. However, there have been considerable changes in use practices in the wake of the Montreal Protocol and emission factors are subject to continual review. Revised emission factors, as described in McCulloch *et al.* (2001 and 2003), have helped improve the situation and have been used in the most recent calculation of emissions (AFEAS, 2004). Nonetheless, the view is that variation and uncertainties in emission factors per se are less significant overall than uncertainties in overall use pattern distributions between rapid-release applications and the delayed-release applications represented by refrigeration and foams.

This implies that, in the near future, further study is needed of the use patterns for the total production forecast for the period 2002-2015 in both the developed and the developing countries. Further study will also be needed to make more precise estimates of future emissions from banks in refrigeration and foams, given the accuracy of calculations of the size of the banks and the emissions derived from them, as well as servicing practices, and issues relating to recovery and recycling and end-of-life..

References

AFEAS (Alternative Fluorocarbons Environmental Acceptability Study), 2004: Alternative Fluorocarbons Environmental Acceptability Study – Production, Sales and Atmospheric Releases of Fluorocarbons through 2002. RAND ES and P Center, Arlington, VA, USA (available online at www.afeas.org).

McCulloch, A., P. Ashford and P.M. Midgley, 2001: Historic emissions of Fluorotrichloromethane (CFC-11) based on a market survey. *Atmos. Environ.*, **35**(26), 4387-4397.

McCulloch, A., P.M. Midgley and P. Ashford, 2003: Releases of refrigerant gases (CFC-12, HCFC-22 and HFC-134a) to the atmosphere. *Atmos. Environ.*, **37**(7), 889-902.

Mizuno, K., 2003: Estimation of Total Amount of CFCs Banked in Building Insulation Foams in Japan. Proceedings of the 14th Annual Earth Technologies Forum, April 22-24, 2003, Washington, D.C., USA.

Molina, M. and F.S. Rowland, 1974: Stratospheric sink for chlorofluoromethanes: Chlorine atom catalysed destruction of ozone. *Nature* 249, 810-812.

Montzka, S.A., P.J. Fraser (Lead Authors), J.H. Butler, P.S. Connell, D.M. Cunnold, J.S. Daniel, R.G. Derwent, S. Lal, A. McCulloch, D.E. Oram, C.E. Reeves, E. Sanhueza, L.P. Steele, G.J.M. Velders, R.F. Weiss and R.J. Zander, 2003: Controlled Substances and Other Source Gases. Chapter 1 in *Scientific Assessment of Ozone Depletion: 2002.* Global Ozone Research and Monitoring Project, Report No. 47, World Meteorological Organization, Geneva, Switzerland.

Appendix 11C Destruction technologies and capacities

The decomposition or destruction of ozone-depleting substances, HFCs and PFCsSGG can be achieved using various technologies. The commercially developed and successfully demonstrated technologies are described briefly in this section. Commercially available operations are also listed by region. The main source documents upon which this section is based are as follows:

- Proceedings of the International Workshop on the Disposal of Ozone-depleting Substances (UNEP, 2000).
- Guidance Document on Disposal Technologies for Ozone-depleting Substances (ODS) in Canada (Environment Canada, 2001)
- Refrigerant Recovery, Recycling, Reclamation and Disposal – an International Assessment, (Snelson and Bouma, 2003, Part 2)

The Canadian study reported, in their Guidance Document, 16 available and potentially efficacious technologies for the effective destruction of ODSs. The selected technologies all had stack emissions of less than 0.1 ng/m3 toxic equivalence for dioxins and furans, destruction or conversion efficiency not less than 99.7%, and the potential to be commercially available in 2003.

Plasma technologies (non-incineration)
- Argon Plasma Arc (Plascon)
- Inductively Coupled Radio Frequency Plasma (ICRF)
- Alternating Current Plasma (ACP)

Other non-incineration technologies
- Solvated Electron
- Gas Phase Chemical Reduction
- Catalytic Dehalogenation
- Liquid Phase Chemical Conversion
- Vitrification

UV photolysis incineration technologies
- Liquid Injection
- Rotary Kiln
- Gaseous/Fume Oxidation
- Internally Circulated Fluidized Bed (ICFB)
- Cement Kiln
- Reactor Cracking
- High Performance Incineration

Table 11C shows technical data for the evaluation of ODS-destruction technologies.

Commercially available technologies

Argon Plasma Arc

The "in-flight" argon plasma process, which has the waste mixing directly with the argon plasma, was developed in Australia and has been operating commercially since 1996. The technology and plant is owned and operated by BCD Technologies. Approximately 1000 tonnes of halons and 800 tonnes of fluorocarbons have been destroyed since commissioning. Destruction efficiencies of 99.9999% have been achieved. Destruction rates

Table 11C. Technical data for the evaluation of ODS-destruction technologies.

Technology	Destruction efficiency (%)	Dioxins & furans (ng/m³)	Other effluents	Energy consumption (kWh/kg CFC)	Estimated costs US$/kg	Availability
Incineration						
High Performance	99.99990	0.100	salt/GHG	1.71	6.00	commercial
Liquid Injection	99.99000	0.100	salt/GHG	1.30	5.00	commercial
Rotary Kiln	99.99000	0.100	salt/GHG	1.54	5.00	commercial
Gas/Fume	99.99000	0.100	salt/GHG	1.30	5.00	commercial
ICFB	99.99900	0.100	salt/GHG	1.30	5.00	demo
Cement Kiln	99.99000	0.100	GHG	1.86	4.00	commercial
Reactor Cracking	99.99900	0.100	wastewater	1.55	3.75	commercial
Plasma						
IC RF	99.99000	0.025	salt	3.70	3.00	Demo
Argon	99.99990	0.025	salt	2.30	2.75	Commercial
AC	99.99000	0.025	salt	2.10	2.75	Condition
Other						
Solvated Electron	99.99000	0.010	salt	10.00	8.00	Demo
UV Photolysis	99.99990	0.010	spent liners	10.00	11.00	Commercial
Gas Phase Chemical Reduction	99.99000	0.060	salt	1.38	6.00	Condition
Catalytic Dehalogenation	99.99000	0.010	salt	0.79	3.60	Demo
Liquid Phase Chemical Conversion	99.99990	0.010	salt	3.00	5.00	Commercial
Vitrification	99.99000	0.100	glass frit	4.88	3.80	Condition

vary from 60 kg/hr for fluorocarbons to 120 kg/hr for halons. Waste refrigerants are rapidly heated to 3000°C where pyrolysis occurs. This process is followed by rapid alkaline quenching that prevents the formation of dioxins and furans. Neutralization occurs with alkaline liquor. The waste gases, CO_2 and argon are vented, whilst the resulting halide salt solution consisting of NaCl and NaF is discharged to wastewater treatment.

Inductively Coupled Radio Frequency Plasma (ICRF)
This process does not require the electrodes needed for the DC process as the energy coupling to the plasma is achieved through the electromagnetic field of the induction coil. A demonstration plant operated from 1993 to 1996, during which time 2443 tonnes of CFC-12 was destroyed. Gaseous CFCs and steam are fed through the plasma torch where they are heated and enter directly into the destruction reactor maintained at approximately 2000°C. The subsequent gases are cooled and scrubbed with a caustic solution to remove acid gases. This process has been commercialized in Japan by Ichikawa Kankyo Engineering. Neutralization occurs with the use of $Ca(OH)_2$, resulting in solid waste CaF_2 and $CaCl_2$. An average of 50 tonnes per annum was destroyed during the second half of the 1990s.

Reactor Cracking
The reactor cracking process uses a cylindrical water-cooled reactor made of graphite, and an oxygen-hydrogen burner system. The reactor is flanged directly to an absorber. Waste refrigerants are fed into the reaction chamber where the temperature is maintained at between 2000°C and 2600°C. The waste is broken down into HF, HCl, Cl_2, CO_2, and H_2O. The cracked products are cooled in the absorber and the acid gases purified and recovered at technical grade quality. This is the most widely used commercial process for the destruction of fluorocarbons in Europe. The plant and technology are owned and operated by Solvay. In mid-2000, approximately 7000 Mt of fluorocarbons had been decomposed with a reactor capacity of 200 kg/hr.

Gaseous/Fume Oxidation
The gaseous/fume process uses refractory-lined combustion chambers for the thermal destruction of waste vapour streams. Waste fluorocarbons may be destroyed by feeding directly from their pressurized storage into the incinerator. Fume incinerators are typically found installed with manufacturing plants and are seldom used for commercial hazardous waste incineration. A version of this process known as High Temperature Steam Decomposition is operated by Ineos Fluor at Mihara in Japan. The rated throughput for CFCs is up to 165 kg/hr. Neutralization occurs with the use of $Ca(OH)_2$, resulting in solid waste CaF_2 and $CaCl_2$.

Liquid Injection
Liquid injection incinerators are usually single chamber units with one or more waste burners into which liquid waste is injected, atomized into fine droplets, and burned in suspension. Problems of flame stability may result with high concentrations

of fluorocarbons. This process has not been applied commercially to waste refrigerants. In Japan, Asahi Glass operates a liquid injection incinerator. The process involves two-stage neutralization using NaOH and $Ca(OH)_2$.

Cement Kiln
Existing cement kilns can destroy organic compounds as the temperature in the burning zone is over 1500°C with a relatively long residence time. The disadvantage is that fluorine and chlorine input rates need to be carefully monitored. Fluorine can be beneficial to the cement process whereas chlorine is generally regarded as an unwanted constituent.

Rotary Kiln
Rotary kiln incinerators are refractory-lined rotating cylindrical steel shells mounted on a slight incline from the horizontal. Most rotary kilns are equipped with an after-burner which ensures complete destruction of exhaust gases. Waste gases can be fed into the rotary kiln or directly into the after-burner. Rotary kilns are most frequently incorporated into the design of commercial incinerator facilities. Because of the production of acid byproducts, there are generally severe restrictions on the amount of halocarbons in the raw material feed.

Demonstrated and pilot processes

Alternating Current Plasma (ACP)
AC plasma is produced directly with 60 Hz high-voltage power but in other respects is similar to ICRF. CFCs have been effectively destroyed in a demonstration plant but the process is not commercially available for this application.

Solvated Electron
The solvated electron process is a batch process operated at atmospheric pressure using two vessels: one heated reaction vessel, and the other a refrigerated ammonia recycle vessel. Refrigerants are decomposed in the reaction vessel with liquid ammonia and metallic sodium. The resulting wastes are chiefly halide salts and biodegradable organic compounds. On a pilot scale, the process has been shown to destroy various fluorocarbons at an efficiency of 99.99% but a commercial facility for this application has not been established.

Gas Phase Chemical Reduction
This process involves the preheating of liquid or gas with boiler steam before injection into the reactor. The atomized gas mixture is heated by vertical radiant tubes with internal electric heating elements to approximately 850°C. Organic compounds are ultimately reduced to methane, hydrochloric acids and low-molecular-weight hydrocarbons. The formation of some dioxins and furans is possible. Hydrochloric acid is neutralized by caustic soda, and a scrubbing system removes inorganics. Hydrogen and methane are recovered for energy use. This process has successfully destroyed other wastes on a commercial basis but not fluorocarbons.

Catalytic Dehalogenation

In this process, CFCs are destroyed over a proprietary metal oxide catalyst at 400°C at atmospheric pressure. The resulting HF and HCL are absorbed in a lime solution. This process has been commercialized for other chlorinated compounds but not fluorocarbons.

Liquid Phase Chemical Conversion

This technology uses a liquid-phase chemical conversion process operating between 80°C and 120°C, where waste is reacted with a blend of potassium hydroxide and polyethylene glycol. Pilot-scale tests have proven a capability for fluorocarbons and the process is used commercially for other chlorinated substances.

Vitrification

This process fixes the products of chlorinated chemical dissociation and hydrolysis into chemically durable glass frit, which can be processed into glass products. The process has not been applied to fluorocarbons but has been proven effective for other chlorinated substances.

Internally Circulating Fluidized Bed (ICFB)

This process involves the waste refrigerant and air being blown through the incinerator fluidized bed, where the refrigerant is broken down by the presence of methane and hydrogen in the reducing atmosphere. The process has been proven to be successful but has not been commercialized.

References

Environment Canada, 2001: Guidance Document on Disposal Technologies for Ozone-depleting Substances (ODS) in Canada. Environment Canada, EPS 1/RA/4, March 2001, Canada.

Snelson, K. and J. Bouma, 2003: Refrigerant Recovery, Recycling, Reclamation and Disposal – An International Assessment, Parts 1 and 2. IEA Heat Pump Centre, Sittard, Netherlands 2003.

UNEP (United Nations Environment Programme), 2000: Proceedings of the International Workshop on the Disposal of Ozone-Depleting Substances, July 10, 2000, Geneva, Switzerland. UNEP Ozone Action Programme, Paris, France (CD-ROM).

Annex I

Authors and Reviewers

I.1 Co-ordinating Lead Authors, Lead Authors, Contributing Authors and Review Editors[1]

Technical Summary

Co-ordinating Lead Authors

D. de Jager	TSU IPCC Working Group III, The Netherlands
L. Kuijpers	Co-chair TEAP, Technical University Eindhoven, The Netherlands
M. Manning	TSU IPCC Working Group I, USA (New Zealand)

Lead Authors

S.O. Andersen	Co-chair TEAP, US Environmental Protection Agency, USA
P.K. Ashford	Caleb Management Services, United Kingdom
P. Atkins	Oriel Therapeutics, USA
N.J. Campbell	Arkema Group, France (United Kingdom)
D.F. Clodic	Ecole des Mines de Paris, France
S. Devotta	National Environmental Engineering Research Institute, India
J. Harnisch	Ecofys GmbH, Germany
M.K.W. Ko	NASA Langley Research Center, USA
S. Kocchi	US Environmental Protection Agency, USA
S. Madronich	National Center for Atmospheric Research, USA
A. McCulloch	Marbury Technical Consulting, United Kingdom
B. Metz	Co-chair IPCC Working Group III, Netherlands Environmental Assessment Agency (MNP), The Netherlands
L.A. Meyer	TSU IPCC Working Group III, Netherlands Environmental Assessment Agency (MNP), The Netherlands
J.R. Moreira	University of Sao Paolo, Brazil
J.G. Owens	3M, USA
R. de Aguiar Peixoto	Mauá Institute of Technology (IMT), Brazil
J.I. Pons	Co-chair TEAP, Spray Quimica, Venezuela
J.A. Pyle	European Ozone Research Coordinating Unit, United Kingdom
S.D. Rand	US Environmental Protection Agency, USA
R.M.S. Shende	United Nations Environment Programme, France (India)

[1] Country in parenthesis: citizenship of contributor

T.G. Shepherd	University of Toronto, Canada
S. Sicars	Multilateral Fund Secretariat, Canada (Germany)
S. Solomon	Co-chair IPCC Working Group I, NOAA Aeronomy Laboratory, USA
G.J.M. Velders	Netherlands Environmental Assessment Agency (MNP), The Netherlands
D.P. Verdonik	Hughes Associates Inc., USA
R.T. Wickham	Wickham Associates, USA
A. Woodcock	Wythenshawe Hospital, United Kingdom
P. Wright	AstraZeneca R&D Charnwood, United Kingdom
M. Yamabe	AIST, Nat. Inst. Advanced Industrial Science and Technology, Japan

Review Editors

O.R. Davidson	Co-chair IPCC Working Group III, University of Sierra Leone, Sierra Leone
M. McFarland	DuPont Fluoroproducts, USA
P.M. Midgley	University of Stuttgart, Germany (United Kingdom)

PART A: Ozone Depletion and the Climate System

Chapter 1: Ozone and Climate: A Review of Interconnections

Co-ordinating Lead Authors

| J.A. Pyle | European Ozone Research Coordinating Unit, United Kingdom |
| T.G. Shepherd | University of Toronto, Canada |

Lead Authors

G. Bodeker	National Institute of Water and Atmospheric research, New Zealand
P.O. Canziani	Universidad de Buenos Aires, Argentina
M. Dameris	Deutschen Zentrum für Luft- und Raumfahrt (DLR), Germany
P.M. de F. Forster	University of Reading, United Kingdom
A.N. Gruzdev	Russian Academy of Sciences, Russia
R. Müller	Forschungszentrum Jülich, Germany
N.J. Muthama	University of Nairobi, Kenya
G. Pitari	Università degli Studi L'Aquila, Italy
W.J. Randel	National Center for Atmospheric Research, USA

Contributing Authors

V.E. Fioletov	Meteorological Service of Canada, Canada
J.-U. Grooß	Forschungszentrum Jülich GmbH, Germany
S.A. Montzka	NOAA Climate Monitoring and Diagnostics Laboratory, USA
P.A. Newman	NASA Goddard Space Flight Center, USA
L.W. Thomason	NASA Langley Research Center, USA
G.J.M. Velders	Netherlands Environmental Assessment Agency (MNP), The Netherlands

Review Editor

| M. McFarland | DuPont Fluoroproducts, USA |

Chapter 2: Chemical and Radiative Effects of Halocarbons and Their Replacement Compounds

Co-ordinating Lead Authors

| S. Madronich | National Center for Atmospheric Research, USA |
| G.J.M. Velders | Netherlands Environmental Assessment Agency (MNP), The Netherlands |

Lead Authors

| C. Clerbaux | Centre National de la Recherche Scientifique, France and Université Libre de Bruxelles, Belgium (Belgium) |

R.G. Derwent	RDScientific, United Kingdom
M. Grutter de la Mora	National Autonomous University of Mexico, Mexico (Swiss)
D.A. Hauglustaine	Centre National de la Recherche Scientifique, France
S. Incecik	Istanbul Technical University, Turkey
M.K.W. Ko	NASA Langley Research Center, USA
J.-M. Libre	Arkema Group, France
O.J. Nielsen	University of Copenhagen, Denmark
F. Stordal	University of Oslo, Norway
T. Zhu	Peking University, China

Contributing Authors

D. Blake	University of California at Irvine, USA
D. Cunnold	Georgia Institute of Technology, USA
J. Daniel	NOAA Aeronomy Laboratory, USA
P.M. de F. Forster	University of Reading, United Kingdom
P. Fraser	CSIRO Atmospheric Research, Australia
P. Krummel	CSIRO Atmospheric Research, Australia
A. Manning	Met Office, United Kingdom
S.A. Montzka	NOAA Climate Monitoring and Diagnostics Laboratory, USA
G. Myhre	University of Oslo, Norway
S. O'Doherty	University of Bristol, United Kingdom
D. Oram	University of East Anglia, United Kingdom
M.J. Prather	University of California at Irvine, USA
R. Prinn	Massachusetts Institute of Technology, USA
S. Reimann	Swiss Federal Laboratories for Materials Testing and Research (EMPA), Switzerland
P. Simmonds	University of Bristol, United Kingdom
T. Wallington	Ford Motor Co., USA
R. Weiss	Scripps Institution of Oceanography, USA

Review Editors

I.S.A. Isaksen	University of Oslo, Norway
B.P. Jallow	Meteorology Division, Gambia

PART B: Options for ODS Phase-out and Reducing GHG Emissions

Chapter 3:	**Methodologies**

Co-ordinating Lead Authors

J. Harnisch	Ecofys GmbH, Germany
J.R. Moreira	University of Sao Paolo, Brazil

Lead Authors

P. Atkins	Oriel Therapeutics, USA
D. Colbourne	Calor Gas, United Kingdom
M. Dieryckx	European Partnership for Energy and the Environment, Belgium
H.S. Kaprwan	Defense Fire Research Institute, India
F.J. Keller	Carrier Corporation, USA
A. McCulloch	Marbury Technical Consulting, United Kingdom
S. Sicars	Multilateral Fund Secretariat, Canada (Germany)
B. Tulsie	Ministry of Physical Development, Environment and Housing, St. Lucia
J.W. Wu	Arkema Group, USA (China)

Contributing Authors

P.K. Ashford	Caleb Management Services, United Kingdom
K. Halsnaes	Risø National Laboratory, Denmark

A. Inaba   Research Center for Life Cycle Assessment, Japan

Review Editors
P.M. Midgley   University of Stuttgart, Germany (United Kingdom)
M. Sideridou   Greenpeace European Unit, Belgium (Greece)

Chapter 4: Refrigeration

Co-ordinating Lead Authors
S. Devotta   National Environmental Engineering Research Institute, India
S. Sicars   Multilateral Fund Secretariat, Canada (Germany)

Lead Authors
R.S. Agarwal   Indian Institute of Technology Delhi, India
J. Anderson   Institute for European Environmental Policy, Belgium (USA)
D.B. Bivens   DuPont Fluoroproducts, USA
D. Colbourne   Calor Gas, United Kingdom
A.T. El-Talouny   UNEP Regional Office of West Asia, Bahrain (Egypt)
G. Hundy   United Kingdom
H. König   Axima Refrigeration, Germany
P.G. Lundqvist   Royal Institute of Technology, Sweden
E.J. McInerney   GE Consumer Products, USA
P. Nekså   SINTEF Energy Research, Norway

Review Editors
E. Calvo   Universidad Nacional de San Marcos, Peru
I. Elgizouli   Higher Council for Environment and Natural Resources, Sudan

Chapter 5: Residential and Commercial Air Conditioning and Heating

Co-ordinating Lead Author
R. de Aguiar Peixoto   Mauá Institute of Technology (IMT), Brazil

Lead Authors
D.J. Butrymowicz   Polish Academy of Sciences, Poland
J.G. Crawford   American Standard Companies, USA
D.S. Godwin   US Environmental Protection Agency, USA
K.E. Hickman   York International, USA
F.J. Keller   Carrier Corporation, USA
H. Onishi   DAIKIN Industries, Japan

Review Editors
M. Kaibara   Japan
A.D. Pasek   Institut Teknologi Bandung, Indonesia

Chapter 6: Mobile Air Conditioning

Co-ordinating Lead Author
D.F. Clodic   Ecole des Mines de Paris, France

Lead Authors
J.A. Baker   Delphi Corporation, USA
J. Chen   Shanghai Jiao-Tong University, China

T.H. Hirata	Japan Refrigeration and Air Conditioning Industry Association, Japan
R.J. Hwang	Natural Resources Defense Council, USA
J. Köhler	Technical University Braunschweig, Germany
C. Petitjean	Valeo Climate Control, France
A. Suwono	Inter University Research Center for Engineering Sciences, Indonesia

Contributing Authors

W. Atkinson	USA
M. Ghodbane	Delphi Harrison Thermal Systems, USA
R. Mager	BMW AG. Germany
G. Picker	Australian Greenhouse Office, Australia
F. Rinne	Sanden Technical Centre Europe GmbH, Germany
J. Wertenbach	DaimlerChrysler AG, Germany
F. Wolf	Obrist Engineering GmbH, Austria

Review Editors

S.O. Andersen	US Environmental Protection Agency, USA
T.A. Cowell	United Kingdom

Chapter 7: Foams

Co-ordinating Lead Author

P.K. Ashford	Caleb Management Services, United Kingdom

Lead Authors

A. Ambrose	A-Gas International, Australia
G.M. Jeffs	ISOPA, European Isocyanate Producers, Belgium (United Kingdom)
R.W. Johnson	RWJ Consulting, USA
S. Kocchi	US Environmental Protection Agency, USA
S.P. Lee	Dow Deutschland GmbH & Co., Germany, (United Kingdom)
D. Nott	Honeywell Chemicals, Belgium (USA)
P. Vodianitskaia	Multibras SA Eletrodomesticos, Brazil
J.W. Wu	Arkema Group, USA (China)

Contributing Authors

T.J. Maine	Dow Chemical Company, USA
J. Mutton	Dow Chemical Canada, Canada
B. Veenendaal	RAPPA Inc., USA

Review Editors

J.A. Leiva Valenzuela	Chile
L.B. Singh	India

Chapter 8: Medical Aerosols

Co-ordinating Lead Authors

P. Atkins	Oriel Therapeutics, USA
A. Woodcock	Wythenshawe Hospital, United Kingdom

Lead Authors

O. Blinova	Russian Scientific Center "Applied Chemistry", Russia
J. Khan FRCP (Edin)	The Aga Khan University, Pakistan
R. Stechert	Boehringer Ingelheim (Schweiz) Gmbh, Switzerland (Germany)
P. Wright	AstraZeneca R&D Charnwood, United Kingdom

Y. You   Journal of Aerosol Communication, China

Review Editors
D. Fakes   United Kingdom
M. Seki   United Nations Environment Programme, Kenya (Japan)

Chapter 9: Fire Protection

Co-ordinating Lead Author
D.P. Verdonik   Hughes Associates Inc., USA

Lead Authors
H.S. Kaprwan   Defense Fire Research Institute, India
E.Th.Morehouse   Institute for Defense Analyses, USA
J.G. Owens   3M, USA
M. Stamp   Great Lakes Chemical Corporation, United Kingdom
R.T. Wickham   Wickham Associates, USA

Review Editors
S.O. Andersen   US Environmental Protection Agency, USA
M. Seki   United Nations Environment Programme, Kenya (Japan)

Chapter 10: Non-Medical Aerosols, Solvents and HFC-23

Co-ordinating Lead Authors
S.D. Rand   US Environmental Protection Agency, USA
M. Yamabe   National Institute Advanced Industrial Science and Technology, Japan

Lead Authors
N.J. Campbell   Arkema Group, France (United Kingdom)
J. Hu   Peking University, China
P.M. Lapin   Falcon Safety Products, USA
A. McCulloch   Marbury Technical Consulting, United Kingdom
A. Merchant   DuPont Fluoroproducts, USA
K.M. Mizuno   National Institute of Advanced Industrial Science and Technology, Japan
J.G. Owens   3M, USA
P. Rollet   AVANTEC, France

Contributing Author
W. Bloch   Siemens, Germany

Review Editors
Y. Fujimoto   Japan
J.I. Pons   Spray Quimica, Venezuela

PART C: Future Estimation and Availability of HFCs and PFCs

Chapter 11: Current and Future Supply, Demand and Emissions of HFCs and PFCs, plus Emissions of CFCs, HCFCs and Halons

Co-ordinating Lead Authors
N.J. Campbell   Arkema Group, France (United Kingdom)
R.M.S. Shende   United Nations Environment Programme, France (India)

Lead Authors

M.E. Bennett	Refrigerant Reclaim Australia, Australia
O. Blinova	Russian Scientific Center "Applied Chemistry", Russia
R.G. Derwent	RDScientific, United Kingdom
A. McCulloch	Marbury Technical Consulting, United Kingdom
J. Shevlin	Australian Greenhouse Office, Australia
T.G.A. Vink	Honeywell Fluorine Products Europe, The Netherlands
M. Yamabe	National Institute Advanced Industrial Science and Technology, Japan

Contributing Authors

P.K. Ashford	Caleb Management Services, United Kingdom
M. McFarland	DuPont Fluoroproducts, USA
P.M. Midgley	University of Stuttgart, Germany (United Kingdom)

Review Editors

G.D. Hayman	AEA Technology, United Kingdom
L. Kuijpers	Technical University Eindhoven, The Netherlands

I.2 Expert Reviewers

Government and National Research and Educational Organisations

Argentina

G. Argüello	Universidad Nacional de Córdoba
A. Bianchi	Government of Argentina
S. Diaz	CADIC/CONICET

Australia

P. Fraser	CSIRO Atmospheric Research
G. Picker	Australian Greenhouse Office
I. Plumb	CSIRO Telecommunications and Industrial Physics
H. Tope	Environment Protection Authority Victoria

Canada

P. Edwards	Government of Canada
N. Gillett	University of Victoria
D. Harvey	University of Toronto

Chile

L. da Silva	Universidad Tecnica F. Santa Maria
M. Préndez	Universidad de Chile
E. Sanhueza	Instituto Venezolano de Investigacioned Ceintíficas Centro de Química

China

S. Liu	Academica Sinica
D. Qin	China Meteorological Administration
S. Zhang	Peking University

Columbia

A. Durán C.	Ministerio de Relaciones Exteriores

Denmark

P. Pedersen	Danish Technological Institute

Estonia
K. Eerme   Tartu Observatory

Fiji Islands
M. Lal   The University of the South Pacific

Finland
T. Oinonen   Finnish Environment Institute

France
M. Gillet   Observatoire National sur les Effets du Réchauffement Climatique

Germany
F. Dettling   German Federal Environmental Agency
R. Engelhardt   German Federal Environment Ministry
U. Fuentes   Federal Environment Ministry
J. Lelieveld   Max Planck Institute for Chemistry
M. Rex   Alfred Wegener Institute for Polar and Marine Research
R. Sartorius   Federal Environmental Agency
M. Schultz   Max Planck Institute for Meteorology
K. Schwaab   German Federal Environmental Agency

Greece
A. Bais   Aristotle University of Thessaloniki
C. Zerefos   National and Kapodistrian University of Athens

India
S. Ramachandran   Physical Research Laboratory
A. Mofidi   Tarbiat Moallem University

Italy
A. Cavallini   University of Padova
S. Girotto   COSTAN S.p.A.
E. Manzini   Istituto Nazionale di Geofisica e Vulcanologia

Japan
H. Nakane   National Institute for Environmental Studies
A. Sekiya   National Institute of Advanced Industrial Science and Technology
K. Hara   Japan Polyurethane Industries Institutes
Y. Matsumoto   Kyoto University

Morocco
A. Allali   Ministry of Agriculture and Rural Development

New Zealand
D. Cleland   Massey University
R. McKenzie   National Institute of Water and Atmospheric Research
A. Reisinger   Ministry for the Environment
H. Struthers   National Institute of Water and Atmospheric Research

Norway
T. Asphjell   Norwegian Pollution Control Authority
H. Haukås   Consulant
G. Myhre   University of Oslo
J. Pettersen   Norwegian University of Science and Technology

Oman
A. bin Saeed AI-Kharoosi   Ministry of Regional Municipalities, Environment and Water Resources

Romania
L. Manea   National Meteorological Administration

South Africa
P.J. Aucamp   North-West University

Spain
J. Saiz Jabardo   Universidad de la Coruña

Sri Lanka
W. Sumathipala   Ministry of Environment and Natural Resources/Open University of Sri Lanka

Sweden
M. Lilliesköld   Swedish Environmental Protection Agency
H. Rodhe   Stockholm University

Switzerland
T. Peter   ETH, Zurich
J. Romero   Swiss Agency for the Environment, Forests and Landscape

The Netherlands
B. Bregman   Royal The Netherlands Meteorological Institute (KNMI)
G. Komen   Royal The Netherlands Meteorological Institute (KNMI)
M. Krol   Institute for Marine and Atmospheric Research Utrecht
J. Olivier   Netherlands Environmental Assessment Agency (MNP)
J. Potting   University of Groningen
P. Siegmund   Royal The Netherlands Meteorological Institute (KNMI)
P. van Velthoven   Royal The Netherlands Meteorological Institute (KNMI)

Togo
A. Ajavon   FDS/Université de Lomé

Tunesia
N. Aïssa   Government of Tunesia

United Kingdom
K. Carslaw   University of Leeds
M. Chipperfield   University of Leeds
R. Critoph   University of Warwick
D. Liddy   UK Ministry of Defense
S. Oliver   UK Department for Environment, Food and Rural Affairs
K. Shine   The University of Reading
W. Sturges   University of East Anglia

United States of America
J. Austin   NOAA Geophysical Fluid Dynamics Laboratory
J. Banks   US Environmental Protection Agency
V. Baxter   Oak Ridge National Laboratory
J. Briskin   US Environmental Protection Agency
J.S. Brown   The Catholic University of America
D. Cunnold   Georgia Institute of Technology
J. Daniel   NOAA Aeronomy Laboratory
P. DeCola   National Aeronautics and Space Administration
C. Delhotal   US Environmental Protection Agency

D. Dokken	US Global Change Research Program
P.A. Domanski	National Institute of Standards and Technology
D. Fahey	NOAA Aeronomy Laboratory
D. Fratz	Consumer Specialty Products Association
D. Hofmann	NOAA Climate Monitoring and Diagnostics Laboratory
L. Hood	University of Arizona
C. Jackman	NASA Goddard Space Flight Center
D. Karoly	University of Oklahoma
B. Maranion	US Environmental Protection Agency
S. McCormick	US Army Tank-automotive Research, Development and Engineering Center
M. Menzer	Air Conditioning and Refrigeration Institute
S. Monroe	US Environmental Protection Agency
S.A. Montzka	NOAA Climate Monitoring and Diagnostics Laboratory
V. Orkin	National Institute of Standards and Technology
D. Ottinger Schaefer	US Environmental Protection Agency
G. Pugh	US Department of State
R. Prinn	Massachusetts Institute of Technology
V. Ramaswamy	NOAA Geophysical Fluid Dynamics Laboratory
A. Ravishankara	NOAA Aeronomy Laboratory
D. Reifsnyder	Office of Global Change
J. Rodriguez	NASA Goddard Space Flight Center
K. Rosenlof	NOAA Aeronomy Laboratory
R. Salawitch	NOAA Jet Propulsion Laboratory
J. Samenow	US Environmental Protection Agency
A.-M. Schmoltner	National Science Foundation
M. Sheppard	US Environmental Protection Agency
D. Shindell	NASA Goddard Institute for Space Studies
R. Soulen	(retired)
W. Stockwell	Desert Research Institute
R.S. Stolarski	NASA Goddard Space Flight Center
K. Taddonio	US Environmental Protection Agency
D. Thompson	Colorado State University
K. Thundiyil	US Environmental Protection Agency
G. Valasek	Intercontinental Chemical Corporation
R. Weiss	Scripps Institution of Oceanography

Uzbekistan
| V. Chub | Uzhydromet |

United Nations Organisations and Specialised Agencies
| F. Dentener | EC Joint Research Centre |
| G. Jimenez Blasco | United Nations Industrial Development Organization |

Non-Governmental Organisations
R. Albach	BayerMaterialScience AG
A. Aloisi	De'Longhi S.p.A.
D. Berchowitz	Global Cooling BV
S. Bernhardt	Honeywell International
F. Billiard	International Institute of Refrigeration
N. Candelori	Alliance for the Polyurethanes Industry
C. Carling	AstraZeneca
M. Cartmell	Cartmell Consulting
M. Christmas	DuPont Fluoroproducts
A. Cohr Pachai	York International
T. Cortina	Halon Alternatives Research Corporation
A. d'Haese	European Aerosol Federation
W. Dietrich	York International

M. Donahue Hardwick	International Pharmaceutical Aerosol Consortium
J.-L. Dupont	International Institute of Refrigeration
R. Frischknecht	ESU-services
S. Haaf	Linde Kältetechnik GmbH
H.S. Hammel	DuPont Fluoroproducts
M. Hardwick	International Pharmaceutical Aerosol Consortium
W. Hare	Greenpeace International
R. Heap	Cambridge Refrigeration Technology
W. Hill	General Motors Corporation
B. Hoare	Greenchill Technology Association Inc.
S. Japar	Ford Motor Company Research Laboratory (retired)
E. Johnson	Atlantic Consulting
J. Judge	York International
A. Lindley	INEOS Fluor
J. Mandyck	Carrier Corporation
H. Mori	Otsuka Pharmaceutical Co. Ltd.
A. Nicoletti	Solvay Solexis
J.J. O'Sullivan	British Airways
A. Pearson	Star Refrigeration Ltd.
E. Preisegger	Solvay Fluor GmbH
E. Rameckers	European Federation of Allergy and Airways Diseases Patient Associations
R. Rubenstein	ICF Consulting, Inc.
D. Stirpe	Alliance for Responsible Atmospheric Policy
R. van Gerwen	Unilever
M. Weick	The Dow Chemical Company
T. Werkema	Arkema Inc.
K. Werner	3M
L. Wethje	Association of Home Appliance Manufacturers
G. William	Boehringer Ingelheim
J. Wolf	American Standard Companies

Annex II

Glossary

Note: The definitions in this glossary refer to the use of the terms in the context of this report. A '→' indicates that the following term is also contained in this glossary. The glossary provides an explanation of specific terms as the lead authors intend them to be interpreted in this report.

Absorption (Refrigeration)
A process by which a material (the absorbent) extracts one or more substances (absorbates) from a liquid or gaseous medium that it is in contact with and changes chemically, physically or both. The process is accompanied by a change in entropy, which makes it a useful mechanism for a refrigeration cycle. Water-lithium bromide and ammonia-water →chillers are examples of absorption chillers.

Adjustment Time
See: →Lifetime in relation to atmospheric concentrations, or → response time in relation to the climate system.

Aerosol
A suspension of very fine solid or liquid particles in a gas. Aerosol is also used as a common name for a spray (or 'aerosol') can, in which a container is filled with a product and a propellant and is pressurized so as to release the product in a fine spray.

Age of Air
The length of time that a stratospheric air mass has been out of contact with the well-mixed →troposphere. The content of a unit element of air at a particular location and particular time of year in the →stratosphere can be thought of as a mixture of different air parcels that have taken different routes from the → tropopause to arrive at that location. The mean age of air at a specific location is defined as the average transit times of the elements since their last contact with the tropopause.

Alcohols
→Hydrocarbon derivatives in which at least one hydrogen atom has been replaced by an -OH group. Alcohols are sometimes used as solvents.

Annex B Countries/Parties (Kyoto Protocol)
The group of countries included in Annex B in the →Kyoto Protocol that have agreed to a target for their →greenhouse-gas emissions. It includes all the →Annex I countries (as amended in 1998) except Turkey and Belarus. See also: →Non-Annex I countries/parties.

Annex I Countries/Parties (Climate Convention)
The group of countries included in Annex I (as amended in 1998) to the →United Nations Framework Convention on Climate Change (UNFCCC). It includes all the developed countries in the Organisation of Economic Co-operation and Development (OECD), and →countries with economies in transition. By default, the other countries are referred to as →Non-Annex I countries. See also: →Annex B countries/parties.

Anthropogenic
Resulting from or produced by human beings.

Aqueous Cleaning
Cleaning parts of equipment with water, to which may be added suitable detergents, saponifiers or other additives.

Article 5(1) Countries (Montreal Protocol)
Developing countries that are →Party to the →Montreal Protocol. These countries are permitted a ten-year grace period in the phase-out schedule in the →Montreal Protocol compared with developed countries.

Atmosphere

The gaseous envelope surrounding the Earth. The dry atmosphere consists almost entirely of nitrogen (N_2) (78.1% volume →mixing ratio) and oxygen (O_2) (20.9% volume mixing ratio). The remaining 1% consists of trace gases, such as argon (Ar) (0.93% volume mixing ratio), helium (He), and radiatively active →greenhouse gases such as →carbon dioxide (CO_2) (0.037% volume mixing ratio in 2004) and →ozone (O_3). In addition, the atmosphere contains water vapour, whose amount is highly variable, clouds and both liquid and particulate aerosols. Most of the matter in the atmosphere occurs in the →troposphere immediately above the Earth's surface and the overlying →stratosphere.

Atmospheric Lifetime

A measure of the average time that a molecule remains intact once released into the →atmosphere. See also: →Lifetime.

Azeotrope (Refrigeration)

A →blend consisting of one or more →refrigerants of different volatilities that does not appreciably change in composition or temperature as it evaporates (boils) or condenses (liquefies) under constant pressure. Refrigerant blends assigned an R-500 series number designation by ANSI/ASHRAE 34 are azeotropes. Compare with: →Zeotrope.

Banks

Banks are the total amount of substances contained in existing equipment, chemical stockpiles, foams and other products not yet released to the atmosphere.

Baseline

A non-intervention →scenario used as a base in the analysis of intervention scenarios. See also: →Business-As-usual (BAU) Scenario.

Best Practice

For this Report, best practice is considered the lowest achievable value of halocarbon emission at a given date, using commercially proven technologies in the production, use, substitution, recovery and destruction of halocarbon or halocarbon-based products.

Blends/Mixtures (Refrigeration)

A mixture of two or more pure fluids. Blends are used to achieve properties that fit many refrigeration purposes. For example, a mixture of flammable and nonflammable components can result in a nonflammable blend. Blends can be divided into three categories: →azeotropic, →non-azeotropic and near-azeotropic blends.

Blowing Agent (Foams)

A gas, volatile liquid or chemical that generates gas during the foaming process. The gas creates bubbles or cells in the plastic structure of a foam.

Bottom-Up Models

A modelling approach that aggregates information from diverse sources, often including technological and engineering details in the analysis. Compare with: →Top-down models.

Business-As-Usual (BAU) Scenario (2015, This Report)

A →baseline scenario for the use of →halocarbons and their alternatives, which assumes that all existing regulation and → phase-out measures, including the →Montreal Protocol and relevant national regulations, continue to 2015. The usual practices (including end-of-life recovery) and emission rates are kept unchanged up to 2015.

Capital Costs

Costs associated with capital or investment expenditure on land, plant, equipment and inventories. Unlike labour and operating costs, capital costs are independent of the level of output for a given capacity of production.

Carbon Dioxide (CO_2)

A naturally occurring gas which occurs as a byproduct of burning →fossil fuels and biomass, as well as other industrial processes and land-use changes. It is the principal →anthropogenic →greenhouse gas that affects the Earth's radiative balance and is the reference gas against which other greenhouse gases are generally measured.

Catalyst

A chemical that acts to speed up or facilitate a chemical reaction, but is not physically changed or used up in the reaction.

Chiller

A cooling system that removes heat from one medium (water) and deposits it into another (ambient air or water).

Chlorine Loading

The total amount of chlorine (generally expressed as a →mixing ratio, or fraction of all air molecules), accounting for the amount of all chlorine-bearing substances and the number of atoms of chlorine in each substance.

Chlorocarbons

→Halocarbons containing carbon and chlorine atoms, but no other →halogen atoms.

Chlorofluorocarbons (CFCs)

→Halocarbons containing only chlorine, fluorine and carbon atoms. CFCs are both →ozone-depleting substances (ODSs) and →greenhouse gases.

Class A Fire (Fire Protection)

Fire in ordinary combustible materials, such as wood, cloth, paper, rubber and many plastics.

Class B Fire (Fire Protection)
Fire in flammable liquids, oils, greases, tars, oil-base paints, lacquers and flammable gases.

Class C Fire (Fire Protection)
Fire that involves energized electrical equipment where the electrical resistivity of the extinguishing media is of importance.

Clean Agent (Fire Protection)
An electrically non-conducting, volatile or gaseous fire-extinguishing agent that does not leave a residue upon evaporation.

Clean Development Mechanism (CDM)
Defined in Article 12 of the →Kyoto Protocol, the Clean Development Mechanism is intended to meet two objectives: (1) to assist →non-Annex I Parties in achieving sustainable development and in contributing to the ultimate objective of the convention; and (2) to assist →Annex I Parties in achieving compliance with their quantified emission limitation and reduction commitments. Certified emission reductions from Clean Development Mechanism projects undertaken in non-Annex I countries that limit or reduce →greenhouse-gas emissions, when certified by operational entities designated by the→Conference of the Parties/Meeting of the Parties, can be accrued to the investor (government or industry) from Parties in →Annex B. A share of the proceeds from the certified project activities is used to cover administrative expenses as well as to assist developing country Parties that are particularly vulnerable to the adverse effects of →climate change to meet the costs of adaptation.

Climate
Climate in a narrow sense is usually defined as the 'average weather', or more rigorously, as the statistical description in terms of the mean and variability of relevant quantities over a period of time ranging from months to thousands or millions of years. The classical period is 30 years, as defined by the → World Meteorological Organization (WMO). These quantities are most often surface variables such as temperature, precipitation, and wind. Climate in a wider sense is the state, including a statistical description, of the climate system.

Climate Change
Climate change refers to a statistically significant variation in either the mean state of the →climate or in its variability, persisting for an extended period (typically decades or longer). Climate change may be due to natural internal processes or external forcings, or to persistent →anthropogenic changes in the composition of the atmosphere or in land use.

Note that Article 1 of the →Framework Convention on Climate Change (UNFCCC) defines 'climate change' as 'a change of climate which is attributed directly or indirectly to human activity that alters the composition of the global atmosphere and which is in addition to natural climate variability, observed over comparable time periods'. The UNFCCC thus makes a distinction between 'climate change' attributable to human activities altering the atmospheric composition, and 'climate variability' attributable to natural causes.

Climate Feedback
An interaction mechanism between processes in the →climate system occurring when the result of an initial process triggers changes in a second process that in turn influences the initial one. A positive feedback intensifies the original process, whereas a negative feedback weakens it.

Climate Scenario
A plausible and often simplified representation of the future climate, based on a coherent and internally consistent set of driving forces and key relations, that has been constructed for use in investigating the potential consequences of →anthropogenic →climate change, and often serves as input to impact models. Climate →projections often serve as the raw material for constructing climate scenarios, but climate scenarios usually require additional information, such as the baseline current climate. A *climate-change scenario* is the difference between a climate scenario and the current climate.

Climate Sensitivity
In IPCC Reports, *equilibrium climate sensitivity* refers to the equilibrium change in global mean surface temperature following a doubling of the atmospheric (→equivalent) →carbon dioxide (CO_2) concentrations. More generally, *equilibrium climate sensitivity* refers to the equilibrium change in surface air temperature following a unit change in →radiative forcing (in units of °C per (W m^{-2})).

In practice, the evaluation of the equilibrium climate sensitivity requires very long simulations with coupled general circulation models. The *effective climate sensitivity* is a related measure that circumvents this requirement. It is evaluated from model output for evolving non-equilibrium conditions. It is a measure of the strengths of the →feedbacks at a particular time and may vary with forcing history and climate state.

Climate System
The highly complex system that consists of five major components: the →atmosphere, the hydrosphere, the cryosphere, the land surface and the biosphere, as well as of the interactions between them. The climate system evolves over time under the influence of its own internal dynamics and because of external forcings, such as volcanic eruptions, solar variations and human-induced forcings (such as the changing composition of the atmosphere and land-use change).

Climate Variability
Variations in the mean state and other statistics (such as the standard deviation and the occurrence of extremes) of the → climate on all temporal and spatial scales beyond that of individual weather events. Climate variability may be caused by natural internal processes within the →climate system (internal variability), or by variations in natural or anthropogenic external forcings (external variability). See also: →Climate change.

CO_2-Equivalent

The amount of →carbon dioxide (CO_2) that would cause the same amount of →radiative forcing as a given amount of another →greenhouse gas. When used with concentrations this refers to the instantaneous radiative forcing caused by the greenhouse gas or the equivalent amount of CO_2. When used with emissions this refers to the time-integrated radiative forcing over a specified time horizon caused by the change in concentration produced by the emissions. See also: →Global warming potential.

Coefficient of Performance (COP) (Refrigeration)

A measure of the energy efficiency of a refrigerating system. It is defined as the ratio between the refrigerating capacity and the electric power consumed by the system. The COP is primarily dependant on the working cycle and the temperature levels (evaporating/condensing temperature) as well as on the properties of the →refrigerant, system design and size.

Column Ozone

The total amount of →ozone in a vertical column above the Earth's surface. Column ozone is measured in →Dobson units (DU).

Commercialization

A sequence of actions necessary to achieve market entry and general market competitiveness of new technologies, processes and products.

Compressor Discharge Temperature (Refrigeration)

The temperature of a gas at the high-pressure outlet from the compressor (superheated gas). The gas temperature is typically 30°C to 40°C higher than the condensing temperature at saturation pressure, mainly depending on the evaporating/condensing temperature, →refrigerant properties and the compressor energy efficiency.

Conference of the Parties (COP) (Climate Convention)

The supreme body of the →United Nations Framework Convention on Climate Change (UNFCCC), comprising countries that have ratified or acceded to the UNFCCC. See also: → Conference of the Parties/Meeting of the Parties and →Meeting of the Parties.

Conference of the Parties/Meeting of the Parties (COP/MOP) (Climate Convention)

The →Conference of the Parties of the →United Nations Framework Convention on Climate Change (UNFCCC) will serve as the →Meeting of the Parties (MOP), the supreme body of the →Kyoto Protocol, but only Parties to the Kyoto Protocol may participate in deliberations and make decisions.

Containment (Refrigeration)

The application of service techniques or special equipment designed to preclude or reduce loss of →refrigerant from equipment during installation, operation, service or disposal of refrigeration and air-conditioning equipment.

Controlled Substance

Under the →Montreal Protocol, any →ozone-depleting substance (ODS) that is subject to control measures, such as a → phase-out requirement.

Cost Effective

A criterion that specifies that a technology or measure delivers a good or service at equal or lower cost than current practice, or the least-cost alternative for the achievement of a given target.

Countries with Economies in Transition (CEITs)

Countries with national economies in the process of changing from a planned economic system to a market economy.

Destruction

Destruction of →ozone-depleting substances (ODSs) by approved destruction plants, in order to avoid their emissions.

Detergent

A product designed to render, for example, oils and greases soluble in water; usually made from synthetic →surfactants.

Dobson Unit (DU)

A unit to measure total →column ozone. The number of Dobson units is the thickness, in units of 10^{-5} m, that the ozone column would occupy if compressed into a layer of uniform density at a pressure of 1013 hPa and a temperature of 0°C. One DU corresponds to a column of ozone containing 2.69×10^{20} molecules per square meter. Although column ozone can vary greatly, 300 DU is a typical value.

Drop-In Replacement (Refrigeration)

The procedure for replacing →CFC refrigerants with non-CFC refrigerants in existing refrigerating, air-conditioning and heat-pump plants without doing any plant modifications. However, drop-in procedures are normally referred to as →retrofitting because plants need minor modifications, such as the change of lubricant, and the replacement of the expansion device and the desiccant material.

Dry Powder Inhaler (DPI) (Medical Aerosols)

An alternate technology to →metered dose inhalers (MDIs) that can be used if the medication being dispensed can be satisfactorily formulated as microfine powder, thus eliminating the use of a chemical propellant.

Emission Factor

The coefficient that relates actual →emissions to activity data as a standard rate of emission per unit of activity.

Emissions

The release of gases or →aerosols into the →atmosphere over a specified area and period of time.

Emission Scenario

A plausible representation of the future development of →emissions of substances that are potentially radiatively active (e.g., →greenhouse gases and →aerosols), based on a coherent and internally consistent set of assumptions about driving forces (such as demographic and socio-economic development, and technological change) and their key relationships. See also: → Scenario (generic) and →climate scenario.

The *IPCC Special Report on Emission Scenarios* (2000) presented emission scenarios, known as the →SRES scenarios, which have been used as a basis for the climate projections in the *IPCC Third Assessment Report* (2001) and in this report.

Energy Balance

Averaged over the globe and over long time periods, the energy budget of the →climate system must be in balance. Because the climate system derives all its energy from the Sun, this balance implies that, globally, the amount of incoming →solar radiation must on average be equal to the sum of the outgoing reflected solar radiation and the outgoing →thermal infrared radiation emitted by the climate system. A perturbation of this global radiation balance, be it →anthropogenic or natural, is called → radiative forcing.

Equivalent-CO_2

See →CO_2-Equivalent.

Equivalent Effective Stratospheric Chlorine (EESC)

An index of the amount of chlorine (Cl) and bromine (Br) that is present in the →stratosphere in forms that can contribute to the depletion of →ozone. The EESC value takes into account different fractional releases of chlorine or bromine from different halocarbons and the much higher efficiency of bromine in the catalytic removal of ozone. See also: →Chlorine loading.

Ethers

Organic compounds with formula R-O-R, where O is an oxygen atom and R is not a hydrogen atom (H).

Expansion Control Devices (Refrigeration)

A device, such as an expansion valve, expansion orifice, turbine or capillary tube, that is used to control the mass flow of a → refrigerant from the high-pressure side to the low-pressure side of a refrigeration system.

External Costs

The costs arising from any human activity when the agent responsible for the activity does not take full account of the negative impacts on others of his or her actions. Similarly, when the impacts are positive and not accounted for in the actions of the agent responsible, they are referred to as *external benefits*. Emissions of particulate pollution from an industrial installation affect the health of people in the vicinity, but this is not often considered, or is given inadequate weight, in private decision making and there is no market for such impacts. Such a phenomenon is referred to as an *externality*, and the costs it imposes are referred to as the external costs.

External Forcing

See: →Climate system.

Externality

See: →External costs.

Feedback

See: →Climate feedback.

Fluorinated Ethers

→Ethers in which one or more hydrogen atoms have been replaced by fluorine.

Fluorocarbons

→Halocarbons containing fluorine atoms, including →chlorofluorocarbons (CFCs), →hydrochlorofluorocarbons (HCFCs), →hydrofluorocarbons (HFCs), and →perfluorocarbons (PFCs).

Fluoroketones (FKs)

Organic compounds in which two fully fluorinated alkyl groups are attached to a carbonyl group (C=O).

Fossil Fuels

Carbon-based fuels derived from geological (fossil) carbon deposits. Examples include coal, oil and natural gas.

Framework Convention on Climate Change

See:→United Nations Framework Convention on Climate Change (UNFCCC).

Global Warming Potential (GWP)

An index comparing the climate impact of an emission of a greenhouse gas relative to that of emitting the same amount of →carbon dioxide. GWPs are determined as the ratio of the time-integrated →radiative forcing arising from a pulse emission of 1 kg of a substance relative to that of 1 kg of carbon dioxide, over a fixed time horizon. See also: →Radiative forcing.

Greenhouse Effect

→Greenhouse gases in the →atmosphere effectively absorb the →thermal infrared radiation that is emitted by the Earth's surface, by the atmosphere itself, and by clouds. The atmosphere emits radiation in all directions, including downward to the Earth's surface. Greenhouse gases trap heat within the surface-troposphere system and raise the temperature of the Earth's surface. This is called the *natural greenhouse effect*.

An increase in the concentration of greenhouse gases leads to increased absorption of infrared radiation and causes a →radiative forcing, or energy imbalance, that is compensated for by an increase in the temperature of the surface-troposphere system. This is the *enhanced greenhouse effect*.

Greenhouse Gases (GHGs)

The gaseous constituents of the →atmosphere, both natural and →anthropogenic, that absorb and emit radiation within the spectrum of the →thermal infrared radiation that is emitted by the Earth's surface, by the atmosphere and by clouds. This property causes the →greenhouse effect. The primary greenhouse gases in the Earth's atmosphere are water vapour (H_2O), →carbon dioxide (CO_2), nitrous oxide (N_2O), methane (CH_4) and →ozone (O_3). Moreover, there are a number of entirely →anthropogenic greenhouse gases in the atmosphere, such as the →halocarbons and other chlorine- and bromine-containing substances that are covered by the →Montreal Protocol. Some other trace gases, such as sulphur hexafluoride (SF_6), hydrofluorocarbons (HFCs), and perfluorocarbons (PFCs), are also greenhouse gases.

Halocarbons

Chemical compounds containing carbon atoms, and one or more atoms of the →halogens chlorine (Cl), fluorine (F), bromine (Br) or iodine (I). *Fully halogenated halocarbons* contain only carbon and halogen atoms, whereas *partially halogenated halocarbons* also contain hydrogen (H) atoms. Halocarbons that release chlorine, bromine or iodine into the →stratosphere cause →ozone depletion. Halocarbons are also →greenhouse gases. Halocarbons include →chlorofluorocarbons (CFCs), →hydrochlorofluorocarbons (HCFCs), →hydrofluorocarbons (HFCs), →perfluorocarbons (PFCs) and →halons.

Halogens

A family of chemical elements with similar chemical properties that includes fluorine (F), chlorine (Cl), bromine (Br) and iodine (I).

Halons

Fully halogenated →halocarbons that contain bromine and fluorine atoms.

Hermetic

An airtight sealed system.

Hermetic Compressors (Refrigeration)

Compressors whose motors are sealed within the →refrigerant loop and are often cooled by the flow of the →refrigerant-lubricant mixture directly over the motor windings.

Hydrocarbons (HCs)

Chemical compounds consisting of one or more carbon atoms surrounded only by hydrogen atoms.

Hydrochlorofluorocarbons (HCFCs)

→Halocarbons containing only hydrogen, chlorine, fluorine and carbon atoms. Because HCFCs contain chlorine, they contribute to →ozone depletion. They are also →greenhouse gases.

Hydrofluorocarbons (HFCs)

→Halocarbons containing only carbon, hydrogen and fluorine atoms. Because HFCs contain no chlorine, bromine or iodine,

they do not deplete the →ozone layer. Like other halocarbons they are potent →greenhouse gases.

Hydrofluoroethers (HFEs)

Chemicals composed of hydrogen, fluorine and →ether, which have similar performance characteristics to certain →ozone-depleting substances (ODSs) that are used as solvents.

Implementation Costs

Costs involved in the implementation of →mitigation options. These costs are associated with the necessary institutional changes, information requirements, market size, opportunities for technology gain and learning, and economic incentives (grants, subsidies and taxes).

Intergovernmental Panel on Climate Change (IPCC)

The Intergovernmental Panel on Climate Change (IPCC) was jointly established by the →World Meteorological Organization (WMO) and the →United Nations Environment Programme (UNEP) in 1988 to assess scientific, technical and socio-economic information relevant for the understanding of climate change, its potential impacts and options for adaptation and mitigation. It is open to all Members of the UN and of WMO. The IPCC provides, on request, scientific, technical and socio-economic advice to the →Conference of the Parties (COP) to the →United Nations Framework Convention on Climate Change (UNFCCC). The IPCC has produced a series of Assessment Reports, Special Reports, Technical Papers, methodologies, and other products that have become standard works of reference and that are widely used by policymakers, scientists, and other experts.

Kyoto Protocol

The Kyoto Protocol to the →United Nations Framework Convention on Climate Change (UNFCCC) was adopted at the Third Session of the →Conference of the Parties (COP) to the UNFCCC in 1997 in Kyoto, Japan. It contains legally binding commitments, in addition to those included in the UNFCCC. Countries included in →Annex B of the Protocol (most OECD countries and →countries with economies in transition) agreed to reduce their →anthropogenic →greenhouse-gas emissions (specifically →carbon dioxide (CO_2) methane (CH_4) nitrous oxide (N_2O) →hydrofluorocarbons (HFCs), →perfluorocarbons (PFCs) and sulfur hexafluoride (SF_6)) by at least 5% below 1990 levels in the commitment period 2008 to 2012. The Kyoto Protocol entered into force on 16 February 2005.

Life Cycle Assessment (LCA)

An assessment of the overall environmental impact of a product over its entire life cycle (manufacture, use and recycling or disposal).

Life Cycle Climate Performance (LCCP)

A measure of the overall global-warming impact of equipment based on the total related →emissions of →greenhouse gases over its entire life cycle. LCCP is an extension of the →total equivalent warming impact (TEWI). LCCP also takes into account the direct fugitive emissions arising during manufacture, and the greenhouse gas emissions associated with their embodied energy.

Lifetime

Lifetime is a general term used for various time scales characterizing the rates of processes affecting the concentration of trace gases. The following lifetimes may be distinguished:

Turnover time (T) is the ratio of the mass (*M*) of a reservoir (e.g., a gaseous compound in the →atmosphere) and the total rate of removal (*S*) from the reservoir: *T* = *M*/*S*. Separate turnover times can be defined for each removal process.

Adjustment time or *response time* (T_a) is a time scale characterizing the decay of an instantaneous pulse input into the reservoir. The term *adjustment time* is also used to characterize the adjustment of the mass of a reservoir following a step change in the source strength. *Half-life* or *decay constant* is used to quantify a first-order exponential decay process. See: →Response time, for a different definition pertinent to climate variations. The term *lifetime* is sometimes used, for simplicity, as a surrogate for *adjustment time*.

In simple cases, such as CFC-11, where the global removal rate of the compound is proportional to the total mass of the reservoir, the adjustment time equals the turnover time: *T* = T_a. In more complex cases removal rates are not proportional to the reservoir mass – for example because of feedback effects – and this equality no longer holds.

Longwave Radiation

See: →Thermal infrared radiation.

Lower Flammability Limit (LFL)

'The minimum concentration of a combustible substance that is capable of propagating a flame through a homogeneous mixture of the combustible and gaseous oxidizer under the specified conditions of test' (ASTM Standard E 681-85). The conditions of test usually reported for →refrigerants are in dry air in ambient temperature and pressure.

Lubricant

Typically a substance introduced between moving surfaces to reduce the friction and wear between them.

Materials Safety Data Sheet (MSDS)

A safety advisory bulletin prepared by chemical producers for a specific →refrigerant or compound.

Meeting of the Parties (to the Kyoto Protocol) (MOP)

→Conference of the Parties to the →United Nations Framework Convention on Climate Change serving as the Meeting of the Parties (MOP) to the →Kyoto Protocol. It is the supreme body of the Kyoto Protocol. See also: →Conference of the Parties/ Meeting of the parties (COP/MOP).

Meeting of the Parties (to the Montreal Protocol) (MOP)

The supreme body of the Montreal Protocol.

Metered Dose Inhalers (MDIs) (Medical Aerosols)

A method of dispensing inhaled pulmonary drugs. See also: → Dry powder inhaler (DPI).

Miscible

The ability of two liquids or gases to uniformly dissolve into each other. Immiscible liquids will separate into two distinguishable layers.

Mitigation

A human intervention to reduce the sources or enhance the sinks of →greenhouse gases.

Mixing Ratio

Mixing ratio, or *mole fraction,* is the ratio of the number of moles of a constituent in a given volume to the total number of moles of all constituents in that volume. It is usually reported for dry air. Typical values for long-lived →greenhouse gases range from μmol/mol (parts per million: ppm), nmol/mol (parts per billion: ppb), to fmol/mol (parts per trillion: ppt). Correcting the mixing ratio for the non-ideality of gases gives the *volume mixing ratio* (sometimes expressed in ppmv, etc.).

Mole Fraction

See: →Mixing ratio

Montreal Protocol

The Montreal Protocol on Substances that Deplete the Ozone Layer was adopted in Montreal in 1987 and subsequently adjusted and amended in London (1990), Copenhagen (1992), Vienna (1995), Montreal (1997) and Beijing (1999). It controls the consumption and production of chlorine- and bromine-containing chemicals (known as →ozone-depleting substances, ODSs) that destroy the stratospheric →ozone layer.

Multilateral Fund

Part of the financial mechanism under the →Montreal Protocol, established by the →Parties to provide financial and technical assistance to →Article 5 Parties.

Non-Annex I Parties/Countries (Climate Convention)

The countries that have ratified or acceded to the →United Nations Framework Convention on Climate Change (UNFCCC) that are not included in →Annex I of the Climate Convention.

Non-Azeotropic (Refrigeration)

A →blend or mixture where the compositions of coexisting liquid and vapour differ and condensation and evaporation occur over a range of temperatures. This effect can in some applications give improved performance in plants with heating/cooling demand with gliding temperatures. Heating of hot tap water is one example. Equipment has to be modified to use a non-azeotropic blend. See also: →zeotropic, →azeotropic.

Non-Condensable Gases (Refrigeration)

Gases with very low temperature boiling points, which are not easily condensed. Nitrogen and oxygen are the most common ones found in →chillers.

Nonlinearity

A process is called 'nonlinear' when there is no simple proportional relation between cause and effect. The →climate system contains many nonlinear processes, resulting in a system with potentially very complex behaviour. Such complexity may lead to rapid climate change.

Normal Boiling Point (NBP)

The boiling point of a compound at atmospheric pressure (1013 hPa).

Not-in-Kind Technologies (NIK)

Not-in-kind technologies achieve the same product objective without the use of →halocarbons, typically with an alternative approach or unconventional technique. Examples include the use of stick or spray pump deodorants to replace CFC-12 aerosol deodorants; the use of mineral wool to replace CFC, HFC or HCFC insulating foam; and the use of dry powder inhalers (DPIs) to replace CFC or HFC metered dose inhalers (MDIs).

One-Component Foam (OCF)

A foam in which the →blowing agent acts both as a frothing agent and as a propellant. These foams are primarily used for gap filling (to prevent air infiltration) rather than for thermal insulation per se. As such the use of blowing agent is fully emissive.

Open Drive (Refrigeration)

A compressor drive motor that is outside the →refrigerant loop and therefore not directly exposed to the circulating refrigerant.

Ozone

The triatomic form of oxygen (O_3), which is a gaseous →atmospheric constituent. In the →troposphere it is created by photochemical reactions involving gases occurring naturally and resulting from →anthropogenic activities (→'smog'). Tropospheric ozone acts as a →greenhouse gas. In the →stratosphere ozone is created by the interaction between solar → ultraviolet radiation and molecular oxygen (O_2). Stratospheric ozone plays a major role in the stratospheric radiative balance. Its concentration is highest in the →ozone layer.

Ozone-Depleting Substances (ODSs)

Substances known to deplete the stratospheric →ozone layer. The ODSs controlled under the →Montreal Protocol and its Amendments are →chlorofluorocarbons (CFCs), →hydrochlorofluorocarbons (HCFCs), →halons, methyl bromide (CH_3Br), carbon tetrachloride (CCl_4), methyl chloroform (CH_3CCl_3), hydrobromofluorocarbons (HBFCs) and bromochloromethane (CH_2BrCl).

Ozone Depletion

Accelerated chemical destruction of the stratospheric →ozone layer by the presence of substances produced by human activities.

Ozone Depletion Potential (ODP)

A relative index indicating the extent to which a chemical product may cause →ozone depletion compared with the depletion caused by CFC-11. Specifically, the ODP of an →ozone-depleting substance (ODS) is defined as the integrated change in total ozone per unit mass emission of that substance relative to the integrated change in total ozone per unit mass emission of CFC-11.

Ozone Layer

The layer in the →stratosphere where the concentration of → ozone is greatest. The layer extends from about 12 to 40 km. This layer is being depleted by →anthropogenic emissions of chlorine and bromine compounds. Every year, during the Southern Hemisphere spring, a very strong depletion of the ozone layer takes place over the Antarctic region. This depletion is caused by anthropogenic chlorine and bromine compounds in combination with the specific meteorological conditions of that region. This phenomenon is called the *Antarctic ozone hole*.

Party

A country that signs and/or ratifies an international legal instrument (e.g., a protocol or an amendment to a protocol), indicating that it agrees to be bound by the rules set out therein. Parties to the →Montreal Protocol or →Kyoto Protocol are countries that have signed and ratified these Protocols.

Perfluorocarbons (PFCs)

Synthetically produced →halocarbons containing only carbon and fluorine atoms. They are characterized by extreme stability, non-flammability, low toxicity, zero →ozone depleting potential and high →global warming potential.

Phase-Out

The ending of all production and consumption of a chemical controlled under the →Montreal Protocol.

Phase-Out Plan

The part of the Country Programme under the →Montreal Protocol that describes the strategy statement of a government defining the →phase-out time schedule for each controlled substance and the government actions to be taken for achieving phase-out. It contains a prioritized list of projects to be undertaken and takes into account the specific industrial, political and legislative situation in the country.

Polar Stratospheric Clouds (PSCs)

A class of clouds composed of particles, including nitric acid hydrates and ice, that occur at high altitudes (of about 15 to 30 km) in the polar →stratosphere. They occur when the temperature is very low (below about –95°C), such as in the Antarctic →polar vortex, and have been observed mainly over Antarctica in the winter and spring, and occasionally over the Arctic. PSCs play a major role in →ozone depletion because chlorine is converted to forms that are highly reactive with ozone through chemical reactions on or within the cloud particles.

Polar Vortex

A dynamical structure that occurs during the polar winter in which →stratospheric air acquires a cyclonic circulation about the pole, with an area of relatively still air in its centre. The vortex core air (above 16 km in altitude) becomes effectively isolated from mid-latitude air. The polar vortex over Antarctica is usually colder and lasts longer (throughout the austral spring) than the polar vortex over the Arctic.

ppm, ppb, ppt

See: →Mixing ratio.

Precursors

Atmospheric compounds which themselves are not →greenhouse gasses or →aerosols, but which have an effect on greenhouse-gas or aerosol concentrations by taking part in physical or chemical processes regulating their production or destruction rates.

Present Value Cost

The sum of all costs over all time periods, with future costs discounted.

Projection (Generic)

A potential future evolution of a quantity or set of quantities, often computed with the aid of a model. Projections are distinguished from *predictions* in order to emphasize that projections involve assumptions concerning, for example, future socio-economic and technological developments that may or may not be realized, and are therefore subject to substantial uncertainty.

Propellant

The component of an →aerosol spray that acts as a forcing agent to expel the product from the aerosol canister.

Purge System (Refrigeration)

A device used on low-pressure chillers to expel air and other non-condensables from the circulating →refrigerant.

Push-Pull Method (Refrigeration)

A method for →recovering and →recycling →refrigerant from a system using a negative pressure (suction) on one side to pull the old refrigerant out and pumping recycled refrigerant vapour to the other side to push the old refrigerant through the system.

Radiative Efficiency

A measure of the efficiency of a gas in changing →radiative forcing. It is calculated as the marginal change in radiative forcing per unit increase in gas concentration and typically given in units of $W\ m^{-2}\ ppb^{-1}$.

Radiative Forcing

Radiative forcing is the change in the net irradiance (expressed in Watts per square meter: $W\ m^{-2}$) at the →tropopause due to an internal change or a change in the external forcing of the →climate system, such as a change in the concentration of → carbon dioxide (CO_2) in the atmosphere or in the output of the Sun. Usually radiative forcing is computed after allowing for stratospheric temperatures to readjust to radiative equilibrium, but with all tropospheric properties held fixed at their unperturbed values. Radiative forcing is called *instantaneous* if no change in stratospheric temperature is accounted for. See also: →Global warming potential.

Radiative Forcing Scenario

A plausible representation of the future development of →radiative forcing associated, for example, with →anthropogenic changes in atmospheric composition or in land-use, or with natural factors such as variations in →solar activity. Radiative forcing scenarios can be used as input into simplified climate models to compute climate projections.

Radical

A molecular entity possessing an unpaired electron.

Reclamation

Reprocessing and upgrading of a recovered controlled substance through mechanisms such as filtering, drying, distillation and chemical treatment in order to restore the substance to a specified standard of performance. Chemical analysis is required to determine that appropriate product specifications are met. It often involves processing off-site at a central facility.

Recovery

The collection and storage of controlled substances from machinery, equipment, containment vessels, etc., during servicing or prior to disposal without necessarily testing or processing it in any way.

Recycling

Reuse of a recovered controlled substance following a basic cleaning process such as filtering and drying. For →refrigerants, recycling normally involves recharge back into equipment and it often occurs 'on-site'.

Refrigerant (Refrigeration)

A heat transfer agent, usually a liquid, used in equipment such as refrigerators, freezers and air conditioners.

Relief Valve (Refrigeration)

A device that vents refrigerant when the pressure in a →chiller becomes dangerously high. Newer relief valves have a resealing mechanism so that when the pressure of the chiller returns to a normal level they reseal and prevent further refrigerant loss.

Research, Development and Demonstration

Scientific or technical research and development of new production processes or products, coupled with analysis and measures that provide information to potential users regarding the application of the new products or processes, such demonstration tests and studies of the feasibility of pilot plants and other pre-commercial applications.

Response Time (Climate System)

The response time or *adjustment time* is the time needed for the →climate system or its components to re-equilibrate to a new state, following a forcing resulting from external or internal processes. Different components of the climate system can have very different response times. The response time of the → troposphere is relatively short, from days to weeks, whereas the response time of the →stratosphere is typically a few months. The oceans have much longer response times, of decades to millennia, because of their large heat capacity. The response time of the strongly coupled surface-troposphere system is mainly determined by the oceans, and is therefore slow compared with that of the stratosphere. The biosphere can respond quickly, for example to droughts, but it can also respond very slowly to other imposed changes.

See: →Lifetime, for the definition of response time in relation to atmospheric concentrations.

Retrofit

The upgrading or adjustment of equipment so that it can be used under altered conditions; for example, of refrigeration equipment to be able to use a non-ozone depleting refrigerant in place of a →chlorofluorocarbon (CFC).

Saturated Vapour Pressure

The maximum vapour pressure of a substance at a given temperature when accumulated over its liquid or solid state in a confined space.

Scenario (Generic)

A plausible and often simplified description of how the future may develop, based on a coherent and internally consistent set of driving forces and key relationships. Scenarios may be derived from →projections, but are often based on additional information from other sources, sometimes combined with a 'narrative storyline'. See also: →SRES scenarios, →climate scenario and →emission scenarios.

Semi-Aqueous Cleaning

Cleaning with a non-water-based cleaner, followed by a water rinse.

Servicing (Refrigeration)

In the refrigeration sector, all kinds of work that may be performed by a service technician, from installation, operations, inspection, repair, retrofitting, redesign and decommissioning of refrigeration systems to handling, storage, recovery and recycling of refrigerants, as well as record-keeping.

Shortwave Radiation

See: →Solar radiation.

Smog

The buildup of high levels of pollution, generally in association with urban areas. Photochemical smog occurs in the →troposphere where sunlight causes chemical reactions in polluted air, one effect of which is the generation of →ozone.

Solar Radiation

Radiation emitted by the Sun, most of which is shortwave radiation at wavelengths less than about 1 μm and is determined by the temperature of the Sun. See also: →Ultraviolet radiation; compare with: →thermal infrared radiation.

Solvent

Any product (aqueous or organic) designed to clean a component or assembly by dissolving the contaminants present on its surface.

Specific Costs (of Abatement Options)

The difference in costs of an abatement option as compared with a reference case, expressed in relevant specific units. In this Report the specific costs of →greenhouse gas emission reduction options are generally expressed in US$ per tonne of avoided →CO_2-equivalents (US$/t$CO_2$-eq).

SRES Scenarios

→Emission scenarios developed by the IPCC Special Report on Emission Scenarios (2000).

Stratosphere

The highly stratified region of the →atmosphere above the → troposphere. It extends from an altitude of about 8 km in high latitudes and 16 km in the tropics to an altitude of about 50 km. This region is characterized by increasing temperature with altitude.

Stratospheric Polar Vortex

See: →Polar vortex

Surfactant

A product designed to reduce the surface tension of water. Also referred to as a tension-active agent/tenside. Detergents are made up principally from surfactants.

Technology and Economic Assessment Panel (TEAP)

A standing subsidiary body of the Parties to the →Montreal Protocol, which was established in 1988 under Article 6 of the Montreal Protocol and is coordinated by the →United Nations Environment Programme (UNEP) Ozone Secretariat. It comprises hundreds of experts from around the world. TEAP is responsible for conducting assessments and for reporting to the Parties on (a) the state of art of production and use technology, options to →phase-out the use of →ozone-depleting substances (ODSs), recycling, reuse and destruction techniques; and (b) the economic effects of ozone layer modification and the economic aspects of technology.

Thermal Infrared Radiation
Radiation emitted by the Earth's surface, the →atmosphere and the clouds, with wavelengths longer than the wavelength of the red colour in the visible part of the spectrum. It is also known as terrestrial or longwave radiation. The spectrum of infrared radiation is distinct from that of →solar or shortwave radiation because of the large difference in temperature between the surface of the Sun and the Earth.

Thermoplastic
A material that can repeatedly become plastic on heating and harden on cooling. Compare with: →Thermosetting.

Thermosetting
A material that sets permanently on heating. Compare with: → Thermoplastic.

Threshold Limit Values (TLVs)
Exposure safety guidelines established by the American Conference of Governmental and Industrial Hygienists (ACGIH) based on an inhalation →time-weighted average. TLVs 'represent conditions under which it is believed that nearly all workers can be repeatedly exposed day after day without adverse effects.' For volatile substances, such as →refrigerants, TLVs are expressed as parts per million volume concentrations in air (ppm).

Time-Weighted Average (TWA)
A technique used to measure the average exposure of workers to a chemical over a given period of time.

Top-Down Models
A modelling approach that evaluates a system from aggregate variables. An example of a top-down model is that of applied macroeconomic theory and econometric techniques applied to historical data on consumption, prices, incomes and factor costs to model final demand for goods and services, and supply from main sectors, like the energy sector, transportation, agriculture and industry. Compare with: →Bottom-up models.

Total Equivalent Warming Impact (TEWI)
A measure of the overall global-warming impact of equipment based on the total related →emissions of →greenhouse gases during the operation of the equipment and the disposal of the operating fluids at the end-of-life. TEWI takes into account both direct fugitive emissions, and indirect emissions produced through the energy consumed in operating the equipment. TEWI is measured in units of mass of →CO_2 equivalent. See also: → Life cycle climate performance (LCCP).

Transitional Substance (Montreal Protocol)
Under the →Montreal Protocol, a chemical whose use is permitted as a replacement for →ozone-depleting substances (ODSs),

but only temporarily because the substance's →ozone depletion potential (ODP) is non-zero.

Tropopause
The boundary between the →troposphere and the →stratosphere.

Troposphere
The lowest part of the →atmosphere above the Earth's surface, where clouds and 'weather' phenomena occur. The thickness of the troposphere is on average 9 km in high latitudes, 10 km in mid-latitudes, and 16 km in the tropics. Temperatures in the troposphere generally decrease with height.

Ultraviolet Radiation (UV)
Radiation from the Sun with wavelengths between visible light and X-rays. UV-B (280–320 nm), one of three bands of UV radiation, is harmful to life on the Earth's surface and is mostly absorbed by the →ozone layer.

United Nations Environment Programme (UNEP)
Established in 1972, UNEP is the specialized agency of the United Nations for environmental protection.

United Nations Framework Convention on Climate Change (UNFCCC)
An international convention whose ultimate objective is the 'stabilization of →greenhouse-gas concentrations in the atmosphere at a level that would prevent dangerous →anthropogenic interface with the →climate system'. The Convention was adopted on 9 May 1992 in New York and signed at the 1992 Earth Summit in Rio de Janeiro by more than 150 countries and the European Community. It contains commitments for all → Parties and entered into force in March 1994. See also: →Kyoto Protocol.

Venting (Refrigeration)
A service practice where the →refrigerant vapour is allowed to escape into the →atmosphere after the refrigerant liquid has been recovered.

Volatile Organic Compounds (VOCs)
Organic compounds that evaporate at their temperature of use. Many VOCs contribute to the formation of →tropospheric → ozone and →smog.

Voluntary Agreement
An agreement between a government authority and one or more private parties, as well as a unilateral commitment that is recognized by the public authority, to achieve environmental objectives or to improve environmental performance beyond compliance.

Voluntary Measures
Measures to reduce →greenhouse-gas emissions that are adopted by firms or other actors in the absence of government man-

dates. Voluntary measures help make climate-friendly products or processes more readily available or encourage consumers to incorporate environmental values in their market choices.

Well-Mixed Greenhouse Gases

→Greenhouse gases with lifetimes that are long compared with the mixing time between the two hemispheres (about 1 year), so that their mixing ratios do not have large gradients except, possibly, close to source regions.

World Meteorological Organization (WMO)

Established in 1950, WMO is the specialized agency of the United Nations for meteorology (weather and climate), operational hydrology and related geophysical sciences.

Zeotrope (Refrigeration)

A blend consisting of several →refrigerants of different volatilities that appreciably changes in composition or temperature as it evaporates (boils) or condenses (liquefies) at a given pressure. A refrigerant blend assigned an R-400 series number designation in ANSI/ASHRAE 34 is a zeotrope. Compare with: → Azeotrope

Annex III

Acronyms and Abbreviations

This is a list of acronyms and abbreviations as they are used in the report. An arrow (→) denotes acronyms or abbreviations that are also glossary items. For a list of chemical substances, see Annex V: Major Chemical Formulae and Nomenclature.

1-D	One-dimensional
2-D	Two-dimensional
3-D	Three-dimensional
AAP	American Academy of Pediatrics
ABNT	Brazilian Association for Technical Standards
AC	Air conditioning
ACGIH	American Conference of Governmental and Industrial Hygienists
ACRIB	Air-Conditioning and Refrigeration Industry Board
ADL	Arthur D. Little, Inc.
AEAT	AEA Technology
AEL	Allowable exposure limit
AFEAS	Alternative Fluorocarbons Environmental Acceptability Study
AFFF	Aqueous film forming foam
AGAGE	Advanced GAGE
AGWP	Absolute GWP
AHAM	Association of Home Appliance Manufacturers
ALE	Atmospheric Lifetime Experiment
ANSI	American National Standards Institute
APME	Association of Plastics Manufacturers in Europe
AR4	IPCC Fourth Assessment Report
ARAP	Alliance for Responsible Atmospheric Policy
ARI	Air-Conditioning and Refrigeration Institute
ASHRAE	American Society of Heating, Refrigerating and Air-Conditioning Engineers
ATM	Automatic teller machine
ATOC	UNEP Aerosols and Miscellaneous Uses Options Committee
BA	→Blowing agent
BAU	→Business As Usual
BCFCs	Bromochlorofluorocarbons

BFCs	Bromofluorocarbons
BING	The Federation of European Polyurethane Rigid Foam Associations
BNF	British National Formulary
BRA	British Refrigeration Association
BRE	DETR Building Research Establishment
BTU	British thermal unit
CAFE	Clean Air For Europe
CAOFSMI	China Association of Organic Fluorine and Silicone Material Industry
CCM	Chemistry-climate model
CCR	China Chemical Reporter
CDHS	California Department of Health Services
CDM	→Clean Development Mechanism
Cefic	Conseil Européen de l'Industrie Chimique (European Chemical Industry Council)
CEITs	→Countries with economies in transition
CEN	Comité Européen de Normalisation (European Committee for Standardization)
CFCs	→Chlorofluorocarbons
CHCP	Combined heating, cooling and power generation
ChCst	Change in cost
ChEU	Change in energy usage
CICADs	Concise International Chemical Assessment Documents
CLASP	Collaborative Labeling and Appliance Standards Program
CMDL	NOAA Climate Monitoring and Diagnostics Laboratory
COP	→Coefficient of performance (Refrigeration)
COP	→Conference of the Parties (Climate Convention)

COPD	Chronic obstructive pulmonary disease		GDP	Gross domestic product
CORINAIR	Core Inventory of Air Emissions in Europe		GEF	Global Environment Facility
CP	→Country Programme		GEIA	Global Emissions Inventory Activity
CPSC	Consumer Products Safety Commission		GEISA	Gestion et Etude des Informations
CPU	Central processing unit			Spectroscopiques Atmosphériques
CTI	Cryo-Trans, Inc.		GHG	→Greenhouse gas
D&T	Deloitte & Touche Consulting Group		GISS	NASA Goddard Institute for Space Studies
DDT	Dichloro-diphenyl-trichloroethane		gpm	Gallon (US) per mile
DEFRA	UK Department of Environment, Food and		GWP	→Global warming potential
	Rural Affairs		HadRT	Hadley Centre Radiosonde Temperature base
DETR	UK Department of the Environment, Trade and		HALOE	Halogen Occultation Experiment
	the Regions		HAPs	Hydrocarbon Aerosol Propellants
DHW	Domestic hot water		HCFCs	→Hydrochlorofluorocarbons
DME	Dimethyl ether		HCs	→Hydrocarbons
DPI	→Dry powder inhaler		HEEP	HFC Emissions Estimating Program
DTU	Danish Technical University		HFCs	→Hydrofluorocarbons
DU	→Dobson unit		HFEs	→Hydrofluoroethers
DX	Direct expansion		HITRAN	High-Resolution Transmission Molecular
EC	European Commission			Absorption
ECMWF	European Centre for Medium-Range Weather		HPC	IEA Heat Pump Centre
	Forecasting		HRBSC	Halon Recycling and Banking Support
EDGAR	Emission Database for Global Atmospheric			Committee
	Research		HTF	Heat transfer fluid
EER	Energy efficiency ratio		HTOC	Halons Technical Options Committee
EESC	Equivalent effective stratospheric chlorine		ICSC	International Chemical Safety Cards
EFCTC	European Fluorocarbon Technical Committee		IEA	International Energy Agency
EHC	Environmental health criteria		IEC	International Electrotechnical Commission
EITs	Economies in transition		IIR	International Institute of Refrigeration
EmR	Direct emission reduction		ILO	International Labour Organisation
EN	European Standards		IMDG	International Maritime Dangerous Goods Codes
EOL	End-of-life		IMO	International Maritime Organization
EPA	Environmental Protection Agency		IPA	Isopropyl alcohol
EPS	Expanded polystyrene		IPAC	International Pharmaceutical Aerosol
ETF	Earth Technology Forum			Consortium
EtO	Ethylene oxide		IPACT-I	International Pharmaceutical Aerosol
ETSU	Renewable and Energy Efficiency Organisation			Consortium for Toxicity Testing of HFC-134a
EU	European Union		IPACT-II	International Pharmaceutical Aerosol
EuroACE	European Alliance for Companies for Energy			Consortium for Toxicity Testing of HFC-227
	Efficiency		IPCC	→Intergovernmental Panel on Climate Change
EXIBA	European Extruded Polystyrene Insulation		IPCS	International Programme on Chemical Safety
	Board Association		IPLV	Integrated part load value
EXPORT	European Export of Precursors and Ozone by		ISAAC-SC	International Study of Asthma and Allergies in
	Long-Range Transport			Childhood Steering Committee
FAO	Food and Agriculture Organization of the United		ISO	International Organization for Standardization
	Nations		ISOPA	European Isocyanate Producers Association
FDH	Fixed dynamical heating		ITH	Integration time horizon
FEA	Fédération Européenne des Aérosols (European		JARN	Japan Air Conditioning, Heating, and
	Aerosol Federation)			Refrigeration News
FEMA	Fire Equipment Manufacturers Association		JPL	NASA Jet Propulsion Laboratory
F-Gases	Fluorinated greenhouse gases		JRAIA	Japan Refrigeration and Air conditioning
FIC	Fire Industry Confederation			Industries Association
FKs	→Fluoroketones		JTCCM	Japan Technical Centre for Construction
FM	Factory Mutual Research Corporation			Materials
FTP 75	US Federal Test Procedure 75		LCA	→Life cycle assessment
GAGE	Global Atmospheric Gases Experiment		LCC	Life cycle cost
GCM	General circulation model		LCCP	→Life cycle climate performance

LCD	Liquid carbon dioxide
LCVs	Light commercial vehicles
LFL	→Lower flammability limit
LKS	Lanzante-Klein-Seidel radiosonde database
LNG	Liquefied natural gas
LPG	Liquefied petrol gas
MAC	Mobile air conditioning
MACS	Mobile Air-Conditioning Society
MDI	→Metered dose inhalers
METI	Japan Ministry of Economy, Trade and Industry
MINOS	Mediterranean Intensive Oxidant Study
MIR	Maximum incremental reactivity
MLF	Multilateral Fund
MOP	→Meeting of the Parties
MSDS	Materials safety data sheet
MSU	Microwave Sounding Unit
MSWI	Municipal Solid Waste Incinerators
NAM	Northern Annular Mode
NAO	North Atlantic Oscillation
NASA	National Aeronautics and Space Administration
NBP	→Normal boiling point
NCEP	National Centers for Environmental Prediction
NEDC	New European Driving Cycle
NESCCAF	Northeast States Center for a Clean Air Future
NFPA	National Fire Protection Association
NH	Northern Hemisphere
NIK	→Not-in-Kind
NIOSH	US National Institute for occupational Safety and Health
NIST	US National Institute of Standards and Technology
NMHC	Non-methane hydrocarbon
NOAA	US National Oceanic and Atmospheric Administration
NODA	US EPA Notice of Data Availability
nPB	n-Propyl Bromide
NPV	Net present value
NREL	National Renewable Energy Laboratory
OCF	→One-component foams
ODP	→Ozone depletion potential
ODSs	→Ozone-depleting substances
OECD	Organisation for Economic Co-operation and Development
OEMs	Original equipment manufacturers
OJEC	Official Journal of the European Community
OORG	Ozone Operations Resource Group
PAG	Polyalkylene glycols
PAO	Polyalphaolefin
PFCAs	Perfluorinated carboxylic acids
PFCs	→Perfluorocarbons
PIR	Polyisocyanurate
POCP	Photochemical ozone creation potential
PS	Polystyrene
PSCs	→Polar stratospheric clouds
PTB	Persistent, toxic and can accumulate in the biosphere

PTC	Positive temperature coefficient
PTFE	Polytetrafluoroethylene
PU	Polyurethane
PVC	Poly(vinyl chloride)
PWD	Planetary wave drag
R&D	Research and development
RF	→Radiative forcing
rpm	Rotations per minute
RSW	Refrigerated sea water
SAC	Stationary air conditioning
SAE	Society of Automotive Engineers
SAGE	Stratospheric Aerosol and Gas Experiment
SAM	Southern Annular Mode
SAR	IPCC Second Assessment Report
SCANVAC	Scandinavian Federation of Heating, Ventilating and Sanitary Engineering Associations
SEER	Seasonal energy efficiency ratio
SH	Southern Hemisphere
SMEs	Small and medium enterprises
SNAP	Significant new alternatives policy
SOLAS	International Convention for the Safety of Life at Sea
SPARC	Stratospheric Processes and their Role in Climate
SPM	Summary for Policymakers
SRES	→Special Report on Emission Scenarios
SRI	Stanford Research Institute, International
SST	Sea-surface temperature
SSU	Stratospheric Sounding Unit
STE	Stratosphere-troposphere exchange
STEK	Association for the Recognition of Refrigeration Engineering Firms (The Netherlands)
STOC	UNEP Solvents Technical Options Committee
STT	Stratosphere into the troposphere
SUV	Sport utility vehicle
TAR	IPCC Third Assessment Report
TCCC	The Coca Cola Company
TEAP	→Technology and Economic Assessment Panel
TEU	Twenty-foot equivalent unit
TEWI	→Total equivalent warming impact
TFA	Trifluoroacetic acid
TFCRS	Task Force on Collection, Recovery and Long-Term Storage
TLV	→Threshold limit value
TOMS	Total Ozone Mapping Spectrometer
TS	Technical Summary
TST	Troposphere into the stratosphere
TTL	Tropical tropopause layer
TWA	→Time-weighted average
UCI	University of California at Irvine
UK	United Kingdom
UL	Underwriters Laboratories
UN	United Nations
UN-ECE	United Nations Economic Commission for Europe
UNEP	→United Nations Environment Programme

UNEP-DTIE	UNEP Division of Technology, Industry, and Economics	UV	→Ultraviolet
UNEP-TOC	UNEP Technical Options Committee	VDA	German Association of the Automotive Industry
UNFCCC	→United Nations Framework Convention on Climate Change	VOCs	→Volatile organic compounds
		VSL	Very short-lived
US	United States	WHO	World Health Organisation
US EPA	US Environmental Protection Agency	WMGHGs	→Well-mixed greenhouse gases
US FDA	US Food and Drug Administration	WMO	→World Meteorological Organization
USA	United States of America	WOUDC	World Ozone and UV Data Centre
USCG	US Coast Guard	WTA	Willingness to accept payment
USD / US$	United States Dollars	WTP	Willingness to pay
		XPS	Extruded polystyrene

Annex IV

Units

IV.1 SI (Systeme Internationale) Units

Table IV.1. Basic SI units.

Physical Quantity	Unit Name	Unit Symbol
Length	meter	m
Mass		kg
Time	second	s
Thermodynamic temperature	kelvin	K
Amount of substance	mole	mol

Table IV.2. Multiplication factors

	Prefix		Multiple	Prefix	Symbol
10^{-1}	deci	d	10	deca	da
10^{-2}	centi	c	10^2	hecto	h
10^{-3}	milli	m	10^3	kilo	k
10^{-6}	micro	μ	10^6	mega	M
10^{-9}	nano	n	10^9	giga	G
10^{-12}	pico	p	10^{12}	tera	T
10^{-15}	femto	f	10^{15}	peta	P

Table IV.3. Special Names and Symbols for Certain SI–Derived Units

Physical Quantity	Unit Name		Definition
Force		N	$kg\ m\ s^{-2}$
Pressure	pascal	Pa	$kg\ m^{-1}\ s^{-2}$ ($= N\ m^{-2}$)
Energy	joule	J	$kg\ m^2\ s^{-2}$
Power	watt	W	$kg\ m^2\ s^{-3}$ ($= J\ s^{-1}$)
Frequency	hertz	Hz	s^{-1} (cycles per second)

Table IV.4. Decimal Fractions and Multiples of SI Units having Special Names

Physical Quantity	Unit Name		Definition
Length	Ångstrom	Å	10^{-10} m = 10^{-8} cm
Length	micron	µm	10^{-6} m
Area	hectare	ha	10^{4} m^2
Volume	litre	L	10^{-3} m^3
Force	dyne	dyn	10^{-5} N
Pressure	bar	bar	10^{5} N m^{-2} = 10^{5} Pa
Pressure	millibar	mb	10^{2} N m^{-2} = 1 hPa
Mass	tonne	t	10^{3} kg
Mass	gram	g	10^{-3} kg
Column density	Dobson units[a]	DU	2.687×10^{16} molecules cm^{-2}
Streamfunction	Sverdrup	Sv	10^{6} m^3 s^{-1}

[a] See 'Dobson units' in glossary.

IV.2 Other Units

Table IV.5. Other units.

Symbol	Description
ºC	Degree Celsius (0°C = 273 K approximately)
	Temperature differences are also given in ºC (= K) rather than the more correct form of 'Celsius degrees'
ppm	Parts per million (10^{6}), mixing ratio[a] (µmol mol^{-1})
ppb	Parts per billion (10^{9}), mixing ratio[a] (nmol mol^{-1})
ppt	Parts per trillion (10^{12}), mixing ratio[a] (fmol mol^{-1})
yr	Year
MtCO$_2$-eq	Megatonnes (1 Mt = 10^{6} kg = 1 Gg) CO$_2$-equivalent[b]
GtCO$_2$-eq	Gigatonnes (1 Gt = 10^{12} kg = 1 Pg) CO$_2$-equivalent[b]
MtN	Megatonnes of nitrogen

[a] See 'mixing ratio' in glossary.
[b] See 'CO$_2$-equivalent' in glossary.

IV.3 Costs

Unless stated otherwise, specific costs are calculated or reported using 5% per year as the default discount rate. The expected lifetime of the equipment is used as the depreciation period.

Costs are expressed in US$$_{(2002)}$, unless stated otherwise.

To correct cost data for the effect of inflation, the deflator for the gross domestic product (GDP) is applied for years other than 2002.

The conversion of currencies is based on the exchange rate on 31 July of the respective year.

Annex V

Major Chemical Formulae and Nomenclature

This annex presents the formulae and nomenclature for halogen-containing species and other species that are referred to in this report (Annex V.1). The nomenclature for refrigerants and refrigerant blends is given in Annex V.2.

V.1   Substances by Groupings

V.1.1   *Halogen-Containing Species*

V.1.1.1   Inorganic Halogen-Containing Species

Atomic chlorine	Cl
Molecular chlorine	Cl_2
Chlorine monoxide	ClO
Chlorine radicals	ClO_x
Chloroperoxy radical	ClOO
Dichlorine peroxide (ClO dimer)	$(ClO)_2$, Cl_2O_2
Hydrogen chloride (Hydrochloric acid)	HCl
Inorganic chlorine	Cl_y
Antimony pentachloride	$SbCl_5$

Atomic bromine	Br
Molecular bromine	Br_2
Bromine monoxide	BrO
Bromine radicals	BrO_x
Bromine nitrate	$BrONO_2$, $BrNO_3$
Potassium bromide	KBr

Atomic fluorine	F
Molecular fluorine	F_2
Hydrogen fluoride (Hydrofluoric acid)	HF
Sulphur hexafluoride	SF_6
Nitrogen trifluoride	NF_3

Atomic iodine	I
Molecular iodine	I_2

V.1.1.2   Halocarbons

For each halocarbon the following information is given in columns:
- Chemical compound [Number of isomers][1] (or common name)
- Chemical formula
- CAS number[2]
- Chemical name (or alternative name)

V.1.1.2.1   Chlorofluorocarbons (CFCs)

CFC-11	CCl_3F	75-69-4	Trichlorofluoromethane
CFC-12	CCl_2F_2	75-71-8	Dichlorodifluoromethane
CFC-13	$CClF_3$	75-72-9	Chlorotrifluoromethane
CFC-113 [2]	$C_2Cl_3F_3$		Trichlorotrifluoroethane
CFC-113	CCl_2FCClF_2	76-13-1	1,1,2-Trichloro-1,2,2-trifluoroethane
CFC-113a	CCl_3CF_3	354-58-5	1,1,1-Trichloro-2,2,2-trifluoroethane
CFC-114 [2]	$C_2Cl_2F_4$		Dichlorotetrafluoroethane
CFC-114	$CClF_2CClF_2$	76-14-2	1,2-Dichloro-1,1,2,2-tetrafluoroethane
CFC-114a	CCl_2FCF_3	374-07-2	1,1-Dichloro-1,2,2,2-tetrafluoroethane
CFC-115	$CClF_2CF_3$		Chloropentafluoroethane

V.1.1.2.2   Hydrochlorofluorocarbons (HCFCs)

HCFC-21	$CHCl_2F$	75-43-4	Dichlorofluoromethane
HCFC-22	$CHClF_2$	75-45-6	Chlorodifluoromethane
HCFC-123 [3]	$C_2HCl_2F_3$		Dichlorotrifluoroethane
HCFC-123	$CHCl_2CF_3$	306-83-2	2,2-Dichloro-1,1,1-trifluoroethane
HCFC-123a	$C_2HCl_2F_3$	354-23-4	1,2-Dichloro-1,1,2-trifluoroethane
HCFC-123b	$C_2HCl_2F_3$	812-04-4	1,1-Dichloro-1,2,2-trifluoroethane
HCFC-124 [2]			Chlorotetrafluoroethane
HCFC-124	$CHClFCF_3$	2837-89-0	2-Chloro-1,1,1,2-tetrafluoroethane
HCFC-124a	C_2HClF_4	354-25-6	1-Chloro-1,1,2,2-tetrafluoroethane
HCFC-141b	CH_3CCl_2F	1717-00-6	1,1-Dichloro-1-fluoroethane
HCFC-142b	CH_3CClF_2	75-68-3	1-Chloro-1,1-difluoroethane
HCFC-225ca	$CHCl_2CF_2CF_3$	422-56-0	3,3-Dichloro-1,1,1,2,2-pentafluoropropane
HCFC-225cb	$CHClFCF_2CClF_2$	507-55-1	1,3-Dichloro-1,1,2,2,3-pentafluoropropane

---

[1] The number of isomers is shown between brackets for figures larger than unity.
[2] The Chemical Abstracts Service (CAS) Registry Number for this substance. The CAS number is an internationally-recognised unique numeric identifier that designates only one chemical substance. The CAS is a division of the American Chemical Society.

V.1.1.2.3 Hydrofluorocarbons (HFCs)

HFC-23		CHF_3	75-46-7	Trifluoromethane
HFC-32		CH_2F_2	75-10-5	Difluoromethane (Methylene fluoride)
HFC-41		CH_3F	593-53-3	Fluoromethane (Methyl fluoride)
HFC-125		CHF_2CF_3	354-33-6	Pentafluoroethane
HFC-134	[2]	$C_2H_2F_4$		Tetrafluoroethane
HFC-134		CHF_2CHF_2	359-35-3	1,1,2,2-Tetrafluoroethane
HFC-134a		CH_2FCF_3	811-97-2	1,1,1,2-Tetrafluoroethane
HFC-143	[2]	$C_2H_3F_3$		Trifluoroethane
HFC-143		CH_2FCHF_2	430-66-0	1,1,2-Trifluoroethane
HFC-143a		CH_3CF_3	420-46-2	1,1,1-Trifluoroethane
HFC-152	[2]	$C_2H_4F_2$		Difluoroethane
HFC-152		CH_2FCH_2F	624-72-6	1,2-Difluoroethane
HFC-152a		CHF_2CH_3	75-37-6	1,1-Difluoroethane
HFC-161		CH_3CH_2F	353-36-6	Monofluoroethane (Ethyl fluoride)
HFC-227	[2]	C_3HF_7		Heptafluoropropane
HFC-227ca		$CF_3CF_2CHF_2$	2252-84-8	1,1,1,2,2,3,3-Heptafluoropropane
HFC-227ca		CF_3CHFCF_3	431-89-0	1,1,1,2,3,3,3-Heptafluoropropane
HFC-236	[4]	$C_3H_2F_6$		Hexafluoropropane
HFC-236ca		$CHF_2CF_2CHF_2$	27070-61-7	1,1,2,2,3,3-Hexafluoropropane
HFC-236cb		$CH_2FCF_2CF_3$	677-56-5	1,1,1,2,2,3-Hexafluoropropane
HFC-236ea		CHF_2CHFCF_3	431-63-0	1,1,1,2,3,3-Hexafluoropropane
HFC-236fa		$CF_3CH_2CF_3$	690-39-1	1,1,1,3,3,3-Hexafluoropropane
HFC-245	[5]	$C_3H_3F_5$		Pentafluoropropane
e.g. HFC-245ce		$CH_2FCF_2CHF_2$	679-86-7	1,1,2,2,3-Pentafluoropropane
HFC-245fa		$CHF_2CH_2CF_3$	460-73-1	1,1,1,3,3-Pentafluoropropane
HFC-365mfc		$CH_3CF_2CH_2CF_3$	406-58-6	1,1,1,3,3-Pentafluorobutane
HFC-c-447ef		$c\text{-}C_5H_3F_7$	15290-77-4	Heptafluorocyclopentane

V.1.1.2.4 Halons

Halon-1202	CBr_2F_2	75-61-6	Dibromodifluoromethane
Halon-1211	$CBrClF_2$	353-59-3	Bromochlorodifluoromethane (Chlorodifluorobromomethane), R-12B1
Halon-1301	$CBrF_3$	75-63-8	Bromotrifluoromethane, R-13B1
Halon-2402	$CBrF_2CBrF_2$	124-73-2	1,2-Dibromotetrafluoroethane (1,1,2,2-Tetrafluoro-1,2-dibromoethane, 1,2-Dibromo-1,1,2,2-tetrafluoroethane)

V.1.1.2.5 Perfluorocarbons (PFCs)

PFC-14	CF_4	75-73-0	Tetrafluoromethane (Carbon tetrafluoride)
PFC-116	C_2F_6 (CF_3CF_3)	76-16-4	Perfluoroethane (Hexafluoroethane)
PFC-218	C_3F_8 ($CF_3CF_2CF_3$)	76-19-7	Perfluoropropane (Octafluoropropane)
PFC-318 or PFC-c318	$c\text{-}C_4F_8$ ($\text{-}(CF_2)_4\text{-}$)	115-25-3	Perfluorocyclobutane (Octafluorocyclobutane)
PFC-3-1-10	C_4F_{10}	355-25-9	Perfluorobutane
PFC-5-1-14	C_6F_{14}	355-42-0	Perfluorohexane
PFC-6-1-16	C_7F_{16}	335-57-9	Perfluoroheptane
PFC-7-1-18	C_8F_{18}	307-34-6	Perfluorooctane

*V.1.1.2.6	Fluorinated Ethers*

HFE-449s1	$C_5H_3F_9O$		
	$CF_3(CF_2)_3OCH_3$	163702-07-6	Methyl nonafluorobutyl ether
	$(CF_3)_2CFCF_2OCH_3$	163702-08-7	Perfluoroisobutyl methyl ether
HFE-569sf2	$C_6H_5F_9O$		
	$CF_3(CF_2)_3OCH_2CH_3$	163702-05-4	Ethyl perfluorobutyl ether
	$(CF_3)_2CFCF_2OCH_2CH_3$		
		163702-06-5	Ethyl perfluoroisobutyl ether
HFE-347pcf2	$C_4H_3F_7O$ $(CF_3CH_2OCF_2CHF_2)$		1,1,2,2-Tetrafluoroethyl 2,2,2-trifluoroethyl ether

*V.1.1.2.7	Chlorocarbons*

Carbon tetrachloride	CCl_4	56-23-5	R-10, (Halon 104)
Chloroform	$CHCl_3$	67-66-3	Trichloromethane, R-20
Methylene chloride	CH_2Cl_2	75-09-2	Dichloromethane, R-30, freon 30
Methyl chloride	CH_3Cl	74-87-3	Chloromethane, R-40, freon 40
Trichloroethene	C_2HCl_3 $(CHCl=CCl_2)$	79-01-6	1,1,2-Trichloroethylene, TCE
Perchloroethene	C_2Cl_4 $(CCl_2=CCl_2)$	127-18-4	Perchloroethylene, tetrachloroethene
Ethyl chloride	CH_3CH_2Cl	75-00-3	Chloroethane
Methyl chloroform	CH_3CCl_3	71-55-6	1,1,1-Trichloroethane
Isopropyl chloride	$CH_3CHClCH_3$	75-29-6	2-Chloropropane

*V.1.1.2.8	Bromocarbons*

Methyl bromide	CH_3Br	74-83-9	Bromomethane (Halon-1001)
Bromoform	$CHBr_3$	75-25-2	Tribromomethane
n-Propyl bromide	$CH_3CH_2CH_2Br$ $(n\text{-}C_3H_7Br)$	106-94-5	1-Bromopropane, n-PB

V.1.1.2.9 Other Halocarbons

FK-5-1-12	$C_6F_{12}O$	756-13-8	Nonafluoro-4- (trifluoromethyl)-3- pentanone
	$(CF_3CF_2C(O)CF(CF_3)_2)$		
	CH_2BrCl	74-97-5	Bromochloromethane (Halon-1011)
	$C_2HF_3O_2$	76-05-1	Trifluoroacetic acid (TFA), Perfluoric acid
	(CF_3COOH)		
	COF_2	353-50-4	Carbonyl fluoride, Carbonic difluoride
	$C_2F_4O\ (CF_3COF)$	354-34-7	Trifluoroacetyl fluoride
	$C_8HF_{15}O_2$	335-67-1	Perfluorooctanoic acid (FOA), Pentadecafluorooctanoic acid
	CH_2ClI	593-71-5	Chloroiodomethane
	CH_2BrI	557-68-6	Bromoiodomethane
	CF_3I	2314-97-8	Trifluoromethyl iodide, trifluoroiodomethane
	CF_3CF_2I	354-64-3	Iodopentafluoroethane
	$COClF$	353-49-1	Chlorofluorocarbonyl
	CF_3COCl	354-32-5	Trifluoroacetyl chloride
	SF_5CF_3	373-80-8	Trifluoromethylsulphur pentafluoride

V.1.2 Other Species

V.1.2.1 Inorganic Species

Atomic oxygen	O
Atomic oxygen (first excited state)	$O(^1D)$
Molecular oxygen (R-732)	O_2
Ozone	O_3
Odd oxygen (O, $O(^1D)$, O_3)	O_x
or oxidant ($O_3 + NO_2$)	

Atomic hydrogen	H
Molecular hydrogen (R-702)	H_2
Hydroxyl radical	OH
Hydroperoxyl radical	HO_2
Water (R-718)	H_2O
Hydrogen peroxide	H_2O_2
Odd hydrogen (H, OH, HO_2, H_2O_2)	HO_x

Atomic sulphur	S
Sulphur dioxide (R-764)	SO_2
Sulfate	SO_4
Sulphuric acid	H_2SO_4
Hydrogen sulfide	H_2S
Sulphur oxides (SO + SO_2 + SO_3)	SO_x

Carbon atom	C
Carbon monoxide	CO
Carbon dioxide (R-744)	CO_2
Sodium bicarbonate	$NaHCO_3$

Atomic nitrogen	N
Molecular nitrogen (R-728)	N_2
Nitrous oxide (R-744A)	N_2O
Nitric oxide	NO
Nitrogen dioxide	NO_2
Nitrogen trioxide, nitrate radical	NO_3
Dinitrogen pentoxide	N_2O_5
Nitrogen oxides (NO + NO_2)	NO_x
Total reactive nitrogen (usually includes NO, NO_2, NO_3, N_2O_5, $ClONO_2$, HNO_4, HNO_3)	NO_y
Amidogen radical	NH_2
Ammonia (R-717)	NH_3
Ammonium	NH_4
Ammonium sulfate	$(NH_4)_2SO_4$
Ammonium nitrate	NH_4NO_3
Ammonium phosphate	$(NH_4)H_2PO_4$
Nitric acid	HNO_3
Hydrogen cyanide	HCN

Helium (R-704)	He
Argon (R-740)	Ar
Radon	Rn

V.1.2.2   Non-Halogenated Hydrocarbons and Other Organic Species

Common or Industrial Designation	Formula	CAS-number	Other Names
Methane	CH_4	74-82-8	R-50
Ethane	C_2H_6 (CH_3CH_3)	74-84-0	R-170
Propane	C_3H_8 $(CH_3CH_2CH_3)$	74-98-6	R-290
Butane	C_4H_{10} $(CH_3CH_2CH_2CH_3)$	106-97-8	R-600, n-Butane
Isobutane	C_4H_{10} $((CH_3)_2CHCH_3)$	75-28-5	R-600a, i-Butane, 2-Methylpropane
Pentane	C_5H_{12} $(CH_3(CH_2)_3CH_3)$	109-66-0	R-601, n-Pentane
Isopentane	C_5H_{12} $((CH_3)_2CHCH_2CH_3)$	78-78-4	R-601a, i-Pentane, 2-Methylbutane
Methyl ether	C_2H_6O (CH_3OCH_3)	115-10-6	R-E170, Dimethyl ether
Cyclopropane	$c\text{-}C_3H_6$ $(\ \text{-}(CH_2)_3\text{-}\ )$	75-19-4	C-270
Cyclopentane	$c\text{-}C_5H_{10}$ $(\ \text{-}(CH_2)_5\text{-}\ )$	287-92-3	
Ethene	C_2H_4 $(CH_2{=}CH_2)$	74-85-1	R-1150, Ethylene
Propene	C_3H_6 $(CH_3CH{=}CH_2)$	115-07-1	R-1270, Propylene
Benzene	C_6H_6	71-43-2	
Toluene	C_7H_8	108-88-3	Methylbenzine
Xylene	C_8H_{10}	several isomers	Dimethylbenzene
Trimethylbenzene	C_9H_{12}	several isomers	
Isoprene	C_5H_8	78-79-5	2-Methyl-1,3-butadiene
Ethyl ether	$C_4H_{10}O$ $(CH_3CH_2OCH_2CH_3)$	60-29-7	R-610, Diethyl ether
n-Octanol	$C_8H_{18}O$ $(CH_3(CH_2)_7OH)$	111-87-5	1-Octanol, 1-Octyl alcohol
Methyl formate	$C_2H_4O_2$ $(HCOOCH_3)$	107-31-3	R-611, Formic acid methyl ester
Isopropanol	C_3H_8O $(CH_3CHOHCH_3)$	67-63-0	Isopropyl alcohol
Methyl amine	CH_5N (CH_3NH_2)	74-89-5	R-630
Ethyl amine	C_2H_7N $(CH_3CH_2(NH_2)\,)$	75-04-7	R-631
Ethyne	C_2H_2 $(CH{\equiv}CH)$	74-86-2	Acetylene
Formaldehyde	CH_2O $(HCHO)$	50-00-0	Oxomethane, Methylene oxide
Acetone	C_3H_6O (CH_3COCH_3)	67-64-1	2-Propanone, Methyl ketone
Methyl peroxide	CH_4O_2 (CH_3OOH)	3031-73-0	Methyl hydroperoxide
Methyl peroxy radical	CH_3OO		
Acetyl peroxy radical	$CH_3C(O)OO$		
Alkoxy radicals	RO		
Organic peroxy radicals	RO_2		

V.2 Refrigerant Nomenclature

V.2.1 Refrigerant Designations for Compounds

Refrigerant Number	CFC	HCFC	HFC	PFC	HC	Other	Chemical name
10						x	Carbon tetrachloride
11	x						Trichlorofluoromethane
12	x						Dichlorodifluoromethane
12B1						x	Bromochlorodifluoromethane (Halon-1211)
13	x						Chlorotrifluoromethane
13B1						x	Bromotrifluoromethane (Halon-1301)
14				x			Carbon tetrafluoride
20						x	Chloroform
21		x					Dichlorofluoromethane
22		x					Chlorodifluoromethane
23			x				Trifluoromethane
30						x	Dichloromethane (Methylene chloride, Freon 30)
31		x					Chlorofluoromethane
32			x				Difluoromethane (Methylene fluoride)
40						x	Chloromethane (Methyl chloride, Freon 40)
41			x				Fluoromethane (Methyl fluoride)
50					x		Methane
113	x						1,1,2-Trichloro-1,2,2-trifluoroethane
114	x						1,2-Dichloro-1,1,2,2-tetrafluoroethane
115	x						Chloropentafluoroethane
116				x			Hexafluoroethane
123		x					2,2-Dichloro-1,1,1-trifluoroethane
124		x					2-Chloro-1,1,1,2-tetrafluoroethane
125			x				Pentafluoroethane
134a			x				1,1,1,2-Tetrafluoroethane
141b		x					1,1-Dichloro-1-fluoroethane
142b		x					1-chloro-1,1-Difluoroethane
143a			x				1,1,1-Trifluoroethane
152a			x				1,1-Difluoroethane
170					x		Ethane
E170					x		Methyl ether (Dimethyl ether)
218				x			Octafluoropropane
227ea			x				1,1,1,2,3,3,3-Heptafluoropropane
236fa			x				1,1,1,3,3,3-Hexafluroropropane
245fa			x				1,1,1,3,3-Pentafluoropropane
C270					x		Cyclopropane
290					x		Propane
C318					x		Octafluorocyclobutane
600					x		Butane
600a					x		2-Methyl propane (Isobutane)
610					x		Ethyl ether
611					x		Methyl formate
630					x		Methyl amine
631					x		Ethyl amine

Examples:

R-11 is also referred to as	CFC-11
R-12B1	,,	Halon-1211
R-22	,,	HCFC-22
R-23	,,	HFC-23
R-116	,,	PFC-116
R-600	,,	HC-600 (or butane)

Refrigerant Number	CFC	HCFC	HFC	PFC	HC	Other	Chemical name
702						x	Hydrogen
704						x	Helium
717						x	Ammonia
718						x	Water
720						x	Neon
728						x	Nitrogen
729						x	Air
732						x	Oxygen
740						x	Argon
744						x	Carbon dioxide
744A						x	Nitrous oxide
764						x	Sulphur dioxide
1132a			x				1,1-Difluoroethene (Vinylidene fluoride)
1150					x		Ethene (Ethylene)
1270					x		Propene (Propylene)

V.2.2 Refrigerant Designations for Blends of Compounds (R-400 and R-500 Blends)
Nominal composition (mass%)

	CFC-12	CFC-13	CFC-114	CFC-115	HCFC-22	HCFC-31	HCFC-124	HCFC-142b	HFC-23	HFC-32	HFC-125	HFC-134a	HFC-143a	HFC-152a	PFC-116	PFC-218	PFC-C318	R-E170	HC-290	HC-600	HC-600a	HC-1270
R-400[1]	x%		y%																			
R-401A					53%		34%							13%								
R-401B					61%		28%							11%								
R-401C					33%		52%							15%								
R-402A					38%						60%								2%			
R-402B					60%						38%								2%			
R-403A					75%											20%			5%			
R-403B					56%											39%			5%			
R-404A											44%	4%	52%									
R-405A					45%			5.5%						7%			42.5%					
R-406A					55%			41%													4%	
R-407A										20%	40%	40%										
R-407B										10%	70%	20%										
R-407C										23%	25%	52%										
R-407D										15%	15%	70%										
R-407E										25%	15%	60%										
R-408A					47%						7%		46%									
R-409A					60%		25%	15%														
R-409B					65%		25%	10%														
R-410A										50%	50%											
R-410B										45%	55%											
R-411A					87.5%									11%								1.5%
R-411B					94%									3%								3%
R-412A					70%			25%								5%						
R-413A												88%				9%					3%	
R-414A					51%		28.5%	16.5%													4%	
R-414B					50%		39%	9.5%													1.5%	
R-415A					82%									18%								
R-415B					25%									75%								
R-416A							39.5%					59%								1.5%		
R-417A											46.6%	50%								3.4%		
R-418A					96%									2.5%					1.5%			
R-419A											77%	19%						4%				
R-500	73.8%													26.2%								
R-501	25%				75%																	

[1] R-400 can have various proportions of CFC-12 and CFC-114. The exact composition needs to be specified, e.g. R-400 (60/40).

	CFC-12	CFC-13	CFC-114	CFC-115	HCFC-22	HCFC-31	HCFC-124	HCFC-142b	HFC-23	HFC-32	HFC-125	HFC-134a	HFC-143a	HFC-152a	PFC-116	PFC-218	PFC-C318	R-E170	HC-290	HC-600	HC-600a	HC-1270
R-502				51.2%	48.8%																	
R-503		59.9%							40.1%													
R-504				51.8%						48.2%												
R-505	78%					22%																
R-506			44.9%			55.1%																
R-507A											50%		50%									
R-508A									39%						61%							
R-508B									46%						54%							
R-509A					44%											56%						

Annex VI

List of Major IPCC Reports

Climate Change – The IPCC Scientific Assessment
The 1990 report of the IPCC Scientific Assessment Working Group

Climate Change – The IPCC Impacts Assessment
The 1990 report of the IPCC Impacts Assessment Working Group

Climate Change – The IPCC Response Strategies
The 1990 report of the IPCC Response Strategies Working Group

Emissions Scenarios
Prepared by the IPCC Response Strategies Working Group, 1990

Assessment of the Vulnerability of Coastal Areas to Sea Level Rise – A Common Methodology, 1991

Climate Change 1992 – The Supplementary Report to the IPCC Scientific Assessment
The 1992 report of the IPCC Scientific Assessment Working Group

Climate Change 1992 – The Supplementary Report to the IPCC Impacts Assessment
The 1992 report of the IPCC Impacts Assessment Working Group

Climate Change: The IPCC 1990 and 1992 Assessments
IPCC First Assessment Report Overview and Policymaker Summaries, and 1992 IPCC Supplement

Global Climate Change and the Rising Challenge of the Sea
Coastal Zone Management Subgroup of the IPCC Response Strategies Working Group, 1992

Report of the IPCC Country Study Workshop, 1992

Preliminary Guidelines for Assessing Impacts of Climate Change, 1992

IPCC Guidelines for National Greenhouse Gas Inventories (3 volumes), 1994

Climate Change 1994 – Radiative Forcing of Climate Change *and* an Evaluation of the IPCC IS92 Emission Scenarios

IPCC Technical Guidelines for Assessing Climate Change Impacts and Adaptations, 1995

Climate Change 1995 – The Science of Climate Change
– Contribution of Working Group I to the Second Assessment Report

Climate Change 1995 – Scientific-Technical Analyses of Impacts, Adaptations and Mitigation of Climate Change
– Contribution of Working Group II to the Second Assessment Report

Climate Change 1995 – The Economic and Social Dimensions of Climate Change – Contribution of Working Group III to the Second Assessment Report

The IPCC Second Assessment Synthesis of Scientific-Technical Information Relevant to Interpreting Article 2 of the UN Framework Convention on Climate Change, 1995

Revised 1996 IPCC Guidelines for National Greenhouse Gas Inventories (3 volumes), 1996

Technologies, Policies and Measures for Mitigating Climate Change – IPCC Technical Paper 1, 1996

An Introduction to Simple Climate Models Used in the IPCC Second Assessment Report – IPCC Technical Paper 2, 1997

Stabilisation of Atmospheric Greenhouse Gases: Physical, Biological and Socio-Economic Implications – IPCC Technical Paper 3, 1997

Implications of Proposed CO$_2$ Emissions Limitations – IPCC Technical Paper 4, 1997

The Regional Impacts of Climate Change: An Assessment of Vulnerability
IPCC Special Report, 1997

Aviation and the Global Atmosphere
IPCC Special Report, 1999

Methodological and Technological Issues in Technology Transfer
IPCC Special Report, 2000

Emissions Scenarios
IPCC Special Report, 2000

Land Use, Land Use Change and Forestry
IPCC Special Report, 2000

Good Practice Guidance and Uncertainty Management in National Greenhouse Gas Inventories
IPCC National Greenhouse Gas Inventories Programme, 2000

Climate Change and Biodiversity - IPCC Technical Paper 5, 2002

Climate Change 2001: The Scientific Basis – Contribution of Working Group I to the Third Assessment Report

Climate Change 2001: Impacts, Adaptation & Vulnerability – Contribution of Working Group II to the Third Assessment Report

Climate Change 2001: Mitigation – Contribution of Working Group III to the Third Assessment Report

Climate Change 2001: Synthesis Report

Good Practice Guidance for Land Use, Land-Use Change and Forestry
IPCC National Greenhouse Gas Inventories Programme, 2003

Enquiries: IPCC Secretariat, c/o World Meteorological Organization, 7 bis, Avenue de la Paix, Case Postale 2300, 1211 Geneva 2, Switzerland